Ecology, Evolution and Diversity of Plants

Ecology, Evolution and Diversity of Plants

Editors

Hong-Hu Meng
Yi-Gang Song

MDPI

Basel • Beijing • Wuhan • Barcelona • Belgrade • Novi Sad • Cluj • Manchester

Editors

Hong-Hu Meng
Center for Integrative Conservation
Xishuangbanna Tropical Botanical Garden, Chinese Academy of Science
Kunming
China

Yi-Gang Song
Eastern China Conservation Centre for Wild Endangered Plant Resources
Shanghai Chenshan Botanical Garden
Shanghai
China

Editorial Office
MDPI
St. Alban-Anlage 66
4052 Basel, Switzerland

This is a reprint of articles from the Special Issue published online in the open access journal *Diversity* (ISSN 1424-2818) (available at: www.mdpi.com/journal/diversity/special_issues/biodiversity_plants).

For citation purposes, cite each article independently as indicated on the article page online and as indicated below:

Lastname, A.A.; Lastname, B.B. Article Title. *Journal Name* **Year**, *Volume Number*, Page Range.

ISBN 978-3-7258-0398-9 (Hbk)
ISBN 978-3-7258-0397-2 (PDF)
doi.org/10.3390/books978-3-7258-0397-2

Cover image courtesy of Hong-Hu Meng

Contents

About the Editors

Hong-Hu Meng

Dr. Hong-Hu Meng is currently working on plant evolution, distribution, and conservation. He is devoting himself to the evolutionary and biogeographic patterns in tropical and subtropical Asia. He is exploring a few groups, i.e., Engelhardia (Juglandaceae), Rhoiptelea (Rhoipteleaceae), and Trigonobalanus (Fagaceae), to explore the geographic distribution and the related geo-events and to build on biodiversity conservation and sustainability in the Anthropocene.

Yi-Gang Song

Dr. Yi-Gang Song is now working on plant evolution, conservation, and seed ecology at Shanghai Chenshan Botanical Garden. He mainly focuses on the families Fagaceae, Juglandaceae, Ulmaceae, and Styracaceae. Based on their phylogeny, he wishes to explore the conservation gap analysis, conservation genomics, trait evolution, and niche evolution of these taxa.

diversity

MDPI

Editorial

Understanding Plant Diversity from Ecological and Evolutionary Perspectives

Hong-Hu Meng [1,*] and Yi-Gang Song [2,*]

1 Plant Phylogenetics and Conservation Group, Center for Integrative Conservation & Yunnan Key Laboratory for Conservation of Tropical Rainforests and Asian Elephants, Xishuangbanna Tropical Botanical Garden, Chinese Academy of Sciences, Mengla 666303, China
2 Eastern China Conservation Centre for Wild Endangered Plant Resources, Shanghai Chenshan Botanical Garden, Shanghai 201602, China
* Correspondence: menghonghu@xtbg.ac.cn (H.-H.M.); ygsong@cemps.ac.cn (Y.-G.S.)

Citation: Meng, H.-H.; Song, Y.-G. Understanding Plant Diversity from Ecological and Evolutionary Perspectives. *Diversity* **2023**, *15*, 1165. https://doi.org/10.3390/d15121165

Received: 16 November 2023
Accepted: 20 November 2023
Published: 22 November 2023

Nowadays, we are living in a world that is benefiting from biodiversity, although environmental change is dramatic and biodiversity has been influenced by climate changes and human activities. As is known, biodiversity, as an important pillar of ecosystem balance, is playing a crucial role in human well-being [1]. In particular, it is too difficult to use plant diversity to evaluate the related indirect value, except the direct value of food, clothing, and shelter. Also, our life is closely bound up with various plants from their functions and primary productivity. Hence, plant diversity is important for the ecosystem service to human beings, playing a significant role in sustaining the equilibrium of global biosphere. However, biodiversity loss has continued to impact people throughout human history, and a sixth mass extinction in Earth's history is coming [2,3].

It is worth noting that plants, as one of the most important globally distributed types of biodiversity, occupy a major proportion of land on our planet. This means that evolutionary and ecological mechanisms are incredibly successful during evolutionary processes, and such mechanisms really determine the plant diversity. In recent decades, there have been increasing numbers of studies on ecological and evolutionary mechanisms for plant diversity, like mushrooms after rain, and scientists from different fields have tried to explore the relationship of plant diversity between ecological and evolutionary approaches. Here, we aim to understand how plant diversity has been enhanced by the establishment of ecological and evolutionary approaches, which had, has, and will continue to have a far-reaching effect on many fields of the basic and applied research within natural science.

In such a context, the Special Issue "*Ecology, Evolution and Diversity of Plants*" has been launched to explore plant diversity from ecological and evolutionary viewpoints. Thus, the published articles have focused on broad key topics in plant diversity, from species to populations and vegetation; as well as from past to current taxa [4,5]. Also, different types of reviews, research articles, and commentaries have been collected to explore the ecology, evolution, and biodiversity conservation of plants [4–6]. Therefore, this Special Issue has published more than 30 articles to understand plant diversity from many angles using ecological and evolutionary approaches.

Plant diversity is primarily recognized using the species, and species delimitation and taxonomic revision are fundamental issues [7]. In the renowned book, *The Origin of Species* by Charles Darwin, published in 1859 [8], the concept of species was proposed and the related approaches to recognize species, i.e., the relationships of species or systematics, were advanced with an unique illustration in such a famous book. As is known, such a simple illustration is also called the Tree of Life (TOL). Accordingly, phylogeny has been become a discipline and even a powerful tool in recognizing plant diversity, which is constantly popular, and many analyses and references have been put forward [9]. Therefore, the topic of phylogeny in plant diversity in relation to species, taxonomy, and geographic distribution, from tropical to subtropical temperatures, as well as alpine plants, has boomed in recent years [10–15]. The ensuing rapid development of phylogeny is accompanied by the development of sequencing technology, transformed from DNA spacers to genomes in recent years.

Here, we considered plant diversity from an ecological point of view on the basis of the ecological processes of past, current, and future. In particular, climate change and human activity were included. In such a context, the concept that habitat suitability has been altered by climate change at the regional scale or global scale, which is causing local extinctions of biodiversity, seems possible [1]. Thus, the notion of whether adaptive genetic variation can keep pace with future climate change is a very critical factor in developing conservation management strategies when the genetic vulnerability of species is assessed [16]. And now, climate change and human activity have been considered to be major drivers of biodiversity loss and species range shifts [17]. However, the response of mountain plant communities has not kept up with climate warmings, especially in the background of global change [18]. Thus, the warming climate may benefit tropical plants, but will be significantly disadvantageous to the alpine plants, and so mountaintops were proposed as climate refugia in tropical rainforests for cold-adapted plants because the warming climate has triggered alpine plants to climb higher altitudes [19]. Different taxa will respond individually to environmental changes, so the identification of refugia from spatial dimensions is important, due to the large number of taxa that are identified in a similar manner, and these factors will be affected by various environmental parameters [20]. Thus, the firm historical insights will provide a better understanding of future climate refugia based on robust predictions [19], especially taxa from regions with dense populations [15]. Notably, the modification of precipitation should also be considered as determinants of plant diversity under the background of climate change. For example, enhanced precipitation has influenced the evolution of subtropical evergreen broad-leaved forests since the early Miocene in East China [21].

At present, the Earth is entering a new geological era, the Anthropocene. Biodiversity will face an unprecedented challenge when the Anthropocene comes, although biodiversity brings essential benefits of nature to human beings and enhances the value of eco-service to human beings [17]. In recent decades, the conservation of plant diversity is an urgent issue, and the related studies of ecological and conservation have developed significantly, due to the large-scale use of DNA spacers, EST-SSR, single-nucleotide polymorphisms (SNPs), and even whole genomes, in order to illuminate the conservation implications of different taxa (especially rare and endangered taxa) from their natural habitats and to evaluate the conservation managements of anthropogenic disturbances [22–24]. Regardless of the different approaches, i.e., genetics or genomic analyses, even the Data Portal (e.g., the *Paris polyphylla* Data Portal, PPDP, which integrates related functional genomics analyses, molecular data resources, and morphological identification points [25]), have provided useful information, both genetic and genomic, for the further molecular breeding and improvement of resources in economic plants.

Although we are living under the conditions of global change, the warming climate and regional poverty are still urgent issues, so the topic of economic forests has boomed in recent decades [26]. It seems that booming afforestation can bring benefits to eliminate poverty from economic growth, and is the most effective strategy for mitigating climate change [26,27]. However, afforestation is causing the disappearance of species, but this fact tends to be ignored [27]. Thus, it can be argued that economic values to eliminate regional poverty using economic forests and photosynthetic carbon captured by the afforestation for climate change are blinding the shifts of biodiversity patterns, which even change the distribution pattern of biodiversity, such as the latitudinal diversity gradient [27,28].

Generally, ecology and evolution have played, are playing, and will continue to play significant roles in shaping plant diversity. However, specific studies on ecology or evolution suggest that plant diversity has two separate parts, and some research topics are not linked to each other. Actually, everything is connected in the natural world. Thus, studies on plant diversity will be facilitated by the flourishing development of ecology and evolution disciplines, just as plant diversity has boomed on a global scale throughout the endless history of the Earth.

Author Contributions: Writing—original draft preparation, H.-H.M. and Y.-G.S.; writing—review and editing, H.-H.M. and Y.-G.S. All authors have read and agreed to the published version of the manuscript.

Funding: This study was supported by the Special Fund for Scientific Research of Shanghai Landscaping & City Appearance Administrative Bureau (G242414, G242416); the "Yunnan Revitalization Talent Support Program" in Yunnan Province (XDYC-QNRC-2022-0028); and the CAS "Light of West China" Program.

Conflicts of Interest: The authors declare no conflict of interest.

References

1. Wiens, J.J. Climate-related local extinctions are already widespread among plant and animal species. *PLoS Biol.* **2016**, *14*, e2001104. [CrossRef] [PubMed]
2. Isbel, F.; Gonzalez, A.; Loreau, M.; Cowles, J.; Diaz, S.; Hector, A.; Mace, M.M.; Wardle, D.A.; O'Connor, M.I.; Duffy, J.M.; et al. Linking the influence and dependence of people on biodiversity across scales. *Nature* **2017**, *546*, 65–72. [CrossRef] [PubMed]
3. Johnson, C.N.; Balmford, A.; Brook, B.W.; Buettel, J.C.; Galetti, M.; Lei, G.; Wilmshurst, J.M. Biodiversity losses and conservation responses in the Anthropocene. *Science* **2017**, *356*, 270–275. [CrossRef]
4. Zhu, H.; Tan, Y. Flora and Vegetation of Yunnan, Southwestern China: Diversity, Origin and Evolution. *Diversity* **2022**, *14*, 340. [CrossRef]
5. Su, T.; Spicer, R.A.; Zhou, Z.K. Tracing the Evolution of Plant Diversity in Southwestern China. *Diversity* **2022**, *14*, 434. [CrossRef]
6. Hu, L.; Le, X.G.; Zhou, S.S.; Zhang, C.Y.; Tan, Y.H.; Ren, Q.; Meng, H.H.; Cun, Y.; Li, J. Conservation Significance of the Rare and Endangered Tree Species, *Trigonobalanus doichangensis* (Fagaceae). *Diversity* **2022**, *14*, 666. [CrossRef]
7. Meng, H.H.; Zhang, C.Y.; Low, S.L.; Li, L.; Shen, J.Y.; Nurainas; Zhang, Y.; Huang, P.H.; Zhou, S.S.; Tan, Y.H.; et al. Two new species from Sulawesi and Borneo facilitate phylogeny and taxonomic revision of *Engelhardia* (Juglandaceae). *Plant Divers.* **2022**, *44*, 552–564. [CrossRef]
8. Darwin, C. *On The Origin of Species by Means of Natural Selection or the Preservation of Favoured Races in the Struggle for Life*; John Murray: London, UK, 1859.
9. Liu, G.Q.; Lian, L.; Wang, W. The Molecular Phylogeny of Land Plants: Progress and Future Prospects. *Diversity* **2022**, *14*, 782. [CrossRef]
10. Song, Y.G.; Li, Y.; Meng, H.H.; Fragnière, Y.; Ge, B.J.; Sakio, H.; Yousefzadeh, H.; Bétrisey, S.; Kozlowski, G. Phylogeny, taxonomy, and biogeography of *Pterocarya* (Juglandaceae). *Plants* **2020**, *9*, 1524. [CrossRef]
11. Zhang, C.Y.; Low, S.L.; Song, Y.G.; Nurainas; Kozlowski, G.; Li, L.; Zhou, S.S.; Tan, Y.H.; Cao, G.L.; Zhou, Z.; et al. Shining a light on species delimitation in the tree genus *Engelhardia* Leschenault ex Blume (Juglandaceae). *Mol. Phylogenet. Evol.* **2020**, *152*, 106918. [CrossRef]
12. Gao, X.Y.; Meng, H.H.; Zhang, M.L. Diversification and vicariance of desert plants: Evidence inferred from chloroplast DNA sequence variation of *Lagochilus ilicifolius* (Lamiaceae). *Biochem. Syst. Ecol.* **2014**, *55*, 93–100. [CrossRef]
13. Jiang, X.L.; Gardner, E.M.; Meng, H.H.; Deng, M.; Xu, G.B. Land bridges in the Pleistocene contributed to flora assembly on the continental islands of South China: Insights from the evolutionary history of *Quercus championii*. *Mol. Phylogenet. Evol.* **2019**, *132*, 36–45. [CrossRef] [PubMed]
14. Meng, H.; Su, T.; Gao, X.; Li, J.; Jiang, X.; Sun, H.; Zhou, Z. Warm-cold colonization: Response of oaks to uplift of the Himalaya–Hengduan Mountains. *Mol. Ecol.* **2017**, *26*, 3276–3294. [CrossRef] [PubMed]
15. Meng, H.H.; Zhang, C.Y.; Song, Y.G.; Yu, X.Q.; Cao, G.L.; Li, L.; Cai, C.N.; Xiao, J.H.; Zhou, S.S.; Tan, Y.H.; et al. Opening a door to the spatiotemporal history of plants from tropical Indochina Peninsula to subtropical China. *Mol. Phylogenet. Evol.* **2022**, *171*, 107458. [CrossRef] [PubMed]
16. Wang, T.R.; Meng, H.H.; Wang, N.; Zheng, S.S.; Jiang, Y.; Lin, D.Q.; Song, Y.G.; Kozlowski, G. Adaptive divergence and genetic vulnerability of relict species under climate change: A case study of *Pterocarya macroptera*. *Ann. Bot.* **2023**, *132*, 241–254. [CrossRef] [PubMed]
17. Meng, H.H.; Gao, X.Y.; Song, Y.G.; Cao, G.L.; Li, J. Biodiversity arks in the Anthropocene. *Reg. Sustain.* **2021**, *2*, 109–115. [CrossRef]
18. Alexander, J.M.; Chalmandrier, L.; Lenoir, J.; Burgess, T.I.; Essl, F.; Haider, S.; Kueffer, C.; McDougall, K.; Milbau, A.; Nunez, M.A.; et al. Lags in the response of mountain plant communities to climate change. *Glob. Change Biol.* **2018**, *24*, 563–579. [CrossRef] [PubMed]
19. Meng, H.H.; Zhou, S.S.; Jiang, X.L.; Gugger, P.F.; Li, L.; Tan, Y.H.; Li, J. Are mountaintops climate refugia for plants under global warming? A lesson from highemountain oaks in tropical rainforest. *Alpine Bot.* **2019**, *129*, 175–183. [CrossRef]

20. Keppel, G.; Van Niel, K.P.; Wardell-Johnson, G.W.; Yates, C.J.; Byrne, M.; Mucina, L.; Schut, A.G.T.; Hopper, S.D.; Franklin, S.E. Refugia: Identifying and understanding safe havens for biodiversity under climate change. *Glob. Ecol. Biogeogr.* **2012**, *21*, 393–404. [CrossRef]
21. Jin, D.; Yuan, Q.; Dai, X.; Kozlowski, G.; Song, Y. Enhanced precipitation has driven the evolution of subtropical evergreen broad-leaved forests in eastern China since the early Miocene: Evidence from ring-cupped oaks. *J. Syst. Evol.* **2023**. *early view*. [CrossRef]
22. Gao, X.Y.; Zhang, X.; Meng, H.H.; Li, J.; Zhang, D.; Liu, C.N. Comparative chloroplast genomes of *Paris* Sect. *Marmorata*: Insight into repeat regions and evolutionary implications. *BMC Genom.* **2018**, *19*, 878.
23. Gao, X.Y.; Su, Q.X.; Yao, B.; Yang, W.; Ma, W.; Yang, B.; Liu, C.N. Development of EST-SSR Markers Related to Polyphyllin Biosynthesis Reveals Genetic Diversity and Population Structure in *Paris polyphylla*. *Diversity* **2022**, *14*, 589. [CrossRef]
24. Fu, X.; Wu, J.; Ma, X.; Li, K.; Zhang, H.; Wu, S.; Sun, K. The Chromosome-Level Genome of *Elaeagnus moorcroftii* Wall., an Economically and Ecologically Important Tree Species in Drylands. *Diversity* **2022**, *14*, 468. [CrossRef]
25. Su, Q.; Zhang, X.; Li, J.; Yang, W.; Ren, Q.; Gao, X.Y.; Liu, C.N. PPDP: A Data Portal of Paris polyphylla for Polyphyllin Biosynthesis and Germplasm Resource Exploration. *Diversity* **2022**, *14*, 1057. [CrossRef]
26. Meng, H.H.; Zhou, S.S.; Li, L.; Tan, Y.H.; Li, J.W.; Li, J. Conflict between biodiversity conservation and economic growth: Insight into rare plants in tropical China. *Biodivers. Conserv.* **2019**, *28*, 523–537. [CrossRef]
27. Zhang, Y.; Duan, M.G.; Huang, P.H.; Li, M.; Meng, H.H.; Qiao, H.J.; Li, J. Blinded by the bright—Afforestation is affecting widespread sampling deficiency in plant collections. *For. Ecol. Manag.* **2023**, *530*, 120765. [CrossRef]
28. Zhang, Y.; Song, Y.G.; Zhang, C.Y.; Wang, T.R.; Su, T.H.; Huang, P.H.; Meng, H.H.; Li, J. Latitudinal diversity gradient in the changing world: Retrospectives and perspectives. *Diversity* **2022**, *14*, 334. [CrossRef]

diversity

MDPI

Commentary

Tracing the Evolution of Plant Diversity in Southwestern China

Tao Su [1,2,*], Robert A. Spicer [1,3] and Zhe-Kun Zhou [1,4]

1 CAS Key Laboratory of Tropical Forest Ecology, Xishuangbanna Tropical Botanical Garden, Chinese Academy of Sciences, Mengla 666303, China; r.a.spicer@open.ac.uk (R.A.S.); zhouzk@xtbg.ac.cn (Z.-K.Z.)
2 University of Chinese Academy of Sciences, Beijing 100049, China
3 School of Environment, Earth and Ecosystem Sciences, The Open University, Milton Keynes MK7 6AA, UK
4 Key Laboratory of Biodiversity and Biogeography, Kunming Institute of Botany, Chinese Academy of Sciences, Kunming 650204, China
* Correspondence: sutao@xtbg.org.cn

Biodiversity hotspots are key regions for understanding the evolutionary history of biodiversity as well as the processes initiating and maintaining it [1,2]. As a biodiversity hotspot, southwestern China is known worldwide for its modern taxonomic richness with numerous endemic and relic plants [3–5]. However, it is still unclear how this high regional plant diversity evolved, largely due to a lack of well-dated fossil evidence. Southwestern China is rich in Cenozoic sedimentary basins, preserving numerous fossil floras [6], and recent paleobotanical progress on these floras sheds new light on this topic.

One of the most important questions on the evolution of plant diversification concerns when modern plant diversity first appeared and, specifically, the pinpointing of the earliest occurrence of a flora that is floristically comparable to present vegetation in the same region. Until recently, it was thought that many taxa in southwestern China originated during the Miocene, evidenced by molecular analyses using fossil records for calibration [7]. Previously, most fossil floras in southwestern China were dated using stratigraphic or floristic comparison [8], which were limited by their low resolution and reasoning circularity. Recently, radiometric dating of volcanic ash materials from the plant-fossil-bearing deposits have substantially revised the ages of some important Cenozoic plant fossil-bearing basins in southwestern China, e.g., Jianchuan Basin [9], Lühe Basin [10], Mangkang Basin [11], and Wenshan Basin [12]. The Lühe flora from central Yunnan Province is dominated by Fagaceae and Betulaceae, and has long been considered to be late Miocene in age. However, recent U-Pb and $^{40}Ar/^{39}Ar$ radiometric dating of tuff layers in the fossil-bearing outcrops indicates an age of ~32 Ma for the flora [10]. Even though some taxa in this flora, such as *Sequoia* and *Metasequoia*, have disappeared from southwestern China, most taxa still survive there, e.g., *Betula*, *Cryptomeria*, *Dipteronia*, *Fraxinus*, *Quercus*, and *Tsuga* [10,13,14]. Besides, the Kajun flora in Mangkang Basin, eastern Xizang and Wenshan flora in southeastern Yunnan also show high floristic similarity to modern vegetation in southwestern China [11,12]. The ages of both these floras were previously thought to be late Miocene [8,15], but radiometric dating has now confirmed these two floras to be 34.6–33.4 Ma and 33 Ma, respectively [11,12]. It seems that regional tectonics near the end of the Eocene created many of the fossil-bearing basins in southwestern China.

The floras mentioned above are floristically different from older floras in southwestern China. For example, the middle Eocene (~47 Ma by U-Pb dating) Jianglang flora in the central Qinghai-Tibetan Plateau bears around 70 morphotypes and some important components in modern vegetation of southwestern China, such as Fagaceae and Betulaceae, are absent from the current collection of more than 2000 fossil specimens. Several extinct taxa, such as *Cedrelospermum*, *Lagokarpos*, and *Limnobiophyllum*, occurred in the flora [16]. The middle Eocene Relu flora in western Sichuan Province is characterized by *Comptonia*, *Hemiptelea*, Myrtaceae, and *Palibinia* [17], which are quite different from those in the modern

Citation: Su, T.; Spicer, R.A.; Zhou, Z.-K. Tracing the Evolution of Plant Diversity in Southwestern China. *Diversity* **2022**, *14*, 434. https://doi.org/10.3390/d14060434

Academic Editor: Mario A. Pagnotta

Received: 12 April 2022
Accepted: 28 May 2022
Published: 29 May 2022

regional flora. Therefore, the modernization of plant diversity in southwestern China should have been taking place in the late Eocene to early Oligocene.

Both climate and topography may be major factors stimulating the modernization of the flora. Current evidence suggests that there was a prolonged regional dry climate during the early and middle Eocene, and the transition from an arid to a humid climate occurred before the Eocene–Oligocene transition (between 45 and 42 Ma), associated with the rise of eastern Tibet to near present elevations and the reorganization of the monsoon system [18]. Currently, the prevailing monsoon climate of southwestern China is characterized by wet summers and dry winters [19], but the humid climate during the late Eocene did not necessarily represent the establishment of modern monsoonal climate. Modelling [20] and leaf-form analysis [11,16] suggest wet winters and dry summers in some parts of the region during the Eocene, and we know that winters and early springs have become drier since the Miocene, with inevitable impacts on plant diversity. Some genera such as *Cedrus*, *Metasequoia*, and *Sequoia* cannot survive under the modern monsoonal climate due to the poor drought resistance of young shoots [21–23], whereas many other genera, e.g., *Betula*, *Elaeagnus*, *Quercus*, *Rosa*, and *Ulmus* have thrived and diversified since then. Diversification within the region is complex and linked to interplay between orography and climate [24].

The topography is another important factor in shaping the modernization of plant diversity. The landscape of southwestern China was mainly established before the Neogene. Phytopaleoaltimetric reconstruction of the Mangkang basin suggests an uplift of ~900 m during the latest Eocene (~34.6 Ma) to the earliest Oligocene (~33.4 Ma) when it achieved its present elevation of 3900 m [10]. Another work in Gongjue Basin with carbonate clumped isotope thermometry suggested an even earlier elevational change from 700 m in the early Eocene to 3800 m in the middle Eocene [18]. All these rises were closely associated with the clockwise rotation and extrusion of Indochina in southeastern Qinghai-Tibetan Plateau initiated as early as the late Paleogene [25].

Generally, both tectonism and the development/fluctuations in monsoonal climate have played important roles in shaping modern plant diversity in southwestern China, but much is still unclear about how the modernization of plant diversity in southwestern China took place, as well as its underlying driving mechanisms. In the future, paleobotanical studies in Paleogene sediments within a high resolution absolute chronostratigraphic framework are needed to establish the sequence and spatial patterns of floristic change. By combining paleoclimate signals derived from fossil evidence, computed tomography (CT) analysis of undisturbed sedimentary core samples, and geochemical analyses and modelling, as well as past landscape reconstructions, the evolutionary history of modern plant diversity can be better understood in this globally important biodiversity center and provide critical processes for bringing together fossil and molecular phylogenetic data.

Author Contributions: All authors discussed and wrote the manuscript.

Funding: This work is supported by National Natural Science Foundation of China (nos. 42072024, 41922010).

Institutional Review Board Statement: No institutional review board statement.

Data Availability Statement: No data statement.

Conflicts of Interest: The authors have no conflict of interest to declare.

References

1. Willis, K.J.; Gillson, L.; Knapp, S. Biodiversity hotspots through time: An introduction. *Philos. Trans. R. Soc. Lond. B Biol. Sci.* **2007**, *362*, 169–174. [CrossRef] [PubMed]
2. Bennett, K.D.; Provan, J. What do we mean by 'refugia'? *Quat. Sci. Rev.* **2008**, *27*, 2449–2455. [CrossRef]
3. Wu, Z.-Y.; Zhu, Y.-C. *Vegetation of Yunnan*; Science Press: Beijing, China, 1987. (In Chinese)
4. Gao, Z.-W.; Sun, H. *Species List of Yunnan (2016)*; Yunnan Science and Techonology Press: Kunming, China, 2017. (In Chinese)
5. Myers, N.; Mittermeier, R.; Mittermeier, C.; da Fonseca, G.A.B.; Kent, J. Biodiversity hotspots for conservation priorities. *Nature* **2000**, *403*, 853–858. [CrossRef] [PubMed]

6. Ge, H.-R.; Li, D.-Y. *Cenozoic Coal-Bearing Basins and Coal-Forming Regularity in West Yunnan*; Yunnan Science and Technology Press: Kunming, China, 1999. (In Chinese)
7. Lu, L.-M.; Mao, L.-F.; Yang, T.; Ye, J.-F.; Liu, B.; Li, H.-L.; Sun, M.; Miller, J.T.; Mathews, S.; Hu, H.-H.; et al. Evolutionary history of the angiosperm flora of China. *Nature* **2018**, *554*, 234–238. [CrossRef]
8. Bureau of Geology and Mineral Resources of Yunnan Province. *The Lithofacies Paleogeography Atlas of Yunnan*; Yunnan Science and Techology Publishing Press: Kunming, China, 1995. (In Chinese)
9. Gourbet, L.; Leloup, P.H.; Paquette, J.-L.; Sorrel, P.; Maheo, G.; Wang, G.; Xu, Y.; Cao, K.; Antoine, P.-O.; Eymard, I.; et al. Reappraisal of the Jianchuan Cenozoic basin stratigraphy and its implications on the SE Tibetan Plateau evolution. *Tectonophysics* **2017**, *700*, 162–179. [CrossRef]
10. Linnemann, U.; Su, T.; Kunzmann, L.; Spicer, R.A.; Ding, W.-N.; Spicer, T.E.V.; Zieger, J.; Hofmann, M.; Moraweck, K.; Gärtner, A.; et al. New U-Pb dates show a Paleogene origin for the modern Asian biodiversity hot spots. *Geology* **2018**, *46*, 3–6. [CrossRef]
11. Su, T.; Spicer, R.A.; Li, S.-H.; Xu, H.; Huang, J.; Sherlock, S.; Huang, Y.-J.; Li, S.-F.; Wang, L.; Jia, L.-B.; et al. Uplift, climate and biotic changes at the Eocene-Oligocene transition in south-eastern Tibet. *Natl. Sci. Rev.* **2019**, *6*, 495–504. [CrossRef]
12. Tian, Y.; Spicer, R.A.; Huang, J.; Zhou, Z.; Su, T.; Widdowson, M.; Jia, L.; Li, S.; Wu, W.; Xue, L.; et al. New early Oligocene zircon U-Pb dates for the 'Miocene' Wenshan Basin, Yunnan, China: Biodiversity and paleoenvironment. *Earth Planet. Sci. Lett.* **2021**, *565*, 116929. [CrossRef]
13. Ding, W.-N.; Huang, J.; Su, T.; Xing, Y.-W.; Zhou, Z.-K. An early Oligocene occurrence of the palaeoendemic genus *Dipteronia* (Sapindaceae) from Southwest China. *Rev. Palaeobot. Palynol.* **2018**, *249*, 16–23. [CrossRef]
14. Ding, W.-N.; Kunzmann, L.; Su, T.; Huang, J.; Zhou, Z.-K. A new fossil species of *Cryptomeria* (Cupressaceae) from the Rupelian of the Lühe Basin, Yunnan, East Asia: Implications for palaeobiogeography and palaeoecology. *Rev. Palaeobot. Palynol.* **2018**, *248*, 41–51. [CrossRef]
15. Tao, J.-R. *The Evolution of the Late Cretaceous-Cenozoic Flora in China*; Science Press: Beijing. China, 2000.
16. Su, T.; Spicer, R.A.; Wu, F.-X.; Farnsworth, A.; Huang, J.; Del Rio, C.; Deng, T.; Ding, L.; Deng, W.-Y.-D.; Huang, Y.-J.; et al. A Middle Eocene lowland humid subtropical "Shangri-La" ecosystem in central Tibet. *Proc. Natl. Acad. Sci. USA* **2020**, *117*, 32989–32995. [CrossRef]
17. Guo, S.-X. An Eocene flora from the Relu Formation in Litang County and the history of Eucalyptus. In *Team of Comprehensive Scientific Expedition to the Qinghai-Xizang Plateau*; Chinese Academy of Sciences, Ed.; Beijing Publishing House of Sciences and Technology: Beijing, China, 1986; (In Chinese with English abstract).
18. Xiong, Z.; Ding, L.; Spicer, R.A.; Farnsworth, A.; Wang, X.; Valdes, P.J.; Su, T.; Zhang, Q.; Zhang, L.; Cai, F.; et al. The early Eocene rise of the Gonjo Basin, SE Tibet: From low desert to high forest. *Earth Planet. Sci. Lett.* **2020**, *543*, 116312. [CrossRef]
19. Wang, B.; Trenberth, K.; Hurrell, J.; Stepaniak, D. *The Asian Monsoon*; Springer: Berlin/Heidelberg, Germany, 2006.
20. Farnsworth, A.; Lunt, D.J.; Robinson, S.A.; Valdes, P.J.; Roberts, W.H.G.; Clift, P.D.; Markwick, P.; Su, T.; Wrobel, N.; Bragg, F.; et al. Past East Asian monsoon evolution controlled by paleogeography, not CO_2. *Sci. Adv.* **2019**, *5*, eaax1697. [CrossRef] [PubMed]
21. Su, T.; Liu, Y.-S.; Jacques, F.M.B.; Huang, Y.-J.; Xing, Y.-W.; Zhou, Z.-K. The intensification of the East Asian winter monsoon contributed to the disappearance of *Cedrus* (Pinaceae) in southwestern China. *Quat. Res.* **2013**, *80*, 316–325. [CrossRef]
22. Zhang, J.-W.; D'Rozario, A.; Adams, J.M.; Li, Y.; Liang, X.-Q.; Jacques, F.M.; Su, T.; Zhou, Z.-K. *Sequoia maguanensis*, a new Miocene relative of the coast redwood, *Sequoia sempervirens*, from China: Implications for paleogeography and paleoclimate. *Am. J. Bot.* **2015**, *102*, 103–118. [CrossRef] [PubMed]
23. Wang, L.; Kunzmann, L.; Su, T.; Xing, Y.-W.; Zhang, S.-T.; Wang, Y.-Q.; Zhou, Z.-K. The disappearance of *Metasequoia* (Cupressaceae) after the middle Miocene in Yunnan, Southwest China: Evidences for evolutionary stasis and intensification of the Asian monsoon. *Rev. Palaeobot. Palynol.* **2019**, *264*, 64–74. [CrossRef]
24. Ding, W.-N.; Ree, R.H.; Spicer, R.A.; Xing, Y.-W. Ancient orogenic and monsoon-driven assembly of the world's richest temperate alpine flora. *Science* **2020**, *369*, 579–581. [CrossRef] [PubMed]
25. Li, S.; Su, T.; Spicer, R.A.; Xu, C.; Sherlock, S.; Halton, A.; Hoke, G.; Tian, Y.; Zhang, S.; Zhou, Z.; et al. Oligocene deformation of the Chuandian Terrane in the SE margin of the Tibetan Plateau related to the extrusion of Indochina. *Tectonics* **2020**, *39*, e2019TC005974. [CrossRef]

diversity

MDPI

Review

The Molecular Phylogeny of Land Plants: Progress and Future Prospects

Guo-Qing Liu [1,2], Lian Lian [1] and Wei Wang [1,2,*]

1 State Key Laboratory of Systematic and Evolutionary Botany, Institute of Botany, Chinese Academy of Sciences, Beijing 100093, China
2 University of Chinese Academy of Sciences, Beijing 100049, China
* Correspondence: wangwei1127@ibcas.ac.cn

Abstract: Phylogenetics has become a powerful tool in many areas of biology. Land plants are the most important primary producers of terrestrial ecosystems and have colonized various habitats on Earth. In the past two decades, tremendous progress has been made in our understanding of phylogenetic relationships at all taxonomic levels across all land plant groups by employing DNA sequence data. Here, we review the progress made in large-scale phylogenetic reconstructions of land plants and assess the current situation of phylogenetic studies of land plants. We then emphasize directions for future study. At present, the phylogenetic framework of land plants at the order and familial levels has been well built. Problematic deep-level relationships within land plants have also been well resolved by phylogenomic analyses. We pointed out five major aspects of molecular phylogenetics of land plants, which are nowadays being studied and will continue to be goals moving forward. These five aspects include: (1) constructing the genus- and species-level phylogenies for land plant groups, (2) updating the classification systems by combining morphological and molecular data, (3) integrating fossil taxa into phylogenies derived from living taxa, (4) resolving deep-level and/or rapidly divergent phylogenetic relationships using phylogenomic data, and (5) building big trees using the supermatrix method. We hope that this review paper will promote the development of plant molecular phylogenetics and other related areas.

Keywords: land plants; molecular systematics; phylogenomics; phylogeny; morphology; tree of life

Citation: Liu, G.-Q.; Lian, L.; Wang, W. The Molecular Phylogeny of Land Plants: Progress and Future Prospects. *Diversity* **2022**, *14*, 782. https://doi.org/10.3390/d14100782

Academic Editors: Hong-Hu Meng and Yi-Gang Song

Received: 29 August 2022
Accepted: 18 September 2022
Published: 21 September 2022

1. Introduction

In his famous book, *The Origin of Species*, Charles Darwin [1] put forward the concept of the Tree of Life (TOL), which is a metaphor for presenting relationships of organisms in space and time. All organisms on Earth originated from a common ancestor and each can be found in the TOL. As a large lineage of TOL, land plants (embryophytes) are the most important primary producers of terrestrial ecosystems. Living land plants are an important source of aliments, timbers, fibers, pharmaceuticals, and other vital resources for human survival and health, and fossilized land plants become one of the sources of fossil fuels, particularly coal, with driving global economy [2]. By an array of innovations, including embryos, sperms, and eggs protected in multicellular structures, alternating generations of diploid sporophytes, and haploid gametophytes, land plants have become the most diverse group of green plants, have dominated modern terrestrial environments, and are the foundation of the vast majority of terrestrial ecosystems [3].

The rise of land plants is one of the major events in the history of life, which irreversibly changed the environments on Earth, including altering atmospheric composition by enhancing photosynthesis and influencing carbon fixation and carbon storage, and affecting evolutionary trajectories of other organisms by promoting the formation of soil and soil microbiota and the establishment of new food chains and new habitats [4,5]. The origin and evolution of land plants is a central theme in evolutionary biology and ecology [3,6]. In the past two decades, tremendous progress has been made in reconstruction of land plant TOL by employing molecular data [4,7–9]. The order- and family-level phylogenetic framework of land plants has been well built [10–12]. In particular, problematic deep-level relationships within land plants have been largely resolved by phylogenomic analyses [4,8,9].

Here, we review the progress made in large-scale phylogenetic reconstructions of land plants and assess the current situation of phylogenetic studies of land plants. We then point out five major aspects of molecular phylogenetics of land plants, which are nowadays being studied and will continue to be goals moving forward. We hope that this review paper will promote the development of plant molecular phylogenetics and other related areas.

2. Large-Scale Phylogenetic Framework of Land Plants

With recent developments in extracting DNA, sequencing, methods of analytical technologies, and computing power, tremendous progress has been made in our understanding of phylogenetic relationships at all taxonomic levels across all land plant groups by employing DNA sequence data. At present, the large-scale phylogenetic framework of land plants has been established and the most likely sister group of land plants has also been identified. Land plants comprise five major clades: bryophytes, lycophytes, monilophytes, gymnosperms, and angiosperms (Figure 1).

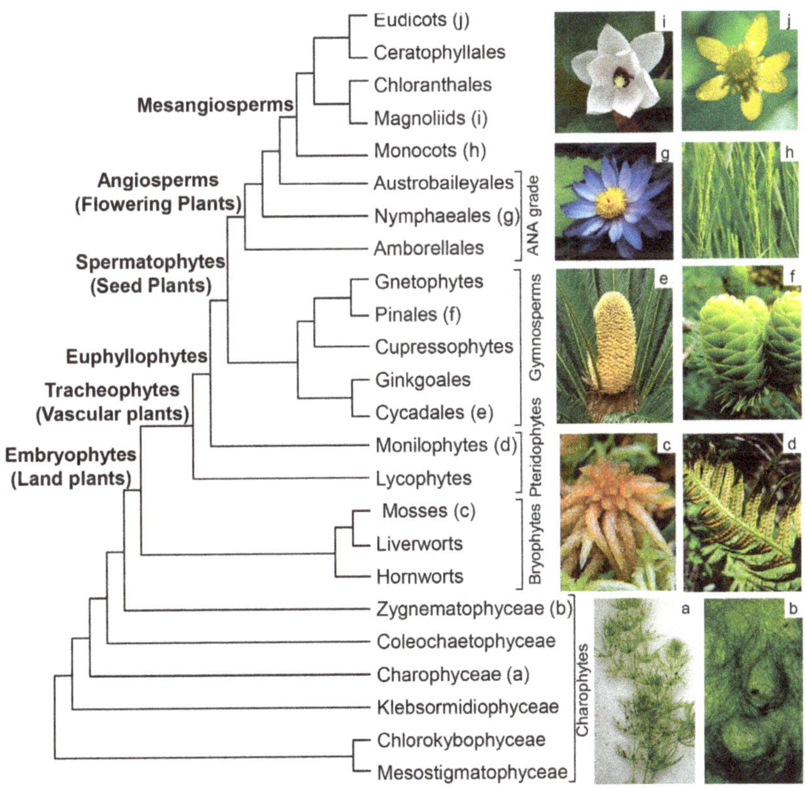

Figure 1. Summary of phylogenetic relationships among major clades of land plants. (**a**) *Chara braunii*; (**b**) *Spirogyra communis*; (**c**) *Sphagnum palustre*; (**d**) *Goniophlebium chinense*; (**e**) *Cycas revoluta*; (**f**) *Larix kaempferi*; (**g**) *Nymphaea gigantea*; (**h**) *Oryza sativa*; (**i**) *Oyama sieboldii*; (**j**) *Ranunculus sieboldii*.

2.1. *Sister Group of Land Plants*

Land plants originated from streptophyte algae in fresh water [13], but which green algal lineage gave rise to land plants has been a long-standing dispute. Phylogenetic studies indicate that streptophyte algae is a paraphyletic group, referred to as charophytes [14], and contains six major clades: Chlorokybophyceae, Mesostigmatophyceae, Klebsormidiophyceae, Charophyceae, Coleochaetophyceae, and Zygnematophyceae (Figure 1). Many clades of streptophytes were once suggested as sisters to land plants, such as Charophyceae [14], Zygnematophyceae [15–17], Coleochaetophyceae [18], and the clade consisting of Zygnematophyceae and Coleochaetophyceae [16,19]. Recently, phylogenomic analyses strongly support Zygnematophyceae as sister to land plants [7,8].

2.2. *Bryophytes*

The simplest morphological structures among land plants, bryophytes have still evolved a range of characters to adapt terrestrial habitats, such as stomata, cuticle, water-conducting cells, and embryos [20–23]. Due to a lack of seeds and a vascular system, life cycles of bryophytes are highly dependent on water [24]. Moreover, the haploid gametophyte phase of bryophytes is dominant in life history, which is distinguished from the remaining land plants. Bryophytes consist of three major monophyletic groups: hornworts, liverworts, and mosses. The relationships among these three groups and the relationships between bryophytes and vascular plants have been energetically debated, and various hypotheses have been proposed. Based on the distribution of three group II introns in the mitochondrial genes *nad1* and *cox2*, Qiu et al. [25] suggested bryophytes were paraphyletic, i.e., liverworts were sister to all other land plants and hornworts were sister to vascular plants, in agreement with the results of the multi-locus phylogenetic analyses [26,27]. Mitochondrial phylogenomic analyses by mitigating the effects of saturation, compositional heterogeneity, and codon-usage bias also supported these relationships [28], whereas plastid phylogenomic analyses by accounting for composition biases among synonymous substitutions supported bryophytes to be monophyletic [29]. Maximum likelihood analysis based on concatenated alignments of first and second codon positions for 674 nuclear genes also suggested bryophytes as a paraphyletic group, but hornworts were identified as the earliest-diverging lineages within land plants and liverworts and mosses formed a clade, sister to vascular plants, whereas the coalescent-based analysis based on 424 gene trees estimated from first and second codon position strongly recognized bryophytes monophyletic [3]. Recently, nuclear genomic phylogenetic analyses further supported bryophytes consist of a monophyletic group and hornworts are sister to liverworts and mosses [4,30,31]. These analyses highlight the negative effects of substitutional saturation and synonymous substitutions on phylogenetic references based on nuclear genomic data.

2.3. *Pteridophytes*

Pteridophytes consist of lycophytes and monilophytes (ferns). Traditionally, pteridophytes were considered to be monophyletic due to lycophytes and monilophytes having a similar life cycle. Phylogenetic analyses based on the 18S rDNA sequences first found that lycophytes were the earliest-diverging lineage within vascular plants and monilophytes were the sister of seed plants [32]. Subsequent multi-locus and phylogenomic analyses consistently supported these hypotheses [3,8,33,34].

2.4. *Gymnosperms*

Gymnosperms and angiosperms are collectively referred to as seed plants because their reproductive organ is seeds, instead of spores. The monophyly of gymnosperms has been unambiguously authenticated by molecular sequence data from nuclear and plastid genomes [35,36]. Gymnosperms consist of four groups: cycads, conifers, *Ginkgo*, and Gnetales. A phylotranscriptomic analysis with a sampling of all 13 families of gymnosperms and main lineages of angiosperms indicates that cycads plus *Ginkgo* as sister to the remaining gymnosperms and Gnetales is embedded within conifers, sister to Pinaceae [9].

Nuclear genomic data also support cycads plus *Ginkgo* as sister to the remaining gymnosperms [37,38]. Convergent molecular evolution or homoplasy is partially responsible for the phylogenetic conflicts in seed plants [9,38].

2.5. *Angiosperms*

As a plant group that is the most closely related to human production and life, angiosperms have spectacular morphological and species diversity, and play an irreplaceable role in global terrestrial ecosystems. Angiosperms are characterized by flowers, ovules covered by carpels, double fertilization, nutritious triploid endosperm, and vessel elements. Monophyly of angiosperms is supported by various phylogenetic analyses. Amborellales, Nymphaeales, and Austrobaileyales are successive sisters to all other angiosperms, referred to as the ANA grade [39–43]. Mesangiospermae is a monophyletic group and comprises five major lineages: monocots, magnoliids, Chloranthales, Ceratophyllales, and eudicots. The relationships among these five lineages have long been an open question because they diverged rapidly within a 3-million-year time window [44]. One Thousand Plant Transcriptomes support monocots as sister to the remaining mesangiosperms [8], and eudicots plus Ceratophyllales is sister to magnoliids plus Chloranthales (Figure 1), in agreement with the results of the nuclear genomic data [45,46]. Guo et al. [46] further found that ancient hybridization may account for the incongruent phylogenetic placements of Chloranthales + magnoliids relative to monocots and eudicots in nuclear and chloroplast trees.

3. The Potential Research Focuses on Reconstructing the TOL of Land Plants in the Long Period Future

In 1993, the article entitled "*Phylogenetics of seed plants: An analysis of nucleotide sequences from the plastid gene rbcL*" is a landmark work in plant molecular systematics [47]. Since then, various molecular data have been widely used to reconstruct the phylogenetic relationships of land plant groups. Development of plant molecular phylogenetics is very rapid, with a process using single locus, multi-locus, to genomic data. It is safe to say that due to the emergence of molecular phylogenetics, the achievements of plant phylogenetics made in the past twenty years have far exceeded the sum of the previous two hundred years. So, Soltis et al. [48] allege that the current period is a "golden era" in plant phylogenetics, as well as organismal phylogenetics in general. Molecular systematics is not only a subdiscipline, but has also become a powerful tool in many areas of biology, such as physiology, ecology, biogeography, paleobiology, genomics, and developmental genetics. Here, we summarized five important aspects of molecular phylogenetics of land plants, which are nowadays being studied and will continue to be goals moving forward.

3.1. *Reconstructing Genus- and Species-Level Phylogenies for Land Plant Groups*

With the establishment of the large-scale phylogenetic framework of angiosperms, a DNA phylogeny-based angiosperm classification system at the order and familial levels was proposed by the Angiosperm Phylogeny Group (APG) in 1998 and has been updated three times [11,49–51]. Other land plants, such as gymnosperms [12], ferns [52], and mosses [53] also have a relatively robust phylogenetic framework. The establishment of the order and familial framework is only the first step in reconstruction the land plant TOL of land plants. Altogether, 500,000 species of green plants occur on Earth, and only less than 30% of the species have reported molecular sequences so far [54,55]. In particular, taxon sampling of large-scale phylogenetic studies is sparse. For example, large-scale phylogenetic analyses in angiosperms or eudicots only included one to three species for each family; thus, the circumscriptions of many families cannot be resolved due to the relatively limited taxon sampling. Recently, some heterogeneous families have redelimited and several new families have been established, such as Arthropteridaceae (ferns) [56], Pteridryaceae (ferns) [57], Borthwickiaceae (angiosperms) [58], and Wightiaceae (angiosperms) [59]. To clarify familial circumscription, a relatively dense taxon sampling at the generic level, especially the inclusion of segregate genera, is necessary.

For delimiting a generic circumscription, species-level taxon sampling is essential. Many traditionally recognized genera have been re-delimited by employing molecular data. For example, Ranunculaceae, a basal eudicot family, contains approximately 60 genera of which 23 were re-delimited or adjusted [60]. Kadereit et al. [61] evaluated the monophyly of all genera of vascular plants in Germany, and identified that *c.* 140 genera are not monophyletic among the 840 genera examined, and the monophyly of *c.* 20 genera is ambiguous. We conducted a statistical survey of new genera of extant land plants published from 2015 to 2021 on the ISI Web of Science (www.webofscience.com/wos/alldb/basic-search, accessed on 12 June 2022) and International Plant Name Index (IPNI, www.ipni.org/, accessed on 10 June 2022) and found that more than 70 genera were described each year since 2015, most of which belongs into angiosperms, with the largest proportion of eudicots (Figure 2). At present, genus- and species-level phylogenetic studies of land plants are reported in almost every issue of molecular systematic journals, such as *Molecular Phylogenetics and Evolution*, *Taxon*, and *Systematic Botany*. Comprehensive understanding of genus- and species-level relationships across land plants is still a great challenge, especially for large and widespread genera. Phylogenetic research at the generic and species level will be the focus of reconstructing the land plant TOL for a long time in the future.

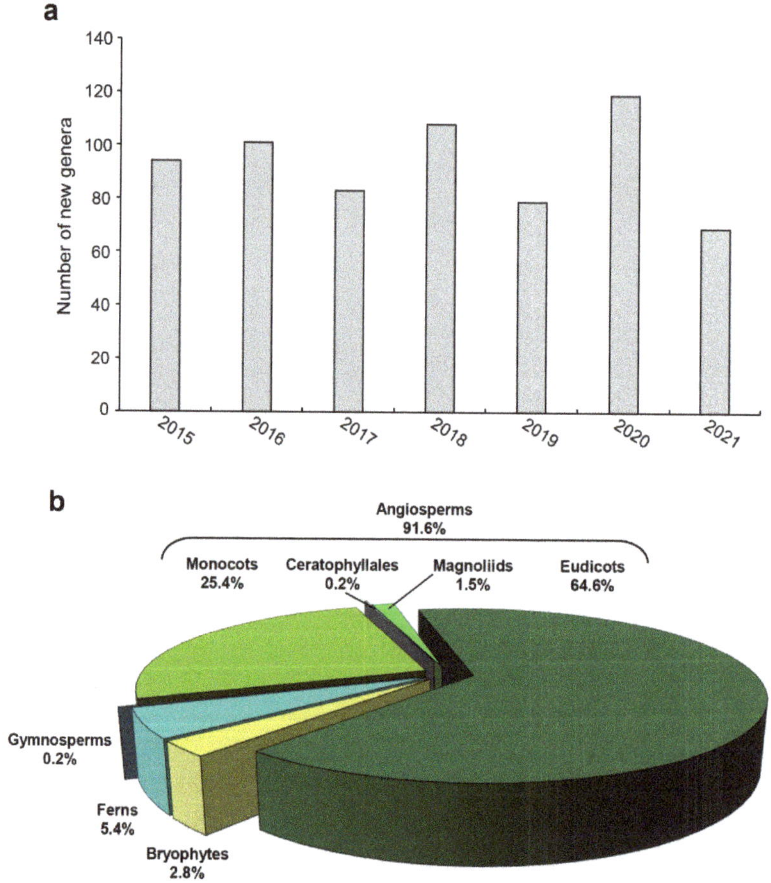

Figure 2. Cumulative number of new genera between 2015 and 2021. (**a**) The number of new genera published annually. (**b**) Proportion of new genera of major groups of land plants.

3.2. Updating the Classification Systems by Combining Morphological and Molecular Data

The APG system has been widely accepted, but it was established based only on molecular data. The orders and families in the APG system were circumscribed using only "monophyly" as criterion. From a morphological viewpoint, some of the currently recognized orders and families in the APG system are highly heterogeneous, such as Malpighiales, Cucurbitales, and Resedaceae [62]. How to determine morphological synapomorphies of those orders is a large challenge. Moreover, a key to distinguish the orders and families of angiosperms is lacking [62]. These prevent the wide implications of the APG system. In particular, the phylogenetic relationships and taxonomic status of many East Asian taxa, including Acanthochlamydaceae, Aceraceae, Bretschneideraceae, Cornaceae *sensu lato*, Hippocastanaceae, Illiciaceae, Leeaceae, Rhoipteleaceae, and Tetracentraceae, need to re-evaluated by integrating morphological and molecular data [62].

Systematics and evolutionary biology have developed for more than one century and have accumulated a wealth of morphological characters through gross morphology, anatomy, embryology, palynology, cytology, and ontogeny. Combining molecular and morphological data can greatly increase the ability to retrospect phylogenetic relationships of organisms and the process of evolution [63,64]. Yet, compared to molecular data, morphological data has been greatly ignored in reconstructing the TOL. At present, more than 95% of systematics articles use only DNA sequence data [65]. Bayesian inference, which can analyze morphological data or combined morphological and molecular data, has greatly improved the application of morphological characters [66,67]. By combing morphological and molecular data, we can determine the diagnostic characters for each lineage in the phylogenetic framework and thereby put forward a natural and reasonable classification system [68,69].

3.3. Integrating Fossil Taxa into Phylogenies Derived from Living Taxa

A complete TOL includes not only living taxa but also extinct taxa. However, more than 95% of current phylogenetic analyses include only extant taxa [65]. With current technology, it is almost impossible to obtain nucleotide sequences from extinct plant species. Therefore, morphological characters have become the only option for phylogenetic analyses of fossil plants. The widespread applications of scanning electron microscopy and synchrotron radiation X-ray tomographic microscopy in fossil taxa have enabled paleobotanists to obtain numerous accurate morphological data from fossils [70]. In contrast, these new techniques are rarely used in studies of extant plant species.

The inclusion of fossil taxa not only affects or improves topology, it is also important for a correct understanding of character evolution because fossils may have a combination of characters unlike those of any extant taxon [71]. Another role of fossil data in combination with molecular phylogenetics is to serve as calibration points based on the assumed position of fossil taxa in extant trees [72]. When multiple fossils from different periods are assigned to the same clade, the oldest fossil is usually selected as the calibration point. For a group with rich fossil records, the inclusion of fossil taxa can heavily influence its biogeographic reconstruction, particularly when the fossil taxa are outside its modern distribution [73]. If multiple fossils belong to a contemporaneous clade, directly placing them in the phylogenetic trees will result in a zero-length branch. The recently developed tip-dating (TD) Bayesian method can make use of all available fossil taxa as terminal tips to generate a timetree containing extinct and extant groups and to overcome zero-length branches to a certain extent [74]. However, this method has a wide application in zoological research. Thus, it has broad prospects in plant biogeographic studies, including living and extinct taxa.

3.4. *Resolve Deep-Level and/or Rapidly Divergent Phylogenetic Relationships using Phylogenomic Data*

When using a few DNA regions cannot resolve phylogenetic relationships well, more loci need to be sampled. Currently, phylogenomic analyses have been widely used in solving recalcitrant phylogenetic relationships, including relationships within rapidly radiating taxa and positions of relict taxa. The application of next-generation sequencing has greatly reduced the cost of sequencing. The plastid genomes in plants are usually a single, non-recombining locus [75] and have been widely applied in plant phylogenetic studies. Currently, most of plastid phylogenomic studies directly use protein-coding gene sequences, which may mislead phylogenetic references. Goremykin et al. [76] found that Amborellales and Nymphaeales formed a clade after removing "fast sites," which is contradicted with the prevalent view that only *Amborella* is the earliest-diverging lineage in angiosperms. Nonetheless, Drew et al. [77] revealed the "noisy" data actually supports *Amborella* as sister to all other angiosperms. To account for "noisy sites," some new nucleotide substitution models or analytic methods have also been proposed. For example, Goremykin et al. [76] proposed the CAT + GTR + Γ + covext model that considered base compositional heterogeneity. When using 82 plastid genes, Xi et al. [78] proposed posteriori data partitioning based on the Bayesian mixture mode, which resolved well the phylogenetic relationships in Malpighiales, the most recalcitrant clade in angiosperms.

Compared to plastid genome, biparental inheritance nuclear genome can not only provide more characters but can also reveal reticular evolution processes, so it has greater potential in phylogenetic studies and may be a key direction of plant phylogeny in the future. Especially, the developments of the restriction-site associated DNA sequencing, target enrichment, and genome skimming technique [79] have reduced sequencing costs and have greatly promoted nuclear phylogenomic studies of land plants, as well as other organisms. However, nuclear genomes have a more complicated evolutionary history, and may contain more evolutionary "noise," such as evolutionary saturation, base compositional heterogeneity, and synonymous codon bias. Incomplete lineage sorting is a common evolutionary phenomenon, and it may cause wrong results based on concatenated alignments. Currently, the coalescent-based method has been widely used in nuclear phylogenomic analyses, alleviating the influence of incomplete lineage sorting to a certain extent.

3.5. *Building "Big Trees" Using the Supermatrix Method*

With unprecedented increase of molecular data available in public databases, such as NCBI (www.ncbi.nlm.nih.gov/ accessed on 15 August 2022), TreeBASE (www.treebase.org/, accessed on 15 August 2022), and Dryad (www.datadryad.org/, accessed on 15 August 2022), massive amounts of molecular data have been generated and can be downloaded free, which provide the possibility to construct mega-phylogenies. There are two methods for constructing a mega-phylogeny: supertree and supermatrix. The supertree method compiles source trees with partially overlapping taxa into a single comprehensive tree, and the supermatrix method assembles numerous matrices with overlapping taxa into a super matrix and then reconstructs a phylogenetic tree [80]. Since the supermatrix contains genetic information of species, it has broader application for downstream analyses, such as estimating divergence times, referring ancestral ranges, and calculating diversification rates. The mega-phylogeny is widely used to explore the evolutionary dynamics of biodiversity in which the inclusion of as many taxa as possible is more important than the improvement of bootstrap values. Smith et al. [81] constructed a phylogeny of 5036 species of Caryophyllales and found that a series of diversification rate shifts occurred more recently than whole genome duplication events. Folk et al. [55] used rosids as a case to illustrate how to construct a comprehensively sampled phylogeny at the species-level on a global-scale and to discuss difficulties and opportunities of associated geographic and phenotypic resources. There were also some studies using the supermatrix approach to construct the TOL and then to explore its phylogenetic diversity, community structure, and biogeographic patterns. Thornhill et al. [82] constructed a phylogeny for California flora and discussed relative phylogenetic diversity, phylogenetic endemism, and neo- and paleo-endemism. Wu et al. [83] reconstructed a phylogenetic tree for 157 species of Zygophyllaceae based on four DNA markers using the supermatrix method and then integrated phylogenetic, molecular dating, biogeographic, and diversification rate methods to investigate the diversity dynamics of the family through time.

As phylogenetic trees become larger, more computer power is required, especially for generating a "big" timetree. Graphics processing unit (GPU) multi-core resources with computer clusters, cloud computing platforms, and parallel version upgrades of analysis software have started to be adopted to improve computer speed. How to visualize constructing a "big" tree is also an urgent problem [48]. In addition, providing an integrated, open, and real-time renewable TOL of land plants for science and society will be a great challenge in the future [84].

Author Contributions: Conceptualization, W.W.; writing—original draft, G.-Q.L.; writing—review and editing, G.-Q.L., L.L. and W.W.; funding acquisition, W.W. All authors have read and agreed to the published version of the manuscript.

Funding: This research was funded by the National Natural Science Foundation of China (32011530072, 32170210, 31770231, and 31770233), the Strategic Priority Research Program of the Chinese Academy of Sciences (XDB31030000), and the K.C. Wong Education Foundation (GJTD-2020-05).

Institutional Review Board Statement: Not applicable for studies not involving humans or animals.

Data Availability Statement: Not applicable.

Acknowledgments: We thank the Special Issue Editor, Hong-Hu Meng, for the invitation to write this review.

Conflicts of Interest: The authors declare no conflict of interest.

References

1. Darwin, C. *The Origin of Species by Means of Natural Selection*; John Murray: London, UK, 1859.
2. Davis, C.C.; Matthews, S. *Evolution of Land Plants*; Oxford University Press: New York, NY, USA, 2019. [CrossRef]
3. Wickett, N.J.; Mirarab, S.; Nguyen, N.; Warnow, T.; Carpenter, E.; Matasci, N.; Ayyampalayam, S.; Barker, M.S.; Burleigh, J.G.; Gitzendanner, M.A.; et al. Phylotranscriptomic analysis of the origin and early diversification of land plants. *Proc. Natl. Acad. Sci. USA* **2014**, *111*, E4859–E4868. [CrossRef]
4. Su, D.; Yang, L.; Shi, X.; Ma, X.; Zhou, X.; Hedges, S.B.; Zhong, B. Large-scale phylogenomic analyses reveal the monophyly of bryophytes and neoproterozoic origin of land plants. *Mol. Biol. Evol.* **2021**, *38*, 3332–3344. [CrossRef] [PubMed]
5. Donoghue, P.C.J.; Harrison, C.J.; Paps, J.; Schneider, H. The evolutionary emergence of land plants. *Curr. Biol.* **2021**, *31*, R1281–R1298. [CrossRef] [PubMed]
6. Cheng, S.; Xian, W.; Fu, Y.; Marin, B.; Keller, J.; Wu, T.; Sun, W.; Li, X.; Xu, Y.; Zhang, Y.; et al. Genomes of subaerial Zygnematophyceae provide insights into land plant evolution. *Cell* **2019**, *179*, 1057–1067.e14. [CrossRef]
7. Gitzendanner, M.A.; Soltis, P.S.; Wong, G.K.; Ruhfel, B.R.; Soltis, D.E. Plastid phylogenomic analysis of green plants: A billion years of evolutionary history. *Am. J. Bot.* **2018**, *105*, 291–301. [CrossRef]
8. One Thousand Plant Transcriptomes Initiative. One thousand plant transcriptomes and the phylogenomics of green plants. *Nature* **2019**, *574*, 679–685. [CrossRef] [PubMed]
9. Ran, J.H.; Shen, T.T.; Wang, M.M.; Wang, X.Q. Phylogenomic resolves the deep phylogeny of seed plants and indicates partial convergent or homoplastic evolution between Gnetales and angiosperms. *Proc. R. Soc. B* **2018**, *285*, 20181012. [CrossRef]
10. The Pteridophyte Phylogeny Group. A community-derived classification for extant lycophytes and ferns. *J. Syst. Evol.* **2016**, *54*, 563–603. [CrossRef]
11. The Angiosperm Phylogeny Group. An update of the Angiosperm Phylogeny Group classification for the orders and families of flowering plants: APG IV. *Bot. J. Linn. Soc.* **2016**, *181*, 1–20. [CrossRef]
12. Yang, Y.; Ferguson, D.K.; Liu, B.; Mao, K.S.; Gao, L.M.; Zhang, S.Z.; Wan, T.; Rushforth, K.; Zhang, Z.X. Recent advances on phylogenomics of gymnosperms and a new classification. *Plant Divers.* **2022**, *44*, 340–350. [CrossRef]
13. McCourt, R.M.; Delwiche, C.F.; Karol, K.G. Charophyte algae and land plant origins. *Trends Ecol. Evol.* **2004**, *19*, 661–666. [CrossRef]
14. Karol, K.G.; McCourt, R.M.; Cimino, M.T.; Delwiche, C.F. The closest living relatives of land plants. *Science* **2001**, *294*, 2351–2353. [CrossRef] [PubMed]
15. Timme, R.E.; Bachvaroff, T.R.; Delwiche, C.F. Broad phylogenomic sampling and the sister lineage of land plants. *PLoS ONE* **2012**, *7*, e29696. [CrossRef] [PubMed]

16. Wodniok, S.; Brinkmann, H.; Glöckner, G.; Heidel, A.J.; Philippe, H.; Melkonian, M.; Becker, B. Origin of land plants: Do conjugating green algae hold the key? *BMC Evol. Biol.* **2011**, *11*, 104. [CrossRef]
17. Turmel, M.; Otis, C.; Lemieux, C. The chloroplast genome sequence of Chara vulgaris sheds new light into the closest green algal relatives of land plants. *Mol. Biol. Evol.* **2006**, *23*, 1324–1338. [CrossRef]
18. Finet, C.; Timme, R.E.; Delwiche, C.F.; Marlétaz, F. Multigene phylogeny of the green lineage reveals the origin and diversification of land plants. *Curr. Biol.* **2012**, *22*, 1456–1457. [CrossRef]
19. Laurin-Lemay, S.; Brinkmann, H.; Philippe, H. Origin of land plants revisited in the light of sequence contamination and missing data. *Curr. Biol.* **2012**, *22*, R593–R594. [CrossRef]
20. Pires, N.D.; Dolan, L. Morphological evolution in land plants: New designs with old genes. *Philos. Trans. R. Soc. B* **2012**, *367*, 508–518. [CrossRef]
21. Jill Harrison, C. Development and genetics in the evolution of land plant body plans. *Philos. Trans. R. Soc. B* **2017**, *372*, 20150490. [CrossRef]
22. Ishizaki, K. Evolution of land plants: Insights from molecular studies on basal lineages. *Biosci. Biotechnol. Biochem.* **2017**, *81*, 73–80. [CrossRef] [PubMed]
23. Szövényi, P.; Waller, M.; Kirbis, A. Evolution of the plant body plan. *Curr. Top. Dev. Biol.* **2019**, *131*, 1–34. [CrossRef]
24. Wang, Q.H.; Zhang, J.; Liu, Y.; Jia, Y.; Jiao, Y.N.; Xu, B.; Chen, Z.D. Diversity, phylogeny, and adaptation of bryophytes: Insights from genomic and transcriptomic data. *J. Exp. Bot.* **2022**, *73*, 4306–4322. [CrossRef]
25. Qiu, Y.L.; Cho, Y.; Cox, C.; Palmer, J.D. The gain of three mitochondrial introns identifies liverworts as the earliest land plants. *Nature* **1998**, *394*, 671–674. [CrossRef]
26. Qiu, Y.L.; Li, L.B.; Wang, B.; Chen, Z.D.; Knoop, V.; Groth-Malonek, M.; Dombrovska, O.; Lee, J.; Kent, L.; Rest, J.; et al. The deepest divergences in land plants inferred from phylogenomic evidence. *Proc. Natl. Acad. Sci. USA* **2006**, *103*, 15511–15516. [CrossRef]
27. Chang, Y.; Graham, S.W. Bryophyte-specific primers for retrieving plastid genes suitable for phylogenetic inference. *Am. J. Bot.* **2011**, *98*, e109–e113. [CrossRef]
28. Liu, Y.; Cox, C.J.; Wang, W.; Goffinet, B. Mitochondrial phylogenomics of early land plants: Mitigating the effects of saturation, compositional heterogeneity, and codon-usage bias. *Syst. Biol.* **2014**, *63*, 862–878. [CrossRef]
29. Cox, C.J.; Li, B.; Foster, P.G.; Embley, T.M.; Civan, P. Conflicting phylogenies for early land plants are caused by composition biases among synonymous substitutions. *Syst. Biol.* **2014**, *63*, 272–279. [CrossRef]
30. Zhang, J.; Fu, X.X.; Li, R.Q.; Zhao, X.; Liu, Y.; Li, M.H.; Zwaenepoel, A.; Ma, H.; Goffinet, B.; Guan, Y.L.; et al. The hornwort genome and early land plant evolution. *Nat. Plants* **2020**, *6*, 107–118. [CrossRef]
31. De Sousa, F.; Foster, P.G.; Donoghue, P.C.J.; Schneider, H.; Cox, C.J. Nuclear protein phylogenies support the monophyly of the three bryophyte groups (Bryophyta Schimp.). *New Phytol.* **2019**, *222*, 565–575. [CrossRef]
32. Kranz, H.D.; Huss, V.A.R. Molecular evolution of pteridophytes and their relationship to seed plants: Evidence from complete 18S rRNA gene sequences. *Plant Syst. Evol.* **1996**, *202*, 1–11. [CrossRef]
33. Pryer, K.M.; Schneider, H.; Smith, A.R.; Cranfill, R.; Wolf, P.G.; Hunt, J.S.; Sipes, S.D. Horsetails and ferns are a monophyletic group and the closest living relatives to seed plants. *Nature* **2001**, *409*, 618–622. [CrossRef] [PubMed]
34. Rai, H.S.; Graham, S.W. Utility of a large, multigene plastid data set in inferring higher-order relationships in ferns and relatives (monilophytes). *Am. J. Bot.* **2010**, *97*, 1444–1456. [CrossRef] [PubMed]
35. Bowe, L.M.; Coat, G.; DePamphilis, C.W. Phylogeny of seed plants based on all three genomic compartments: Extant gymnosperms are monophyletic and Gnetales' closest relatives are conifers. *Proc. Natl. Acad. Sci. USA* **2000**, *97*, 4092–4097. [CrossRef]
36. Chaw, S.M.; Parkinson, C.L.; Cheng, Y.C.; Vincent, T.M.; Palmer, J.D. Seed plant phylogeny inferred from all three plant genomes: Monophyly of extant gymnosperms and origin of Gnetales from conifers. *Proc. Natl. Acad. Sci. USA* **2000**, *97*, 4086–4091. [CrossRef]
37. Liu, Y.; Wang, S.; Li, L.; Yang, T.; Dong, S.; Wei, T.; Wu, S.; Liu, Y.; Gong, Y.; Feng, X.; et al. The Cycas genome and the early evolution of seed plants. *Nat. Plants* **2022**, *8*, 389–401. [CrossRef]
38. Wan, T.; Liu, Z.M.; Li, L.F.; Leitch, A.R.; Leitch, I.J.; Lohaus, R.; Liu, Z.J.; Xin, H.P.; Gong, Y.B.; Liu, Y.; et al. A genome for gnetophytes and early evolution of seed plants. *Nat. Plants* **2018**, *4*, 82–89. [CrossRef]
39. Parkinson, C.L.; Adams, K.L.; Palmer, J.D. Multigene analyses identify the three earliest lineages of extant flowering plants. *Curr. Biol.* **1999**, *9*, 1485–1491. [CrossRef]
40. Saarela, J.M.; Rai, H.S.; Doyle, J.A.; Endress, P.K.; Mathews, S.; Marchant, A.D.; Briggs, B.G.; Graham, S.W. Hydatellaceae identified as a new branch near the base of the angiosperm phylogenetic tree. *Nature* **2007**, *446*, 312–315. [CrossRef]
41. Qiu, Y.L.; Dombrovska, O.; Lee, J.; Li, L.; Whitlock, B.A.; Bernasconi-Quadroni, F.; Rest, J.S.; Davis, C.C.; Borsch, T.; Hilu, K.W.; et al. Phylogenetic analyses of basal angiosperms based on nine plastid, mitochondrial, and nuclear genes. *Int. J. Plant Sci.* **2005**, *166*, 815–842. [CrossRef]
42. Amborella Genome Project; Albert, V.A.; Barbazuk, W.B.; Depamphilis, C.W.; Der, J.P.; Leebens-Mack, J.; Ma, H.; Palmer, J.D.; Rounsley, S.; Sankoff, D.; et al. The Amborella genome and the evolution of flowering plants. *Science* **2013**, *342*, 1241089. [CrossRef]
43. Zhang, L.; Chen, F.; Zhang, X.; Li, Z.; Zhao, Y.; Lohaus, R.; Chang, X.; Dong, W.; Ho, S.Y.W.; Liu, X.; et al. The water lily genome and the early evolution of flowering plants. *Nature* **2020**, *577*, 79–84. [CrossRef] [PubMed]
44. Moore, M.J.; Bell, C.D.; Soltis, P.S.; Soltis, D.E. Using plastid genome-scale data to resolve enigmatic relationships among basal angiosperms. *Proc. Natl. Acad. Sci. USA* **2007**, *104*, 19363–19368. [CrossRef] [PubMed]
45. Ma, J.; Sun, P.; Wang, D.; Wang, Z.; Yang, J.; Li, Y.; Mu, W.; Xu, R.; Wu, Y.; Dong, C.; et al. The Chloranthus sessilifolius genome provides insight into early diversification of angiosperms. *Nat. Commun.* **2021**, *12*, 6929. [CrossRef] [PubMed]

46. Guo, X.; Fang, D.; Sahu, S.K.; Yang, S.; Guang, X.; Folk, R.; Smith, S.A.; Chanderbali, A.S.; Chen, S.; Liu, M.; et al. Chloranthus genome provides insights into the early diversification of angiosperms. *Nat. Commun.* **2021**, *12*, 6930. [CrossRef]
47. Chase, M.; Soltis, D.E.; Olmstead, R.G.; Morgan, D.; Les, D.H.; Mishler, B.D.; Duvall, M.R.; Price, R.A.; Hills, H.G.; Qiu, Y.L.; et al. Phylogenetics of seed plants: An analysis of nucleotide sequences from the plastid gene rbcL. *Ann. Mo. Bot. Gard.* **1993**, *80*, 528–580. [CrossRef]
48. Soltis, D.E.; Moore, M.J.; Burleigh, G.; Soltis, P.S. Molecular markers and concepts of plant evolutionary relationships: Progress, promise, and future prospects. *Crit. Rev. Plant Sci.* **2009**, *28*, 1–15. [CrossRef]
49. The Angiosperm Phylogeny Group. An ordinal classification for the families of flowering plants. *Ann. Mo. Bot. Gard.* **1998**, *85*, 531–553. [CrossRef]
50. The Angiosperm Phylogeny Group. An update of the Angiosperm Phylogeny Group classification for the orders and families of flowering plants: APG II. *Bot. J. Linn. Soc.* **2003**, *141*, 399–436. [CrossRef]
51. The Angiosperm Phylogeny Group. An update of the Angiosperm Phylogeny Group classification for the orders and families of flowering plants: APG III. *Bot. J. Linn. Soc.* **2009**, *161*, 105–121. [CrossRef]
52. Lehtonen, S. Towards resolving the complete fern tree of life. *PLoS One* **2011**, *6*, e24851. [CrossRef]
53. Liu, Y.; Johnson, M.G.; Cox, C.J.; Medina, R.; Devos, N.; Vanderpoorten, A.; Hedenäs, L.; Bell, N.E.; Shevock, J.R.; Aguero, B.; et al. Resolution of the ordinal phylogeny of mosses using targeted exons from organellar and nuclear genomes. *Nat. Commun.* **2019**, *10*, 1485. [CrossRef] [PubMed]
54. Hinchliff, C.E.; Smith, S.A.; Allman, J.F.; Burleigh, J.G.; Chaudhary, R.; Coghill, L.M.; Crandall, K.A.; Deng, J.; Drew, B.T.; Gazis, R.; et al. Synthesis of phylogeny and taxonomy into a comprehensive tree of life. *Proc. Natl. Acad. Sci. USA* **2015**, *112*, 12764–12769. [CrossRef] [PubMed]
55. Folk, R.A.; Sun, M.; Soltis, P.S.; Smith, S.A.; Soltis, D.E.; Guralnick, R.P. Challenges of comprehensive taxon sampling in comparative biology: Wrestling with rosids. *Am. J. Bot.* **2018**, *105*, 433–445. [CrossRef]
56. Liu, H.-M.; Jiang, R.-H.; Guo, J.; Hovenkamp, P.; Perrie, L.R.; Shepherd, L.; Hennequin, S.; Schneider, H. Towards a phylogenetic classification of the climbing fern genus Arthropteris. *Taxon* **2013**, *62*, 688–700. [CrossRef]
57. Zhou, X.M.; Zhang, L.; Lu, N.T.; Gao, X.F.; Zhang, L.B. Pteridryaceae: A new fern family of Polypodiineae (Polypodiales) including taxonomic treatments. *J. Syst. Evol.* **2018**, *56*, 148–173. [CrossRef]
58. Su, J.X.; Wang, W.; Zhang, L.B.; Chen, Z.D. Phylogenetic placement of two enigmatic genera, Borthwickia and Stixis, based on molecular and pollen data, and the description of a new family of Brassicales, Borthwickiaceae. *Taxon* **2012**, *61*, 601–611. [CrossRef]
59. Liu, B.; Tan, Y.H.; Liu, S.; Olmstead, R.G.; Min, D.Z.; Chen, Z.D.; Joshee, N.; Vaidya, B.N.; Chung, R.C.K.; Li, B. Phylogenetic relationships of Cyrtandromoea and Wightia revisited: A new tribe in Phrymaceae and a new family in Lamiales. *J. Syst. Evol.* **2019**, *58*, 1–17. [CrossRef]
60. Wang, W.; Liu, Y.; Yu, S.X.; Gao, T.G.; Chen, Z.D. Gymnaconitum, a new genus of Ranunculaceae endemic to the Qinghai–Tibetan Plateau. *Taxon* **2013**, *62*, 713–722. [CrossRef]
61. Kadereit, J.W.; Albach, D.C.; Ehrendorfer, F.; Galbany-Casals, M.; Garcia-Jacas, N.; Gehrke, B.; Kadereit, G.; Kilian, N.; Klein, J.T.; Koch, M.A.; et al. Which changes are needed to render all genera of the German flora monophyletic? *Willdenowia* **2016**, *46*, 39–91. [CrossRef]
62. Wang, W.; Zhang, X.X.; Chen, Z.D.; Lu, A.M. Comments on the APG's classification of angiosperms. *Biodivers. Sci.* **2017**, *25*, 418–426. [CrossRef]
63. Nandi, O.; Chase, M.W.; Endress, P.K. A combined cladistic analysis of angiosperms using rbcL and non-molecular data sets. *Ann. Mo. Bot. Gard.* **1998**, *85*, 137–214. [CrossRef]
64. Doyle, J.A.; Endress, P.K. Morphological phylogenetic analysis of basal angiosperms: Comparison and combination with molecular data. *Int. J. Plant Sci.* **2000**, *161*, S121–S153. [CrossRef]
65. Soltis, D.E.; Moore, M.J.; Sessa, E.B.; Smith, S.A.; Soltis, P.S. Using and navigating the plant tree of life. *Am. J. Bot.* **2018**, *105*, 287–290. [CrossRef] [PubMed]
66. Glenner, H.; Hansen, A.J.; Sørensen, M.V.; Ronquist, F.; Huelsenbeck, J.P.; Willerslev, E. Bayesian inference of the metazoan phylogeny: A combined molecular and morphological approach. *Curr. Biol.* **2004**, *14*, 1644–1649. [CrossRef] [PubMed]
67. O'Reilly, J.E.; Puttick, M.N.; Pisani, D.; Donoghue, P.C.J. Probabilistic methods surpass parsimony when assessing clade support in phylogenetic analyses of discrete morphological data. *Palaeontology* **2018**, *61*, 105–118. [CrossRef] [PubMed]
68. Wang, W.; Lu, A.M.; Ren, Y.; Endress, M.E.; Chen, Z.D. Phylogeny and classification of Ranunculales: Evidence from four molecular loci and morphological data. *Perspect. Plant Ecol. Evol. Syst.* **2009**, *11*, 81–110. [CrossRef]
69. Yu, S.X.; Janssens, S.B.; Zhu, X.Y.; Lidén, M.; Gao, T.G.; Wang, W. Phylogeny of Impatiens (Balsaminaceae): Integrating molecular and morphological evidence into a new classification. *Cladistics* **2016**, *32*, 179–197. [CrossRef] [PubMed]
70. Friis, E.M.; Pedersen, K.R.; Crane, P.R. The emergence of core eudicots: New floral evidence from the earliest Late Cretaceous. *Proc. R. Soc. B* **2016**, *283*, 20161325. [CrossRef]
71. Christiansen, P. Phylogeny of the great cats (Felidae: Pantherinae), and the influence of fossil taxa and missing characters. *Cladistics* **2008**, *24*, 977–992. [CrossRef]
72. Parham, J.F.; Donoghue, P.C.; Bell, C.J.; Calway, T.D.; Head, J.J.; Holroyd, P.A.; Inoue, J.G.; Irmis, R.B.; Joyce, W.G.; Ksepka, D.T.; et al. Best practices for justifying fossil calibrations. *Syst. Biol.* **2012**, *61*, 346–359. [CrossRef]
73. Xiang, X.; Xiang, K.; Ortiz, R.D.C.; Jabbour, F.; Wang, W. Integrating palaeontological and molecular data uncovers multiple ancient and recent dispersals in the pantropical Hamamelidaceae. *J. Biogeogr.* **2019**, *46*, 2622–2631. [CrossRef]

74. Pyron, R.A. Divergence time estimation using fossils as terminal taxa and the origins of Lissamphibia. *Syst. Biol.* **2011**, *60*, 466–481. [CrossRef] [PubMed]
75. Palmer, J.D.; Jansen, R.K.; Michaels, H.J.; Chase, M.W.; Manhart, J.R. Chloroplast DNA variation and plant phylogeny. *Ann. Mo. Bot. Gard.* **1988**, *75*, 1180–1206. [CrossRef]
76. Goremykin, V.V.; Nikiforova, S.V.; Biggs, P.J.; Zhong, B.; Delange, P.; Martin, W.; Woetzel, S.; Atherton, R.A.; McLenachan, P.A.; Lockhart, P.J. The evolutionary root of flowering plants. *Syst. Biol.* **2013**, *62*, 50–61. [CrossRef]
77. Drew, B.T.; Ruhfel, B.R.; Smith, S.A.; Moore, M.J.; Briggs, B.G.; Gitzendanner, M.A.; Soltis, P.S.; Soltis, D.E. Another look at the root of the angiosperms reveals a familiar tale. *Syst. Biol.* **2014**, *63*, 368–382. [CrossRef]
78. Xi, Z.; Ruhfel, B.R.; Schaefer, H.; Amorim, A.M.; Sugumaran, M.; Wurdack, K.J.; Endress, P.K.; Matthews, M.L.; Stevens, P.F.; Mathews, S.; et al. Phylogenomics and a posteriori data partitioning resolve the Cretaceous angiosperm radiation Malpighiales. *Proc. Natl. Acad. Sci. USA* **2012**, *109*, 17519–17524. [CrossRef]
79. McKain, M.R.; Johnson, M.G.; Uribe-Convers, S.; Eaton, D.; Yang, Y. Practical considerations for plant phylogenomics. *Appl. Plant Sci.* **2018**, *6*, e1038. [CrossRef]
80. Smith, S.A.; Beaulieu, J.M.; Donoghue, M.J. Mega-phylogeny approach for comparative biology: An alternative to supertree and supermatrix approaches. *BMC Evol. Biol.* **2009**, *9*, 37. [CrossRef]
81. Smith, S.A.; Brown, J.W.; Yang, Y.; Bruenn, R.; Drummond, C.P.; Brockington, S.F.; Walker, J.F.; Last, N.; Douglas, N.A.; Moore, M.J. Disparity, diversity, and duplications in the Caryophyllales. *New Phytol.* **2018**, *217*, 836–854. [CrossRef]
82. Thornhill, A.H.; Baldwin, B.G.; Freyman, W.A.; Nosratinia, S.; Kling, M.M.; Morueta-Holme, N.; Madsen, T.P.; Ackerly, D.D.; Mishler, B.D. Spatial phylogenetics of the native California flora. *BMC Biol.* **2017**, *15*, 96. [CrossRef]
83. Wu, S.D.; Zhang, L.J.; Lin, L.; Yu, S.X.; Chen, Z.D.; Wang, W. Insights into the historical assembly of global dryland floras: The diversification of Zygophyllaceae. *BMC Evol. Biol.* **2018**, *18*, 166. [CrossRef]
84. Eiserhardt, W.L.; Antonelli, A.; Bennett, D.J.; Botigué, L.R.; Burleigh, J.G.; Dodsworth, S.; Enquist, B.J.; Forest, F.; Kim, J.T.; Kozlov, A.M.; et al. A roadmap for global synthesis of the plant tree of life. *Am. J. Bot.* **2018**, *105*, 614–622. [CrossRef]

diversity MDPI

Review

Flora and Vegetation of Yunnan, Southwestern China: Diversity, Origin and Evolution

Hua Zhu * and Yunhong Tan

Center for Integrative Conservation, Xishuangbanna Tropical Botanical Garden, Chinese Academy of Sciences, Xishuangbanna 666303, China; tyh@xtbg.org.cn
* Correspondence: zhuh@xtbg.ac.cn

Abstract: Yunnan has a complicated geological history, a particular geography, and a complex topography, which have influenced the formation of various habitats of high biodiversity: 245 families; 2140 genera; 13,253 species and varieties of seed plants; more than 12 types of vegetation; and 167 plant formations, including tropical rain forests, tropical dry forests, subtropical evergreen broad-leaved forest, cold temperate coniferous forests, and alpine bushes and meadows. An analysis of the geographic elements to the current Yunnan flora shows that the tropical distribution contributed to 51% of all families and to 57.5% of all genera, of which the genera from the tropical Asian distribution make up the highest proportion among all geographical elements. During the late evolution of Yunnan, its flora was strongly affected by the tropical Asian flora. The complicated patterns and diversity in Yunnan flora and vegetation have been shaped mainly by its particular geological histories, which include the differential uplifts in topography, the clock-wise rotation of the Simao-Lanping geoblock, and the extrusion of the Indochina geoblock by the Himalayan uplift. The flora and vegetation of Yunnan were possibly derived from tropical-subtropical Tertiary flora before later diverging. Northwestern Yunnan flora likely evolved due to rapid speciation from families and genera from cosmopolitan and northern temperate distributions during the uplift of the Himalayas and climatic oscillations after the late Tertiary. Southern Yunnan flora likely evolved into tropical Asian flora following the southeastward extrusion of the Indochina block, which brought along tropical Asian elements. Central Yunnan flora inherited most of the elements of the Tertiary flora from East Asia. The formation and strengthening of the southwest monsoon by the uplift of the Himalayas was also a direct factor in the formation of the tropical rain forests found in southern Yunnan. The flora from southern and southeastern Yunnan also diverged, with the former being more closely related to Indo-Malaysian flora and the latter being more closely related to Eastern Asian flora. This floristic divergence is well supported by the geological history of these regions: that is, the tropical flora of southeastern Yunnan derived from the South China geoblock, whereas the flora of southern and southwestern Yunnan mainly derived from the Shan-Thai geoblock.

Keywords: flora; vegetation; origin; evolution; Yunnan; SW China

Citation: Zhu, H.; Tan, Y. Flora and Vegetation of Yunnan, Southwestern China: Diversity, Origin and Evolution. *Diversity* **2022**, *14*, 340. https://doi.org/10.3390/d14050340

Academic Editor: Mario A. Pagnotta

Received: 13 March 2022
Accepted: 7 April 2022
Published: 26 April 2022

1. Introduction

Yunnan is biogeographically located at the junction where tropical Asia, East Asia, and the Himalayan subtropical-temperate zones meet (Figure 1). With a complex topography consisting of deep valleys, plateaus, and mountains, the change in elevation is huge (from the 76 m altitude at the southeastern most boundary to the 6740 m altitude at Kavagbo Peak of the Meili Mountains in the northwestern part), decreasing from north to south (Figure 2). Geologically, it is a suture zone between Gondwana and Laurasia and its territory mainly formed when the Himalayas rose during the Tertiary, parting from the ancient south China geoblock. The Gondwana and Laurasia flora contributed to the evolution of the flora, and the current flora shows a mixture of elements from tropical Southeast Asia, East Asia, and the temperate Himalayas. Yunnan was also affected by the Himalayan uplift and the accompanying geological events and large river captures. Due to its exceptional geological, geographical and topographical features, Yunnan has evolved to have high biodiversity.

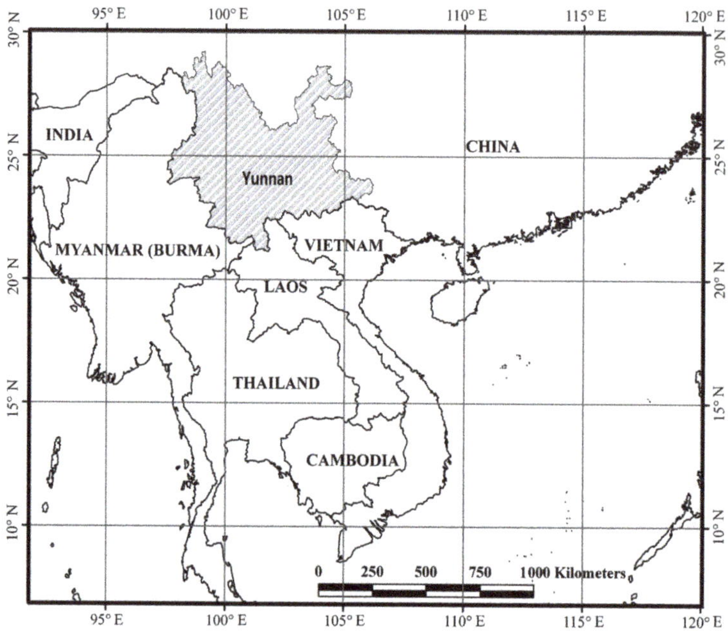

Figure 1. Geographical location of Yunnan (this figure was made by the Landscape Ecology Lab., Xishuangbanna Tropical Botanical Garden, CAS).

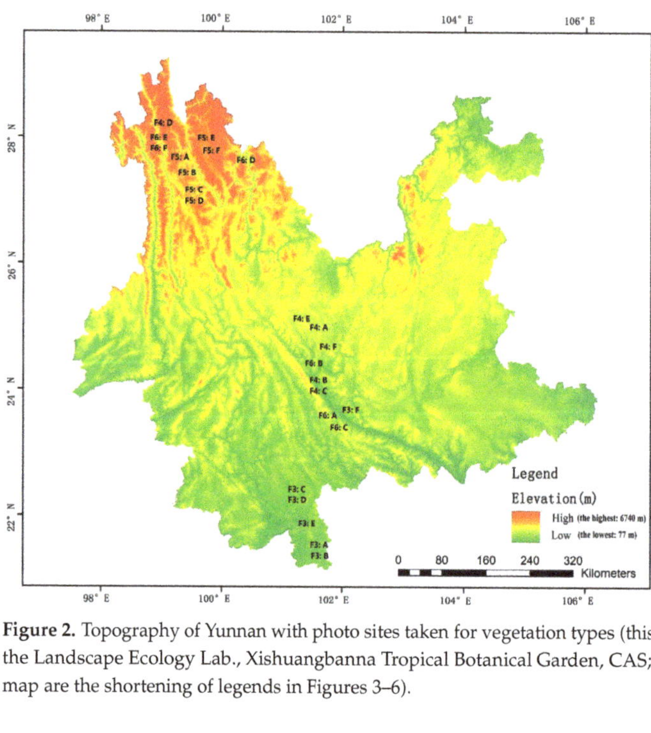

Figure 2. Topography of Yunnan with photo sites taken for vegetation types (this figure was made by the Landscape Ecology Lab., Xishuangbanna Tropical Botanical Garden, CAS; F3, F4, F5, F6 in the map are the shortening of legends in Figures 3–6).

Figure 3. (**A**), Tropical rain forest in southern Yunnan; (**B**), understory of the tropical rain forest in southern Yunnan; (**C**), tropical lower montane evergreen broad-leaved forest in southern Yunnan; (**D**), understory of the tropical lower montane evergreen broad-leaved forest in southern Yunnan; (**E**), tropical seasonal moist forest on limestone in southern Yunnan; (**F**), tropical deciduous forest in southern Yunnan.

Yunnan takes up only 4.1% of the total land area of China but includes more than half of the total plant species found and has various types of vegetation, from tropical to cold temperate forests. Based on an analysis of herbarium specimens collected, 245 families, 2140 genera, and 13,253 species and varieties of seed plants were confirmed to be found in Yunnan [1]. Additionally, from the data recorded from field surveys, the main types of vegetation found in Asia are also found in Yunnan [2]. For example, forests similar in floristic composition and physiognomy to the tropical rain forests of SE Asia were found in southwestern, southern, and southeastern Yunnan [3–16]; tropical deciduous forests and savanna-like vegetation found in hot dry valleys [17,18], and the subtropical evergreen broad-leaved forests of eastern Asia were found also in the central plateau of Yunnan [19–21]; cold temperate coniferous forests dominated by spruce, fir, and larch trees and temperate deciduous forests dominated by birch wood, maple, and poplar trees were found in high northern mountains; and sclerophyllous evergreen broad-leaved forests dominated by the *Quercus* species were commonly found in diverse habitats throughout Yunnan, from hot, dry, and deep valleys to cold temperate mountains. The monograph "Vegetation of Yunnan" by Wu recorded 12 main natural types of vegetation [22], which includes all of the main types of vegetation found in China.

Figure 4. (**A**), Semi-evergreen broad-leaved forest in central Yunnan; (**B**), mid-montane wet evergreen broad-leaved forest in central Yunnan; (**C**), understory of the mid-montane wet evergreen broad-leaved forest, showing abundant epiphytes; (**D**), sclerophyllous oak forest in northern Yunnan; (**E**), temperate coniferous forest in central Yunnan; (**F**), temperate coniferous and evergreen broad-leaved mixed forest in central Yunnan.

Figure 5. (**A**), Temperate coniferous and deciduous broad-leaved mixed forest in northern Yunnan; (**B**), cold temperate coniferous and sclerophyllous oak mixed forest in northern Yunnan; (**C**), cold temperate coniferous forest in northern Yunnan; (**D**), understory of the cold temperate coniferous forest in northern Yunnan; (**E**), alpine Rhododendron shrubs in northern Yunnan; (**F**), alpine meadows in northern Yunnan.

Geological events that have occurred since the Tertiary period—such as the differential uplift of Yunnan; the clock-wise rotation of the Simao-Lanping geoblock; the southeastward extrusion of the Indochina geoblock caused by the collision of India with Asia; and the strengthening of the southwestern Asian monsoon climate triggered by an uplift in the Himalayas as well as several large rivers that flow southward, which were once tributaries to a single, southward flowing system (the paleo-Red River since at least the late-Miocene period [23])—influenced the origin and evolution of Yunnan flora and vegetation [1,9–11,17,24–29]. These geological events also deeply influenced the divergence of flora in Yunnan, such as the formation of some suggested biogeographical demarcation lines [24,26,30–34].

Figure 6. (**A**), Savanna-like vegetation in hot dry valleys in southeastern Yunnan; (**B**), savanna-like shrubs in hot dry valleys in southeastern Yunnan; (**C**), succulent shrubs in hot dry valleys in southeastern Yunnan; (**D**), savanna-like vegetation in hot dry valleys in northern Yunnan; (**E**), maquis-like vegetation in warm dry valleys in northern Yunnan; (**F**), close up of the maquis-like vegetation in warm dry valleys in northern Yunnan.

However, the origin and evolution of the flora and vegetation of Yunnan are still somewhat unclear. Paleobotanical studies offer good clues about their origin and evolution. The divergence of the evergreen broad-leaved forests of Yunnan was suggested to have started in the late Miocene [35–37]. During the Miocene, the vegetation of southwestern Yunnan was suggested to be tropical in nature, while that of eastern and northern Yunnan were suggested to be subtropical [38]. The preliminary composition and distribution of the present flora and vegetation of Yunnan formed during the Miocene [27,37]. Based on paleobotanical and molecular phylogeny studies of the Hengduan Mountains, psychrophytes grew during the early Oligocene [39]; some current genera and species, such as *Quercus, Alnus, Betula, Carpinus, Carya, Pterocaryae*, etc., appeared in the Lühe basin of central Yunnan in the early Oligocene [40]; and subtropical forests supposedly formed in the central Qinghai-Tibetan Plateau 47 million years ago (Ma) [41]. Paleobotanical data can serve as evidence of the origin and evolution of the flora and vegetation of Yunnan.

Most studies on the flora of Yunnan focus on local flora in nature reserves, while research of the regional and provincial floras is rare: only found in the studies by Li, Li and Walker, and Zhu [1,42,43] and in the monograph by Wu [44]. Although some studies on the origin and evolution of Yunnan have been published [11,27,45,46], synthetic studies combining its paleobotanical, geological, and climatic histories are still needed.

A descriptive work on the vegetation in Yunnan was first published by Wang [47]. Later, Wu published the monograph *Vegetation of Yunnan* [22], which laid the foundation for studies of the vegetation in Yunnan. Zeng published a study on the natural forests of Yunnan based on vegetation classification and distributions [48]. Zhu discussed the main types of vegetation in Yunnan, and their origin and evolution [2]. However, more work on the extremely diverse vegetation of Yunnan is further needed.

Combining community ecology and floristic plant geography to study types of vegetation could give a better overall understanding of vegetation: not only its ecological and physiognomic features, but also its biogeographical sources. This article aims to review and clarify the diversity, origin, and evolution of the flora and vegetation of Yunnan and to provide information useful for biodiversity conservation.

2. Composition and Characteristics of the Flora and Vegetation of Yunnan

2.1. Floristic Composition and Geographical Elements

We identified 245 families, 2140 genera, and 13,253 species and varieties of seed plants in Yunnan [1]. The species-rich families included Poaceae (874 species), Asteraceae (787), Orchidaceae (774), Fabaceae (637), Rosaceae (460), Lamiaceae (446), Rubiaceae (365), Ericaceae (360), and more. The families with the highest species richness were cosmopolitan ones. Of the geographical elements, the tropical distribution contributed to 52.7%, and the pantropic distribution obtained the highest ratio, making up 34.4% of all families of flora in Yunnan, such as Acanthaceae, Anacardiaceae, Annonaceae, Apocynaceae, Araceae, Arecaceae, and Clusiaceae. The northern temperate distribution contributed to 12.5% of all families, such as Adoxaceae, Betulaceae, Caprifoliaceae, Cornaceae, Fagaceae and Hamamelidaceae. The flora of Yunnan included 12 families from the tropical Asian distribution, such as Crypteroniaceae, Iteaceae, Ixonanthaceae, Pentaphragmataceae, Pentaphylacaceae, Sabiaceae and Sladeniaceae, and 8 families from the eastern Asian distribution, such as Cercidiphyllaceae, Circaeasteraceae, Eupteleaceae and Stachyuraceae. The families from the tropical distribution were found mainly in southwestern to southeastern Yunnan and in hot dry valleys of several large rivers and their tributaries, while those of a temperate distribution were found mainly on the plateau, and in northwestern and northern Yunnan.

Floristic evolutionary history could be obtained from geographical elements at the generic level in flora. Of the 2140 genera in Yunnan flora, the tropical distribution contributed to 57.4%, of which those of the tropical Asian distribution made up the highest ratio among all geographical elements, contributing 22.2% of all genera. The temperate distribution contributed to 33%, of which the north temperate contributed 10.9%, followed by the eastern Asian distribution, contributing 9.9% of all genera [1]. Tropical elements contributed the most to the flora in Yunnan, showing its tropical affinity.

2.2. Vegetation Types and Distribution

From the monograph *Vegetation of Yunnan* [22], 12 vegetation types and 167 formations were identified, including tropical rain forests, subtropical evergreen broad-leaved forests, sclerophyllous oak forests, warm temperate deciduous forests, temperate coniferous and broad-leaved mixed forests, cold temperate coniferous forests, subalpine and alpine shrubs and meadows, and savanna-like vegetation from hot dry valleys, etc. (see Figures 3–6).

Due to the particular landform and topography of Yunnan, the climate and natural vegetation change dramatically at very short distances. Various types of vegetation display complicated distribution patterns, not only at the provincial but also at the regional levels. Generally, tropical rain forests and tropical deciduous forests are distributed in southwestern, southern, and southeastern Yunnan; subtropical evergreen broad-leaved forests, coniferous and broad-leaved mixed forests, and warm temperate deciduous forests are mainly distributed on the central plateau; and cold temperate coniferous forests are distributed in northern Yunnan. Savanna-like vegetation can be found in the hot dry valleys of several large rivers, and sclerophyllous and microphyllous shrubs and maquis-like vegetation with a Mediterranean floristic affinity can be found in the warm dry valleys in northern Yunnan. Sclerophyllous oak forests dominated by *Quercus franchetii* and *Q. cocciferoides* are found mainly in the hot dry valleys, while those dominated by *Q. guajavifolia* and *Q. semecarpifolia* are found mainly in the central plateaus and in the high mountains of northwestern Yunnan.

Some arguments have been made about vegetation zonation and vegetation regionalization in Yunnan. Liu et al. argued that the vegetation distribution in Yunnan should be regarded as a large vertical distribution based on altitude and that areas below 1500 m in altitude should be considered tropical regions because the topography of Yunnan is generally sloped to the south and most of the climate in Yunnan is affected by the Indian Ocean [49]. Ren and Xiang also suggested that Yunnan is a big tropical mountain with clearly vertical zonality [50]. We support these points of view on vegetation and flora in Yunnan because tropical climates and tropical vegetation occur below 1300 m in altitude in Yunnan regardless of the latitude [51]. However, the vegetation distribution in Yunnan can also be regarded based on latitude [52,53], as was accepted in the monograph *Vegetation of Yunnan* [22].

2.3. Biogeographical Divergence of the Flora and Vegetation of Yunnan

On floristic regionalization, Wu and Wu raised the "Eastern Asiatic floristic region" delineated by Takhtajan [54] to the kingdom level, "the Eastern Asiatic Kingdom", and further divided it into the Sino-Himalaya and Sino-Japanese Subkingdoms [55]. The demarcation line between the Sino-Himalaya and Sino-Japanese Subkingdoms runs north–southward in eastern Yunnan. This line was supposedly of significance in the divergence of Yunnan flora.

Tanaka suggested a demarcation line between the citrus "*Archicitrus*" and "*Metacitrus*" groups, which extends in the northwest–southeast direction from Yunnan, China, to northern Vietnam [30]. This geographical line was given the name "Tanaka line of Citrus distribution" or the "Tanaka line" [30]. Later, this line was suggested to be the rough demarcation line between Sino-Himalaya and Sino-Japanese flora-based research on some genera of the Sino-Himalaya and Sino-Japanese distributions [56]. Combined with some orchid plants and their distributions, this line was renamed the "Tanaka–Kaiyong line" [57]. This line has been studied and discussed by many researchers [13,32,33,58–62]. However, Pang et al. studied the phylogenetic tree of *Citrus* based on an AFLP marker analysis, which did not support the "Tanaka line of Citrus distribution" [63]. We consider that the "Tanaka line" has no practical significance for the floristic divergence in Yunnan [64].

We also found geographic differences in the plants between the southwest, south, and southeast of Yunnan. Although dominated by tropical components and having the highest proportion of tropical Asian components, the flora of southeastern Yunnan has subtropical, temperate families with relatively abundant species numbers, such as Magnoliaceae, Theaceae, Cornaceae, Symplocaceae, Caprifoliaceae, and Aquifoliaceae. In particular, the characteristic families of East Asia and the Himalayas, such as Diapensiaceae, Dipentodontaceae, Eupteleaceae, Grossulariaceae, and Torricelliaceae, are present in the tropical flora of southeast Yunnan but absent in the flora of southern Yunnan [28]. Furthermore, 349 genera, including 57 genera of East Asian distribution, 53 genera of northern temperate distribution, 22 genera endemic to China, and 17 genera of East Asian–Northern American disjuncted distribution, were found in southeast Yunnan but were absent in southern Yunnan. On the other hand, 237 genera in southern Yunnan were not found in southeast Yunnan. Although the southern and southeastern tropical regions of Yunnan have similar tropical monsoon climates and tropical rain forest vegetation, the floristic difference between them implicates that they possibly went through different evolutionary histories on their flora [24,34]. This is supported by the geological histories of these regions. Southeastern Yunnan was mainly derived from the South China geoblock, while southwestern and southern Yunnan were derived from the Shan-Thai geoblock [65,66] (Figure 7). However, a clear floristic divergence between the flora of southwestern Yunnan and southern Yunnan has not been found.

Based on physiognomic and floristic studies of the tropical and subtropical evergreen forests in Yunnan, two conspicuous ecotones were also found: at the 800–1200 m altitude, the tropical lowland seasonal rain forest changes to a tropical lower montane evergreen forest; at the 1800–2100 (2200) m altitude, the tropical lower montane evergreen forest changes to a subtropical (warm temperate) evergreen forest, almost completely replaced by subtropical (warm temperate) Himalayan and Chinese endemic tree flora [67].

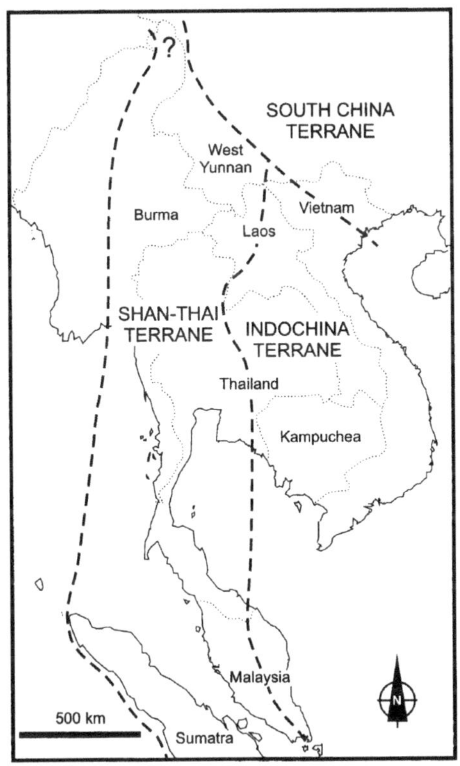

Figure 7. The main geoblocks in Southeast Asia (from Fortey et al., 1998).

3. Uplift in the Himalayan-Qinghai-Tibet Plateau and the Following Monsoon Climate Formation Affected the Evolution of Vegetation and Flora in Yunnan

The Indian plate moved northward and collided with the Eurasian plate in the early Cenozoic period (about 50 Ma (million years) ago), leading to the formation and uplift of the Himalayas [68]. With the Himalayan uplift, the Indochina geological plate was extruded toward Southeast Asia [69–73], ending in the late Miocene (10 Ma) [74]. Meanwhile, the Simao-Lanping geoblock in Yunnan moved toward the southeast with the Indochina plate and underwent a clockwise rotation (about 30°) [75,76] and moved southeast by about 800 km, and its co-instantaneous clockwise rotation continued into the Miocene [71,77] (Figure 8). These geological events affected the origin and evolution of the flora and vegetation of Yunnan [11,27]. The Indian plate collided with the Asian plate, pushing away part of the Myanmar geological plate to the north by about 1000 km [78]. The northern movement of western Yunnan also occurred correspondingly, leading to the northwest–southeast inclined distribution pattern of tropical flora and tropical vegetation in Yunnan [17,26,79].

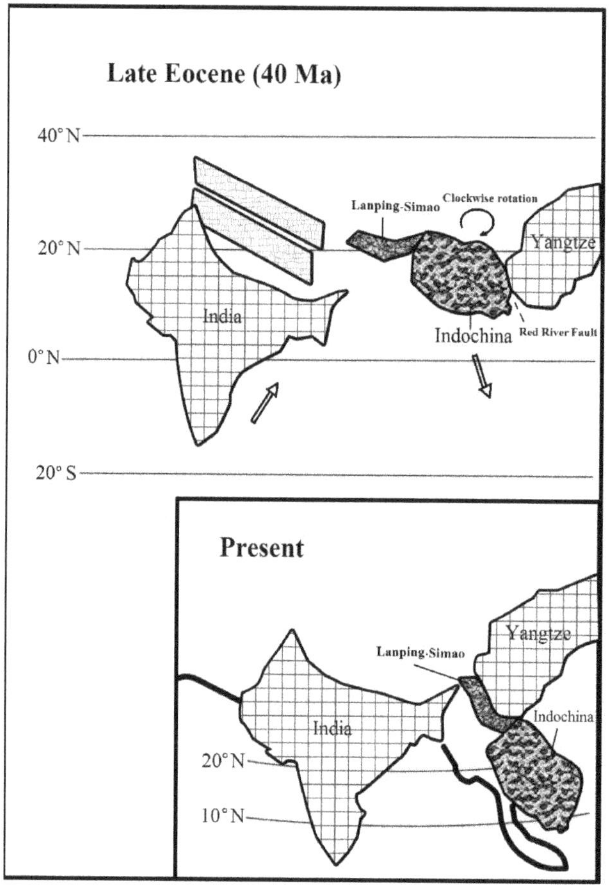

Figure 8. Clockwise rotation and southeastward extrusion of the Lanping-Simao and Indochina geoblocks during the late Eocene (redrawn from Sato et al., 2001, Figure 7).

The uplift in the Himalayan-Tibetan Plateau affected the global climate and widespread environmental changes since the late Cenozoic [80–82]. Due to the uplift of the Tibetan Plateau, the southwest monsoon occurring in South Asia emerged, which plays a decisive role in the development of tropical vegetation in India, mainland SE Asia and southwest China [83]. The formation and evolution of vegetation in Yunnan, especially tropical rain forests, are directly affected by the southwest monsoon, while the formation and evolution of Yunnan flora are obviously affected by historical geological events.

Much debate surrounds the timing of the monsoon climate in Southeast and East Asia. The mainstream view has been that, in the late Eocene, about 45–50 Ma ago, the Indian plate and the Eurasian plate collided and became integrated; however, the Himalayan-Tibetan

Plateau did not rise strongly but rather experienced a long process of uplift and deplanation at low altitudes (1000–2000 m alt.). The Himalayas were not very high until the Quaternary before 3.4 Ma or 2.5 Ma [81,84]. However, according to Su et al. and Liu et al., the Qinghai-Tibet Plateau should have risen earlier based on paleobotanical research [85,86]. When the Himalayas reached a considerable altitude (above 6000 m), the warm and humid air flow in the south was blocked by the high mountains, causing an abundance of precipitation on the southern slope, forming a warm and humid subtropical and tropical climate at lower altitudes and playing a decisive role in the development of tropical vegetation in SE Asia and southwest China [81,83]. Based on the simulation of late Oligocene paleogeographic data, Li et al. believe that the uplift of the northern Qinghai-Tibet Plateau during the Paleogene enhanced the East Asian monsoon climate system and drove the formation of humid and semi-humid vegetation types dominated by the evergreen broad-leaved forests in East Asia [87]. Undoubtedly, the formation of the Himalayas greatly influenced the strength of the southwest monsoon [88].

The geological and climatic histories of a region directly affect the formation and evolution of its flora and vegetation. The findings of paleobotanical research not only provide a basis for exploring the period when the Himalayan uplift occurred and the monsoon climate formed but also are a key factor in piecing together the evolutionary history of regional flora and vegetation. Jacques et al. revealed that ancient vegetation in central and south-central Yunnan during the Miocene were similar to the current subtropical forests [35], but in southern Fujian in southeastern China, during the Miocene, a large number of dipterocarp fossils were found [36,89], which show that the vegetation in southern Fujian was similar to those of the tropical rain forests of Southeast Asia [90]. The paleobotanical information for southwest and southeast China during the Miocene revealed different vegetation, meaning that southwestern and southeastern China should have different geological, climatic, and vegetational histories.

Our studies on the flora and vegetation in Yunnan suggest that their formation and evolution are closely related to the Himalayan uplift, the monsoon climate formation, and the various accompanying geological events [11,24,25,27,37]. The clockwise rotation and displacement of the Simao-Lanping geoblock to the southeast, and the extrusion of the Indochina geological plate toward Southeast Asia are supposedly the main geological events that directly affected the formation and evolution of flora in Yunnan, while the formation and strengthening of the southwest monsoon is a direct factor in the evolution of the evergreen broad-leaved forests and tropical rain forests found in Yunnan.

The modern rivers draining into the plateau at the border of SW China, i.e., the Jinshajiang, Nujiang, and Lancanjiang (the upper reaches of the Mekong River), were once tributaries into a single, southward flowing system—the paleo-Red River, which drained into the South China Sea since at least the late Miocene [23]. A disruption of the paleo-drainage was caused by river capture and reversal prior to or coeval with the initiation of the Miocene uplift of the Himalayas [23]. River captures interrupted these large rivers, which led to species isolation and differentiation, and further excited new taxa evolution in these river drainage areas.

4. Origin and Evolution of Yunnan Biodiversity

Plant diversity is closely related to the evolution of flora and vegetation. Yunnan flora and vegetation are considered to have evolved from the Tertiary tropical and subtropical flora and vegetation. With the uplift in the Himalayas and the extrusion of the Indochina geological plate into Southeast Asia, the cosmopolitan and northern temperate floristic components quickly speciated and proliferated in the high mountains in the north, evolving into the present-day temperate flora and vegetation, while in the southern lowland region, the tropical Asian floristic elements influenced and evolved into the current tropical flora and vegetation. However, the original ancient floristic components in the central region of Yunnan remained and were inherited [11,27].

Taking southern Yunnan as an example, tropical elements account for 78.3% of all genera, among which the proportion of tropical Asian distributions is the highest, accounting for 30.19%, showing that the flora of southern Yunnan is now tropical Asian flora in nature. Furthermore, regarding the floristic composition of the tropical rain forest in southern Yunnan, 80% of families, 94% of genera, and more than 90% of species are from a tropical distribution, of which about 38% of genera and 74% of species are from a tropical Asian distribution [7]. The large proportion of tropical Asian genera and species in the tropical rain forest in southern Yunnan, are supposedly closely related to the formation and strengthening of the southwest monsoon, and the extrusion of the Indochina geological plate into Southeast Asia. Obviously, the formation and strengthening of the southwest monsoon have led to the emergence of the tropical wet climate locally in southern Yunnan, which enabled tropical rain forests to emerge. According to paleobotanical data, from the late Cretaceous to early Tertiary in southern Yunnan, the representative vegetation was speculated to be dry subtropical or southern subtropical montane evergreen broad-leaved forests [91]; from the Eocene to the Oligocene, the main vegetation may have still shared characteristics from the previous period; from the Miocene to the Pliocene, the forest vegetation was speculated to mainly be an evergreen broad-leaved forest based on the southern subtropical or subtropical features in southern Yunnan [92–95]; and the tropical flora and rain forest in southern Yunnan likely evolved after the Pliocene [9,10].

Studies on the flora in northwestern Yunnan have revealed that the worldwide and northern temperate families and genera had proliferated and experienced rapid speciation during the Tertiary during the Himalayan uplift and under climate fluctuations, where they evolved into the current temperate flora [11,25]. Paleobotany and paleovegetation studies revealed that the accumulation of alpine plant diversity in Hengduan Mountain began during the early Oligocene [39]. Therefore, the flora and vegetation in the northwestern region of Yunnan may have begun in the early Oligocene.

Apart from the southern-most and the northern Yunnan, most of Yunnan, especially the central part originally covered by evergreen broad-leaved forests on plateau, lower and upper montanes, which were further divided into three major vegetation sub-types—monsoon evergreen broad-leaved forest on the lower montane in southern and southwestern Yunnan (with a southern subtropical climate); the semi-wet evergreen broad-leaved forest mainly on plateau (with a subtropical climate), both on limestone and acid soil habitats; and the mid-montane wet evergreen broad-leaved forest on the upper montanes (with a subtropical to temperate climate) [22]. At vegetation level, the three evergreen broad-leaved forest types diverged considerably in species composition, physiognomy, and biogeography, although they are commonly dominated by species of the families Fagaceae, Lauraceae, and Theaceae. The monsoon evergreen broad-leaved forest is extremely rich in species and is characterized by a tropical physiognomy. It is dominated by tropical Asian species, which is similar to the tropical lower montane evergreen forest in Southeast Asia. The semi-wet evergreen broad-leaved forest and the mid-montane wet evergreen broad-leaved forest were characterized by a subtropical physiognomy and are dominated by Sino-Himalayan and Chinese endemic species, which are unique in southwestern China [21]. All of the three forest types commonly have species-rich families, which tend to have cosmopolitan distributions, but the families with fewer species have other distribution types. At generic and specific levels, the semi-wet evergreen broad-leaved forest and the mid-montane wet evergreen broad-leaved forest showed similar biogeographical patterns in the proportions of tropical (45% and 44%, respectively) and temperate (46% and 48%) elements, with northern temperate distributions comprising the highest percentage (18% in the semi-wet evergreen broad-leaved forest and 20% in the mid-montane wet evergreen broad-leaved forest) of total genera, while in the monsoon evergreen broad-leaved forest in southern Yunnan, tropical elements comprised 79% of the total genera, with elements of tropical Asian distributions contributing the highest percentage (27%). The tropical species in the monsoon evergreen broad-leaved forest make up 70.52% of its total species, of which the tropical Asian distribution contribute 64.73% of the total species, while both of the semi-wet evergreen broad-leaved forest and mid-montane wet evergreen broad-leaved forest have the majority of temperate distribution species, 72.88% on limestone and 78.67% on non-limestone habitats in the former, and 80.27% in the latter. Of their temperate distribution species, both of them have the Chinese endemic and Chinese-Himalayan species contributing to the majority (Zhu, 2021). The three vegetation subtypes of evergreen broad-leaved forest in Yunnan are commonly dominated by species of the families Fagaceae, Lauraceae, and Theaceae, which reveals their common early origin. Later their divergence took place, with events in the geological history of Yunnan caused by the uplift of the Himalayas.

Comparing the flora of northern, central, and southern Yunnan, they commonly include families of a tropical distribution, which make up the highest ratio, but in the genera and species, northern Yunnan is distinct from southern Yunnan [11]. This revealed that the Yunnan flora is identical to the tropical families that dominated before the uplift of the Himalayas. A floristic divergence took place with the uplift of Himalayas and with the formation of the monsoon climate. The northern region evolved into temperate flora and vegetation dominated by worldwide and northern temperate families and genera, while the southern region evolved with the extrusion of the geological plate to Southeast Asia, with tropical Asian element infiltration causing the evolution into tropical Asian flora [11,25]. East Asia is a relatively stable region in geological history, where the continuous evolution of flora and vegetation has occurred. Therefore, the flora and vegetation of central Yunnan are mainly from the East Asian flora and vegetation.

Deciduous forests with the same ecological physiognomy and structure as the ones in the Indo-Myanmar region appeared disjointedly in parts of the deep valleys of the Yuanjiang, Nujiang, Jinsha, and Lancang rivers and in some of the open valley basins most strongly affected by the monsoon. Regarding geographical elements, deciduous forests consist of 80% of the genera and 70% of the species that belong to tropical components, showing that they are tropical forests by nature. They are believed to have extensive associations with the tropical deciduous forests in the Indo-Myanmar region and occurred in a much larger area during some period during or before the Pliocene to Pleistocene, when a drier climate appeared in most parts of Yunnan. These forests have survived as refuges in hot dry valleys and basins possibly since the Pleistocene [18]. The higher rainfall in Yunnan during the later Pleistocene and early Holocene [96] likely explains their survival in these hot dry rain-shadow valleys. The tropical deciduous forest shrank to isolated dry habitats and valleys with expansion of evergreen broad-leaved forests with the increasing rainfall during the later Pleistocene.

Savanna-like vegetation is also commonly found in deep, hot and dry valleys within several large rivers strongly affected by the monsoon and by the foehn effect in Yunnan. Its flora is dominated by tropical families and genera and is fundamentally tropical in nature. An endemic genus *Tsaiodendron* (Euphorbiaceae), a newly published genus [97], is found locally in the savanna-like vegetation in the hot dry valley of Yuanjiang, SE Yunnan, and the compositae tree genus *Nouelia*, which is a mono-typical (*N. insignis*), is endemic to the hot dry valley of Jinshajiang, Northern Yunnan, offering evidence that these vegetations have existed in hot dry valleys for a long time. Some studies, such as those on *Musella lasiocarpa* (Musaceae) [98] and *Terminalia franchetii* (Combretaceae) [99,100], showed that their phylogenetic evolution could have been caused by the rapid uplift of Himalaya and concomitant river captures.

The geological events that have occurred since the Cenozoic, especially the river capture caused by the rise of the Himalayas and the climatic fluctuation, have influenced the evolution and divergence of the flora of the savanna-like vegetation in Yunnan [17]. The flora in hot dry valleys of the Yuanjiang, Nujiang, and Jinshajiang were compared. At the family level, the flora from each were almost the same, with a similarity coefficient of more than 87% (Similarity coefficient between A and B = the number of taxa shared by both A and B divided by the lowest number of taxa of A or B, multiplied by 100%). At the generic level, the highest similarity was between the Yuanjiang and Jinshajiang (73.84%), and all have a similarity of more than 62%, showing that close floristic affinity remains. However, at the species level, the similarity coefficients were generally low, with the highest (53.76%) being between the Yuanjiang and Jinshajiang and the lowest (35.03%) being between the Yuanjiang and Nujiang. The flora of the hot dry valleys in Yunnan has diverged at the specific level [17].

The current flora in Yunnan can be categorized into five historical phytogeographical elements: East Asian subtropical floristic element (mainly in central Yunnan), Southeast Asian tropical floristic element (mainly in the southwest, south and southeast of Yunnan), the Himalayan floristic element (mainly in the northern region), the Mediterranean residual flora (widespread in Yunnan), and the Gondwana flora (mainly in the dry hot river valleys). Morley suggested that the Indian plate drifted to low latitudes in Asia during the Eocene, bringing the floristic components of Gondwana to Southeast Asia, which successfully evolved in Southeast Asia after 41 Ma [101]. Later, these Gondwana floristic elements also entered the dry and hot river valley regions of Yunnan province [17]. In paleogeography, Yunnan was located on the southeast edge of Eurasia. After the Indian plate and Asian plate collided, the Neo-Tethys closed and the Himalayan Plateau was uplifted. The sclerophyllous and microphyllous shrubs and maquis-like vegetation with Mediterranean floristic affinity in warm dry valleys in northern Yunnan remained, and the wide-spread sclerophyllous oak forests in Yunnan evolved from the Mediterranean-related flora [102,103].

In summary, the formation and evolution of plant diversity in Yunnan are closely related to the geological events and the formation and strengthening of the monsoon climate.

5. Conclusions

In the eyes of a biologist, the flora of Yunnan of SW China make up a mysterious kingdom. Yunnan only takes up 4.1% of the total land area but includes more than half of all plant species found in China and has various types of vegetation, including tropical rainforests to cold temperate coniferous forests. The complex topography, landforms, and huge difference in altitudes have created diverse habitats in Yunnan. In physical geography and biogeography, Yunnan is a transitional region located in the tropical SE Asia and subtropical-temperate East Asia and Himalaya. Located at the particular intersection between SE Asia, East Asia, and Himalaya, the possibilities for its biota origins are plentiful, because SE Asia, East Asia, and Himalaya offer it ample biodiversity sources. In geological history, it has experienced geological events such as the extrusion of the Indochina geoblock, the fracture and dislocation of fragments of smaller geoblocks, and large rivers' captures due to the uplift in the Himalayas. It also experienced the formation of a monsoon climate due to the Himalayan uplift. These particular natural and historical conditions serve as the basis for the evolution of such high biodiversity in Yunnan.

In this unique geographical area, these geological events, monsoon climate formation and strengthening, and climate fluctuation evidently led to frequent species migration, isolation, and divergence, and to quicker-than-usual speciation. Based on paleobotanical studies, the majority of the vegetation and flora in Yunnan may have formed during the period from the Oligocene to the Miocene. With a further uplift in the Himalayas, a topographically differential uplift, the strengthening of monsoon climates, and climate fluctuation after the Pliocene, the south–north vegetation and the flora differences increased, and west–east biogeographical divergence became clear. Abundant biodiversity has arisen and evolved accordingly in Yunnan.

Author Contributions: Conceptualization, H.Z.; Methodology, H.Z. & Y.T.; Software, H.Z. & Y.T.; Validation, H.Z. & Y.T.; Formal analysis, H.Z. & Y.T.; Investigation, H.Z. & Y.T.; Resources, H.Z. & Y.T.; Data curation, H.Z. & Y.T.; Writing—original draft preparation, H.Z. & Y.T.; Writing—review and editing, H.Z. & Y.T.; Visualization, H.Z. & Y.T.; supervision, H.Z.; Project administration, H.Z. & Y.T.; Funding acquisition, H.Z. & Y.T. All authors have read and agreed to the published version of the manuscript.

Funding: This project was funded by The National Natural Science Foundation of China (41471051, 31970223).

Institutional Review Board Statement: Not applicable.

Data Availability Statement: Not applicable.

Conflicts of Interest: The authors declare no conflict of interest.

References

1. Zhu, H. A biogeographical comparison between Yunnan, Southwest China, and Taiwan, Southeast China, with implications for the evolutionary history of the East Asian Flora. *Ann. Mo. Bot. Gard.* **2016**, *101*, 750–771. [CrossRef]
2. Zhu, H. Vegetation diversity of Yunnan. *J. Southwest For. Univ.* **2022**, *42*, 1–12.
3. Zhu, H. The tropical rainforest vegetation in Xishuangbanna. *Chin. Geogr. Sci.* **1992**, *2*, 64–73. [CrossRef]
4. Zhu, H. Floristic plant geography on the dipterocarp forest of Xishuangbanna. *Acta Bot. Yunnan* **1993**, *15*, 233–253.
5. Zhu, H. The floristic characteristics of the tropical rainforest in Xishuangbanna. *Chin. Geogr. Sci.* **1994**, *4*, 174–185. [CrossRef]
6. Zhu, H. Floristic relationships between dipterocarp forest of Xishuangbanna and forests of tropical Asia and S China. *Acta Bot. Yunnan* **1994**, *16*, 97–106.
7. Zhu, H. Ecological and biogeographical studies on the tropical rain forest of south Yunnan, SW China with a special reference to its relation with rain forests of tropical Asia. *J. Biogeogr.* **1997**, *24*, 647–662. [CrossRef]
8. Zhu, H. Forest vegetation of Xishuangbanna, south China. *For. Stud. China* **2006**, *8*, 1–58.
9. Zhu, H. The tropical flora of southern Yunnan, China, and its biogeographical affinities. *Ann. Mo. Bot. Gard.* **2008**, *95*, 661–680.
10. Zhu, H. Advances in biogeography of the tropical rainforest in southern Yunnan, southwestern China. *Trop. Conserv. Sci.* **2008**, *1*, 34–42.
11. Zhu, H. Biogeographical divergence of the flora of Yunnan, southwestern China initiated by the uplift of Himalaya and extrusion of Indochina block. *PLoS ONE* **2012**, *7*, e45601.
12. Zhu, H. The tropical forests of southern China and conservation of biodiversity. *Bot. Rev.* **2017**, *83*, 87–105. [CrossRef]
13. Zhu, H. A biogeographical study on tropical flora of southern China. *Ecol. Evol.* **2017**, *7*, 10398–10408. [CrossRef]
14. Zhu, H.; Shi, J.P.; Zhao, C.J. Species composition, physiognomy and plant diversity of the tropical montane evergreen broad-leaved forest in southern Yunnan. *Biodivers. Conserv.* **2005**, *14*, 2855–2870. [CrossRef]
15. Zhu, H.; Cao, M.; Hu, H.B. Geological history, flora, and vegetation of Xishuangbanna, southern Yunnan, China. *Biotropica* **2006**, *38*, 310–317. [CrossRef]
16. Zhu, H.; Chai, Y.; Zhou, S.S.; Wang, H.; Yan, L.C. Vegetation, floristic composition and species diversity in a tropical mountain nature reserve in southern Yunnan, SW China with implications to conservation. *Trop. Conserv. Sci.* **2015**, *8*, 528–546. [CrossRef]
17. Zhu, H.; Tan, Y.H.; Yan, L.C.; Liu, F.Y. Flora of the savanna-like vegetation in hot dry valleys, southwestern China with implications to their origin and evolution. *Bot. Rev.* **2020**, *86*, 281–297. [CrossRef]
18. Zhu, H.; Ashton, P.; Gu, B.J.; Zhou, S.S.; Tan, Y.H. Tropical deciduous forest in Yunnan, southwestern China: Implications for geological and climatic histories from a little-known forest formation. *Plant Divers.* **2021**, *34*, 444–451. [CrossRef]
19. Zhu, H.; Chai, Y.; Zhou, S.S.; Yan, L.C.; Shi, J.P.; Yang, G.P. Combined community ecology and floristics, a synthetic study on the upper montane evergreen broad-leaved forests in Yunnan, southwestern China. *Plant Divers.* **2016**, *38*, 295–302. [CrossRef]
20. Zhu, H.; Zhou, S.S.; Yan, L.C.; Shi, J.P.; Shen, Y.X. Studies on the evergreen broad-leaved forests of Yunnan, southwestern China. *Bot. Rev.* **2019**, *85*, 131–148. [CrossRef]
21. Zhu, H. Vegetation geography of evergreen broad-leaved forests in Yunnan, southwestern China. *Chin. J. Plant Ecol.* **2021**, *45*, 224–241. [CrossRef]
22. Wu, Z.Y. *Vegetation of Yunnan*; Science Press: Beijing, China, 1987; pp. 143–163.
23. Clark, M.K.; Schoenbohm, L.M.; Royden, L.H.; Whipple, K.X.; Burchfiel, B.C.; Zhang, X.; Tang, W.; Wang, E.; Chen, L. Surface uplift, tectonics, and erosion of eastern Tibet from large-scale drainage patterns. *Tectonics* **2004**, *23*, TC1006. [CrossRef]
24. Zhu, H. The floras of southern and tropical southeastern Yunnan have been shaped by divergent geological histories. *PLoS ONE* **2013**, *8*, e64213.
25. Zhu, H. Biogeography of Shangri-la flora in southwestern China. *Phytotaxa* **2015**, *203*, 231–244.
26. Zhu, H. Geographical patterns of Yunnan seed plants may be influenced by the Clockwise Rotation of the Simao-Indochina Geoblock. *Front. Earth Sci.* **2015**, *3*, 53. [CrossRef]
27. Zhu, H. On the origin and evolution of the flora of Yunnan. *Plant Sci. J.* **2018**, *36*, 32–37.
28. Zhu, H.; Yan, L.C. Biogeographical affinities of the flora of southeastern Yunnan, China. *Bot. Stud.* **2009**, *50*, 467–475.
29. Zhu, H.; Wang, H.; Li, B.G.; Sirirugsa, P. Biogeography and floristic affinity of the limestone flora in southern Yunnan, China. *Ann. Mo. Bot. Gard.* **2003**, *90*, 444–465. [CrossRef]
30. Tanaka, T. *Species Problem in Citrus*; Japanese Society for the Promotion of Science: Tokyo, Japan, 1954; pp. 58–69.
31. Li, H.; He, D.M.; Bartholomew, B.; eLong, C.-L. Re-examination of the biological effect of plate movement—Impact of Shan-Malay Plate displacement (the movement of Burma-Malay Geoblock) on the biota of the Gaoligong Mountains. *Acta Bot. Yunnan* **1999**, *21*, 407–425.
32. Zhu, H.; Yan, L.C. A discussion on biogeographical lines of the tropical-subtropical Yunnan. *Chin. Geogr. Sci.* **2002**, *12*, 90–96. [CrossRef]
33. Zhu, H.; Yan, L.C. Note on the realities and significances of the"Tanaka line"and the "Ecogeographical Diagonal line"in Yunnan. *Adv. Earth Sci.* **2003**, *18*, 871–877.

34. Zhu, H. A new biogeographical line between south Yunnan and southeast Yunnan. *Adv. Earth Sci.* **2011**, *26*, 916–925.
35. Jacques, F.M.B.; Su, T.; Spicer, R.A.; Xing, Y.W.; Huang, Y.J.; Zhou, Z.K. Late Miocene southwestern Chinese floristic diversity shaped by the southeastern uplift of the Tibetan Plateau. *Palaeogeogr. Palaeocl.* **2014**, *411*, 208–215. [CrossRef]
36. Jacques, F.M.B.; Shi, G.L.; Su, T.; Zhou, Z.K. A tropical forest of the middle Miocene of Fujian (SE China) reveals Sino-Indian biogeographic affinities. *Rev. Palaeobot. Palynol.* **2015**, *216*, 76–91. [CrossRef]
37. Zhu, H. Floristic divergence of the evergreen broad-leaved forests in Yunnan, southwestern China. *Phytotaxa* **2019**, *393*, 001–020. [CrossRef]
38. Huang, Y.J.; Jia, L.B.; Wang, Q.; Mosbrugger, V.; Utescher, T.; Su, T.; Zhou, Z.K. Cenozoic plant diversity of Yunnan: A review. *Plant Divers.* **2016**, *38*, 271–282. [CrossRef]
39. Ding, W.N.; Ree, R.H.; Spicer, R.A.; Xing, Y.W. Ancient orogenic and monsoon-driven assembly of the world's richest temperate alpine flora. *Science* **2020**, *369*, 578–581. [CrossRef]
40. Tang, H.; Li, F.S.; Su, T.; Spicer, R.A.; Zhang, S.; Li, S.; Liu, J.; Lauretano, V.; Witkowski, C.R.; Spicer, T.E.V.; et al. Early Oligocene vegetation and climate of southwestern China inferred from palynology. *Palaeogeogr. Palaeoclimatol. Palaeoecol.* **2020**, *560*, 109988. [CrossRef]
41. Su, T.; Spicer, R.A.; Wu, F.X.; Farnsworth, A.; Huang, J.; del Rio, C.; Deng, T.; Ding, L.; Deng, W.; Huang, Y.; et al. Middle Eocene lowland humid subtropical "Shangri-La" ecosystem in central Tibet. *Proc. Natl. Acad. Sci. USA* **2020**, *117*, 32989–32995. [CrossRef]
42. Li, X.W. Floristic study of Yunnan province. *Acta Bot. Yunnan* **1985**, *7*, 361–382.
43. Li, X.W.; Walker, D. The plant geography of Yunnan Province, southwest China. *J. Biogeogr.* **1986**, *13*, 367–397.
44. Wu, Z.Y. *Flora Yunnanica*; Science Press: Beijing, China, 1977–2005. (In Chinese)
45. Liu, S.Y.; Zhu, H.; Yang, J. A phylogenetic perspective on biogeographical divergence of the flora in Yunnan, Southwestern China. *Sci. Rep.* **2017**, *7*, 43032. [CrossRef]
46. Li, R.; Yang, J. A phylogenetic perspective on the evolutionary processes of floristic assemblages within a biodiversity hotspot in eastern Asia. *J. Syst. Evol.* **2020**, *58*, 413–442. [CrossRef]
47. Wang, C.W. A preliminary study of the vegetation of Yunnan. *Bull. Fan. Mem. Inst. Bot.* **1939**, *9*, 65–135.
48. Zeng, J.M. The classification system of natural forest and its geographical distribution in Yunnan. *J. Southwest For. Univ.* **2018**, *38*, 1–18.
49. Liu, S.E.; Feng, Z.W.; Zhao, D.C. Some principles of vegetation zoning in China. *J. Integr. Plant Biol.* **1959**, *8*, 87–105. (In Chinese)
50. Ren, M.E.; Xiang, R.Z. Some theoretical problems of natural zoning in China from a contradictory perspective—On the issue of natural zoning again in China. *J. Nanjing Univ. Nat. Sci. Ed.* **1963**, *3*, 1–12. (In Chinese)
51. Zhu, H. Distribution patterns of genera of Yunnan seed plants with references to their biogeographical significances. *Adv. Earth Sci.* **2008**, *23*, 830–839.
52. Jiang, H.Q. Distribution features and zonal regularity of vegetation in Yunnan. *Acta Bot. Yunnan* **1980**, *2*, 22–32.
53. Jiang, H.Q. Distribution features and zonal regularity of vegetation in Yunnan (continue). *Acta Bot. Yunnan* **1980**, *2*, 142–151.
54. Takhtajian, Y. *Floristic Region of the World*, English ed.; Cronquist, A., Ed.; Crovello, T.S., Translator; Soviet Sciences Press: Leningrad Branch, 1986; p. 50. (In Russian)
55. Wu, Z.Y.; Wu, S.G. A Proposal for a new floristic kingdom (realm)—The Asiatic kingdom, its delineation and characteristics. In *Floristic Characteristics and Diversity of East Asian Plants*; Zhang, A.L., Wu, S.G., Eds, Eds.; China Higher Education & Springer Press: Beijing, China, 1996; pp. 3–42.
56. Li, X.W.; Li, J. On the validity of Tanaka Line and its significance viewed from the distribution of eastern Asiatic genera in Yunnan. *Acta Bot. Yunnan* **1992**, *14*, 1–12. (In Chinese with English Abstract)
57. Li, X.W.; Li, J. The Tanaka-Kaiyong Line—An important floristic line for the study of the flora of East Asia. *Ann. Mo. Bot. Gard.* **1997**, *84*, 888–892.
58. Feng, J.M.; Zhu, Y.Y. Tanaka line and its biogeographical significance: A further discussion. *Chin. J. Ecol.* **2010**, *29*, 1–7.
59. Fan, D.M.; Yue, J.P.; Nie, Z.L.; Li, Z.M.; Comes, H.P.; Sun, H. Phylogeography of *Sophora davidii* (Leguminosae) across the "Tanaka-Kaiyong Line", an important phytogeographic boundary in southwest China. *Mol. Ecol.* **2013**, *22*, 4270–4288. [CrossRef] [PubMed]
60. Tian, B.; Zhou, Z.; Du, F.K.; He, C.; Xin, P.; Ma, H. The Tanaka line shaped the phylogeographic pattern of the cotton tree (Bombax ceiba) in southwest China. *Biochem. Syst. Ecol.* **2015**, *60*, 150–157. [CrossRef]
61. Ye, J.W.; Zhang, Y.; Wang, X.J. Phylogeographic breaks and the mechanisms of their formation in the Sino-Japanese floristic region. *Chin. J. Plant Ecol.* **2017**, *41*, 1003–1019.
62. Ju, M.M.; Fu, Y.; Zhao, G.F.; He, C.Z.; Li, Z.H.; Tian, B. Effects of the Tanaka Line on the genetic structure of *Bombax ceiba* (Malvaceae) in dry- hot valley areas of southwest China. *Ecol. Evol.* **2018**, *8*, 3599–3608. [CrossRef]
63. Pang, X.M.; Hu, C.G.; Deng, X.X. Phylogenetic relationships within Citrus and its related genera as inferred from AFLP markers. *Genet. Resour. Crop Evol.* **2007**, *54*, 429–436. [CrossRef]
64. Zhu, H. Problems of "Tanaka line" and its application on biogeography as a floristic line. *Plant Sci. J.* **2018**, *36*, 761–766.
65. Fortey, R.A.; Cocks, L.R.M. Biogeography and palaeogeography of the Sibumasu terrene in the Ordovician: A review. In *Biogeography and Geological Evolution of SE Asia*; Hall, R., Holloway, J.D., Eds.; Backbuys Publishers: Leiden, The Netherlands, 1998; pp. 43–56.
66. Metcalfe, I. Tectonic framework and Phanerozoic evolution of Sundaland. *Gondwana Res.* **2011**, *19*, 3–21. [CrossRef]

67. Zhu, H.; Ashton, P. Ecotones in the tropical-subtropical vegetation transition at the tropical margin of southern China. *Chin. Sci. Bull.* **2021**, *66*, 3732–3743. [CrossRef]
68. Jain, A.K. When did India-Asia collide and make the Himalaya? *Curr. Sci. India* **2014**, *106*, 254–266.
69. Tapponnier, P.; Pelter, G.; Armijo, R.; Le Dain, A.Y.; Cobbold, P. Propagation extrusion tectonics in Asia: New insights from simple experiments with plasticine. *Geology* **1982**, *10*, 611–616. [CrossRef]
70. Tapponnier, P.; Lacassin, R.; Leloup, P.H.; Schärer, U.; Zhong, D.L.; Liu, X.H.; Ji, S.C.; Zhang, L.S.; Zhong, J.Y. The Ailao Shan/Red River metamorphic belt: Tertiary left-lateral shear between Indochina and South China. *Nature* **1990**, *343*, 431–437. [CrossRef]
71. Schärer, U.; Tapponnier, P.; Lacassin, R.; Leloup, P.H.; Zhong, D.-L.; Ji, S.-C. Intraplate tectonics in Asia, a precise age for large-scale Miocene movements along the Ailao Shan-Red River shear zone, China. *Earth Planet. Sci. Lett.* **1990**, *97*, 65–77. [CrossRef]
72. Lee, T.Y.; Lawver, L.A. Cenozoic plate reconstruction of Southeast Asia. *Tectonophysics* **1995**, *251*, 85–138. [CrossRef]
73. Leloup, P.H.; Laeassin, R.; Tapponnier, P.; Scharer, U.; Zhong, D.L.; Liu, X.; Zhang, L.; Ji, S.C.; Phan, T.T. The Ailao Shan—Red River shear zone (Yunnan, China), Tertiary transform boundary of Indochina. *Tectonophysics* **1995**, *251*, 3–8. [CrossRef]
74. Hall, R. The plate tectonics of Cenozoic SE Asia and the distribution of land and sea. In *Biogeography and Geological Evolution of SE Asia*; Hall, R., Holloway, J.D., Eds.; Backbuys Publishers: Leiden, The Netherlands, 1998; pp. 99–131.
75. Sato, K.; Liu, Y.Y.; Zhu, Z.C.; Yang, Z.Y.; Otofuji, Y. Tertiary paleomagnetic data from northwestern Yunnan, China: Further evidence for large clockwise rotation of the Indochina block and its tectonic implications. *Earth Planet. Sci. Lett.* **2001**, *185*, 185–198. [CrossRef]
76. Sato, K.; Liu, Y.Y.; Wang, Y.B.; Yokoyam, M.; Yoshioka, S.; Yang, Z.Y.; Otofuji, Y. Paleomagnetic study of Cretaceous rocks from Pu'er, western Yunnan, China: Evidence of internal deformation of the Indochina block. *Earth Planet. Sci. Lett.* **2007**, *258*, 1–15. [CrossRef]
77. Chen, H.H.; Dobson, J.; Heller, F.; Hao, J. Paleomagnetic evidence for clockwise rotation of the Simao region since the Cretaceous, A consequence of India-Asia collision. *Earth Planet. Sci. Lett.* **1995**, *134*, 203–217.
78. Mitchell, A.H.G. Cretaceous-Cenozoic tectonic events in the western Myanmar (Burma)-Assam region. *J. Geol. Soc.* **1993**, *150*, 1089–1102. [CrossRef]
79. Zhu, H. Suggestions for the northern boundary of the tropical zone in China. *Plant Sci. J.* **2018**, *36*, 893–898. (In Chinese)
80. Raymo, M.; Ruddimen, W. Tectonic forcing of late Cenozoic climate. *Nature* **1992**, *359*, 117–122. [CrossRef]
81. Shi, Y.F. Plateau uplift and environmental evolution. In *Formation, Evolution and Development of Qinghai-Xizang (Tibetan)*; Plateausun, H.L., Zheng, D., Eds.; Guangdong Science and Technology Press: Guangzhou, China, 1998; pp. 73–138.
82. Shi, Y.F.; Li, J.Y.; Li, B.Y. Uplift of the Qinghai-Xizang (Tibetan) plateau and east Asia environmental change during late Cenozoic. *Acta Geogr. Sin.* **1999**, *54*, 10–21.
83. Liu, D.S.; Zhang, X.S.; Yan, B.Y. The impact of plateau uplifting on surrounding areas. In *Formation, Evolution and Development of Qinghai-Xizang (Tibetan) Plateau*; Sun, H.L., Zheng, D., Eds.; Guangdong Science and Technology Press: Guangzhou, China, 1998; pp. 179–230.
84. Pan, Y.S. Plateau lithosphere structure, evolution, and dynamics. In *Formation, Evolution and Development of Qinghai-Xizang (Tibetan) Plateau*; Sun, H.L., Zheng, D., Eds.; Guangdong Science and Technology Press: Guangzhou, China, 1998; pp. 1–72.
85. Su, T.; Farnsworth, A.; Spicer, R.A.; Huang, J.; Wu, F.-X.; Liu, J.; Li, S.-F.; Xing, Y.-W.; Huang, Y.-J.; Deng, W.-Y.-D.; et al. No high Tibetan Plateau until the Neogene. *Sci. Adv.* **2019**, *5*, eaav2189. [CrossRef]
86. Liu, J.; Su, T.; Spicera, R.A.; Tang, H.; Deng, W.-Y.; Wu, F.-X.; Srivastava, G.; Spicer, T.; Van Do, T.; Deng, T.; et al. Biotic interchange through lowlands of Tibetan Plateau suture zones during Paleogene. *Palaeogeogr. Palaeocl.* **2019**, *524*, 33–40. [CrossRef]
87. Li, S.F.; Valdes, P.J.; Farnsworth, A.; Davies-Barnard, T.; Su, T.; Lunt, D.J.; Spicer, R.A.; Liu, J.; Deng, W.-Y.; Huang, J.; et al. Orographic evolution of northern Tibet shaped vegetation and plant diversity in eastern Asia. *Sci. Adv.* **2021**, *7*, eabc7741. [CrossRef]
88. Spicer, R.A.; Farnsworth, A.; Su, T. Cenozoic topography, monsoons and biodiversity conservation within the Tibetan Region: An evolving story. *Plant Divers.* **2020**, *42*, 229–254. [CrossRef]
89. Shi, G.; Li, H. A fossil fruit wing of Dipterocarpus from the Middle Miocene of Fujian, China, and its peoclimatic significance. *Rev. Palaeobot. Palyn.* **2010**, *162*, 599–608. [CrossRef]
90. Wang, B.; Shi, G.; Xu, C.; Spicer, R.A.; Perrichot, V.; Schmidt, A.R.; Feldberg, K.; Heinrichs, J.; Chény, C.; Pang, H.; et al. The mid-Miocene Zhangpu biota reveals an outstandingly rich rainforest biome in East Asia. *Sci. Adv.* **2021**, *7*, eabg0625. [CrossRef]
91. Sun, X.J. Palynofloristical investigation on the Late Cretaceous and Paleocene of China. *J. Syst. Evol.* **1979**, *17*, 8–21.
92. Liu, J.L.; Tan, L.Y.; Qiao, Y.; Head, M.J.; Walker, D. Late Quaternary vegetation history at Menghai, Yunnan province, southwest China. *J. Biogeogr.* **1986**, *13*, 399–418.
93. Song, Z.Z.; Li, H.M.; Zheng, Y.H. Miocene floristic region of China. In *Editorial Board of Basic Palaeobiogeographic Theory Series. Palaeobiogeographic Provinces of China*; Science Press: Beijing, China, 1983; pp. 178–184.
94. Song, Z.Z. Miocene Phytogeographical Area of East Asia. In *Stratum and Paleontology*; Nanjing Institute of Geology and Paleontology, Ed.; Nanjing Institute of Geology and Paleontology: Nanjing, China, 1984; Volume 13, pp. 63–69.
95. Wang, W.M. A palynological survey of Neocene strata in Xiaolongtan Basin, Yunnan province of south China. *Acta Bot. Sin.* **1996**, *38*, 743–748.
96. Walker, D. Late Pleistocene—Early Holocene vegetational and climatic changes in Yunnan Province, southwest China. *J. Biogeogr.* **1986**, *13*, 477–486. [CrossRef]

97. Zhou, Z.; Gu, B.G.; Sun, H.; Zhu, H.; Tan, Y.H. Molecular phylogenetic analyses of Euphorbiaceae tribe Epiprineae, with the description of a new genus, Tsaiodendron gen. nov., from south-western China. *Bot. J. Linn. Soc.* **2017**, *184*, 167–184. [CrossRef]
98. Ma, H.; Wang, D.X.; Li, T.Q.; Li, Z.H. Phylogeographic study of Musella lasiocarpa (Musaceae): Providing insightinto the historical river capture events. *Pak. J. Bot.* **2019**, *51*, 191–199. [CrossRef]
99. Zhang, T.C.; Sun, H. Phylogeographic structure of *Terminalia franchetii* (Combretaceae) in southwest China and its implications for drainage geological history. *J. Plant Res.* **2011**, *124*, 63–73. [CrossRef]
100. Zhang, T.C.; Comes, H.P.; Sun, H. Chloroplast phylogeography of *Terminalia franchetii* (Combretaceae) from the eastern Sino-Himalayan region and its correlation with historical river capture events. *Mol. Phylogenetics Evol.* **2011**, *60*, 1–12. [CrossRef]
101. Morley, R.J. Assembly and division of the South and South East Asian flora in relation to tectonics and climate change. *J. Trop. Ecol.* **2018**, *34*, 209–234. [CrossRef]
102. Mai, D.H. Development and regional differentiation of the European vegetation during the Tertiary. *Plant Syst. Evol.* **1989**, *162*, 79–91. [CrossRef]
103. Sun, H.; Li, Z.M. Qinghai-Tibet plateau uplift and its act on Tethys flora. *Adv. Earth Sci.* **2003**, *18*, 852–862.

diversity

MDPI

Review

Latitudinal Diversity Gradient in the Changing World: Retrospectives and Perspectives

Yu Zhang [1,2], Yi-Gang Song [3], Can-Yu Zhang [4], Tian-Rui Wang [3], Tian-Hao Su [2,5], Pei-Han Huang [1,2], Hong-Hu Meng [1,6,*] and Jie Li [1,7,*]

1 Plant Phylogenetics and Conservation Group, Center for Integrative Conservation, Xishuangbanna Tropical Botanical Garden, Chinese Academy of Sciences, Kunming 650223, China; zhangyu@xtbg.ac.cn (Y.Z.); peihanhuang@foxmail.com (P.-H.H.)
2 University of Chinese Academy of Sciences, Beijing 100049, China; sutianhao@xtbg.ac.cn
3 Eastern China Conservation Centre for Wild Endangered Plant Resources, Shanghai Chenshan Botanical Garden, Shanghai 201602, China; ygsong@cemps.ac.cn (Y.-G.S.); wtianrui@163.com (T.-R.W.)
4 Yunnan Normal University, Kunming 650500, China; zhangcanyu11@outlook.com
5 CAS Key Laboratory of Tropical Forest Ecology, Xishuangbanna Tropical Botanical Garden, Chinese Academy of Sciences, Kunming 650223, China
6 Southeast Asia Biodiversity Research Institute, Chinese Academy of Sciences, Naypyidaw 05282, Myanmar
7 Center of Conservation Biology, Core Botanical Gardens, Chinese Academy of Sciences, Mengla 666303, China
* Correspondence: menghonghu@xtbg.ac.cn (H.-H.M.); jieli@xtbg.ac.cn (J.L.)

Abstract: The latitudinal diversity gradient (LDG) is one of the most extensive and important biodiversity patterns on the Earth. Various studies have established that species diversity increases with higher taxa numbers from the polar to the tropics. Studies of multicellular biotas have supported the LDG patterns from land (e.g., plants, animals, forests, wetlands, grasslands, fungi, and so forth) to oceans (e.g., marine organisms from freshwater invertebrates, continental shelve, open ocean, even to the deep sea invertebrates). So far, there are several hypotheses proposed to explore the diversity patterns and mechanisms of LDG, however, there has been no consensus on the underlying causes of LDG over the past few decades. Thus, we reviewed the progress of LDG studies in recent years. Although several explanations for the LDG have been proposed, these hypotheses are only based on species richness, evolution and the ecosystems. In this review, we summarize the effects of evolution and ecology on the LDG patterns to synthesize the formation mechanisms of the general biodiversity distribution patterns. These intertwined factors from ecology and evolution in the LDG are generally due to the wider distribution of tropical areas, which hinders efforts to distinguish their relative contributions. However, the mechanisms of LDG always engaged controversies, especially in such a context that the human activity and climate change has affected the biodiversity. With the development of molecular biology, more genetic/genomic data are available to facilitate the estimation of global biodiversity patterns with regard to climate, latitude, and other factors. Given that human activity and climate change have inevitably impacted on biodiversity loss, biodiversity conservation should focus on the change in LDG pattern. Using large-scale genetic/genomic data to disentangle the diversity mechanisms and patterns of LDG, will provide insights into biodiversity conservation and management measures. Future perspectives of LDG with integrative genetic/genomic, species, evolution, and ecosystem diversity patterns, as well as the mechanisms that apply to biodiversity conservation, are discussed. It is imperative to explore integrated approaches for recognizing the causes of LDG in the context of rapid loss of diversity in a changing world.

Keywords: latitudinal diversity gradient; species richness; biodiversity pattern; climate change; conservation; diversification; speciation; extinction

Citation: Zhang, Y.; Song, Y.-G.; Zhang, C.-Y.; Wang, T.-R.; Su, T.-H.; Huang, P.-H.; Meng, H.-H.; Li, J. Latitudinal Diversity Gradient in the Changing World: Retrospectives and Perspectives. *Diversity* **2022**, *14*, 334. https://doi.org/10.3390/d14050334

Academic Editor: Michael Wink

Received: 30 March 2022
Accepted: 20 April 2022
Published: 25 April 2022

1. Introduction

In 1807, Alexander von Humboldt (1769–1859; Figure 1A) proposed the embryonic framework of latitudinal diversity gradient (LDG) and wrote that "The nearer we approach the tropics, the greater the increase in the variety of structure, grace of form, and mixture of colors, as also in perpetual youth and vigor of organic life" [1,2] (Notes: von Humboldt published the first edition of a series of essays that entitled 'Ansichten der Natur' in Berlin, 1807. The essays initiated while he was in South and Central America that was translated variously as 'Aspects of Nature' or 'Views of Nature'. One of the four essays was composed the first edition, 'Ideas for a physiognomy of plants' that contains the following paragraph was translated by Otté and Bohn in 1850). Subsequently, the LDG has been recognized and studied by many biologists, ecologists, and geographers for over 200 years [3–12]. The pattern of LDG is responsible for the broadest and most notable of biodiversity patterns globally (Figure 1B). It has been well documented on the land, in the open ocean, and even discernible in deep sea, which distribution patterns have been characterized for plants, animals, fungi, and marine organisms [13,14]. However, distribution patterns of organisms are not balanced around the globe so that many naturalists and scientists have tried to understand the cause of LDG for centuries [15]. Notably, the LDG is mostly consistent, regardless the geographic context, taxonomic affiliation, or time scale of the biota [16,17]. Previous studies have attested and cataloged many hypotheses to explain the underlying mechanisms that increase species diversity and taxon numbers from high latitudes to the tropics [18]. However, the mechanisms of LDG are not very clear, even with little consensus. It is necessary to review the LDG in the context of a changing world wherein human activity and climate warming are affecting the patterns of biodiversity in evolution and ecology.

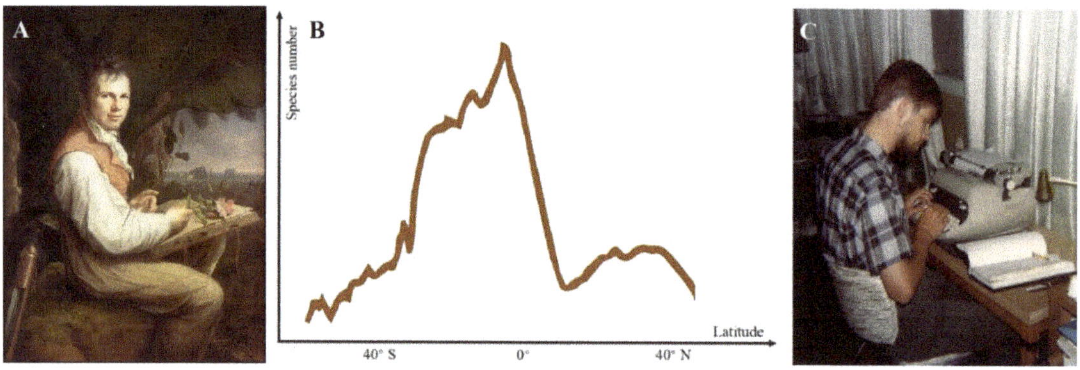

Figure 1. (**A**), Alexander von Humboldt (1769–1859) is widely regarded as the "father of phytography". He was the first to state observations about the LDG (photo reproduces from Biogeography [19]); (**B**), diagram of the LDG, with count of species by latitude (diagram adapted from Spatio-temporal climate change contributes to latitudinal diversity gradients [9]); (**C**), Eric Pianka provided insights into the concept and major hypotheses for the LDG in 1966 (Photo courtesy of E. Pianka on 23 February 2022 via Email; this photo has also been used in *Am. Nat.* 2017 [10]).

The investigation of the LDG has drawn on in ecological and evolutionary studies, i.e., some species survive whereas others die out in the process [15]. Almost six decades ago, Eric Pianka (Figure 1C) proposed the first comprehensive review on LDG and the six major hypotheses that compiled a wide range of ideas to explain and address possible causes for patterns in diversity [6,10,15]. Subsequently, LDG explanations have been focused on evolutionary mechanisms [5], such as differences in the time and area available for diversification in tropical and temperate biomes, latitudinal differences in the rates of diversification, speciation and/or extinction in combination with tropical energy, and niche conservatism [10,15]. However, evolutionary biology and ecology must be combined to explain why larger numbers of taxa are distributed in certain areas of the planet [15]. There is still no consensus on the drivers of LDG that elevate tropical diversity [20].

Alternative LDG hypotheses can be tested with the help of a rich body of biodiversity database, including data concerning phylogeny and biogeography [21]. However, the primary cause of LDG at a global scale is unexplained; and a new synthesis has emerged to determine the geographic ranges of species and their concentration of species within regions, based on evolutionary, biogeographical, and contemporary (i.e., climate and environmental variables) factors [15]. Understanding the underlying mechanisms of LDG is a major goal in conservation biology, biogeography, and ecology [22]. Multiple dimensions of LDG have been described, from intraspecific genetic variation to species richness and phylogenetic diversity, all of which are vital for assessing the underlying processes that shape the distribution of life on Earth and for providing maximum support for global biodiversity conservation [23]. Fascination with the pattern of higher biodiversity in tropical regions has stimulated increasing interest in community ecology [17]. Similarly, the erosion of biodiversity from the local/regional level to the global scale has catalyzed many studies in conservation biology [24]. Thus, the study of the LDG provided a unique opportunity to comprehensively understand latitude-associated patterns in ecology, biogeographical origin, and maintenance of species diversity [25].

Given the potential value of biodiversity conservation efforts, it is critical to examine the mechanisms that create the patterns in biodiversity that produce the LDG. In this review, we aim to identify and discuss the hypotheses for LDG mechanisms, i.e., whether the distribution patterns of different species are consistent with LDG, and explain the factors that cause the number of species to change with latitude. We also discuss the effects and formation of LDG in various periods during which species have been restricted to different latitudes. Additionally, the role of LDG in evolution and ecology will be discussed in relation to global warming and human activity that are affecting the biodiversity. Finally, we concentrate on the relevance of the LDG to biodiversity conservation in the hope that a deeper understanding and reasonable scientific approach (e.g., the big data of genetic/genomic sources) to it can be obtained by studying the LDG.

2. Status of LDG from Previous Studies

Pianka's comprehensive review on LDG aided the organization of the myriad hypotheses regarding it in 1966 [6]. Since then, many major journals have published numerous studies about the LDG, such as the flagship journals, *Global Ecology and Biogeography*, and *Journal of Biogeography*, have been the main sources on LDG articles in the past few decades; other mainstream journals including *Ecography*, *Ecology*, *Ecological Letters*, *American Naturalist*, and *Proceedings of the Royal Society B: Biological Sciences*, have also published many papers on LDG (Figure 2). In addition, the number of articles on LDG about land-based flora/fauna, marine organisms, and micro-organisms has increased since 1995, although the trend from 2019 and 2020 has slowed (Figure 3). Overall, the body of LDG-related literatures has grown substantially in recent decades.

All taxa, regardless of the land or the ocean, are conformed to the association between latitude gradients and taxonomic richness, e.g., terrestrial arthropods and terrestrial plants, mangrove trees, birds, mollusks, mammals, corals, freshwater arthropods, marine protists, marine arthropods, and reptiles. The existing distribution patterns of LDG of land-based flora, fauna, micro-organisms, and marine organisms are useful benchmarks to explore the distribution mechanisms so that they are stated briefly as follows.

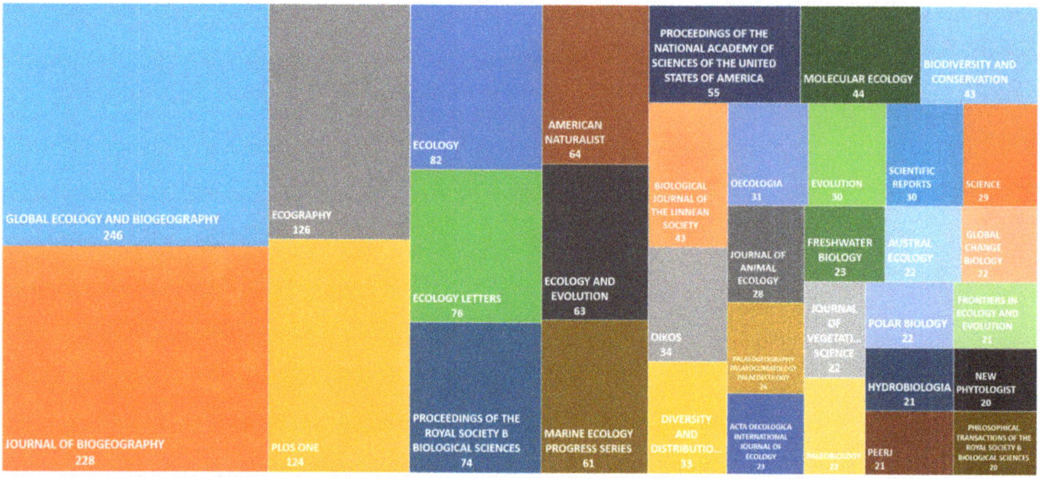

Figure 2. Number of documents per journal relating to the LDG research using the keywords "latitudinal diversity gradient" from Web of Science (Data accessed on 22 February 2022); only journals with more than 20 published papers are shown. The size of a block is proportional to the number of relevant papers in the given journal.

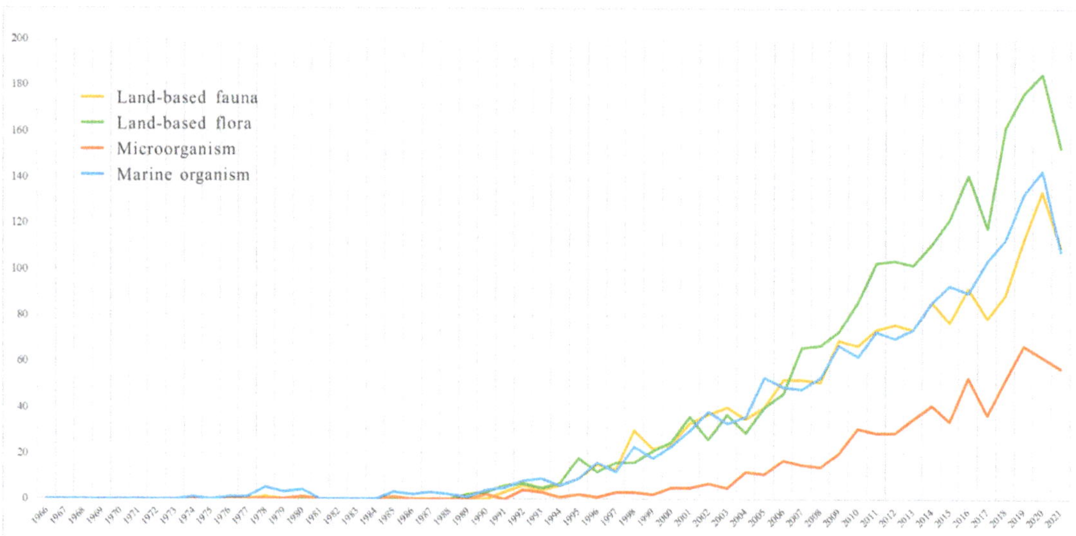

Figure 3. Search results using Web of Science (Data accessed on 22 February 2022) from 1966 to 2021, with the numbers of the published papers in each year; results include all documents and trends obtained using the keywords, "latitudinal diversity gradient" and "Land-based flora" (including plant, forest, flora, vegetation, algae, bryophyte, pteridophyte, spermatophyte, or lichen); "Land-based fauna" (including animal, zoology, fauna, vertebrate, invertebrate, bird, fish, mammal, amphibians, reptiles, or insect); "Microorganism" (including microbes, microorganism, bacteria, or germ); "Marine organism" (including marine, ocean, or sea), respectively.

2.1. *Land-Based Flora*

The LDG is most strongly visible in phytogeographical composition. A previous study has observed a similar distribution pattern of all seed plants in China, considering all geographical and climatic variables [26]. Huang et al. revealed that the endemic seed plants show a clear distribution of LDG from north to south, indicating that large-scale phytogeography of endemic flora that is strongly related to latitude, e.g., tropical genera account for approximately 75% of flora species at the southernmost tip (i.e., Hainan), which decreases to nearly 0 at latitude 45–50° N [25]. In Australia, the differences in plant reproductive strategies between communities at high and low latitudes indicate the importance of climate diversity patterns [27]. A global species density map for liverworts showed a significant pattern of LDG in species richness [28]. And latitudinal gradients of the richness of shrub and liana showed the same trends, i.e., the latitudes occupied by shrubs ranged nearly from 18–50° N, a greater range than that occupied by shrubs (18–45° N) and lianas (18–40° N) of latitude in species range size [29]. However, not all plants conform to the obvious LDG, i.e., the latitudinal trend in species richness is weakly negative in some liverworts and woody plants [22]. Notably, climate change is already altering the biogeographical patterns of flora, and the substantially diminish the extent and richness of Europe's high-latitude mountain flora has demonstrated that climate change is predicted to the lost at high latitudes [30]. Therefore, biogeographic histories of flora affected their vulnerability to the climate change, especially the climate warming; and the vulnerability of the endemic plants, implying high significance for conservation decisions in the shifts of LDG.

2.2. *Land-Based Fauna*

The LDG pattern of animals is significant in a broad sweep of taxa, e.g., in the tropics, the squamate lineages originating in situ reflect it, and the patterns are driven primarily by the dispersal and diversification rates [14,31]. Ant species richness also peaks in the tropics, which is consistent with that of many other taxa [32]. Extant birds are globally distributed, although the ubiquity of birds and their penchant for dispersal, and avian diversity studies have shown that both species numbers and the presence of higher taxonomic groups are skewed towards the tropical environments [33]. Approximately 92% of all mammalian diversity peaks in the tropical regions, with the exception of Lagomorpha, which shows maximum diversity in the northern and temperate regions [34]. LDG patterns are strikingly consistent with the diversification rates, wherein the peaks for species richness are always associated with low extinction rates and/or high speciation rates [34].

2.3. *Microorganisms*

The pattern of diversity in microorganisms with a small body size, fast population growth, high abundance, and high dispersal rates, which characteristics are contrary to that of macroorganisms. However, microorganisms unexpectedly showed the LDG pattern existed in bacteria as well as in marine protists (i.e., planktonic foraminifera) on a global scale [16,35,36]. Similarly, the beta diversity and phylogenetic diversity of *Streptomyces* strains showed an LDG pattern [37]. The LDG patterns were speculated to be involved in the geographic distribution of the host organisms. Surprisingly, both host richness and parasite abundance increased across 20° of latitude despite assumptions about diversity in parasites suggesting that their parasites exhibited no pattern or reverse latitudinal gradients to their hosts [38]. The reason for the reverse latitudinal gradients or lack of pattern in LDG in microorganisms is that the greater areas of wetlands at higher latitudes provide many habitats for larval amphibians to enhance host density, contributing positively to parasite richness [39]. In different organism groups (e.g., meiofauna, zooplankton and unicellular taxa), there is a significant difference in the LDG pattern between species richness and latitude [35].

2.4. Marine Organisms

The existence of LDG patterns in the sea is surprisingly controversial, especially when land organisms show pervasive LDG with maximum species richness in the tropical regions [40]. However, the LDG patterns for known marine organisms are well-studied in many taxa. Although the extant data from the deep seas are insufficient to analyze the LDG for sparsely distributed deep-sea organisms, some marine epifauna resident on the surface of the substratum in the ocean show a typical LDG [40]. The diversity of nematodes shows a positive LDG in the deep sea in the Atlantic, with a decline from 0° to 40° S [41]. The eastern Pacific bivalvia show a strong LDG within increasing numbers of species from the tropics to the southern tropical boundary in the northern Hemisphere; species numbers outside of the tropics show a stepwise decline toward the poles from 5° S to 8–9° N [40]. Interestingly, the fossil records of marine bivalves from the three successive slices in the late Cenozoic showed that species with tropical origins tended to expand from the tropics to higher latitudes [13]. These results support LDG in the marine shelf benthos, which is congruent with the diversity trends of gastropods along both northeastern Pacific shelves and the northwestern Atlantic [13,42]. For the coastal plants, e.g., mangrove, the mixing-isolation-mixing cycles can potentially generate species at an exponential rate, thus combining speciation and biodiversity in a unified framework by permitting intermittent gene flow during speciation [43]. But, the LDG pattern is still unclear for this.

2.5. LDG and Biodiversity Conservation

Biodiversity enhances many natural resources that are essential for human well-being, yet human activity has resulted in rapid biodiversity loss [44]. The sixth mass extinction in Earth's history has been driven by human activity [45,46] and global warming [44,47,48]. Accelerated biodiversity loss is a hallmark of the Anthropocene, in which large declines in population size, habitat loss, fragmentation, biological invasions, pollution, and climate change have been widely observed requiring effective and efficient conservation managements [45,48–51]. LDG, as a wide-scale diversity pattern on Earth, has inevitably been affected by the changing environment. The spatial model indicates that forests and jungles are exposed to anthropogenic threats (e.g., changes in fire regimes and deforestation) in Amazonia, Central America, the Eastern Arc Mountains, the Northern Andes, the Brazilian Atlantic, southeastern Asia, and Sub-Saharan Africa [23,52]. The pattern of LDG will inevitably be influenced in this dynamic environment; hence, the changing trends of biodiversity loss to human activity and climate change in the Anthropocene require more attention.

However, our understanding of the full impact of humanity on biodiversity, as well as of the links between the processes occurring in natural ecosystems, is incomplete [51]. As an important indicator of biodiversity patterns, the LDG patterns will be affected by climate warming and human activity over the global biodiversity framework. Over the past few decades, scientific studies have played important roles in verifying and identifying explicit goals for plant conservation, which are used in assessments of extinction risk, the prediction of range changes under climate change, and adaptation measures [30,53–56]. Therefore, we suggest that the cohesive nature of species richness and ecosystem diversity, particularly the trends of LDG in the changing world, will provide important insights into prioritizing conservation efforts.

3. Formation Mechanisms of the LDG

3.1. LDG Hypotheses

For the formation mechanism of LDG, there are six main hypotheses in a comprehensive review of LDG have been proposed previously, and these mainly focus on ecology and evolution [6,10]. The hypotheses from 1966 to 2021 for the latitudinal diversity gradient and the other sources are listed in Table 1. With the increase in the knowledge about LDG, more hypotheses have been proposed, but the late-comers are mainly deduced from the six hypotheses. In this review, we simply elaborate the relationships between the six main hypotheses and the others; and mainly focus on the widely embraced hypotheses. Thus, in this review the six broadly accepted hypotheses are revisited.

Table 1. Hypotheses from 1966 to 2021 for the latitudinal diversity gradient, P1–P6 from Pianka (1966), F1–F5 from Fine (2015), and O1–O7 from the other sources.

Hypothesis	Primary Focus	References
P1. The time theory	Ecology and evolution	[6]
P2. The theory of spatial heterogeneity	Ecology	[6]
P3. The competition hypothesis	Ecology	[6]
P4. The predation hypothesis	Ecology	[6]
P5. The theory of climatic stability	Ecology and evolution	[6]
P6. The productivity hypothesis	Ecology	[6]
F1. Time-integrated area, energy, and tropical niche conservatism	Evolution	[15]
F2. Climate stability	Evolution	[15]
F3. Temperature and evolutionary speed	Evolution	[15]
F4. Biotic interactions and speciation rate	Evolution	[15]
F5. Biotic interactions and finer niches	Ecology	[15]
O1.The ecological regulation hypothesis	Ecology	[32]
O2.The "diversification rate hypothesis	Evolution	[57]
O3.The out of the tropics hypothesis	Ecology and evolution	[58]
O4.The out-of-the-extratropics hypothesis	Ecology and evolution	[59]
O5.The evolutionary time hypothesis	Evolution	[32]
O6.The time-for-speciation hypothesis	Evolution	[60]
O7. The tropical niche conservatism hypothesis	Ecology	[61]

As known, the other hypotheses originate from the six main hypotheses and make differing predictions about the spatial distribution of organisms. For example, the out of the tropics (OTT) hypothesis describes how the combination of evolutionary dynamics and dispersal may have shaped the LDG of marine species [58]. Later, some groups may have also been shaped by dispersal towards the tropics, as in the out of the extratropics (OET) hypothesis [59]. Similarly, the evolutionary time hypothesis (ETH) suggests that tropical areas have been occupied for longer than temperate region, and thus have had more time to accumulate species [32]. Alternatively, under the diversification-rate hypothesis (DRH), higher richness in some clades is explained by faster rates of net diversification, and high species richness is associated with clades that have accumulated many species in a relatively short period of time [57]. The ecological regulation hypothesis (ERH) posits that there are equilibrated the ecological limits to species numbers, which vary systematically with latitude, perhaps due to the direct influence of climate and/or available energy [32]. The tropical conservatism hypothesis (TCH) suggests that the tropics have been occupied for longer, dispersal out of the tropics is rare, and the greater past area of the tropics yielded more present-day tropical clades [61]. From the above-mentioned examples of hypotheses, each of them is related with one or two of the six main hypotheses. Thus, the review briefly elaborates on the possible causes and/or hypothesis of LDG as follows.

The time hypothesis is perhaps the most widely accepted and the oldest of the six hypotheses. It can be dated back to the time of Alfred Russel Wallace, who proposed that evolutionary (speciation) and ecological (immigration) factors drove the increase in species richness of communities over time [6]. In tropical regions where have been occupied for longer periods, yielding more present-day tropical clades, and dispersal out of the tropics is rare [62]. The evolutionary time hypothesis suggests that the tropics provide more time for lineages to accumulate species [21]. This can be explained by geological events, i.e., the tropical regions remained relatively undisturbed, whereas the diversity in northern latitudes was reduced due to multiple glaciations, which led to the formation of the current LDG patterns [10].

The environmental heterogeneity hypothesis suggests that habitat heterogeneity promotes species richness [63], and the probability of species coexistence augments different niches [64]. Heterogeneity is beneficial in case of adverse environmental conditions. For the environmental heterogeneity, it has been shown that communities with increasing species capability and speciation withstood isolation or adaptation to diverse environmental conditions [65,66]. In addition, lower species ecological tolerance in the tropics may result in denser speciation and spatial heterogeneity at lower latitudes [54].

The competition hypothesis was proposed based on natural selection in the temperate zone, which was controlled by abiotic more than biotic factors with stronger biotic interactions and increased speciation rates for positive feedback; niches are narrower, competition is stronger, and co-evolutionary rates across the geographic mosaic have reduced the extinction rates [6,10,15]. Therefore, competition in the tropical regions is lower, as intense predation in the tropical environments results in reduced populations.

As an alternative to the competition hypothesis, the predation hypothesis suggests that competition in the tropics is lower due to higher predation in tropical environments, which causes a reduction in population size [6,10]. From the polar to tropical regions, more diversification was observed in the community composition and structure that are the greater probability that a larger proportion of predators can be maintained. Predators can then effectively control the number of prey and producer populations, such that there are more predators, parasites, and prey in the tropics than in other regions [67]. In tropical forests, trees attract their consumers, so that the consumption of seeds and seedlings by animals/herbivores reduces the number/survival rate of seeds/seedlings, and consequently the density of the population. In this way, herbivores increase the space available for the invasion of seeds and seedlings of other plant species, concomitantly increasing the diversity of tree species in the tropical forests [68].

The climatic stability hypothesis predicts that the tropical regions have a higher species richness, greater specialization, and narrower niches due to their stable climate [6,9,10]. The stable climate in tropical regions increases species richness; for example, climate oscillations have affected species diversity globally especially during the Late Quaternary period [69]. However, in the tropical regions with relatively stable climate trends that are likely to have prevented large demographic fluctuations, thus promoting the maintenance of species richness and intraspecific genetic diversity [69,70]. By contrast, rapid climate change in temperate and cold and/or arid regions has led to more profound effects on precipitation and temperature trends, which has directly or indirectly affected species demographics over time [69,71–73]. LDG patterns of increasing latitudinal ranges in animal and plant species richness from low to high are related to the tolerance to seasonal temperature variability [11,74] and Ice Age temperature fluctuations [75].

The productivity hypothesis suggests that species richness increases because of the greater productivity in tropical regions, allowing tighter species packing and narrower niches with a greater overlap of niches [6,10]. Energy input may enhance mutation and physiological rates, which then increase speciation by decreasing generation times [76]; thus population sizes should also correlate positively with productivity [77]. The net primary productivity and constraint on species richness due to limited resources may lead to geographical variation in species diversity [7,78]. Plant richness is primarily limited by water availability and solar energy at the base of the global food web [77,78]. For example, higher amounts of energy lead to faster evolutionary rates and more species richness in flowering plants [54]. In turn, predator richness is limited by the secondary production of herbivores in the food chain, whereas herbivore richness is limited by the net primary production of plants [78]. A positive relationship between richness and productivity could be responsible for the LDG pattern; however, the productivity hypothesis has not been accepted as an important cause of the LDG [17].

3.2. Climate Change, Temperature, and Precipitation

There is an equilibrium ecological limit to species numbers that varies systematically with latitude in LDG patterns due to the direct influence of available energy and/or climate [79,80]. The LDG mechanisms are difficult to determine due to the strong correlations between related ecological parameters, including climate, latitude, and temperature [37]. The species richness often varies with elevation and latitude or between geographic regions and temperature, precipitation, or other factors [7]. That is, the effects of species distribution on different species reflect different climatic tolerance among them. Thus, the explanations of the climatic variability hypothesis for variation along the latitudinal gradient favor the evolution of broader climatic tolerance of species at high latitudes [29,30].

During the geological ages, i.e., between the Tertiary and early Eocene period increased, global atmospheric temperatures increased and taxa evolved in tropical regions (low latitudes), and dispersed to higher latitudes, showing latitudinal patterns through tropical areas to higher latitudes [26]. And the current day species richness is also affected by climate oscillations on species demographic during the Late Quaternary [69,81]. Trends have been observed in the changes in the relative frequencies of tropical and temperate genera along the temperature, precipitation, and radiation gradients along the LDG [25]. For most plants, climate cooling (i.e., freezing tolerance) created an evolutionary barrier or survival limit, which determined the distribution range of many flowering plant taxa. For example, cold-intolerant plants of the boreotropical and evergreen flora were forced southward by climate cooling in the northern hemisphere [82]. Thus, this evolutionary process for cold adaptation is reasonable, and the tropical genera in flora descend along latitudes from low to high [26]; even some genera colonized from warm to cold [83]. An understanding of the environmental aspects that influence the traits underpinning adaptive resilience to changing climates could help in assessing the vulnerability of populations to climate change [84,85].

Intraspecific genetic diversity and high levels of species richness are also correlated with the past inter-annual precipitation variability [23]. For example, frequent variations in precipitation during the Late Quaternary and the resultant fluctuations are proposed to have driven population isolation and adaptive divergence in suitable habitats at lower latitudes [86,87]. Precipitation can be used to explain the patterns of selection of local and regional variations in climate regimes [88].

Model selection and hierarchical partitioning of species richness in liverworts showed that water-related variables are dominant [89]. For precipitation seasonality and availability, the spatial precipitation heterogeneity in mosses is considered an important predictor of species richness, but the temperature variables generally have higher explanatory power than water variables in woody plants [22]. Also, hierarchical partitioning in the species richness of liverworts and mosses indicated that the independent effects of temperature variables are higher than those of water variables and that water variables have more variation in these families than in woody plants [29]. This is consistent with the evolution of terrestrial plants, which involves their ability to adapt to water deficiencies [89]. Additionally, population persistence due to long-term climate stability results in a higher accumulation of species richness in the tropics than at higher latitudes, and climate-driven processes at lower latitudes always result in higher population divergence due to frequent precipitation variability [23].

4. Evolutionary Responses for LDG

Generally, evolutionary responses are prerequisites for the long-term persistence of biodiversity during ongoing and projected scenarios of the extreme climate events [85,90]. Many LDG results from ecological, historical geology, and demographic events that influence dispersal and diversification can be explained by the evolutionary responses based on the historical contingency proposed [62,91]. Likewise, the diversification rate suggested that extinction and/or speciation rates vary systematically with LDG [32]. Evolutionary responses (i.e., speciation, extinction, diversification, and dispersal rates) are inevitable in

LDG patterns. The evolutionary forces generated by LDG have been debated for many years [5,62]. Here, related evolutionary responses are listed to address the LDG pattern.

4.1. Speciation Rate

The key to explaining LDG patterns is to understand the variation in speciation and extinction rates with latitude; however, it is difficult to estimate the diversification rates associated with specific geographic locations [34]. The speciation rates differ due to a range of factors and in different geographic regions [7]. In tropical clades, speciation is relatively high, and extinction appears to be very low, whereas in temperate lineages, speciation and extinction are very high [31,92]; the speciation rate is always linked to the diversification rate. For example, the relative stability in tropical climes led to older, slower-evolving but still species-rich communities, and long-term cooling had a disproportionate effect on the non-tropical diversification rates, leading to dynamic young communities outside of the tropics [93]. Thus, nonequilibrium explanations for LDG have been proposed based on different speciation rates in the tropics [35]. For example, in tropical regions where the ambient conditions are warmer, environmental factors could increase the mutation rates and lead to faster rates of evolution [8,94]. Thus, the speciation rate affects LDG patterns.

4.2. Extinction Rate

The extinction rates have led to geographic patterns of species richness; for example, a geographic region with climatic stability may play a key role in extinction levels [7]. In addition, species with larger distributions are less prone to extinction, and regions experiencing intense climatic fluctuations always experience increased extinction rates [70,95,96]. In the phylogeny, the older crown groups from the phylogenetic tree always accumulate more species as expected, whereas the stem groups have fewer species so that older temperate taxa have an apparently attenuated extinction [31,92]. And in the tropics, the mammals, e.g., amphibians, and squamates, also have lower extinction rates that contributed to more net diversification than speciation [31,34,97]. The dependence of extinction and speciation on diversity is a general process that regulates the shapes of taxonomic diversity curves [5]. Together, the association between a high species richness and low extinction rate suggested that extinction may play a more important role in driving differences in net diversification rates than speciation along latitudinal gradients [7].

4.3. Net Diversification Rate

The outcome of both speciation and extinction is the net diversification rate, which is typically higher in tropical clades and represents the difference between speciation and extinction [31]. The rates of diversification are higher in tropical latitudes; a strong, consistent role of net diversification in driving latitudinal species richness gradients has been proposed [7,8,71]. Net diversification rates suggest that higher temperatures or energy fluxes promote more rapid evolution; and a more equitable climate favors habitat specialization, leading to the dispersal and gene flow reduced and/or more intense biological interactions drive the adaptive changes [6,11,54,98]. Notably, the areas with high net diversification rates are more likely to be evolutionary cradles [14]. In the tropics, the squamate lineages suggested the greater species richness due to speciation, but the LDG appears to be driven primarily by the diversification rates [31]. Nevertheless, there are exceptions; e.g., the clades of passerines and swallowtails showed a significant latitudinal effect on the relative diversification rates [99].

4.4. Dispersal Rate

Dispersal always promotes gene flow wherein the colonization of habitats among isolated patches occurs. It has played a key role in species and population persistence in fragmented systems under rapid climatic change [81]. Two main hypotheses on dispersal dominance have been proposed. Firstly, the "out of the tropics" hypothesis suggests that lineages originate and diversify in the tropics and disperse from the tropics to the temperate regions. Secondly, the "tropical niche conservatism" hypothesis suggests that the lineages originate in the tropics and accumulate in tropical regions, because it is difficult to disperse and adapt in temperate regions [13,34]. Most of the present-day diversity in marine bivalves in extratropical regions is from taxa shared with the tropics, which has supported dispersal and the persistence of an LDG [40,58]. The role of dispersal in generating diversity gradients has been determined by studying the frequency of shifts from temperate to tropical biomes and vice versa [9]. More lineages are dispersed from the tropical to temperate regions than from the temperate to tropical regions, and the dominance of tropical to temperate dispersal is statistically significant in all scenarios [9,12,13]. In tropical regions, the newly evolved taxa might have extended northward into the temperate regions [100]. However, the range of extension in warm regions was curbed by frost tolerance; rapid and large-scale range contractions and expansions may have resulted in population extirpation and the subsequent loss of species richness at these latitudes [80,81].

5. Future Perspectives

The most unique feature of the Earth is the existence of life, and the most extraordinary feature of life is its diversity (i.e., genes, species, and ecosystem diversity), which provides numerous essential services to a human-dominated society [48,49]. Considering that LDG is the broadest and most notable biodiversity pattern [15], the conservation efforts should be linked to the mechanisms of ecology and evolution from genetic/genomic diversity, species diversity, and ecosystem diversity in the future projects.

In recent decades, the studies of LDG have evolved due to the promotion of studies on land-based flora/fauna, microorganisms, and marine organisms (Figure 3). Using the comprehensive review written by Pianka as a milestone [6,10], the myriad of hypotheses for LDG has been organized into a manageable framework for future studies. Many studies on LDG have developed scientific tools for the elucidation of diversity and the related causes. However, most biodiversity patterns of LDG have mainly been explored based on ecosystem and species numbers [10,15]. As known, species is the evolutionary unit, and species diversity is the basic element for evolutionary change. Also, the ecosystem diversity of specific areas preserves many species and the subsequent genetic diversity [48]. According to the Convention on Biological Diversity (CBD, www.cbd.int, accessed on 23 January 2022), genetic diversity is recognized as one of the three basic elements of biodiversity and is the focus of many conservation genetics studies. Thus, we suggest that LDG patterns are useful for investigating the ecological and evolutionary mechanisms from genetic/genomic and phylogenetic diversity to facilitate the estimation of clade-specific diversification rates in relation to latitude, climate, and other factors. Estimating the contribution of these factors from ecology and evolution in promoting the LDG patterns from phylogenetic and/or genetic/genomic data is challenging, but the availability of big data is gradually reality. Also, it is worth noting that the development of Geographic Information System (GIS) mapping and satellite imagery has provided unprecedented resources to study LDG from the perspectives of global patterns of climate, productivity, landform, and species richness [10]. In addition, LDG can be explored with technological achievements from both paleo- and modern studies [18].

With the coming of new era of the Anthropocene, the biodiversity crisis is closely connected to the human activity. Predicting the influence of human-induced climatic change on a short and/or long-term organismal distribution is imperative in contemporary biology [101]. Over the course of a century, humans have markedly altered the planet, causing various effects, including an increase in ocean acidity to landscape fragmentation and climate change [48]. The far-reaching influence of human activity has contributed to a loss of biodiversity, changing land use, habitat loss, plant extinction, predatory fish, defaunation, and a reduction in species abundance [48,102]. Therefore, investigations of the LDG pattern must consider the knock-on effects on biodiverse communities from the concomitant influence of human activity and climate change. According to the shifts or trends of LDG patterns, some protected areas can be identified. Protected areas are safeguarded from human activity to a certain extent through conservation planning and prioritization in a human-dominated and fast-changing world [48]. Now the solution of human activity to biodiversity is easily found to these issues, i.e., the establishments of related legislations and conservation areas to decrease the impacts from human activity. However, little is known about the effects of climate change on biodiversity, and thus, significant research on this aspect is crucial. Previous research on well-studied large organisms showed that biodiversity may be more sensitive to climate, such that the impact of ongoing Anthropocene climatic change may be much more serious than previously thought [58]. In particular, wide-scale extinctions and population decline across taxonomic groups have been caused by human activity over the past 500 years [103]. However, most studies on LDG have mainly focused on ecological factors, i.e., species and ecosystems. Therefore, the impact of human activity or imprint and climate change (e.g., the warming climate) on the diversity of ecosystems along latitudinal gradient has been generally disregarded.

In addition, the explanations of LDG patterns and/or mechanisms from genetics and genomics are rare. The reasonable utilization and benefit of sharing of genetic resources have been inferred from the CBD to ensure the conservation of biodiversity [104]. Based on whole genomes, a strategy for cataloging adaptive genetic diversity to climate change across a range of ecologically important non-model species has improved population datasets and provided a high-resolution record of variation in structural information and genomes [88]. The models linking genomics with eco-evolution provide unique opportunities for predicting and tracking vulnerability and adaptive responses to climate change, which will benefit the biodiversity issues along the LDG distribution. Thus, LDG taken from large-scale genetic/genomic data (e.g., the DNA data from NCBI) or the genetic variation in specific taxa in wild populations with a vast distribution will reveal the distribution patterns and mechanisms to assist the CBD in meeting targets to halt the acceleration in biodiversity loss. In the face of environmental change, the ever-increasing availability of ecological studies is going on, integrating the big data from evolution across large-scale distributions and taxa, will provide important and novel opportunities to enhance our understanding of the adaptive potential of LDG globally. Hence, ecologists, evolutionists, and related scholars are called upon to rethink the LDG in view of the available genetic resources in the changing world.

6. Conclusions

Half a century ago, the review from Pianka is believed to be a milestone in understanding LDG patterns, which is and will be the manageable framework for future studies [6,10]. Subsequently, a time-integrated biogeographic analysis of geographical diversification suggests that both time-integrated and stable climate areas will determine the baseline of marine diversity and terrestrial patterns at the global scale. Patterns of the LDG mainly focused on the ecology and evolution in tropics from the species richness to mechanism. The tropics are geologically older, and have had more time for diversification, which is consistent with the time hypothesis, in which biotic interactions likely augment coexistence and diversification [15]. Thus, the related studies from tropical regions to warm and/or cold regions are the key points we have to take into account, e.g., the different taxa and/or communities along with latitudinal gradients. According to our own experience, *Engelhardia* is distributed from 10° S to 30° N on a latitudinal distribution in Southeast Asia, where the plants are typical to unique substrates to disentangle the LDG from integrated disciplines [100,105].

Here, this review presents a systematic overview of the broad and comprehensive literatures on the state of LDG in a changing world where ecology and evolution have been applied to its distribution patterns. In the future, we suggest that genetic/genomic-based approaches on LDG should be integrated for the understanding of biodiversity conservation. This facet is still underdeveloped, and the numbers of studies that elaborate biodiversity patterns of LDG based on the genetic/genomic data are still scarce, especially regarding the relationship between the variations in genetic/genomic data and the pre-existing data from ecosystems and speciation (e.g., phylogeny and genetic diversity). Moreover, the use of molecular studies (i.e., genetic/genomic diversity) is important to highlight the potential for the establishment of theories on LDG that do not have sufficient support from genetic/genomic implementation, because the dataset for a considerably large-scale distribution is difficult to obtain, especially the vast regions across oceans. However, genetic/genomic criteria of specific taxa with a vast distribution are useful and possible in some plants with a large latitudinal distribution if the field works is comprehensive. The large-scale taxa collection combined the big data DNA information, will enable the CBD to ensure biodiversity conservation through the sharing and utilization of benefits from genetic/genomic resources. Thus, exploring the LDG from large-scale DNA data to disentangle the diversity mechanisms and patterns to inform biodiversity conservation and management measures may be a valid approach. Comprehensive biodiversity patterns, together with the determination of ecological and evolutionary mechanisms should therefore be used to understand the mechanisms and causes of LDG in a changing environment where biodiversity is rapidly declining and disappearing.

Author Contributions: Conceptualization, Y.Z. and H.-H.M.; Funding acquisition, H.-H.M. and Y.-G.S.; Project administration, H.-H.M.; Visualization, Y.Z. and H.-H.M.; Formal Analysis, Y.Z.; Writing—original draft, Y.Z., Y.-G.S., C.-Y.Z., T.-R.W., T.-H.S., P.-H.H., J.L. and H.-H.M.; Writing—review and editing, Y.Z., Y.-G.S., C.-Y.Z., T.-R.W., T.-H.S., P.-H.H., J.L. and H.-H.M. All authors have read and agreed to the published version of the manuscript.

Funding: This research was funded by National Natural Science Foundation of China (No. 42171063); Southeast Asia Biodiversity Research Institute, Chinese Academy of Sciences (No. Y4ZK111B01); the Special Fund for Scientific Research of Shanghai Landscaping & City Appearance Administrative Bureau (Nos. G192422 and G212406); Youth Innovation Promotion Association, Chinese Academy of Sciences (No. 2018432); and the CAS "Light of West China" Program.

Institutional Review Board Statement: Not applicable.

Data Availability Statement: Not applicable.

Acknowledgments: The authors are grateful to Chong-Rui Ai (Xishuangbanna Tropical Botanical Garden, Chinese Academy of Sciences) and Hui-Jie Qiao (Institute of Zoology, Chinese Academy of Sciences) for their kind assistances and suggestions in the references & literatures investigation for this paper.

Conflicts of Interest: The authors declare no conflict of interest.

References

1. von Humboldt, A. *Views of Nature: Or Contemplations on the Sublime Phenomena of Creation with Scientific Illustrations*; (Transl. Otté, E.C., Bohn, H.G., 1st edition in 1850); Higher Education Press: Beijing, China, 2016.
2. Hawkins, B.A. Ecology's oldest pattern? *Trends Ecol. Evol.* **2001**, *16*, 470. [CrossRef]
3. Hillebrand, H. On the generality of the latitudinal diversity gradient. *Am. Nat.* **2004**, *163*, 192–211. [CrossRef]
4. Kinlock, N.L.; Prowant, L.; Herstoff, E.M.; Foley, C.M.; Akin-Fajiye, M.; Bender, N.; Umarani, M.; Ryu, H.Y.; Sen, B.; Gurevitch, J. Explaining global variation in the latitudinal diversity gradient: Meta-analysis confirms known patterns and uncovers new ones. *Glob. Ecol. Biogeogr.* **2017**, *27*, 125–141. [CrossRef]
5. Mittelbach, G.G.; Schemske, D.W.; Cornell, H.V.; Allen, A.P.; Brown, J.M.; Bush, M.B.; Harrison, S.P.; Hurlbert, A.H.; Knowlton, N.; Lessios, H.; et al. Evolution and the latitudinal diversity gradient: Speciation, extinction and biogeography. *Ecol. Lett.* **2007**, *10*, 315–331. [CrossRef]
6. Pianka, E.R. Latitudinal Gradients in Species Diversity: A Review of Concepts. *Am. Nat.* **1966**, *100*, 33–46. [CrossRef]
7. Pulido–Santacruz, P.; Weir, J.T. Extinction as a driver of avian latitudinal diversity gradients. *Evolution* **2016**, *70*, 860–872. [CrossRef]
8. Rohde, K. Latitudinal gradients in species diversity: The search for the primary cause. *Oikos* **1992**, *65*, 514–527. [CrossRef]
9. Saupe, E.E.; Myers, C.E.; Townsend Peterson, A.; Soberon, J.; Singarayer, J.; Valdes, P.; Qiao, H. Spatio–temporal climate change contributes to latitudinal diversity gradients. *Nat. Ecol. Evol.* **2019**, *3*, 1419–1429. [CrossRef]
10. Schemske, D.W.; Mittelbach, G.G. "Latitudinal gradients in species diversity": Reflections on Pianka's 1966 article and a look forward. *Am. Nat.* **2017**, *189*, 599–603. [CrossRef]
11. Stevens, G.C. The latitudinal gradient in geographical range: How so many species coexist in the tropics. *Am. Nat.* **1989**, *133*, 240–256. [CrossRef]

12. Tomašových, A.; Jablonski, D.; Sandel, B. Decoupling of latitudinal gradients in species and genus geographic range size: A signature of clade range expansion. *Glob. Ecol. Biogeogr.* **2016**, *26*, 288–303. [CrossRef]
13. Jablonski, D.; Belanger, C.L.; Berke, S.K.; Huang, S.; Krug, A.Z.; Roy, K.; Tomašových, A.; Valentine, J.W. Out of the tropics, but how? Fossils, bridge species, and thermal ranges in the dynamics of the marine latitudinal diversity gradient. *Proc. Natl. Acad. Sci. USA* **2013**, *110*, 10487–10494. [CrossRef] [PubMed]
14. Jablonski, D.; Roy, K.; Valentine, J.W. Out of the tropics: Evolutionary dynamics of the latitudinal diversity gradien. *Science* **2006**, *314*, 102–106. [CrossRef]
15. Fine, P.V.A. Ecological and evolutionary drivers of geographic variation in species diversity. *Annu. Rev. Ecol. Evol. Syst.* **2015**, *46*, 369–392. [CrossRef]
16. Fuhrman, J.A.; Steele, J.A.; Hewson, I.; Schwalbach, M.S.; Brown, M.V.; Green, J.L.; Brown, J.H. A latitudinal diversity gradient in planktonic marine bacteria. *Proc. Natl. Acad. Sci. USA* **2008**, *105*, 7774–7778. [CrossRef]
17. Willig, M.R.; Kaufman, D.M.; Stevens, R.D. Latitudinal gradients of biodiversity: Pattern, process, scale, and synthesis. *Annu. Rev. Ecol. Evol. Syst.* **2003**, *34*, 273–309. [CrossRef]
18. Jablonski, D.; Huang, S.; Roy, K.; Valentine, J.W. Shaping the latitudinal diversity gradient: New perspectives from a synthesis of paleobiology and biogeography. *Am. Nat.* **2017**, *189*, 1–12. [CrossRef]
19. Lomolino, M.V.; Riddle, B.R.; Whittaker, R.J.; Brown, J.H. *Biogeography*, 4th ed.; Sinauer Associates, Inc.: Sunderland, MA, USA, 2010.
20. Hanly, P.J.; Mittelbach, G.G.; Schemske, D.W. Speciation and the latitudinal diversity gradient: Insights from the global distribution of endemic fish. *Am. Nat.* **2017**, *189*, 604–615. [CrossRef]
21. Jansson, R.; Rodriguez–Castaneda, G.; Harding, L.E. What can multiple phylogenies say about the latitudinal diversity gradient? A new look at the tropical conservatism, out of the tropics, and diversification rate hypotheses. *Evolution* **2013**, *67*, 1741–1755. [CrossRef]
22. Chen, S.B.; Ferry Slik, J.W.; Gao, J.; Mao, L.F.; Bi, M.J.; Shen, M.W.; Zhou, K.X. Latitudinal diversity gradients in bryophytes and woody plants: Roles of temperature and water availability. *J. Syst. Evol.* **2015**, *53*, 535–545. [CrossRef]
23. Theodoridis, S.; Fordham, D.A.; Brown, S.C.; Li, S.; Rahbek, C.; Nogues-Bravo, D. Evolutionary history and past climate change shape the distribution of genetic diversity in terrestrial mammals. *Nat. Commun.* **2020**, *11*, 2557. [CrossRef] [PubMed]
24. Chown, S.L.; Gaston, K.J. Areas, cradles and museums: The latitudinal gradient in species richness. *Trends Ecol. Evol.* **2000**, *15*, 311–315. [CrossRef]
25. Huang, J.; Ma, K.; Huang, J. Phytogeographical patterns of genera of endemic flora in relation to latitudinal and climatic gradients in China. *Plant Syst. Evol.* **2017**, *303*, 689–698. [CrossRef]
26. Qian, H.; Song, J.S.; Krestov, P.; Guo, Q.; Wu, Z.; Shen, X.; Guo, X. Large–scale phytogeographical patterns in East Asia in relation to latitudinal and climatic gradients. *J. Biogeogr.* **2003**, *30*, 129–141. [CrossRef]
27. Smith, S.; Brauer, C.J.; Sasaki, M.; Unmack, P.J.; Guillot, G.; Laporte, M.; Bernatchez, L.; Beheregaray, L.B. Latitudinal variation in climate–associated genes imperils range edge populations. *Mol. Ecol.* **2020**, *29*, 4337–4349. [CrossRef] [PubMed]
28. von Konrat, M.; Renner, M.; Soederstroem, L.; Hagborg, A.; Mutke, J. Chapter nine: Early land plants today: Liverwort species diversity and the relationship with higher taxonomy and higher plants. *Fieldiana. Bot.* **2008**, *47*, 91–104. [CrossRef]
29. Liu, H.; Yu, R.; Huang, J.; Liu, Y.; Zang, R.; Guo, Z.; Ding, Y.; Lu, X.; Li, Q.; Chen, H.Y.H. Latitudinal diversity gradients and Rapoport effects in Chineseendemic woody seed plants. *Forests* **2020**, *11*, 1029. [CrossRef]
30. Niskanen, A.K.J.; Niittynen, P.; Aalto, J.; Väre, H.; Luoto, M.; Diez, J. Lost at high latitudes: Arctic and endemic plants under threat as climate warms. *Divers. Distrib.* **2019**, *25*, 809–821. [CrossRef]
31. Pyron, R.A. Temperate extinction in squamate reptiles and the roots of latitudinal diversity gradients. *Glob. Ecol. Biogeogr.* **2014**, *23*, 1126–1134. [CrossRef]
32. Economo, E.P.; Narula, N.; Friedman, N.R.; Weiser, M.D.; Guenard, B. Macroecology and macroevolution of the latitudinal diversity gradient in ants. *Nat. Commun.* **2018**, *9*, 1778. [CrossRef]
33. Saupe, E.E.; Farnsworth, A.; Lunt, D.J.; Sagoo, N.; Pham, K.V.; Field, D.J. Climatic shifts drove major contractions in avian latitudinal distributions throughout the Cenozoic. *Proc. Natl. Acad. Sci. USA* **2019**, *116*, 12895–12900. [CrossRef] [PubMed]
34. Rolland, J.; Condamine, F.L.; Jiguet, F.; Morlon, H. Faster speciation and reduced extinction in the tropics contribute to the mammalian latitudinal diversity gradient. *PLoS Biol.* **2014**, *12*, e1001775. [CrossRef] [PubMed]
35. Hillebrand, H.; Azovsky, A.I. Body size determines the strength of the latitudinal diversity gradient. *Ecography* **2001**, *24*, 251–256. [CrossRef]
36. Wilhelm, F. Biogeography and dispersal of micro-organisms: A review emphasizing protists. *Acta Protozool.* **2006**, *45*, 111–136.
37. Andam, C.P.; Doroghazi, J.R.; Campbell, A.N.; Kelly, P.J.; Choudoir, M.J.; Buckley, D.H. A latitudinal diversity gradient in Terrestrial Bacteria of the genus *Streptomyces*. *mBio* **2016**, *7*, e02200-15. [CrossRef]
38. Johnson, P.T.J.; Haas, S.E. Why do parasites exhibit reverse latitudinal diversity gradients? Testing the roles of host diversity, habitat and climate. *Glob. Ecol. Biogeogr.* **2021**, *30*, 1810–1821. [CrossRef]
39. McDonald, C.P.; Rover, J.A.; Stets, E.G.; Striegl, R.G. The regional abundance and size distribution of lakes and reservoirs in the United States and implications for estimates of global lake extent. *Limnol. Oceanogr.* **2012**, *57*, 597–606. [CrossRef]
40. Roy, K.; Jablonski, D.; Valentine, J.W. Dissecting latitudinal diversity gradients: Functional groups and clades of marine bivalves. *Proc. R. Soc. Lond. B-Biol. Sci.* **2000**, *267*, 293–299. [CrossRef]

41. Lambshead, P.J.D.; Tietjen, J.; Timothy, F.; Jensen, P. Latitudinal diversity gradients in the deep sea with special reference to North Atlantic nematodes. *Mar. Ecol. Prog. Ser.* **2000**, *194*, 159–167. [CrossRef]
42. Roy, K.; Jablonski, D.; Valentine, J.W.; Rosenberg, G. Marine latitudinal diversity gradients: Tests of causal hypotheses. *Proc. Natl. Acad. Sci. USA* **1998**, *95*, 3699–3702. [CrossRef]
43. He, Z.W.; Li, X.N.; Yang, M.; Wang, X.F.; Zhong, C.R.; Duke, N.C.; Wu, C.I.; Shi, S.H. Speciation with gene flow via cycles of isolation and migration: Insights from multiple mangrove taxa. *Natl. Sci. Rev.* **2019**, *6*, 275–288. [CrossRef] [PubMed]
44. Meng, H.H.; Zhou, S.S.; Li, L.; Tan, Y.H.; Li, J.W.; Li, J. Conflict between biodiversity conservation and economic growth: Insight into rare plants in tropical China. *Biodivers. Conserv.* **2019**, *28*, 523–537. [CrossRef]
45. Johnson, C.N.; Balmford, A.; Brook, B.W.; Buettel, J.C.; Galetti, M.; Guangchun, L.; Wilmshurst, J.M. Biodiversity losses and conservation responses in the Anthropocene. *Science* **2017**, *356*, 270–275. [CrossRef] [PubMed]
46. Isbel, F.; Gonzalez, A.; Loreau, M.; Cowles, J.; Diaz, S.; Hector, A.; Mace, M.M.; Wardle, D.A.; O'Connor, M.I.; Duffy, J.M.; et al. Linking the influence and dependence of people on biodiversity across scales. *Nature* **2017**, *546*, 65–72. [CrossRef]
47. Meng, H.H.; Zhou, S.S.; Jiang, X.L.; Gugger, P.F.; Li, L.; Tan, Y.H.; Li, J. Are mountaintops climate refugia for plants under global warming? A lesson from high–mountain oaks in tropical rainforest. *Alp. Bot.* **2019**, *129*, 175–183. [CrossRef]
48. Meng, H.H.; Gao, X.Y.; Song, Y.G.; Cao, G.L.; Li, J. Biodiversity arks in the Anthropocene. *Reg. Sustain.* **2021**, *2*, 109–115. [CrossRef]
49. Cardinale, B.J.; Duffy, J.E.; Gonzalez, A.; Hooper, D.U.; Perrings, C.; Venail, P.; Narwani, A.; Mace, G.M.; Tilman, D.; Wardle, D.A.; et al. Biodiversity loss and its impact on humanity. *Nature* **2012**, *486*, 59–67. [CrossRef]
50. Li, H.P.; XiangYu, J.G.; Dai, G.Y.; Gu, Z.L.; Ming, C.; Yang, Z.F.; Ryder, O.A.; Li, W.H.; Fu, Y.X.; Zhang, Y.P. Large numbers of vertebrates began rapid population decline in the late 19th century. *Proc. Natl. Acad. Sci. USA* **2016**, *113*, 14079–14084. [CrossRef]
51. Tucker, C.M.; Aze, T.; Cadotte, M.W.; Cantalapiedra, J.L.; Chisholm, C.; Diaz, S.; Grenyer, R.; Huang, D.W.; Mazel, F.; Pearse, W.D.; et al. Assessing the utility of conserving evolutionary history. *Biol. Rev.* **2019**, *94*, 1740–1760. [CrossRef]
52. Barlow, J.; Franca, F.; Gardner, T.A.; Hicks, C.C.; Lennox, G.D.; Berenguer, E.; Castello, L.; Economo, E.P.; Ferreira, J.; Guenard, B.; et al. The future of hyperdiverse tropical ecosystems. *Nature* **2018**, *559*, 517–526. [CrossRef]
53. Bonebrake, T.C.; Brown, C.J.; Bell, J.D.; Blanchard, J.L.; Chauvenet, A.; Champion, C.; Chen, I.C.; Clark, T.D.; Colwell, R.K.; Danielsen, F.; et al. Managing consequences of climate–driven species redistribution requires integration of ecology, conservation and social science. *Biol. Rev.* **2018**, *93*, 284–305. [CrossRef] [PubMed]
54. Davies, T.J.; Barraclough, T.G.; Savolainen, V.; Chase, M.W. Environmental causes for plant biodiversity gradients. *Philos. Trans. R. Soc. Lond. B-Biol. Sci.* **2004**, *359*, 1645–1656. [CrossRef] [PubMed]
55. Lenoir, J.; Svenning, J.C. Climate–related range shifts a global multidimensional synthesis and new research directions. *Ecography* **2015**, *38*, 15–28. [CrossRef]
56. Margules, C.R.; Pressey, R.L. Systematic conservation planning. *Nature* **2000**, *405*, 243–253. [CrossRef] [PubMed]
57. Scholl, J.P.; Wiens, J.J. Diversification rates and species richness across the Tree of Life. *Proc. R. Soc. Lond. B-Biol. Sci.* **2016**, *283*. [CrossRef]
58. Chiu, W.T.R.; Yasuhara, M.; Cronin, T.M.; Hunt, G.; Gemery, L.; Wei, C.L. Marine latitudinal diversity gradients, niche conservatism and out of the tropics and Arctic: Climatic sensitivity of small organisms. *J. Biogeogr.* **2020**, *47*, 817–828. [CrossRef]
59. Raja, N.B.; Kiessling, W. Out of the extratropics: The evolution of the latitudinal diversity gradient of Cenozoic marine plankton. *Proc. R. Soc. B-Biol. Sci.* **2021**, *288*, 20210545. [CrossRef]
60. Laenen, B.; Patiño, J.; Hagborg, A.; Désamoré, A.; Wang, J.; Shaw, A.J.; Goffinet, B.; Vanderpoorten, A. Evolutionary origin of the latitudinal diversity gradient in liverworts. *Mol. Phylogenet. Evol.* **2018**, *127*, 606–612. [CrossRef]
61. Qian, H.; Jin, Y.; Ricklefs, R.E. Patterns of phylogenetic relatedness of angiosperm woody plants across biomes and life-history stages. *J. Biogeogr.* **2017**, *44*, 1383–1392. [CrossRef]
62. Wiens, J.J.; Donoghue, M.J. Historical biogeography, ecology and species richness. *Trends Ecol. Evol.* **2004**, *19*, 639–644. [CrossRef]
63. Stein, A.; Gerstner, K.; Kreft, H. Environmental heterogeneity as a universal driver of species richness across taxa, biomes and spatial scales. *Ecol. Lett.* **2014**, *17*, 866–880. [CrossRef] [PubMed]
64. Tews, J.; Brose, U.; Grimm, V.; Tielbörger, K.; Wichmann, M.C.; Schwager, M.; Jeltsch, F. Animal species diversity driven by habitat heterogeneity/diversity: The Importance of Keystone Structures. *J. Biogeogr.* **2004**, *31*, 79–92. [CrossRef]
65. Hughes, C.; Eastwood, R. Island radiation on a continental scale: Exceptional rates of plant diversification after uplift of the Andes. *Proc. Natl. Acad. Sci. USA* **2006**, *103*, 10334–10339. [CrossRef] [PubMed]
66. Kallimanis, A.S.; Bergmeier, E.; Panitsa, M.; Georghiou, K.; Delipetrou, P.; Dimopoulos, P. Biogeographical determinants for total and endemic species richness in a continental archipelago. *Biol. Conserv.* **2009**, *19*, 1225–1235. [CrossRef]
67. Harms, K.E.; Wright, S.J.; Calderón, O.; Hernández, A.; Herre, E.A. Pervasive density–dependent recruitment enhances seedling diversity in a tropical forest. *Nature* **2000**, *404*, 493–495. [CrossRef]
68. Janzen, D.H. Herbivores and the Number of Tree Species in Tropical Forests. *Am. Nat.* **1970**, *104*, 501–528. [CrossRef]
69. Hewitt, G.M. Genetic consequences of climatic oscillations in the Quaternary. *Philos. Trans. R. Soc. Lond. B-Biol. Sci.* **2004**, *359*, 183–195. [CrossRef]
70. Dynesius, M.; Jansson, R. Evolutionary consequences of changes in species' geographical distributions driven by Milankovitch climate oscillations. *Proc. Natl. Acad. Sci. USA* **2000**, *97*, 9115–9120. [CrossRef]
71. Fischer, A.G. Latitudinal variations in organic diversity. *Evolution* **1960**, *14*, 64–81. [CrossRef]

72. Gao, X.Y.; Meng, H.H.; Zhang, M.L. Diversification and vicariance of desert plants: Evidence inferred from chloroplast DNA sequence variation of *Lagochilus ilicifolius* (Lamiaceae). *Biochem. Syst. Ecol.* **2014**, *55*, 93–100. [CrossRef]
73. Meng, H.H.; Gao, X.Y.; Huang, J.F.; Zhang, M.L. Plant phylogeography in arid Northwest China: Retrospectives and perspectives. *J. Syst. Evol.* **2015**, *53*, 33–46. [CrossRef]
74. Brown, J.H.; Stevens, G.C.; Kaufman, D.M. The geographic range: Size, shape, boundaries, and internal structure. *Annu. Rev. Ecol. Evol. Syst.* **1996**, *27*, 597–623. [CrossRef]
75. Rohde, K. Robust Phylogenies and adaptive radiations: A critical examination of methods used to identify key Innovations. *Am. Nat.* **1996**, *148*, 481–500. [CrossRef]
76. Rohde, K. Latitudinal gradients in species diversity and Rapoport's rule revisited: A review of recent work and what can parasites teach us about the causes of the gradients? *Ecography* **1999**, *22*, 593–613. [CrossRef]
77. Wright, D.H. Speciesenergy theory: An extension of speciesarea theory. *Oikos* **1983**, *41*, 496–506. [CrossRef]
78. Hawkins, B.A.; Field, R.; Cornell, H.V.; Currie, D.J.; Guégan, J.F.; Kaufman, D.M.; Kerr, J.T.; Mittelbach, G.G.; Oberdorff, T.; O'Brien, E.M.; et al. Energy, water, and broadscale geographic patterns of species richness. *Ecology* **2003**, *84*, 3105–3117. [CrossRef]
79. Rabosky, D.L. Ecological limits and diversification rate: Alternative paradigms to explain the variation in species richness among clades and regions. *Ecol. Lett.* **2009**, *12*, 735–743. [CrossRef]
80. Rabosky, D.L.; Hurlbert, A.H. Species richness at continental scales is dominated by ecological limits. *Am. Nat.* **2015**, *185*, 572–583. [CrossRef]
81. Theodoridis, S.; Randin, C.; Szovenyi, P.; Boucher, F.C.; Patsiou, T.S.; Conti, E. How do coldadapted plants respond to climatic cycles? Interglacial expansion explains current distribution and genomic diversity in *Primula farinosa* L. *Syst. Biol.* **2017**, *66*, 715–736. [CrossRef]
82. Tiffney, B.H. erspectives on the origin of the floristic similarity between Eastern Asia and Eastern North America. *J. Arn. Arbor.* **1985**, *66*, 73–94.
83. Meng, H.H.; Su, T.; Gao, X.Y.; Li, J.; Jiang, X.L.; Sun, H.; Zhou, Z.K. Warm-cold colonization: Response of oaks to uplift of the Himalaya-Hengduan Mountains. *Mol. Ecol.* **2017**, *26*, 3276–3294. [CrossRef] [PubMed]
84. Siepielski, A.M.; Morrissey, M.B.; Buoro, M.; Carlson, S.M.; Caruso, C.M.; Clegg, S.M.; Coulson, T.; DiBattista, J.; Gotanda, K.M.; Francis, C.D.; et al. Precipitation drives global variation in natural selection. *Science* **2017**, *355*, 959–962. [CrossRef] [PubMed]
85. Waldvogel, A.M.; Feldmeyer, B.; Rolshausen, G.; Exposito-Alonso, M.; Rellstab, C.; Kofler, R.; Mock, T.; Schmid, K.; Schmitt, I.; Bataillon, T.; et al. Evolutionary genomics can improve prediction of species' responses to climate change. *Evol. Lett.* **2020**, *4*, 4–18. [CrossRef] [PubMed]
86. Haffer, J. Speciation in Amazonian Forest Birds. *Science* **1969**, *165*, 131–137. [CrossRef]
87. Jon, F.; Jon, C.L. Geographical patterns of old and young species in African forest biota: The significance of specific montane areas as evolutionary centres. *Biol. Conserv.* **1997**, *6*, 325–346.
88. Parker, L.D.; Hawkins, M.T.R.; Camacho-Sanchez, M.; Campana, M.G.; West-Roberts, J.A.; Wilbert, T.R.; Lim, H.C.; Rockwood, L.L.; Leonard, J.A.; Maldonado, J.E. Little genetic structure in a Bornean endemic small mammal across a steep ecological gradient. *Mol. Ecol.* **2020**, *29*, 4074–4090. [CrossRef]
89. Oliver, M.J.; Tuba, Z.; Mishler, B.D. The Evolution of Vegetative Desiccation Tolerance in Land Plants. *Plant Ecol.* **2000**, *151*, 85–100. [CrossRef]
90. Scheffers, B.R.; De Meester, L.; Bridge, T.C.; Hoffmann, A.A.; Pandolfi, J.M.; Corlett, R.T.; Butchart, S.H.; Pearce-Kelly, P.; Kovacs, K.M.; Dudgeon, D.; et al. The broad footprint of climate change from genes to biomes to people. *Science* **2016**, *354*, aaf7671. [CrossRef]
91. Wiens, J.J.; Ackerly, D.D.; Allen, A.P.; Anacker, B.L.; Buckley, L.B.; Cornell, H.V.; Damschen, E.I.; Jonathan Davies, T.; Grytnes, J.A.; Harrison, S.P.; et al. Niche conservatism as an emerging principle in ecology and conservation biology. *Ecol. Lett.* **2010**, *13*, 1310–1324. [CrossRef]
92. Pyron, R.A.; Burbrink, F.T. Extinction, ecological opportunity, and the origins of global snake diversity. *Evolution* **2012**, *66*, 163–178. [CrossRef]
93. Sun, M.; Folk, R.A.; Gitzendanner, M.A.; Soltis, P.S.; Chen, Z.D.; Soltis, D.E.; Guralnick, R.P. Recent accelerated diversification in rosids occurred outside the tropics. *Nat. Commun.* **2020**, *11*, 3333. [PubMed]
94. Gillman, L.N.; Wright, S.D. Species richness and evolutionary speed: The influence of temperature, water and area. *J. Biogeogr.* **2014**, *41*, 39–51.
95. Claramunt, S.; Cracraft, J. A new time tree reveals Earth history's imprint on the evolution of modern birds. *Sci. Adv.* **2015**, *1*, e1501005. [PubMed]
96. Gaston, K.J. Species–range–size distributions: Patterns, mechanisms and implications. *Trends Ecol. Evol.* **1996**, *11*, 197–201. [CrossRef]
97. Pyron, R.A.; Wiens, J.J. Large–scale phylogenetic analyses reveal the causes of high tropical amphibian diversity. *Proc. R. Soc. B Biol. Sci.* **2013**, *280*, 20131622. [CrossRef]
98. Allen, A.P.; Brown, J.H.; Gillooly, J.F. Global biodiversity, biochemical kinetics, and the energetic–equivalence rule. *Science* **2002**, *297*, 1545–1548. [CrossRef]
99. Cardillo, M. Latitude and rates of diversification in birds and butterflies. *Proc. R. Soc. Lond. Ser. B Biol. Sci.* **1999**, *266*, 1221–1225.

100. Meng, H.H.; Zhang, C.Y.; Song, Y.G.; Yu, X.Q.; Cao, G.L.; Li, L.; Cai, C.N.; Xiao, J.H.; Zhou, S.S.; Tan, Y.H.; et al. Opening a door to the spatiotemporal history of plants from the tropical Indochina Peninsula to subtropical China. *Mol. Phylogenet. Evol.* **2022**, *171*, 107458. [CrossRef]
101. Pecl, G.T.; Araujo, M.B.; Bell, J.D.; Blanchard, J.; Bonebrake, T.C.; Chen, I.C.; Clark, T.D.; Colwell, R.K.; Danielsen, F.; Evengard, B.; et al. Biodiversity redistribution under climate change: Impacts on ecosystems and human well-being. *Science* **2017**, *355*, eaai9214.
102. Braje, T.J.; Erlandson, J.M. Human acceleration of animal and plant extinctions: A Late Pleistocene, Holocene, and Anthropocene continuum. *Anthropocene* **2013**, *4*, 14–23.
103. Newbold, T.; Hudson, L.N.; Hill, S.L.; Contu, S.; Lysenko, I.; Senior, R.A.; Borger, L.; Bennett, D.J.; Choimes, A.; Collen, B.; et al. Global effects of land use on local terrestrial biodiversity. *Nature* **2015**, *520*, 45–50. [CrossRef] [PubMed]
104. Locke, H.; Ellis, E.C.; Venter, O.; Schuster, R.; Ma, K.; Shen, X.; Woodley, S.; Kingston, N.; Bhola, N.; Strassburg, B.B.N.; et al. Three global conditions for biodiversity conservation and sustainable use: An implementation framework. *Natl. Sci. Rev.* **2019**, *6*, 1080–1082. [CrossRef] [PubMed]
105. Zhang, C.Y.; Low, S.L.; Song, Y.G.; Nurainas Kozlowski, G.; Do, T.V.; Li, L.; Zhou, S.S.; Tan, Y.H.; Cao, G.L.; Zhou, Z.; et al. Shining a light on species delimitation in the tree genus *Engelhardia* Leschenault ex Blume (Juglandaceae). *Mol. Phylogenet. Evol.* **2020**, *152*, 106918. [CrossRef] [PubMed]

diversity

MDPI

Article

Genome-Wide Analysis of SPL Gene Families Illuminate the Evolution Patterns in Three Rubber-Producing Plants

Renping Su [1], Boxuan Yuan [1], Yang Yang [2], Guoen Ao [1] and Juanying Wang [1,*]

[1] The Key Laboratory of Plant Resource Conservation and Germplasm Innovation in Mountainous Region (Ministry of Education), Institute of Agro-Bioengineering, College of Life Sciences, Guizhou University, Guiyang 550025, China; ls.rpsu20@gzu.edu.cn (R.S.); yuanboxuan111@163.com (B.Y.); aoguoen@163.com (G.A.)

[2] College of Life Sciences, Hainan Normal University, Haikou 571158, China; 15776580499@163.com

* Correspondence: juanying020@163.com; Tel.: +86-133-3842-2948

Abstract: Transcription factors SQUAMOSA Promoter-binding Protein-like (SPL) play a crucial role in regulating plant response to stress, root development, and flower production. However, analysis of SPL gene families in the three rubber-producing plants *Taraxacum kok-saghyz*, *Hevea brasiliensis*, and *Eucommia ulmoides*, renowned for their natural rubber production, has not yet been conducted. In this study, we utilized reference genomes to perform genome-wide analysis, and obtained new insights on the evolution of SPL gene families in these three rubber-producing plants. Our results revealed the following: (1) *T. kok-saghyz*, *H. brasiliensis*, and *E. ulmoides* harbored 25, 16, and 13 SPL genes, respectively, containing conserved structural domains of SBP. (2) A phylogenetic analysis categorized 90 SPL proteins from 25 TkSPLs, 16 HbSPLs, 13 EuSPLs, 17 AtSPLs, and 19 OsSPLs into eight groups. (3) Analysis of cis-acting elements demonstrated that the promoters of EuSPLs contained a significant number of light response elements, hormone regulatory elements, and stress response elements. (4) Transcriptome data analysis revealed that the EuSPL8 gene had strong expression in bark, as well as TkSPL4 and TkSPL8 exhibit high expression levels specifically in roots and latex. This study provides valuable insights into the biological functions of the SPL gene family in the three rubber plants and might serve as a reference for identifying efficient genes.

Keywords: *Taraxacum kok-saghyz*; *Hevea brasiliensis*; *Eucommia ulmoides*; genome-wide; SPLs; expression pattern

Citation: Su, R.; Yuan, B.; Yang, Y.; Ao, G.; Wang, J. Genome-Wide Analysis of SPL Gene Families Illuminate the Evolution Patterns in Three Rubber-Producing Plants. *Diversity* **2023**, *15*, 983. https://doi.org/10.3390/d15090983

Academic Editors: Hong-Hu Meng and Yi-Gang Song

Received: 13 July 2023
Revised: 22 August 2023
Accepted: 23 August 2023
Published: 31 August 2023

1. Introduction

Transcription factors (TFs), a crucial subclass of regulatory proteins that control gene expression, are vital for plant growth and development because they bind to cis-acting areas to either activate or inhibit the production of downstream genes. SQUAMOSA Promoter-binding Protein-like (SPL) proteins are a family of plant-specific TFs with a highly conserved DNA-binding SBP domain made up of two distinct zinc finger architectures, Zn-1 (Cys3Hes or Cys4) and Zn-2 (Cys2HisCys) [1] Since the SBP genes AmSBP1 and AmSBP2 were discovered for the first time in *Antirrhinum majus* [2], the genome-wide analysis of the SPL gene family has now been performed for numerous species, including 12 SPLs in sweet cherry (*Prunus avium* L.) [3], 23 SPLs in quinoa (*Chenopodium quinoa* Willd.) [4], and 18 SPLs in foxtail millet (*Setaria italica*) [5]. As more plant genomic resources become available, more SPL genes will probably be found.

SPL transcription factors are crucial regulators of plant growth, development, and stress response. The microRNA-targeted transcription factor SPL3, which initiates and subsequently activates the transcription factors LEAFY, FRUITFULL, and APETALA1, controls the time of flower development in *Arabidopsis* [6]. Early blooming in *Arabidopsis* was reportedly stopped by AtSPL3 [7]. AtSPL8 appears to be implicated in gibberellin signaling, according to an overexpression study [8]. SPL genes control the homeostasis of copper (Cu) and the response to Cu deficiency [9,10], and cadmium (Cd) tolerance [11,12]; they take part in hormone signaling as well as reactions to a variety of biotic and abiotic stimuli, including as heat, cold, salt, thirst, and injury [13–16]. AtSPL9 and AtSPL15 play complementary roles in controlling plastochron length and shoot maturation [17,18], and AtSPL9-controlled cell elongation and the transition from the vegetative phase through the BR (brassinosteroid) signaling pathway [19]. SPL genes are essential for phase transitions in addition to the roles they perform in the aforementioned processes [20,21], latitudinal root growth [22], trichome development [23–25], embryogenesis [26], and seedling growth [27]. SPL genes control rice grain shape, size, quality, and yield in addition to plant architecture [28–31].

Natural rubber, with distinct physical characteristics [32,33], is a valuable industrial raw material used extensively in the transportation, medicinal, and defense sectors. Currently, the prevalent plants that produce rubber are *Hevea brasiliensis*, *Taraxacum kok-saghyz*, and *Eucommia ulmoides* [33,34]. Several TFs participate in the control of the expression of the genes involved in rubber biosynthesis in *H. brasiliensis*. As an example, the expression of HbSRPP was decreased by the three genes HbWRKY14, HbWRKY1, and HbMADS4 [35,36]. HbFPS1 expression was increased by HblMYB19, HblMYB44, and HbWRKY27 [37,38]. HbRZFP1 decreased the expression of HRT2 [39]. HbMYC2b controls the expression of HbSRPP [40]. While TkWRKY21 was up-regulated under heat stress, TkWRKY18, TkWRKY23, and TkWRKY38 were all considerably up-regulated during cold stress [41]. The majority of CBF genes in TKS seedlings that had undergone cold acclimation took longer to react to the cold signal than those that had not [42]. Most of the EuWRKYs genes were highly expressed in leaf buds and involved in leaf development [43]. The genes EuMADS39 and EuMADS65 were highly expressed in male individuals, whether in flower or leaf tissues [44]. These findings demonstrated that TFs are essential for rubber biosynthesis.

However, the precise role of SPL genes in rubber biosynthesis remains incompletely understood in these three rubber-producing plants. Despite that, it is believed that they may regulate the expression of genes involved in either the mevalonate (MVA) pathway or the methylerythritol phosphate (MEP) pathway, both of which contribute to the synthesis of the isoprenoid precursor for rubber molecules. Therefore, we selected three significant natural rubber producers, *H. brasiliensis*, *T. kok-saghyz*, and *E. ulmoides*. In this study, a thorough analysis of the gene structure, conserved domain, chromosome location, cis-acting element, and expression pattern of the SPL genes in three rubber-producing plants was conducted in this study. Through analyzing the SPL gene family in these plants, our objective was to gain potential insights into the involvement of SPL genes in the biosynthesis of natural rubber.

2. Materials and Methods

2.1. SPL Identification across the Genome and Phylogenetic Analysis

The genomic sequences of *T. kok-saghyz*, *H. brasiliensis*, and *E. ulmoides* were retrieved from the NCBI and Genomic Warehouse databases (https://ngdc.cncb.ac.cn, accessed on 23 April 2023). This study used the Phytozome (https://phytozome-next.jgi.doe.gov, accessed on 23 April 2023) database to download the AtSPLs and OsSPLs protein sequences that are used to retrieve the protein sequences of *T. kok-saghyz*, *H. brasiliensis*, and *E. ulmoides*. The SBP domain sequences (ID: PF03110) from the Pfam database were used to scan the three genome sequences of rubber-producing plants for putative SPL genes using HMMER (http://hmmer.org, accessed on 23 April 2023). The local BLAST and hidden Markov models were utilized for comparison in order to locate the SPL family members in all three species, and the findings were combined. Delete domain-less or domain-incomplete sequences manually to guarantee the accuracy of the results. The first transcript is chosen as the typical sequence in situations when there are numerous transcripts of the same gene. The protein domain was analyzed by NCBI Batch CD-search and plotted by Chiplot's website (https://www.ncbi.nlm.nih.gov/Structure/bwrpsb/bwrpsb.cgi and https://www.chiplot.online/, accessed on 23 April 2023). Gene mapping was performed using TBtools software (https://github.com/CJ-Chen/TBtools/releases, accessed on 23 April 2023).

The amino acid sequences of *T. kok-saghyz*, *H. brasiliensis*, *E. ulmoides*, *Oryza sativa* and *Arabidopsis thaliana* SPLs were compared in multiple sequences using ClustalW 2.0 software, while the phylogenetic tree was built using the neighbor-joining method with MEGA 11.0 software, with the number of replications set at 1000 and all parameters left at their default settings, and figures with confidence levels below 60 were selected to be hidden in Figure 1 and, finally, touched up in the iTQL website (https://itol.embl.de, accessed on 23 April 2023).

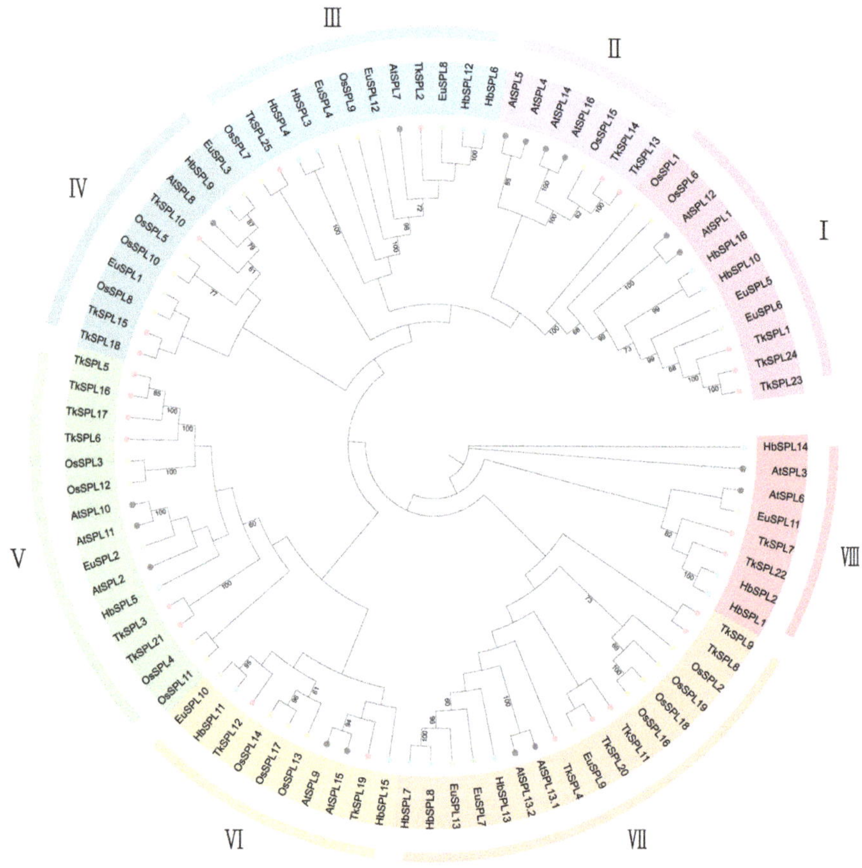

Figure 1. Phylogenetic tree includes 90 SPLs from 5 different plants. Using MEGA 11.0 and the neighbor-joining (NJ) method, a phylogenetic tree based on the SBP domain was created. The bootstrap value is represented by the number of branches.

2.2. Physicochemical Characterization and Structural Analysis of the Protein

The amino acid count, relative molecular weight, theoretical isoelectric point, and instability index of TkSPLs, HbSPLs, and EuSPLs were examined using the ExPASy Prot Param program (https://web.expasy.org/protparam/, accessed on 23 April 2023). By comparing SPL gene sequences to the matching genomic DNA sequences, the exons, introns, CDS, and UTRs of the genes were identified. WoLF PSORT (https://wolfpsort.hgc.jp, accessed on 23 April 2023) is used to predict protein subcellular localization. Expasy and TMHMM-2.0 were used to assess the hydrophobicity and transmembrane characteristics of proteins (https://web.expasy.org/protscale/ and https://ser-vices.healthtech.dtu.dk/Services/TMHMM-2.0/, accessed on 23 April 2023). The secondary and tertiary structures of proteins can be predicted using the SOPMA and Phyre2 programs (https://npsa-prabi.

ibcp.fr/cgi-bin/npsa_automat.pl?page=npsa_sopma.html and http://www.sbg.bio.ic.ac.uk/phyre2/html/page.cgi?id=index, accessed on 23 April 2023).

2.3. Analysis of Protein Domains and Conserved Motifs

Using ClustalW 2.0, the protein sequences of AtSPL, OsSPL, and the 54 SPL sequences discovered in this investigation were aligned to identify the SBP domain [45]. TBtools was used to examine the exon–intron structure [46], using the genome and cDNA sequences of the SPL gene retrieved from the genomes of four rubber-producing plants. Three rubber-producing plant gene family members were subjected to protein motif analysis using the web program MEME (https://meme-suite.org/meme/tools/meme, accessed on 25 April 2023) [47]. This analysis used up to 10 motlets and other default parameters. Using GFF files of three rubber-producing plants and two model plants, as well as the files of the previous construction of the evolutionary tree, the gene structure and motif combination map were drawn on the TBtools software. The PlantCare website (http://bioinformatics.psb.ugent.be/webtools/plantcare/html/, accessed on 25 April 2023) was used to predict the cis-acting elements in the promoter region of the three rubber-producing plant SPLs using the 2000 bp sequence upstream of the translation start codon (ATG), and the TBtools software was then used for visual analysis.

2.4. Collinearity Analysis

The genome sequences of Oryza sativa and *Arabidopsis* thaliana were downloaded from Phytozome (https://phytozome-next.jgi.doe.gov, accessed on 23 April 2023) to create a chromosome location and collinearity plot in order to analyze the replication patterns and evolutionary mechanisms of SPL genes.

2.5. Analysis of Expression Patterns across Several Developmental Stages and Tissue Types

The NCBI database was used to download the transcriptome information for various tissues and time periods in *E. ulmoides*, the relative expression abundance of SPLs of the three rubber-producing plants was expressed by FPKM value, the logarithm of the value was statistically analyzed, and the TBtools software generated a heat map of gene expression, where the highest expression was denoted by a red box and the lowest expression by a blue box.

3. Results

3.1. SPL Identification across the Genome and Phylogenetic Analysis

Local BLAST and the hidden Markov model were utilized for screening purposes, resulting in the identification of 25 TkSPLs, 16 HbSPLs, and 13 EuSPLs (Table 1). The Supplementary Materials provide information on the SBP domain and gene localization of these three species following the identification and screening process (Figures S1 and S2).

To enhance our understanding of the evolutionary trajectory of the SPL gene across various species, a phylogenetic analysis was conducted on a set of 90 SPL proteins, including those from *T. kok-saghyz* (25), *H. brasiliensis* (16), *E. ulmoides* (13), *Arabidopsis thaliana* (17) and *Oryza sativa* (19). The analysis categorized these proteins into eight distinct groups, as shown in Figure 1.

The results showed that in the phylogenetic tree (I–VIII), 90 SPL genes were divided into 8 subfamilies. Their agreement with the AtSPL and OsSPL protein classification groups implies that SPL genes have been highly conserved throughout molecular evolution. SPL genes of *T. kok-saghyz* and *Arabidopsis thaliana* are distributed in all eight subfamilies, while SPL genes of *E. ulmoides* and *H. brasiliensis* are lacking in subfamily II and SPL genes of *Oryza sativa* are lacking in subfamily VIII. Subfamily VII had the most individuals (17 SPLs) out of the eight subfamilies, while subfamily II only had seven SPLs. The phylogenetic tree also revealed that a number of SPL genes from the five species clustered together, with a support rating of $\geq$70.

Table 1. Analysis of basic physical and chemical properties of 54 SPLs.

Gene	Gene ID	Chr ID	Gene Range	Exon	Intron	CDS	UTR	Size/aa	MW/kD	PI	Instability Index
TkSPL1	GWHPBCHF001261	GWHBCHF00000001	14822969:14828316	12	11	11	3	927	103,394.48	6.2	53.89
TkSPL2	GWHPBCHF002675	GWHBCHF00000002	2073280:2078116	11	10	10	3	748	83,959.66	5.95	58.08
TkSPL3	GWHPBCHF006473	GWHBCHF00000002	100814431:100818025	4	3	4	0	261	29,108.21	7.73	74.85
TkSPL4	GWHPBCHF019170	GWHBCHF00000004	830539:832919	3	2	2	3	302	34,271.32	9.08	68.14
TkSPL5	GWHPBCHF020579	GWHBCHF00000004	30786520:30796342	5	4	5	2	357	39,818.25	9	51.09
TkSPL6	GWHPBCHF020580	GWHBCHF00000004	30786520:30796342	5	4	4	3	349	38,902.15	9.1	51.1
TkSPL7	GWHPBCHF022201	GWHBCHF00000004	84269226:84270914	4	3	4	2	298	33,372.53	8.41	56.09
TkSPL8	GWHPBCHF023484	GWHBCHF00000004	107063807:107065169	2	1	2	2	198	22,079.54	9.24	65.76
TkSPL9	GWHPBCHF024887	GWHBCHF00000004	129182786:129185573	2	1	2	2	192	22,005.9	8.36	70.5
TkSPL10	GWHPBCHF025127	GWHBCHF00000004	132569671:132571972	3	2	3	2	298	33,497.02	8.76	53.76
TkSPL11	GWHPBCHF026791	GWHBCHF00000004	155862873:155865148	4	3	3	3	279	30,897.47	9.74	67.09
TkSPL12	GWHPBCHF028134	GWHBCHF00000005	14701253:14704746	4	3	3	3	351	37,175.07	7.61	52.58
TkSPL13	GWHPBCHF029402	GWHBCHF00000005	33473738:33478697	11	10	10	3	1042	115,521.09	8.36	56.28
TkSPL14	GWHPBCHF029413	GWHBCHF00000005	33656280:33661229	11	10	10	3	1039	115,165.69	8.36	57.09
TkSPL15	GWHPBCHF035319	GWHBCHF00000006	28696570:28698241	2	1	2	2	183	20,547.1	9.49	44.27
TkSPL16	GWHPBCHF039801	GWHBCHF00000007	11469584:11479658	5	4	5	2	357	39,883.37	9.1	53.96
TkSPL17	GWHPBCHF039802	GWHBCHF00000007	11469584:11479658	5	4	4	3	349	38,967.27	9.19	54.04
TkSPL18	GWHPBCHF040059	GWHBCHF00000007	13895402:13897881	3	2	2	3	171	19,026.37	8.94	60.4
TkSPL19	GWHPBCHF044300	GWHBCHF00000008	4487440:4488774	2	1	2	2	296	32,163.23	9.48	57.94
TkSPL20	GWHPBCHF044302	GWHBCHF00000008	4501786:4503817	3	2	2	3	261	29,845.17	7.03	73.8
TkSPL21	GWHPBCHF050461	GWHBCHF00000009	9840573:9843749	5	4	4	2	275	30,724.02	9.55	73.95
TkSPL22	GWHPBCHF052188	GWHBCHF00000009	28195236:28196956	3	2	3	2	316	35,456.72	9.36	68.57
TkSPL23	GWHPBCHF053518	GWHBCHF00000009	42852554:42856744	10	9	9	3	928	104,212.96	6.82	46.08
TkSPL24	GWHPBCHF053519	GWHBCHF00000009	42852554:42856744	9	8	9	2	924	103,871.69	6.95	45.77
TkSPL25	GWHPBCHF056162	GWHBCHF00000009	103280194:103281999	2	1	2	2	137	15,812.68	8.56	63.94
EuSPL1	GWHPAAAL001077	GWHAAAL00000030	988480:992976	11	10	11	0	178	20,179.44	9.03	55.87
EuSPL2	GWHPAAAL003913	GWHAAAL00000058	2037987:2040046	3	2	3	0	547	60,636.72	6.91	54.8
EuSPL3	GWHPAAAL004171	GWHAAAL00000080	203789:215644	2	1	2	0	369	41,404.83	8.07	72.6
EuSPL4	GWHPAAAL007895	GWHAAAL00000112	735958:741319	5	4	5	0	133	15,322.89	7.69	84.16
EuSPL5	GWHPAAAL008461	GWHAAAL00000115	1029716:1033122	5	4	5	0	968	107,979.44	7.33	57.38
EuSPL6	GWHPAAAL008647	GWHAAAL00000172	475295:476266	2	1	2	0	978	109,120.78	8.36	51.07

Table 1. *Cont.*

Gene	Gene ID	Chr ID	Gene Range	Exon	Intron	CDS	UTR	Size/aa	MW/kD	PI	Instability Index
EuSPL7	GWHPAAAL012748	GWHAAAL00000184	1290342:1300388	11	10	11	0	347	37,321.69	8.1	64.57
EuSPL8	GWHPAAAL012866	GWHAAAL00000249	1634971:1649669	12	11	12	0	875	98,038.82	6.26	58.57
EuSPL9	GWHPAAAL016673	GWHAAAL00006095	779825:852698	9	8	9	0	259	28,558.46	8.79	61.98
EuSPL10	GWHPAAAL020125	GWHAAAL00014124	1676:4668	3	2	3	0	363	38,927.58	9.01	64.09
EuSPL11	GWHPAAAL025240	GWHAAAL00017881	529023:533014	3	2	3	0	354	40,444.43	8.94	48.71
EuSPL12	GWHPAAAL026381	GWHAAAL00020293	536331:540436	3	2	3	0	696	77,869.21	8.88	53.91
EuSPL13	GWHPAAAL026606	GWHAAAL00026424	227394:229146	3	2	3	0	131	15,166.23	10.47	72.74
HbSPL1	GT021460.t1	CM021228.1	13875757:13877545	2	1	2	0	531	57,107.57	7.27	46.45
HbSPL2	GT021495.t1	CM021228.1	13970213:13972002	2	1	2	0	442	48,140.64	8.2	48.31
HbSPL3	GT010583.t1	CM021228.1	86118185:86120916	4	3	4	0	194	21,651.15	9.06	64.25
HbSPL4	GT010615.t1	CM021233.1	98164232:98172531	10	9	10	0	194	21,651.15	9.06	64.25
HbSPL5	GT017984.t1	CM021234.1	94639220:94645994	4	3	4	0	458	49,852.23	8.87	56.52
HbSPL6	GT003116.t1	CM021238.1	65041899:65062959	24	23	24	0	707	79,038.55	7.23	55.88
HbSPL7	GT030712.t1	CM021238.1	65225362:65248838	24	23	24	0	1065	117,893.71	8.75	46.86
HbSPL8	GT030809.t1	CM021239.1	12363240:12365845	3	2	3	0	1005	110,811.41	8.33	48.37
HbSPL9	GT019690.t1	CM021239.1	23400072:23405078	10	9	10	0	293	32,788.22	9.21	64.77
HbSPL10	GT039310.t1	CM021239.1	26622170:26632180	12	11	12	0	995	110,256.79	6.4	43.85
HbSPL11	GT033052.t1	CM021239.1	39504948:39506565	3	2	3	0	339	37001.37	9.5	59.73
HbSPL12	GT019861.t1	CM021240.1	72675955:72677659	3	2	3	0	686	77,405.9	5.86	55.24
HbSPL13	GT025439.t1	CM021241.1	2028142:2034204	10	9	10	0	368	41,070.02	9.44	51.55
HbSPL14	GT042310.t1	CM021241.1	27863168:27864903	3	2	3	0	862	97,241.22	6.29	47.94
HbSPL15	GT040479.t1	JAAGAX010000040.1	559530:565521	4	3	4	0	398	44,823.71	8.64	49.13
HbSPL16	GT028737.t1	JAAGAX010000040.1	620658:626675	4	3	4	0	404	44,567.07	8.41	55.27

3.2. Physicochemical Properties and Secondary and Three-Dimensional Structural Analysis

In the three rubber-producing plants, 54 SPL genes in all were found. The protein's amino acid composition ranged from 131 to 1065, its molecular weight from 15,166.23 to 117,893.71 kD, and its isoelectric point from 5.86 to 10.47 as a basic protein, and the instability index ranged from 43.85 to 84.16. With only 131 amino acids, EuSPL13 was the smallest of the 54 SPL proteins. HbSPL7, in contrast, had the most amino acids (1065), making it the biggest.

The secondary structure analyses predicted that all SPLs of the three species comprised alpha helices, extended strands, beta turns, and random coils (Table S1). The alpha helices and the random coils were the main secondary structural elements of SPLs. It was projected that TkSPL1 would localize to the plasma membrane in *T. kok-saghyz,* while TkSPL2 would localize to the cytoplasm, TkSPL9 to the chlo, and TkSPL18 to the cytoplasm. EuSPL8 was predicted to localize to the peroxisome in *E. ulmoides*. The remaining 49 SPL genes, excluding the aforementioned 5 SPL genes, are found in the nucleus. The Phyre2 sever was used to model the three-dimensional structure of the 54SPLs (Figures S9–S11). The results showed that the SPLs were mainly composed of alpha helices and random coils, and the structure ratios were consistent with the predicted secondary structures. The tertiary structures of all three rubber-producing plant SPLs proteins were very similar, and the results are shown in the accompanying figure, with the largest proportion of irregular curls, while the different protein spatial structures will determine the differences in function. In addition, 54 SPL proteins were predicted for hydrophilicity and transmembrane conditions (Figures S3–S8).

3.3. Analysis of Protein Domains and Conserved Motifs

Exon–intron distribution and conserved motifs were studied in order to further evaluate the structural characteristics of the five species (Figure 2). We selected SPL genes from two model plants (*Arabidopsis thaliana* and *Oryza sativa*) for comparison with SPL genes from the three rubber-producing plants found in this study. A phylogenetic tree with ten conserved motifs was built using sequence data from the 54 SPLs and the SPL genes from the two additional plants using the NJ method; 10 conservative motifs were designated Motif1–Motif10.

Motifs are shared by genes in the same subfamily, which causes them to group together and generate an unequal distribution of the TkSPL, HbSPL, and EuSPL genes in the evolutionary tree. The most varied motifs were found in subfamilies V and VII, while motifs 8 and 10 were frequently seen near the start and conclusion of the patterning, respectively. Additionally, we discovered that in subfamilies I and II, motif 9 was consistently distributed towards the start of the pattern, and pattern 7 was consistently distributed at the very end of patterning.

The protein sequences of AtSPL, OsSPL, and the 54 SPL sequences found during this work were aligned to determine the SBP domain; 14 conserved amino acids are found in the basic region, which is made up of 70–80 amino acids. The Zn-fingers (Zn-1 and Zn-2) are shown in green, while the bidirectional nuclear localization signal (NLS) structures are highlighted in red (Figure 3).

To define the protein structure, the multiple sequence alignment of full-length proteins was created using MEGA 11.0 software. The length of the conserved SPL domains has been aligned many times. With roughly 74 amino acid residues, the SBP domains at the SCR, RRR, and CQQC sequences were largely preserved. Both the nuclear localization signal (NLS) and the Zn-1 and Zn-2 zinc finger configurations are retained in these SPL domains. The amino acid distribution of the structural domains of the SPL genes of the three rubber-producing plants was very similar to that of *Arabidopsis* and rice, with the BASIC region consisting of 19 amino acids, 9 of which were highly conserved, and the HLH region containing 5 highly conserved amino acids.

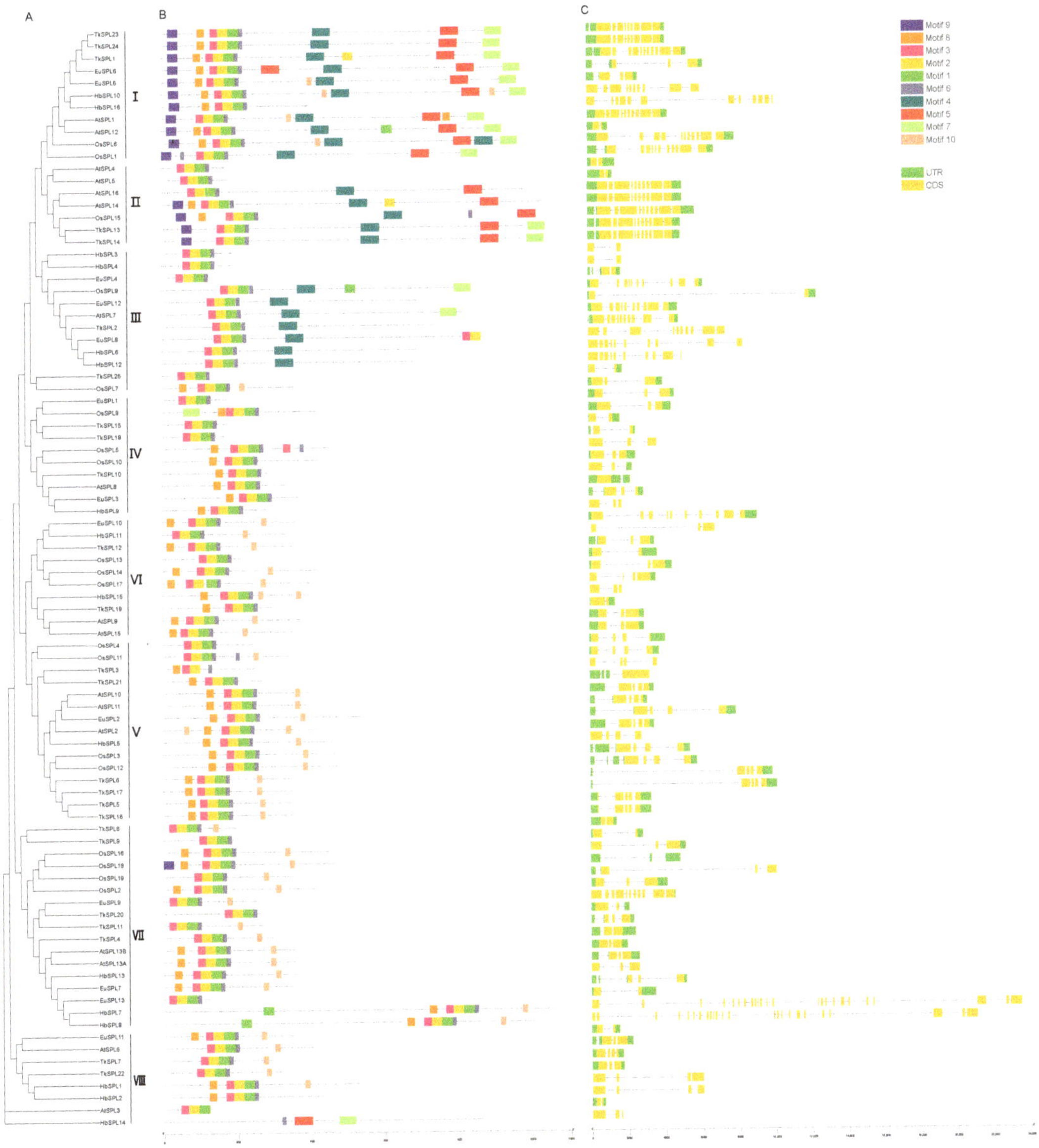

Figure 2. Analyses of protein domains and conserved motifs. (**A**) Phylogenetic tree. (**B**) Conserved motifs. (**C**) Gene structure.

3.4. *Cis-Acting Elements and Collinearity Analysis*

To investigate the gene functions and expression regulation patterns of EuSPLs, Plant CARE online analysis software was used to search for cis-acting elements in the 2000 bp sequences upstream of the start codon of 13 EuSPL proteins (Figure 4).

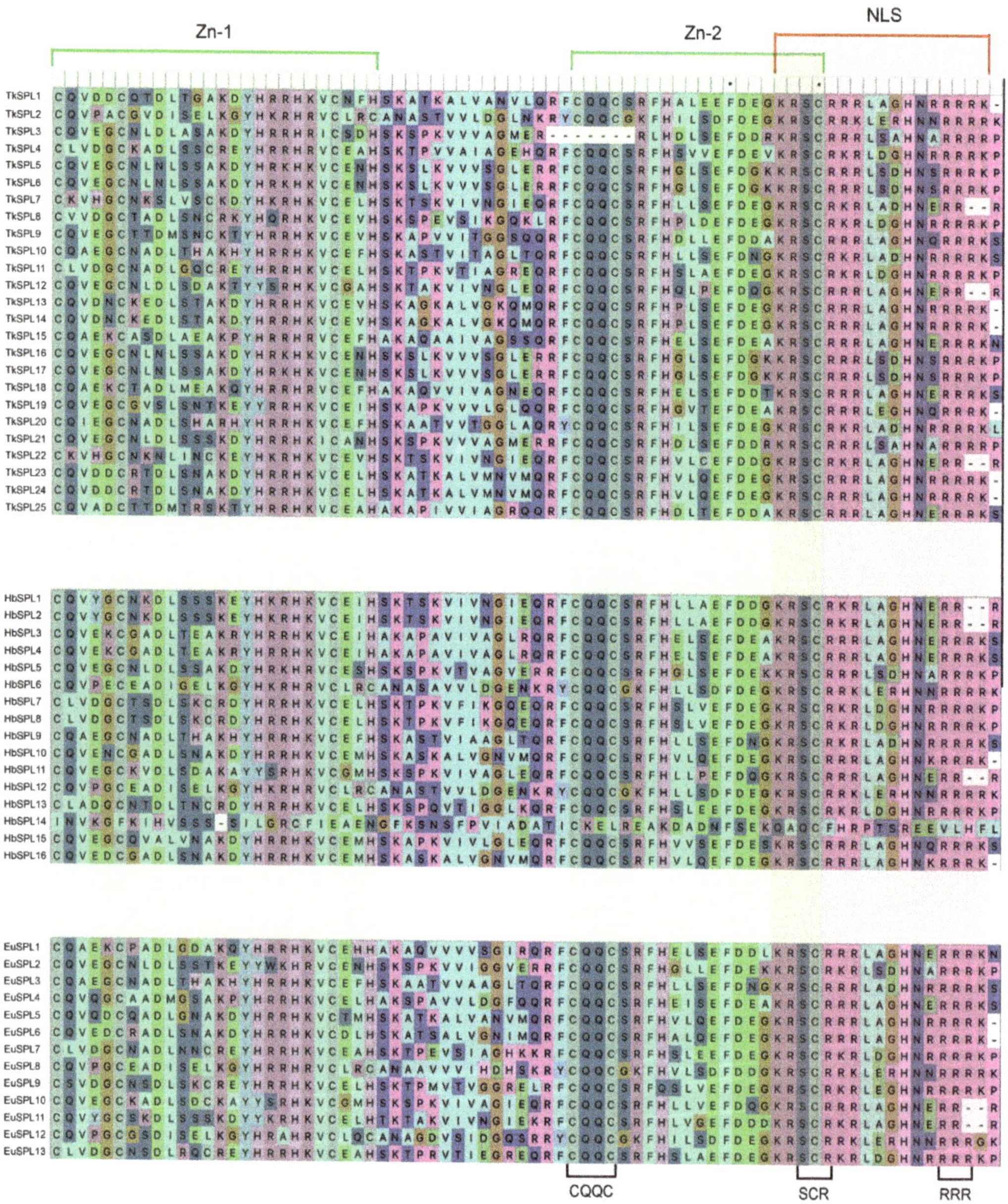

Figure 3. SBP domains from the SPL protein family with multiple sequence alignments. The symbol '*' represents a residue that is fully preserved.

Upstream of the gene coding sequence, the promoter of the gene contains a large number of cis-acting elements that can control how the gene expresses itself. The SPL promoter contains not only basic cis-acting elements but also four types of elements: (1) hormone regulatory elements, such as gibberellin response element ABRE, growth element AuxRR-core, and salicylic acid response element CGTCA-motif; (2) light response elements, such as G-Box, G-box, Box 4, GATA-motif, etc.; (3) stress response elements, like the MBS drought stress response elements and the low-temperature response elements; and (4) physiological response elements, such as O2-site, CAT-box, etc. It is speculated that SPL genes may perform crucial functions in stress response, growth, hormone regulation, and photoperiodic regulation in *E. ulmoides*. The promoter region of EuSPLs contains 19 ABRE elements and 26 ARE elements (Figure 4B), suggesting that EuSPLs may be involved in ABA regulation and anaerobic regulation.

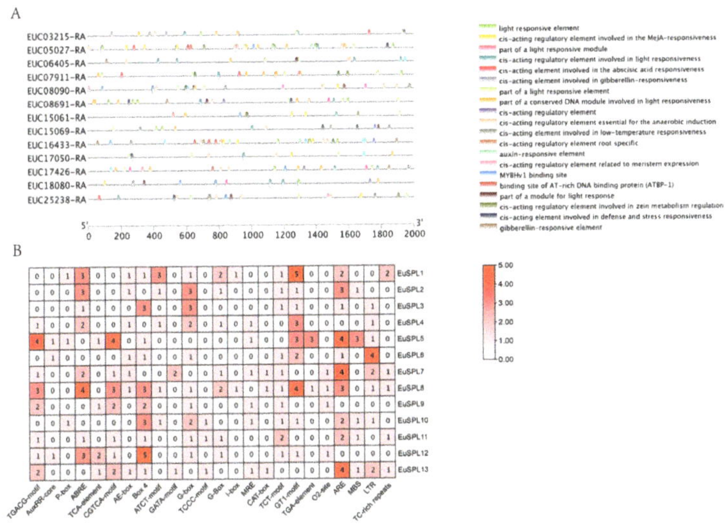

Figure 4. Cis-acting elements of the 13 EuSPLs in *E. ulmoides*: (**A**) the examination of cis-elements in SPL genes was conducted, where different cis-elements were illustrated using colored boxes; (**B**) the number of sequential elements varies, with a greater quantity of elements in darker colors.

To gain deeper insights into the evolutionary relationships, replication patterns, and evolutionary processes of the SPL genes, we created a chromosome location and collinearity plot for the 54 SPL genes (Figure 5).

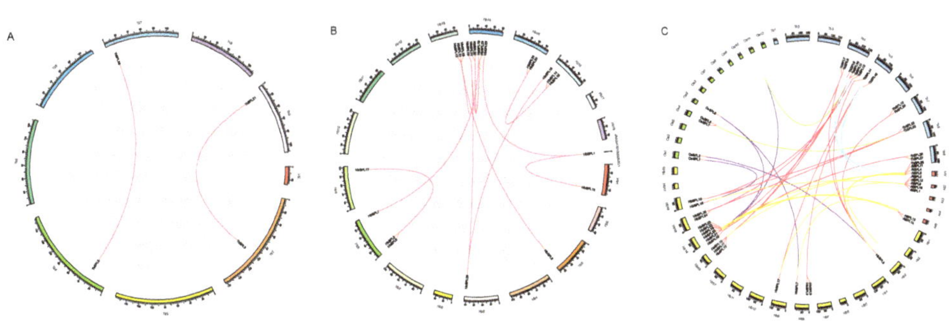

Figure 5. Collinearity analysis of rubber-producing plants. The gray line depicts all the covariates, and the colorful line depicts the pairwise replication. (**A**) *T. kok-saghyz*; (**B**) *H. brasiliensis*; (**C**) The collinearity among *H. brasiliensis*, *T. kok-saghyz*, *Oryza sativa* and *Arabidopsis thaliana*.

The tandem relationship between two species, *H. brasiliensis* and *T. kok-saghyz*, was confirmed and five and two pairs were found in their own genomes, respectively. The number of homologous SPLs of AtSPLs and OsSPLs in HbSPLs was more than that in TkSPLs. Between AtSPLs and HbSPLs, there were nine lines, but only five lines between

AtSPLs and TkSPLs. Between OsSPLs and HbSPLs, there are five lines, yet just two lines separate OsSPLs from TkSPLs. There were the most connections between *T. kok-saghyz* and *H. brasiliensis* (15).

3.5. Analysis of Expression Patterns across Several Developmental Stages and Tissue Types

According to transcriptome data, the expression pattern of the EuSPLs gene was found in order to examine the role of the SPL gene in various developmental stages and tissues of *E. ulmoides*.

Based on transcriptome data, the expression patterns were analyzed in order to analyze the roles of the EuSPLs and TkSPLs genes throughout various developmental phases and rubber production, and the gene numbers shown in the figure are consistent with the gene naming in the article (Figures 6 and 7). Low levels of expression of the majority of the EuSPLs genes were observed, but two EuSPLs genes (EuSPL5 and EuSPL6) were expressed in high abundance in all parts of *E. ulmoides* at all stages of development. Otherwise, EuSPL3 and EuSPL7 were significantly expressed in fruit, EuSPL4 was highly expressed in fruit and leaf, and EuSPL8 was highly expressed in bark.

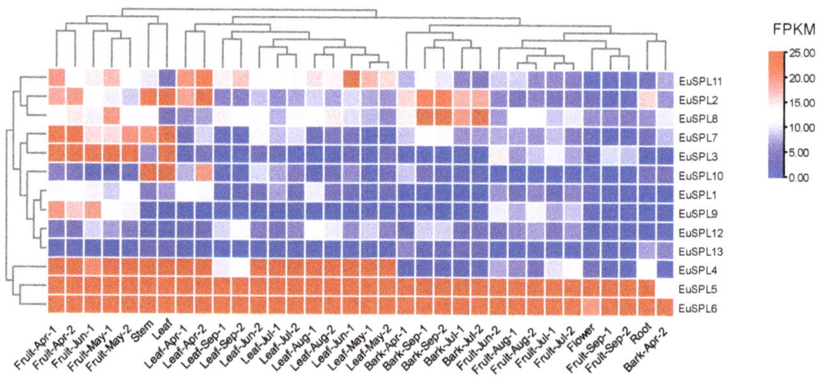

Figure 6. Patterns of gene expression for EuSPLs in various *E. ulmoides* tissue sections at various developmental stages.

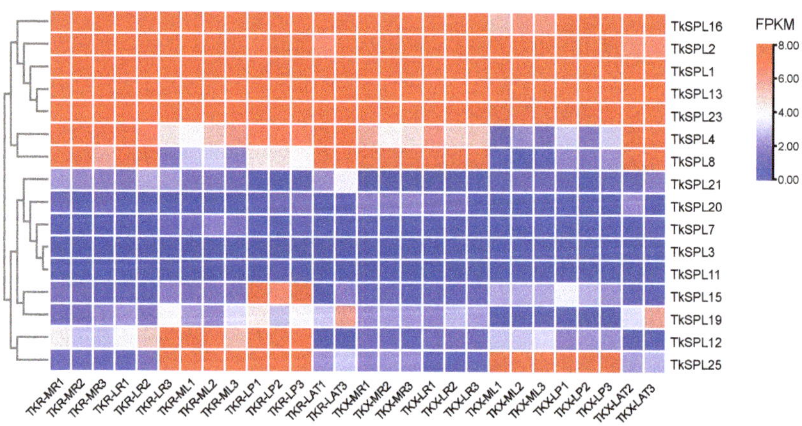

Figure 7. Patterns of gene expression for TkSPLs in various *T. kok-saghyz* tissue sections. TKX: varieties with lower rubber yield; TKR: varieties with intermediate rubber yield; MR: main roots; LR: lateral roots; ML: mature leaf; LP: leaf petiole; LAT: latex.

In *T. kok-saghyz*, several genes including TkSPL16, TkSPL2, TkSPL1, TkSPL13, and TkSPL23 show relatively high expression levels in multiple tissues. TkSPL4 and TkSPL8 exhibit high expression levels specifically in main roots, lateral roots, and latex. TkSPL25 demonstrates higher expression levels in mature leaves and leaf petioles. Furthermore, TkSPL12 shows higher expression levels in mature leaves and leaf petioles of TKR.

4. Discussion

This is the first account of a simultaneous investigation of three SPL gene families from rubber-producing plants at the genome-wide level. These findings advance our comprehension of the biological role of the SPL gene as well as the underlying molecular mechanisms.

4.1. SPL Identification across the Genome and Phylogenetic Analysis

Plant-specific TFs called SPL proteins have a structural domain of the SBP that is highly conserved [48,49], and it plays a critical role in stress response and plant growth and development [7,50,51]. Natural rubber has important economic and strategic values, and the in-depth study of rubber-producing plants is of great significance. The identification and characterization of SPL genes in numerous plants, including *Arabidopsis thaliana* [49], *Oryza sativa* [52], and *Camellia sinensis* [53], has been made possible by the quick advancement of genome sequencing technologies. However, the association of the SPL gene family in these three rubber-producing plants is not so clear that the related research is necessary and urgent.

The number of SPL members is unaffected by the size of the genome [54]. In the study, 54 SPL sequences, including 25 TkSPLs, 16 HbSPLs, and 13 EuSPLs were identified. The prediction of the physicochemical properties of SPLs showed that most of the members were basic hydrophobic proteins. The majority of the SPL proteins from the three rubber-producing plants are high in basic amino acids, which may be important in acidic subcellular settings. By bioinformatic analysis of SPL genes from three rubber-producing plants, we found that the number of SPL members in *T. kok-saghyz* was higher than that in *H. brasiliensis* and *E. ulmoides*. However, the species *T. kok-saghyz* and *E. ulmoides* shared greater similarities in terms of the number of exons, introns, and CDS as well as the number of amino acids and molecular weight.

4.2. Analysis of Protein Domains and Conserved Motifs

The predicted motifs, conserved domains, and tertiary structures show that all members have typical SBP domains and that the tertiary structures are similar. Eight groups were formed from the evolutionary examination of three rubber-producing plants, and each group had at least one gene from each of the other two plant species (*Arabidopsis thaliana* and *Oryza sativa*). It suggests that the AtSPLs and OsSPLs shared a high homology of similarity with the SPLs of three rubber-producing plants. In addition, the phylogenetic tree revealed that several SPL genes from the five species clustered together tightly (bootstrap support 70) (Figure 1). This suggests that SPL genes are functionally conserved across multiple plant species and these proteins may be orthologous [49,55], and as a result, may share similar biological roles. Sequence alignment showed that the 54 SPL genes of the three rubber-producing plants had high similarity to the structural domains of *Arabidopsis* and rice SPL proteins. According to Motif analysis, it was found that the Motif 1 motif was found to be present in all SPL transcription factors, indicating that they all belong to the SPL transcription factor family and are relatively conserved during gene evolution. Analysis of cis-acting elements revealed that EuSPL contains a large number of response elements related to hormone, light response, stress and physiology; therefore, SPL genes may play important roles in growth and development, stress response, hormone regulation and photoperiodic regulation in *E. ulmoides*.

4.3. Analysis of Responses and Expression Patterns

Plant growth and development depend heavily on hormonal cues and environmental changes [56–58]. SPL genes may play a substantial role in the control of photoperiodic, hormone, growth, and responses to stress in three rubber-producing plants. The majority of the members of this family respond to a range of hormones and stresses, which in our study's investigation of cis-acting regions upstream of the EuSPLs promoter revealed (Figure 6) that PavSPLs may be regulated by light, stresses, and phytohormones. As with studies in other species, there have been cases that have not yet been assembled at the chromosomal level (Figure 5A) [59]. Previous studies have shown that transcriptional-level analysis can identify genes involved in plant regulation [60]. Therefore, this study aims to obtain key SPL genes involved in the growth regulation of rubber-producing plants through transcriptome analysis. Based on transcriptome data, low levels of expression of the majority of the EuSPLs genes were observed, but two EuSPLs genes (EuSPL5 and EuSPL6) were expressed in high abundance in all parts of *E. ulmoides* at all stages of development. Due to the latex production primarily occurring in the roots of *T. kok-saghyz*, it is observed that TkSPL4 and TkSPL8 show higher expression levels in the main roots, lateral roots, and latex. Therefore, it is speculated that these two genes, TkSPL4 and TkSPL8, may have a strong correlation with the latex production process in *T. kok-saghyz*. To better understand how these EuSPL and TkSPL genes regulate the response, the analyses and research of other taxa will provide tests of our perspective that are essential to the molecular mechanism of the related gene families in rubber-producing plants.

5. Conclusions

This work combines genome-wide data from three rubber-producing plants to conduct a comprehensive bioinformatics investigation of the SPL transcription factor family. A total of 54 SPL transcription factor family genes were identified and classified into eight groups based on evolutionary tree analysis. The members of the SPL family exhibit the typical SBP structural domains. Furthermore, a thorough analysis of the gene structure, conserved domain, chromosome location, cis-acting element, and expression pattern of the SPL genes in the three rubber-producing plants was conducted. Additionally, the expression patterns of the EuSPLs gene family were analyzed, revealing that the EuSPL8 gene was highly expressed in bark, EuSPL3 and EuSPL7 were highly expressed in fruit, and EuSPL4 was highly expressed in both fruit and leaf. This article focuses on the genome-wide investigation of SPL gene families in rubber-producing plants. These findings will facilitate a better understanding of the biological function and molecular mechanisms of SPL genes in rubber-producing plants.

Supplementary Materials: The following supporting information can be downloaded at: https://www.mdpi.com/article/10.3390/d15090983/s1, Table S1: predicted secondary structure and subcellular location of SPLs of three species; Figure S1: SPL domains of three rubber-producing plants; Figure S2: locations of SPL genes in three rubber-producing plants; Figure S3–S5: transmembrane situation of 13 EuSPLs, 16 HbSPLs, and 25 TkSPLs; Figure S6–S8: hydrophilicity index plot of 13 EuSPLs, 16 HbSPLs, and 25 TkSPLs; Figure S9–S11: three-dimensional structure of 13 EuSPLs, 16 HbSPLs, and 25 TkSPLs.

Author Contributions: J.W. conceived and designed the experiment. R.S. carried out bioinformatics analysis. R.S., G.A. and Y.Y. co-wrote the manuscript. J.W. and B.Y. revised the manuscript. All authors have read and agreed to the published version of the manuscript.

Funding: This work was supported by the National Key Research and Development Program of China (No. 2021YFA1300401), Guizhou University Introduction of Talents Research Project (2021, No. 6), and the specific research fund of the Innovation Platform for Academicians of Hainan Province (2022YSCXTD0005).

Institutional Review Board Statement: Not applicable.

Data Availability Statement: The data and code supporting the current study are available from the first author or corresponding author upon request.

Acknowledgments: We are very grateful to Xuchu Wang for his helpful suggestions and funding support of this research.

Conflicts of Interest: The authors declare no conflict of interest.

References

1. Yamasaki, K.; Kigawa, T.; Inoue, M.; Tateno, M.; Yamasaki, T.; Yabuki, T.; Aoki, M.; Seki, E.; Matsuda, T.; Nunokawa, E.; et al. A novel zinc-binding motif revealed by solution structures of DNA-binding domains of *Arabidopsis* SBP-family transcription factors. *J. Mol. Biol.* **2004**, *337*, 49–63. [CrossRef]
2. Klein, J.; Saedler, H.; Huijser, P. A new family of DNA binding proteins includes putative transcriptional regulators of the *Antirrhinum majus* floral meristem identity gene *SQUAMOSA*. *Mol. Gen. Genet.* **1996**, *250*, 7–16. [CrossRef]
3. Sun, Y.; Wang, Y.; Xiao, Y.; Zhang, X.; Du, B.; Turupu, M.; Wang, C.; Yao, Q.; Gai, S.; Huang, J.; et al. Genome-Wide Identification of the SQUAMOSA Promoter-Binding Protein-like (SPL) Transcription Factor Family in Sweet Cherry Fruit. *Int. J. Mol. Sci.* **2023**, *24*, 2880. [CrossRef]
4. Zhao, H.; Cao, H.; Zhang, M.; Deng, S.; Li, T.; Xing, S. Genome-Wide Identification and Characterization of SPL Family Genes in *Chenopodium quinoa*. *Genes* **2022**, *13*, 1455. [CrossRef]
5. Lai, D.; Fan, Y.; Xue, G.; He, A.; Yang, H.; He, C.; Li, Y.; Ruan, J.; Yan, J.; Cheng, J. Genome-wide identification and characterization of the SPL gene family and its expression in the various developmental stages and stress conditions in foxtail millet (*Setaria italica*). *BMC Genom.* **2022**, *23*, 389. [CrossRef]
6. Yamaguchi, A.; Wu, M.F.; Yang, L.; Wu, G.; Poethig, R.S.; Wagner, D. The microRNA-regulated SBP-Box transcription factor SPL3 is a direct upstream activator of LEAFY, FRUITFULL, and APETALA1. *Dev. Cell* **2009**, *17*, 268–278. [CrossRef]
7. Gandikota, M.; Birkenbihl, R.P.; Hohmann, S.; Cardon, G.H.; Saedler, H.; Huijser, P. The miRNA156/157 recognition element in the 3′ UTR of the *Arabidopsis* SBP box gene SPL3 prevents early flowering by translational inhibition in seedlings. *Plant J.* **2007**, *49*, 683–693. [CrossRef]
8. Zhang, Y.; Schwarz, S.; Saedler, H.; Huijser, P. SPL8, a local regulator in a subset of gibberellin-mediated developmental processes in *Arabidopsis*. *Plant Mol. Biol.* **2007**, *63*, 429–439. [CrossRef]
9. Yamasaki, H.; Hayashi, M.; Fukazawa, M.; Kobayashi, Y.; Shikanai, T. SQUAMOSA Promoter Binding Protein-Like7 Is a Central Regulator for Copper Homeostasis in *Arabidopsis*. *Plant Cell* **2009**, *21*, 347–361. [CrossRef]
10. Kropat, J.; Tottey, S.; Birkenbihl, R.P.; Depege, N.; Huijser, P.; Merchant, S. A regulator of nutritional copper signaling in *Chlamydomonas* is an SBP domain protein that recognizes the GTAC core of copper response element. *Proc. Natl. Acad. Sci. USA* **2005**, *102*, 18730–18735. [CrossRef]
11. Chen, W.W.; Jin, J.F.; Lou, H.Q.; Liu, L.; Kochian, L.V.; Yang, J.L. LeSPL-CNR negatively regulates Cd acquisition through repressing nitrate reductase-mediated nitric oxide production in tomato. *Planta* **2018**, *248*, 893–907. [CrossRef] [PubMed]
12. Zhang, L.; Ding, H.; Jiang, H.; Wang, H.; Chen, K.; Duan, J.; Feng, S.; Wu, G. Regulation of cadmium tolerance and accumulation by miR156 in *Arabidopsis*. *Chemosphere* **2020**, *242*, 125168. [CrossRef] [PubMed]
13. Cui, L.G.; Shan, J.X.; Shi, M.; Gao, J.P.; Lin, H.X. The miR156-SPL9-DFR pathway coordinates the relationship between development and abiotic stress tolerance in plants. *Plant J.* **2014**, *80*, 1108–1117. [CrossRef]
14. Wang, J.; Ye, Y.; Xu, M.; Feng, L.; Xu, L.A. Roles of the SPL gene family and miR156 in the salt stress responses of tamarisk (*Tamarix chinensis*). *BMC Plant Biol.* **2019**, *19*, 370. [CrossRef] [PubMed]
15. Dong, H.; Yan, S.; Jing, Y.; Yang, R.; Zhang, Y.; Zhou, Y.; Zhu, Y.; Sun, J. MIR156-Targeted SPL9 Is Phosphorylated by SnRK2s and Interacts With ABI5 to Enhance ABA Responses in *Arabidopsis*. *Front. Plant Sci.* **2021**, *12*, 708573. [CrossRef] [PubMed]
16. Ma, Y.; Xue, H.; Zhang, F.; Jiang, Q.; Yang, S.; Yue, P.; Wang, F.; Zhang, Y.; Li, L.; He, P.; et al. The miR156/SPL module regulates apple salt stress tolerance by activating MdWRKY100 expression. *Plant Biotechnol. J.* **2021**, *19*, 311–323. [CrossRef]
17. Schwarz, S.; Grande, A.V.; Bujdoso, N.; Saedler, H.; Huijser, P. The microRNA regulated SBP-box genes SPL9 and SPL15 control shoot maturation in *Arabidopsis*. *Plant Mol. Biol.* **2008**, *67*, 183–195. [CrossRef]
18. Wang, J.W.; Schwab, R.; Czech, B.; Mica, E.; Weigel, D. Dual effects of miR156-targeted SPL genes and CYP78A5/KLUH on plastochron length and organ size in *Arabidopsis thaliana*. *Plant Cell* **2008**, *20*, 1231–1243. [CrossRef]
19. Wang, L.; Yu, P.; Lyu, J.; Hu, Y.; Han, C.; Bai, M.Y.; Fan, M. BZR1 physically interacts with SPL9 to regulate the vegetative phase change and cell elongation in *Arabidopsis*. *Int. J. Mol. Sci.* **2021**, *22*, 10415. [CrossRef]
20. Wu, G.; Poethig, R.S. Temporal regulation of shoot development in *Arabidopsis thaliana* by miR156 and its target SPL3. *Development* **2006**, *133*, 3539–3547. [CrossRef]
21. Wu, G.; Park, M.Y.; Conway, S.R.; Wang, J.W.; Weigel, D.; Poethig, R.S. The sequential action of miR156 and miR172 regulates developmental timing in *Arabidopsis*. *Cell* **2009**, *138*, 750–759. [CrossRef] [PubMed]
22. Yu, N.; Niu, Q.W.; Ng, K.H.; Chua, N.H. The role of miR156/SPLs modules in *Arabidopsis* lateral root development. *Plant J.* **2015**, *83*, 673–685. [CrossRef] [PubMed]
23. Yu, N.; Cai, W.J.; Wang, S.; Shan, C.M.; Wang, L.J.; Chen, X.Y. Temporal control of trichome distribution by microRNA156-targeted SPL genes in *Arabidopsis thaliana*. *Plant Cell* **2010**, *22*, 2322–2335. [CrossRef] [PubMed]
24. Xue, X.Y.; Zhao, B.; Chao, L.M.; Chen, D.Y.; Cui, W.R.; Mao, Y.B.; Wang, L.J.; Chen, X.Y. Interaction between two timing microRNAs controls trichome distribution in *Arabidopsis*. *PLoS Genet.* **2014**, *10*, e1004266. [CrossRef]
25. Ioannidi, E.; Rigas, S.; Tsitsekian, D.; Daras, G.; Alatzas, A.; Makris, A.; Tanou, G.; Argiriou, A.; Alexandrou, D.; Poethig, S.; et al. Trichome patterning control involves TTG1 interaction with SPL transcription factors. *Plant Mol. Biol.* **2016**, *92*, 675–687. [CrossRef]

26. Nodine, M.D.; Bartel, D.P. MicroRNAs prevent precocious gene expression and enable pattern formation during plant embryogenesis. *Genes Dev.* **2010**, *24*, 2678–2692. [CrossRef]
27. Martin, R.C.; Asahina, M.; Liu, P.-P.; Kristof, J.R.; Coppersmith, J.L.; Pluskota, W.E.; Bassel, G.W.; Goloviznina, N.A.; Nguyen, T.T.; Martínez-Andújar, C.; et al. The regulation of post-germinative transition from the cotyledon- to vegetative-leaf stages by microRNA-targeted Squamosa Promoter-Binding Protein LIKE13 in *Arabidopsis*. *Seed Sci. Res.* **2010**, *20*, 89–96. [CrossRef]
28. Miura, K.; Ikeda, M.; Matsubara, A.; Song, X.J.; Ito, M.; Asano, K.; Matsuoka, M.; Kitano, H.; Ashikari, M. OsSPL14 promotes panicle branching and higher grain productivity in rice. *Nat. Genet.* **2010**, *42*, 545–549. [CrossRef]
29. Jiao, Y.; Wang, Y.; Xue, D.; Wang, J.; Yan, M.; Liu, G.; Dong, G.; Zeng, D.; Lu, Z.; Zhu, X.; et al. Regulation of OsSPL14 by OsmiR156 defines ideal plant architecture in rice. *Nat. Genet.* **2010**, *42*, 541–544. [CrossRef]
30. Wang, S.; Wu, K.; Yuan, Q.; Liu, X.; Liu, Z.; Lin, X.; Zeng, R.; Zhu, H.; Dong, G.; Qian, Q.; et al. Control of grain size, shape and quality by OsSPL16 in rice. *Nat. Genet.* **2012**, *44*, 950–954. [CrossRef]
31. Wang, S.; Li, S.; Liu, Q.; Wu, K.; Zhang, J.; Wang, S.; Wang, Y.; Chen, X.; Zhang, Y.; Gao, C.; et al. The OsSPL16-GW7 regulatory module determines grain shape and simultaneously improves rice yield and grain quality. *Nat. Genet.* **2015**, *47*, 949–954. [CrossRef] [PubMed]
32. Backhaus, R.A. Rubber formation in plants. *Isr. J. Bot.* **1985**, *34*, 283–293.
33. van Beilen, J.B.; Poirier, Y. Establishment of new crops for the production of natural rubber. *Trends Biotechnol.* **2007**, *25*, 522–529. [CrossRef] [PubMed]
34. Cornish, K. Alternative natural rubber crops: Why should we care? *Technol. Innov.* **2017**, *18*, 244–255. [CrossRef]
35. Wang, Y.; Guo, D.; Li, H.L.; Peng, S.Q. Characterization of HbWRKY1, a WRKY transcription factor from *Hevea brasiliensis* that negatively regulates HbSRPP. *Plant Physiol. Biochem.* **2013**, *71*, 283–289. [CrossRef] [PubMed]
36. Li, H.L.; Wei, L.R.; Guo, D.; Wang, Y.; Zhu, J.H.; Chen, X.T.; Peng, S.Q. HbMADS4, a MADS-box transcription factor from *Hevea brasiliensis*, negatively regulates HbSRPP. *Front. Plant Sci.* **2016**, *7*, 1709. [CrossRef]
37. Wang, Y.; Zhan, D.F.; Li, H.L.; Guo, D.; Zhu, J.H.; Peng, S.Q. Transcriptome-wide identification and characterization of MYB transcription factor genes in the laticifer cells of *Hevea brasiliensis*. *Front. Plant Sci.* **2017**, *8*, 1974. [CrossRef]
38. Qu, L.; Li, H.L.; Guo, D.; Zhu, J.H.; Wang, Y.; Yin, L.Y.; Peng, S.Q. HbWRKY27, a group IIe WRKY transcription factor, positively regulates HbFPS1 expression in *Hevea brasiliensis*. *Sci. Rep.* **2020**, *10*, 20639. [CrossRef]
39. Guo, D.; Yang, Z.P.; Li, H.L.; Wang, Y.; Zhu, J.H.; Peng, S.Q. The 14-3-3 protein HbGF14a interacts with a RING zinc finger protein to regulate expression of the rubber transferase gene in *Hevea brasiliensis*. *J. Exp. Bot.* **2018**, *69*, 1903–1912. [CrossRef]
40. Guo, D.; Li, H.L.; Wang, Y.; Zhu, J.H.; Peng, S.Q. A myelocytomatosis transcription factor from *Hevea brasiliensis* positively regulates the expression of the small rubber particle protein gene. *Ind. Crops Prod.* **2019**, *133*, 90–97. [CrossRef]
41. Cheng, Y.; Luo, J.; Li, H.; Wei, F.; Zhang, Y.; Jiang, H.; Peng, X. Identification of the WRKY Gene Family and Characterization of Stress-Responsive Genes in *Taraxacum kok-saghyz* Rodin. *Int. J. Mol. Sci.* **2022**, *23*, 10270. [CrossRef]
42. Zhang, H.; Gong, Y.; Sun, P.; Chen, S.; Ma, C. Genome-wide identification of CBF genes and their responses to cold acclimation in *Taraxacum kok-saghyz*. *PeerJ* **2022**, *10*, e13429. [CrossRef]
43. Liu, J.; Wang, X.; Chen, Y.; Liu, Y.; Wu, Y.; Ren, S.; Li, L. Identification, evolution and expression analysis of WRKY gene family in *Eucommia ulmoides*. *Genomics* **2021**, *113*, 3294–3309. [CrossRef]
44. Zhang, X.; Wang, X.; Pan, L.; Guo, W.; Li, Y.; Wang, W. Genome-wide identification and expression analysis of MADS-box transcription factors reveal their involvement in sex determination of hardy rubber tree (*Eucommia ulmoides* oliv). *Front. Genet.* **2023**, *14*, 1138703. [CrossRef] [PubMed]
45. Larkin, M.A.; Blackshields, G.; Brown, N.P.; Chenna, R.; McGettigan, P.A.; McWilliam, H.; Valentin, F.; Wallace, I.M.; Wilm, A.; Lopez, R.; et al. Clustal W and clustal X version 2.0. *Bioinformatics* **2007**, *23*, 2947–2948. [CrossRef] [PubMed]
46. Chen, C.; Chen, H.; Zhang, Y.; Thomas, H.R.; Frank, M.H.; He, Y.; Xia, R. TBtools: An integrative toolkit developed for interactive analyses of big biological data. *Mol. Plant* **2020**, *13*, 1194–1202. [CrossRef]
47. Bailey, T.L.; Boden, M.; Buske, F.A.; Frith, M.; Grant, C.E.; Clementi, L.; Ren, J.; Li, W.W.; Noble, W.S. MEME SUITE: Tools for motif discovery and searching. *Nucleic Acids Res.* **2009**, *37*, W202–W208. [CrossRef] [PubMed]
48. Birkenbihl, R.P.; Jach, G.; Saedler, H.; Huijser, P. Functional Dissection of the Plant-specific SBP-Domain: Overlap of the DNA-binding and Nuclear Localization Domains. *J. Mol. Biol.* **2005**, *352*, 585–596. [CrossRef]
49. Yang, Z.; Wang, X.; Gu, S.; Hu, Z.; Xu, H.; Xu, C. Comparative study of SBP-box gene family in *Arabidopsis* and rice. *Gene* **2008**, *407*, 1–11. [CrossRef]
50. Hyun, Y.; Richter, R.; Vincent, C.; Martinez-Gallegos, R.; Porri, A.; Coupland, G. Multi-layered regulation of SPL15 and cooperation with SOC1 integrate endogenous flowering pathways at the *Arabidopsis* shoot meristem. *Dev. Cell* **2015**, *37*, 254–266. [CrossRef]
51. Gao, R.; Wang, Y.; Gruber, M.Y.; Hannoufa, A. miR156/SPL10 modulates lateral root development, branching and leaf morphology in *Arabidopsis* by silencing AGAMOUS-LIKE 79. *Front. Plant Sci.* **2017**, *8*, 2226. [CrossRef]
52. Kabin, X.; Congqing, W.; Lizhong, X. Genomic organization, differential expression, and interaction of SQUAMOSA promoter-binding-like transcription factors and microRNA156 in rice. *Plant Physiol.* **2006**, *142*, 280–293. [CrossRef]
53. Zhang, D.; Han, Z.; Li, J.; Qin, H.; Zhou, L.; Wang, Y.; Zhu, X.; Ma, Y.; Fang, W. Genome-wide analysis of the SBP-box gene family transcription factors and their responses to abiotic stresses in tea (*Camellia sinensis*). *Genomics* **2020**, *112*, 2194–2202. [CrossRef] [PubMed]

54. Song, J.; Ma, D.; Yin, J.; Yang, L.; Gao, C. Genome-wide characterization and expression profiling of squamosa promoter binding protein-like (SBP) transcription factors in wheat (*Triticum aestivum* L.). *Agronomy* **2019**, *9*, 527. [CrossRef]
55. Cai, C.; Guo, W.; Zhang, B. Genome-wide identification and characterization of SPL transcription factor family and their evolution and expression profiling analysis in cotton. *Sci. Rep.* **2018**, *8*, 762. [CrossRef] [PubMed]
56. Santner, A.; Calderon-Villalobos, L.I.; Estelle, M. Plant hormones are versatile chemical regulators of plant growth. *Nat. Chem. Biol.* **2009**, *5*, 301–307. [CrossRef] [PubMed]
57. Leng, P.; Yuan, B.; Guo, Y. The role of abscisic acid in fruit ripening and responses to abiotic stress. *J. Exp. Bot.* **2014**, *65*, 4577–4588. [CrossRef]
58. Guo, J.; Wang, S.; Yu, X.; Dong, R.; Li, Y.; Mei, X.; Shen, Y. Polyamines regulate strawberry fruit ripening by abscisic acid, auxin, and ethylene. *Plant. Physiol.* **2018**, *177*, 339–351. [CrossRef]
59. Duan, Q.; Li, G.-R.; Qu, Y.-P.; Yin, D.-X.; Zhang, C.-L.; Chen, Y.-S. Genome-Wide Identification, Evolution and Expression Analysis of the Glutathione S-Transferase Supergene Family in Euphorbiaceae. *Front. Plant Sci.* **2022**, *13*, 808279. [CrossRef]
60. Jaiswal, P.; Cheruku, J.R.; Kumar, K.; Yadav, S.; Singh, A.; Kumari, P.; Dube, S.C.; Upadhyaya, K.C.; Verma, P.K. Differential transcript accumulation in chickpea during early phases of compatible interaction with a necrotrophic fungus Ascochyta rabiei. *Mol. Biol. Rep.* **2012**, *39*, 4635–4646. [CrossRef]

diversity

MDPI

Article

Diverse Host Spectrum and the Parasitic Process in the Pantropical Hemiparasite *Cassytha filiformis* L. (Lauraceae) in China

Zhi-Fang Liu [1,2,3,4,*], Xiu-Qin Ci [4,5], Shi-Fang Zhang [4,6], Xiao-Yan Zhang [4,6], Xue Zhang [7], Li-Na Dong [8], John G. Conran [9] and Jie Li [4,5,*]

1 Institute of Leisure Agriculture, Shandong Academy of Agricultural Sciences, Jinan 250131, China
2 Key Laboratory of East China Urban Agriculture, Ministry of Agriculture and Rural Affairs, Jinan 250131, China
3 Shandong Engineering Research Center of Ecological Horticultural Plant Breeding, Jinan 250131, China
4 Plant Phylogenetics and Conservation Group, Center for Integrative Conservation, Xishuangbanna Tropical Botanical Garden, Chinese Academy of Sciences, Kunming 650201, China
5 Center of Conservation Biology, Core Botanical Gardens, Chinese Academy of Sciences, Mengla 666300, China
6 University of Chinese Academy of Sciences, Beijing 100049, China
7 Key Laboratory of Chemical Biology (Ministry of Education), School of Pharmaceutical Sciences, Cheeloo College of Medicine, Shandong University, Jinan 250012, China
8 Guangxi Institute of Botany, Guangxi Zhuang Autonomous Region and Chinese Academy of Sciences, Guilin 541006, China
9 Australian Centre for Evolutionary Biology and Biodiversity & Sprigg Geobiology Centre, School of Biological Sciences, University of Adelaide, Adelaide, SA 5005, Australia
* Correspondence: sghliuzhifang2010@163.com (Z.-F.L.); jieli@xtbg.ac.cn (J.L.)

Abstract: Many hemiparasites attach to a range of different host species, resulting in complex parasite–host interactions. Comprehensive molecular phylogenies allow the investigation of evolutionary relationships between these host plants. We surveyed the hosts of the laurel dodder (*Cassytha filiformis*, Lauraceae) in China, representing 184 species from 146 genera, 67 families, and spanning flowering plants, conifers, and ferns, using host phylogenetic relationships to investigate the susceptibility to attack by this hemiparasitic plant among the vascular plants. The process of produced well-formed haustoria by *C. filiformis* was also observed in detail for six different hosts. Our results show that *C. filiformis* grows mainly on trees and shrubs from phylogenetically divergent members of the rosid and asterid eudicot clades, often attacking multiple adjacent hosts simultaneously, and forming extensive colonies. However, whether and to what extent transitions between *C. filiformis* and host plants occur remain unclear. Physiological evidence for the complex parasite–host species interactions need to be studied in the future.

Keywords: hemiparasite–host species; phylogenetic; *Cassytha*; Lauraceae; woody plants

Citation: Liu, Z.-F.; Ci, X.-Q.; Zhang, S.-F.; Zhang, X.-Y.; Zhang, X.; Dong, L.-N.; Conran, J.G.; Li, J. Diverse Host Spectrum and the Parasitic Process in the Pantropical Hemiparasite *Cassytha filiformis* L. (Lauraceae) in China. *Diversity* **2023**, *15*, 492. https://doi.org/10.3390/d15040492

Academic Editors: Michael Wink and Mario A. Pagnotta

Received: 2 February 2023
Revised: 20 March 2023
Accepted: 24 March 2023
Published: 27 March 2023

1. Introduction

Parasitic plants are a diverse group of 292 genera: approximately 4750 species (1% of all flowering plants) with parasitism evolving at least 12 times [1,2]. Parasitic plants negatively influence host growth and development [3]. They have had increased attention by researchers over the past three decades [1]. Parasitic plants can obtain water, mineral nutrients, and carbon from other plants using the haustorium [1,4]. Parasitic plants can either be photosynthetic hemiparasites or achlorophyllous holoparasites [5], but the majority are hemiparasites [2]. They occur in most major ecosystems across the world from subpolar to tropical latitudes, as well as agricultural ecosystems [6].

Generalist hemiparasitic plants as a group are morphologically diverse with a broad range of host interactions [2], often attacking multiple co-occurring plant species [7]. Parasite performance can be influenced by host characteristics, including biomass [8], carbon content [9], resistance [10], genetic constitution [11], rate of growth [12], host condition [13], secondary compounds [14], etc., complicating research into the evolution of host range [7]. Hemiparasite fitness has often been linked to host functional plant groups such as legumes, grasses, or forbs [15], but is influenced by life history strategies and resource-linked attributes [16]. However, some functional groups are monophyletic, while others have more complex evolutionary relationships and hemiparasite performance may sometimes be predicted better by host phylogenetic relationships [7]. Host identity may be an important factor in determining the success of the parasitic species.

The literature on parasitic plant host biology covers a range of taxa, including the root hemiparasites in Orobanchaceae [17] and Santalaceae [18], mistletoes of Loranthaceae [19] and Viscaceae [20], and the stem holoparasite *Cuscuta* L. (dodder, Convolvulaceae) [21] and hemiparasite *Cassytha* L. (laurel dodder, Lauraceae) [22]. Parasitic plants also interact with other plants, influencing competition, community biodiversity, and nutrient cycling [23–25]. Although molecular phylogenetic methods have been used to address several long-standing issues in parasitic plant taxonomy and evolutionary biology [1], few studies have examined the evolutionary relationships of host species.

There is a long tradition of the human use of parasitic plants for medicinal and cultural purposes worldwide [26]. *Cassytha filiformis* L. (laurel dodder) is a pantropical hemiparasite which has also been used variously for medicine, cosmetics, cushion, and rope-making in many regions [6,27], as well as being regarded as a serious weed in some tropical regions [7,28]. The species has been reported as having a very wide, possibly indiscriminate host range [28–30], although with some apparent host preferences [30,31]. Identifying its host spectrum and phylogenetic relationships may help with the use of this parasite to control invasive weeds by understanding their interactive biology, as has been suggested for *C. pubescens* R.Br. in Australia [32–34]. Parasitic plants develop haustoria for absorbing water and nutrients from the host plant, so understanding the seed germination to host selection and haustorial attachment process is critical [35].

Cassytha is the only parasitic genus Lauraceae, but it is also morphologically similar to *Cuscuta* L. (Convolvulaceae), with both being leafless, haustorial stem parasites [35]. *Cassytha* is hemiparasitic with white or light green flowers borne in spicate, capitate, or racemose inflorescences and 1-seeded drupes included in a dilated, fleshy, free perianth tube. In contrast, *Cuscuta* is holoparasitic, with white, orange, or purplish-red stems; white, yellow, or pink flowers in globular inflorescences; and two–four seeded capsules [36]. *Cassytha* parasitizes a wide range of mainly woody plants, including plants of agricultural and economic value, and although most species occur in Australia, *C. filiformis* occurs across the tropics worldwide [22,28]. According to Nickrent (2002), *C. filiformis* is indiscriminate in host choice, often attacking multiple co-occurring hosts simultaneously [29], and surveys by Li et al. (1992) in Nanning, Tianlin, Yulin, and other places in Guangxi recorded 137 host species from 113 genera and 58 families [37].

In China, *C. filiformis* causes serious damage to crop and forests in some regions, with Huang et al. (1957) finding that the severe parasitism of oil tea by *C. filiformis* caused a massive yield reduction in Shangsi and Luchuan, Guangxi, China [38]. Similarly, Zhang (1988) found that heavy infestation of *Pinus massoniana* Lamb by *C. filiformis* led to acute deciduous needle disease caused by *Lophodermium pinastri* Chev. in Fuzhou, Fujian [39]. These studies show the importance of investigating the host–parasite relationships of *C. filiformis* and their effects, but to date no systematic and comprehensive studies have been undertaken on host species occurring in China.

Accordingly, we here investigate and summarize the hosts of *C. filiformis* in China, with a particular focus on its host phylogenetic relationships and host–parasite species interactions. Our aim is to understand the evolutionary relationships of the diverse range of host species in China and the potential growth responses of the hosts, specifically: (1) to determine the host range and responses to attack for *C. filiformis* in China and analyze the parasitic progress of *C. filiformis*; and (2) examine the evolutionary relationships among the hosts of *C. filiformis*.

2. Materials and Methods

2.1. Plant Materials

Collection sites in China were selected using data from the Chinese Virtual Herbarium (CVH, https://www.cvh.ac.cn, accessed on 5 May 2019), Plant Photo Bank of China (PPBC, http://ppbc.iplant.cn, accessed on 5 May 2019), and Global Biodiversity Information Facility (GBIF, https://www.gbif.org, accessed on 6 May 2019). We conducted a series of targeted field trips to the geographical range areas of particular interest in 2019–2021 (Figure S1). Efforts were made to ensure sampling from localities across most of the known geographical ranges of the species (144 sample sites; for details see Table S1 and Figure S1). Inflorescences and fruits of *C. filiformis* were gathered (with reproductive characters collected where possible) from Guangzhou (Guangdong), Lingshui (Hainan), and Xishuangbanna (Yunnan) (Figure 1), but the Guangzhou locality (Zhongkai University of Agriculture and Engineering) was the only one with ripe fruit, despite repeated site visits. Host plants were recorded for samples from six provinces: Fujian, Guangdong, Guangxi, Hainan, Yunnan, and Zhejiang (Figure S1; Table S1), with hosts defined as plants having direct haustorial connections to *C. filiformis* (Figure 2). Host identification was performed by experts at PE (Herbarium, Institute of Botany, CAS), Bing Liu, XTBG (Xishuangbanna Tropical Botanical Garden, Chinese Academy of Sciences), Yunhong Tan and CATAS (Chinese Academy of Tropical Agricultural Sciences), and Shengzhuo Huang, using the reproductive or vegetative characters available on the vouchers. All vouchers were stored at the Herbarium of Xishuangbanna Tropical Botanical Garden (HITBC). In addition, we summarized the host plants of the published literature, together with data from our surveys (40 specimens by the previous published literature, 104 specimens by our surveys; details can be seen in Table S1). In total, the Chinese host range assessment was based on 184 specimens of *C. filiformis*.

Figure 1. Representative inflorescence and fruits of *Cassytha filiformis*. (**a**) Inflorescence collected in Guangzhou (Guangdong, longitude: 113°17′ E, latitude: 23°0′ N, altitude: 10 m). (**b**) Inflorescence collected in Xishuangbanna (Yunnan, longitude: 100°45′ E, latitude: 22°28′ N, altitude: 1129 m). (**c**) Inflorescence and unripe fruits collected in Lingshui (Hainan, longitude: 109°54′ E, latitude: 18°34′ N, altitude: 76 m). (**d**) Ripe fruit collected in Guangzhou (Guangdong, longitude: 113°17′ E, latitude: 23°0′ N, altitude: 10 m). Scale bars = 5 mm.

Figure 2. The host *Bridelia balansae* and *Cassytha filiformis* with its haustoria. (**a**) Host and *C. filiformis* collected in Lingshui, Hainan (longitude: 109°55′ E, latitude: 18°33′ N, altitude: 42 m). (**b**) Vines and fruits. (**c**) The haustorium. Scale bars = 1 mm.

2.2. Growth Conditions of C. filiformis

The seeds of *C. filiformis* displayed physical dormancy [40], which was broken by soaking the seeds in warm water at 60 °C for 1 h. The treated seeds were then grown in the soils that collected in Guangzhou (Zhongkai University of Agriculture and Engineering, Guangdong, China) under controlled conditions in a PHCbi CO_2 Incubator (PHC Holdings Corporation, Tokyo, Japan) with a 24–27 °C, 18 h light and 6 h dark cycle. Germination time was measured as the total number of days to seedling emergence after planting. Host seeds (*Bidens pilosa* L. (Asteraceae) herb; *Chromolaena odorata* (L.) R.M.King and H.Rob. (Asteraceae), herb; *Cleome rutidosperma* DC. (Cleomaceae), herb; *Dimocarpus longan* Lour. (Sapindaceae), tree; *Mikania micrantha* Kunth (Asteraceae), liana; and *Scoparia dulcis* L. (Plantaginaceae), herb) were also collected from the sites of *C. filiformis* seed collection, in Guangzhou (Zhongkai University of Agriculture and Engineering, Guangdong, China). These six species were confirmed as hosts of *C. filiformis*, based on field observations of haustorial attachment. We filled pots (18 cm diameter, 16 cm height) with soil collected from these field sites and sowed a minimum of five seeds of a single species per pot, with three replicate pots per host species for a total of eighteen pots. Hosts were observed at the same time each day (10 a.m.) in each pot once germination had commenced.

To investigate the interaction between *C. filiformis* and hosts, we observed and recorded the growth and germination progress. Three replicates were used, each of which was pooled from at least five seeds. To obtain laurel dodder and hosts from host–laurel dodder parasitization systems, we put different host plants with five seeds of *C. filiformis* together to form the host–laurel dodder parasitization system. This system was cultivated under controlled conditions in PHCbi CO_2 Incubator (18 h light, 6 h dark) at 24–27 °C. From germination to growth conditions, *C. filiformis* as well as hosts were observed at the same time each day (10 a.m.) in each pot.

2.3. Phylogenic Analysis

To examine the host–parasite relationship, a host frequency analysis was determined using two methods of classifying species: (1) life history (woody or herbaceous) and (2) taxonomic levels (family, genus, and species). We recorded the percentage of the number of instances *C. filiformis* was found on each of the hosts from the total host occurrences. The lists of family, genus, and species' hosts were clustered using the R packages APE [41] and V.PhyloMaker2 [42], based on the botanical nomenclature of the World Plants (WP) Database for Pteridophytes and Gymnosperms (https://www.worldplants.de, accessed on 20 September 2022), the Angio-sperm Phylogeny Website (http://www.mobot.org/MOBOT/research/Apweb/, accessed on 20 September 2022), and the Angiosperm Phylogeny Group IV (APG IV) [43] classification for angiosperms. The percentage of different types was imported into Excel to summarize the hosts' frequency.

3. Results

3.1. Biology of Laurel Dodder and Its Hosts

Cassytha filiformis and the host seeds germinated under the soil surface without light, with *C. filiformis* producing a rudimentary primary root about 0.5 mm long (Figure 3a). This root grows for 2–4 weeks, degenerating when the embryonic axis starts to grow and three–four adventitious roots 2–6 mm long and 0.2–0.4 mm thick develop at the base of the axis (Figure 3a). This free-living phase can last up to two months (Figure 3b,c), but unless a suitable host is found during this period, the seedling dies (Figure 3c).

Figure 3. Images of *Cassytha filiformis* growth and parasitic processes on host species in laboratory conditions. (**a**) Germinated seed with a rudimentary and short-lived root. (**b**) Independent growth stage of seedlings. (**c**) The dead *C. filiformis* seedlings unsuccessfully parasitized a viable host; the red arrow indicates the dead stem of *C. filiformis*. (**d**) The survived *Cassytha filiformis* seedlings successfully parasitized the viable host *Mikania micrantha* with the haustoria. (**e–h**) *Cassytha filiformis–Cleome rutidosperma* host associations via the haustoria. The thin stem is *C. filiformis*, and the thick stem is *C. rutidosperma*. The red arrows indicate infected parts.

For the laurel dodder as well as the hosts, the percent germination exceeded 80% (Table S2). The average time to germination of *C. filiformis* was 15 days (±10–20 SD) and for the host species: *M. micrantha* (3 ± 3–4), *B. pilosa* (5 ± 4–6), *C. odorata* (5 ± 4–6), *C. rutidosperma*

(7 ± 6–8), *S. dulcis* (8 ± 7–9), and *D. longan* (21 ± 20–22). The host species belongs to the family Asteraceae germinated firstly (*M. micrantha*, three days), then *B. pilosa* and *C. odorata* (five days), followed by *C. rutidosperma* (Cleomaceae, seven days), *S. dulcis* (Scrophulariaceae, eight days), and *D. longan* (Sapindaceae, three weeks). It took eight weeks for parasitism to occur following germination of the host liana species *M. micrantha* (Figure 3d). Then, the same *C. filiformis* vine that was attacking *M. micrantha* (Asteraceae) was also connecting to herb species *C. rutidosperma* (Cleomaceae) a month later (Figure 3e–h). When the *C. rutidosperma* (Cleomaceae) plant died, some parts of the host in contact with the haustoria survived for a further two months (Figure 3f–h), apparently deriving water and nutrients from the parasite and even producing new shoots (Figure 3h). The signal transduction mechanism for infection is unclear yet. Except tree species *D. longan* (Sapindaceae), the other host species were all successfully parasitized. Under the controlled conditions, it seemed easier to infect herbs than woody plants.

Flowering and fruiting in wild Chinese *C. filiformis* occur from May to December, with considerable overlap (Figure 1a,c). The inflorescence is arranged into a spicate, capitate, or racemose (Figure 1a,b), and the ripe fruit is a drupe with a white translucent, fleshy pericarp (Figure 1d). The cultivated *C. filiformis* plants did not flower under the laboratory conditions used here, but this may be because at least several Australian *Cassytha* species take >5 years to flower from seed in the wild (J.G. Conran, pers. obs.).

3.2. Host Range of *C. filiformis*

After three years of field investigation, we found *C. filiformis* preferred to grow on trees and shrubs in well-lit, open, well-watered habitats, especially along roadsides. Based on our surveys and the published literature [37,44,45], we found that *C. filiformis* produced well-formed haustoria onto diverse vascular plants, including angiosperms, conifers, and ferns (Figures 4 and 5; Table S1). The overall host range of Chinese *C. filiformis* included 184 species, which belonged to 146 genera, 67 families, and 32 orders (Table S1). In total, 80.80% of the host species in this study, defined by the presence of haustoria, were woody taxa, including trees (82/184, 44.57%), shrubs (58/184, 31.52%), and occasionally lianas (9/184, 4.91%), but herbaceous plants were also affected (35/184, 19.20%) (Figure S2). The most common hosts at order-, family-, genus-, and species levels are summarized in Table S1. Nine preferred orders had hosts with more than ten species (Asterales, Ericales, Fabales, Gentianales, Laurales, Malpighiales, Myrtales, Rosales, and Sapindales. All contain 30 families, 96 genera, and 126 species; Figure 5; Table S1 noted with different colors). To our knowledge, there is currently no comprehensive survey of the host range of laurel dodder for such a broad geographic range of China.

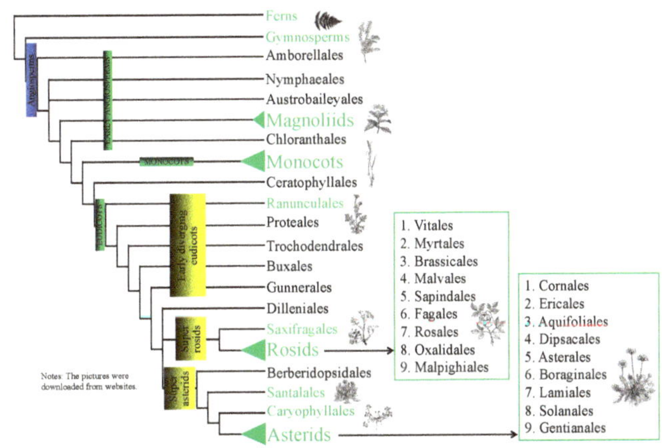

Figure 4. A simplified phylogenetic tree showing major orders that include *Cassytha filiformis* hosts. The green labels indicate the host positions among the vascular phylogenetic relationships.

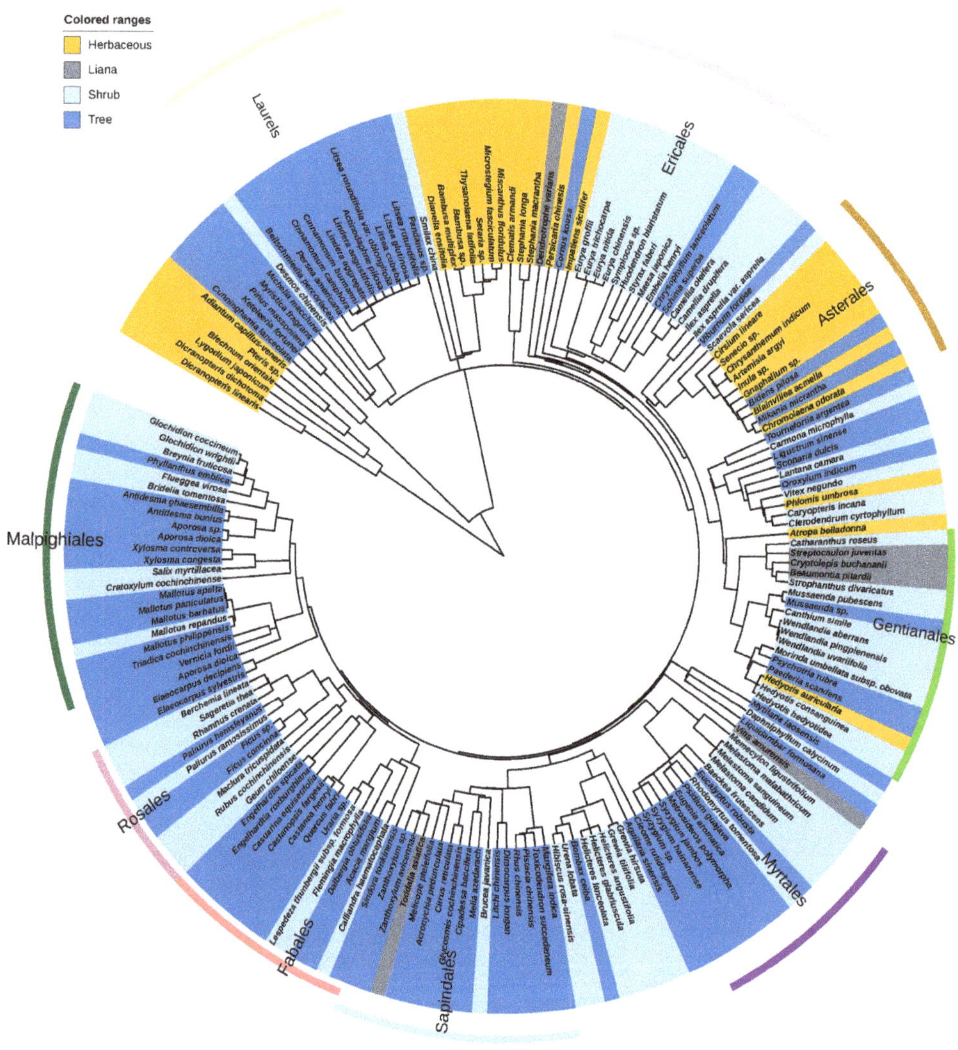

Figure 5. The host plants' phylogenetic tree for *Cassytha filiformis* in China. Yellow notes herbs. Grey notes lianas. Light blue notes shrubs; and darker blue notes trees. Hosts belonging to ferns, gymnosperms, and angiosperms are coded with different colors corresponding to Table S1.

The host range of *C. filiformis* spanned most of the vascular plants (Figure 4), with six ferns (*Adiantum capillus-veneris* L., *Blechnopsis orientalis* C. Presl, *Dicranopteris pedata* (Houttuyn) Nakaike, *D linearis* (Burm.) Underw., *Lygodium japonicum* (Thunb.) Sw., and *Pteris* sp. L.) and three conifers (*Cunninghamia lanceolata* (Lamb.) Hook., *Keteleeria fortunei* (Murr.) Carr., and *Pinus massoniana* Lamb.) parasitized by *C. filiformis* (Figure 5).

In the flowering plants, rosids were common hosts, with Malpighiales containing the largest number of host species (twenty-two species), and Sapindales containing the largest number of infected genera (fifteen genera), while the asterid order Ericales had the most host families (seven) (Figures 4 and 5; Table S1). Euphorbiaceae was the largest single family with eight genera and thirteen species (Figure 5; Table S1). The order Gentianales represented the widest array of host plant growth forms, including the tree *Psychotria rubra* (Lour.) Poir., shrub *Wendlandia aberrans* Cowan, herb *Hedyotis auricularia* L., and liana *Cryptolepis buchananii* Roem. et Schult (Figure 5). Four angiosperm families had at least 10 host species (Table S1): Rubiaceae (12), Lauraceae (11), Phyllanthaceae (11), and Asteraceae (10). These results differ from the study by Zhang et al. [6], where they reported seven families at the world level with at least 10 host species: Anacardiaceae, Asteraceae, Euphorbiaceae, Fabaceae, Myrtaceae, Phyllanthaceae, and Rubiaceae. This confirms that *C. filiformis* is a generalist hemiparasite that is associated with many host species. These species are relatively common (Figures 4 and 5), with no obvious evolutionary relationships, as observed by Zhang et al. [6].

4. Discussion

4.1. Laurel Dodder Prefers Woody Hosts

Eighty-one percent of the hosts noted in this study were woody (Figures 5 and S2), but numerous herbaceous plants were also affected, agreeing with the conclusions of previous studies such as those by Zhang et al. [6] and Parra-Tabla et al. [44]. However, the percentage of host life form preferences varies considerably between countries, with trees most widely affected in Tanzania (69.7%), China (44.6%), India (36.1%), and down to Japan (19.2%). Shrub hosts percentages were similar for China (36.9%) and India (36.1%), but lower for Japan and Tanzania (both ca. 20%) (Table 1). In contrast, herbaceous host ratios for the Japanese Ryukyu Islands were much higher than these other three reported regions (46.2%).

Table 1. Host plants' life forms for *Cassytha filiformis* infestation in four countries.

Country	Herb	Shrub	Tree	Liana	Study
China	15.51%	36.90%	44.39%	3.20%	[37,46]; Present study
India	13.89%	36.11%	36.11%	13.89%	[6,47]
Japan	46.15%	19.23%	19.23%	15.39%	[48]
Tanzania	0.09%	21.21%	69.70%	/	[45]

In addition to *C. filiformis*, the Australian species *C. melantha* and *C. pubescens* also seem to prefer woody hosts [46–48] and this preference for woody hosts may accord with the perennial life form and hemiparasitic nature of *Cassytha*. However, the networks created by a parasitic plant attached simultaneously to multiple hosts may trigger transferring systemic signals between the hosts [26]. Herbaceous species might also act as bridging hosts to allow *Cassytha* seedlings to survive long enough to grow onto nearby shrubs or trees [6]. It may also reflect the speciose nature of many woody host families, combined with the problem that as many herbaceous plants are annuals this could be problematic for their use by perennial *Cassytha* species [6]. The reasons why *Cassytha* seem to prefer certain plant families as hosts are unclear, though there is some evidence from Australia of a preference for nitrogen-fixing taxa [33,49]. However, the factors that might contribute to host susceptibility are currently unknown.

Although unrelated, *Cuscuta* and *Cassytha* are morphologically similar rootless stem parasites that spread by developing haustoria along their stems. Similarly, *Cuscuta* different species can vary in host specificity from a single species to hundreds of taxa covering diverse genera and families [35], with 237 species, 120 genera, and 32 families reported as hosts in one study [49]. *Cassytha* and *Cuscuta* mainly occupy different habitats and have different ecologies, which may indicate that the hemiparasitic *Cassytha* represent a less specialized parasite than the holoparasitic *Cuscuta*. It is not known what factors might contribute to the susceptibility of various hosts in *Cassytha* [50], and the location of hosts in *Cuscuta* occurs by a range of methods, including host chemistry [51,52]. Host choice in these two genera may also be influenced in part by the availability suitable host-derived resources, based on the degree to which the parasites can function independently [6,53]. Long-term host range divergence also helps to drive speciation in these two parasitic genera, but further study is needed.

4.2. The Host Phylogenetic Relationships of Laurel Dodder and Its Potential Application

The mechanisms by which *C. filiformis* selects the suitable hosts are complex. *C. filiformis* parasitizes plants throughout the vascular plant phylogeny. Vascular plants play a major role in global carbon cycling and are of fundamental importance to life on earth. Phylogenetic studies have led to tremendous progress in our understanding of the origin, phylogeny, and evolution of the plants [54]. The wide variability of the host cannot predict hemiparasite performance as the host species is scattered on vascular plant phylogenetic clades (Figure 4). Asterids (more than 80,000 species) and rosids (70,000 species) comprise more than half of the hosts (Figures 4 and 5), as they are the two largest core eudicot clades (>50% of the total angiosperm species diversity) [55,56]. Some hosts' clades do not comprise as many species as the above two clades, such as the ferns (about 10,500 to 15,000), gymnosperms (only 900 living species), and some angiosperm orders (Ranunculales, Saxifragales, and Santalales have about 2000 species, respectively) [43,56–58]. We found these host species have wide ranges and high population densities. So, the parasitism of the hosts seems likely related to the species' richness and distributions. That is different from Parra-Tabla et al. who studied the host of *C. filiformis* distributed in the coastal dunes of Yucatan. They claimed the frequency of parasitized plants by *C. filiformis* was not dependent on host plant abundance [44].

Parasitic plants have been documented repeatedly to play the role of keystone species in the ecosystems [59]. Many parasitic plants also parasitize multiple hosts simultaneously; thus, they may serve as a common network connecting multiple individuals in a plant community [26]. Harmful effects on community dominants, including invasive species, may facilitate species coexistence and thus increase biodiversity [27]. In our experiment, we found *C. filiformis* may act as a trigger transferring systemic signal to connect *M. micrantha* (Asteraceae) and *C. rutidosperma* (Cleomaceae) (Figure 3). Haustorial connections very likely allow the flow or even exchange of various molecules between *C. filiformis* and hosts, and it has been long known that viruses [60] and phytoplasmas [61] can be transmitted between *Cuscuta* and its hosts. The study on *Cassytha* has been relatively neglected, leading to it being less well-characterized compared to its companion *Cuscuta* [62]. We would like to take this opportunity to encourage researchers to explore the detailed relations of *C. filiformis*–host associations because understanding the host associations will likely provide an insight into the transition between laurel dodder and their host plants.

5. Conclusions

This study investigated the host range of *C. filiformis* over major areas of its geographic range in China and reported observations of the germination and parasitism of the selected host species. *Cassytha filiformis* was found to attach to many different host species, displaying complex parasite–host interactions. The results confirm that *C. filiformis* grows mainly on diverse woody species with divergent phylogenetic relationships. However, evidence for more complex parasite–host species interactions is unclear, and the physiological basis for these associations requires further study.

Supplementary Materials: The following supporting information can be downloaded at: https://www.mdpi.com/article/10.3390/d15040492/s1, Figure S1. The collection geographical range of *Cassytha filiformis* and its hosts in China; Figure S2. The percentage of different host life forms parasitized by *Cassytha filiformis* in China; Table S1. A summary of the hosts of *Cassytha filiformis*; Table S2. The hosts' germination rates and parasitizing situation by laurel dodder.

Author Contributions: Conceived and designed the research: Z.-F.L., J.L. and X.-Q.C.; performed the sampling: Z.-F.L., S.-F.Z. and L.-N.D.; conducted the experiments, analyzed the data: Z.-F.L., X.-Y.Z. and X.Z.; wrote the manuscript: Z.-F.L.; contributed to the revision of the manuscript: J.G.C., J.L. and X.-Q.C. All authors have read and agreed to the published version of the manuscript.

Funding: This work is funded by 1. the National Natural Science Foundation of China (31370245, 31770569, 31970222); 2. the Natural Science Foundation of Shandong (ZR2022QC214); 3. the Biodiversity Conservation Program of the Chinese Academy of Sciences (ZSSD-013); 4. the Science and

Technology Basic Resources Investigation Program of China: Survey and Germplasm Conservation of Plant Species with Extremely Small Populations in South-West China (2017FY100102); 5. the 135 programs of the Chinese Academy of Sciences (2017XTBG-T03).

Institutional Review Board Statement: Not applicable.

Informed Consent Statement: Not applicable.

Data Availability Statement: Not applicable.

Conflicts of Interest: The authors have declared that no competing interests exist.

References

1. Nickrent, D.L. Parasitic angiosperms: How often and how many? *Taxon* **2020**, *69*, 5–27. [CrossRef]
2. Twyford, A.D. Parasitic plants. *Curr. Biol.* **2018**, *28*, R857–R859. [CrossRef]
3. Zhang, J.; Xu, Y.; Xie, J.; Zhuang, H.; Liu, H.; Shen, G.; Wu, J. Parasite dodder enables transfer of bidirectional systemic nitrogen signals between host plants. *Plant Physiol.* **2021**, *185*, 1395–1410. [CrossRef] [PubMed]
4. Yoder, J.I.; Scholes, J.D. Host plant resistance to parasitic weeds; recent progress and bottlenecks. *Curr. Opin. Plant Biol.* **2010**, *13*, 478–484. [CrossRef] [PubMed]
5. Irving, L.J.; Cameron, D.D. You are what you eat: Interactions between root parasitic plants and their hosts. *Adv. Bot. Res.* **2009**, *50*, 87–138.
6. Zhang, H.; Florentine, S.; Tennakoon, K.U. The angiosperm stem hemiparasitic genus *Cassytha* (Lauraceae) and its host interactions: A review. *Front. Plant Sci.* **2022**, *13*, 864110. [CrossRef] [PubMed]
7. Brown, M.R.; Moore, P.G.P.; Twyford, A.D. Performance of generalist hemiparasitic *Euphrasia* across a phylogenetically diverse host spectrum. *New Phytol.* **2021**, *232*, 2165–2174. [CrossRef]
8. Matthies, D. Interactions between a root hemiparasite and 27 different hosts: Growth, biomass allocation and plant architecture. *Perspect. Plant Ecol.* **2017**, *24*, 118–137. [CrossRef]
9. Tesitel, J.; Tesitelova, T.; Fisher, J.P.; Leps, J.; Cameron, D.D. Integrating ecology and physiology of root-hemiparasitic interaction: Interactive effects of abiotic resources shape the interplay between parasitism and autotrophy. *New Phytol.* **2015**, *205*, 350–360. [CrossRef]
10. Bize, P.; Jeanneret, C.; Klopfenstein, A.; Roulin, A. What makes a host profitable? Parasites balance host nutritive resources against immunity. *Am. Nat.* **2008**, *171*, 107–118. [CrossRef]
11. Rowntree, J.K.; Cameron, D.D.; Preziosi, R.F. Genetic variation changes the interactions between the parasitic plant-ecosystem engineer *Rhinanthus* and its hosts. *Philos. Trans. R. Soc. B Biol. Sci.* **2011**, *366*, 1380–1388. [CrossRef] [PubMed]
12. Hautier, Y.; Hector, A.; Vojtech, E.; Purves, D.; Turnbull, L.A. Modelling the growth of parasitic plants. *J. Ecol.* **2010**, *98*, 857–866. [CrossRef]
13. Bickford, C.P.; Kolb, T.E.; Geils, B.W. Host physiological condition regulates parasitic plant performance: *Arceuthobium vaginatum* subsp. *cryptopodum on Pinus ponderosa. Oecologia* **2005**, *146*, 179–189. [CrossRef]
14. Adler, L.S. Alkaloid uptake increases fitness in a hemiparasitic plant via reduced herbivory and increased pollination. *Am. Nat.* **2000**, *156*, 92–99. [CrossRef] [PubMed]
15. Matthies, D. Interactions between the root hemiparasite *Melampyrum arvense* and mixtures of host plants: Heterotrophic benefit and parasite-mediated competition. *Oikos* **1996**, *75*, 118–124. [CrossRef]
16. Garnier, E. Growth analysis of congeneric annual and perennial grass species. *J. Ecol.* **1992**, *80*, 665–675. [CrossRef]
17. Phoenix, G.K.; Press, M.C. Linking physiological traits to impacts on community structure and function: The role of root hemiparasitic Orobanchaceae (ex-Scrophulariaceae). *J. Ecol.* **2005**, *93*, 67–78. [CrossRef]
18. Tennakoon, K.U.; Pate, J.S.; Arthur, D. Ecophysiological aspects of the woody root hemiparasite *Santalum acuminatum* (R Br) A DC and its common hosts in south western Australia. *Ann. Bot.* **1997**, *80*, 245–256. [CrossRef]
19. Le, Q.V.; Tennakoon, K.U.; Metali, F.; Lim, L.B.L.; Bolin, J.F. Ecophysiological responses of mistletoe *Dendrophthoe curvata* (Loranthaceae) to varying environmental parameters. *J. Trop. For. Sci.* **2016**, *28*, 59–67.
20. Glatzel, G.; Geils, B.W. Mistletoe ecophysiology: Host–parasite interactions. *Botany* **2009**, *87*, 10–15. [CrossRef]
21. Furuhashi, T.; Furuhashi, K.; Weckwerth, W. The parasitic mechanism of the holostemparasitic plant *Cuscuta*. *J. Plant Interact.* **2011**, *6*, 207–219. [CrossRef]
22. Weber, J. A taxonomic revision of *Cassytha* (Lauraceae) in Australia. *J. Adel. Bot. Gard.* **1981**, *3*, 187–262.
23. Press, M.C.; Phoenix, G.K. Impacts of parasitic plants on natural communities. *New Phytol.* **2005**, *166*, 737–751. [CrossRef] [PubMed]
24. Bardgett, R.D.; Smith, R.S.; Shiel, R.S.; Peacock, S.; Simkin, J.M.; Quirk, H.; Hobbs, P.J. Parasitic plants indirectly regulate below-ground properties in grassland ecosystems. *Nature* **2006**, *439*, 969–972. [CrossRef] [PubMed]
25. Hatcher, M.J.; Dick, J.T.A.; Dunn, A.M. Diverse effects of parasites in ecosystems: Linking interdependent processes. *Front. Ecol. Environ.* **2012**, *10*, 186–194. [CrossRef]
26. Tesitel, J.; Li, A.R.; Knotkova, K.; McLellan, R.; Bandaranayake, P.C.G.; Watson, D.M. The bright side of parasitic plants: What are they good for? *Plant Physiol.* **2021**, *185*, 1309–1324. [CrossRef]

27. Adamu, A.A.; Garba, F.N.; Ahmed, T.M.; Abubakar, A. Pharmacognostic studies and elemental analysis of *Cassytha filiformis* Linn. *J. Pharmacogn. Phytother.* **2017**, *9*, 131–137. [CrossRef]
28. Nelson, S.C. Cassytha filiformis. In *Plant Disease PD-42*; University of Hawaii: Honolulu, HI, USA, 2008; pp. 1–10.
29. Nickrent, D.L. Plantas parásitas en el mundo. Capitulo 2. In *Plantas Parásitas de la Península Ibérica e Islas Baleares, López-Sáez, J.A., Catalán, P., Sáez, L., Eds.*; Mundi-Prensa Libros, S.A.: Madrid, Spain, 2002; pp. 7–27.
30. Werth, C.R.; Pusateri, W.P.; Eshbaugh, W.H.; Wilson, T.K. Field observations on the natural history of *Cassytha filiformis* L. (Lauraceae) in the Bahamas. In *The 2nd International Symposium on Parasitic Weeds*; Musselman, L.J., Worsham, A.D., Eplee, R.E., Eds.; North Carolina State University College: Raleigh, NC, USA, 1979; pp. 94–102.
31. Thorogood, C.J.; Rumsey, F.J.; Hiscock, S.J. Host-specific races in the holoparasitic angiosperm Orobanche minor: Implications for speciation in parasitic plants. *Ann. Bot.* **2009**, *103*, 1005–1014. [CrossRef]
32. Cirocco, R.M.; Facelli, J.M.; Watling, J.R. High water availability increases the negative impact of a native hemiparasite on its non-native host. *J. Exp. Bot.* **2016**, *67*, 1567–1575. [CrossRef]
33. Cirocco, R.M.; Facelli, J.M.; Watling, J.R. Does nitrogen affect the interaction between a native hemiparasite and its native or introduced leguminous hosts? *New Phytol.* **2017**, *213*, 812–821. [CrossRef]
34. Prider, J.; Watling, J.; Facelli, J.M. Impacts of a native parasitic plant on an introduced and a native host species: Implications for the control of an invasive weed. *Ann. Bot.* **2009**, *103*, 107–115. [CrossRef] [PubMed]
35. Furuhashi, K.; Iwase, K.; Furuhashi, T. Role of light and plant hormones in stem parasitic plant (*Cuscuta* and *Cassytha*) twining and haustoria induction. *Photochem. Photobiol.* **2021**, *97*, 1054–1062. [CrossRef]
36. Li, H.W.; Li, J.; Huang, P.H.; Wei, F.N.; Cui, H.B.; van der Werff, H. Lauraceae. In *Flora of China, Calycanthaceae–Schisandraceae*; Wu, Z.Y., Raven, P.H., Eds.; Science Press: Beijing, China; Missouri Botanical Garden Press: St. Louis, MI, USA, 2008; Volume 7, pp. 102–254.
37. Li, Q.; Yao, D.; Cai, J.; Huang, L. A field survey of *Cassytha filiformis* host range. *Guangxi Plant Prot.* **1992**, *4*, 21–24.
38. Huang, Z.J. Some observations on the host of *Cassytha filiformis* and the exploration of its harmfulness. *J. Sci. Res. Repert. Guangxi Agric. Coll.* **1957**, 54–60.
39. Shao, S.Z. Records of the detriment of *Cassytha filiformis* on *Pinus massoniana*. *For. Pest Commun.* **1988**, *4*, 3.
40. Mahadevan, N.; Jayasuriya, K. Water-impermeable fruits of the parasitic angiosperm *Cassytha filiformis* (Lauraceae): Confirmation of physical dormancy in Magnoliidae and evolutionary considerations. *Aust. J. Bot.* **2013**, *61*, 322–329. [CrossRef]
41. Paradis, E.; Claude, J.; Strimmer, K. APE: Analyses of phylogenetics and evolution in R language. *Bioinformatics* **2004**, *20*, 289–290. [CrossRef] [PubMed]
42. Jin, Y.; Qian, H.V. PhyloMaker2: An updated and enlarged R package that can generate very large phylogenies for vascular plants. *Plant Divers.* **2022**, *44*, 335–339. [CrossRef]
43. The Angiosperm Phylogeny Group. An update of the angiosperm phylogeny group classification for the orders and families of flowering plants: APG IV. *Bot. J. Linn. Soc.* **2016**, *181*, 12385. [CrossRef]
44. Parra-Tabla, V.; Tun-Garrido, J.; García-Franco, J.; Martínez, M. The recent expansion of the invasive hemiparasitic plant *Cassytha filiformis* and the reciprocal effect with its main hosts. *Res. Square* **2023**. [CrossRef]
45. Buriyo, A.S.; Kasuga, L.; Moshi, H.N.; Nene, W.A. Ecological distribution and abundance of the parasitic weed, *Cassytha filiformis* L. (Lauraceae) in major cashew, *Anacardium occidentale* L. growing regions in Tanzania. *Int. J. Basic Appl. Sci.* **2015**, *5*, 109–116.
46. Zhang, C.; Ma, H.; Sanchez-Puerta, M.V.; Li, L.; Xiao, J.; Liu, Z.; Ci, X.; Li, J. Horizontal gene transfer has impacted cox1 gene evolution in *Cassytha filiformis*. *J. Mol. Evol.* **2020**, *88*, 361–371. [CrossRef] [PubMed]
47. Rajanna, L.; Shivamurthy, G.R. Occurrence of graniferous tracheary elements in the haustorium of *Cassytha filiformis* Linn., a stem parasite of Lauraceae. *Taiwania* **2001**, *46*, 40–48.
48. Kokubugata, G.; Nakamura, K.; Forster, P.I.; Wilson, G.W.; Holland, A.E.; Hirayama, Y.; Yokota, M. *Cassytha pubescens* and *C. glabella* (Lauraceae) are not disjunctly distributed between Australia and the Ryukyu Archipelago of Japan—Evidence from morphological and molecular data. *Aust. Syst. Bot.* **2012**, *25*, 364–373. [CrossRef]
49. Costea, M.; ElMiari, H.; Farag, R.; Fleet, C.; Stefanović, S. *Cuscuta* sect. *Californicae* (Convolvulaceae) revisited:'cryptic' speciation and host range differentiation. *Syst. Bot.* **2020**, *45*, 638–651. [CrossRef]
50. Facelli, E.; Wynn, N.; Tsang, H.T.; Watling, J.R.; Facelli, J.M. Defence responses of native and invasive plants to the native generalist vine parasite *Cassytha pubescens*—Anatomical and functional studies. *Aust. J. Bot.* **2020**, *68*, 300. [CrossRef]
51. Kelly, C.K. Resource choice in *Cuscuta europaea*. *Proc. Natl. Acad. Sci. USA* **1992**, *89*, 12194–12197. [CrossRef]
52. Runyon, J.B.; Mescher, M.C.; De Moraes, C.M. Plant defenses against parasitic plants show similarities to those induced by herbivores and pathogens. *Plant Signal. Behav.* **2010**, *5*, 929–931. [CrossRef]
53. Machado, M.A.; Zetsche, K. A structural, functional and molecular analysis of plastids of the holoparasites *Cuscuta reflexa* and *Cuscuta europaea*. *Planta* **1990**, *181*, 91–96. [CrossRef]
54. Liu, Z.-F.; Ma, H.; Ci, X.-Q.; Li, L.; Song, Y.; Liu, B.; Li, H.-W.; Wang, S.-L.; Qu, X.-J.; Hu, J.-L.; et al. Can plastid genome sequencing be used for species identification in Lauraceae? *Bot. J. Linn. Soc.* **2021**, *197*, 1–14. [CrossRef]
55. Schonenberger, J.; von Balthazar, M. Asterids. *Bot. J. Linn. Soc.* **2013**, *173*, 321–324. [CrossRef]
56. Fay, M.F. Rosids. *Bot. J. Linn. Soc.* **2013**, *172*, 399–403. [CrossRef]

57. Soltis, D.E.; Mort, M.E.; Latvis, M.; Mavrodiev, E.V.; O'Meara, B.C.; Soltis, P.S.; Burleigh, J.G.; Rubio de Casas, R. Phylogenetic relationships and character evolution analysis of Saxifragales using a supermatrix approach. *Am. J. Bot.* **2013**, *100*, 916–929. [CrossRef]
58. Su, H.-J.; Hu, J.-M.; Anderson, F.E.; Der, J.P.; Nickrent, D.L. Phylogenetic relationships of Santalales with insights into the origins of holoparasitic Balanophoraceae. *Taxon* **2015**, *64*, 491–506. [CrossRef]
59. Tesitel, J.; Fibich, P.; de Bello, F.; Chytry, M.; Leps, J. Habitats and ecological niches of root-hemiparasitic plants: An assessment based on a large database of vegetation plots. *Preslia* **2015**, *87*, 87–108.
60. Birschwilks, M.; Haupt, S.; Hofius, D.; Neumann, S. Transfer of phloem-mobile substances from the host plants to the holoparasite *Cuscuta* sp. *J. Exp. Bot.* **2006**, *57*, 911–921. [CrossRef]
61. Kaminska, M.; Korbin, M. Graft and dodder transmission of phytoplasma affecting lily to experimental hosts. *Acta Physiol. Plant.* **1999**, *21*, 21–26. [CrossRef]
62. Lanini, W.T.; Kogan, M. Biology and management of *Cuscuta* in crops. *Cienc. Investig. Agrar.* **2005**, *32*, 127–141. [CrossRef]

diversity

MDPI

Article

Population Survey Combined with Genomic-Wide Genetic Variation Unravels the Endangered Status of *Quercus gilva*

Yi-Gang Song [1,*,†], Tian-Rui Wang [1,†], Zi-Jia Lu [1,2], Bin-Jie Ge [1], Xin Zhong [1], Xiao-Chen Li [1], Dong-Mei Jin [1], Quan Yuan [1,2], Yu Li [1,3], Yi-Xin Kang [1,4], Xin Ning [1], Si-Si Zheng [1], Li-Ta Yi [3], Xi-Ling Dai [2], Jian-Guo Cao [2], Jung-Hyun Lee [5] and Gregor Kozlowski [1,6,7]

1 Eastern China Conservation Centre for Wild Endangered Plant Resources, Shanghai Chenshan Botanical Garden, Shanghai 201602, China
2 College of Life Sciences, Shanghai Normal University, 100 Guilin Rd., Shanghai 200234, China
3 College of Forestry and Biotechnology, Zhejiang A&F University, Hangzhou 311300, China
4 School of Ecology and Nature Conservation, Beijing Forestry University, 35 Qinghua East Road, Beijing 100080, China
5 Department of Biology Education, Chonnam National University, Gwangju 61186, Republic of Korea
6 Department of Biology and Botanic Garden, University of Fribourg, Chemin du Musée 10, 1700 Fribourg, Switzerland
7 Natural History Museum Fribourg, Chemin du Musée 6, 1700 Fribourg, Switzerland
* Correspondence: ygsong@cemps.ac.cn
† These authors contributed equally to this work.

Abstract: Since the Anthropocene, biodiversity loss owing to human activity and climate change has worsened. *Quercus gilva* is an evergreen oak species native to China, Japan, and South Korea and is threatened by a long history of human impact. The purpose of this study was to (1) reassess the threatened category of *Q. gilva* based on a detailed survey, and (2) identify the genetic structure and diversity of *Q. gilva* based on genomic data. First, we conducted a detailed survey of the populations in China. Second, we collated all the literature and information. Finally, genome-wide genetic variation was analyzed based on 65 individuals from 22 populations. We found that *Q. gilva* has suffered rapid population decline, and at present, most populations are very small. The evolutionary path of *Q. gilva* was from the southwest to east of China and then to Japan and South Korea. *Quercus gilva* showed no distinct genetic structure and had a relatively low genetic diversity. Among the 22 populations, most populations in southwestern China, South Korea, and Japan had high genetic diversity. The populations in Jingning (Zhejiang province; ZJN), Wuyuan (Jinaxi province; JWY), and Zherong (Fujian province; FZR) suffered a strong bottleneck. In conclusion, *Q. gilva* is an endangered species native to East Asia. Because of the very low genetic diversity of *Q. gilva* and most populations are small, we need to (1) strengthen the protection of this species, (2) conduct conservation actions with *in-situ* reinforcement populations, and (3) select populations with high genetic diversity as provenances for afforestation efforts. Finally, we suggest that in the future, genetic diversity should be considered as the sixth criterion for IUCN to evaluate the threatened category.

Keywords: biodiversity loss; conservation genomics; endangered species; Fengshui/shrine/temple forests; genetic diversity; human impact

Citation: Song, Y.-G.; Wang, T.-R.; Lu, Z.-J.; Ge, B.-J.; Zhong, X.; Li, X.-C.; Jin, D.-M.; Yuan, Q.; Li, Y.; Kang, Y.-X.; et al. Population Survey Combined with Genomic-Wide Genetic Variation Unravels the Endangered Status of *Quercus gilva*. *Diversity* **2023**, *15*, 230. https://doi.org/10.3390/d15020230

Academic Editor: Julio Peñas de Giles

Received: 29 November 2022
Revised: 31 January 2023
Accepted: 2 February 2023
Published: 6 February 2023

1. Introduction

Trees form the principal components of forests and serve as immense support for terrestrial ecosystems and are of vital importance ecologically, economically, and culturally [1–3]. *Quercus* (oaks), predominantly in the Northern Hemisphere, is the largest genus of the family Fagaceae and one of the largest genera of all tree families [4]. Unquestionably, oaks are among the most successful, widely distributed, and valuable hardwood trees ecologically, economically, and culturally [5]. As keystone species in many ecosystems, oaks play pivotal roles in shaping biodiversity, creating healthy ecosystems, and carbon sequestration [6,7]. During the Anthropocene, oaks have also been a valuable source of food, housing components, and materials [7].

Since the Anthropocene, biodiversity loss owing to human activity and climate change has worsened, and more attention should be paid to biodiversity conservation [8]. Through the global tree assessment, we know that currently, 30% of tree species are threatened with extinction [1]. Forty-one percent of oaks are of conservation concern, and 31% are estimated to be threatened with extinction [4]. Although the percentage of threatened species is already high, the assessment of many species of least concern (LC) is very rough (with only the area of occupancy (AOO) and extent of occurrence (EOO) calculated based on the occurrence data). A detailed population survey and genetic diversity estimation can help us to reassess the conservation status of these species of LC.

Genetic diversity is recognized as one of the three basic elements of biodiversity [9]. Current approaches to biodiversity conservation are largely based on geographic areas, ecosystems, ecological communities, and species, with less attention paid to genetic diversity and the evolutionary continuum from population to species [10,11]. Genetic diversity within all species, not just domesticated species and their wild relatives, must be conserved and monitored using appropriate metrics [11]. Thus, genetic diversity should be recognized as one of the main targets for biodiversity conservation under the international agreements on the "post-2020" framework [12].

Quercus gilva Blume is an ecologically important large tree of evergreen broad-leaved forests in China, Japan, and South Korea [13,14]. It is a precious tree species with hard and reddish-brown timber [15]. Because of the long history of large-scale regional development and excessive logging, many populations of *Q. gilva* have limited habitats and a small population size in their entire distribution range [14,16,17]. Most of the natural populations of *Q. gilva* are threatened with extinction, and local governments have classified this species as endangered or critically endangered [13–15,18]. *Quercus gilva* has recently been assessed as LC by the Botanical Gardens Conservation International (BGCI) and IUCN SSC Global Tree Specialist Group [19]. Thus, the reassessment of this species based on detailed population survey data is urgently needed.

Forest and landscape restoration are approaches that aim to regain ecological functionality and enhance human well-being in deforested or degraded landscapes [20]. Well planned and executed for reforestation with selected species and populations of selected provenances could maximize carbon sequestration, biodiversity, and livelihood benefits [21]. During the last 10 years, *Q. gilva* has been recognized as an important tree species and has been used for forest restoration in Zhejiang, Fujian, Jiangxi, and Hunan province of China. In the future, focus needs to be on the conservation of natural populations, germplasm evaluation, and utilization of excellent germplasm. Genetic diversity has been recognized as an important criterion to consider the prioritizing populations for protection [22] and as the basis for excellent germplasm selection [23]. The rapid expansion of genomic information will transform our understanding of the amount, distribution, and functional significance of genome-wide genetic variation in natural populations to guide conservation and reforestation [24,25].

The main aim of this study was to understand the endangered and conservation status of *Q. gilva* based on a detailed population survey and genetic diversity. The following specific aspects were explored: (1) the size, age composition, and main threats to the natural populations of *Q. gilva*, (2) the phylogeny and population structure of *Q. gilva* based on the genomic data, and (3) the patterns of genetic diversity at the genomic level.

2. Materials and Methods

2.1. Data Collection and Population Survey

Occurrence data with geographical coordinates of *Q. gilva* were compiled from the Chinese Virtual Herbarium [26], IUCN Red List of Threatened Species 2019 [19], and other publications related to *Q. gilva*. We then collected the population status (size, age composition, and main threats, if possible) of *Q. gilva* in Japan and South Korea based on the publications. Additionally, between 2020 and 2022, an intensive field survey was conducted to explore the size, age composition, and main threats to each population in China. Finally, we surveyed 40 mainland Chinese populations. The populations were then divided into four categories based on the number of individuals in each population: large (>500 individuals), medium (100–500 individuals), small (30–100 individuals), and very small (<30 individuals) populations. The AOO and EOO were calculated using the GeoCAT online browser (http://geocat.kew.org/ (accessed on 23 November 2022)) [27]. We also collected the main threats, population trends within three generations, and habitats for each population of *Q. gilva*. Finally, we reassessed the status of *Q. gilva* across its distribution, following the "IUCN Red List Categories" [28].

2.2. Plant Material Samples, Resequencing, Control, and Mapping

A total of 65 individuals from 22 populations (three individuals for each population, except one population (only two individuals for the population of Jingning, Zhejiang province (ZJN)) were carefully selected to represent most of the natural populations of *Q. gilva* in East Asia (Figure 1 and Table S1). For each sample, genomic DNA was extracted from mature leaves using a cetyltrimethylammonium bromide (CTAB)-based protocol [29]. The concentration and quality of the total genomic DNA were determined using a NanoDrop 2000 spectrophotometer (Thermo Fisher Scientific, Waltham, MA, USA). DNA libraries (350 bp) for Illumina sequencing were constructed for each accession according to the manufacturer's specifications. After DNA library construction, sequencing was performed on an Illumina NovaSeq 6000 platform by a commercial service (Biomarker Technologies, Beijing, China) with 150 bp paired-end reads. Raw reads were filtered based on the following criteria: paired-end reads with >10% 'N' bases, reads on which more than 50% of the bases had a quality score of less than 20 (Phred-like score), and sequencing adapter. Finally, high-quality clean reads were obtained for subsequent analysis.

2.3. SNP and Insertion/Deletion (InDels) Calling

All clean reads for each individual were mapped to the reference genome using the MEM algorithm of the Burrows–Wheeler Aligner (bwa-mem2 v2.2). The average mapping rate was 89.5%, and the average coverage rate was 10-fold for the reference genome. The mapping results were sorted, and duplicate reads were removed using SAMtools rmdup (version 1.9) [30]. SNPs and InDels were called using the HaplotypeCaller module in the Genome Analysis Toolkit (GATK) (version 3.8) [31] and were filtered with the following parameters: QD < 2.0 | | MQ < 40.0 | | FS > 60.0 | | QUAL < 30.0 | | MQrankSum < −12.5 | | ReadPosRankSum < −8.0-clusterSize 2-clusterWindowSize 5. The SNPs identified above were subjected to a second round of filtering to improve the accuracy and efficiency of subsequent analyses. Only SNPs with a minor allele frequency greater than 5% and less than 20% of missing data were considered as high-quality SNPs. Transition (Ti), transversion (Tv), Ti/Tv, heterozygosity, homozygosity, and heterozygosity ratio were further identified using GATK. We used *Fagus sylvatica* [32] as an outgroup for phylogenetic analysis. Finally, 4,020,695 SNPs containing outgroup and 2,993,608 SNPs without outgroup was identified and used for subsequent downstream analysis.

Figure 1. Geographic distribution (black dotted line) and the sampling locations (black dots) of *Quercus gilva* (**A**). The forests and selected old trees of *Q. gilva* (**B–H**). Population code abbreviations in Figure 1A are the same as in Table S1.

2.4. Phylogenetic Inference and Population Genomic Analysis

A neighbor-joining (NJ) phylogenetic tree was constructed using MEGAX [33] under the p-distances model with the 4,020,695 SNPs. We also used IQ-TREE [34] with self-estimated best substitution models to generate a maximum likelihood (ML) phylogenetic tree. The two phylogenetic trees were run with 1,000 bootstrap repetitions, using *Fagus sylvatica* as the outgroup.

To visualize the genetic relationships among the samples, principal component analysis (PCA) was performed using the smartpca program in EIGENSOFT version 6.0 based on 2,993,608 SNPs [35]. The initial three eigenvectors were plotted in three dimensions. ADMIXTURE version 1.22 [36] was used to infer historical ancestor clusters showing clusters of similar genotypes. The membership of each genotype was run for a range of genetic clusters from a value of K = 1 to 10 by using the admixture model.

2.5. Population Genetic Diversity and Linkage Disequilibrium (LD) Analyses

The observed heterozygosity (H_O), expected heterozygosity (H_E), polymorphism information content (*PIC*), Nei diversity index (*H*), and Shannon–Wiener index (*I*) were calculated using the "PopGenome" package in the R project [37,38]. Nucleotide diversity (π) was calculated within a non-overlapping 100-kb window using VCFtools (version 0.1.13) [39]. The LD was calculated using PLINK version 1.9, within a 1000 kb window, and a maximum of 999,999 SNPs for each window [40]. The squared correlation coefficient (r^2) of each chromosome was calculated using SNP pairs only from the corresponding chromosome. Pairwise r^2 values within and between different chromosomes were averaged across the entire genome. We compared the LD patterns among different populations using the LD decay distance, indicated by the r^2 decreased to half of the maximum.

3. Results

3.1. Reassessment of Q. gilva

After collating all the distribution data of *Q. gilva* from different resources, there were a total of 108 known populations in East Asia (68 populations in China, 35 in Japan, and 5 in South Korea). According to the information gathered from indigenous people, almost all old trees (except individuals located in Fengshui forests and temples) have been deforested during the last 100 years. Based on this information, *Q. gilva* can be listed as endangered as per the EN-A4ad criteria.

Although the EOO of *Q. gilva* was very high, the area in South Korea was very small. The AOO of *Q. gilva* was less than 500 km^2 with the largest being in China (272 km^2) and the smallest in South Korea (20 km^2). More than half of the known populations have been surveyed in China. Only one large population had more than 500 individuals, and most of the surveyed populations were very small, with fewer than 30 individuals. There were even some occurrences with only one individual (Tables 1 and S2). Based on the population status, we estimated that more than 40% of the AOO after three generations (future-AOO) would be lost. Considering the populations without information, we inferred that more than 50% of the AOO would be lost in the next three generations. Finally, we estimated that there were fewer than 10,000 individuals within the distribution area of *Q. gilva*. During the last 100 years, the main threat to *Q. gilva* in natural populations has been logging and wood harvesting, as it is used as a biological resource. To support the rapid development of society, expansion of land under agriculture, residential use, and transportation infrastructure has also led to the destruction of natural *Q. gilva* populations. According to our survey, most of the current populations were conserved in the Fengshui forests near the villages, and forests surrounding shrines and temples with severe fragmentation. According to the IUCN Red List categories and criteria, the conservation status of *Q. gilva* is also determined to be endangered as per the EN-A4c criteria.

Table 1. Summary of the current status of *Quercus gilva*.

Country	China	Japan	South Korea	Total/Summary
Number of populations	68	35	5	108
AOO (km^2)	272	140	20	432
Future-AOO (km^2)	148	92	8	248
EOO (km^2)	873,462	161,420	84	1,921,293
NLP	1	0	0	1
NMP	2	0	0	2
NSP	12	0	1	13
NVSP	31	12	4	47
NISS	23	23	0	46
Total individuals	<5000	<2000	<600	<10,000
Main threats	Logging and wood harvesting; Agriculture and development	Logging; Agriculture and development	Human-mediated disturbance	Agriculture and Biological resource use
PTTG	Decrease noticeable	Decrease noticeable	No information	Decrease noticeable
Main area conserved	Fengshui forests and temples	Forests surrounding shrines and temples	Gotjawal (conserved area)	Protected Trees

AOO, area of occupancy; Future-AOO, AOO after three generations; NLP, Number of large populations: >500 individuals; NMP, Number of medium populations: 100–500 individuals; NSP, Number of small populations: 30–100 individuals; NVSP, Number of very small populations: <30 individuals; PTTG, Population trends within Three Generations; NISS, No information about population size and structure.

3.2. Detection of Genome-Wide Variant

We re-sequenced 65 individuals (22 populations) of *Q. gilva* collected from its main distribution area in East Asia: 19 populations from China, one population from South Korea, and two populations from Japan. A total of 706 Gb of high-quality clean reads were obtained. Among the 65 individuals, seven of them had 20 Gb clean reads and for all other individuals, clean reads were between 8.9 Gb and 11.2 Gb. We obtained an average of 36,206,470 reads, with an average Q20 value of 95.72%, Q30 of 89.46%, and average GC content of 37.01%. The average sequencing depth was 10.23. The 1× coverage of all individuals was higher than 80% with an average of 84.72%, except for one individual with a coverage of 56.75%. These high-quality sequences were aligned to the chromosome-level high-precision genome with the average mapping rate of 91.25%; alignment and proper mapping reached 83.17% (Table S1).

Among the 65 individuals of *Q. gilva*, 15,377,234 SNPs and 4,405,966 InDels were identified. The number of SNPs for each population was between 2,172,504 and 4,293,739, while for each individual, the number of SNPs for each individual was between 1,477,213 and 2,560,913 (Tables 2 and S2). Transitions and transversions accounted for 71.87% and 28.12% of the total number of SNPs, respectively, with an average transition/transversion (Ti/Tv) ratio of 2.56. The number of heterozygosities in different samples varied from a lowest of 712,570 to a highest of 1,513,312, with an average of 1,097,305 (Tables 2 and S2). The number of homozygosities in different samples varied between 761,543 and 2,136,876, with an average of 906,786 (Tables 2 and S1).

Table 2. Summary of genetic variation in *Quercus gilva* populations.

Population Code (Location)	SNPs	Indels	Transition	Transversion	Ti/Tv	Heterozygosity	Homozygosity	Het-Ratio
GLP (Liping, Guizhou)	3,507,918	1,115,687	1,403,828	547,870	2.56	1,106,329	845,370	0.5666
GJK (Jiangkou, Guizhou)	3,470,289	1,150,966	1,510,648	589,730	2.56	1,194,235	906,143	0.5636
GCS (Changshun, Guizhou)	3,448,941	1,055,047	1,507,376	579,442	2.597	1,132,727	954,091	0.5425
HXX (Xiangxiang, Hunan)	4,293,739	1,378,636	1,596,453	626,624	2.54	1,343,567	879,510	0.6023
HXS (Xinshao, Hunan)	3,561,533	1,162,826	1,514,130	589,450	2.56	1,247,186	856,394	0.5928
HDK (Dongkou, Hunan)	3,932,356	1,236,055	1,451,795	562,065	2.58	1,159,165	854,695	0.5755
HCN (Changning, Hunan)	3,819,985	1,222,766	1,477,122	574,678	2.57	1,179,174	872,625	0.5746
HSZ (Sangzhi, Hunan)	3,083,571	983,796	1,351,905	522,997	2.58	1,027,765	847,137	0.5453
HPJ (Pingjiang, Hunan)	3,149,545	1,019,904	1,324,570	516,136	2.56	931,328	909,378	0.5036
HYL (Yanling, Hunan)	2,445,579	850,489	1,370,498	535,637	2.56	1,046,839	859,296	0.5485
JWY (Wuyuan, Jiangxi)	3,129,945	1,056,778	1,451,178	566,446	2.56	1,043,045	974,579	0.5157
FCT (Changting, Fujian)	3,581,645	1,144,872	1,448,578	563,974	2.56	1,123,022	889,529	0.5525
FZR (Zherong, Fujian)	5,183,429	1,750,582	1,501,902	719,470	2.21	908,988	1,312,384	0.4290
FMQ (Minqing, Fujian)	3,486,120	1,157,688	1,578,398	614,521	2.56	1,296,886	896,033	0.5890
FJO (Jian'ou, Fujian)	3,361,907	1,112,365	1,471,386	573,316	2.56	1,128,122	916,580	0.5518
ZYZ (Yinzhou, Zhejiang)	2,428,260	858,412	1,382,278	544,668	2.54	986,717	940,225	0.5115
ZZS (Zhoushan, Zhejiang)	3,229,541	1,053,082	1,416,464	550,877	2.57	1,088,961	878,380	0.5522
ZNH (Ninghai, Zhejiang)	3,732,811	1,179,976	1,421,104	540,260	2.63	1,075,951	885,413	0.5442
ZJN (Jingning, Zhejiang)	2,172,504	699,612	1,112,899	413,417	2.69	714,120	812,196	0.4684
JGU (Gueok-ri, Jeju)	3,876,609	1,232,391	1,518,845	590,571	2.57	1,218,079	891,338	0.5774
MMY (Miyakonojo-shi, Miyazaki)	3,852,953	1,223,121	1,477,301	572,304	2.58	1,146,487	903,117	0.5564
MNB (Nobeoka-shi, Miyazaki)	3,551,342	1,109,444	1,400,624	533,912	2.62	1,069,663	864,873	0.5527
Total/Average	15,377,234	4,405,966	1,440,422	533,912	2.56	1,097,305	906,786	0.5462

3.3. Phylogenetic and Population Structure Analyses of *Q. gilva*

The NJ and ML phylogenetic trees were constructed using 4,020,695 SNPs in the single-copy genes. The NJ and ML trees consistently showed that individuals from the Zherong, Fujian (FZR), Dongkou, Hunan (HDK), and Xiangxiang, Hunan (HXX) populations did not cluster into one lineage. According to the NJ and ML trees, all the *Q. gilva* individuals could be divided into three major groups: West, Central, and East groups. Generally, the populations from Guizhou and western Hunan provinces comprised the western group. The populations from Eastern Hunan, Jiangxi, Fujian, and most of Zhejiang provinces formed the central group. Populations from South Korea, Japan, and ZZS (Zhoushan, Zhejiang) formed the main part of eastern group (Figure 2). There were three main differences in the phylogenetic structures between the NJ and ML trees of *Q. gilva* populations (Figures 2 and S1). First, the GLP population (Liping, Guizhou) was nested into the central group on the ML tree, whereas the western group was nested in the NJ tree. Second, the FMQ population (Minqing, Fujian) was nested into the central group on the ML tree, whereas the eastern group was nested in the NJ tree. Finally, compared to the NJ tree, the ML tree had four clear clades for the Central and East groups (Figures 2 and S1).

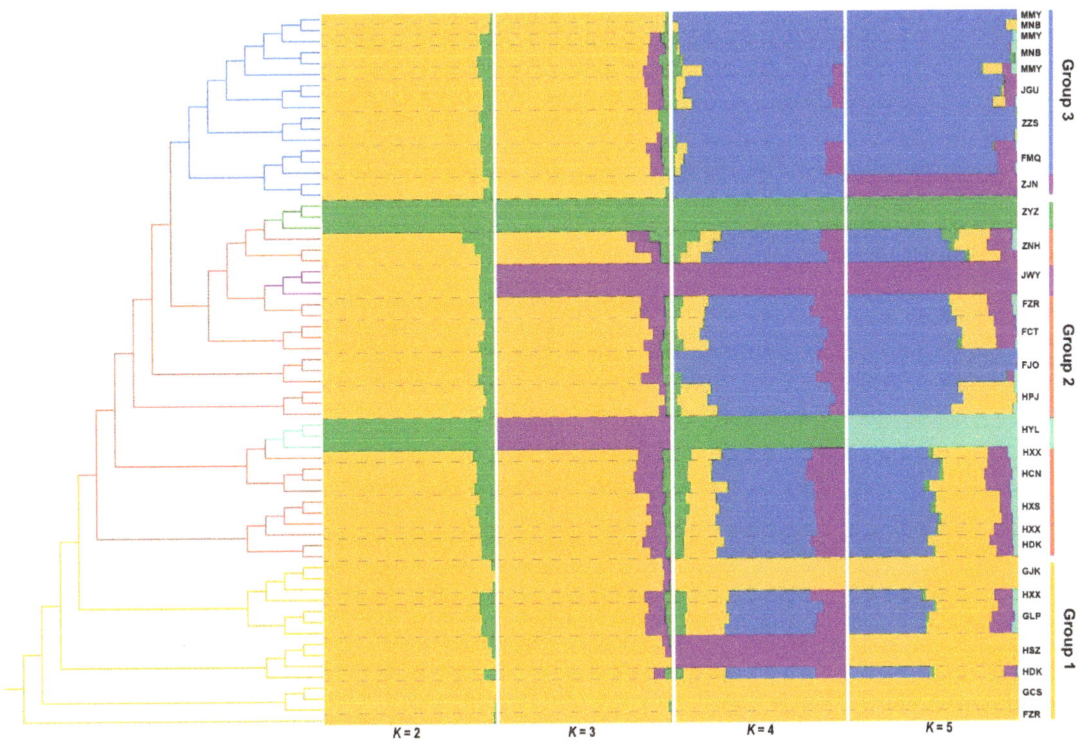

Figure 2. Neighbor-joining phylogenetic tree and population structure of *Quercus gilva*. *Fagus sylvatica* was used as the outgroup for the phylogenetic analysis. The figure does not show the outgroup. Population codes abbreviations are the same as in Table 2.

The results of the cross-validation (CV) provided by admixture analysis showed that the CV error rate had a minimum value when K = 1. The CV error rate was relatively low value when K = 2–5 (Figure S2). When K = 2, the populations of ZYZ and HYL formed one group, and the remaining populations formed the second group. When K = 3, the two populations in Jiangxi province formed one group, the ZYZ population was identified as the second group, and the remaining populations were classified into the third group.

When K = 4, the two most western populations (GCS and GJK) formed the first group, the HSZ and JWY populations formed the second group, the ZYZ and HYL populations formed the third group, and the remaining populations were classified into the fourth group. When K = 5, the minor change observed as that the HSZ population merged into the western group and the HYL population separated again (Figure 2). Based on the PCA results, we found that the ZYZ, HYL, and JWY populations were the most distinct. The remaining populations were clustered together (Figure S3).

3.4. Genome-Wide Patterns of Nucleotide Diversity and LD Analyses

Among the 22 populations, the values for observed heterozygosity (H_O) and expected heterozygosity (H_E) ranged between 0.1506 and 0.2441 and between 0.1156 and 0.2199, respectively. The polymorphism information content (*PIC*) values were between 0.089 and 0.1765, indicating that all the *Q. gilva* populations had a low level of polymorphism. Moreover, the Nei diversity index (*H*: ranged between 0.1399 and 0.265), Shannon–Wiener index (*I*: between 0.1646 and 0.328), and nucleotide diversity ($\pi \times 10^{-3}$: between 0.522 and 0.973) were calculated to evaluate the genetic diversity of different populations. The nucleotide diversity of *Q. gilva* was found to be 0.994. The ZYZ, FZR, ZJN, and HYL populations showed substantially lower diversity than the HXX, JGU, HDK, HXS, and MMY populations (Table 3).

Table 3. Genetic diversity of *Quercus gilva* populations.

Population	H_O	H_E	*PIC*	*H*	*I*	$\pi \times 10^{-3}$
GLP	0.2083	0.1876	0.15	0.2269	0.2787	0.863
GJK	0.2197	0.1807	0.1437	0.2182	0.2666	0.834
GCS	0.1867	0.165	0.131	0.1999	0.2431	0.735
HXX	0.2401	0.2199	0.1765	0.265	0.328	0.887
HXS	0.2322	0.1904	0.1516	0.2297	0.2815	0.964
HDK	0.216	0.2096	0.168	0.2534	0.3122	0.965
HCN	0.2192	0.2048	0.1639	0.2472	0.3045	0.727
HSZ	0.1965	0.1641	0.1301	0.1994	0.2413	0.727
HPJ	0.1793	0.1681	0.1337	0.2038	0.248	0.757
HYL	0.2077	0.1256	0.0968	0.1521	0.179	0.568
JWY	0.1975	0.1651	0.1307	0.1992	0.2425	0.762
FCT	0.2132	0.1945	0.1553	0.2354	0.2884	0.892
FZR	0.1913	0.1761	0.1389	0.2248	0.257	0.546
FMQ	0.2441	0.1897	0.1506	0.2285	0.2795	0.884
FJO	0.2126	0.1764	0.1409	0.2128	0.2618	0.815
ZYZ	0.1936	0.1156	0.089	0.1399	0.1646	0.522
ZZS	0.2113	0.1773	0.1408	0.2144	0.2612	0.812
ZNH	0.205	0.2021	0.1616	0.2449	0.3003	0.905
ZJN	0.1506	0.1207	0.0948	0.1609	0.175	0.549
JGU	0.2294	0.21	0.168	0.2532	0.3122	0.973
MMY	0.2161	0.2068	0.1655	0.25	0.3075	0.956
MNB	0.2079	0.1973	0.1576	0.2386	0.2928	0.891
Total						0.994

H_O, observed heterozygosity; H_E, expected heterozygosity; *PIC*, polymorphism information content; *H*, Nei diversity index; *I*, Shannon-Wiener index; π, nucleotide diversity. Population code abbreviations are the same as in Table 2.

Half of the maximum squared correlation coefficients (r^2) between pairwise SNPs ranged from 0.319 to 0.463. Linkage disequilibrium decayed to half among different populations in the range of 0.26 to 685.23 kb. The LD decay measured by physical distance, at which the pairwise correlation dropped to half of its maximum value, occurred at 685.23 kb in the GJK population ($r^2 = 0.368$) and 0.27 kb in the HYL population ($r^2 = 0.451$). There are three populations (ZJN, FZR, and JWY) that did not reach the half of the maximum r^2 (Figure 3 and Table S3).

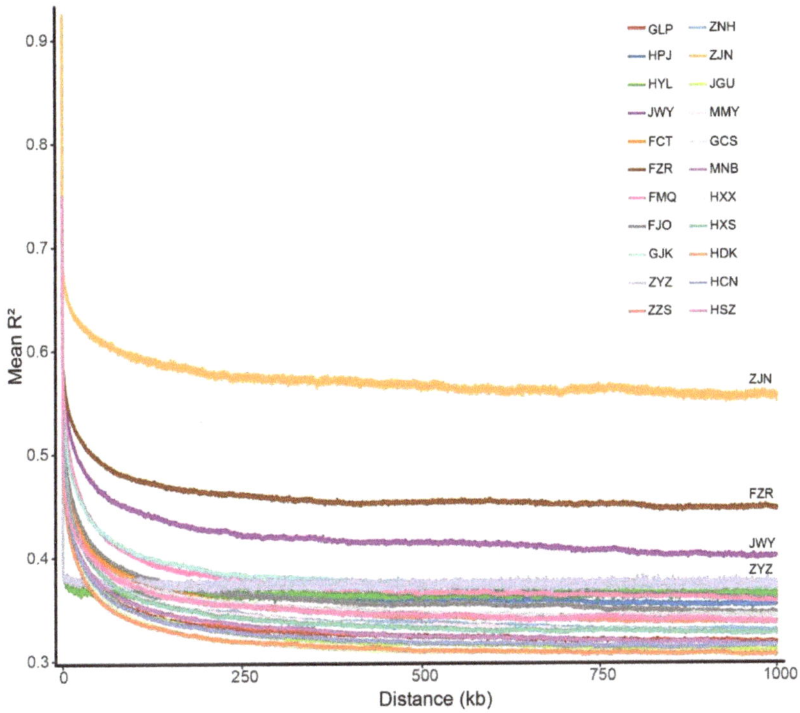

Figure 3. Linkage disequilibrium decay measured by r^2 in *Quercus gilva* species and each of the 22 populations. Population code abbreviations are the same as in Table 2.

4. Discussion

Our assessment showed that *Q. gilva* is an endangered (EN) species as per the EN-A4acd criteria. According to our extensive field survey and more than 30 literature sources on *Q. gilva*, we found that this species has suffered massive population decline and will be facing accelerated declines in the future. During the last 100 years, many natural populations have been logged for industrial timber, agriculture, and economic development. Currently, natural communities dominated by *Q. gilva* are rare, and most of the existing *Q. gilva* are scattered in other forest communities with ancient trees. More than 80% of *Q. gilva* populations occurred in the Fengshui forests or forests surrounding shrines and temples. Most of these populations were very small, or even just individual ancient trees. These populations have no natural regeneration of young adults and seedlings and thus seem to have no future. Therefore, legislation is required to protect this endangered species, and actively assist in the restoration of small populations. To date, *Q. gilva* has been listed as vulnerable (VU) in the Korea Red Data Book [41], endangered (EN) or critically endangered (CR) in several districts of Japan [14] and has also been described as a rare and endangered tree species in China [15]. The assessment results show a large disparity between the local government and the IUCN. Based on our global study, we suggested that the IUCN elevates the threatened category of *Q. gilva* from LC to EN.

In this study, we analyzed the genome sequences of 65 individuals representing the entire distributional range of *Q. gilva*. More than 15 million SNPs were identified, from which we determined the phylogeny, population structure, and genetic diversity of *Q. gilva*. Although the NJ and ML analyses showed considerable differences, both phylogenetic trees showed that *Q. gilva* has a strong evolutionary path from southwestern China to Central China, then to East China, and finally from the east coast of China to Japan and/or South Korea (Figures 2 and S1). The same pattern has been detected in many taxa native to the Sino-Japanese Forest sub-kingdom, such as *Cercidiphyllum japonicum* [42], *Quercus glauca* [43], and Asian butternuts (*Juglans* section *Cardiocaryon*) [44]. The characterized genetic relationships among all individuals based on structure and PCA showed that the populations of Yinzhou, Zhejiang (ZYZ), Yanling, Hunan (HYL), and Wuyuan, Jiangxi (JWY) had the most distinctive genetic composition.

Quercus gilva exhibited a substantially lower genetic diversity (0.994×10^{-3}) than *Q. acutissima* ($\pi = 8.7 \times 10^{-3}$), *Q. variabilis* ($\pi = 9.0 \times 10^{-3}$), and *Q. chenii* ($\pi = 7.2 \times 10^{-3}$) at the genome-wide level, which are species that belong to *Quercus* in East Asia [45]. Compared with tree species from other genera in East Asia, *Q. gilva* exhibited genetic diversity of a level similar to that of *C. japonicum* (mean $\pi = 1.00 \times 10^{-3}$) [42], and two or three times lower than the living fossil *Ginkgo biloba* ($\pi = 2.11 \times 10^{-3}$) [46] and an endangered maple *Acer yangbiense* ($\pi = 3.13 \times 10^{-3}$) [47].

Among the 22 populations, the very low level of genetic diversity in the populations of Yinzhou, Zhejiang (ZYZ), Yanling, Hunan (HYL), Zherong, Fujian (FZR), and Jingning, Zhejiang (ZJN) indicates a possibility of different demographical dynamics. The LD decay was very slow for the FZR and ZJN populations, which did not decay to half of their maximum value at the end of the distance. In contrast, the HYL and ZYZ populations exhibited the fastest decay rates. The highest r^2 of the HYL and ZYZ populations ($r^2 = 0.9$) suggested that these two populations are artificial cultivation populations, and the seeds maybe from one individual. According to the genetic diversity and LD, a strong bottleneck was detected in the small populations of FZR, ZJN, and JWY. Overall, the populations with relatively high genetic diversity and large populations are suggested as the provenance of seeds for artificial breeding, such as the populations from southwest China, Jeju Island of South Korea, and Kyushu in Japan. It is important to highlight the limitations and risks of using seeds from areas with different environmental conditions for restoration purposes. Thus, we will continue to study the adaptive evolution of *Q. gilva* under the climate change in the future to provide more detailed guidance on provenance applications.

5. Conclusions

Genetic diversity is the basis for evolutionary change and is critical for species to adapt to changing climates and biotic interactions, including novel diseases [11]. Human-mediated destruction and environmental changes disrupt population and community dynamics, resulting in the loss of population genetic diversity and species extinction [48,49]. Based on this study, we confirmed that *Q. gilva* is an endangered (EN) species, regardless of population survey or genetic evidence. In the future, we need to uncover the evolutionary history, population vulnerability, and adaptive capacity under climate change for *Q. gilva*.

Based on a detailed survey of population status and the study of genetic diversity, we could provide a more accurate assessment of the endangered status of species. This study helps initiate the assessment of threatened categories of species combined with population field survey data on genetic diversity. We suggested that in the future, a sixth criterion regarding genetic diversity should be added to the IUCN criteria used to evaluate the threatened category.

Supplementary Materials: The following supporting information can be downloaded at https://www.mdpi.com/article/10.3390/d15020230/s1, Figure S1: The maximum-likelihood (ML) phylogenetic tree of *Quercus gilva*; Figure S2: PCA (principal component analysis) of 65 individuals of *Q. gilva*. The cycles with different colors represent the different populations. The details of abbreviation codes for populations showed in Tables 2 and S1; Table S1: Information of each individual and population used in our study, and the quality of sequencing. Table S2: All the information of population status of *Q. gilva*; Table S3: Linkage disequilibrium decay measured by r^2 in each population and their position when LD decayed to half of their maximum value. References [14,19,50] are cited in the Supplementary Materials.

Author Contributions: Conceptualization, Y.-G.S., J.-H.L. and G.K.; methodology, Y.-G.S. and T.-R.W.; software, T.-R.W. and S.-S.Z.; validation, Y.-G.S.; formal analysis, T.-R.W. and S.-S.Z.; investigation,

Y.-G.S., T.-R.W., Z.-J.L., B.-J.G., X.Z., X.-C.L., D.-M.J., Q.Y., Y.L., Y.-X.K. and X.N.; resources, Y.-G.S. and X.Z.; writing—original draft preparation, Y.-G.S., T.-R.W., S.-S.Z., L.-T.Y., X.-L.D., J.-G.C., J.-H.L. and G.K.; writing—review and editing, L.-T.Y., X.-L.D., J.-G.C., J.-H.L. and G.K.; visualization, T.-R.W., X.Z., D.-M.J. and B.-J.G.; supervision, G.K.; project administration, Y.-G.S.; funding acquisition, Y.-G.S. and X.Z. All authors have read and agreed to the published version of the manuscript.

Funding: This research was funded by the Special Fund for Scientific Research of Shanghai Landscaping and City Appearance Administrative Bureau, grant number G192422; the National Natural Science Foundation of China, grant number 31901217; Science and Technology Development Center, National Forestry and Grassland Administration (KJZXSA202214); and the National Wild Plant Germplasm Resource Center for Shanghai Chenshan Botanical Garden, grant number ZWGX2202.

Institutional Review Board Statement: Not applicable.

Informed Consent Statement: Not applicable.

Data Availability Statement: Not applicable.

Acknowledgments: The authors are grateful for the help of the local guide for sampling collection. Special thanks to Zhi-Yong Wen and Rui Zou, who works for Hui-Chang Shan National Forest Park in Huichang County; Jian-Sheng Shen, who works for Xiangjiang Yuan Provincial Nature reserve.

Conflicts of Interest: The authors declare that they have no conflict of interest.

References

1. Botanic Gardens Conservation International (BGCI). *State of the World's Trees*; BGCI: Richmond, UK, 2021; pp. 3–6.
2. Crowther, T.W.; Glick, H.B.; Covey, K.R.; Bettigole, C.; Maynard, D.S.; Thomas, S.M.; Smith, J.R.; Hintler, G.; Duguid, M.C.; Amatulli, G.; et al. Mapping tree density at global scale. *Nature* **2015**, *525*, 201–205. [CrossRef] [PubMed]
3. Fazan, L.; Song, Y.G.; Kozlowski, G. The woody planet: From past triumph to manmade decline. *Plants* **2020**, *9*, 1593. [CrossRef] [PubMed]
4. Carrero, C.; Jerome, D.; Beckman, E.; Byrne, A.; Coombes, A.J.; Deng, M.; González-Rodríguez, A.; Hoang, V.S.; Khoo, E.; Nguyen, N.; et al. *The Red List of Oaks 2020*; The Morton Arboretum: Lisle, IL, USA, 2020; pp. 6–7.
5. Cavender-Bares, J. Diversification, adaptation, and community assembly of the American oaks (*Quercus*), a model clade for integrating ecology and evolution. *New Phytol.* **2019**, *221*, 669–692. [CrossRef]
6. Song, Y.G.; Petitpierre, B.; Deng, M.; Wu, J.P.; Kozlowski, G. Predicting climate change impacts on the threatened *Quercus arbutifolia* in montane cloud forests in southern China and Vietnam: Conservation implications. *For. Ecol. Manag.* **2019**, *444*, 269–279. [CrossRef]
7. Sork, V.L.; Cokus, S.J.; Fitz-Gibbon, S.T.; Zimin, A.V.; Puiu, D.; Garcia, J.A.; Gugger, P.F.; Henriquez, C.L.; Zhen, Y.; Lohmueller, K.E.; et al. High-quality genome and methylomes illustrate features underlying evolutionary success of oaks. *Nat. Commun.* **2022**, *13*, 2047. [CrossRef] [PubMed]
8. Meng, H.H.; Gao, X.Y.; Song, Y.G.; Cao, G.L.; Li, J. Biodiversity arks in the Anthropocene. *Reg. Sustain.* **2021**, *2*, 109–115. [CrossRef]
9. Convention on Biological Diversity. In *Strategic Plan for Biodiversity 2011–2020, Including Aichi Biodiversity Targets*; UNEP: Nagoya, Japan, 2010.
10. Coates, D.J.; Byrne, M.; Moritz, C. Genetic diversity and conservation units: Dealing with the species-population continuum in the age of genomics. *Front. Ecol. Evol.* **2018**, *6*, 165. [CrossRef]
11. Hoban, S.; Bruford, M.; Jackson, J.D.; Lopes-Fernandes, M.; Heuertz, M.; Hohenlohe, P.A.; Paz-Vinas, I.; Sjogren-Gulve, P.; Segelbacher, G.; Vernesi, C.; et al. Genetic diversity targets and indicators in the CBD post-2020 Global Biodiversity Framework must be improved. *Biol. Conserv.* **2020**, *248*, 108654. [CrossRef]
12. Open-Ended Working Group. Zero Draft of the Post-2020 Global Biodiversity Framework. In Proceedings of the Convention on Biological Diversity, Kunming, China, 24-29 February 2020.
13. Han, E.K.; Cho, W.B.; Park, J.S.; Choi, I.S.; Kwak, M.; Kim, B.Y.; Lee, J.H. A disjunctive marginal edge of evergreen broad-leaved oak (*Quercus gilva*) in East Asia: The high genetic distinctiveness and unusual diversity of Jeju Island populations and insight into a massive, independent postglacial colonization. *Genes* **2020**, *11*, 1114. [CrossRef]
14. Sugiura, N.; Tang, D.Q.; Kurokochi, H.; Saito, Y.; Ide, Y. Genetic structure of *Quercus gilva* Blume in Japan as revealed by chloroplast DNA sequences. *Botany* **2015**, *93*, 873–880. [CrossRef]
15. Ouyang, Z.Y.; Ouyang, S.; Wu, J.Y.; Zhou, Z.C.; Li, Z.H. Research progress of precious commercial tree species *Cyclobalanopsis gilva*. *Hunan For. Sci. Technol.* **2021**, *48*, 74–79.
16. Sugiura, N.; Kurokochi, H.; Tan, E.; Asakawa, S.; Sato, N.; Saito, Y.; Ide, Y. Development of 13 polymorphic chloroplast DNA markers in *Quercus gilva*, a regionally endemic species in Japan. *Conserv. Genet. Resour.* **2014**, *6*, 961–965. [CrossRef]
17. Lin, W.J.; Liu, Q.Y.; Luo, X.; Zeng, Q.M.; Xie, J.; Ni, B.Y.; Chen, S.P. Species diversity and niche of small populations in the natural forest *Cyclobalanopsis gilva*. *Fujian Linye* **2017**, *2017*, 38–41.

18. Kim, H.R.; You, Y.H. Ecophysiological responses of *Quercus gilva*, endangered species and *Q. glauca* to long-term exposure to elevated CO^2 concentration and temperature. *J. Ecol. Field Biol.* **2012**, *35*, 203–212.
19. Botanic Gardens Conservation International (BGCI) & IUCN SSC Global Tree Specialist Group. *Quercus gilva*. The IUCN Red List of Threatened Species 2019. e.T18749666A136775665. Available online: https://www.iucnredlist.org/species/18749666/136775665 (accessed on 17 October 2022).
20. Besseau, P.; Graham, S.; Christophersen, T. Restoring forests and landscapes: The key to a sustainable future. In *Global Partnership on Forest and Landscape Restoration*; International Union of Forest Research Organizations: Vienna, Austria, 2018.
21. Di Sacco, A.; Hardwick, K.A.; Blakesley, D.; Brancalion, P.H.S.; Breman, E.; Rebola, L.C.; Chomba, S.; Dixon, K.; Elliott, S.; Ruyonga, G.; et al. Ten golden rules for reforestation to optimize carbon sequestration, biodiversity recovery and livelihood benefits. *Glob. Change Biol.* **2021**, *27*, 1328–1348. [CrossRef] [PubMed]
22. Bonin, A.; Nicole, F.; Pompanon, F.; Miaud, C.; Taberlet, P. Population adaptive index: A new method to help measure intraspecific genetic diversity and prioritize populations for conservation. *Conserv. Biol.* **2007**, *21*, 697–708. [CrossRef]
23. Grant, V. *The Evolutionary Process, a Critical Study of Evolutionary Theory*, 2nd ed.; Columbia University Press: New York, NY, USA, 1991; pp. 121–135.
24. Allendorf, F.W.; Hohenlohe, P.A.; Luikart, G. Genomics and the future of conservation genetics. *Nat. Rev. Genet.* **2010**, *11*, 697–709. [CrossRef]
25. Kardos, M.; Armstrong, E.E.; Fitzpatrick, S.W.; Hauser, S.; Hedrick, P.W.; Miller, J.M.; Tallmon, D.A.; Funk, W.C. The crucial role of genome-wide genetic variation in conservation. *Proc. Natl. Acad. Sci. USA* **2021**, *118*, e2104642118. [CrossRef]
26. Chinese Virtual Herbarium (CVH). Available online: http://www.cvh.ac.cn/ (accessed on 8 September 2022).
27. Bachman, S.; Moat, J.; Hill, A.W.; De la Torre, J.; Scott, B. Supporting Red List threat assessment with GeoCAT: Geospatial conservation assessment tool. *ZooKeys* **2011**, *150*, 117–126. [CrossRef]
28. IUCN Standards and Petitions Committee. Guidelines for Using the IUCN Red List Categories and Criteria. Version 15.1. Prepared by the Standards and Petitions Committee. 2022. Available online: https://www.iucnredlist.org/documents/RedListGuidelines.pdf (accessed on 9 September 2022).
29. Doyle, J. A rapid DNA isolation procedure for small quantities of fresh leaf tissue. *Phytochem. Bull.* **1987**, *19*, 11–15.
30. Li, H.; Handsaker, B.; Wysoker, A.; Fennell, T.; Ruan, J.; Homer, N.; Marth, G.; Abecasis, G.; Durbin, R. The sequence alignment/Map format and SAMtools. *Bioinformatics* **2009**, *25*, 2078–2079. [CrossRef] [PubMed]
31. McKenna, A.; Hanna, M.; Banks, E.; Sivachenko, A.; Cibulskis, K.; Kernytsky, A. The genome analysis toolkit: A mapReduce framework for analyzing next generation DNA sequencing data. *Genome Res.* **2010**, *20*, 1297–1303. [CrossRef] [PubMed]
32. Mishra, B.; Ulaszewski, B.; Meger, J.; Aury, J.M.; Bodénès, C.; Lesur-Kupin, I.; Pfenninger, M.; Da Silva, C.; Gupta, D.K.; Guichoux, E.; et al. A Chromosome-level genome assembly of the European Beech (*Fagus sylvatica*) reveals anomalies for organelle DNA integration, repeat Content and distribution of SNPs. *Front. Genet.* **2022**, *12*, 691058. [CrossRef]
33. Kumar, S.; Stecher, G.; Li, M.; Knyaz, C.; Tamura, K. MEGA X: Molecular evolutionary genetics analysis across computing platforms. *Mol. Biol. Evol.* **2018**, *35*, 1547–1549. [CrossRef] [PubMed]
34. Nguyen, L.T.; Schmidt, H.A.; Von Haeseler, A.; Minh, B.Q. IQ-TREE: A fast and effective stochastic algorithm for estimating maximum-likelihood phylogenies. *Mol. Biol. Evol.* **2015**, *32*, 268–274. [CrossRef]
35. Patterson, N.; Price, A.L.; Reich, D. Population structure and eigenanalysis. *PLoS Genet.* **2006**, *2*, e190. [CrossRef]
36. Alexander, D.H.; Lange, K. Enhancements to the ADMIXTURE algorithm for individual ancestry estimation. *BMC Bioinform.* **2011**, *12*, 1–6. [CrossRef]
37. R Core Team. R: A Language and Environment for Statistical Computing. Vienna, Austria, 2019. R Foundation for Statistical Computing. Available online: https://www.R-project.org/ (accessed on 6 May 2022).
38. Pfeifer, B.; Wittelsbürger, U.; Ramos-Onsins, S.E.; Lercher, M.J. PopGenome: An efficient Swiss army knife for population genomic analyses in R. *Mol. Biol. Evol.* **2014**, *31*, 1929–1936. [CrossRef]
39. Danecek, P.; Auton, A.; Abecasis, G.; Albers, C.A.; Banks, E.; DePristo, M.A.; Handsaker, R.E.; Lunter, G.; Marth, G.T.; Sherry, S.T.; et al. The variant call format and VCFtools. *Bioinformatics* **2011**, *27*, 2156–2158. [CrossRef]
40. Purcell, S.; Neale, B.; Todd-Brown, K.; Thomas, L.; Ferreira, M.A.R.; Bender, D.; Maller, J.; Sklar, P.; de Bakker, P.I.W.; Daly, M.J.; et al. PLINK: A tool set for whole-genome association and population-based linkage analyses. *Am. J. Hum. Genet.* **2007**, *81*, 559–575. [CrossRef]
41. Ministry of the Environment of Korea. *Korean Red List of Threatened Species*; National Institute of Biological Resources: Incheon, Republic of Korea, 2012.
42. Zhu, S.S.; Chen, J.; Zhao, J.; Comes, H.P.; Li, P.; Fu, C.X.; Xie, X.; Lu, R.S.; Xu, W.Q.; Feng, Y.; et al. Genomic insights on the contribution of balancing selection and local adaptation to the long-term survival of a widespread living fossil tree, *Cercidiphyllum japonicum*. *New Phytol.* **2020**, *228*, 1674–1689. [CrossRef] [PubMed]
43. Xu, J.; Deng, M.; Jiang, X.L.; Westwood, M.; Song, Y.G.; Turkington, R. Phylogeography of *Quercus glauca* (Fagaceae), a dominant tree of East Asian subtropical evergreen forests, based on three chloroplast DNA interspace sequences. *Tree Genet. Genomes.* **2015**, *11*, 805. [CrossRef]
44. Bai, W.N.; Wang, W.T.; Zhang, D.Y. Phylogeographic breaks within Asian butternuts indicate the existence of a phytogeographic divide in East Asia. *New Phytol.* **2016**, *209*, 1757–1772. [CrossRef]

45. Liang, Y.Y.; Shi, Y.; Yuan, S.; Zhou, B.F.; Chen, X.Y.; An, Q.Q.; Ingvarsson, P.K.; Plomion, C.; Wang, B.S. Link selection shapes the landscape of genomic variation in three oak species. *New Phytol.* **2022**, *233*, 555–568. [CrossRef] [PubMed]
46. Zhao, Y.P.; Fan, G.Y.; Yin, P.P.; Sun, S.; Li, N.; Hong, X.N.; Hu, G.; Zhang, H.; Zhang, F.M.; Han, J.D.; et al. Resequencing 545 ginkgo genomes across the world reveals the evolutionary history of the living fossil. *Nat. Commun.* **2019**, *10*, 4201. [CrossRef]
47. Ma, Y.P.; Liu, D.T.; Wariss, H.M.; Zhang, R.G.; Tao, L.D.; Milne, R.I.; Sun, W.B. Demographic history and identification of threats revealed by population genomic analysis provide insights into conservation for an endangered maple. *Mol. Ecol.* **2021**, *31*, 767–779. [CrossRef]
48. Ceballos, G.; Ehrlich, P.R. Mammal population losses and the extinction crisis. *Science* **2002**, *296*, 904–907. [CrossRef]
49. De Kort, H.; Prunier, J.G.; Ducatez, S.; Honnay, O.; Baguette, M.; Stevens, V.M.; Blanchet, S. Life history, climate and biogeography interactively affect worldwide genetic diversity of plant and animal populations. *Nat. Commun.* **2021**, *12*, 516. [CrossRef]
50. Kim, G.U.; Jang, K.S.; Lim, H.; Kim, E.H.; Lee, K.H. Genetic diversity of *Quercus gilva* in Je-ju Island. *J. Korean Soc. For. Sci.* **2018**, *107*, 151–157.

diversity

MDPI

Article

Numerical Ecology and Social Network Analysis of the Forest Community in the Lienhuachih Area of Taiwan

Tung-Yu Hsieh [1,2,3,4,*], Chun-Jheng Yang [5], Feng Li [2,3,4] and Chyi-Rong Chiou [5,*]

1 School of Food and Bioengineering, Fujian Polytechnic Normal University, Fuqing 350300, China
2 Fujian Universities and Colleges Engineering Research Center of Modern Facility Agriculture, Fujian Polytechnic Normal University, Fuqing 350300, China
3 Lanxi Agricultural Technology Co., Ltd., Fuqing 350300, China
4 Wuyou Ecological Agriculture Co., Ltd., Fuqing 350300, China
5 School of Forestry and Resource Conservation, National Taiwan University, No. 1, Sec. 4, Roosevelt Rd., Taipei 10617, Taiwan
* Correspondence: dyxie@sibs.ac.cn (T.-Y.H.); esclove@ntu.edu.tw (C.-R.C.)

Abstract: In this study, the integration of useful statistical methods from different disciplines for analyzing the forest community of the Lienhuachih area of central Taiwan was attempted. We employed a seriated heat map to confirm the presence of multiple community patterns in the area and used the gap statistics and a clustplot to confirm the number and structure of the patterns, respectively. A minimum spanning tree was used to display a succession series among the quadrats, and Renyi diversity was used to indicate the species composition of these patterns. The results confirmed the existence of six patterns with different biodiversity structures in which pattern C was the succession prototype of the local community patterns. Next, we used the patterns as nodes of a social network to perform bipartite network analysis. The results showed that a community network consisted of 108 taxa and six syntaxa. The syntaxa were highly vulnerable to extinction; therefore, the optimal conservation strategy for local species would be to first protect the syntaxa. The random forest method and bipartite modularity were used to analyze the dominant characteristic species of the six syntaxa. The results showed that these two methods are useful for detecting characteristic species of the syntaxa. Therefore, this study proposed a new nomenclature system, namely the Mathematic Code of Syntaxonomic Nomenclature, to support the results of numerical vegetation analysis. Finally, the potential for an apparently competitive network was examined, the role of an apparently competitive network in the local structuring community was explored, and six new associations in the Lienhuachih area were described.

Keywords: numerical taxonomy; random forest; bipartite networks; modularity analysis; apparent competition; minimum spanning tree; Renyi diversity; new syntaxa; CART; MCSN; taxonomy modeling; classifier modeling

Citation: Hsieh, T.-Y.; Yang, C.-J.; Li, F.; Chiou, C.-R. Numerical Ecology and Social Network Analysis of the Forest Community in the Lienhuachih Area of Taiwan. *Diversity* **2023**, *15*, 60. https://doi.org/10.3390/d15010060

Academic Editor: Michael Wink

Received: 12 October 2022
Revised: 11 December 2022
Accepted: 13 December 2022
Published: 3 January 2023

1. Introduction

Humboldt, who first proposed the basic concepts of vegetation appearance and physiognomy, originally propounded vegetation ecology, also known as plant community ecology [1]. However, the titles, concepts, definitions, methods, and focus areas of early studies on vegetation ecology varied considerably; consequently, multiple terms, such as phytocoenology, symmorphology, synecology, symphysiology, syntaxonomy, and syndynamics, were used to indicate vegetation ecology. Since the early 19th century, some ecologists have studied plots of vegetation, which they considered samples of a plant community to study as integrated units which can be classified, described, and analyzed. Early vegetation classification efforts were driven largely by a desire to understand the natural diversity of vegetation and the factors that create and sustain it. Vegetation classification is critical to basic scientific research as a tool for organizing and interpreting information and placing that information in context. The use of vegetation classification has increased over the past few decades. Vegetation description and classification provide units critical for the inventory and monitoring of natural communities, planning and managing conservation programs, documenting the requirements of individual species, monitoring the use of natural resources such as forest and range lands, and providing targets for restoration. Vegetation types are even achieving legal status where they are used to define endangered habitats and where their protection is mandated [2].

By far the most widely applied approach to vegetation classification is that developed by Josias Braun-Blanquet [3]. The method centers on recording the species composition of the basic unit (plot or relevé) and the association as a vegetation unit that could be classified in a similar way to a species. Associated with the plot are records of its location, size, physical setting, and vegetation composition. The advantages of the Braun-Blanquet system include the consistency of the approach, the enormous number of plots that have been recorded, and a large number of published descriptions of vegetation types. Weaknesses include a seemingly arbitrary definition of units, the lack of a requirement that new units be integrated with established units, and the lack of any formal registry of published units [2]. After a long period of continuous development, vegetation ecology became a sub-discipline of modern ecology and now mainly involves studies on the species composition, function, dynamics, succession, classification, and distribution of plant communities [4,5].

Early classification of vegetation was subjective. Numerical methods were developed to provide objective procedures. Hence, the most common approach nowadays to vegetation classification is by numerical means [4]. Typically, this requires defining a similarity or dissimilarity matrix among all of the vegetation plots and then clustering the plots into types. Numerous methods of classification were developed with various similarity measures and different strategies for grouping plots together [6]. These methods of numerical ecology were introduced from the field of numerical taxonomy at the end of the past century and have been recently used for advancements in the study of vegetation ecology [5,7]. With the availability of textbooks and software packages, there has been a massive expansion in the use of numerical methods in vegetation science and other areas of ecology [6,7]. The development of numerical methods in vegetation ecology has led to considerable advancements in the discipline of analogy. Similarly, research in the social sciences has been supported by social network analysis in the past decades [8]. However, the methods of social network analysis are rarely applied to vegetation ecology. Interdisciplinary studies are of high importance and are needed to solve the complex problems of vegetation classification. Progress in vegetation science depends on the development of explicit theory and numerical methods capable of discriminating between rival theories [4]. Therefore, in the present study, we used the plots in the Lienhuachih area of central Taiwan as a case study, attempted to synthesize methods originating from two disciplines with similar theories, and explored the ecological implications of the results obtained by combining these methods. We also hope that this will help vegetation ecologists progressively improve the statistical methods used in vegetation ecology through this early interdisciplinary work.

2. Materials and Methods

2.1. Study Areas

Because previous related studies [9–12] have provided extremely detailed information about the Lienhuachih area, herein we provide only a summary of relevant information. The study plots of Lienhuachih were set up in 2008. The geographical location is approximately 23°54′49″ N and 120°52′43″ E; the elevation ranges from 667 to 845 m, and the average slope is 35.3°. The mean annual temperature is 20.8 °C; the rainy season in this region lies between May and September, and the annual precipitation is approximately 2285 mm in the Lienhuachih area.

2.2. Experimental Design

The vegetation survey in the Lienhuachih area was conducted in the year 2012. Abundance was recorded as the number of tree species, and sampling and data collection were conducted according to the standardized protocols adopted by the Center for Tropical Forest Science plots network. In total, 25 quadrats, each measuring 20×20 m^2, were systematically sampled within the study site. Each quadrat was spaced at 100 m and was arranged in a square grid with a projected area of 25 ha of evergreen forest. The codes used to identify the quadrats and species were according to those used in a previous study [11]. The species code accompanies the scientific name at the first mention in this article.

2.3. Statistical Methods

Multiple methods have been applied to numerical taxonomy/ecology and social network analysis in recent years [5,7,8,13]. In this study, we carefully applied methods from different disciplines. All data analysis and statistical methods in this study were conducted using R software packages [14] according to the following steps:

Vegetation classification:

1. The seriated heat map is a novel method that has been recently applied to numerical taxonomy [13]. The advantage of this method is that it simultaneously implements clustering and ordination visualizations in a plot [15,16]. The Q-Q-type seriated heat map [13,15,16], which employs the Bray–Curtis dissimilarity index [17–19], was used to detect possible community patterns in this study.
2. The gap statistic, Gapk, is a goodness of clustering measure obtained after running 10,000 Monte Carlo samples [20,21] and was employed to estimate the number of community patterns hidden in the seriated heat map. Next, the detected community patterns were coded in capital letters for subsequent analyses.
3. We used the results from the seriated heat map and gap statistics and applied the clustplot by using the partitioning around medoids (PAM) algorithm [20,22,23] to display a clustering ordination of the community patterns.
4. A minimum spanning tree was embedded into a PCoA ordination [5,18]; this was used to determine the shortest path of succession among the quadrats.
5. Renyi diversity [24] profiles were used to determine the biodiversity structure of the community patterns [7,18,25–28].

Social network analysis:

1. The community patterns and their species were used as nodes in a social network; they were then converted into a taxon–syntaxon bipartite web for social network analysis [29–32]. All of the selected indices were calculated [33] to describe network properties [34,35] in this study.
2. The characteristic value of a species (IncNodePurity) was determined using the random forest model for each community pattern consisting of 100,000 trees. The characteristic species of each syntaxon was determined [36–39].
3. We conducted bipartite modularity by applying Newman's modularity measure in a weighted web to detect the modules and compute their information in the bipartite network [33,40–42].
4. The potential for apparent competition (PAC) among or within community patterns [43–45] was calculated using a previously reported formula [46–49].
5. This study adopted the classification and regression tree [50] of the dominant/characteristic species as the key to syntaxa in the Lienhuachih area of Taiwan.

3. Results

3.1. Vegetation Classification:

Figure 1 is the Q-Q type seriated heat map of the 25 quadrats used in this study; the map appears to have several inconspicuous pattern blocks, but the blocks are vague. Therefore, we calculated the Gapk [20,21] to confirm the number and validity of the patterns in this study. The result (Figure 2) showed that a steep gap appeared when the samples were divided into 6 or 23 clusters. Based on the fact that, only between 5–6 or 22–23 clusters, there is a bigger difference in the average of clusters than the standard deviations of 5–6 or 22–23 clusters, dividing the samples into 6 or 23 clusters was reasonable. However, when the samples were divided into 23 clusters, the clusters were very similar to the individual quadrats, which would not be useful for pattern detection. Hence, the 25 quadrats were divided into 6 clusters, which was the only reasonable pattern structure apart from the individual structure. The two results, Figures 1 and 2, consistently indicated that six community patterns were hidden in the 25 squares. The detected community patterns were then randomly coded using capital letters (from A to F).

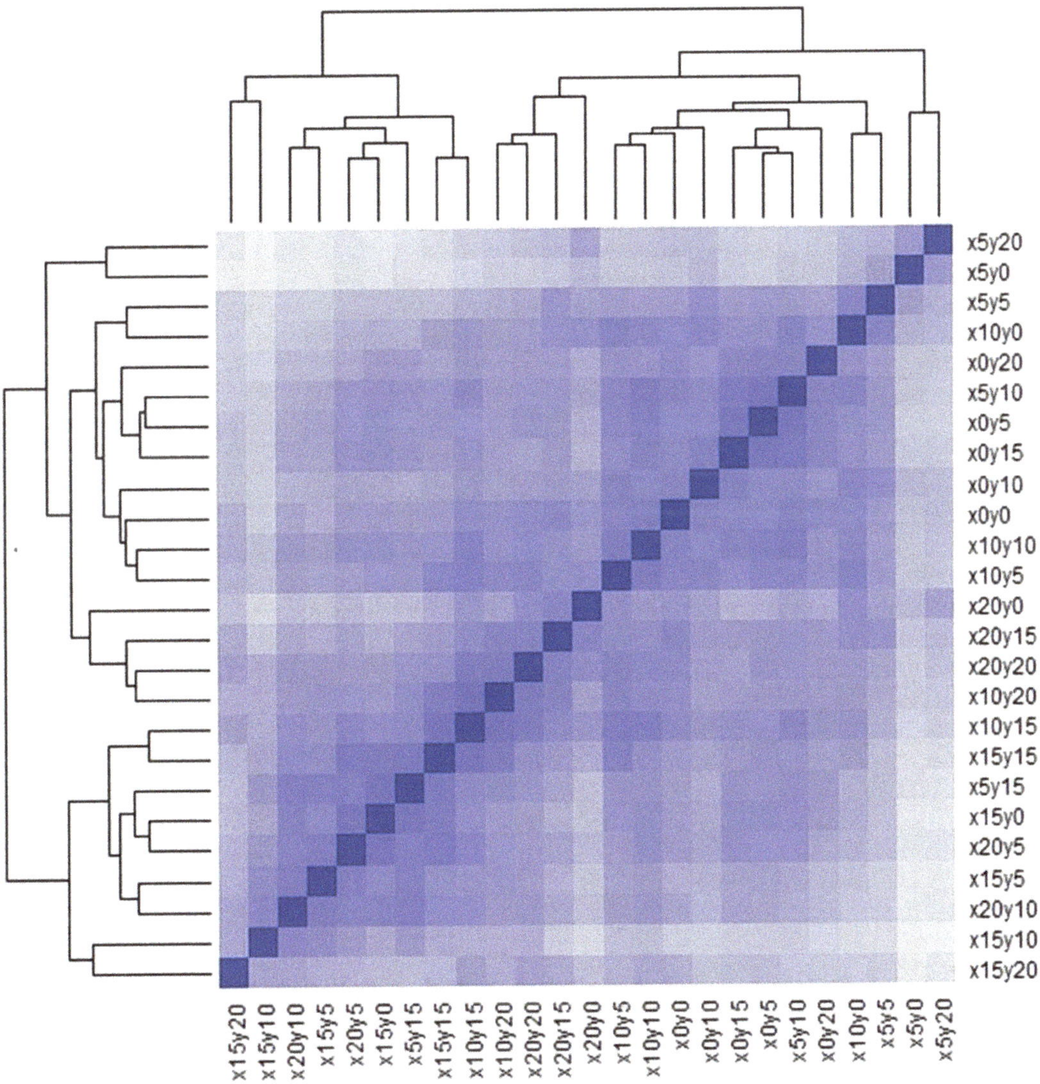

Figure 1. Seriated heat map of 25 quadrats. The blue-scaled color as a meter indicates the value of similarity.

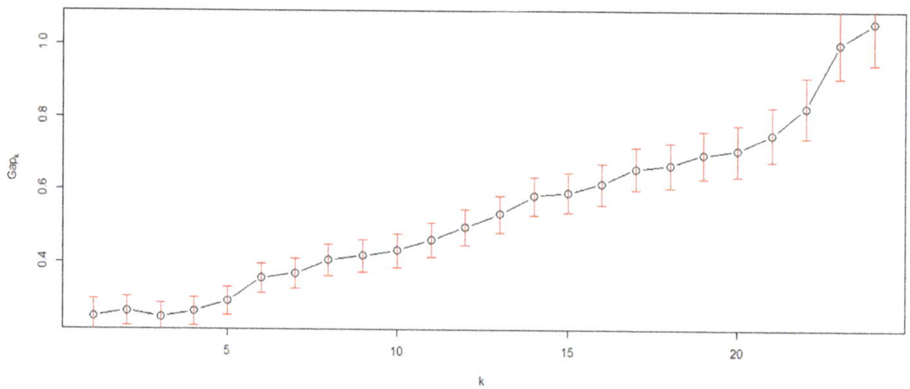

Figure 2. Gap_k and its standard deviations came from 10,000 Monte Carlo samples that were calculated to estimate the number of clusters k, k from 1 to 24.

To reveal the clear structure of hidden patterns, a clustplot (bivariate cluster plot), made using the PAM algorithm [20,22,23] was used to visualize the distinct community patterns as well as the community members. The result showed that the forest community of the study site could be divided into six patterns, and an ellipse was drawn around each pattern to indicate its members. The relationships among the patterns were indicated using red lines (Figure 3). The quadrats x15y20, x15y5, x20y20, x5y10, x10y0, and x5y20 were the six medoids of patterns A–F, respectively. In the next step, a minimum spanning tree embedded into a PCoA ordination [5,18] was conducted to determine the shortest path of succession among the quadrats. The graph shows the successional series in the form of a tree. Figure 4 shows that the center and five main branches of the minimum spanning tree correspond to the six patterns in Figure 3. The two aforementioned methods were complementary. Figure 3 reveals the relationship between the patterns, and Figure 4 reveals the relationship between the members of the patterns. Combining the results of these two methods, the internal and external structures of the community patterns were completely revealed. Figures 3 and 4 show that C is the center pattern in the forest community of the study site. A is a unique pattern that contains only one quadrat (x15y20).

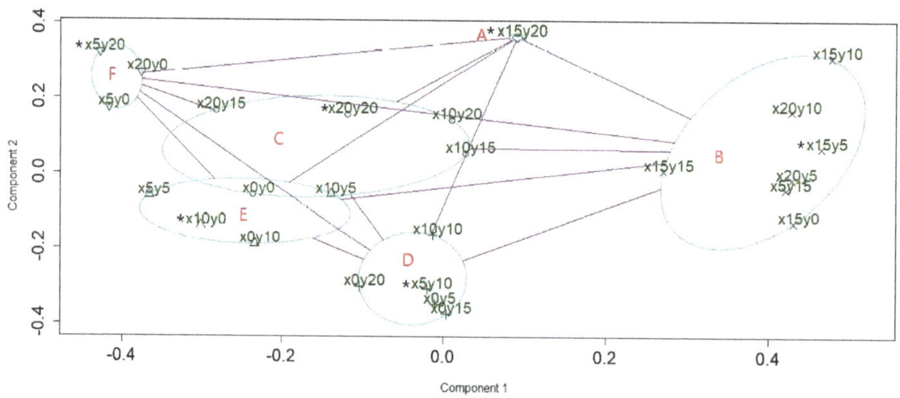

Figure 3. The clustplot shows the relationship of six community patterns with their members. The first two principal components explained 43.21% of the point variability. The medoid of a cluster is indicated with a star.

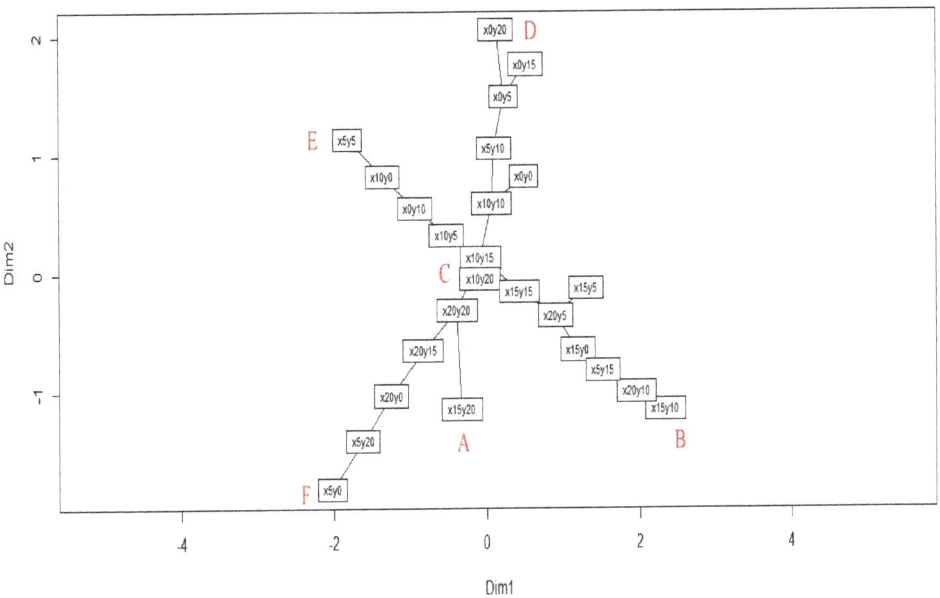

Figure 4. The minimum spanning tree of succession series from 25 quadrats in the forest community of Lienhuachih area.

Renyi entropy [24] was adopted to reveal and compare the biodiversity structures of these community patterns by using the Renyi diversity profiles [7,18,25–28]. The result (Figure 5) demonstrated that biodiversity structures differed among the six patterns; D exhibited the highest diversity along the upper line, and A exhibited the lowest diversity along the lower line. The median lines of Renyi diversity (red lines) in Figure 5 show that patterns B and E have relatively high species richness (scale = 0), but the Shannon (scale = 1) and Simpson (scale = 2) diversities of patterns B and E are similar to the median lines. The species richness of pattern E is higher than that of F, but the Shannon, Simpson, and Berger–Parker (scale = ∞/Inf) diversities of these two patterns are similar and along the median lines. The ordering, in terms of Berger–Parker diversity, of these patterns in descending order is D, C, F, E, B, and A.

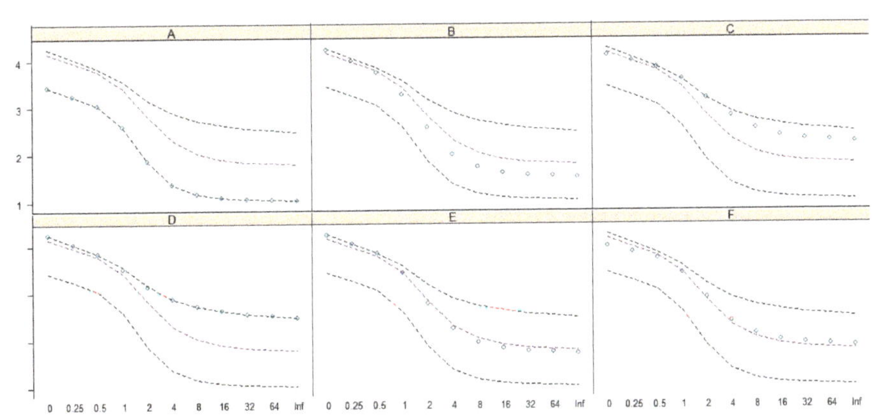

Figure 5. The Renyi diversity profiles show the biodiversity structure of patterns (**A–F**). The *y*-axis and *x*-axis are the values and scales of Renyi diversity, respectively.

3.2. Social Networks Analysis

We used the taxa and syntaxa as nodes in social network analysis and created a taxa–syntaxa bipartite ecological web for social network analysis [29–32]. Hence, the tree species were the higher taxa, whereas the patterns were the lower syntaxa in the bipartite web. Multiple indices can be used to describe the network properties or topography, but some of them do not apply to ecological networks [32]. Hence, listing all of the indices for the bipartite network in this study is unnecessary. We selected some useful indices to reveal network properties and their ecological implications in this study. The results (Table 1) revealed that the web size was 108 taxa and six syntaxa. The connectance index is the standardized number of species combinations often used in co-occurrence analyses [49]. A connectance index of 0.54 indicated that approximately half the tree species in a web could coexist in the same syntaxon. If the number of compartments was one, it indicated that the web was not separated into sub-sets [51]. A web asymmetry of 0.89 indicated that the number of taxa was considerably higher than that of syntaxa [52]. The extinction slopes for the taxa and syntaxa were 77.41 and 3.52, respectively. That extinction slope values indicated that the syntaxa were more vulnerable than the taxa because of high slope estimates, indicating relatively few effects of extinction on the network [35]. The C-scores of the taxa and syntaxa were 0.38 and 0.2, respectively, which indicated that the syntaxa were more aggregated than the taxa [53]. The V-ratios showed that the syntaxa and taxa exhibited positive aggregations [54]. High values of the togetherness of the syntaxa indicated that the taxa exhibited more disaggregation than the syntaxa [53]. The mean number of shared syntaxa per taxon indicated that an average of 1.85 syntaxa was shared by a taxon. The mean number of shared taxa per syntaxon indicated that an average of 37.27 tree species was recorded in a syntaxon [53]. The values of robustness indicated that the syntaxa were more vulnerable to secondary extinction [35].

Table 1. All selected indices for the taxon-syntaxon bipartite network of study site.

Indices	Values
number of syntaxa	6
number of taxa	108
connectance	0.54
number of compartments	1
web asymmetry	0.89
extinction slopes of taxon	77.41
extinction slopes of syntaxon	3.52
C-Score of taxon	0.38
C-Score of syntaxon	0.2
V-Ratio of taxon	2.14
V-Ratio of syntaxon	10.39
togetherness of taxa	0.17
togetherness of syntaxa	0.35
mean number of shared syntaxon per taxon	1.85
mean number of shared taxa per syntaxon	37.27
robustness of taxa	0.99
robustness of syntaxa	0.76

A random forest with 100,000 trees was used to measure the characteristic values of species for each syntaxon. Table 2 lists the first five species with high characteristic values for each syntaxon. The characteristic species of syntaxon A are GORDAXP (*Gordonia axillaris* (Roxb.) Dietr.), MELACA (*Melastoma candidum* D. Don), and ILEXLO (*Ilex lonicerifolia* Hayata), with a characteristic value of approximately 0.16. Thus, this community exhibited a high abundance of GORDAX, ILEXLO, and MELACA. The number of GORDAX in this syntaxon was as high as 57, but it did not exceed 12 in the other syntaxa, while MELACA in this syntaxon was 7, but it did not exceed 3 in the other syntaxa. The characteristic species of syntaxon B were ILEXGO (*Ilex goshiensis* Hayata), SYZYBU (*Syzygium buxifolium*

Hook. and Arn.), and RANDCO (*Randia cochinchinensis* (Lour.) Merr.). They were rare in the other syntaxa, but syntaxon B was rich in these species. The characteristic species of syntaxon C was MELISQ (*Meliosma squamulata* Hance), with a characteristic value of 0.23. The characteristic species for syntaxa D, E, and F were PSYCRU (*Psychotria rubra* (Lour.) Poir.), HELIFO (*Helicia formosana* Hemsl.), and CINNSU (*Cinnamomum subavenium* Miq.), respectively. Notably, in syntaxon F, CINNSU and ORMOFO (*Ormosia formosana* Kanehira) were negatively characteristic species for F; therefore, CINNSU and ORMOFO existed in all syntaxa except for F. In this case, we only employed the third-order species, FICUFI (*Ficus fistulosa* Reinw. ex Bl.) or SAURTR (*Saurauia tristyla* DC. var. *oldhamii* (Hemsl.) Finet and Gagnep.), but these two species had low characteristic values, which indicated that these were not suitable characteristic species for syntaxon F. In conclusion, two highly suitable negatively characteristic species were determined; however, suitably characteristic species were not found in syntaxon F.

Table 2. The first 5 high values of species for each syntaxon are listed with the species code (sc) and characteristic value (cv).

Syntaxon / Species	A		B		C		D		E		F	
	sc	cv	sc	cv	sc	cv	sc	cv	sc	cv	sc	cv
1	GORDAX	0.16	ILEXGO	1.03	MELISQ	0.23	PSYCRU	0.51	HELIFO	0.97	CINNSU	0.55
2	ILEXLO	0.16	SYZYBU	1.02	ENGERO	0.18	SCHEOC	0.48	STYRSU	0.69	ORMOFO	0.54
3	MELACA	0.15	RANDCO	1.02	BLASCO	0.18	LITSAC	0.43	BLASCO	0.22	FICUFI	0.15
4	SCHISU	0.15	EUONLA	0.57	HELIRE	0.18	MICHCO	0.26	BEILER	0.13	SAURTR	0.15
5	ILEXMI	0.06	RHAPIN	0.23	TRICDU	0.17	HELICO	0.25	CLERCY	0.12	ARDIQU	0.13

We implemented bipartite modularity by applying Newman's modularity measure in a weighted web to compute modules and their information in the network [33,40–42]. Consequently (Figure 6), the bipartite network can be grouped into five modules with the corresponding species group, also called the group of dominant characteristic species, in which syntaxa E and F share a similar pattern of species group in one module. The corresponding dominant species for syntaxon A happened to be its characteristic species, GORDAX. The dominant species for syntaxon B are RANDCO, SYZYBU, and EUONLA (*Euonymus laxiflorus* Champion ex Bentham), in which RANDCO and SYZYBU are the species with highly characteristic values. The dominant species for syntaxon C are DIOSMO (*Diospyros morrisiana* Hance.), HELIRE (*Helicia rengetiensis* Masam.), RANDCO, and BLASCO (*Blastus cochinchinensis* Lour.), in which BLASCO is not included in its species group. For syntaxon D, the dominant species are TRICDU (*Tricalysia dubia* (Lindl.) Ohwi), ARDIQU (*Ardisia quinquegona* Blume), CRYPCH (*Cryptocarya chinensis* (Hance) Hemsl.), MALLPA (*Mallotus paniculatus* Lam. Muell.-Arg.), PSYCRU, and SCHEOC. The syntaxa E and F have the same dominant species, BLASCO and HELIFO, but they are not included in the species group of F.

The PAC [43–45] of each syntaxa pair was calculated using species composition and visualized in the PAC network as a circular graph in which syntaxa were represented by circles and shared tree species were represented by connecting lines [33,46,47]. The results (Figure 7 and Table 3) revealed that syntaxon B had the highest PAC (0.54) within the syntaxon (i.e., B vs. B). In comparison with syntaxon B, the PACs of syntaxon A, C, D, E, and F were 0.24, 0.31, 0.25, 0.17, and 0.09, respectively. Syntaxon A exhibited a high PAC within the syntaxon, and syntaxon F exhibited the smallest PAC among other syntaxa. Syntaxon C exhibited the lowest PAC within the syntaxon. This study revealed the PAC network of syntaxa in the Lienhuachih forest area for the first time. As shown in Figures 3–7 and Table 3, the species composition of syntaxon A was unique; hence, the PAC within the community was the highest. The species compositions of the dominant species of syntaxon B and the other syntaxa were similar; consequently, the PAC relationship outside the syntaxon was the highest. Although syntaxon C exhibited relatively high species richness, no species was distinctly dominant; consequently, the PAC inside and outside the syntaxon was very low. Figures 3 and 4 show that syntaxon C was also the succession prototype of all of the syntaxa in the area. Syntaxa D and E exhibited similar PAC structures, which showed that the common species structure of these two communities was similar. For syntaxon F, the external PAC with E was higher than that with others; this indicated the species composition structure and common species were similar between syntaxa E and F. These results indicate that PAC plays a crucial role in structuring community [46] in the forest of the Lienhuachih area, but the results do not clarify how the apparent competition govern the spatial distribution, species composition, biodiversity structure, and the succession of syntaxa. Additional studies are necessary to answer these questions.

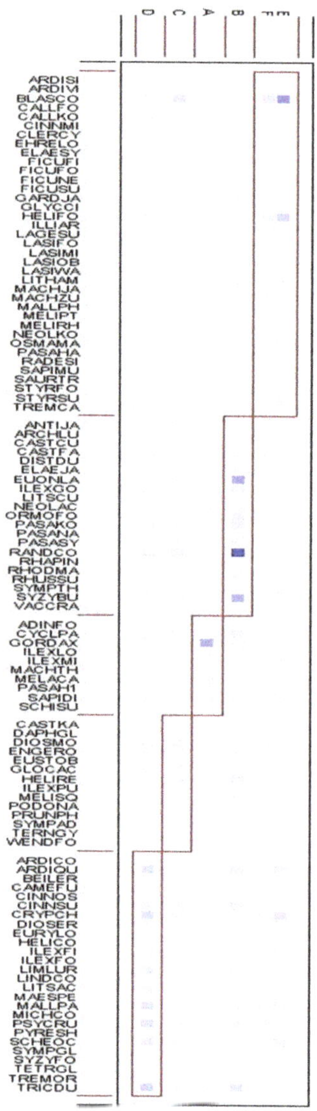

Figure 6. Modularity plot shows 5 modules with the abundance of the corresponding species group by the blue-scale.

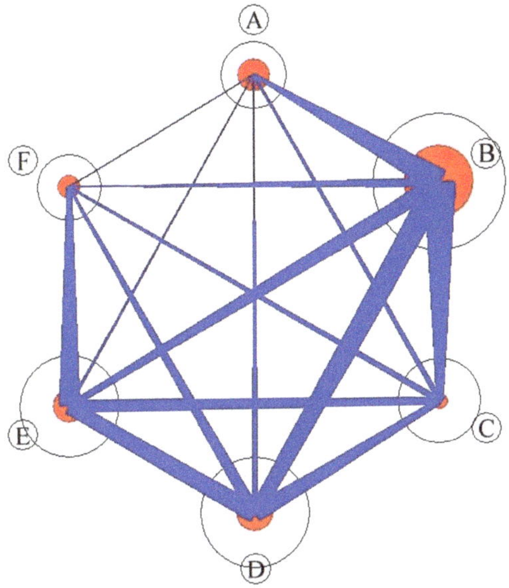

Figure 7. Quantitative circular graph of PAC network in the forest community of the study site. The circle's size is proportional to the species' abundance in the syntaxon; the red circles represent the apparent competition within syntaxon, and blue line presents the apparent competition among the syntaxa.

Table 3. PAC values in the network of syntaxa.

Syntaxon	A	B	C	D	E	F
A	0.52	0.24	0.09	0.09	0.05	0.01
B	0.06	0.54	0.12	0.17	0.09	0.02
C	0.05	0.31	0.19	0.2	0.18	0.06
D	0.03	0.25	0.11	0.36	0.19	0.05
E	0.02	0.17	0.13	0.22	0.34	0.12
F	0.02	0.09	0.1	0.15	0.28	0.36

Figure 8 shows the algorithm results from using the classification and regression tree of the dominant/characteristic species. The result demonstrated that RANDCO was used first to distinguish syntaxon B from the other species, in which syntaxon B was recorded more/equal to 53 and others were less than 53 RANDCO individuals. Then, PSYCRU helped distinguish syntaxon D (more/equal to 14) from the others (less than 14); HELIFO distinguished syntaxon A, C (less than 6) from the E, F (more/equal to 6), in which GORDAX was used to distinguish syntaxa A (more/equal to 30) from D (less than 30), and TRICDU was used to distinguish syntaxa E (more/equal to 8) from F (less than 8). In this study, 28%, 20%, 20%, 16%, 12%, and 4% of the quadrats belong to syntaxon B, C, D, E, F, and A, respectively. The rate of classification accuracy of the key to the six syntaxa was 100%.

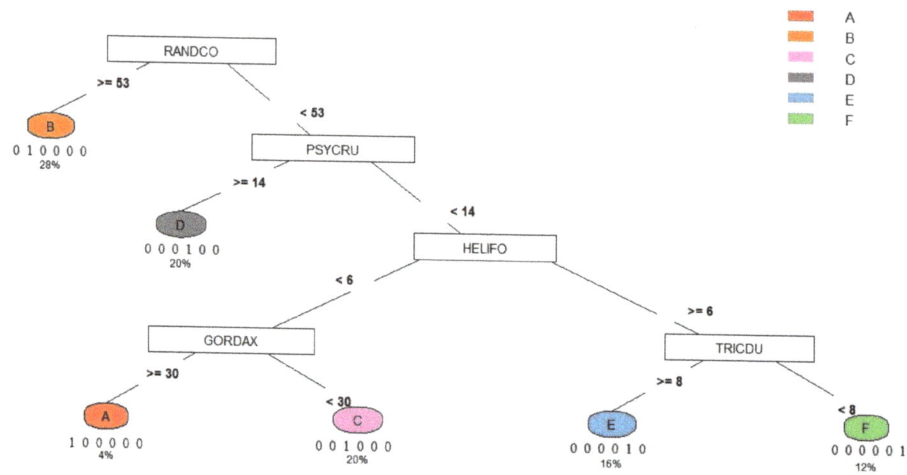

Figure 8. The key to syntaxa in the Lienhuachih area of Taiwan. The numbers below each syntaxon were the rate of classification accuracy and the rate of quadrats that belong to the syntaxon, respectively.

4. Discussion

Vegetation surveys have been conducted using small dispersed sample plots for several decades in Taiwan. Since 1990, to understand dynamic changes in forests, several large-forest dynamics plots, such as Nanjenshan, Nantzuhsienhsi, Fushan, and Lienhuachih, have been established [9]. We selected the Lienhuachih area as our study site because of its numerous advantages, such as having been extensively investigated for over 10 years, being exposed to low levels of human disturbance, exhibiting high biodiversity, and possessing a large cover of primeval forest area. Many of these are the characteristics of ideal study sites [54]. Consequently, many vegetation studies have been conducted at this experimental site in the past decade; however, none of these investigations have involved social networks. The social relationships among forest communities in this area have not been elucidated. Most vegetation analysts over the past few decades have preferred to conduct two-way indicator species analysis (TWINSPAN) or COCKTAIL [55]. Although the two methods yielded more objective results than traditional methods, they provided different results when applied to the same data set. Therefore, vegetation researchers urgently need to find more useful methods for solving the many problems in the field of vegetation ecology [56,57]. Accordingly, in this study, a combination of numerical taxonomy and network analysis techniques was used to study the forest community in the Lienhuachih area. Furthermore, the feasibility of applying these methods to vegetation analysis was examined. The results show that these methods have their own functions and are complementary to each other. The module analysis method could be used to identify modules and the dominant characteristic species group; hence, it is somewhat similar to the COCKTAIL method [58]. However, a starting species or species group must be preselected while using COCKTAIL. This pre-selection, to some extent, determines the final composition of the species group [57,58]. These problems have been reported by many vegetation ecologists. The results of the present study revealed the presence of six patterns or syntaxa in the forest community in the study site of the Lienhuachih area. The syntaxa E and F were similar; consequently, the module analysis and cluster analysis yielded different results. Figure 5 shows that the two syntaxa differed in rare species, while the common species were similar. Therefore, cluster analysis, which focus on overall similarities, and modular analysis, which focuses on dominant species, yielded different results. In a study by [9], four plant communities were identified and represented using dominant and indicator species based on TWINSPAN at the study site of the Lienhuachih area. The communities were as follows: Type I, *Pasania nantoensis—Randia cochinchinensis*; Type II, *Mallotus paniculatus—Engelhardtia roxburghiana*; Type III, *Diospyros morrisiana—Cryptocarya chinensis*; Type IV, *Machilus japonica* var. *kusanoi—Helicia formosana*. Among the aforementioned communities, no community type corresponded with any syntaxa in the present study, and the four types were also highly different from those identified in the present study. Type I is most likely to correspond to syntaxon B in this study because the *Randia cochinchinensis* was the most-dominant characteristic species of pattern B, but *Pasania nantoensis* was the fourteenth-most dominant characteristic species of pattern B (Figure 6). Type II may correspond to syntaxon C or D, because *Mallotus paniculatus* was the fifth-most dominant characteristic species of syntaxon D, and *Engelhardtia roxburghiana* was the tenth-most dominant characteristic species of syntaxon C, but neither of these two species was the most-dominant characteristic species of syntaxa C and D. Type III may also correspond to syntaxa C or D, because *Diospyros morrisiana* was the most-dominant characteristic species of syntaxon C, and *Cryptocarya chinensis* was the second-most dominant characteristic species of syntaxon D. Type IV probably corresponds to the syntaxa E and F in the present study, because both *M. japonica* var. *kusanoi* and *Helicia formosana* were among the dominant characteristic species of syntaxa E and F, but neither was the most-dominant characteristic species. *Machilus japonica* var. *kusanoi* was the thirty-seventh-most dominant characteristic species of syntaxon E and the fifth of syntaxon F. *Helicia formosana* was the second-most dominant characteristic species of syntaxa E and F. Therefore, the results of this study differed substantially from those of a previous study conducted using TWINSPAN.

The detected community patterns or syntaxa in the present study were consistent with the definitions of an association in all aspects [59,60]. Therefore, the next question was the nomenclature of these associations. The nomenclature of syntaxa has evolved over a long time with many controversies [4,57,60]; for example, the scientists of the Clements school prefer to use nomenclature coming from associations defined by dominant species, whereas scientists of the Floristic school [3] prefer nomenclature based on the characteristic species, and they proposed the International Code of Phytosociological Nomenclature (ICPN) [61]. Scientists from many countries, such as the United Kingdom [62], China [58], Taiwan [63,64], and the United States [59], have proposed principles of vegetation classification and nomenclature. Although numerical classification has enabled many useful statistical methods in recent years, a nomenclature system supporting numerical methodology has not been proposed, and the nomenclature of plant communities has not yet been unified globally [56,58,65]. Therefore, we referred to the literature and case studies [13,55,58,61,66–72] and proposed in this study a binomial nomenclature system for numerical syntaxonomy that meets the requirements of statistical analysis results. This nomenclature system, the Mathematic Code of Syntaxonomic Nomenclature (MCSN), adopts the most-dominant characteristic species in a module as the generic name and the most-characteristic species as the specific epithet for an association. In the middle of the generic name and the specific epithet is a mathematical symbol, plus or minus, which represents the fact that the association is named by the MCSN. The PAM algorithm is used to designate the nomenclature type relevé of an association. According to the MCSN, association A can be named Assoc. *Gordonia axillaris + Gordonia axillaris* Tung-Yu Hsieh et al., 2022, and the nomenclature type relevé is x15y20, which indicates both that the association is published by Hsieh et al. (2022) and that *Gordonia axillaris* is the most-dominant characteristic species. Association B can be named Assoc. *Randia cochinchinensis + Ilex goshiensis* Tung-Yu Hsieh et al., 2022, the nomenclature type relevé is x15y5, which represents an association with *R. cochinchinensis* as the most-dominant characteristic species and *Ilex goshiensis* as the most-characteristic species. Association C can be named Assoc. *Diospyros morrisiana + Meliosma squamulata* Tung-Yu Hsieh et al., 2022, and the type relevé is x20y20. Association D can be named Assoc. *Tricalysia dubia + Psychotria rubra* Tung-Yu Hsieh et al., 2022, and the type relevé is x5y10. Association E can be named Assoc. *Blastus cochinchinensis + Helicia formosana* Tung-Yu Hsieh et al., 2022, and the type relevé is x10y0. Association F can be named Assoc. *Blastus cochinchinensis-Cinnamomum subavenium* Tung-Yu Hsieh et al., 2022, and the middle symbol of the scientific name is a minus, which indicates that the most-dominant characteristic species is *Blastus cochinchinensis* and that *Cinnamomum subavenium* is the most negatively characteristic species of the association, and the type is x5y20. The characteristic species data of nomenclature types relevés A–F are presented in Table 4. In total, five genera and six species of associations were identified in the study site of the Lienhuachih area. Among the associations, E and F are associations of the same genus. Syntaxa belonging to the higher or lower ranks that are not mentioned in this study can also be applied to this nomenclature system. In addition, the MCSN retains valid publications of authors and year and the codes of priority in the ICPN and discards the suffix changes and tautonymous names in the list of scientific names to improve compatibility with the properties of the syntaxon. Because binomial nomenclature and this form of naming and presenting associations have been used by many biologists in previous studies on vegetation ecology, we are only redefining them now. The prefix word "Assoc." identifies the rank of a syntaxon; hence, the suffix of the scientific name need not be changed to indicate its rank, making the MCSN easier to understand. If some scholars cannot accept such a nomenclature system, they can also rename the syntaxon according to their own preference. For example, according to the Braun-Blanquet approach (ICPN), these associations would be renamed as follows:

Table 4. Characteristic species data of the nomenclature type relevés.

Type Releve'	ARD IQU	BLA SCO	CIN NSU	CRY PCH	DIO SMO	EUO NLA	FIC UFI	GOR DAX	HEL IFO	HEL IRE	ILE XLO	ILE XGO	MAL LPA	MEL ACA	MEL ISQ	ORM OFO	PSY CRU	RAN DCO	SAU RTR	SCH EOC	SYZ YBU	TRI CDU	Syntaxon Code
x15y20	3	0	2	1	1	10	0	57	0	1	6	0	0	7	0	3	0	4	0	1	3	2	A
x15y5	31	0	7	0	20	62	0	8	0	30	0	25	21	3	0	3	2	318	0	2	52	33	B
x20y20	13	30	6	6	21	12	0	3	0	3	0	0	1	0	1	9	1	21	0	12	2	4	C
x5y10	49	14	21	14	9	0	0	0	3	0	0	0	3	0	0	9	18	33	0	24	2	61	D
x10y0	17	121	6	16	10	0	0	0	28	0	0	0	2	0	0	1	8	4	0	28	0	10	E
x5y20	2	25	0	7	2	0	6	0	10	0	0	0	3	0	0	0	1	0	1	4	0	0	F

Syntaxonomical Synopsis

A Assoc. *Gordonietum axillare* Tung-Yu Hsieh et al., 2022 ass. nova hoc loco

Nomenclature type releve': x15y20 (holotypus hoc loco designatus).

B Assoc. *Randio cochinchinensis-Iletum goshiensis* Tung-Yu Hsieh et al., 2022, ass. nova hoc loco

Nomenclature type releve': x15y5 (holotypus hoc loco designatus).

C Assoc. *Diospyro morrisianae-Meliosmetum squamulatae* Tung-Yu Hsieh et al., 2022, ass. nova hoc loco

Nomenclature type releve´: x20y20 (holotypus hoc loco designatus).

D Assoc. *Tricalysio dubiae-Psychotretum rubrae* Tung-Yu Hsieh et al., 2022, ass. nova hoc loco

Nomenclature type releve´: x5y10 (holotypus hoc loco designatus).

E Assoc. *Blasto cochinchinensis-Helicetum formosanae* Tung-Yu Hsieh et al., 2022, ass. nova hoc loco

Nomenclature type releve´: x10y0 (holotypus hoc loco designatus).

F Assoc. *Blasto cochinchinensis-Cinnamometum subaveniae* Tung-Yu Hsieh et al., 2022, ass. nova hoc loco

Nomenclature type releve': x5y20 (holotypus hoc loco designatus).

In general, the names of associations A–E were sufficiently compatible between the ICPN and MCSN. Only the properties of the association F cannot be adequately described by the ICPN. However, when named as follows, the characteristics of the association F were clearer.

Assoc. *Blasto cochinchinensis-Ficetum fistulosae* Tung-Yu Hsieh et al., 2022, ass. nova hoc loco
Nomenclature type releve': x5y20 (holotypus hoc loco designatus).

In this study, a classification and regression tree was adopted to replace the man-made key. Compared with the conventional key, the classification and regression tree had great processing power and could carry out more precise algorithms for many numerical characteristics that are excellent in classification. For large-scale and complicated data, this advantage is especially evident. Another advantage of the classification and regression tree is its great flexibility of application. Not only can it offer professional biologists a method of classification through an observable, comprehensible, and accessible presentation like the conventional key, but it can also be used as a computer algorithm program that accesses the database from the internet and the computer programs to offer ordinary people quick species identification online. This is what conventional keys find hard to achieve.

Taxonomy and vegetation classification have a similar theoretical background, so the problems they encountered in the development process were also very similar. Early research on taxonomy and vegetation classification was often questioned because the research was subjective and lacked the use of objective methods. Although numerical taxonomy methods have improved some of the issues, many problems remain unsolved, such as type designation, the discovery of new taxa/syntaxa, revision of scientific names, and the production of a key, these important works in taxonomy/vegetation classification still lack objective analysis methods. After many years of case studies, the authors have gradually proposed appropriate numerical analysis methods for the above problems [13,72] and call them "taxonomy modeling" or "classifier modeling".

5. Conclusions

In recent years, statistical technology has made significant progress in taxonomy and social network research. However, the methods from social network analysis are still rarely used in vegetation research. Therefore, this study takes the plots of the Lienhuachih area as a case study and tries to combine some new methods from different disciplines into vegetation ecology.

In this study, the results of the heat map and gap analysis showed that a total of 108 tree species of six different syntaxa were investigated in the lotus pond area, and A–F codes were assigned to six different syntaxa. Through clustplot and a minimum spanning tree, it can be found that B is the most common syntaxa in the area, A is the rarest, and C is the succession prototype of the local community patterns; the Renyi index shows that D is the highest diversity syntaxon, and A is the lowest one; the results of Newman's modularity measure showed that there are five different modules in the Lienhuachih area. The results of this study showed that the numerical taxonomy methods are good at discriminating the species structure, and the modular method derived from social network analysis is good at distinguishing the generic structure. These two methods complement each other and apply to the analysis of the vegetation ecology; they could solve the rank problems that are not easily discernible in past vegetation research. The network of PAC showed that the most common syntaxon B in Lienhuachih had the greatest PAC, while C had the least PAC among other syntaxa. This indicates that, when B is adjacent to other syntaxa, they will have a larger PAC and be more susceptible to the common influence factor.

In the past, in the syntaxonomic nomenclature system, as with the taxonomic nomenclature system of species, the lack of objective analytical techniques often led to many inconsistencies in the nomenclature of the syntaxa in different studies. Therefore, this study redefines some nomenclature codes based on the properties of numerical analysis methods. The Mathematic Code of Syntaxonomic Nomenclature (MCSN) is proposed for the syntaxon naming system, but this does not mean that the traditional naming system must be abandoned. We just proposed a naming system that is more in line with the numerical analysis methods.

From the case study of Lienhuachih, we can find that these new methods, which are derived from numerical taxonomy and social network analysis, can be well combined with the research of vegetation analysis. This study, combined with several of our relevant published research cases, has solved the problem of lacking objective analysis methods in many key works of these two disciplines and has completed, unifying taxonomy and vegetation classification in methodology. In the future, these methods can be used in taxa/syntaxa classification but also have great application potential in the research of germplasm resources, such as the discovery of new cultivars or the identification of cultivars. The authors believe that there will be more new methods in different fields that will be applied to the field of vegetation ecology/taxonomy in the future. This research is only the beginning of the application of social network methods in these two fields.

Author Contributions: T.-Y.H. contributed significantly to data analysis and wrote the manuscript; C.-R.C. contributed to the conception of the study; C.-J.Y. performed the vegetation survey; F.L. helped perform the analysis with constructive discussions. All authors have read and agreed to the published version of the manuscript.

Funding: This study was supported by the Chinese Academy of Sciences (Grant no.: 2013 TW2SA 0003, 2015TW1SA0001), the Forestry Bureau, Council of Agriculture, Executive Yuan (Grant no.: 101 agriculture-13.5.4-forestry-e1), and the National Science Council (Grant no.: NSC 102-2313-B-002-038).

Institutional Review Board Statement: "Not applicable" for studies not involving humans or animals.

Data Availability Statement: The data presented in this study are available on request from the corresponding author.

Acknowledgments: We are thankful to the Forestry Research Institute for providing the data on the Lienhuachih forest dynamic samples and the Forestry Bureau for providing the national flora map.

Conflicts of Interest: The authors declare no conflict of interest.

References

1. Von Humboldt, A.; Aimé, B. *Ideen zu einer Geographi der Pflanzen nebst einem Naturgemalde der Tropenlander*; Akad: Leipzig, Germany, 1807; p. 139.
2. Waterton, C. From field to fantasy: Classifying nature, constructing Europe. *Soc. Stud. Sci.* **2002**, *32*, 177–204. [CrossRef]
3. Braun-Blanquet, J. *Pflanzensoziologie: Grundzüge der Vegetationskunde*; Springer: Vienna, Austria, 1964; p. 865.
4. Maarel, E.v.d.; Franklin, J. *Vegetation Ecology*, 2nd ed.; Wiley: Chichester, UK, 2012.
5. Wildi, O. *Data Analysis in Vegetation Ecology*, 2nd ed.; CABI: Wallingford, UK, 2013.
6. Jongman, R.G.H.; ter Braak, C.J.F.; van Tongeren, O.F. *Data Analysis in Community and Landscape Ecology*; Cambridge University Press: Cambridge, UK, 1995.
7. Legendre, P.; Legendre, L. *Numerical Ecology*; Elsevier: Amsterdam, The Netherlands, 2012.
8. Freeman, L.C. *The Development of Social Network Analysis: A Study in the Sociology of Science*; Empirical Press: Vancouver, BC, Canada, 2004; p. 205.
9. Chang, L.-W.; Chiu, S.-T.; Yang, K.-C.; Wang, H.-H.; Hwong, J.-L.; Hsieh, C.-F. Changes of plant communities classification and species composition along the micro-topography at the Lienhuachih Forest Dynamics Plot in the central Taiwan. *Taiwania* **2012**, *57*, 359–371. [CrossRef]
10. Chang, L.-W.; Hwong, J.-L.; Chiu, S.-T.; Wang, H.-H.; Yang, K.-C.; Chang, H.-Y.; Hsieh, C.-F. Species composition, size-class structure, and diversity of the Lienhuachih Forest Dynamics Plot in a subtropical evergreen broad-leaved forest in central Taiwan. *Taiwan J. For. Sci.* **2010**, *25*, 81–95. [CrossRef]
11. Yang, C.-J. *The Study of Species Selection in Restoration Plan: An Example of Lienhuachih Region*; National Taiwan University: Taipei, Taiwan, 2013.
12. Chang, L.; Huang, J.; Luo, S.; Lee, P. Understory plant composition and its relations with environmental factors of the Lienhuachih forest dynamics plot at a subtropical evergreen broadleaf forest in central Taiwan. *Taiwan J. For. Sci.* **2015**, *30*, 245–257.
13. Hsieh, T.-Y.; Ku, S.-M.; Chien, C.-T.; Liou, Y.-T. Classifier modeling and numerical taxonomy of Actinidia (Actinidiaceae) in Taiwan. *Bot. Stud.* **2011**, *52*, 337–357.
14. R Core Team, R. *A Language and Environment for Statistical Computing, 4.2.0*, R Foundation for Statistical Computing: Vienna, Austria, 2022.
15. Earle, D.; Hurley, C.B. Advances in dendrogram seriation for application to visualization. *J. Comput. Graph. Stat.* **2015**, *24*, 1–25. [CrossRef]
16. Hahsler, M.; Buchta, C.; Hornik, K. Seriation: Infrastructure for Ordering Objects Using Seriation, R Package Version 1.3.6. 2022. Available online: https://CRAN.r-project.org/package=seriation (accessed on 1 January 2022).
17. Bray, J.R.; Curtis, J.T. An ordination of the upland forest communities of southern Wisconsin. *Ecol. Monogr.* **1957**, *27*, 325–349. [CrossRef]

18. Oksanen, J.; Simpson, G.; Blanchet, F.; Kindt, R.; Legendre, P.; Minchin, P.; O'Hara, R.; Solymos, P.; Stevens, M.; Szoecs, E.; et al. Vegan: Community Ecology Package, R Package Version 2.6-2. 2022. Available online: https://CRAN.r-project.org/package=vegan (accessed on 18 June 2022).
19. Faith, D.P.; Minchin, P.R.; Belbin, L. Compositional dissimilarity as a robust measure of ecological distance. *Vegetatio* **1987**, *69*, 57–68. [CrossRef]
20. Maechler, M.; Rousseeuw, P.; Struyf, A.; Hubert, M.; Hornik, K. Cluster: Cluster Analysis Basics and Extensions, R Package Version 2.0.7-1. 2018. Available online: https://CRAN.r-project.org/package=cluster (accessed on 1 July 2018).
21. Tibshirani, R.; Walther, G.; Hastie, T. Estimating the number of clusters in a data set via the gap statistic. *J. R. Stat. Soc. B* **2002**, *63*, 411–423. [CrossRef]
22. Reynolds, A.P.; Richards, G.; de la Iglesia, B.; Rayward-Smith, V.J. Clustering rules: A comparison of partitioning and hierarchical clustering algorithms. *J. Math. Model. Algorithms* **2006**, *5*, 475–504. [CrossRef]
23. Pison, G.; Struyf, A.; Rousseeuw, P.J. Displaying a clustering with CLUSPLOT. *Comput. Stat. Data Anal.* **1999**, *30*, 381–392. [CrossRef]
24. Rényi, A. On measures of entropy and information. In Proceedings of the 4th Berkeley symposium on mathematical statistics and probability, Berkeley; 1961; pp. 547–561.
25. Hill, M.O. Diversity and evenness: A unifying notation and its consequences. *Ecology* **1973**, *54*, 427–432. [CrossRef]
26. Kindt, R.; Van Damme, P.; Simons, A.J. Tree diversity in western Kenya: Using profiles to characterise richness and evenness. *Biodivers. Conserv.* **2006**, *15*, 1253–1270. [CrossRef]
27. Tóthmérész, B. Comparison of different methods for diversity ordering. *J. Veg. Sci.* **1995**, *6*, 283–290. [CrossRef]
28. Caruso, T.; Pigino, G.; Bernini, F.; Bargagli, R.; Migliorini, M. The Berger-Parker index as an effective tool for monitoring the biodiversity of disturbed soils: A case study on Mediterranean oribatid (Acari: Oribatida) assemblages. *Biodivers. Conserv.* **2006**, *16*, 3277–3285. [CrossRef]
29. Lau, M.K.; Borrett, S.R.; Hines, D.E.; Singh, P. EnaR: Tools for Ecological Network Analysis, R Package Version 3.0.0. 2017. Available online: https://CRAN.R-project.org/package=enaR (accessed on 15 January 2017).
30. Borrett, S.R.; Lau, M.K. EnaR: An R package for ecosystem network analysis. *Methods Ecol. Evol.* **2014**, *5*, 1206–1213. [CrossRef]
31. Scott, J. *Social Network Analysis*, 4th ed.; SAGE Publications: London, UK, 2017; p. 227.
32. Dormann, C.F.; Gruber, B.; Fründ, J. Introducing the bipartite Package: Analysing Ecological Networks. *R News* **2008**, *8*, 8–11.
33. Dormann, C.F.; Fruend, J.; Grube, B. Bipartite: Visualising Bipartite Networks and Calculating Some (Ecological) Indices, R Package Version 2.08. 2017. Available online: https://CRAN.R-project.org/package=bipartite (accessed on 22 January 2017).
34. Dormann, C.F.; Fründ, J.; Blüthgen, N.; Gruber, B. Indices, graphs and null models: Analyzing bipartite ecological networks. *Open Ecol. J.* **2009**, *2*, 7–24. [CrossRef]
35. Memmott, J.; Waser, N.M.; Price, M.V. Tolerance of pollination networks to species extinctions. *Proc. R. Soc. Lond. B Biol. Sci.* **2004**, *271*, 2605–2611. [CrossRef]
36. Liaw, A.; Wiener, M. RandomForest: Breiman and Cutler's Random Forests for Classification and Regression. R Package Version 2.6-2. 2018. Available online: https://CRAN.r-project.org/package=randomForest (accessed on 11 June 2018).
37. Breiman, L. Random forests. *Mach. Learn.* **2001**, *45*, 5–32. [CrossRef]
38. Cutler, D.R.; Edwards, T.C., Jr.; Beard, K.H.; Cutler, A.; Hess, K.T.; Gibson, J.; Lawler, J.J. Random forests for classification in ecology. *Ecology* **2007**, *88*, 2783–2792. [CrossRef] [PubMed]
39. Liaw, A.; Wiener, M. Classification and regression by randomForest. *R News* **2002**, *2*, 18–22.
40. Beckett, S.J. Improved community detection in weighted bipartite networks. *R. Soc. Open Sci.* **2016**, *3*, 140536. [CrossRef] [PubMed]
41. Liu, X.; Murata, T. An efficient algorithm for optimizing bipartite modularity in bipartite networks. *J. Adv. Comput. Intell. Intell. Inform.* **2010**, *14*, 408–415. [CrossRef]
42. Dormann, C.F.; Strauss, R. Detecting modules in quantitative bipartite networks: The QuaBiMo algorithm. *Methods Ecol. Evol.* **2013**, *5*, 90–98. [CrossRef]
43. Holt, R.D. Predation, apparent competition, and the structure of prey communities. *Theor. Popul. Biol.* **1977**, *12*, 197–229. [CrossRef]
44. Morris, R.J.; Lewis, O.T.; Godfray, H.C.J. Experimental evidence for apparent competition in a tropical forest food web. *Nature* **2004**, *428*, 310–313. [CrossRef] [PubMed]
45. Carvalheiro, L.G.; Biesmeijer, J.C.; Benadi, G.; Fründ, J.; Stang, M.; Bartomeus, I.; Kaiser-Bunbury, C.N.; Baude, M.; Gomes, S.I.; Merckx, V. The potential for indirect effects between co-flowering plants via shared pollinators depends on resource abundance, accessibility and relatedness. *Ecol. Lett.* **2014**, *17*, 1389–1399. [CrossRef]
46. Morris, R.J.; Lewis, O.T.; Godfray, H.C.J. Apparent competition and insect community structure: Towards a spatial perspective. *Ann. Zool. Fenn.* **2005**, *42*, 449–462.
47. Muller, C.; Adriaanse, I.; Belshaw, R.; Godfray, H. The structure of an aphid–parasitoid community. *J. Anim. Ecol.* **1999**, *68*, 346–370. [CrossRef]
48. Gotelli, N.J.; Graves, G.R. *Null Models in Ecology*; Smithsonian Institution Press: Washington, DC, USA, 1996; p. 368.
49. Pimm, S.L. *Food Webs*; Chapman and Hall: London, UK, 1982; p. 219.

50. Therneau, T.; Atkinson, B. Rpart: Recursive Partitioning and Regression Trees, R Package Version 4.1.16. 2022. Available online: https://CRAN.R-project.org/package=rpart (accessed on 8 October 2022).
51. Blüthgen, N.; Menzel, F.; Hovestadt, T.; Fiala, B.; Blüthgen, N. Specialization, constraints, and conflicting interests in mutualistic networks. *Curr. Biol.* **2007**, *17*, 341–346. [CrossRef] [PubMed]
52. Stone, L.; Roberts, A. Competitive exclusion, or species aggregation? *Oecologia* **1992**, *91*, 419–424. [CrossRef] [PubMed]
53. Schluter, D. A variance test for detecting species associations, with some example applications. *Ecology* **1984**, *65*, 998–1005. [CrossRef]
54. Condit, R. *Tropical Forest Census Plots*; Springer: New York, NY, USA, 1998; p. 211.
55. Li, C.-F.; Zelený, D.; Chytrý, M.; Chen, M.-Y.; Chen, T.-Y.; Chiou, C.-R.; Hsia, Y.-J.; Liu, H.-Y.; Yang, S.-Z.; Yeh, C.-L. Chamaecyparis montane cloud forest in Taiwan: Ecology and vegetation classification. *Ecol. Res.* **2015**, *30*, 771–791. [CrossRef]
56. Bruelheide, H.; Chytrý, M. Towards unification of national vegetation classifications: A comparison of two methods for analysis of large data sets. *J. Veg. Sci.* **2000**, *11*, 295–306. [CrossRef]
57. Bruelheide, H. A new measure of fidelity and its application to defining species groups. *J. Veg. Sci.* **2000**, *11*, 167–178. [CrossRef]
58. Song, Y.-C. Recognition and proposal on the vegetation classification system of China. *Chin. J. Plant Ecol.* **2011**, *35*, 882–892. [CrossRef]
59. Jennings, M.D.; Faber-Langendoen, D.; Loucks, O.L.; Peet, R.K.; Roberts, D. Standards for associations and alliances of the US National Vegetation Classification. *Ecol. Monogr.* **2009**, *79*, 173–199. [CrossRef]
60. Mucina, L. Classification of vegetation: Past, present and future. *J. Veg. Sci.* **1997**, *8*, 751–760. [CrossRef]
61. Weber, H.E.; Moravec, J.; Theurillat, J.P. International code of phytosociological nomenclature. *J. Veg. Sci.* **2000**, *11*, 739–768. [CrossRef]
62. Rodwell, J.S. The UK National Vegetation Classification. *Phytocoenologia* **2018**, *48*, 133–140. [CrossRef]
63. Song, Y.-C.; Xu, G.-S. A scheme of vegetation classification of Taiwan, China. *Acta Bot. Sin.* **2003**, *45*, 883–895.
64. Li, C.F.; Chytrý, M.; Zelený, D.; Chen, M.Y.; Chen, T.Y.; Chiou, C.R.; Hsia, Y.J.; Liu, H.Y.; Yang, S.Z.; Yeh, C.L. Classification of Taiwan forest vegetation. *Appl. Veg. Sci.* **2013**, *16*, 698–719. [CrossRef]
65. Mucina, L.; Bültmann, H.; Dierßen, K.; Theurillat, J.P.; Raus, T.; Čarni, A.; Šumberová, K.; Willner, W.; Dengler, J.; García, R.G. Vegetation of Europe: Hierarchical floristic classification system of vascular plant, bryophyte, lichen, and algal communities. *Appl. Veg. Sci.* **2016**, *19*, 3–264. [CrossRef]
66. Hsieh, T.-Y.; Hatch, K.A.; Chang, Y.-M. Phlegmariurus changii (Huperziaceae), a new hanging firmoss from Taiwan. *Am. Fern J.* **2012**, *102*, 283–288. [CrossRef]
67. Turland, N.J.; Wiersema, J.H.; Barrie, F.R.; Greuter, W.; Hawksworth, D.L.; Herendeen, P.S.; Knapp, S.; Kusber, W.-H.; Li, D.-Z.; Marhold, K.; et al. *International Code of Nomenclature for Algae, Fungi, and Plants (Shenzhen Code) Adopted by the Nineteenth International Botanical Congress Shenzhen, China, July 2017*; Koeltz Botanical Books: Glashütten, Germany, 2018.
68. Hsieh, T.-Y.; Hsu, T.-C.; Kono, Y.; Ku, S.-M.; Peng, C.-I. Gentiana bambuseti (Gentianaceae), a new species from Taiwan. *Bot. Stud.* **2007**, *48*, 349–355.
69. Balpinar, N.; Kavgaci, A.; Bingöl, M.Ü.; Ketenoğlu, O. Diversity and gradients of vegetation of Sivrihisar Mountains (Eskişehir-Turkey). *Acta Bot. Croat.* **2018**, *77*, 18–27. [CrossRef]
70. Peng, C.-I.; Hsieh, T.-Y.; Ngyuen, Q.H. *Begonia kui* (sect. Coelocentrum, Begoniaceae), a new species from Vietnam. *Bot. Stud.* **2007**, *48*, 127–132.
71. Chen, J.-H.; Hsieh, T.-Y.; Chen, W.-Z.; Chen, C.-H.; Chang, Y.-M. A new species of the cicada genus Euterpnosia (Hemiptera: Cicadidae) from Taiwan, with morphometric approaches. *Formos. Entomol.* **2021**, *41*, 192–207.
72. Hsieh, T.-Y. Taxonomy and distribution of indigenous Actinidia in Taiwan. Ph.D. Thesis, National Chung-Hsing University, Taichung, Taiwan, 2011.

diversity

MDPI

Article

PPDP: A Data Portal of *Paris polyphylla* for Polyphyllin Biosynthesis and Germplasm Resource Exploration

Qixuan Su [1,2], Xuan Zhang [3], Jing Li [2], Wenjing Yang [2,4], Qiang Ren [4], Xiaoyang Gao [2,*] and Changning Liu [1,2,5,6,*]

1 School of Life Sciences, University of Science and Technology of China, Hefei 230026, China
2 CAS Key Laboratory of Tropical Plant Resources and Sustainable Use, Xishuangbanna Tropical Botanical Garden, Chinese Academy of Sciences, Kunming 650223, China
3 Center for Molecular Diagnosis and Precision Medicine, The Department of Clinical Laboratory, The First Affiliated Hospital of Nanchang University, Nanchang 330006, China
4 College of Life Sciences, University of Chinese Academy of Sciences, Beijing 100049, China
5 Center of Economic Botany, Core Botanical Gardens, Chinese Academy of Sciences, Xishuangbanna 666303, China
6 The Innovative Academy of Seed Design, Chinese Academy of Sciences, Kunming 650223, China
* Correspondence: gaoxiaoyang@xtbg.ac.cn (X.G.); liuchangning@xtbg.ac.cn (C.L.)

Abstract: *Paris polyphylla* Smith is a perennial medicinal herb with records from around 2000 years ago. Polyphyllins are the main bioactive compounds of this herb, which are found to have remarkable effects on bacteriostatic, antiphlogistic, sedative, and antitumor. However, the market demand for *P. polyphylla* is sharply increasing, and the wild resources are threatened by plundering exploitation. Integrating molecular data of *P. polyphylla* can benefit the sustainable resource exploitation. Here, we constructed PPDP (*Paris polyphylla* Data Portal) to provide a data platform for polyphyllin biosynthesis and germplasm resource research. PPDP integrates related molecular data resources, functional genomics analysis, and morphological identification. The database provides abundant data (transcriptome, CDS, lncRNA, alternative splicing, gene family, SSR, and chloroplast genome) and practical analytical tools (network construction, heatmap of expression profiles, enrichment, and pathway search) with a user-friendly interface. So far, PPDP is the first biomolecular database for the genus *Paris* plants. In the future, we will gradually add genomic data and other necessary molecular biological information to improve the database.

Keywords: *Paris polyphylla*; polyphyllin; germplasm resource; database

Citation: Su, Q.; Zhang, X.; Li, J.; Yang, W.; Ren, Q.; Gao, X.; Liu, C. PPDP: A Data Portal of *Paris polyphylla* for Polyphyllin Biosynthesis and Germplasm Resource Exploration. *Diversity* **2022**, *14*, 1057. https://doi.org/10.3390/d14121057

Academic Editor: Mario A. Pagnotta

Received: 1 November 2022
Accepted: 29 November 2022
Published: 1 December 2022

1. Introduction

Paris polyphylla Smith (Melanthiaceae) has a medicinal history of over 2000 years, it is distributed across Assam, Bangladesh, North-Central China, South-Central China, East Himalaya, Myanmar, Nepal, Qinghai, Tibet, and West Himalaya. *Paris* plants mostly grow in evergreen broad-leaved forests, bamboo forests, shrubs, or grass slopes. Currently, more than 10 varieties have been found in *P. polyphylla* [1,2]. *Paris polyphylla* var. *yunnanensis* (Franch.) Hand.-Mazz. is one of the most significant varieties of *P. polyphylla* recorded in Pharmacopoeia of the People's Republic of China [3]. In a recent monograph published in 2021, it is recognized as a species named *Paris yunnanensis* Franch. [4]. Pharmacological studies have shown that *P. polyphylla* has significant effects on hemostasis, analgesia, bacteriostasis, anti-inflammatory, and tumor cell inhibition [5]. To date, more than 210 compounds have been isolated and identified from *P. polyphylla*, including steroidal saponins, C-21 steroids, fatty acid esters, triterpenes, flavonoids, β-ecdysone, and polysaccharides [6,7]. Steroidal saponins, i.e., polyphyllins, are the most important active ingredients, accounting for approximately 57% of the total number of identified ingredients. Therefore, rhizoma paridis has become the key drug material of more than 80 kinds of patented medicines, such as gongxuening, yunnan baiyao, reduqing, and jidesheng sheyao tablets [8]. However, the long growth cycle of 5–7 years, and the post-embryonic maturation of seeds result in the slow reproduction of *P. polyphylla*. Moreover, sharply increasing market demand and large-scale over-excavation driven by high medical value cause resource shortage, and also endanger the wild resources of *P. polyphylla* and other *Paris* species [9]. Therefore, understanding polyphyllin biosynthesis mechanism is significant for the effective utilization of this medicinal herb.

With the development of high-throughput sequencing, transcriptome sequencing has been widely used to mine functional genes involved in plant secondary metabolism processes, especially the specialized metabolite biosynthesis of non-model plants lacking genome sequences. The exploration of polyphyllin biosynthesis genes (PBGs) has remarkably advanced through transcriptome sequencing. Several candidate genes related to polyphyllin biosynthesis were predicted based on the transcriptome data from rhizomes [10] or leaves (stems) [11] of *P. polyphylla*. Meanwhile, another 25 transcripts encoding 17 key enzymes related to polyphyllin biosynthesis were identified from the transcriptome of *P. polyphylla* tissue mixtures [12]. More specifically, a gene encoding cytochrome P450 monooxygenase (P450) was identified, which catalyzes the oxidative 5,6-spiroketalization of cholesterol to produce diosgenin [13]. A total of 137 PBGs, 74 transcription factor genes, and 1 transporter gene associated with polyphyllin biosynthesis and accumulation were identified in our previous study [14]. Additionally, polyphyllin biosynthesis is found to be in response to tissue-specific combinatorial development cues through RNA-Seq analysis [14,15]. Nonetheless, molecular mechanisms underlying polyphyllin biosynthesis and accumulation are still unclear.

A database that integrates omics data and analytical tools is very beneficial to unravel specialized metabolite biosynthesis mechanism. Some medicinal plant databases such as the Ginseng Genome Database [16] facilitate terpenoid biosynthesis research and molecular breeding [17,18]. However, there is no database available for *P. polyphylla* so far. Genetic information of *P. polyphylla* is primarily from next-generation sequencing (NGS) transcriptome data [19,20] and very few third-generation sequencing (TGS) transcriptome data [21]. In this study, we constructed the first data portal PPDP (http://ppdp.liu-lab.com) for *P. polyphylla*, mainly based on the PacBio SMRT-based RNA-Seq and Illumina-based RNA-Seq data. The current version of the database provides diverse data, including 36,504 genes, 9020 lncRNAs, 7029 alternative splicing (AS) events, 25,767 SSRs and 49 valid SSR markers, 108 chloroplast genomes, 931 genes from 50 transcription factor (TF) families, 93 CYP and 56 UGT gene candidates, and gene expression profiles at different growth stages. A total of 186 PBGs were identified, of which 88 genes contained SSRs. PPDP provides a kit of practical analytical tools, such as network construction, heatmap drawing, pathway search, etc., with a user-friendly interface. In all, PPDP provides new molecular data and helps to gain further insight into polyphyllin biosynthesis.

2. Materials and Methods

2.1. Sampling and PacBio SMRT Sequencing

Four fresh tissue samples were collected from six 7-year-old plants of *P. polyphylla* var. *yunnanensis*, which were planted under the same field management in the greenhouse of Xishuangbanna Tropical Botanical Garden, Chinese Academy of Sciences (Kunming). Each fresh tissue sample was a mixture, which combined the same tissue collected from three plants with similar size and status to obtain credible sequencing data. The *P. polyphylla* var. *yunnanensis* plants in this study were obtained through seed germination and seedling cultivation for years. The seeds were kindly provided by Yunnan Yuxin Agriculture and Forestry Biological Technology Co., Ltd (Yunnan, China). Seven-year-old is an important stage in the agriculture cultivation of *Paris* plants. The rhizomes of *P. polyphylla* plants from seed propagation are often harvested at the seventh year. This age is chosen for providing a representative dataset and a reasonable reference in metabolism biosynthesis research. The species identification was confirmed by Professor Li Heng (Kunming Institute of Botany, CAS). RNA from roots and stems were pooled into one sample for SMRT sequencing.

mRNA was purified using magnetic beads with Oligo (dT), and the library was prepared according to the Isoform Sequencing protocol (Iso-Seq) using the Clontech SMARTer PCR cDNA Synthesis Kit and the BluePippin Size Selection System protocol as described by Pacific Biosciences (PN 100-092-800-03). The sequencing was performed using the PacBio Sequel II platform after confirming library quality. Flowers and leaves were sequenced separately in the same way after RNA extraction. An additional PacBio isoform sequencing dataset of *P. polyphylla* var. *yunnanensis* was downloaded from The National Genomics Data Center (NGDC) (accession number: CRA004081) [21] (Figure 1A).

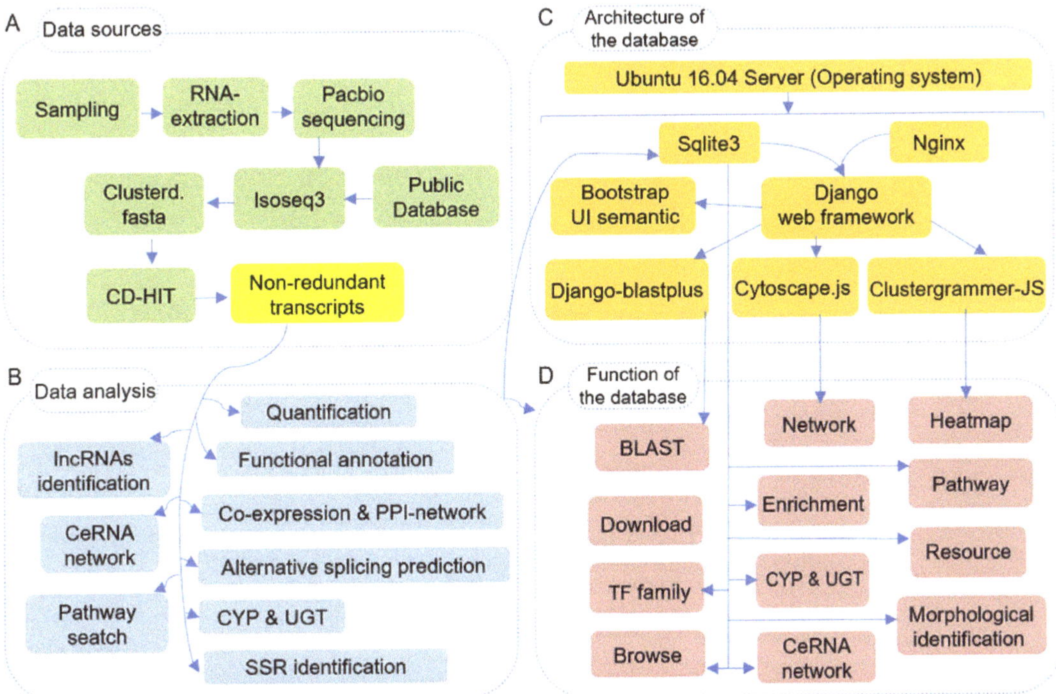

Figure 1. Implementation of PPDP. (**A**) RNA-Seq of sampled tissues and data pre-processing. (**B**) The main data analysis process. (**C**) The basic architecture of PPDP. (**D**) The main functions and related tools of PPDP.

2.2. Transcriptome Determination and Evaluation

Three sets of PacBio raw data of *P. polyphylla* var. *yunnanensis* (two sequenced in this study and one downloaded from the public database) were processed according to the Iso-Seq3 pipeline (https://github.com/PacificBiosciences/IsoSeq (accessed on 10 November 2021)). First, circular consensus sequence (CCS) was generated from subread BAM files with parameter settings: min_length 50, max_length 15,000, min-rq 0.8. Then, CCS reads were defined as full-length (FL) and non-full-length (NFL) isoforms, depending on the presence or absence of 5′ primers, 3′ primers, and poly(A) tails. Primers, poly(A) tails, and rapid concatemers were also removed. Finally, full-length non-chimeric (FLNC) reads were clustered to generate high-quality (HQ) isoforms and low-quality (LQ) isoforms. In addition, the three FLNC reads were merged by clustering in the expectation of a new transcript of higher quality. De-redundancy using CD-HIT v4.8.1 [22] and the removal of transcripts shorter than 200 bp yielded the final non-redundant transcripts, respectively. Together with our previous NGS transcripts based on Illumina sequencing data from 4 developmental stages and 5 tissues in *P. polyphylla* var. *yunnanensis* [13,14], all five transcripts were assessed using BUSCO v5.0.0 [23].

2.3. Functional Annotation and Classification

Functional annotation was performed to gain an insight into the biological context of the transcripts. All the non-redundant transcripts were searched against Nr and KEGG using BLAST v2.2.25 [24] with an E-value threshold of 1×10^{-5}. The annotation of GO, Swiss-Prot, and Pfam were determined using Trinotate v3.2.2 (http://trinotate.github.io/ (accessed on 20 November 2021)). Transcripts were also annotated by eggnog-mapper v2.6.1 [25]. Coding sequences (CDS) and open reading frames (ORFs) were identified with TransDecoder v5.5.0 (https://github.com/TransDecoder/TransDecoder/releases (accessed on 20 November 2021)).

2.4. SSR Analysis

SSR determination was performed using MISA v2.1 [26]. Six types of SSRs including mono-, di-, tri-, tetra-, penta-, and hexa-nucleotide repeats with minimum repeat numbers of 10, 6, 5, 5, 5, and 5 were identified from the transcriptome. The distance between adjacent SSRs < 100 bp was defined as compound SSR. Further, functional annotations of transcripts containing SSR were obtained. In addition, valid SSR markers and chloroplast genomes were collected from the literature and NCBI database, respectively.

2.5. Identification of AS and lncRNAs

The transcripts gtf file raw.gtf was first obtained. Then, the AS events were predicted using SUPPA v2.3 with default parameters [27]. lncRNAs are non-coding transcripts longer than 200 bp that regulate gene expression in various forms. lncRNAs in PPDP were identified by evaluating the coding potential of transcripts using CNCI v2 [28], CPC2 v 0.1 [29], PLEK v1.2 [30], and Pfam. Finally, transcripts at any two intersections of the prediction results of CNCI, CPC2, PLEK, and Pfam were used as candidate lncRNAs.

2.6. Prediction and Construction of ceRNA Network

In plants, miRNA genes are a class of highly conserved gene families. As no whole genome of *P. polyphylla* is available, *Arabidopsis thaliana* mature miRNAs from the PmiREN database [31] were used to predict the ceRNA network. The psRNATarget web server was used to predict targets [32]. The ceRNA score was calculated with our python script to determine whether lncRNA–mRNA pairs were ceRNAs (p-value < 0.05) [33]. The intersection was taken to obtain the final predicted ceRNA regulatory relationship pair and the ceRNA network was acquired. Then the ceRNA network was constructed using Cytoscape v3.9.1 [34].

2.7. Explore the Biosynthesis Pathway of Secondary Metabolites in P. polyphylla

The following biosynthesis pathways, including "fatty acid elongation", "steroid biosynthesis", "starch and sucrose metabolism", "terpenoid backbone biosynthesis", "phenylpropanoid biosynthesis", and "flavonoid biosynthesis" in *P. polyphylla* were explored by defining *A. thaliana* as a template. Protein sequences of *A. thaliana* were obtained from TAIR [35] and gene accession numbers of the above-mentioned pathways were obtained from the KEGG PATHWAY Database. Then, the biosynthetic genes from the pathways of *P. polyphylla* were acquired by BLAST with the *A. thaliana* sequences and data filtration with an E-value threshold of 1×10^{-5}.

2.8. Prediction of Transcription Factor and CYP&UGT

The TF families were identified by mapping the protein sequences to the database PlantTFDB 5.0 [36]. As CYPs and UGTs play very important roles in secondary metabolism, they are also critical in polyphyllin biosynthesis. A total of 288 CYP protein sequences from cytochrome P450 (http://drnelson.uthsc.edu/CytochromeP450.html (accessed on 21 March 2022)); and 122 UGT protein sequences from *Arabidopsis thaliana* cytochromes P450, cytochromes b_5, NADPH-cytochrome P450 reductases, and β-Glucosidases site (http://www.p450.kvl.dk/UGT.shtml (accessed on 21 March 2022)) were used to search for homologous CYPs and UGTs in *P. polyphylla* (E-value < 1×10^{-5}). Hidden Markov models (accession numbers: PF00067 and PF00201) were obtained from the Pfam [37] based on the evolutionary conservation of gene family domains. They were used as the queries to search against the protein sequences of *P. polyphylla* using Hmmsearch v3.3.2 [38]. The protein sequences of screened complete ORFs were further submitted to NCBI–CDD and Pfam to confirm the domain. Finally, candidate CYPs and UGTs were obtained.

2.9. Construction of Co-Expression Network and PPI Network

Quantification of the merged transcripts was achieved by RSEM [39] using the Illumina sequencing data. The co-expression between two genes was estimated by using the traditional Pearson correlation coefficient. Gene pairs with correlation values > 0.8 and adjusted p values < 0.01 were considered to show co-expression [40]. The protein interactions of *A. thaliana* were collected from three databases, namely AtPID v5.0 [41], AtPIN v9.0 [42], and PAIR v3.0 [43] and the literature [44–46]. A total of 18,037 *A. thaliana* genes and 241,468 interactions were obtained. Orthologous groups between the *P. polyphylla* and *A. thaliana* were detected using InParanoid v4.2 [47] with default parameters. The PPI network in *P. polyphylla* was inferred from the *A. thaliana* PPI network by homology mapping (Figure 1B).

2.10. Development of the Morphological Identification

First of all, the description of morphological characteristics of different species of the genus *Paris*, various specimen information of *Paris* in the CHV China Digital Herbarium (https://primulaworld.blogspot.com/2015/12/the-chinese-virtual-herbarium-cvh.html (accessed on 2 May 2020)), expert opinions, and the morphological classification feature data of *Paris* were collected. Then, the random forest classifier [48] was used to train the model of the preprocessed data after comparison and screening. Finally, the trained model was encapsulated and saved for use [49].

2.11. Implementation of PPDP

PPDP was developed using Django web framework (http://www.djangoproject.com/ (accessed on 23 February 2021)) based on Python programming. The website runs on a Linux Server. The Nginx web server (https://www.nginx.com/ (accessed on 10 March 2021)) and SQLite (https://www.sqlite.org/ (accessed on 20 March 2021)) were used as the web server and database server, respectively. The web front end was developed using Bootstrap (https://www.ubuntu.com/ (accessed on 30 March 2021)) and the Semantic UI framework (https://semantic-ui.com/ (accessed on 15 April 2021)). JavaScript libraries, highchart.js (https://www.highcharts.com/ (accessed on 10 May 2022)), and datatables (https://datatables.net/ (accessed on 3 April 2021)) were applied for rendering interactive graphs and tables. The network visualization was implemented using Cytoscape.js v.3.3 [50]. The online BLAST server was built with Django-blastplus (https://pypi.python.org/pypi/django-blastplus/ (accessed on 20 May 2021)) (Figure 1C). Finally, PPDP provides a series of user-friendly functions, such as browse, download, BLAST, heatmap, network, enrichment, pathway search, TF family, etc. (Figure 1D).

3. Results

3.1. The Web Interface of PPDP

3.1.1. Browse PPDP

All the genes were listed on the browse page and the detailed information page can be accessed by clicking the gene ID. For each gene in the browse, comprehensive information was provided, including multiple gene functional annotations, AS events, heatmaps of gene expression, and co-expression sub-networks. The number of co-expressed genes displayed is optional in the gene expression section.

3.1.2. Visualization of Heatmap, Network, and Enrichment

The 'Enrichment' module provides users with GO and KEGG enrichment for genes of interest (Figure 2B). The expression patterns of genes at different growth periods are available in the 'Heatmap' module (Figure 2C). The 'Network' module enables to extract a sub-network of user-specified genes from the global co-expression network (Figure 2D). Additionally, the 'BLAST' is a user-friendly tool for interacting with the NCBI BLAST+ toolkits (Figure 2A).

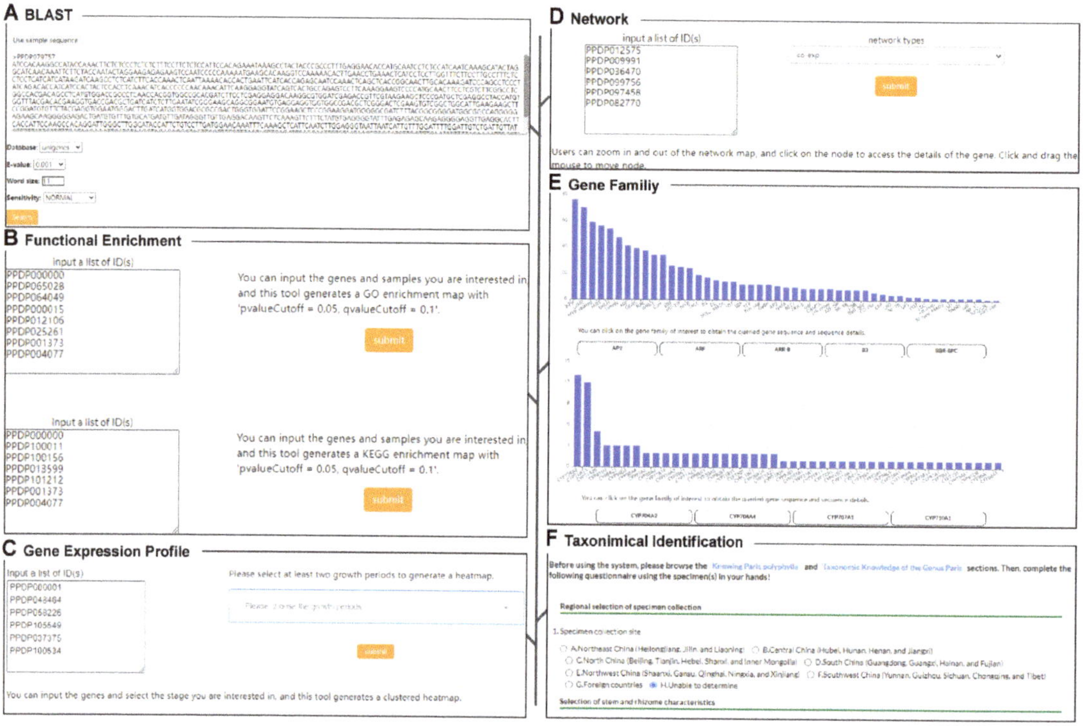

Figure 2. Web interface of PPDP. (**A**) Finding homologous genes of *P. polyphylla* using sequence similarity search with BLAST. (**B**) KEGG and GO enrichment of genes of interest. (**C**) Viewing gene expression profiles at different growth stages. (**D**) Prediction of co-expression network and PPI-network. (**E**) TF and CYP gene family statistics and the specific gene sequences. (**F**) Morphological identification of *P. polyphylla* and other *Paris* species.

3.1.3. Pathway Search

Plant secondary metabolites play an important role in plant growth and development, biotic and abiotic stresses, and mediating interactions with other organisms [51–53]. Particularly, specialized secondary metabolites of medicinal plants are responsible for medicinal activity. A total of 61, 36, 25, 171, 36, and 128 genes involved in terpenoid backbone biosynthesis, steroid biosynthesis, flavonoid biosynthesis, starch and sucrose metabolism, fatty acid elongation, and phenylpropanoid biosynthesis in *A. thaliana* were obtained. Correspondingly, a total of 38, 21, 12, 99, 16, and 49 genes of these pathways in *P. polyphylla* were acquired through BLAST. The reference pathway map, related biosynthetic genes in *A. thaliana* and *P. polyphylla*, and their corresponding KEGG enzymes are presented by selecting a wanted pathway.

3.1.4. TF, CYP, and UGT Family

In the 'TF family' module, the distribution of TF families is represented as a bar chart. The TF family of interest can be clicked to obtain the queried sequences and each sequence is linked to its detail page. The 'CYP & UGT' module demonstrates the CYP and UGT gene families in the same way (Figure 2E).

3.1.5. Other Data Resource

An abundance of *P. polyphylla* datasets can be freely downloaded through the 'Downloads' section, including transcript sequences, CDS sequences, lncRNA sequences, and annotations. The 'Resource' section provides valid SSR markers and chloroplast genomes of *P. polyphylla* and other *Paris* species. In the 'Identification System' section, users can taxonomically identify their samples by the morphological characters of *Paris* specimens collected, such as features of roots, stems, leaves, sepals, flowers, and fruits (Figure 2F). Moreover, pictures of the taxonomic characteristics are supplied for useful reference.

3.2. Database Statistics and Use Case

3.2.1. General Properties of PacBio Sequencing Data and Evaluation of Transcriptome

A total of 110,666 HQ isoforms and 824 LQ isoforms were acquired, and 36,812 transcripts were attained after clustering and de-redundancy. A total of 36,504 high-quality non-redundant transcripts were defined in *P. polyphylla* after filtering transcripts < 200 bp. The average length of the transcriptome determined in this study was 2008 bp, with an N50 of 2330 bp and a GC content of 45.88% (Table 1).

Table 1. Summary of the transcriptome of *P. polyphylla* var. *yunnanensis*.

Item	Value
High-quality isoforms	110,666
Low-quality isoforms	824
Non-reductant transcripts	36,812
High-quality non-redundant transcripts	36,504
Total bases (bp)	73,313,381
N50 (bp)	2330
Average length (bp)	2008
Reads mapping (%)	86.80
Percent GC (%)	45.88

According to BUSCO assessment results, the merged transcriptome has the most complete sequences and single-copy complete sequences, followed by the Illumina-based transcriptome (Figure 3A). In addition, 82.50% of the predicted genes from the merged transcriptome indicated relatively high completeness of the merged transcriptome assembly. Furthermore, 86.8% of the Illumina reads were mapped to the newly merged transcriptome, demonstrating a high level of transcriptome coverage. Because it was of higher quality, it was used for subsequent analyses.

3.2.2. Data Summary

Approximately 90% of the genes were annotated to at least one of the seven public databases. Specifically, 33,100, 29,249, 17,926, 25,855, 26,493, 19,907, and 29,696 transcripts were annotated to NR, GO, KEGG, Pfam, Swiss-prot, Kog, and eggNOG, accounting for 90.67%, 80.1%, 49.1%, 70.8%, 72.6%, 54.53%, and 81.3% of the total transcripts, respectively (Figure 3B). A total of 7029 transcripts had multiple types AS. Among them, those with retained introns (RI) had the largest number (3230), accounting for 46% (Figure 3C). The result is consistent with previous reports indicating that RI is the most frequent type of AS events in plants [54,55]. Additionally, a total of 931 candidate genes from 50 TF families were found (Figure 3D). The top five TF families are bZIP, C2H2, MYB_related,

FAR1, and bHLH. Moreover, 100,201 interactions between miRNAs, lncRNAs, and mRNAs (157 miRNAs, 879 lncRNAs, and 4922 mRNAs) were identified.

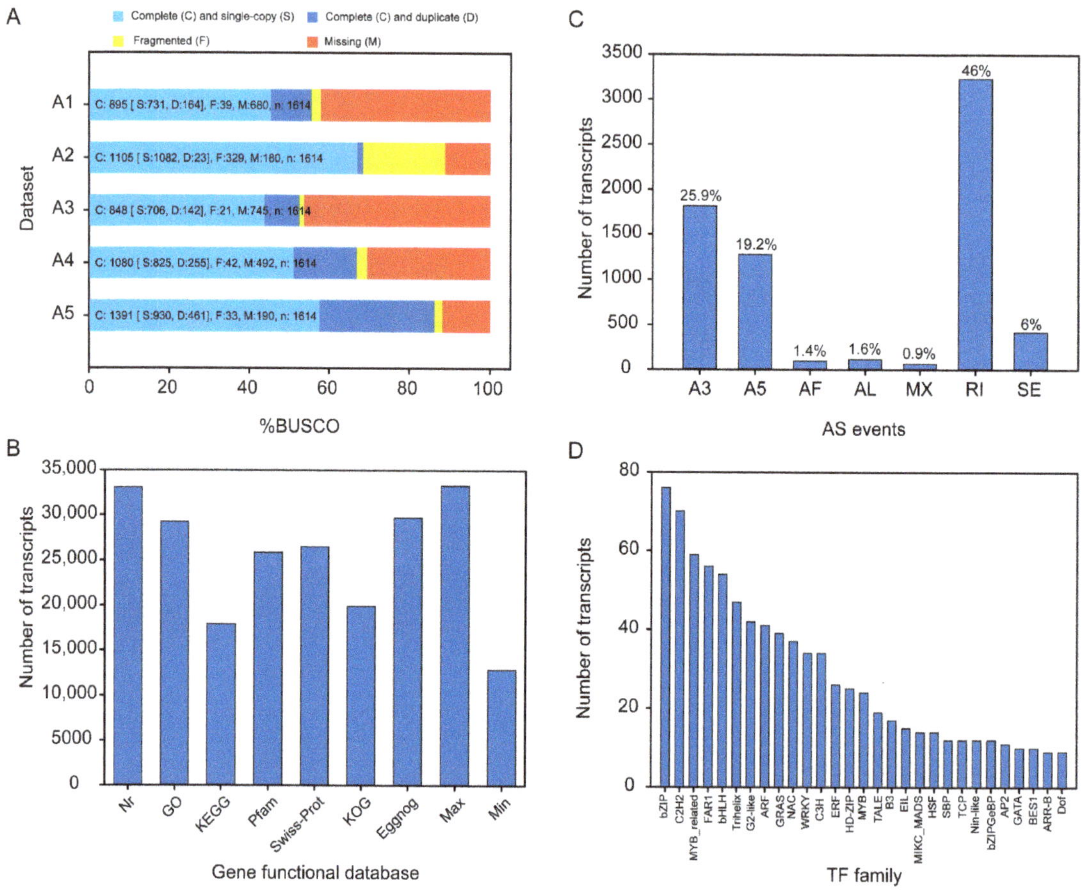

Figure 3. Transcriptome evaluation and annotation. (**A**) BUSCO assessment results for five transcriptomes with different datasets and assembly strategies. A1: PacBio-based RNA-Seq data from public database; A2: Illumina-based RNA-Seq data from our previous work; A3: PacBio-based RNA-Seq data from roots and stems; A4: PacBio-based RNA-Seq data from flowers and leaves; A5: merging all PacBio-based RNA-Seq data; C: complete orthologs; S: single copy orthologs; D: duplicated orthologs; F: fragmented orthologs; M: missing orthologs. (**B**) Statistics of functional annotations. (**C**) Classification of the AS events. A3: alternative 3′ splice sites; A5: alternative 5′ splice sites; AF: alternative first exon; AF: alternative last exon; MX: mutually exclusive exons; RI: retained introns; SE: skipping exon events. (**D**) The top 30 TF families identified in *P. polyphylla*.

3.2.3. Comparison of lncRNAs and mRNAs

A total of 9020 lncRNAs were identified in the *P. polyphylla* transcriptome using CPC2, CNCI, PLEK, and Pfam (Figure 4A). According to the length distribution, the average length of mRNAs (2008 bp) was longer than that of lncRNAs (1613 bp) (Figure 4C). Compared to mRNAs, lncRNAs featured fewer isoforms and AS events (Figure 4B,D).

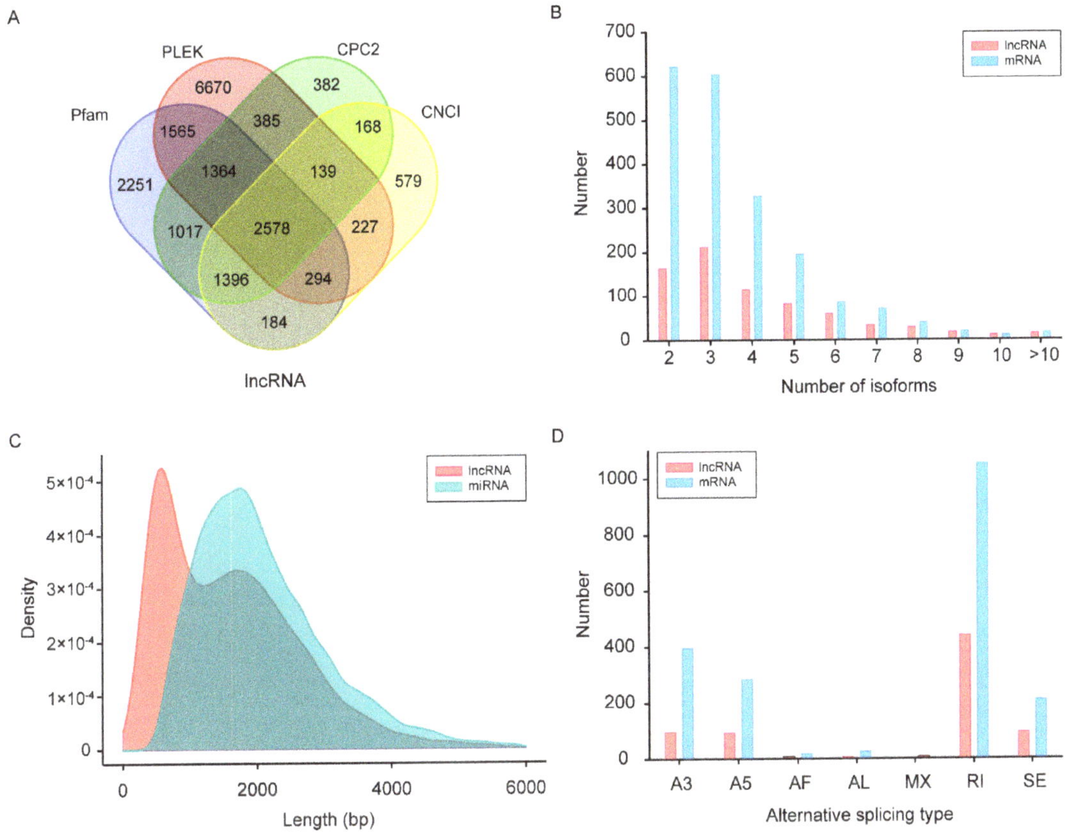

Figure 4. Distribution and features of lncRNA. (**A**) Venn diagram of predicted lncRNAs. (**B**) Distribution of isoform numbers of lncRNA and mRNA. (**C**) Density distribution of lncRNA and mRNA length. (**D**) Distribution of AS events of lncRNA and mRNA.

3.2.4. Exploration of the Polyphyllin Biosynthesis Pathway

A total of 186 candidate genes involved in the upstream of the polyphyllin biosynthesis pathway were identified (Figure 5A). These genes participate in two branch pathways of plant saponin biosynthesis: the cytosolic mevalonate (MVA) pathway and the plastidial 2-C-methyl-D-erythritol 4-phosphate (MEP) pathway. Quantitatively, more genes were related to the MVA pathway than to the MEP pathway. The transcriptional regulation of polyphyllin biosynthesis was also predicted in this study. It showed that a total of 432 TF genes (Figure 5B) may participate in the regulation of polyphyllin biosynthesis (Supplementary Materials Figure S1).

3.2.5. Data Related to Germplasm Resources

A total of 25,767 potential SSRs were identified from 36,504 transcripts. Among them, dinucleotides (46.75%) were the most abundant, followed by mononucleotides (39.45%) and trinucleotides (12.14%). Moreover, 4027 compound SSRs were identified (Table 2). After functional annotation, 16,120 SSR-containing transcripts had 12,626, 7678, 11,106, 11,465, and 14,412 homologous sequences in GO, KEGG, Pfam, Swiss-prot, and NR, respectively. In total, 186 PBGs were identified from the transcriptome, of which 97 contained SSRs. Additionally, valid SSR markers and the related primers of *P. polyphylla* were collected from reference. There were 14 and 35 valid SSR markers for *P. polyphylla* var. *chinensis* [56] and

P. polyphylla var. *yunnanensis* [57–59], respectively. Similarly, *P. polyphylla* var. *chinensis* is also considered as a species, named *Paris chinensis* Franch. in the monograph [4]. As the chloroplast genome is an ideal system for plant phylogeny study and an efficient marker source in germplasm resource investigation, chloroplast genomes of the *Paris* species were also collected from the public database. A total of 108 chloroplast genomes of *P. polyphylla* and other Paris species have been added in PPDP.

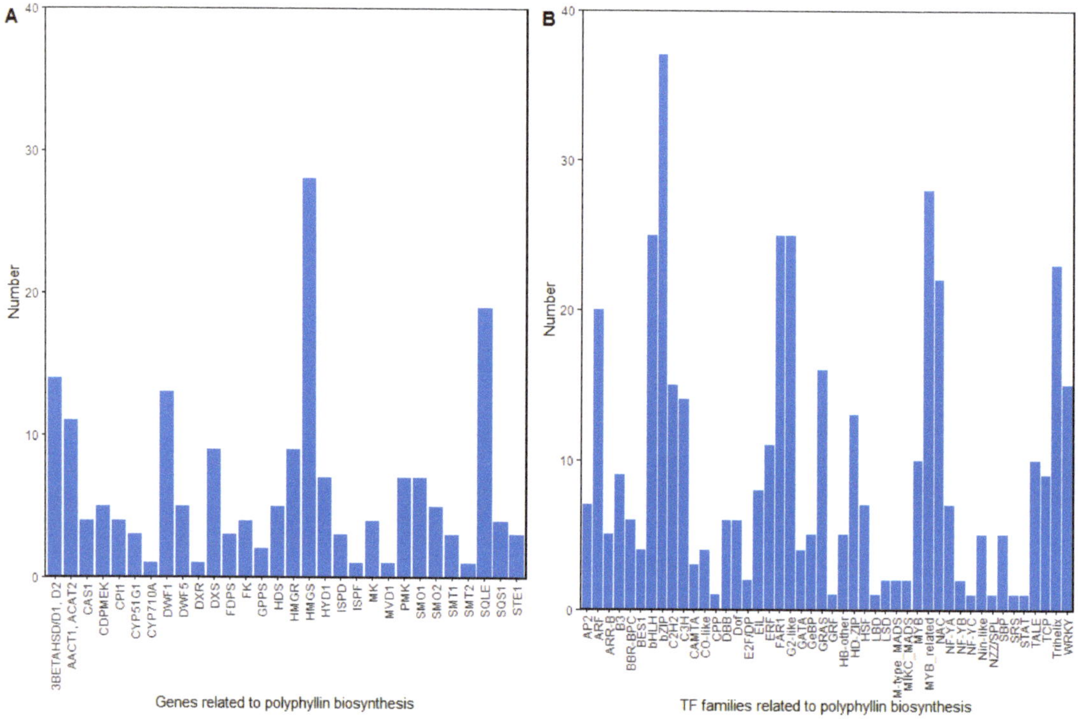

Figure 5. Genes and TF genes involved in polyphyllin biosynthesis. (**A**) Putative polyphyllin biosynthetic genes. (**B**) Putative TF genes related to the biosynthesis.

Table 2. Statistics of SSRs from the transcriptome.

Item	Value
Total number of sequences examined	36,504
Total size of examined sequences (bp)	73,313,381
Total number of identified SSRs	25,767
SSR-containing sequences	16,120
sequences containing more than 1 SSR	6169
SSRs present in compound formation	4027
Mononucleotide repeats	10,164
Dinucleotide repeats	12,045
Trinucleotide repeats	3129
Tetranucleotide repeats	88
Pentanucleotide repeats	71
Hexanucleotide repeats	270

3.2.6. A Case Study of Mining Genes Related to Polyphyllin Synthesis Using PPDP

PPDP is a comprehensive data platform with molecular data and analytical tools, such as BLAST, network construction, enrichment analysis, etc. Here, we demonstrate how to mine genes involved in the polyphyllin biosynthesis pathway. Polyphyllins belong to steroidal saponins, which share a similar biosynthesis backbone to that of terpenoids. From KEGG, 61 terpenoid backbone biosynthesis-related genes of *A. thaliana* were obtained, and their corresponding sequences from TAIR. Then, 38 optimal polyphyllin biosynthesis-related gene candidates in *P. polyphylla* were identified using PPDP BLAST. The expression profiles of these genes in different tissues at different growth stages can be viewed using the 'Heatmap' function (Figure 6A,B). Apparently, polyphyllin biosynthesis-related genes are expressed differentially in tissues across growth stages, especially PPDP043478, PPDP051101, and PPDP057972, which are highly expressed in leaves at pollination and fruit stages, and also in the stem at the pollination stage. A sub-network of polyphyllin backbone biosynthesis can be constructed using the 'Network Construction', indicating the existence of connections between polyphyllin biosynthesis-related genes in *P. polyphylla* (Figure 6C). Additionally, these candidate genes can be examined using 'GO Enrichment'. As expected, there was significant enrichment related to polyphyllin biosynthesis. The most abundant GO entries are isoprenoid biosynthesis processes, followed by sterol biosynthesis and terpenoid biosynthesis processes (Figure 6D). The ceRNA network of 61 genes is queried using the 'ceRNA network'. Only PPDP057217 has eight lncRNAs as its ceRNAs, indicating that these eight lncRNAs may be involved in polyphyllin backbone biosynthesis in *P. polyphylla*. (Figure 6E). Finally, the SSRs distribution of the related genes can be retrieved using the 'SSR search'. The 16 SSRs retrieved have no tetranucleotide and pentanucleotide repeats; however, they are very uniformly distributed in type mononucleotide, dinucleotides, trinucleotides, and compound repeats, which are four, five, three, and four, respectively.

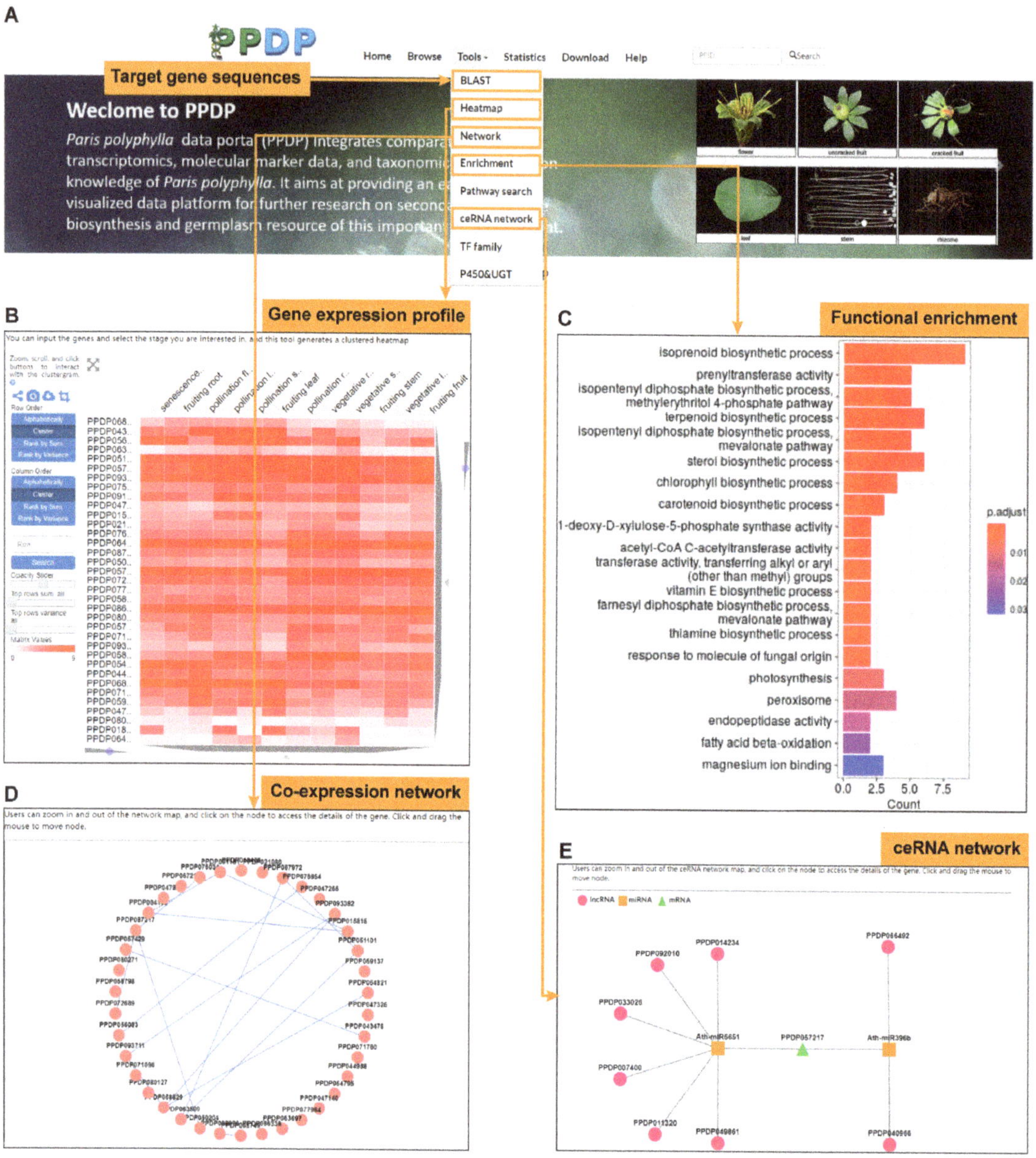

Figure 6. Case study: gene function prediction using PPDP tools. (**A**) Interface of tools in PPDP. (**B**) Heatmap of polyphyllin backbone biosynthesis-related genes in *P. polyphylla*. (**C**) Sub-network of predicted genes related to the biosynthesis. (**D**) GO enrichment of predicted genes related to biosynthesis. (**E**) CeRNA network of polyphyllin backbone biosynthesis-related gene candidates. Triangles represent miRNAs, squares represent mRNAs, and circles represent lncRNAs.

4. Discussion

Databases typically consolidate redundant-rich data, and provide many useful analytical tools [60]. Therefore, databases can greatly facilitate the research of various growth and development processes, genetic research and breeding, and secondary metabolite synthesis pathways of plants. The *Lonicera japonica* functional genomics database (LjaFGD) includes a *Lonicera japonica* genome and 77 sets of transcriptomes, converging multiple tools for the purpose of gene functional analysis and mining of *Lonicera japonica* [61]. GinkgoDB is a comprehensive database with multi-dimensional research resources of ginkgo, which contains two versions of genomes, expression profiles, distribution information, monitoring data, and morphological photos [62]. It assists the research, development, and conservation of the entire community of ginkgo. The Citrus Pan-Genome to Breeding Database (CPBD) covers 23 genomes of 17 citrus species, 4038 sets of transcriptomes of 13 horticultural species, variations of 167 citrus resource materials, and DNA methylome of 44 citrus samples at different tissues and developmental stages [63]. Practical analysis tools are also provided in CPBD, including gene search, BLAST, gene ID conversion, KEGG/GO enrichment, CRISPR design, and genome-wide association analysis (GWAS). PPDP currently lacks phenotype–genotype association data and genomic data. It is because that GWAS/QTL studies in *P. polyphylla* have not been carried out and the large genome size of *P. polyphylla* is unavailable at present. Therefore, we need continued efforts in this area.

So far, PPDP is the first database for the *Paris* plants. Based on PacBio SMRT-based RNA-Seq and Illumina-based RNA-Seq data, PPDP contains a variety of datasets, including 36,504 genes, 9020 lncRNAs, 7029 AS events, 25,767 SSRs and 49 valid SSRs, 108 chloroplast genomes, 931 genes from 50 TF families, 93 CYP and 56 UGT gene candidates, and gene expression profiles at different growth stages. PPDP provides a range of practical analytical tools, such as BLAST online, heatmap drawing, network construction, and pathway search. In conclusion, PPDP provides new molecular data and contributes to further understanding of polyphyllin biosynthesis and germplasm resource research.

In the future, as related research increases and advances, PPDP will continue to be upgraded to collect more data and provide more functions. Our goal is that PPDP can facilitate more aspects of research on *P. polyphylla*.

5. Conclusions

PPDP is a user-friendly data platform that integrates functional genomic analyses, molecular data resources, and morphological identification of *P. polyphylla*. Currently, the analytical tools of PPDP v1.0 are mainly based on transcriptome data. Notably, it offers a kit of necessary tools for functional analyses of databases, including co-expression, PPI network prediction, heatmap, functional enrichment, pathway search, online BLAST, and other general tools. PPDP can contribute to functional genomic analyses and germplasm resource research on *P. polyphylla*, and it will help research of polyphyllin biosynthesis and regulation. In the future, we will add more molecular biology data to progressively improve the database.

Supplementary Materials: The following supporting information can be downloaded at: https://www.mdpi.com/article/10.3390/d14121057/s1, Figure S1: co-expression network of polyphyllins biosynthesis genes (PBGs) and the related TF genes. The blue block indicates the PBG and the yellow block indicates TF gene; Table S1: characteristics of the transcriptomic data in this study; Table S2: information on valid SSRs collected.

Author Contributions: Conceptualization, C.L. and X.G.; software, Q.S., X.Z. and Q.R.; formal analysis, Q.S.; investigation, Q.S. and X.G.; resources, X.Z. and W.Y.; data curation, J.L.; writing—original draft preparation, Q.S.; writing—review and editing, X.G. and C.L.; visualization, Q.S. and X.G. All authors have read and agreed to the published version of the manuscript.

Funding: This research was funded by the National Natural Science Foundation of China (Grant No. 31970609, 31800273); 'Crop Varietal Improvement and Insect Pests Control by Nuclear Radiation'; Yunnan Fundamental Research Projects (Grant No. 202001AT070114); Startup Fund from Xishuangbanna Tropical Botanical Garden; 'Top Talents Program in Science and Technology' from Yunnan Province. The funders had no role in the study design, data collection, analysis and interpretation, or preparation of the manuscript. Publication costs are funded by 'Crop Varietal Improvement and Insect Pests Control by Nuclear Radiation'.

Institutional Review Board Statement: Not applicable.

Data Availability Statement: The data presented in this study are available in the NCBI database with accession number PRJNA893196.

Acknowledgments: We are very grateful to Heng Li (Kunming Institute of Botany, Chinese Academy of Sciences) for her taxonomic consultation. We thank the following people for their kind help in this study: Xiaoke Jiang and Jiazhi Liu (Xishuangbanna Tropical Botanical Garden, Chinese Academy of Sciences). We thank the Institutional Center for Shared Technologies and Facilities of Xishuangbanna Tropical Botanical Garden, CAS for providing the computer resources and technical support.

Conflicts of Interest: The authors declare no conflict of interest.

References

1. Li, H. *The Genus Paris Plants*; Science Press: Beijing, China, 2008; pp. 1–36.
2. Chase, M.W.; Christenhusz, M.J.M.; Fay, M.F.; Byng, J.W.; Judd, W.S.; Soltis, D.E.; Mabberley, D.J.; Sennikov, A.N.; Soltis, P.S.; Stevens, P.F. An Update of the Angiosperm Phylogeny Group Classification for the Orders and Families of Flowering Plants: APG IV. *Bot. J. Linn. Soc.* **2016**, *181*, 1–20.
3. Committee of National Pharmacopoeia. *Pharmacopoeia of the People's Republic of China (I)*; China Medical Science and Technology Press: Beijing, China, 2020.
4. Ji, Y. *A Monograph of Paris (Melanthiaceae): 1–203*; Science Press: Beijing, China, 2021; pp. 1–203.
5. Liu, Z.; Li, N.; Gao, W.; Man, S.; Yin, S.; Liu, C. Comparative study on hemostatic, cytotoxic and hemolytic activities of different species of *Paris* L. *J. Ethnopharmacol.* **2012**, *142*, 789–794. [CrossRef] [PubMed]
6. Wei, J.-C.; Gao, W.-Y.; Yan, X.-D.; Wang, Y.; Jing, S.-S.; Xiao, P.-G. Chemical Constituents of Plants from the Genus *Paris*. *Chem. Biodivers.* **2014**, *11*, 1277–1297. [CrossRef] [PubMed]
7. Wu, X.; Wang, L.; Wang, G.-C.; Wang, H.; Dai, Y.; Yang, X.-X.; Ye, W.-C.; Li, Y.-L. Triterpenoid saponins from rhizomes of *Paris polyphylla* var. *yunnanensis*. *Carbohydr. Res.* **2013**, *368*, 1–7. [CrossRef]
8. Xin, G.; Ruoshi, L.; Duan, B.; Ying, W.; Min, F.A.N.; Shuang, W.; Zhang, H.; Xia, C. Advances in research on chemical constituents and pharmacological effects of *Paris* genus and prediction and analysis of quality markers. *Chin. Tradit. Herb. Drugs* **2019**, *50*, 4838–4852.
9. Cunningham, A.; Brinckmann, J.; Bi, Y.-F.; Pei, S.-J.; Schippmann, U.; Luo, P. *Paris* in the spring: A review of the trade, conservation and opportunities in the shift from wild harvest to cultivation of *Paris polyphylla* (Trilliaceae). *J. Ethnopharmacol.* **2018**, *222*, 208–216. [CrossRef]
10. Liu, T.; Li, X.; Xie, S.; Wang, L.; Yang, S. RNA-seq analysis of *Paris polyphylla* var. *yunnanensis* roots identified candidate genes for saponin synthesis. *Plant Divers.* **2016**, *38*, 163–170. [CrossRef]
11. Li, B.; Peng, L.; Sun, X.; Huang, W.; Wang, N.; He, Y.; Shi, X.; Liu, Y.; Zhang, P.; Yang, X.; et al. Organ-specific transcriptome sequencing and mining of genes involved in polyphyllin biosynthesis in *Paris polyphylla*. *Ind. Crop. Prod.* **2020**, *156*, 112775. [CrossRef]
12. Yang, Z.; Yang, L.; Liu, C.; Qin, X.; Liu, H.; Chen, J.; Ji, Y. Transcriptome analyses of *Paris polyphylla* var. *chinensis*, *Ypsilandra thibetica*, and *Polygonatum kingianum* characterize their steroidal saponin biosynthesis pathway. *Fitoterapia* **2019**, *135*, 52–63. [CrossRef]
13. Christ, B.; Xu, C.; Xu, M.; Li, F.-S.; Wada, N.; Mitchell, A.J.; Han, X.-L.; Wen, M.-L.; Fujita, M.; Weng, J.-K. Repeated evolution of cytochrome P450-mediated spiroketal steroid biosynthesis in plants. *Nat. Commun.* **2019**, *10*, 1–11. [CrossRef]
14. Gao, X.; Su, Q.; Li, J.; Yang, W.; Yao, B.; Guo, J.; Li, S.; Liu, C. RNA-Seq analysis reveals the important co-expressed genes associated with polyphyllin biosynthesis during the developmental stages of *Paris polyphylla*. *BMC Genom.* **2022**, *23*, 1–17. [CrossRef]
15. Gao, X.; Zhang, X.; Chen, W.; Li, J.; Yang, W.; Zhang, X.; Li, S.; Liu, C. Transcriptome analysis of *Paris polyphylla* var. *yunnanensis* illuminates the biosynthesis and accumulation of steroidal saponins in rhizomes and leaves. *Phytochemistry* **2020**, *178*, 112460. [CrossRef] [PubMed]
16. Jayakodi, M.; Choi, B.-S.; Lee, S.-C.; Kim, N.-H.; Park, J.Y.; Jang, W.; Lakshmanan, M.; Mohan, S.V.G.; Lee, D.-Y.; Yang, T.-J. Ginseng Genome Database: An open-access platform for genomics of *Panax ginseng*. *BMC Plant Biol.* **2018**, *18*, 62. [CrossRef] [PubMed]
17. Fan, H.; Li, K.; Yao, F.; Sun, L.; Liu, Y. Comparative transcriptome analyses on terpenoids metabolism in field- and mountain-cultivated ginseng roots. *BMC Plant Biol.* **2019**, *19*, 82. [CrossRef]
18. Tien, N.Q.D.; Ma, X.; Man, L.Q.; Chi, D.T.K.; Huy, N.X.; Nhut, D.-T.; Rombauts, S.; Ut, T.; Loc, N.H. De novo whole-genome assembly and discovery of genes involved in triterpenoid saponin biosynthesis of *Vietnamese ginseng* (Panax vietnamensis Ha et Grushv.). *Physiol. Mol. Biol. Plants* **2021**, *27*, 2215–2229. [CrossRef] [PubMed]
19. Yin, Y.; Gao, L.; Zhang, X.; Gao, W. A cytochrome P450 monooxygenase responsible for the C-22 hydroxylation step in the *Paris polyphylla* steroidal saponin biosynthesis pathway. *Phytochemistry* **2018**, *156*, 116–123. [CrossRef]
20. Qi, J.J.; Zheng, N.; Zhang, B.; Sun, P.; Hu, S.N.; Xu, W.J.; Ma, Q.; Zhao, T.Z.; Zhou, L.L.; Qin, M.J.; et al. Mining genes involved in the stratification of *Paris Polyphylla* seeds using high-throughput embryo Transcriptome sequencing. *BMC Genom.* **2013**, *14*, 358. [CrossRef]
21. Hua, X.; Song, W.; Wang, K.; Yin, X.; Hao, C.; Duan, B.; Xu, Z.; Su, T.; Xue, Z. Effective prediction of biosynthetic pathway genes involved in bioactive polyphyllins in *Paris polyphylla*. *Commun. Biol.* **2022**, *5*, 50. [CrossRef]
22. Li, W.; Godzik, A. Cd-hit: A fast program for clustering and comparing large sets of protein or nucleotide sequences. *Bioinformatics* **2006**, *22*, 1658–1659. [CrossRef]

23. Simão, F.A.; Waterhouse, R.M.; Ioannidis, P.; Kriventseva, E.V.; Zdobnov, E.M. BUSCO: Assessing genome assembly and annotation completeness with single-copy orthologs. *Bioinformatics* **2015**, *31*, 3210–3212. [CrossRef]
24. Altschul, S.F.; Gish, W.; Miller, W.; Myers, E.W.; Lipman, D.J. Basic local alignment search tool. *J. Mol. Biol.* **1990**, *215*, 403–410. [CrossRef] [PubMed]
25. Cantalapiedra, C.P.; Hernández-Plaza, A.; Letunic, I.; Bork, P.; Huerta-Cepas, J. eggNOG-mapper v2: Functional Annotation, Orthology Assignments, and Domain Prediction at the Metagenomic Scale. *Mol. Biol. Evol.* **2021**, *38*, 5825–5829. [CrossRef] [PubMed]
26. Beier, S.; Thiel, T.; Münch, T.; Scholz, U.; Mascher, M. MISA-web: A web server for microsatellite prediction. *Bioinformatics* **2017**, *33*, 2583–2585. [CrossRef]
27. Trincado, J.L.; Entizne, J.C.; Hysenaj, G.; Singh, B.; Skalic, M.; Elliott, D.J.; Eyras, E. SUPPA2: Fast, accurate, and uncertainty-aware differential splicing analysis across multiple conditions. *Genome Biol.* **2018**, *19*, 40. [CrossRef] [PubMed]
28. Sun, L.; Luo, H.; Bu, D.; Zhao, G.; Yu, K.; Zhang, C.; Liu, Y.; Chen, R.; Zhao, Y. Utilizing sequence intrinsic composition to classify protein-coding and long non-coding transcripts. *Nucleic Acids Res.* **2013**, *41*, e166. [CrossRef]
29. Kang, Y.-J.; Yang, D.-C.; Kong, L.; Hou, M.; Meng, Y.-Q.; Wei, L.; Gao, G. CPC2: A fast and accurate coding potential calculator based on sequence intrinsic features. *Nucleic Acids Res.* **2017**, *45*, W12–W16. [CrossRef] [PubMed]
30. Li, A.; Zhang, J.; Zhou, Z. PLEK: A tool for predicting long non-coding RNAs and messenger RNAs based on an improved k-mer scheme. *BMC Bioinform.* **2014**, *15*, 311. [CrossRef]
31. Guo, Z.; Kuang, Z.; Wang, Y.; Zhao, Y.; Tao, Y.; Cheng, C.; Yang, J.; Lu, X.; Hao, C.; Wang, T.; et al. PmiREN: A comprehensive encyclopedia of plant miRNAs. *Nucleic Acids Res.* **2019**, *48*, D1114–D1121. [CrossRef]
32. Dai, X.; Zhuang, Z.; Zhao, P.X. psRNATarget: A plant small RNA target analysis server (2017 release). *Nucleic Acids Res.* **2018**, *46*, W49–W54. [CrossRef]
33. Li, J.H.; Liu, S.; Zhou, H.; Qu, L.-H.; Yang, J.-H. starBase v2. 0: Decoding miRNA-ceRNA, miRNA-ncRNA and protein–RNA interaction networks from large-scale CLIP-Seq data. *Nucleic Acids Res.* **2014**, *42*, D92–D97. [CrossRef]
34. Shannon, P.; Markiel, A.; Ozier, O.; Baliga, N.S.; Wang, J.T.; Ramage, D.; Amin, N.; Schwikowski, B.; Ideker, T. Cytoscape: A software environment for integrated models of Biomolecular Interaction Networks. *Genome Res.* **2003**, *13*, 2498–2504. [CrossRef] [PubMed]
35. Poole, R.L. The TAIR database. *Methods Mol. Biol.* **2007**, *406*, 179–212. [PubMed]
36. Guo, A.-Y.; Chen, X.; Gao, G.; Zhang, H.; Zhu, Q.-H.; Liu, X.-C.; Zhong, Y.-F.; Gu, X.; He, K.; Luo, J. PlantTFDB: A comprehensive plant transcription factor database. *Nucleic Acids Res.* **2007**, *36*, D966–D969. [CrossRef] [PubMed]
37. Finn, R.D.; Bateman, A.; Clements, J.; Coggill, P.; Eberhardt, R.Y.; Eddy, S.R.; Heger, A.; Hetherington, K.; Holm, L.; Mistry, J.; et al. Pfam: The protein families database. *Nucleic Acids Res.* **2013**, *42*, D222–D230. [CrossRef] [PubMed]
38. Finn, R.D.; Clements, J.; Eddy, S.R. HMMER web server: Interactive sequence similarity searching. *Nucleic Acids Res.* **2011**, *39*, W29–W37. [CrossRef]
39. Li, B.; Dewey, C.N. RSEM: Accurate transcript quantification from RNA-Seq data with or without a reference genome. *BMC Bioinform.* **2011**, *12*, 323. [CrossRef]
40. Chen, W.; Li, J.; Huang, S.; Li, X.; Zhang, X.; Hu, X.; Xiang, S.; Liu, C. GCEN: An Easy-to-Use Toolkit for Gene Co-Expression Network Analysis and lncRNAs Annotation. *Curr. Issues Mol. Biol.* **2022**, *44*, 1479–1487. [CrossRef]
41. Li, P.; Zang, W.; Li, Y.; Xu, F.; Wang, J.; Shi, T. AtPID: The overall hierarchical functional protein interaction network interface and analytic platform for *Arabidopsis*. *Nucleic Acids Res.* **2010**, *39*, D1130–D1133. [CrossRef]
42. Brandão, M.M.; Dantas, L.L.; Silva-Filho, M.C. AtPIN: *Arabidopsis* thaliana Protein Interaction Network. *BMC Bioinform.* **2009**, *10*, 454. [CrossRef]
43. Lin, M.; Shen, X.; Chen, X. PAIR: The predicted *Arabidopsis* interactome resource. *Nucleic Acids Res.* **2010**, *39*, D1134–D1140. [CrossRef]
44. Jones, A.M.; Xuan, Y.; Xu, M.; Wang, R.S.; Ho, C.H.; Lalonde, S.; You, C.H.; Sardi, M.I.; Parsa, S.A.; Smith-Valle, E.; et al. Border control—A membrane-linked interactome of *Arabidopsis*. *Science* **2014**, *344*, 711–716. [CrossRef]
45. *Arabidopsis* Interactome Mapping Consortium. Evidence for Network Evolution in an *Arabidopsis* Interactome Map. *Science* **2011**, *333*, 601–607. [CrossRef]
46. Mukhtar, M.S.; Carvunis, A.-R.; Dreze, M.; Epple, P.; Steinbrenner, J.; Moore, J.; Tasan, M.; Galli, M.; Hao, T.; Nishimura, M.T.; et al. Independently Evolved Virulence Effectors Converge onto Hubs in a Plant Immune System Network. *Science* **2011**, *333*, 596–601. [CrossRef] [PubMed]
47. O'Brien, K.P.; Remm, M.; Sonnhammer, E.L.L. Inparanoid: A comprehensive database of eukaryotic orthologs. *Nucleic Acids Res.* **2005**, *33*, D476–D480. [CrossRef] [PubMed]
48. Pedregosa, F.; Varoquaux, G.; Gramfort, A.; Michel, V.; Thirion, B.; Grisel, O.; Blondel, M.; Prettenhofer, P.; Weiss, R.; Dubourg, V.; et al. Scikit-learn: Machine Learning in Python. *J. Mach. Learn. Res.* **2011**, *12*, 2825–2830.
49. Ren, Q.; Sang, S.Y.; Liu, C.N. Design and implementation of *Paris* plants online classification and identification system. *Computer Era* **2020**, *9*, 72–75.
50. Franz, M.; Lopes, C.T.; Huck, G.; Dong, Y.; Sumer, O.; Bader, G.D. Cytoscape.js: A graph theory library for visualisation and analysis. *Bioinformatics* **2015**, *32*, 309–311. [CrossRef]
51. Izhaki, I. Emodin—A secondary metabolite with multiple ecological functions in higher plants. *New Phytol.* **2002**, *155*, 205–217. [CrossRef]
52. Khare, S.; Singh, N.B.; Singh, A.; Hussain, I.; Niharika, K.; Yadav, V.; Bano, C.; Yadav, R.K.; Amist, N. Plant secondary metabolites synthesis and their regulations under biotic and abiotic constraints. *J. Plant Biol.* **2020**, *63*, 203–216. [CrossRef]

53. Bartwal, A.; Mall, R.; Lohani, P.; Guru, S.K.; Arora, S. Role of Secondary Metabolites and Brassinosteroids in Plant Defense Against Environmental Stresses. *J. Plant Growth Regul.* **2012**, *32*, 216–232. [CrossRef]
54. Zhang, G.; Guo, G.; Hu, X.; Zhang, Y.; Li, Q.; Li, R.; Zhuang, R.; Lu, Z.; He, Z.; Fang, X.; et al. Deep RNA sequencing at single base-pair resolution reveals high complexity of the rice transcriptome. *Genome Res.* **2010**, *20*, 646–654. [CrossRef] [PubMed]
55. Shen, Y.; Zhou, Z.; Wang, Z.; Li, W.; Fang, C.; Wu, M.; Ma, Y.; Liu, T.; Kong, L.-A.; Peng, D.-L.; et al. Global Dissection of Alternative Splicing in Paleopolyploid Soybean. *Plant Cell* **2014**, *26*, 996–1008. [CrossRef]
56. Wang, H.M.; Ruan, C.J.; Liang, J.H.; Han, P.; Yan, R.; Dong, X.; Liu, L.Y.; Shuai, X.C.; Liu, H.H. Development of SSR Markers of Fanjingshan *Paris polyphylla* Smith var.*chinensis* Based on High Throughput RNA-seq. *Mol. Plant Breed.* **2019**, *18*, 6059–6065.
57. Gao, X.; Su, Q.; Yao, B.; Yang, W.; Ma, W.; Yang, B.; Liu, C. Development of EST-SSR Markers Related to Polyphyllin Biosynthesis Reveals Genetic Diversity and Population Structure in *Paris polyphylla*. *Diversity* **2022**, *14*, 589. [CrossRef]
58. Chen, Z.S.Z.; Tian, B.; Cai, C.T. Genetic diversity of *Paris polyphylla* var. *yunnanensis* by SSR marker. *Chin. Tradit. Herb. Drugs* **2017**, *9*, 1834–1838.
59. Yang, W.Z.; Xu, Z.L.; Yang, S.B.; Yang, M.Q.; Zuo, Y.M.; Yang, T.M.; Zhang, J.Y. Transferability Analysis of EST-SSR Marker of Three Plants to *Paris polyphylla* Smith var. *yunnanensis (Franch.) Hand Mazz. Southwest China J. Agric. Sci.* **2014**, *4*, 1686–1690.
60. Fayyad, U.; Piatetsky-Shapiro, G.; Smyth, P. From data mining to knowledge discovery in databases. *AI Mag.* **1996**, *17*, 37.
61. Xiao, Q.; Li, Z.; Qu, M.; Xu, W.; Su, Z.; Yang, J. LjaFGD: *Lonicera japonica* functional genomics database. *J. Integr. Plant Biol.* **2021**, *63*, 1422–1436. [CrossRef]
62. Gu, K.J.; Lin, C.F.; Wu, J.J.; Zhao, Y.P. GinkgoDB: An ecological genome database for the living fossil, *Ginkgo biloba*. *Database J. Biol. Databases Curation* **2022**, *2022*, baac046. [CrossRef]
63. Liu, H.; Wang, X.; Liu, S.; Huang, Y.; Guo, Y.-X.; Xie, W.-Z.; Liu, H.; Qamar, M.T.U.; Xu, Q.; Chen, L.-L. *Citrus* Pan-genome to Breeding Database (CPBD): A comprehensive genome database for citrus breeding. *Mol. Plant* **2022**, *15*, 1503–1505. [CrossRef]

diversity MDPI

Article

Geographic Patterns of the Richness and Density of Wild Orchids in Nature Reserves of Jiangxi, China

Qinghua Zhan [1,2,†], Yuelong Liang [3,†], Zhong Zhang [4,†], Feihu Liu [1,2], Liyang Li [1,2], Xiaodong Tang [1,2], Zhongxuan Liang [1,2], Weixing Chen [1,2], Mingtao Hu [1,2], Shaolin Tan [1,2], Huolin Luo [1,2], Yadong Zhou [1,2] and Boyun Yang [1,2,*]

1 School of Life Sciences, Nanchang University, Nanchang 330000, China
2 Key Laboratory of Plant Resources in Jiangxi Province, Nanchang 330000, China
3 Jiulianshan National Nature Reserve of Jiangxi, Ganzhou 341700, China
4 Jinggangshan National Nature Reserve of Jiangxi, Ji'an 343000, China
* Correspondence: yangboyun@163.com
† These authors contributed equally to this work.

Abstract: Orchids have attracted much attention from researchers, because of their richness of species and their great ornamental and medicinal value. Jiangxi Province, which is located in southeastern China and ringed on three sides by mountains, contains many nature reserves and harbors large number of orchids. Here, we conducted field surveys of orchids in 35 nature reserves in Jiangxi, using sampling lines and plots. We also analyzed the relationship between orchid richness and density with environmental variables and studied the relationship among these nature reserves. We found that the mountainous areas of southwestern, southern, and northeastern Jiangxi have a high richness and density of orchids, while the mountainous areas of central and northwestern Jiangxi have low richness and density. Jiulianshan and Jinggangshan are the two most rich-species reserves, with 58 and 55 orchids, respectively. Eight reserves (22% of those surveyed) had fewer than 10 orchids. Compared with soil, climate, and vegetation, topography was more closely related to the richness and density of orchids. Topographical variables explained 19% and 20% of the total variation in SR and SD, respectively. The result of hierarchical clustering analysis showed that the 35 nature reserves of Jiangxi obviously fall into two main clusters, which are separated by the Ganjiang River–Poyang Lake water system. In conclusion, the geographical patterns of richness and the density of orchids in Jiangxi are uneven and are affected by topography and vegetation, while their distribution is affected by the terrain of Jiangxi. Our work explains the richness and density patterns and the assembly mechanism of the orchids in Jiangxi and also provides new ideas for the protection of orchids in this region.

Keywords: Orchidaceae; investigation; diversity; mountains; Poyang Lake basin

Citation: Zhan, Q.; Liang, Y.; Zhang, Z.; Liu, F.; Li, L.; Tang, X.; Liang, Z.; Chen, W.; Hu, M.; Tan, S.; et al. Geographic Patterns of the Richness and Density of Wild Orchids in Nature Reserves of Jiangxi, China. *Diversity* **2022**, *14*, 855. https://doi.org/10.3390/d14100855

Academic Editors: Hong-Hu Meng and Yi-Gang Song

Received: 13 September 2022
Accepted: 1 October 2022
Published: 9 October 2022

1. Introduction

The family of orchids (Orchidaceae) is one of the largest families of flowering plants in the world, with approximately 750 genera and 28,000 species, and contains a large number of critically endangered species [1]. Orchids are widely distributed in various terrestrial ecosystems around the world, especially in tropical regions, except for polar and arid desert regions [1–3]. China is one of the countries with the richest diversity of orchids in the world [4]. Currently, there are approximately 1708 orchids recorded in China, belonging to 181 genera, 17 tribes, and 5 subfamilies [4].

At present, the understanding of biodiversity is still insufficient. Many new species are discovered and published every year [5]. Orchidaceae is one of the families with the richest discovery of new taxa by researchers [5–7]. For example, a total of 20 species, 1 hybrid, 6 varieties, and 1 form of Orchidaceae have been discovered in China in 2020 [6]. The main reasons may be that orchids are highly diverse, with small plants and a short flowering period, and most of them grow in an undisturbed environment, such as epiphytic orchids that grow on trees or rock walls. In the early stages, researchers had difficulties in obtaining complete species inventories.

Environmental variables affect the propagation and growth of plants and could determine the distribution of plant species to a great extent. Previous studies showed that the distribution of orchids was regulated by habitat size and elevation range at large scales, while it was regulated by light availability, soil moisture, and vegetation at small scales [8–11]. Research based on the literature, floras, specimens, and other empirical data showed that net primary productivity, elevation range, and temperature seasonality together explained 34.4% of variance in orchid richness in China [11]. However, research based on field survey data showed that vegetation type, elevation, aspect, slope, mean annual temperature, and precipitation could only explain 3.7% of the variation in the composition of wild orchid plants on Hainan Island, China [12].

Jiangxi is located in southeastern China, with its geographical range roughly overlapping the Poyang Lake basin [13]. There are more 190 nature reserves at different levels in Jiangxi, including 16 national and 39 provincial nature reserves. These nature reserves harbor a large number of wild orchids, but the field investigation for orchids in these reserves is incomplete [14]. For example, there are only 69 orchids of Jiangxi recorded in *Flora of China* [15]. However, the latest statistics showed that there are more than 100 orchids in Jiangxi [14], and this number has increased to 193 after several years of investigation by our team [16–19]. There have been a lot of new records and new taxa reported in recent years, such as *Danxiaorchis yangii* B.Y.Yang & Bo Li [20] and *Calanthe sieboldopsis* B.Y.Yang & Bo Li [21] from Jinggangshan National Nature Reserve. Here, based on the survey data from thousands of plots in 35 nature reserves of Jiangxi, we studied the orchid diversity in this region. Specifically, we intend to address the following questions: (1) What are the geographic patterns of the richness and density of orchids in the nature reserves of Jiangxi? (2) Which environment variables are most important in affecting the orchids in Jiangxi? (3) How are the mountain ranges of Jiangxi affecting the distribution of orchids in this region?

2. Materials and Methods

2.1. Study Area

Jiangxi is located in the southeastern China, bordered by Zhejiang and Fujian in the east, Guangdong in the south, Hunan in the west, and Hubei and Anhui in the north. Jiangxi is surrounded by mountains in the east, west, and south, and Poyang Lake plain in the north [13]. The main mountain ranges in the territory include Wuyi Mountains and Huaiyu Mountains in the east; Luoxiao Mountains, Mufu Mountains, and Jiuling Mountains in the west; and Dayu Mountains and Jiulian Mountains in the south (Figure 1). Ganjiang River flows through Jiangxi from south to north and flows into Poyang Lake. This water system divides the main mountain ranges of Jiangxi into east and west parts and may have geographical impact on the distribution of plants.

2.2. Field Surveys

Our field investigation in 35 nature reserves of Jiangxi lasted for two years, from 2021 to 2022. These reserves almost cover all the main mountains of Jiangxi (Figure 1). The places with rich orchids were chosen for setting up the sampling lines with a length greater than 1000 m; then, 1 to 20 5 m × 5 m plots were set up on both sides of the sampling lines. The criteria of setting plot are that plots should harbor orchids and distance between two plots is greater than 10 m. In total, more than 2400 plots were set up in the 35 nature reserves (Figure 1; Supplementary Material Table S1). Geographical coordinates and orchid species were recorded for each plot. In order to protect orchids, we did not collect specimens from every orchid that was encountered. We just took pictures for each orchid and collected samples and specimens for some species that cannot be identified in the wild. Samples and specimens were used to accurately identify those unknown species in the wild.

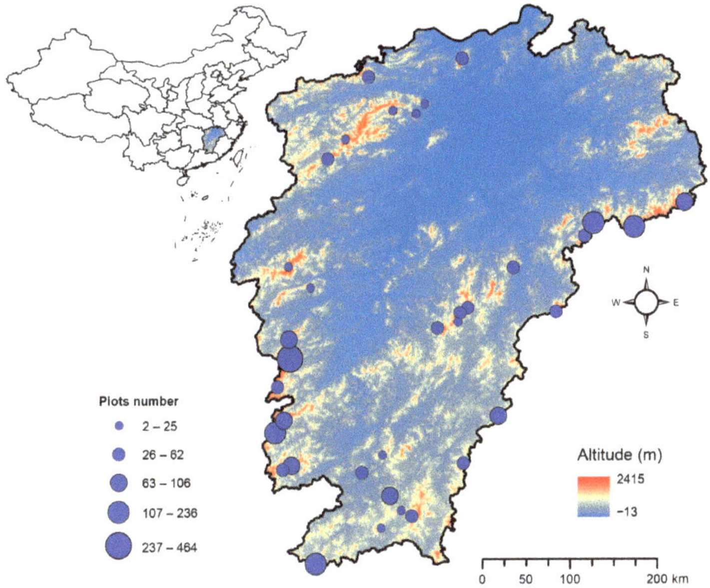

Figure 1. The locality of the 35 nature reserves of Jiangxi.

2.3. Environmental Variables

Ten environmental variables of these plots were used for analysis in this study, including topography (elevation and slope), vegetation (canopy density (CD) and normalized difference vegetation index (NDVI)), soil (pH, total nitrogen (TN), total phosphorus (TP), and total potassium (TK)), and climate (annual mean temperature (AMT) and annual precipitation (AP)). Elevation, slope, and CD were measured in the field, and other variables were extracted from raw data sets using ArcGIS 10.5 [22]. The raw data sets of NDVI during 2011 to 2020 were downloaded from Resource and Environment Science and Data Center (https://www.resdc.cn/, accessed on 20 August 2022), soil variables were downloaded from National Tibetan Plateau/Third Pole Environment Data Center (http://www.tpdc.ac.cn/, accessed on 20 August 2022) [23], and climatic variables were downloaded from WorldClim (http://www.worldclim.org/, accessed on 5 May 2021). The environmental variables of each nature reserve were calculated by the mean values of the variables among all the plots located in the nature reserve (Table S1).

2.4. Data Analysis

Based on the data matrix of orchids' present or absent in the 35 nature reserves of Jiangxi, we calculated the species richness (SR) of orchids in these reserves. There is always a positive correlation between SR and the area of nature reserves, so we also calculated the species density (SD) of orchids for each nature reserve using the formula $SD = SR/\ln(A)$ [24,25], where A is the area of each nature reserve. Linear regressions were used to explore the relationship between SR or SD and all the environmental variables. Variance partitioning analysis was used to examine the contribution of total, shared, and independent effects of the four groups of environmental variables on SR and SD of orchids. Before performing variance partitioning analysis, the correlations of pairwise variables were tested using Pearson correlation, and all these variables did not exhibit strong correlation ($|r| < 0.70$) [26]. In order to find out the relationships among nature reserves, we used Jaccard's index to calculate the taxonomic β-diversity of orchids and used Ward's method to carry out the hierarchical clustering analysis [27]. All the analyses were carried out using the R software (Version 3.3.3) [28].

3. Results

3.1. Richness and Density of Orchids in Nature Reserves of Jiangxi

In total, 125 orchids belonging to 55 genera were found by us in the 35 nature reserves of Jiangxi, including 85 terrestrial (including 10 mycoheterotrophic) and 40 epiphytic species (Table S2). Of these orchids, 50 species belonged to the least concern (LC) category of the IUCN Red List, 13 to the near threatened (NT) category, 17 to the vulnerable (VU) category, 11 to the endangered (EN) category, and 5 to the critically endangered (CR) category (Appendix S2). *Bulbophyllum* Thou. and *Calanthe* R. Br., both with 10 species, are the two most species-rich genera, while over half of these genera only have one species (Figure 2). Based on our field work, there were 23 new records of orchids in Jiangxi (Table S2). Reserves with high SR and SD of orchids are mainly distributed in the reserves of the mountain areas around Jiangxi, while these are low in the reserves of the mountain areas in central and northwestern Jiangxi (Figure 3). Jiulianshan and Jinggangshan are the two most species-rich reserves, with 58 and 55 orchids, respectively. Eight reserves (22% of those surveyed) had fewer than 10 orchids (Figure 3; Table S2).

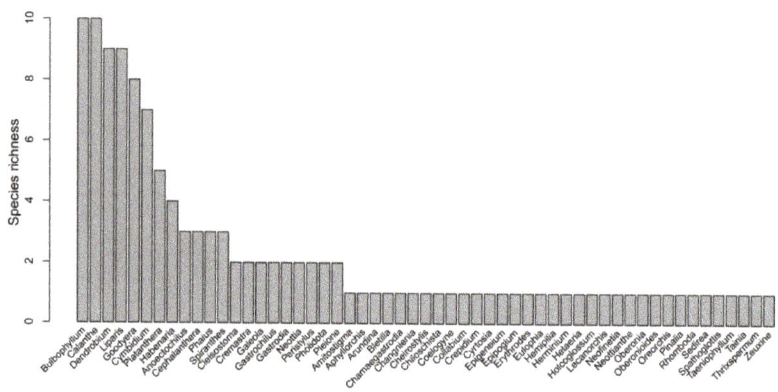

Figure 2. Species richness of each genus of orchids.

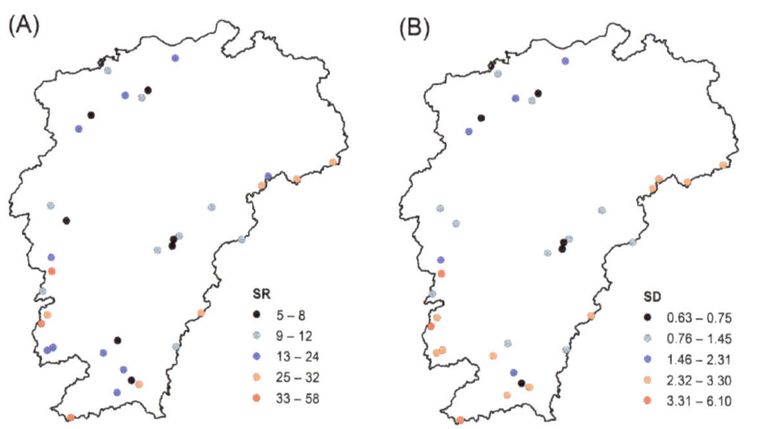

Figure 3. Geographic patterns of (**A**) species richness (SR) and (**B**) species density (SD) of orchids in 35 nature reserves of Jiangxi.

3.2. Effects of Environmental Variables on Richness and Density of Orchids

Except for slope having significant positive correlation with SR and SD (Figure 4A,B), and NDVI having significant negative correlation with SR (Figure 4C), other environmental variables have no relationship with both indexes. The results of variance partitioning analysis showed that these four groups of environmental variables could not explain the richness and density of orchids very well, i.e., they only together explained 22% or 26% of the variation in SR and SD of orchids (Figure 5). Topographical variables explained more variation of both SR and SD than the other three groups of variables. Topographical variables explained 19% and 20% of the total variation in SR and SD, respectively (Figure 5).

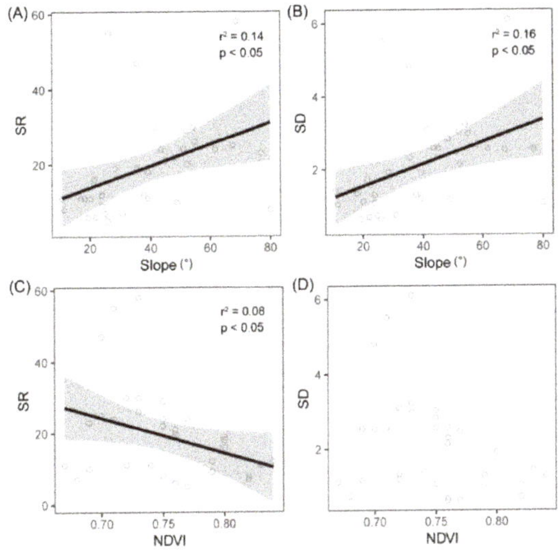

Figure 4. Linear relationships between species richness (SR) and species density (SD) with slope and normalized difference vegetation index (NDVI). (**A**) relationship between SR and slope; (**B**) relationship between SD and slope; (**C**) relationship between SR and NDVI; (**D**) relationship between SD and NDVI.

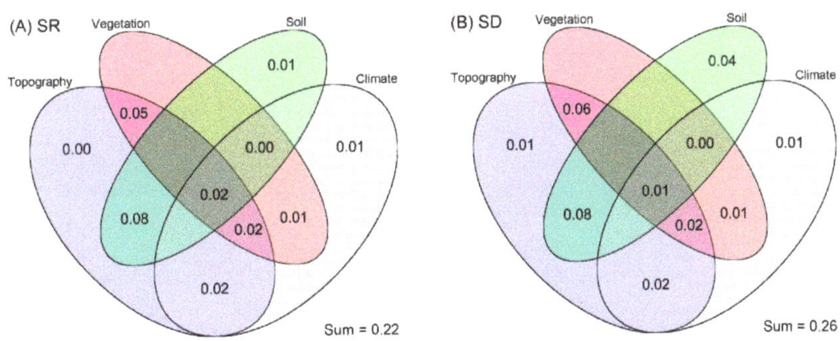

Figure 5. Variation partitioning of four environmental variable groups for (**A**) species richness (SR) and (**B**) species density (SD) of orchids in Jiangxi. The four groups of environmental variables are topography (elevation and slope), vegetation (canopy density and normalized difference vegetation index), soil (pH, total nitrogen, total phosphorus, and total potassium), and climate (annual mean temperature and annual precipitation).

3.3. *Hierarchical Clustering of Nature Reserves*

The result of hierarchical clustering analysis showed all these 35 nature reserves fall into two main clusters (A and B) (Figure 6). Cluster A contained most of the nature reserves from eastern and southern Jiangxi, while cluster B contained the nature reserves from western and northwestern Jiangxi; Ganjiang River and Poyang Lake could separate the two clusters geographically (Figure 6).

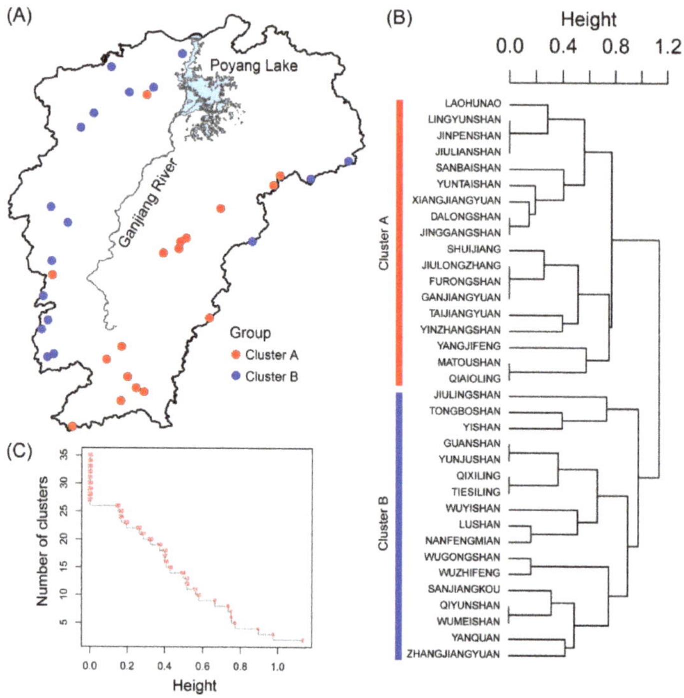

Figure 6. Map (**A**), dendrogram (**B**), and fusion level (**C**) resulting from ward hierarchical clustering of 35 nature reserves of Jiangxi, based on the similarity of the reserves in terms of orchid richness.

4. Discussion

Jiangxi is surrounded by mountains on three sides and basically coincides with the Poyang Lake basin [14]; Jiangxi has a complex terrain, developed water systems, and a warm and humid climate, which breeds rich biodiversity [29]. However, the field survey of plant diversity in Jiangxi is incomplete, especially for some species-rich large families, such as Orchidaceae. Our field surveys showed that there are 125 species of orchids in only 35 nature reserves, including 23 new records of Jiangxi. This work fully showed that the investigation of orchids in Jiangxi is not comprehensive, especially for some mountains with high species richness, such as the Wuyi Mountains and the mountains of southern Jiangxi [14].

Currently, there are more than 1700 orchid species recorded in China, and *Bulbophyllum* Thou. and *Dendrobium* Sw. are the two most species-rich genera, each with more than 100 species [4]. Both of the two genera are also rich in species in Jiangxi, based on our investigation. The areas rich in orchids are concentrated in southwest, south, and northeast Jiangxi, though are relatively low in central and northwest Jiangxi. This is basically consistent with the results of previous studies [18]. For example, a previous study based on a summary of literatures and field surveys showed that there are more than 120 species of orchids in the Luoxiao Mountains in southwest Jiangxi [30].

A positive species-area relationship is one of the most robust summaries in ecology [31]. We also found that larger nature reserves harbor more orchid species. After removing the area effect, we found that the species density and species richness of orchids have similar geographical patterns in Jiangxi. However, some reserves in south Jiangxi have few species, but the species density is relatively high in, for instance, Zhangjiangyuan, Taojiangyuan, and Yuntaishan. Our results showed that environmental variables do not explain orchid richness and density very well, and this result was consistent with that of

Hainan Island [12] and also the whole country of China [11]. However, we found that topography, especially slope, was closely related to orchid richness and density, rather than other environmental variables, such as soil and climate. This was different from the results of previous studies [12] and indicated that the complexity of terrain can affect orchid richness and density more than climate, on small geographical scales [32].

Most previous studies of the orchids in Jiangxi considered the impact of mountains on the diversity pattern, but ignored the role of water systems [13,16,18]. The mountains of Jiangxi generally stretch from north to south, while the Ganjiang River and Poyang Lake divide the main mountain ranges of Jiangxi into east and west groups. The distances between the mountain ranges of Jiangxi from east to west are far greater than those between the mountain ranges from north to south, and this terrain could cause certain obstacles to the interaction between mountain plants on both sides. We conducted cluster analysis on 35 nature reserves based on orchid distribution information. Our results strongly supported this speculation, i.e., the 35 nature reserves could obviously fall into two main clusters, and the boundary between these two clusters roughly coincided with the Ganjiang River–Poyang Lake water system. Southern Jiangxi is close to the Nanling Mountains, which are considered as an important glacial refuge for plants [33,34]. We speculated that orchids took refuge in the warmer regions of the Nanling Mountains during the glacial period, expanding northward along two different mountain groups after the end of the glacial period [18]. This inference has been confirmed by the studies of population genetic differentiation of some species, such as *Loropetalum chinense* (R. Br.) Oliver [33], *Cercis chuniana* Metc. [34], and *Machilus pauhoi* Kanehira [35].

Orchids are of great interest because of their high ornamental and medicinal value, and their survival is also seriously threatened by climate change, habitat degradation and fragmentation, human disturbance and excavation, etc. [4,11,36,37]. In our study, we did not include human disturbance variables for analysis, but we found that larger reserves harbor more orchids, including some rare and endangered mycoheterotrophic orchids, such as *Danxiaorchis yangii* in Jinggangshan National Nature Reserve [20]. Our survey showed that more than 40 orchids appear in only one of the 35 nature reserves. For example, the vulnerable species *Goodyera bomiensis* K.Y. Lang was only found in Sanbai Mountain National Forest Park. Thus, increasing connectivity between small protected areas and habitats may be beneficial to the protection of orchid diversity [38].

5. Conclusions

In this study, we conducted field surveys of orchids in 35 nature reserves in Jiangxi Province, China. We found that the mountainous areas of southwestern, southern, and northeastern Jiangxi have a high richness and density of orchids, while few orchids were found in the central and northwestern Jiangxi. We also found that topography is closely related to the richness and density of orchids, rather than soil, climate, and vegetation. The 35 nature reserves, representing most of the mountain ranges of Jiangxi, were obviously clustered in two main groups, which were also separated by the Ganjiang River–Poyang Lake water system. This indicates that terrain plays an important role in the dispersal of orchids in Jiangxi. Meanwhile, we also put forward a suggestion for the protection of orchid diversity in these regions, i.e., some small protected areas could be integrated to increase the ecological connectivity.

Supplementary Materials: The following supporting information can be downloaded at: https://www.mdpi.com/article/10.3390/d14100855/s1. Table S1: Environmental variables of the 35 nature reserves in Jiangxi Province, China; Table S2: Checklist of orchids in the 35 nature reserves of Jiangxi Province, China.

Author Contributions: Data curation, Q.Z., M.H., Y.Z. and B.Y.; formal analysis, M.H. and Y.Z.; funding acquisition, B.Y. and Y.Z.; investigation, Y.L., Z.Z., F.L., L.L., X.T., Z.L., W.C., M.H., S.T., H.L. and B.Y.; methodology, M.H. and Y.Z.; supervision, B.Y.; writing—original draft, Q.Z. and Y.Z.; writing—review and editing, Q.Z. All authors have read and agreed to the published version of the manuscript.

Funding: This work was supported by grants from the National Forestry and Grassland Administration of China (No. 2021070704) and the National Science Foundation of China (No. 32260046).

Institutional Review Board Statement: Not applicable.

Informed Consent Statement: Not applicable.

Data Availability Statement: Data are available in supplementary files.

Acknowledgments: We would like to thank Antony Njogu Waigwa from East China Normal University for revising the English of this manuscript. Many thanks should also be given to the students from Nanchang University and the staff from the nature reserves of Jiangxi for their kindly help with the field surveys.

Conflicts of Interest: The authors declare no conflict of interest.

References

1. Chase, M.W.; Cameron, K.M.; Freudenstein, J.V.; Pridgeon, A.M.; Salazar, G.; van den Berg, C.; Schuiteman, A. An updated classification of Orchidaceae. *Bot. J. Linn. Soc.* **2015**, *177*, 151–174. [CrossRef]
2. Dressler, R.L. *Phylogeny and Classification of the Orchid Family*; Cambridge University Press: Cambridge, UK, 1993.
3. Gustovo, A.R. The Orchid family. In *Orchids: Status, Survey, and Conservation Action Plan*; Hagsater, E., Dumont, V., Eds.; IUCN: Gland, Switzerland; Cambridge, UK, 1996.
4. Zhou, Z.; Shi, R.; Zhang, Y.; Xing, X.; Jin, X. Orchid conservation in China from 2000 to 2020: Achievements and perspectives. *Plant Divers.* **2021**, *43*, 343–349. [CrossRef] [PubMed]
5. Wan, X.; Zhang, L. Global new taxa of vascular plants published in 2021. *Biodivers. Sci.* **2022**, *30*, 22116. [CrossRef]
6. Du, C.; Liu, J.; Ye, W.; Liao, S.; Ge, B.; Liu, B.; Ma, J. Annual report of new taxa and new names for Chinese plants in 2020. *Biodivers. Sci.* **2021**, *29*, 1011–1020. [CrossRef]
7. Du, C.; Liu, J.; Ye, W.; Liao, S. 2021 annual report on new taxa and nomenclatural changes of Chinese plants. *Biodivers. Sci.* **2022**, *30*, 22207. [CrossRef]
8. Jacquemyn, H.; Micheneau, C.; Roberts, D.L.; Pailler, T. Elevational gradients of species diversity, breeding system and floral traits of orchid species on Reunion Island. *J. Biogeogr.* **2005**, *32*, 1751–1761. [CrossRef]
9. Acharya, K.P.; Vetaas, O.R.; Birks, H.J.B. Orchid species richness along Himalayan elevational gradients. *J. Biogeogr.* **2011**, *38*, 1821–1833. [CrossRef]
10. McCormick, M.K.; Jacquemyn, H. What constrains the distribution of orchid populations? *New Phytol.* **2014**, *202*, 392–400. [CrossRef]
11. Zhang, Z.; Yan, Y.; Tian, Y.; Li, J.; He, J.; Tang, Z. Distribution and conservation of orchid species richness in China. *Biol. Conserv.* **2015**, *181*, 64–72. [CrossRef]
12. Zuo, Y.; Yang, X.; Li, D.; Wu, E.; Yang, N.; Li, L.; Zhang, P.; Chen, L.; Li, C. Effects of environmental variables on the species composition and distribution patterns of wild orchids in Hainan Island. *Chin. J. Plant Ecol.* **2021**, *45*, 1341–1349. [CrossRef]
13. Xie, H.; Peng, D.; Chen, Y.; Luo, H.; Yang, B.; Xiong, D. Genetic structure and differentiation of wild populations of *Cymbidium goeringii* (Rchb. f.) Rchb. f. in the main mountain range of Jiangxi Province, China. *Plant Sci. J.* **2020**, *38*, 123–133.
14. Ji, L.; Xie, Y.; Li, Z.; Xu, T.; Yang, B.; Li, B. A checklist of wild vascular plants in Jiangxi Province, China. *Biodivers. Sci.* **2022**, *30*, 22057. [CrossRef]
15. Chen, X.; Liu, Z.J.; Zhu, G.H.; Lang, K.; Ji, Z.; Luo, Y.; Jin, X.; Cribb, P.J.; Wood, J.J.; Gale, S.W.; et al. *Flora of China: 25. Orchidaceae*; Science Press: Beijing, China, 2009.
16. Kong, L. Wild Orchid Plant Resources of Jiangxi Province: Floristic Composition and Characteristics. Master's Thesis, Nanchang University, Nanchang, China, 2011.
17. Wang, C.; Liang, Y.; Zhang, Z.; Liao, W.; Huang, W.; Yang, B. New records of Orchidaceae in Jiangxi Province. *J. For. Environ.* **2018**, *38*, 367–371.
18. Chen, Y. Phylogenetic Diversity and Its Structure of Wild Orchid Plants in Jiangxi Province and the Change Trend of the Suitable Areas. Master's Thesis, Nanchang University, Nanchang, China, 2019.
19. Liu, H.; Wang, C.; Xiao, H.; Liu, N.; Liang, Y.; Yang, B. Additional records of Orchidaceae in Jiangxi Province. *J. Nanchang Univ. (Nat. Sci.)* **2020**, *44*, 167–171.
20. Yang, B.; Xiao, S.; Jiang, Y.; Luo, H.; Xiong, D.; Zhai, J.; Li, B. *Danxiaorchis yangii* sp. nov. (Orchidaceae: Epidendroideae), the second species of *Danxiaorchis*. *Phytotaxa* **2017**, *306*, 287–295. [CrossRef]

21. Yang, B.; Luo, H.; Huang, W.; Xiong, D.; Tan, S.; Li, B. *Calanthe sieboldopsis* (Orchidaceae, Epidendroideae, Collabieae), a new species from Luoxiao Mountains, eastern China. *PhytoKeys* **2020**, *145*, 37–45. [CrossRef]
22. ESRI. ArcGIS Release 10.5 [Online]. 2016. Available online: https://www.esri.com (accessed on 20 January 2019).
23. Dai, Y.; Shangguan, W. *Dataset of Soil Properties for Land Surface Modeling over China*; National Tibetan Plateau Data Center: Beijing, China, 2019.
24. Li, R.; Kraft, N.J.B.; Yu, H.Y.; Li, H. Seed plant phylogenetic diversity and species richness in conservation planning within a global biodiversity hotspot in eastern Asia. *Conserv. Biol.* **2015**, *29*, 1552–1562. [CrossRef]
25. Zhou, Y.; Chen, S.; Hu, G.; Mwachala, G.; Yan, X.; Wang, Q. Species richness and phylogenetic diversity of seed plants across vegetation zones of Mount Kenya, East Africa. *Eco. Evol.* **2018**, *8*, 8930–8939. [CrossRef]
26. Dormann, C.F.; Elith, J.; Bacher, S.; Buchmann, C.; Carl, G.; Carre, G.; Marquez, J.R.G.; Gruber, B.; Lafourcade, B.; Leitao, P.J.; et al. Collinearity: A review of methods to deal with it and a simulation study evaluating their performance. *Ecography* **2013**, *36*, 27–46. [CrossRef]
27. Wang, S.; Boru, B.H.; Njogu, A.W.; Ochola, A.C.; Hu, G.; Zhou, Y.; Wang, Q. Floristic composition and endemism pattern of vascular plants in Ethiopia and Eritrea. *J. Sys. Evol.* **2020**, *58*, 33–42. [CrossRef]
28. R Core Team. R: A language and environment for statistical computing. In *R Foundation for Statistical Computing*; R Core Team: Vienna, Austria, 2017.
29. Wu, D.; Li, B.; Yang, B. *Checklist of Wild Angiosperm in Jiangxi Province*; China Forestry Publishing House: Beijing, China, 2021.
30. Liu, H. Flora and Its Ecological Geography of the Orchidaceae in the Luoxiao Mountains Range, Jiangxi Province. Master's Thesis, Nanchang University, Nanchang, China, 2019.
31. He, F.; Legendre, P. Species diversity patterns derived from species-area models. *Ecology* **2002**, *83*, 1185–1198.
32. Han, J.; Shen, Z.; Shi, S.; Peng, P. Comparison of plant species diversity and composition in the dry valleys of Yalong River and Dadu River: Evaluating the effects of climate, topography and space. *Biodivers. Sci.* **2016**, *24*, 421–430. [CrossRef]
33. Gong, W.; Liu, W.; Gu, L.; Kaneko, S.; Koch, M.A.; Zhang, D. From glacial refugia to wide distribution range: Demographic expansion of *Loropetalum chinense* (Hamamelidaceae) in Chinese subtropical evergreen broadleaved forest. *Org. Divers. Evol.* **2016**, *16*, 23–38. [CrossRef]
34. Liu, W.; Xie, J.; Zhou, H.; Kong, H.; Hao, G.; Fritsch, P.W.; Gong, W. Population dynamics linked to glacial cycles in Cercis chuniana F. P. Metcalf (Fabaceae) endemic to the montane regions of subtropical China. *Evol. Appl.* **2021**, *14*, 2647–2663. [CrossRef] [PubMed]
35. Zhu, Q.; Liao, B.Y.; Li, P.; Li, J.C.; Deng, X.M.; Hu, X.S.; Chen, X.Y. Phylogeographic pattern suggests a general northeastward dispersal in the distribution of *Machilus pauhoi* in South China. *PLoS ONE* **2017**, *12*, e0184456. [CrossRef]
36. Qin, H.; Zhao, L. Evaluating the threat status of higher plants in China. *Biodivers. Sci.* **2017**, *25*, 689–695. [CrossRef]
37. Liu, H.; Liu, Z.; Jin, X.; Gao, J.; Chen, Y.; Liu, Q.; Zhang, D. Assessing conservation efforts against threats to wild orchids in China. *Biol. Conserv.* **2020**, *243*, 108484. [CrossRef]
38. Zhao, Z.; Wang, P. Significance and key issues of protected area connectivity in China. *Landsc. Archit.* **2022**, *29*, 12–17.

diversity

MDPI

Article

Assessment of Climate Change and Land Use Effects on Water Lily (*Nymphaea* L.) Habitat Suitability in South America

John M. Nzei [1,2,3], Boniface K. Ngarega [4], Virginia M. Mwanzia [5], Joseph K. Kurauka [6], Qing-Feng Wang [1,2,7], Jin-Ming Chen [1,2], Zhi-Zhong Li [1,2,*] and Cheng Pan [1,*]

1 Wuhan Botanical Garden, Chinese Academy of Sciences, Wuhan 430074, China
2 Center of Conservation Biology, Core Botanical Gardens, Chinese Academy of Sciences, Wuhan 430074, China
3 University of Chinese Academy of Sciences, Beijing 100049, China
4 Centre for Integrative Conservation, Xishuangbanna Tropical Botanical Garden, Chinese Academy of Sciences, Menglun 666303, China
5 School of Agriculture and Technical Studies, Lukenya University, Mtito Andei P.O. Box 90-90128, Kenya
6 School of Environmental Studies, Kenyatta University, Nairobi P.O. Box 43844-00100, Kenya
7 Sino-Africa Joint Research Center, Chinese Academy of Sciences, Wuhan 430074, China
* Correspondence: lizhizhong@wbgcas.cn (Z.-Z.L.); pancheng@wbgcas.cn (C.P.)

Abstract: Many aquatic species have restricted dispersal capabilities, making them the most vulnerable organisms to climate change and land use change patterns. These factors deplete *Nymphaea* species' suitable habitats, threatening their populations and survival. In addition, the species are poorly documented, which may indicate how scarce they are or will become. Members of *Nymphaea* are ecologically important as well as having cultural and economic value, making them of conservation interest. Therefore, using the maximum entropy (MaxEnt) approach, climatic variables, land use, and presence points were modeled for seven *Nymphaea* species in South America, using three general circulation models (CCSM4, HADGEM2-AO, and MIROC5) and in two representative concentration pathways (RCPs 4.5 and 8.5) and two scenarios (2050 and 2070). Our results indicated that mean diurnal range (bio2), precipitation of the wettest month (bio13), temperature seasonality (bio15), and land use (dom_lu) were the main influencing factors. For all species, suitable areas were concentrated east of Brazil, and they were variable in northern parts of the continent. Besides, inconsistent expansion and contraction of suitable habitats were noticed among the species. For example, *N. amazonum*, *N. rudgeana*, and *N. lasiophylla* future habitat expansions declined and habitat contraction increased, while for *N. ampla* and *N. jamesoniana*, both future habitat expansion and contraction increased, and for *N. pulchella* and *N. rudgeana* it varied in the RCPs. Moreover, the largest projected suitable habitats were projected outside protected areas, characterized by high human impacts, despite our analysis indicating no significant change between protected and unprotected areas in suitable habitat change. Finally, understanding how climate change and land use affect species distribution is critical to developing conservation measures for aquatic species.

Keywords: climate change; distribution; habitat suitability; *Nymphaea*; land use; conservation

Citation: Nzei, J.M.; Ngarega, B.K.; Mwanzia, V.M.; Kurauka, J.K.; Wang, Q.-F.; Chen, J.-M.; Li, Z.-Z.; Pan, C. Assessment of Climate Change and Land Use Effects on Water Lily (*Nymphaea* L.) Habitat Suitability in South America. *Diversity* **2022**, *14*, 830. https://doi.org/10.3390/d14100830

Academic Editors: Hong-Hu Meng and Yi-Gang Song

Received: 5 September 2022
Accepted: 30 September 2022
Published: 3 October 2022

1. Introduction

Temperature and precipitation fluctuations have been reported in South America over the last decade [1]. For example, in Brazil, the temperature has increased by approximately 0.5 °C [2], while the mean temperature variability across the continent varied between 0.2 and 0.8 °C, with a projected increase of 1 to 4 °C by the end of the century [3]. The average precipitation is projected to increase in the southeast and decrease in other areas, especially between latitudes 5° and 20° south [1]. The effect of climate change on the continent is also seen in the rapid melting of glaciers, which is associated with changes in temperature and humidity [4,5]. As global warming continues to rise, species respond to climate change, causing increased shifts and redistributions in search of suitable habitats.

Ultimately, this has an impact on conservation and management plans in maintaining biodiversity [6].

As the earth becomes much warmer, various land regions become drier, such as parts of the Amazon Forest in Brazil which are being replaced by non-forest environments [7]. Global warming is also linked to the increased savannah land in the Amazon and semi-arid areas northeast of Brazil, which are slowly transforming to deserts [8]. Besides climate warming, the increased human activities pose negative consequences for many parts of the region, such as the Caatinga biome east of Brazil [8]. As climate change affects species community assemblages, the direct effect of human activity in the ecosystem is loss of habitat, which is the primary loss of biodiversity [9]. The unprecedented population growth, urbanization, and need for sustainable food production are among the major humanized factors leading to ecosystem exploitation and loss of biodiversity. These activities also create large disruption to stream flow and wetlands, as well as influencing the hydrological events that lead to floods and droughts, which further threaten extinction of wetland ecosystems and their biodiversity.

The distribution of *Nymphaea* species is considered widespread worldwide [10], however in South America the distribution is poorly explored and documented. Their distribution is particularly prone to climate change and human activities, more so from the characteristic assemblages of the presence data. Most occurrence points were obtained from Brazil compared to the other states in the continent, which might signify inconsistent availability of the species or sampling in the most accessible areas [11]. Much of their restricted ranges threaten the species' habitat suitability with a risk of decline or extinction. This may signify the loss of wild populations that play a key role not only scientifically or culturally but also ecologically in providing food, habitat, water sediments, and turbidity management, and as an indicator for a healthy wetland ecosystem [12,13]. Besides, other studies have indicated climate change to be of concern in aquatic species' distribution [14–16]. Although human influence was not included in those studies, its impact, combined with climate change, poses an unlimited threat to biodiversity and habitat loss for the species [17]. For example, it is approximated that 15% (560,000 km^2) of the forested area in Brazil has so far been lost to ranching and agriculture [18,19]. Considering the sensitivity of the species habitat environment and the species distribution data deficit, their vulnerability to climate change and human influence make them a great choice for this study.

The ecological niche models (ENMs) approach has been employed as a valuable tool for assessing the species habitat suitability [20], enabling insights for conservation measures. These models have been widely used to estimate potentially suitable habitats for a variety of species in temporal and spatial ranges using species occurrence records and environmental variables, in addition to species range shifts, habitat quality, habitat requirements, and to identify potential distribution regions for species. However, maximum entropy (MaxEnt) is much preferred for its ability to accommodate presence only data [21], it performs well and can work with small sample sets [22–24], and avoids commission errors when projecting species distribution [25], thus making it suitable in the assessment of aquatic organism-suitable habitats. It also performs well on both large and narrow geographical distribution scales, such as in the distribution of *Ottelia* and water lily species across Africa and Australia, respectively [14–16].

Using the MaxEnt modeling approach, we assessed the habitat distribution for seven *Nymphaea* species in South America. Our goals were to: (i) model and predict the current distributional range for *Nymphaea* species in South America, (ii) identify the environmental variables shaping the habitat and distribution of *Nymphaea* species, (iii) predict the future habitat suitability of the species, (iv) evaluate human effects on the distribution of the water lilies, and (v) predict the percentage threat of the water lilies' suitable habitats by evaluating the habitat suitability under protected areas (PAs) and unprotected areas (un-PAs).

2. Materials and Methods

2.1. Species Occurrence Data

The species occurrence data for *Nymphaea* were obtained from the Global Biodiversity Information Facility using rgbif package in R (GBIF; https://www.gbif.org/, retrieved on 6 February 2022) and from [26,27]. The downloaded distribution localities were manually filtered to remove duplicate samples and coordinates with ambiguous geographic localities. Google Earth (https://earth.google.com, accessed on 7 February 2022) was used to examine the precision of the coordinates, and those with apparent errors in their geographic coordinates were eliminated. The remaining localities for each species were then rarefied to a spatial distance of 5 km between the points to reduce spatial autocorrelation. This analysis was implemented in R software with the package spThin [28]. The remaining points included *N. amazonum* Mart. and Zucc. (93), *N. ampla* (Salib.) DC. (32), *N. jamesoniana* Planch (23), *N. lasiophylla* Mart. and Zucc. (47), *N. lingulata* Wiersma (47), *N. pulchella* DC (138), and *N. rudgeana* G. Mey. (62) (Figure 1; Table S1).

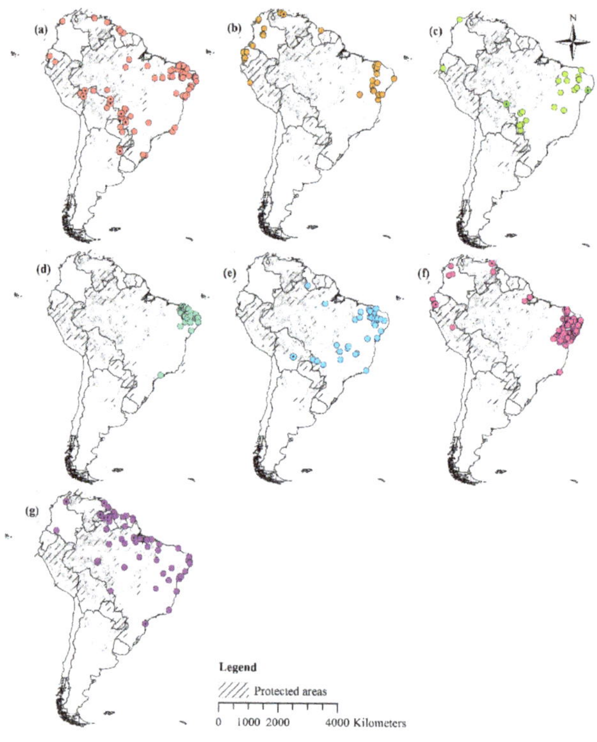

Figure 1. Occurrence points for the seven water lily (*Nymphaea*) species in South America: (**a**) *N. amazonum*, (**b**) *N. ampla*, (**c**) *N. jamesoniana*, (**d**) *N. lasiophylla*, (**e**) *N. lingulata*, (**f**) *N. pulchella*, and (**g**) *N. rudgeana*. The points marked with a black dot inside indicate occurrence points within protected areas.

2.2. Environmental Data

Nineteen bioclimatic variables spanning 1950–2000 were downloaded from WorldClim v1.4 (http://worldclim.org/version2, accessed on 10 April 2021 [29]) at a 2.5 arc-min spatial resolution. These variables define yearly climatic temperature and precipitation trends, as well as seasonality and extreme factors that could exert physiological limits on organisms and influence their geographic distribution. In addition, the land use variable was obtained from the Food and Agricultural Organization (https://www.fao.org/, accessed on: 16 June 2022) at a spatial resolution of 30 arc sec (1 km) and resampled to match the bioclimatic variables' resolution in ArcGis 10.8 [30]. The variables were then masked to the M area of the BAM diagram [31] using South America's freshwater ecoregions as the species' accessible regions [32]. Bioclimatic variables bio8, bio9, bio18, and bio19 were omitted from further analysis as they are known to contain some artifacts and express unrealistic climatic changes between adjacent pixels, although the discontinuities in the variables may not have significant biological meaning [33,34]. Then, we performed the variance inflation factor (VIF: r = 0.7) in the usdm package in the R program to eliminate correlated variables [35] in 16 variables (Table S2). Three general climate models (GCMs): Community Climate System Model version 4 (CCSM4) [36], Hadley Centre Global Environment Model version AO (HADGEM2-AO), and the Model for Interdisciplinary Research on Climate (MIROC5), were selected for the future habitat suitability prediction as they are less biased and have been shown to produce good modeling results [16,22,37]. We selected two representative concentration pathways (RCPs), RCP 4.5 and RCP 8.5, in two scenarios (2050 and 2070) to represent medium- and high-emission scenarios consistent with the IPCC Fifth Assessment Report [3].

2.3. Model Parameterization and Calibration

The climatic niche of *Nymphaea* species was modeled using MaxEnt v.3.4.1. [38]. The model was built with default settings, except for 5000 iterations, 10 replications, cross-validated run-type, and 75% and 25% of occurrence data for training and testing of the model, respectively. To control over-fitting, regularization multiplier (rm) and feature classes were selected using ENMeval in R program in five feature classes (L, LQ, LQH, LQHP, LQHPT, being linear (L), quadratic (Q), hinge (H), product (P), and threshold (T)) at rm values of 0.5 to 4.0 at a 0.5 increment [39]. To evaluate the predictive performance of the models, we utilized the area under the curve of the receiver operating characteristic curve (ROC) [40].

2.4. Predicting Current and Future Range Shifts

Using a threshold criterion of maximum training sensitivity plus specificity (MTSS), the average 'Logistic' outputs were converted to binary maps depicting climatic suitable and unsuitable areas. In this step, the continuous suitability outputs were changed to binary maps so that over 95% of the training occurrence data fell inside the suitable range. Finally, using the binary maps, current and future habitat suitability change were assessed using the SDM-Toolbox extension in ArcGis 10.8 [30,40,41]. The generated maps show habitat suitability changes in (i) stable, (ii) expansion, (iii) contraction, and (iv) unsuitable areas.

2.5. Species Conservation/Threat Area

To assess the current and future possible habitat threat to the species, a map of PAs was obtained from world protected planet (available at: https://www.protectedplanet.net/search?q=natura+2000, accessed on: 24 May 2022) and overlaid with the species' occurrences to assess the percentage of populations inside PAs, the current distribution, and the future projection percentage change of suitable habitats inside and outside PAs. Further, the area of land use in both PAs and un-PAs was used to assess the likely influence of human activities in the species' area of distribution.

3. Results

3.1. Variable Selection and Model Performance

After the correlation analyses, mean diurnal range (bio2), temperature seasonality (bio4), the maximum temperature of the wettest month (bio5), precipitation of the wettest month (bio13), precipitation seasonality (bio15), and Land use (dom_lu) were used in the model building. The selected variables had VIF values of less than three (Table 1). The ENMeval analysis predicted three feature classes (LQ, LQH, and LQHP) and rm values of

2–4 as the best parameters for the models (Table 2). Besides, the AUC values were above 0.8 in all projection scenarios (Table 3), demonstrating high model accuracy. This suggests that the produced models outperformed a random model (AUC = 0.5), indicating that the suitability of the models in the forecast for the distributions of these seven *Nymphaea* species was reliable. In addition, the mean training and testing AUC values had no significant difference and the standard deviation values were close to the probability distribution (Table S3).

Table 1. Climatic variables retained after correlation analysis using variance inflation factors (VIF).

Variable No.	Bioclimatic Variable	Code	VIF
1	Mean diurnal range (mean of monthly (max temp − min temp))	bio2	1.7436
2	Temperature seasonality (standard deviation × 100)	bio4	1.9289
3	Maximum temperature of the warmest month	bio5	1.1728
4	Precipitation of the wettest month	bio13	1.8567
5	Precipitation seasonality (coefficient of variation)	bio15	1.4996
6	Land use cover	dom_lu	1.2866

Table 2. Feature classes selected for the modeling of the *Nymphaea* species in South America. The abbreviations represent linear (L), quadratic (Q), hinge (H), and product (P) at varying regularization multiplier (rm) values.

Species	Features Class	rm Value	Current Habitat Suitability (km^2)
N. amazonum	LQHP	3.5	2,339,884
N. ampla	LQ	2.5	1,558,143
N. jamesoniana	LQH	4	2,790,903
N. lasiophyla	LQH	2	862,217
N. lingulata	LQH	4	1,341,940
N. pulchella	LQH	4	907,696.1
N. rudgeana	LQH	3.5	2,657,233

Table 3. The area under the curve (AUC) and standard deviation values for the seven *Nymphaea* species distribution models in two representative concentration pathways (RCPs 4.5 and 8.5) and two scenarios (2050 and 2070).

		RCP 4.5		RCP 8.5	
Species	Current	2050	2070	2050	2070
N. amazonum	0.847 ± 0.075	0.808 ± 0.116	0.803 ± 0.118	0.815 ± 0.115	0.811 ± 0.111
N. ampla	0.878 ± 0.081	0.887 ± 0.089	0.889 ± 0.090	0.891 ± 0.084	0.879 ± 0.085
N. jamesoniana	0.842 ± 0.083	0.859 ± 0.086	0.848 ± 0.096	0.848 ± 0.090	0.848 ± 0.094
N. lasiophylla	0.963 ± 0.036	0.954 ± 0.045	0.951 ± 0.049	0.953 ± 0.048	0.958 ± 0.035
N. lingulata	0.902 ± 0.036	0.890 ± 0.074	0.889 ± 0.075	0.891 ± 0.078	0.882 ± 0.087
N. pulchella	0.927 ± 0.025	0.936 ± 0.022	0.935 ± 0.021	0.938 ± 0.021	0.933 ± 0.023
N. rudgeana	0.814 ± 0.086	0.794 ± 0.083	0.800 ± 0.089	0.796 ± 0.089	0.806 ± 0.090

3.2. Contribution of Variables

The percentage contributions of the bioclimatic variables to the final *Nymphaea* models are shown in Table 4. The best variables explaining the distribution of all *Nymphaea* species in all scenarios were bio2, bio13, dom_lu, and bio4. Bio2 had a limited contribution for *N. jamesoniana*. Bio4 indicated a greater contribution in the current projection compared to future, except for *N. rudgeana* and *N. lasiophyla*. Bio5 was the less contributing variable among the species in all scenarios except for *N. amazonum* and *N. jamesoniana*. Bio13 contributed greatly among the species in all scenarios except for *N. rudgeana*. Bio15 had much influence on the distribution of *N. lasiophylla*, *N. lingulata*, and *N. rudgeana*. Finally, dom_lu is of influence to all species' suitable habitats.

Table 4. The mean relative contribution for the six bioclimatic variables used for the habitat suitability modeling of the water lily species in South America in two representative concentration pathways (RCPs 4.5 and 8.5) and two scenarios (2050 and 2070).

			RCP 4.5		RCP 8.5	
Species	**Variable**	**Current**	2050	2070	2050	2070
N. amazonum	bio2	**32.6**	**28.3**	**29.8**	**32.5**	**30.0**
	bio4	17.6	3.3	3.3	3.4	2.7
	bio5	16.1	15.3	11.6	12.2	12.0
	bio13	9.5	**19.5**	**18.8**	**20.4**	**20.6**
	bio15	4.3	9.8	12.7	10.4	11.3
	dom_lu	**20.0**	**23.7**	**23.9**	**21.2**	**23.3**
N. ampla	bio2	**23.5**	**35.3**	**35.7**	**40.2**	**37.5**
	bio4	17.8	13.5	13.8	12.2	15.2
	bio5	0.0	1.6	1.6	2.3	3.8
	bio13	**25.8**	**20.3**	**20.2**	**15.9**	**15.9**
	bio15	8.3	10.5	9.5	11.2	8.6
	dom_lu	**24.6**	**18.9**	**19.2**	**18.1**	**19.2**
N. jamesoniana	bio2	0.0	0.3	0.4	0.2	0.6
	bio4	18.6	6.4	6.8	3.9	4.5
	bio5	29.1	16.6	14.3	12.6	8.9
	bio13	2.8	**20.9**	**17.0**	**22.7**	**21.1**
	bio15	**19.3**	**21.3**	**23.5**	**23.2**	**25.1**
	dom_lu	**30.3**	**34.4**	**38.0**	**37.5**	**39.9**
N. lasiophylla	bio2	**24.9**	**27.3**	**26.6**	**30.5**	**31.1**
	bio4	0.2	0.8	0.9	0.3	0.6
	bio5	0.9	0.3	0.5	1.1	2.3
	bio13	7.4	10.3	7.1	7.5	7.1
	bio15	**58.2**	**52.9**	**56.2**	**53.0**	**50.9**
	dom_lu	8.4	8.3	8.6	7.7	8.0
N. lingulata	bio2	**19.4**	**20.6**	**21.6**	**23.0**	**23.0**
	bio4	9.8	4.9	4.1	4.9	3.3
	bio5	4.9	3.8	3.1	1.8	1.2
	bio13	1.8	6.9	6.2	6.0	5.7
	bio15	**43.0**	**41.0**	**42.5**	**40.5**	**42.2**
	dom_lu	**21.1**	**22.7**	**22.5**	**23.8**	**24.6**
N. pulchella	bio2	**27.5**	**27.9**	**29.9**	**31.7**	**30.5**
	bio4	**38.6**	**19.5**	**17.9**	**18.6**	**16.3**
	bio5	2.5	2.9	3.6	4.0	6.0
	bio13	4.6	23.3	21.1	20.6	20.7
	bio15	1.2	2.2	2.2	1.8	1.9
	dom_lu	**25.6**	**24.4**	**25.2**	**23.4**	**24.6**
N. rudgeana	bio2	**70.0**	**66.7**	**61.1**	**62.3**	**60.1**
	bio4	6.9	10.3	9.4	8.6	11.6
	bio5	1.2	0.3	0.2	0.3	0.1
	bio13	0.4	0.1	0.1	0.2	0.6
	bio15	12.7	10.7	16.5	16.0	15.4
	dom_lu	8.8	11.8	12.7	12.6	12.2

3.3. Current Potential Distribution

The sampling of occurrence points indicated that *N. rudgeana* (12.90%), *N. amazonum* (10.75%), and *N. jamesoniana* (8.70%) had the highest percentage of points within the protected area, followed by *N. ampla* (3.13%), *N. lingulata* (2.13%), and *N. pulchella* (1.45%), while for *N. lasiophylla* all points were outside the protected area. The projected distribution maps for the seven *Nymphaea* species indicate Brazil and particularly eastern regions, Venezuela, Colombia, Ecuador, Bolivia, Guyana, and Peru (Figure 2a–g), to contain suitable habitat areas for the species. Some species had partial distribution, such as *N. ampla* in

Bolivia (Figure 2b), *N. pulchella* in Guyana (Figure 2f), and *N. rudgeana* in Bolivia and Peru (Figure 2g). Majority of the species were limited in distribution in states such as Argentina, Paraguay, Suriname, Chile, and Uruguay (Figure 2a–g). Generally, the eastern, northern parts, and some regions northwest of South America had the highest probabilities of suitable habitat, with certain regions in the central part of the continent also indicating potentially appropriate ranges. Among the seven water lily species, *N. jamesoniana* (2,790,903 km^2), *N. rudgeana* (2,657,233 km^2), and *N. amazonum* (2,339,884 km^2) occupied the most ranges in current distribution, followed by *N. ampla* (1,558,143 km^2), *N. lingulata* (1,341,940 km^2), *N. pulchella* (907,696.1 km^2), and *N. lasiophylla* (862,217 km^2), respectively (Table 2).

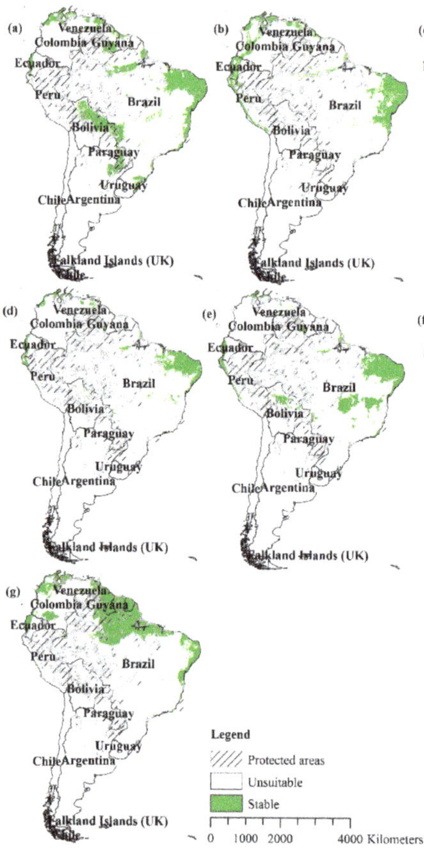

Figure 2. The current distribution of the seven *Nymphaea* species in South America: (**a**) *N. amazonum*, (**b**) *N. ampla*, (**c**) *N. jamesoniana*, (**d**) *N. lasiophylla*, (**e**) *N. lingulata*, (**f**) *N. pulchella*, and (**g**) *N. rudgeana*.

3.4. Future Distribution Changes

The assessment of the species' future distribution is important for conservation measures. The projection change summary for the distribution changes is presented in Figure 3 (Table S4). The future projection for *N. amazonum* indicates increased habitat gain and major habitat loss of suitability in Bolivia in RCP 4.5 (2050). Further loss of habitat suitability is projected consecutively in RCP 4.5 (2070) and both RCP 8.5 scenarios, especially in Bolivia and in most parts north of the continent, such as Venezuela, Guyana, and Suriname (Figure S1). The projection of *N. ampla* indicates a persistent habitat increase in RCP 4.5 (2070) and in both RCP 8.5 scenarios compared to RCP 4.5 (2050). Most habitat gain is projected in Brazil, Venezuela, and Guyana, while most parts of Peru and northwest of the continent indicate habitat loss (Figure S2). Interestingly, *N. ampla* indicated increasing habitat expansion and loss in both RCP 4.5 scenarios and RCP 8.5 (2050), although habitat loss declined in RCP 8.5 (2070). The distribution of *N. jamesoniana* indicated greater changes in Brazil, Bolivia, Paraguay, and Argentina. These regions experience greater habitat loss compared to RCP 4.5 (2050), especially Brazil and Bolivia (Figure S3). However, in both RCP 8.5 scenarios, some regions in Brazil and Peru experienced an increase of suitable habitats, although habitat loss was increasing (Figures 3 and S3). The expansion ranges for *N. lasiophylla* in RCP 4.5 are lost in future scenarios of RCP 8.5, especially in Brazil and Guyana (Figure S4). *N. lasiophylla* indicated reduced habitat gain and increased habitat loss (Figure 3). *N. lingulata* was indicated to have the highest habitat gain in RCP 4.5 (2050) compared to all other species. Its expansion ranges are projected to contract in RCP 4.5 (2070) and both RCP 8.5 scenarios, especially in Brazil and Bolivia, which are characterized by continuous habitat loss (Figures 3 and S5). The projection of *N. pulchella* indicated greater habitat loss in Bolivia, Ecuador, and northeast of Brazil in both RCPs 4.5 and 8.5 (2050) (Figure S6). Although the expansion ranges are not obviously visible in the maps compared to the other species, Figure 3 indicates habitat gain in RCPs 4.5 and 8.5 of scenario 2070 compared to 2050. In scenario 2050 for both RCPs, *N. rudgeana* was projected to have a greater habitat increase compared to scenario 2070 for both RCPs (Figure 3). Many of the changes are noticed in central Brazil, Colombia, and Venezuela, with habitat expansion declining and habitat loss increasing (Figure S7). In extreme conditions of RCP 8.5 (2070), expansion ranges reduced greatly with increasing contractions (Figure 3).

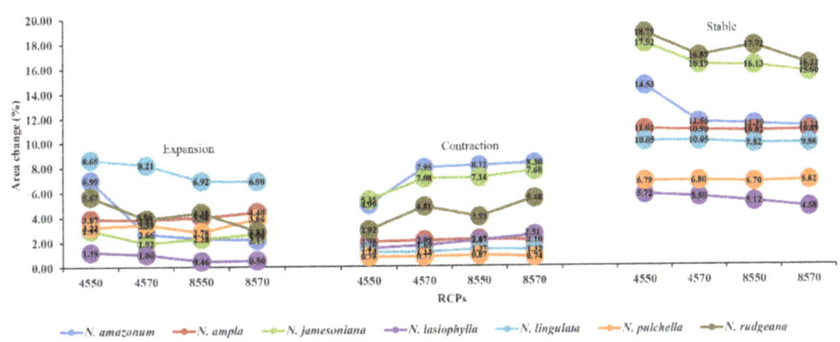

Figure 3. Future distribution changes in expansion (habitat gain), contraction (habitat loss), and stable environment among the seven *Nymphaea* species.

Generally, the species varied considerably in suitable habitat expansion compared to the contraction trend. Species such as *N. ampla*, *N. lasiophylla*, *N. lingulata*, and *N. pulchella* indicated an almost similar trend in habitat loss and the stable environment, although for *N. lasiophylla* the stable environment experienced shrinkage (Figure 3). The loss of suitable habitat expansion, stable environment, and the increase in contraction in the suitable ranges is a call for alarm when it comes to species' conservation. This loss is not only experienced by *N. lasiophylla* but also *N. amazonum* and *N. rudgeana* in RCP 8.5 (2070). *N. jamesoniana* indicates increased habitat expansions and contraction, and loss of stable environment, which indicate possible shifting to new areas. Lastly, many contractions occur in the most central parts of the continent compared to the coastlines, where many habitat growths and stable habitats are projected. These regions are therefore important in providing shifting ranges for suitable habitats.

3.5. Land Use and the Distribution of Water Lilies

The species-accessible area (M) for land use is projected in Figure 4. Most parts of the distribution regions remain unprotected, and this exposes them to human influence and degradation, more so the eastern parts of the continent (Table S5, Figure 4). Much influence is projected in the north, northwest, central, east, and southeast parts of the continent according to the species-accessible area. Most of the areas are characterized by low land cover of between 50% and 75% (Figure 4). Land use was among the main contributing factors for the water lily species' distribution (Table 4). The current projection (Figure 2) and the future projections indicate that large areas of suitable habitats are projected outside PAs, areas acquired by accessible area classifications minus protected area classifications. The areas are characterized by grasslands or shrub areas, crop land areas, and areas composed of mixed activities (Figures S1–S7). Although large areas are covered by forests, especially in Brazil, crops, non-vegetation, and mixed activity areas are of significant influence to the continent (Table S5). Besides PAs playing a key role in the conservation and management of natural resources, they are limited in size compared to the un-PAs (Figure 5; Table S6) and therefore they can sometimes fail in ensuring sustainable refuge and survival for species under threat of extinction, especially when it is out of human control. In this case, the projection of the water lily species indicates an almost similar trend in expansion and contraction in both un-PAs and PAs, with the exception of *N. lingulata* which has higher expansion in un-PAs compared to PAs (Figure 5). Besides, Brazil, Ecuador, Colombia, Venezuela, and Bolivia are among the states with great habitat suitability and among those influenced by land use, more specifically Bolivia and Brazil, where many of the projected areas are declining. Besides, the stable environment had no significant change between un-PAs and PAs (Figure 6). The un-PAs had much of the stable area among the species, except for *N. rudgeana*, for which in both areas the stable environment was almost the same (Figure 6).

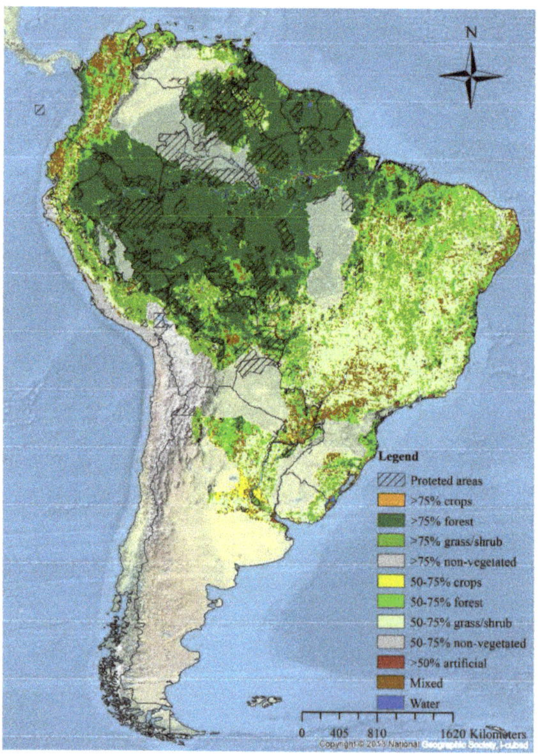

Figure 4. Land use characteristics in South America within the water lily (*Nymphaea*)-accessible area (M), indicating the land use category in protected and unprotected areas.

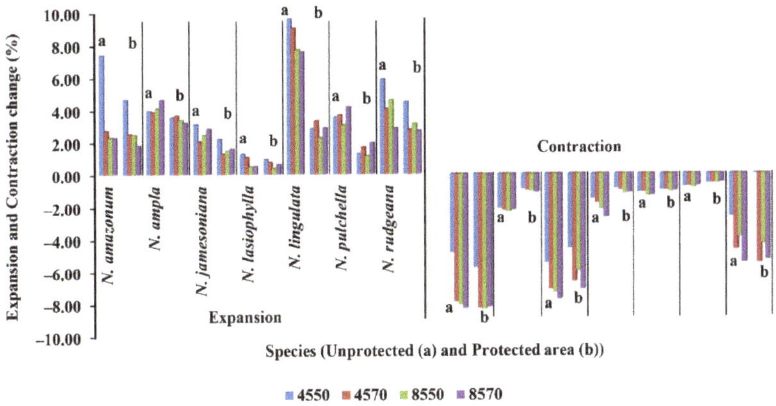

Figure 5. Comparison of habitat suitability change for the water lily species within unprotected areas (a) and protected areas (b). Change area is represented in percentage, expansion in positive values, and contraction in negative values.

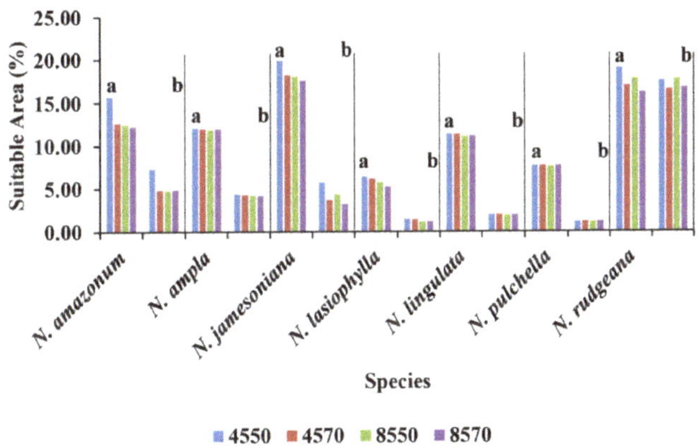

Figure 6. Comparison of the stable habitat change for the water lily species within unprotected areas (a) and protected areas (b). Change area is represented in percentage.

4. Discussion

This study developed a set of potential ecological niche models for the *Nymphaea* species to examine their distribution patterns across South America. We used species occurrence data and bioclimatic variables to simulate species' ecological niches. These factors have been proven fundamental for MaxEnt modeling [38]. As a result of the variable selection procedure, the VIF values were less than three, indicating less collinearity [42]. The rm values in the selected feature classes were higher compared to the MaxEnt default rm of one, thus highly minimizing the overfitting of the models [43]. The mean AUC values obtained from the MaxEnt models reflected the common values expected for fairly fitted models (Table 2) and were not significantly different from previous studies, and were thus considered descriptive [16,24,44].

In this study, temperature and precipitation variables (bio2, bio13, and bio15) influenced *Nymphaea* species' distribution in South America similarly to the African *Nymphaea* species [16]. For instance, bio2 was predicted as the primary temperature variable with the greatest impact in all the studied *Nymphaea* species except *N. jamesoniana* in South America. The mean diurnal range (bio2) indicates the variation in temperature over a day, and the temperature changes may have a significant impact on plant growth, especially photosynthesis and respiration of the plant, contributing to nutrient buildup [45]. Bio4 and bio15 mediate species' suitable distribution by forcing them to shift between favorable seasons. In this case, the species are likely to shift to regions experiencing less dry seasons and cooler temperatures [46]. Unlike bio13, which provides a more conducive environment by increasing precipitation that replenishes the isolated habitats such as ponds and increasing surface runoff, bio5 increases evaporation and transpiration, which significantly affect the plants and their environment by causing dryness. As a result, the plants are forced to seek suitable locations. Lastly, land use directly declines species' habitats depending on the magnitude of land use.

The current projected distribution of the *Nymphaea* species revealed areas of high environmental suitability that corresponded to the observed records. This might indicate that ecological niches are defined by variables affecting the distribution and abundance of resources on which they rely on [47]. Nevertheless, due to continuous habitat destruction on the continent, the projected suitable habitat areas may over-represent habitat suitability where *Nymphaea* species habitats no longer exist or their habitats are under the extreme pressure of sustaining them, such as in the northeast of Brazil where human influence is high [8]. Greater changes in habitat suitability were noticed in the future distribution of *N. amazonum*, *N. jamesoniana*, and *N. rudgeana*, which may have been caused by the reduced moisture, increasing aridity in multiple locations, and precipitation variability [48]. Such situations are propagated by temperature increases, especially in the warm months, which increase evaporation and dryness; for example, the contribution of bio5 for *N. amazonum* and *N. jamesoniana* distribution. As global temperatures rise, the threat of heat stress increases, possibly causing a decrease in species-suitable habitats and richness [3,49]. The rise in temperature may push the species to adjust in habitat-suitable areas, thus causing geographical shifts [50].

Although the environmental requirements for the *Nymphaea* species slightly differed, these species utilized similar habitats in the northeastern parts of the continent. Moreover, it is proposed that in most wetland ecosystems, aquatic species are sympatric [11]. Similarly, in the South American freshwater ecoregions, such relationships between *Nymphaea* species have been documented [47,51]. Furthermore, *Nymphaea* species would display niche partitioning and geographic segregation in South America [52].

The protected areas (PAs) act as shelter for species' survival in this human-dominated world [53]. Species take refuge in the uninfluenced ecosystems, thus ensuring their survival. Although the regions hold biodiversity and conservation value, they are also under extreme pressure from the growing human populations and need for development resources [54]. Nonetheless, with the high magnitude of global warming, protected areas also face adversities in conservation, resulting in declining habitat suitability and vegetation shifts [55]. In this study, a similar change in habitat suitability was observed between PAs and un-PAs in habitat suitability. This could imply a declining capability of PAs to provide a safe refuge for species with declining habitats [56,57]. The only hope with this kind of projection is that the declining habitats in PAs will happen slower and more visibly compared to the non-PAs due to regulated human impacts. However, it is a wakeup call for conservation managers to rethink on the impact of PAs on some species as climate change will cut across all environments.

Surprisingly, the Amazon ecosystem was thought to provide the most suitable habitat for the *Nymphaea* species, however the species' limited distribution in the area was linked to the increasing climatic events such as drought and numerous floods which destroy the species' habitat [2]. Besides, the Amazon's humid environment may not be suitable for numerous growth of the water lily species compared to the savannah areas, where the habitat has been favored [15]. Since habitat suitability distribution in the Amazon might be controlled by many factors, we leave it as an open gap for further studies involving field surveys which can provide the absent data that are capable of improving our model in confirming the water lily species' distribution in the region. Unlike the PAs, much distribution is projected in human-disturbed environments, especially in some parts of the Amazon, which could support the hypothesis that deforestation reduces evapotranspiration and increases stream flow [58]. In addition, human influence through land use changes may have a significant impact on land water resources by affecting the mechanisms at which latent heat flux, surface runoff, discharge from the rivers, and regional and continental precipitation patterns occur [59].

In conclusion, this study provided insight on the implications of climate change and land use effects in assessing the potential habitat suitability for the *Nymphaea* species as the main factors influencing aquatic ecosystems and biodiversity. The study is important for the conservation and management of these species, which are scientifically and ecologically important. The loss of suitable habitat areas always raises concern for possible changes, leading to the reduction of species' populations that may result in species being threatened or becoming extinct [60]. As PAs face increased protection and conservation crises, it is high time for the conservationists to turn to and engage the local indigenous people, civil societies, and private sectors in biodiversity conservation and management [61]. Such measures have been implemented in states such as Bolivia, Peru, Guyana, and other areas outside South America, and they may be of higher significance in the conservation of the South American wetland ecosystem and biodiversity [62].

Supplementary Materials: The following supporting information can be downloaded at: https://www.mdpi.com/article/10.3390/d14100830/s1, Figures S1–S7: The projected distribution change for the seven *Nymphaea* species for the future projection scenarios. (Figure S1) *N. amazonum*, (Figure S2) *N. ampla*, (Figure S3) *N. jamesoniana*, (Figure S4) *N. lasiophylla*, (Figure S5) *N. lingulata*, (Figure S6) *N. pulchella*, and (Figure S7) *N. rudgeana*. Table S1: The occurrence points for the seven water lily (*Nymphaea*) species used for this study. Table S2: The sixteen variables used for correlation in variance inflation factor (VIF). Table S3: Training (75%) and testing (25%) AUC values and their standard deviation values for the *Nymphaea* species' distribution suitability using the maximum training sensitivity plus specificity logistic threshold (MTSS). The AUC values describe the fitness of the model in predicting the species' distribution. Table S4: Distribution change obtained by comparing binary changes between the current and future potential distribution for the seven water lily species (*Nymphaea*) in South America. Table S5: Projected area size between protected areas and unprotected areas for the seven *Nymphaea* species' accessible areas in South America. Table S6: Distribution change obtained by comparing binary changes between the current and future potential distribution for the seven water lily species (*Nymphaea*) in protected areas of South America.

Author Contributions: J.M.N.: conceptualization, investigation, formal analysis, writing—original draft; B.K.N.: formal analysis, writing—review and editing; V.M.M.: conceptualization, writing—review and editing; J.K.K.: writing—review and editing; Q.-F.W.: resources, funding acquisition; C.P. and J.-M.C.: conceptualization, resources, supervision; Z.-Z.L.: supervision, writing—review and editing. All authors have read and agreed to the published version of the manuscript.

Funding: This work was supported by grants from the National Natural Science Foundation of China, China (No. 32070231), and CAS-TWAS President's Ph.D. Fellowship Program University of the Chinese Academy of Sciences, China.

Institutional Review Board Statement: Not applicable.

Data Availability Statement: All data presented in this study are available in the article and in the Supplementary Materials.

Conflicts of Interest: The authors declare no conflict of interest.

References

1. Chou, S.C.; Lyra, A.; Mourão, C.; Dereczynski, C.; Pilotto, I.; Gomes, J.; Bustamante, J.; Tavares, P.; Silva, A.; Rodrigues, D. Assessment of climate change over South America under RCP 4.5 and 8.5 downscaling scenarios. *Am. J. Clim. Chang.* **2014**, *3*, 512. [CrossRef]
2. Marengo, J.; Nobre, C.; Tomasella, J.; Cardoso, M.; Oyama, M. Hydro-climatic and ecological behaviour of the drought of Amazonia in 2005. *Philos. Trans. R. Soc. Lond. B Proc. Biol. Sci.* **2008**, *363*, 1773–1778. [CrossRef] [PubMed]

3. Rowbotham, J. Intergovernmental Panel on Climate Change Fifth Assessment Report: Review of Working Group I Contribution. 2014. Available online: https://publications.parliament.uk/ (accessed on 16 June 2022).
4. Leiva, J. Assessment of climate change impacts on the water resources at the northern oases of Mendoza province, Argentina. In *Global Change in Mountain Regions*; Price, M.F., Ed.; Sapiens Publishing: Kirkmahoe, UK, 2006; pp. 81–83.
5. Vuille, M.; Bradley, R.S.; Werner, M.; Keimig, F. 20th century climate change in the tropical Andes: Observations and model results. In *Climate Variability and Change in High Elevation Regions: Past, Present & Future*; Springer: Berlin/Heidelberg, Germany, 2003; pp. 75–99.
6. Crimmins, S.M.; Dobrowski, S.Z.; Greenberg, J.A.; Abatzoglou, J.T.; Mynsberge, A.R. Changes in climatic water balance drive downhill shifts in plant species' optimum elevations. *Science* **2011**, *331*, 324–327. [CrossRef] [PubMed]
7. Cook, K.H.; Vizy, E.K. Effects of twenty-first-century climate change on the Amazon rain forest. *J. Clim.* **2008**, *21*, 542–560. [CrossRef]
8. Franchito, S.H.; Fernandez, J.P.R.; Pareja, D. Surrogate climate change scenario and projections with a regional climate model: Impact on the aridity in South America. *Am. J. Clim. Chang.* **2014**, *3*, 474. [CrossRef]
9. Sala, O.E.; Stuart Chapin, F.; Armesto, J.J.; Berlow, E.; Bloomfield, J.; Dirzo, R.; Huber-Sanwald, E.; Huenneke, L.F.; Jackson, R.B.; Kinzig, A. Global biodiversity scenarios for the year 2100. *Science* **2000**, *287*, 1770–1774. [CrossRef] [PubMed]
10. Borsch, T.; Hilu, K.W.; Wiersema, J.H.; Löhne, C.; Barthlott, W.; Wilde, V. Phylogeny of *Nymphaea* (Nymphaeaceae): Evidence from substitutions and microstructural changes in the chloroplast *trnT-trnF* region. *Int. J. Plant Sci.* **2007**, *168*, 639–671. [CrossRef]
11. Murphy, K.; Efremov, A.; Davidson, T.A.; Molina-Navarro, E.; Fidanza, K.; Betiol, T.C.C.; Chambers, P.; Grimaldo, J.T.; Martins, S.V.; Springuel, I. World distribution, diversity and endemism of aquatic macrophytes. *Aquat. Bot.* **2019**, *158*, 103127. [CrossRef]
12. Dalziell, E.L.; Lewandrowski, W.; Merritt, D.J. Increased salinity reduces seed germination and impacts upon seedling development in *Nymphaea* L. (Nymphaeaceae) from northern Australia's freshwater wetlands. *Aquat. Bot.* **2020**, *165*, 103235. [CrossRef]
13. Parveen, S.; Singh, N.; Adit, A.; Kumaria, S.; Tandon, R.; Agarwal, M.; Jagannath, A.; Goel, S. Contrasting reproductive strategies of two *Nymphaea* species affect existing natural genetic diversity as assessed by microsatellite markers: Implications for conservation and wetlands restoration. *Front. Plant Sci.* **2022**, *13*, 773572. [CrossRef]
14. Ngarega, B.K.; Nzei, J.M.; Saina, J.K.; Halmy, M.W.A.; Chen, J.-M.; Li, Z.-Z. Mapping the habitat suitability of *Ottelia* species in Africa. *Plant Divers.* **2022**. [CrossRef]
15. Nzei, J.M.; Mwanzia, V.M.; Ngarega, B.K.; Musili, P.M.; Wang, Q.-F.; Chen, J.-M.; Li, Z.-Z. Ecological Niche Modeling of water Lily (*Nymphaea* L.) species in Australia under climate change to ascertain habitat suitability for conservation measures. *Plants* **2022**, *11*, 1874. [CrossRef] [PubMed]
16. Nzei, J.M.; Ngarega, B.K.; Mwanzia, V.M.; Musili, P.M.; Wang, Q.-F.; Chen, J.-M. The past, current, and future distribution modeling of four water lilies (*Nymphaea*) in Africa indicates varying suitable habitats and distribution in climate change. *Aquat. Bot.* **2021**, *173*, 103416. [CrossRef]
17. Willis, S.G.; Hole, D.G.; Collingham, Y.C.; Hilton, G.; Rahbek, C.; Huntley, B. Assessing the impacts of future climate change on protected area networks: A method to simulate individual species' responses. *Environ. Manag.* **2009**, *43*, 836–845. [CrossRef] [PubMed]
18. Macedo, M.N.; DeFries, R.S.; Morton, D.C.; Stickler, C.M.; Galford, G.L.; Shimabukuro, Y.E. Decoupling of deforestation and soy production in the southern Amazon during the late 2000s. *PNAS* **2012**, *109*, 1341–1346. [CrossRef] [PubMed]
19. Morton, D.C.; DeFries, R.S.; Shimabukuro, Y.E.; Anderson, L.O.; Arai, E.; del Bon Espirito-Santo, F.; Freitas, R.; Morisette, J. Cropland expansion changes deforestation dynamics in the southern Brazilian Amazon. *PNAS* **2006**, *103*, 14637–14641. [CrossRef]
20. Papeş, M.; Gaubert, P. Modelling ecological niches from low numbers of occurrences: Assessment of the conservation status of poorly known viverrids (Mammalia, Carnivora) across two continents. *Divers. Distrib.* **2007**, *13*, 890–902. [CrossRef]
21. Elith, J.; Graham, C.H.; Anderson, R.P.; Dudík, M.; Ferrier, S.; Guisan, A.; Hijmans, R.J.; Huettmann, F.; Leathwick, J.R.; Lehmann, A. Novel methods improve prediction of species' distributions from occurrence data. *Ecography* **2006**, *29*, 129–151. [CrossRef]
22. Ngarega, B.K.; Masocha, V.F.; Schneider, H. Forecasting the effects of bioclimatic characteristics and climate change on the potential distribution of *Colophospermum mopane* in southern Africa using Maximum Entropy (Maxent). *Ecol. Inform.* **2021**, *65*, 101419. [CrossRef]
23. Phillips, S.J.; Elith, J. On estimating probability of presence from use availability or presence background data. *Ecology* **2013**, *94*, 1409–1419. [CrossRef]
24. Wan, J.-N.; Mbari, N.J.; Wang, S.-W.; Liu, B.; Mwangi, B.N.; Rasoarahona, J.R.; Xin, H.-P.; Zhou, Y.-D.; Wang, Q.-F. Modeling impacts of climate change on the potential distribution of six endemic baobab species in Madagascar. *Plant Divers.* **2021**, *43*, 117–124. [CrossRef] [PubMed]
25. Townsend Peterson, A.; Papeş, M.; Eaton, M. Transferability and model evaluation in ecological niche modeling: A comparison of GARP and Maxent. *Ecography* **2007**, *30*, 550–560. [CrossRef]
26. De Lima, C.T.; Guilietti, A.M. *Nymphaea vanildae* (Nymphaeaceae): A new species from the Caatinga in Brazil. *Phytotaxa* **2013**, *134*, 42–48. [CrossRef]
27. de Lima, C.T.; Machado, I.C.; Giulietti, A.M. Nymphaeaceae of Brasil. *SITIENTIBUS Série Ciências Biológicas* **2021**, *21*. [CrossRef]
28. Aiello-Lammens, M.E.; Boria, R.A.; Radosavljevic, A.; Vilela, B.; Anderson, R.P. spThin: An R package for spatial thinning of species occurrence records for use in ecological niche models. *Ecography* **2015**, *38*, 541–545. [CrossRef]

29. Fick, S.E.; Hijmans, R.J. WorldClim 2: New 1-km spatial resolution climate surfaces for global land areas. *Int. J. Climatol.* **2017**, *37*, 4302–4315. [CrossRef]
30. Esri, R. *ArcGIS Desktop: Release 10*; Environmental Systems Research Institute: Redlands, CA, USA, 2011.
31. Barve, N.; Barve, V.; Jiménez-Valverde, A.; Lira-Noriega, A.; Maher, S.P.; Peterson, A.T.; Soberón, J.; Villalobos, F. The crucial role of the accessible area in ecological niche modeling and species distribution modeling. *Ecol. Model.* **2011**, *222*, 1810–1819. [CrossRef]
32. Soberon, J.; Peterson, A.T. Interpretation of models of fundamental ecological niches and species' distributional areas. *Biodivers. Inform.* **2005**. [CrossRef]
33. Escobar, L.E.; Medina-Vogel, G.; Peterson, A.T. Potential for spread of the white-nose fungus (*Pseudogymnoascus destructans*) in the Americas: Use of Maxent and NicheA to assure strict model transference. *Geospat. Health* **2014**, *9*, 221–229. [CrossRef]
34. Booth, T.H. Checking bioclimatic variables that combine temperature and precipitation data before their use in species distribution models. *Austral Ecol.* **2022**. [CrossRef]
35. Team, R.C. *A Language and Environment for Statistical Computing*; R Foundation for Statistical Computing: Vienna, Austria, 2014; Available online: www.r-project.org (accessed on 14 February 2019).
36. Gent, P.R.; Danabasoglu, G.; Donner, L.J.; Holland, M.M.; Hunke, E.C.; Jayne, S.R.; Lawrence, D.M.; Neale, R.B.; Rasch, P.J.; Vertenstein, M. The community climate system model version 4. *J. Clim.* **2011**, *24*, 4973–4991. [CrossRef]
37. McSweeney, C.; Jones, R.; Lee, R.W.; Rowell, D. Selecting CMIP5 GCMs for downscaling over multiple regions. *Clim. Dyn.* **2014**, *44*, 3237–3260. [CrossRef]
38. Phillips, S.J.; Anderson, R.P.; Schapire, R.E. Maximum entropy modeling of species geographic distributions. *Ecol. Model.* **2006**, *190*, 231–259. [CrossRef]
39. Muscarella, R.; Galante, P.J.; Soley-Guardia, M.; Boria, R.A.; Kass, J.M.; Uriarte, M.; Anderson, R.P. ENM eval: An R package for conducting spatially independent evaluations and estimating optimal model complexity for Maxent ecological niche models. *Methods Ecol. Evol.* **2014**, *5*, 1198–1205. [CrossRef]
40. Phillips, S.J.; Dudík, M.; Schapire, R. Maxent Software for Modeling Species Niches and Distributions (Version 3.4.1). Available online: http.biodiversityinformatics.amnh.org/open_source/Maxent/ (accessed on 31 July 2017).
41. Brown, J.L. SDM toolbox: A python-based GIS toolkit for landscape genetic, biogeographic and species distribution model analyses. *Methods Ecol. Evol.* **2014**, *5*, 694–700. [CrossRef]
42. Merow, C.; Smith, M.J.; Silander Jr, J.A. A practical guide to MaxEnt for modeling species' distributions: What it does, and why inputs and settings matter. *Ecography* **2013**, *36*, 1058–1069. [CrossRef]
43. Deka, M.A.; Morshed, N. Mapping disease transmission risk of Nipah virus in South and Southeast Asia. *Trop. Med. Infect. Dis.* **2018**, *3*, 57. [CrossRef]
44. Zhao, D.; He, H.S.; Wang, W.J.; Wang, L.; Du, H.; Liu, K.; Zong, S. Predicting wetland distribution changes under climate change and human activities in a mid-and high-latitude region. *Sustainability* **2018**, *10*, 863. [CrossRef]
45. Bachman, G.R.; McMahon, M.J. Day and Night Temperature Differential (DIF) or the absence of far-red light alters cell elongation in celebrity white 'Petunia. *J. Am. Soc. Hortic. Sci.* **2006**, *131*, 309–312. [CrossRef]
46. Deb, J.C.; Phinn, S.; Butt, N.; McAlpine, C.A. The impact of climate change on the distribution of two threatened *Dipterocarp* trees. *Ecol. Evol.* **2017**, *7*, 2238–2248. [CrossRef]
47. Löhne, C.; Yoo, M.-J.; Borsch, T.; Wiersema, J.; Wilde, V.; Bell, C.D.; Barthlott, W.; Soltis, D.E.; Soltis, P.S. Biogeography of Nymphaeales: Extant patterns and historical events. *Taxon* **2008**, *57*, 1123–1146. [CrossRef]
48. Deininger, M.; Ward, B.M.; Novello, V.F.; Cruz, F.W. Late quaternary variations in the South American monsoon system as inferred by speleothems new perspectives using the SISAL database. *Quaternary* **2019**, *2*, 6. [CrossRef]
49. Corlett, R.T. Impacts of warming on tropical lowland rainforests. *Trends Ecol. Evol.* **2011**, *26*, 606–613. [CrossRef] [PubMed]
50. Hewitt, G.M. Genetic consequences of climatic oscillations in the Quaternary. *Philos. Trans. R. Soc. Lond. B Proc. Biol. Sci.* **2004**, *359*, 183–195. [CrossRef] [PubMed]
51. Ritter, N.P.; Crow, G.E.; Wiersema, J.H. *Nymphaea* (Nymphaeaceae) in Bolivia: Notes on several species, three new country records, and a key to species. *Rhodora* **2001**, *103*, 326–331.
52. de Sousa, D.; Matias, L.Q. Synopsis of genus *Nymphaea* L. (Nymphaeaceae) in state of Ceará. *Rev. Caatinga* **2012**, *25*, 72–78.
53. Langdon, J.G.; Lawler, J.J. Assessing the impacts of projected climate change on biodiversity in the protected areas of western North America. *Ecosphere* **2015**, *6*, 1–14. [CrossRef]
54. Thompson, J.; Peepre, J. Economic Benefits of Protected Areas to Community and Tourism 2014, pp. 1–39. Available online: http://nbsapforum.net/sites/default/files/Economic%20Benefits%20of%20Protected%20Areas.pdf (accessed on 18 June 2022).
55. Rowland, E.L.; Fresco, N.; Reid, D.; Cooke, H.A. Examining climate-biome ("cliome") shifts for Yukon and its protected areas. *Glob. Ecol. Conserv.* **2016**, *8*, 1–17. [CrossRef]
56. Hannah, L.; Midgley, G.; Andelman, S.; Araújo, M.; Hughes, G.; Martinez-Meyer, E.; Pearson, R.; Williams, P. Protected area needs in a changing climate. *Front. Ecol. Environ.* **2007**, *5*, 131–138. [CrossRef]
57. Hole, D.G.; Willis, S.G.; Pain, D.J.; Fishpool, L.D.; Butchart, S.H.; Collingham, Y.C.; Rahbek, C.; Huntley, B. Projected impacts of climate change on a continent wide protected area network. *Ecol. Lett.* **2009**, *12*, 420–431. [CrossRef]
58. Scanlon, B.R.; Jolly, I.; Sophocleous, M.; Zhang, L. Global impacts of conversions from natural to agricultural ecosystems on water resources: Quantity versus quality. *Water Resour. Res.* **2007**, *43*. [CrossRef]

59. Coe, M.T.; Costa, M.H.; Soares-Filho, B.S. The influence of historical and potential future deforestation on the stream flow of the Amazon river land surface processes and atmospheric feedbacks. *J. Hydrol.* **2009**, *369*, 165–174. [CrossRef]
60. Urban, M.C. Accelerating extinction risk from climate change. *Science* **2015**, *348*, 571–573. [CrossRef] [PubMed]
61. Butchart, S.H.; Clarke, M.; Smith, R.J.; Sykes, R.E.; Scharlemann, J.P.; Harfoot, M.; Buchanan, G.M.; Angulo, A.; Balmford, A.; Bertzky, B. Shortfalls and solutions for meeting national and global conservation area targets. *Conserv. Lett.* **2015**, *8*, 329–337. [CrossRef]
62. Ellison, K. Renting Biodiversity: The conservation concessions approach: With all the money we spend making conservation pay for itself, we could just pay for conservation. *Conserv. Pract.* **2003**, *4*, 20–29. [CrossRef]

diversity

Article

Characterization of Okra (*Abelmoschus esculentus* L.) Accessions with Variable Drought Tolerance through Simple Sequence Repeat Markers and Phenotypic Traits

Sonto Silindile Mkhabela [1,2,*], Hussein Shimelis [1,3], Abe Shegro Gerrano [2], Jacob Mashilo [1,3] and Admire Shayanowako [3]

1 Discipline of Crop Science, School of Agricultural, Earth and Environmental Sciences, University of KwaZulu-Natal, Pietermaritzburg 3209, South Africa
2 Industrial and Medicinal Plants, Agricultural Research Council—Vegetable, Private Bag X293, Pretoria 0001, South Africa
3 African Centre for Crop Improvement (ACCI), University of KwaZulu-Natal, Pietermaritzburg 3209, South Africa
* Correspondence: slindilentekati@gmail.com

Abstract: Genetic diversity analysis of crop genetic resources is a prerequisite for parental selection with suitable and complementary profiles for breeding. The objectives of this study were to determine genetic diversity present among okra accessions using simple sequence repeat (SSR) and complementary phenotypic markers and to select genetically divergent and superior parental accessions for pre-breeding. Twenty-six preliminarily selected okra accessions were assessed using nine highly polymorphic SSR markers and phenotyped under drought-stressed (DS) and non-stressed (NS) environmental conditions using a 13 × 2 alpha lattice design with two replications. Data were collected on the following eleven phenotypic traits: plant height (PH), days to 50% maturity (DTM), fresh pod length (FPL), dry pod weight (DPW), dry pod length (DPL), number of pods per plant (NPPP), pod yield per plant (PYPP), total above-ground biomass (AGB), harvest index (HI), root weight (RW), and root to shoot ratio (RSR). The SSR markers revealed an expected mean heterozygosity value of 0.54, indicating moderate genetic diversity among the tested okra accessions. Cluster analysis based on phenotypic and SSR markers differentiated the accessions into three distinct genetic groups. Wide phenotypic variation was observed for PH, FPL, NPPP, and PYPP under NS and DS conditions. PYPP was positively and significantly correlated with FPL ($r = 0.81$), ABG ($r = 0.69$), and HI ($r = 0.67$) under DS conditions, and FPL ($r = 0.83$) and AGB ($r = 0.60$) under NS conditions. Genetically complementary accessions such as LS04, LS05, LS06, LS07, LS08, LS10, LS11, LS15, LS18, LS23, LS24, and LS26 were identified for their high yield potential and related yield-improving traits under DS conditions. The identified accessions are recommended as parents for hybridization and selection programs to improve the yield potential of okra under drought-stressed environments.

Keywords: abiotic stress; genotyping; okra; phenotyping; molecular markers; SSR

Citation: Mkhabela, S.S.; Shimelis, H.; Gerrano, A.S.; Mashilo, J.; Shayanowako, A. Characterization of Okra (*Abelmoschus esculentus* L.) Accessions with Variable Drought Tolerance through Simple Sequence Repeat Markers and Phenotypic Traits. *Diversity* **2022**, *14*, 747. https://doi.org/10.3390/d14090747

Academic Editors: Hong-Hu Meng and Yi-Gang Song

Received: 11 August 2022
Accepted: 7 September 2022
Published: 11 September 2022

1. Introduction

Okra (*Abelmoschus esculentus* L., $2n = 130$) is an allotetraploid derived from the natural hybridization of a wild progenitor *A. tuberculatus* ($2n = 58$), with another yet unidentified species with $2n = 72$ chromosomes. Okra is a vegetable crop that is widely cultivated for its fresh and succulent pods [1,2]. Okra is an autogamous species and predominantly a self-pollinating crop. However, varying levels of cross-pollination have also been reported depending on the activity of insect pollinators and the growing environment [1]. The tender green pod is the most economical and vital source of vitamins A, B_1, B_3, B_6, folic acid, C, and K, essential for the human diet [3]. Potassium, magnesium, phosphorus, and calcium are the principal and essential mineral elements present in the green and immature pods of okra [4]. In addition, the pod contains 9.7% carbohydrate, 2.2 % protein, and 1% fibre [5]. Okra grains contain 22.14% protein, rare amino acids (such as lysine and tryptophan), fat, and fibre. The seed oil content varies from 20–40%, and the major fatty acids of the seed are linoleic acid (49.54%), palmitic acid (28.60%), and oleic acid [6]. These nutritional attributes make okra an important food security crop, especially in sub-Saharan Africa (SSA), where malnutrition is the highest. Africa accounts for 32.8% of the world's okra production. West and Central African countries contribute to over 75% of total okra production in SSA [7]. Despite the significant contribution by SSA toward global okra production, the average crop yields are low and variable in the region due to a lack of improved and modern varieties.

Okra is drought-tolerant crop and can successfully grow under water-limited conditions with minimal supplemental irrigation. Despite being relatively drought-tolerant, the crop fails to reach its maximum yield potential, resulting in low marketable pod yields, primarily when drought stress occurs at the flowering and pod development stages. For example, 37 to 83% yield losses attributed to drought stress occurred during the reproductive stage [8,9]. The low yield performance is related with the cultivation of low-yielding and drought-sensitive varieties.

The crop exhibits extensive morphological variation for traits such as plant height, fresh pod length, number of days to 50% flowering and maturity, number of branches, number of pods per plant, and pod yield [1,2]. Phenotypic traits such as plant height, number of branches, fresh pod length, number of pods per plant, total biomass, and seed yield exhibit positive associations with pod yield [4,9,10]. Hence, these traits could serve as useful product profiles for breeding of improved okra accessions for high yield and related traits. Therefore, rigorous phenotypic assessment of okra genetic resources will identify beneficial traits for future breeding. However, phenotypic traits are influenced by the genotype, environment, and genotype-by-environment interaction, confounding trait heritability and genotype performances [10,11]. To complement phenotypic assessment and for detailed genetic analysis, molecular markers are eminent genetic tools [1,6,12]. Molecular markers improve selection efficiency through phenotypic traits and accelerate genetic gains for desired traits.

Markers such as random amplified polymorphic DNA (RAPD) inter-simple sequence repeat (ISSR) [10,12,13], amplified fragment length polymorphism (AFLP) [14,15], sequence-related amplified polymorphism (SRAP), and simple sequence repeats (SSR) [16,17] have been successfully used. The marker systems were applied to explore genetic diversity and relatedness among okra genetic resources. These allowed for delineating heterotic groups among core collections of the crop and assisted in the selection and variety design. Among the molecular markers, the SSRs are highly polymorphic and reproducible markers useful for effective genotyping and selection programs [9,18,19].

Improved varieties of okra are yet to be developed and marketed for food security, better nutrition, and economic gains. There are limited drought-adapted varieties in SSA, which necessitates developing high-yielding and drought-tolerant okra genotypes adapted to the region. Therefore, the objectives of this study were to determine the genetic diversity present among okra accessions using simple sequence repeat and complementary phenotypic markers and to identify heterotic groups to select genetically divergent and superior parental accessions for pre-breeding.

2. Materials and Methods

2.1. Plant Materials

The present study used 25 okra landrace accessions sourced from the Agricultural Research Council, Vegetable, Industrial and Medicinal Plants (ARC-VIMP)/South Africa. Locally adapted and grown okra variety "Clemson Spineless" was also included as a comparative control. The code and number and geographical origin of the okra accessions used in the study are presented in Table 1.

Table 1. Accession code and number, and geographical origin of the okra accessions used in the study.

Accession Code	Accession Number	Geographical Origin
LS01	VI033775	Malaysia
LS02	VI033797	Malaysia
LS03	VI056457	Yugoslavia
LS04	VI039651	Bangladesh
LS05	VI046561	Thailand
LS06	VI047672	Bangladesh
LS07	VI050150	Taiwan
LS08	VI050957	Zambia
LS09	VI050960	Zambia
LS10	VI055110	Malaysia
LS11	VI055119	Myanmar
LS12	VI055219	Malaysia
LS13	VI055220	Malaysia
LS14	VI055421	Viet Nam
LS15	VI056069	Cambodia
LS16	VI056079	Cambodia
LS17	VI056081	Cambodia
LS18	VI056449	United States of America
LS19	VI060131	Mali
LS20	VI060313	Tanzania
LS21	VI060679	India
LS22	VI060803	Turkey
LS23	VI060817	Brazil
LS24	VI060822	Nigeria
LS25	VI060823	Nigeria
LS26	Clemson Spineless	South Africa

2.2. DNA Extraction, Purification, and Quantification

Okra seeds were sent to SciCorp Laboratories (SciCorp-lab, SA Pty Ltd., Pietermaritzburg, South Africa) for SSR analysis. Genomic DNA was extracted from 20 seeds per genotype using modified CTAB method [20]. The quantity and quality of total genomic DNA were determined by 0.7% Tris-Borate-EDTA (TBE) agarose gel electrophoresis and spectrophotometer, respectively. A working concentration of 20 ng μL^{-1} was standardized for all extracted DNA.

2.3. Polymerase Chain Reaction (PCR) and SSR Analysis

Okra seeds were found to be better for DNA sampling and analysis due to the mucilaginous material present in the leaves. Bulked DNA was used for amplification and analysis. SSR sequences were amplified through PCR using 9 selected diagnostic polymorphic SSR markers developed for okra (Table 2). These markers were selected based on their high polymorphic information content (PIC) and that they were developed and recommended for okra genetic diversity studies [18,21–23]. PCR amplification reaction contained 20 µL of PCR mix. The mix contained 1x PCR buffer, 3 mM MgCl, 1.25 U Taq polymerase, 0.2 mM dNTPs, 4pM each primer, and 5 ng genomic DNA (Bioline, Meridian, MI, USA). A PCR profile of initial denaturation for 2 min at 94 °C and 33 cycles of denaturation for 1 min at 94 °C, the annealing temperature of 63 °C for 2 min, and extension for 2 min at 72 °C was used. PCR products were fluorescently labelled and separated by capillary electrophoresis on ABI 3130 automatic sequencer (Applied Biosystems, Johannesburg, South Africa) and analysis was performed using GeneMapper 4.1 (Applied Biosynthesis, Johannesburg, South Africa). A 36 cm capillary and 3130 POP-7 polymer (Applied Biosystem, Johannesburg, South Africa) were used.

Table 2. Description of the SSR primers used for genotyping of 26 okra landrace accessions.

Marker Name	Forward Primer Sequence	Reverse Primer Sequence	PIC
Okra 111	GATGGAATTGAGAAACCAGA	TGTGTTCTTCACTCTCGTCA	0.89
Okra 152	GCTCTATTGATGGCGAGTAA	AAAGTCATCCAAGGTGACAA	0.81
Okra 166	TTCCAGTTGGAGAGGTAAGA	CTTCCATTTCATCGACTTTC	0.82
AVRDC-Okra17	ACGAGAGTGAAGTGGAACTG	CTCCTCTTTCCTTTTTCCAT	0.81
AVRDC-Okra70	GTAGCTGAACCCTTTGCTTA	CTATCATGGCGGATTCTTTA	0.98
AVRDC-Okra39	TGAGGTGATGATGTGAGAGA	TTGTAGATGAGGTTTGAACG	0.99
AVRDC-Okra64	AAGGAGGAGAAAGAGAAGGA	ATTTACTTGAGCAGCAGCAG	0.87
AVRDC-Okra9	ACCTTGAACACCAGGTACAG	TTGCTCTTATGAAGCAGTGA	0.85
AVRDC-Okra57	CGAGGAGACCATGGAAGAAG	ATGAGGAGGACGAGCAAGAA	0.78
Okra137	GAGAGAGATTGCTTCGACTG	TAAACTTTAAACTCAGCGGC	0.80

SSR = simple sequence repeats, PIC = polymorphic information content.

2.4. Marker Data Analysis

2.4.1. Computation of Principal Coordinate Analysis (PCoA) and Genetic Parameters

The GenAlex software version 6.5 [24] was used for data analyses and to summarize PCoA and genetic diversity parameters. Two approaches were adopted to investigate the genetic diversity and structure among the accessions. The first approach treated DNA polymorphisms as binary data (presence or absence). To determine the genetic structure within and among landraces, a second approach was adopted based on the co-dominant nature of the marker. Genetic parameters such as number of alleles per locus (Na), number of effective alleles per locus (Ne), allelic richness (Ar), Shannon's information index (I), observed heterozygosity (Ho), and expected heterozygosity (He) were calculated using GenAlex version 6.5 according to Nei and Li [25]. Polymorphic information content (PIC) was calculated using the formula PIC = $1 - \Sigma Pij^2$, where Pij is the frequency of the jth allele of the ith locus [25]. The number of polymorphic loci was estimated for each pre-determined group based on pedigree information. Online-based ClustVis (https://biit.cs.ut.ee/clustvis_large/, accessed on 22 May 2022) was used to visualize the heatmap, and plots of total genetic variation were analysed using pairwise genetic distance for haploid and co-dominant SSR markers [26].

2.4.2. Cluster Analysis

The binary data were used to obtain a dissimilarity matrix using the Jaccard index. The matrix was used to perform cluster analysis based on the unweighted pair group method using the arithmetic mean algorithm (UPGMA) in DARwin 5.0 software [27]. A dendrogram was then generated on the dissimilarity matrix to determine genetic relationships among the tested accessions. Bootstrap analysis was performed for node construction using 10,000 bootstrap values to estimate the reliability of the clustering pattern. A joint hierarchical cluster was generated to determine the association between genotypic and phenotypic data for both stressed and non-stressed conditions. The clusters were constructed using the "Cluster" package in R software [28].

2.5. Phenotyping Okra Accessions

2.5.1. Experimental Design and Crop Establishment

Two seeds of each genotype were grown in 5 L capacity plastic pots filled with composted pine bark growing media. Two plants were established per pot for each genotype. The day/night temperatures in the greenhouse (GH) were 30 °C/20 °C and the relative humidity ranged between 45 and 55% during the study. Inorganic fertilizers consisting of nitrogen (N), phosphorus (P), and potassium (K) were applied at a rate of 120, 30, and 30 kg ha^{-1}, based on soil fertility recommendations using urea (46-0-0), phosphorus pentoxide (P_2O_5), and potassium oxide (P_2O), respectively. The okra accessions were evaluated using a 13 × 2 alpha lattice design under drought-stressed (DS) and non-stressed (NS) conditions with two replications. DS was imposed at 50% flowering until physiological maturity to mimic terminal drought stress by withholding irrigation until the soil water content reached 30% field capacity. In addition, plots were irrigated at field capacity to allow for continued plant growth and development. The NS conditions involved maintaining soil moisture content at field capacity by supplying water through the dripper irrigation system until physiological maturity under GH environment. Tensiometers (Spectrum Technologies, Inc., Aurora, IL, USA) were used to monitor soil moisture status during the experiment.

2.5.2. Phenotypic Data Collection

Data were collected from three randomly selected and tagged plants for each genotype. At physiological maturity, data were collected on the following phenotypic traits. Plant height (PH) was measured in cm from the ground level to the apex of the plant on the main stem. Pods were harvested when 50% of the pods were 3–5 cm long, which is regarded as a marketable size [9]. Harvesting was conducted every third day by hand. At each harvest, the number of pods per plant (NPPP) were counted, and fresh pod length (FPL) was measured in cm. At the end of the experiment, data were computed on the number of pods per plant (NPPP), fresh pod length (FPL), and pod yield per plant (PYPP). Plants from the second pots were left until maturity to collect data on dry pod length (DPL) which was measured in cm, and mature dry pod weight (DPW) was determined by weighing dry pods harvested per plant and expressed in grams. Yield per plant was determined by weighing fresh pods harvested per plant and expressed in grams. The plants were cut at the soil surface to separate shoots and roots biomass. Total above-ground biomass (AGB) was determined in grams by weighing the stem and the pod of the plants per pot. Root weight (RW) was determined in grams by weighing all roots. Root to shoot ratio (RSR) was calculated as the ratio of shoot to root biomass. Harvest index (%) was calculated as HI = (pod weight/total above − ground biomass) $\times 100$.

2.5.3. Phenotypic Data Analysis

Phenotypic data were subjected to analysis of variance (ANOVA) using a lattice procedure with GenStat 18th Edition (VSN International, Hempstead, UK). Treatment means were separated using the least significant difference (LSD) at the 5% significance level. Pearson's correlation coefficients were calculated using IBM SPSS Statistics 25.0 (SPSS Inc., Chicago, IL, USA, 2008) to determine the magnitude of the relationship among phenotypic traits. Principal component analysis (PCA) was used to identify influential traits under NS and DS conditions using R Studio version 4.0, ggplot2(R Core Team, 2020, Vienna, Austria). Biplots were constructed using R version 4.0, ggplot2 (R Core Team, 2020) to determine relationships between the accessions and the assessed phenotypic traits. Hierarchical clusters were generated using phenotypic data based on the Gower method [29], using the Cluster package in R software [28].

3. Results and Discussion

3.1. Marker Characterization

Understanding the genetic diversity present among diverse okra accessions is useful for identifying and selecting suitable and contrasting parental genotypes for breeding, leading to the accelerated development of improved varieties. The estimated genetic parameters derived using SSR markers are presented in Table 3. The SSR markers amplified 24 putative alleles among the tested okra accessions, ranging from 2 for the markers AVRDC-OKRA64, Okra 152, Okra 162, Okra 111, and AVRDC-Okra57 to 5 for the marker AVRDC-Okra 9 with a mean value of 2.70 alleles per locus. Out of 10 selected SSRs used in this study, 9 SSR primers could amplify successfully. The level of amplification in the present study was higher compared to the 75% reported in okra by Kumar et al. [30]. This indicates the suitability of the sampled SSR markers for the analysis of genetic variation and relationship in okra. A total of 24 alleles were amplified, with an average of 2.70 alleles per locus (Table 3), and this average number of alleles indicates that this genetic diversity would be relatively moderate [31]. This was lower than the value of 71 alleles per locus reported by Mohammed et al. [18] when assessing 32 okra accessions genotypes with 16 SSR markers. The variability in the number of alleles observed could be attributed to the genetic differences in the tested lines and the difference in the sampled SSR markers. Effective allele number (Ne) ranged from 2 for markers AVRDC-OKRA64, Okra 152, Okra 162, Okra 111, and AVRDC-Okra57 to 2.76 for marker AVRDC-Okra9 with a mean of 2.24, indicating that this genetic diversity would be relatively small. However, greater diversity has been reported [16], indicating a mean number of 4.8 effective alleles after evaluating 20 okra accessions using SSR markers. This corroborates with the results of two to seven (mean = four) alleles per locus reported by Kpodo et al. [15]. The mean Shannon information index value of the test population was 0.83, ranging from 0.69 for markers AVRDC-OKRA64, Okra 152, Okra 162, Okra 111, and AVRDC-Okra57 to 1.16 for marker AVRDC-Okra9. The observed heterozygosity value was 1, suggesting that all the accessions reached 100% heterozygosity. The expected heterozygosity ranged from 0.50 to 0.64 with a mean value of 0.54. Marker AVRDC-Okra9 had the highest He of 0.64. The inbreeding coefficient varied from −0.57 to −1.00, with a mean of −0.85. Nine markers (100%) were highly polymorphic with PIC > 0.50, indicating their high discriminating ability and their utility for genetic analysis studies in okra. PIC values ranged from 0.50 to 0.64 with a mean of 0.55, which is relatively higher than the mean PIC value of 0.51 reported by Kpodo et al. [15] and lower than the PIC value of 0.81 reported by Mohammed et al. [18] in okra. The average PIC value of 0.55 indicates that these markers are informative for genetic diversity analysis [22]. High polymorphism values suggest that the selected markers are suitable for distinguishing the genetic diversity among the tested accessions. The high polymorphism values observed when using the sampled SSR markers may be due to the amphipolyploid nature of *Abelmoschus* species. In addition, there is a higher frequency of mutations in polyploids, such as in okra, than diploids [21], leading to increased genetic diversity and genetic plasticity [10].

Table 3. Genetic diversity parameters generated by SSR markers among 26 okra accessions.

Marker	Genetic Parameters						
	Na	Ne	I	Ho	He	F_{IS}	PIC
AVRDC-Okra70	3	2.47	0.97	1.00	0.60	−0.68	0.60
AVRDC-Okra64	2	2.00	0.69	1.00	0.50	−1.00	0.50
Okra 152	2	2.00	0.69	1.00	0.50	−1.00	0.50
Okra 166	2	2.00	0.69	1.00	0.50	−1.00	0.50
AVRDC-Okra9	5	2.76	1.16	1.00	0.64	−0.57	0.64
AVRDC-Okra39	3	2.31	0.91	1.00	0.57	−0.76	0.57
Okra 111	2	2.00	0.69	1.00	0.50	−1.00	0.50
Okra137	3	2.58	1.01	1.00	0.61	−0.63	0.61
AVRDC-Okra57	2	2.00	0.69	1.00	0.50	−1.00	0.50
Average	2.70	2.24	0.83	1.00	0.54	−0.85	0.55
Standard deviation	1.00	0.30	0.18	0.00	0.06	0.19	0.06
Standard error	0.34	0.15	0.10	0.10	0.06	0.06	0.02

Na = total number of alleles per locus; Ne = number of effective alleles per locus; I = Shannon information index; Ho = observed heterozygosity; He = expected heterozygosity; F_{IS} = inbreeding coefficient; PIC = polymorphic information content.

3.2. Principal Coordinate Analysis (PCoA) of 26 Okra Accessions Genotyped Using 9 SSR Markers

The genetic structure of the assessed accessions was inferred with the PCoA based on the genetic matrix (Figure 1). The first principal coordinate (PC) accounted for 25.19% of variation present among accessions. The coordinate analysis indicated higher genetic diversity among the two accessions LS02 and LS11 compared to other genotypes, due to their inherent genetic variation. The second principal component suggests a further separation between LS02, LS11, and LS13, accounting for 18.24% of the total variation. The grouping of LS01, LS03, LS04, LS09, and LS26 into the same cluster may indicate the genetic similarity among these accessions. Hence, the genetic information generated can be useful to design crosses and exploit genetic diversity through selection programs.

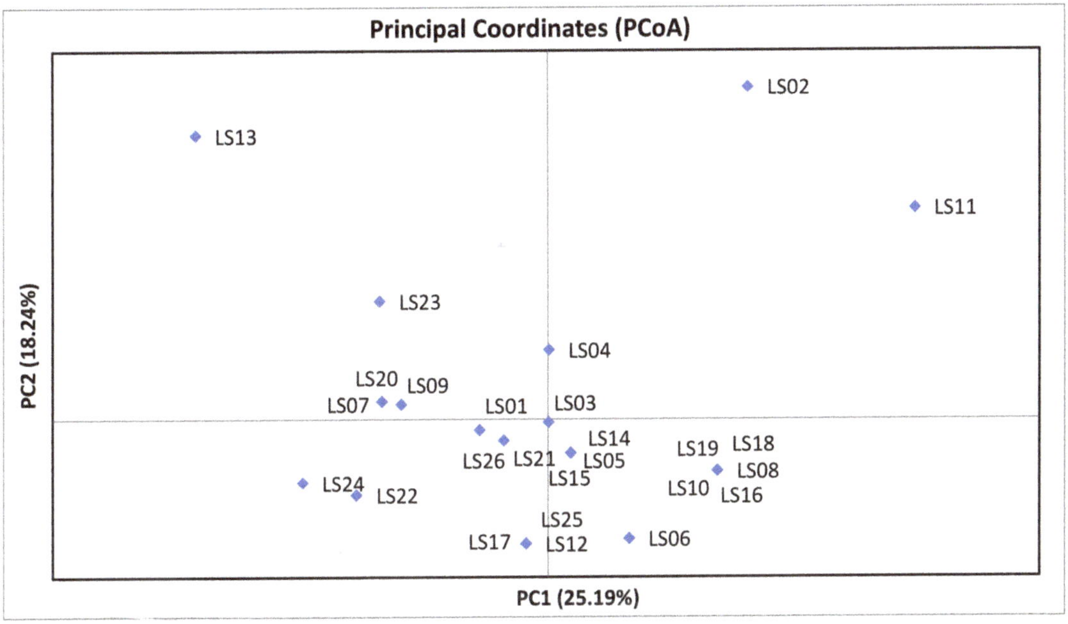

Figure 1. Principal coordinate analysis (PCoA) based on 9 polymorphic SSR markers and 26 okra accessions (coded LS01 to LS26, Table 1). Note PC1 denotes the first principal coordinate and PC2 the second principal coordinate.

3.3. Heatmap Cluster

A heatmap based on SSR marker transcription was constructed using the hierarchical clustering method discerning the genetic relationship of 26 okra accessions based on Jaccard's coefficient (Figure 2). The assayed okra accessions and SSR markers each were grouped into two main clusters. The first cluster had one subcluster consisting of eight accessions, of which four were collected from Malaysia (LS02 and LS10), Myanmar (LS11), and Cambodia (LS16) in Asia and two from Mali (LS19) and Zambia in Africa (LS08). The second cluster contained two subclusters with seven accessions, including LS24, LS13, LS07, LS22, LS23, LS09, and LS20 on the first subcluster, which was dominated by accessions collected from Nigeria, Malaysia, Taiwan, Turkey, Brazil, Zambia, and Tanzania, respectively, and eleven accessions LS06, LS01, LS26, LS25, LS12, LS17, LS04, LS21, LS15, LS05, and LS14 on the second subcluster, of which nine were sampled from Asia and one each from South Africa (LS26) and Nigeria (LS25). The observed genetic dissimilarity indicates that these accessions are related to different geographic locations and most of the cultivated accessions in each geographic region were uniquely differentiated. This may be due to the limited outcrossing rate among the geographic regions of the evaluated okra accessions. Genetic variability among okra accessions was also reported by Massucato et al. [12]. Information on the genetic grouping of the accessions is essential in selecting contrasting parents based on the breeding history and genetic relationship of the assessed population and test environment.

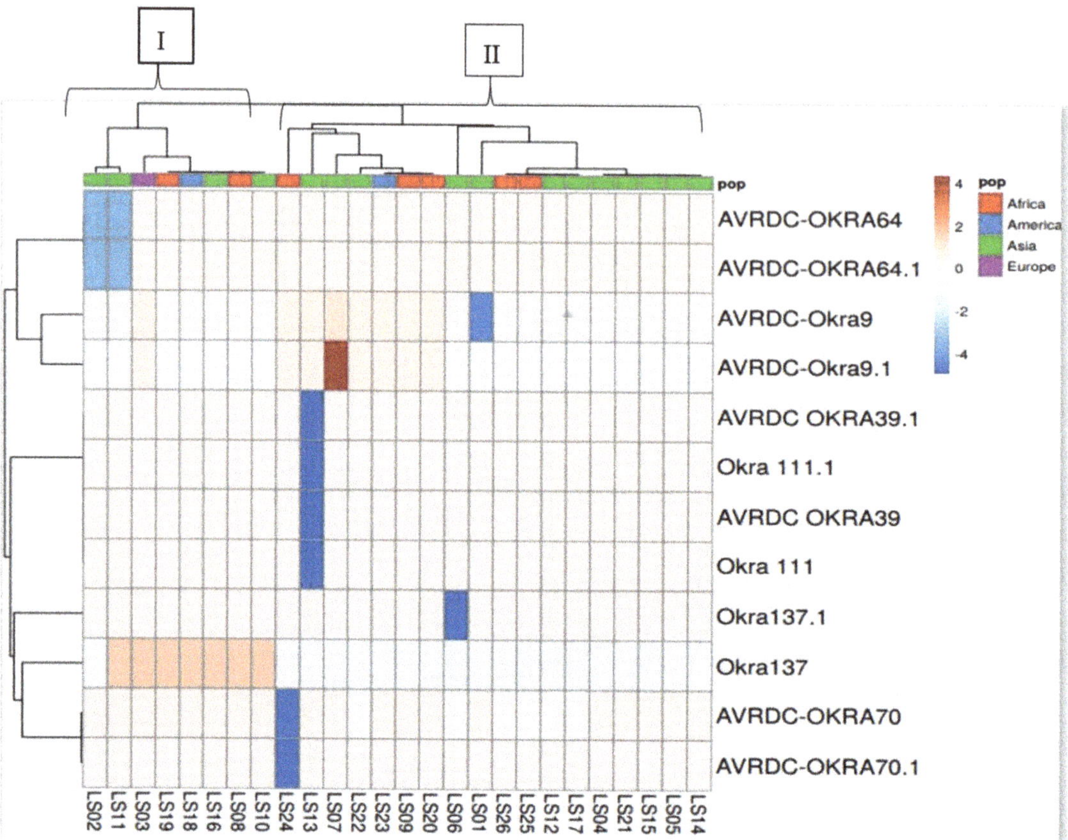

Figure 2. Heatmap showing the genetic relationship among 26 okra accessions using 9 SSR markers. Annotations in the heatmap show grouping of accessions and SSR marker clusters.

3.4. Cluster Analysis

Figure 3 summarizes the unweighted pair group method with the arithmetic mean method using Jaccard's dissimilarity matrix, showing the genetic inter-relationships among the studied okra accessions. Clustering is a multivariate technique that assists in indicating the pattern of genetic relationships among accessions. The accessions were grouped into three distinct major clusters, namely cluster I, consisting of eight accessions, and clusters II and III, consisting of nine accessions each, indicating the presence of a wide genetic variation among the studied okra accessions. This corroborates with the findings of Reddy et al. [7], Pradip et al. [32], and Ravishankar et al. [33], who presented a dendrogram that classified the tested accessions into three major groups. Accessions allocated in different clusters are genetically divergent and may serve as prospective parents for a breeding programme. Most accessions maintained their positions on the dendrogram compared to the heatmap cluster analysis, except for the accessions LS01, LS03, and LS06. These clustering patterns indicate that accessions from different regions were genetically diverse. The high diversity among the accessions makes the assessed genotypes unique genetic resources to develop new breeding populations.

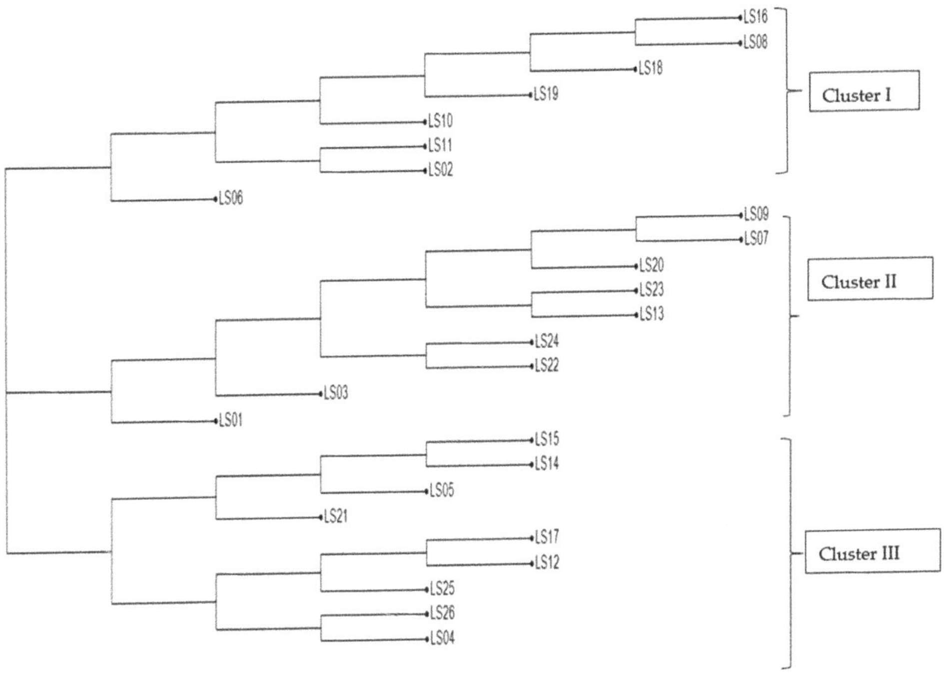

Figure 3. Dendrogram showing genetic relationships among 26 okra accessions assessed via 9 polymorphic SSR markers.

3.5. Accession and Environmental Effects on Phenotypic Traits

Analysis of variance indicated significant ($p < 0.05$) differences among the test accessions under different water treatments and their interactions for the assessed phenotypic traits (Table 4). Accessions had significant ($p < 0.05$) difference for PH, DTM, NPPP, PYPP, AGB, RW, and FPL. The water treatments had a significant ($p < 0.05$) effect on PH, FPL, and PYPP. The genotype × water treatment interaction exerted a significant ($p < 0.05$) effect on DTM, FPL, and PYPP (Table 4).

Table 4. Analysis of variance showing mean square values and significant tests of 26 okra accessions assessed for phenotypic responses in glasshouse environment under drought-stressed (DS) and non-stressed (NS) conditions.

S.O.V.	df	PH	DTM	FPL	DPL	DPW	NPPP	PYPP	AGB	HI	RW	RSR
Replications	1	23.80 ns	11.44 ns	21.61 *	35.41 ns	81.39 *	0.01 ns	55.20 *	1423.50 **	1153 ns	1.50 ns	1.09 **
Incomplete blocks	1	2063.50 **	0.08 ns	9.99 ns	11.22 ns	13.89 ns	3.47 ns	1.11 ns	104.56 ns	790.10 ns	85.87 *	0.09 ns
Genotype (G)	26	336.40 *	225.57 *	15.15 **	13.76 ns	7.54 ns	7.92 *	15.97 *	136.00 *	664.10 ns	17.97 *	0.14 ns
Water regime (WC)	1	2231.00 **	75.84 ns	77.13 **	13.18 ns	10.93 ns	16.56 *	229.47 **	578.52 ns	4736.10 *	82.41 *	0.04 ns
G × WC	25	234.60 ns	89.58 *	6.82 *	10.15 ns	6.99 ns	4.43 ns	12.01 *	55.27 ns	714.90 ns	8.91 ns	0.07 ns
Residual	49	139.80	48.05	3.96	11.88	8.26	3.98	6.96	75.76	429.40	10.19	0.09

S.O.V.: source of variation, PH: plant height, DTM: days to maturity, FPL: fresh pod length, DPL: dry pod length, DPW: dry pod weight, NPPP: number of pods per plant, PYPP: pod yield per plant, AGB: above-ground biomass, HI: harvest index, RW: root weight, RSR: root: shoot ratio, * significant at 5% level of significance, ** significant at 1% level of significance, ns non-significant.

3.5.1. Performance of Okra Accessions for Phenotypic Traits under Drought-Stressed and Non-Stressed Conditions

Phenotypic traits provide useful selection criteria for genotype selection and breeding. Mean values of phenotypic traits recorded among the tested 26 okra accessions evaluated under DS and NS treatments are presented in Table 5. The present study revealed a significant genotype × water regime effect interaction for several traits, including days to maturity, fresh pod length, and fresh pod yield. This allowed the identification and selection of ideal accessions suited for irrigated and drought-prone environments. Highly significant genotypic differences ($p < 0.001$) were observed for PH under NS conditions. Plant height is an important agronomic trait that reflects the vegetative growth behaviour of crop plants in response to drought-stressed conditions. In the present study, drought stress reduced plant height (Table 6), with accessions LS01, LS02, LS08, LS09, LS11, LS13, and LS18 being the tallest under DS conditions. According to Eshiet and Brisibe [34], the height of an okra plant can potentially affect yield, as taller plants are more prone to lodging, thus resulting in a reduced number of pods and yield. Taller plant height was observed for accessions LS01, LS02, LS08, LS09, LS11, LS13, and LS18 under DS than NS that could be attributed to higher biomass portioning in okra as a means for higher yield and drought tolerance [35]. Days to maturity is an important trait in evaluating and selecting drought-tolerant okra genotypes. Under water-limited conditions, plants synthesize phytohormones, which synchronize the transition from vegetative to reproductive phases. Hence, the synthesis allows plants to regulate flowering, reproduction, and maturity periods [36]. Accessions LS01, LS03, LS06, LS08, and LS13 were early maturing under DS conditions (<85 days to maturity). Early maturing accessions could be selected as parents when breeding for drought escape through early maturity. Significant ($p < 0.05$) genotypic differences were recorded for FPL under both DS and NS conditions. Accessions LS06, LS07, and LS10 recorded the highest FPL (>8 cm) under DS conditions, whereas accessions LS10 and LS21 had the highest FLP (>10 cm) under NS conditions. Dry weight showed a reduction from 2.58 to 1.92 g under DS conditions. Chaturvedi et al. [9] reported that the reduction in dry weight is associated with the suppression of cell expansion and cell growth due to lower turgor pressure that occurs when plants are experiencing water shortages. Another record of reduced plant dry weight under drought stress was reported by Komolafe et al. [3]. Significant ($p < 0.05$) genotypic differences were recorded for NPPP under DS conditions. Accessions LS03, LS04, LS11, LS12, LS15, LS17, LS18, and LS21 recorded the highest NPPP (≥5), while LS13, LS19, and LS25 recorded the lowest NPPP (<2) under DS conditions. Drought stress reduced NPPP in okra accessions due to the disturbance in photosynthesis and low carbohydrate production caused by limited water availability [37].

Significant ($p < 0.05$) genotypic differences were recorded for PYPP under DS conditions. Accessions LS05, LS07, LS18 and LS10 recorded the highest PYPP (≥10 g). Under drought stress conditions, okra plants can accumulate sufficient photo-assimilates, resulting in higher YPPP [9]. In addition, Komolafe et al. [3], reported that pod yield could be improved with selection of a higher number of pods per plant and heavier pods as breeding parents. Genotypes LS19, LS20, and LS25 recorded the lowest PYPP (<1 g) under DS conditions. The reduction in PYPP in DS is attributed to low water availability, which reduces cell division, resulting in lower dry matter and pod yield [37]. Significant ($p < 0.05$) genotypic differences were recorded for AGB under both DS and NS conditions. There were significant ($p < 0.05$) differences among accessions for HI under NS conditions only. The highest HI (>70%) was observed for accessions LS12, LS21, and LS26, whereas the lowest HI (<10%) was recorded for LS08 and LS19 under NS conditions. Non-significant differences were recorded for RW under both DS and NS conditions. Significant ($p < 0.05$) genotypic differences were recorded for RSR under DS condition. Genotype LS19 recorded the highest RSR (>1) compared to all other test accessions under DS conditions. Under water-limited conditions, the productivity of a plant depends on some essential processes, such as temporal biomass distribution and dry matter partitioning [9]. Hence, the high fresh and dry weight of plants under restricted water supply is desirable and relates to high conversion efficiency. In the present study, accessions LS08, LS10, LS17, and LS23 indicated higher biomass production. Selecting parents with high biomass expression can help improve genetic gains. Based on this study, it can be indicated that reductions in most studied traits were highly associated with drought stress. These traits can effectively assess the drought tolerance potential of okra accessions and genotype variability for the studied traits and can be used to improve okra through selection.

Table 5. Mean values for phenotypic traits among 26 okra accessions evaluated under drought-stressed (DS) and non-stressed (NS) conditions.

Accession Code	PH (cm)		DTM		FPL (cm)		DPL (cm)		DPW (g Per Plant)		NPPP		PYPP (g Per Plant)		AGB (g Per Plant)		HI (%)		RW (g Per Plant)		RSR	
	DS	NS	DS	NS	DS	NS	DS	NS	DS	NS	DS	NS	DS	NS	DS	NS	DS	NS	DS	NS	DS	NS
LS01	71.50	64.25	80.25	95.17	3.92	7.96	11.59	7.78	6.50	2.00	3.50	8.00	3.92	7.02	13.92	13.02	17.15	54.91	5.25	7.75	0.58	0.59
LS02	65.37	61.62	101.00	95.67	5.19	7.13	9.13	1.75	4.00	0.00	3.50	3.50	2.58	7.83	7.08	17.83	36.67	44.47	6.00	5.25	0.83	0.30
LS03	60.50	68.38	83.00	83.50	3.03	5.96	5.15	6.83	3.50	3.50	8.00	7.00	2.50	8.79	17.50	17.29	13.96	51.86	5.25	6.25	0.33	0.37
LS04	49.87	52.12	86.25	86.75	6.21	7.50	8.29	8.54	1.00	2.50	6.00	5.50	4.19	6.09	9.19	12.09	47.52	56.69	0.75	3.00	0.09	0.29
LS05	54.50	72.50	89.00	80.25	5.73	7.25	6.50	7.08	2.00	8.00	4.00	5.50	6.17	7.33	16.17	18.33	38.30	44.78	5.25	3.75	0.32	0.27
LS06	61.50	82.88	77.00	77.00	8.23	9.06	4.50	8.85	2.50	5.50	4.00	5.00	5.05	8.00	16.55	29.50	62.42	30.64	4.00	7.50	0.50	0.25
LS07	58.38	70.38	86.25	89.00	8.58	4.96	7.60	9.00	3.00	6.50	4.00	3.50	7.92	2.92	16.92	14.92	65.16	29.01	5.00	4.50	0.34	0.37
LS08	73.50	64.00	77.00	95.00	6.00	1.50	3.50	4.67	1.50	1.00	4.50	2.50	6.58	2.63	21.58	17.12	41.57	8.97	8.75	12.25	0.49	0.96
LS09	82.25	69.75	89.67	83.50	3.80	7.83	7.50	4.21	1.00	1.00	3.00	3.00	4.60	7.17	17.10	16.67	21.83	48.07	10.50	7.75	0.85	0.50
LS10	68.75	88.25	95.17	90.00	10.17	11.50	0.00	8.38	0.00	1.50	3.00	3.50	14.00	13.25	32.00	37.75	43.75	35.80	4.00	7.25	0.13	0.20
LS11	79.00	66.50	95.50	86.00	6.51	9.75	8.33	3.63	3.00	0.00	6.00	4.00	6.76	9.23	16.76	19.23	40.34	47.33	7.75	9.00	0.46	0.47
LS12	65.25	82.38	101.00	83.17	3.99	5.51	3.75	5.08	2.50	4.00	6.50	6.50	2.85	7.68	19.85	21.68	19.12	46.84	5.00	7.25	0.33	0.40
LS13	74.25	49.75	80.25	98.25	2.50	7.46	5.71	6.19	2.50	0.50	1.00	5.50	2.00	6.13	17.00	9.13	10.98	79.55	6.50	4.00	0.40	0.48
LS14	65.50	87.25	86.42	83.17	5.93	8.09	6.44	7.38	0.50	3.00	4.50	6.00	4.48	8.56	18.97	23.06	23.56	60.68	8.00	8.25	0.42	0.49
LS15	52.50	57.88	89.00	92.42	6.71	5.47	6.40	6.40	1.50	1.50	5.00	6.00	4.71	4.82	10.71	8.32	47.71	57.66	3.50	3.75	0.40	0.45
LS16	53.50	78.12	95.17	92.25	3.95	7.75	7.40	6.92	3.50	3.00	2.50	4.50	2.63	11.55	13.62	24.55	18.99	47.05	4.25	8.75	0.29	0.36
LS17	67.00	79.00	98.00	83.50	4.23	3.88	4.75	8.08	1.00	5.00	5.50	3.50	3.69	6.00	21.19	20.00	23.54	30.08	7.00	9.25	0.49	0.46
LS18	83.62	75.00	98.25	86.75	7.21	7.33	7.75	10.38	1.50	1.50	6.00	3.50	5.42	6.10	16.42	25.10	33.94	24.68	10.75	6.25	0.74	0.25
LS19	54.12	72.00	95.50	83.50	1.00	0.00	4.00	5.25	0.50	0.00	1.00	1.00	0.50	0.00	6.50	12.50	16.67	0.00	5.50	9.25	1.43	0.73
LS20	60.25	67.12	95.16	89.67	1.81	8.46	1.88	7.63	0.00	2.50	2.50	4.00	0.75	8.08	11.25	18.08	7.89	48.95	4.50	6.25	0.45	0.42
LS21	54.12	66.25	92.17	92.25	8.50	10.08	7.58	7.58	4.00	1.50	6.00	6.00	4.17	9.58	10.17	12.08	41.96	84.21	2.00	2.75	0.27	0.24
LS22	63.50	119.25	89.67	92.75	5.52	8.04	5.83	7.63	0.00	6.00	3.00	6.00	1.75	11.44	10.75	37.94	12.96	30.57	6.25	11.75	0.62	0.31
LS23	69.75	86.75	92.17	89.50	4.94	7.29	6.13	8.79	2.50	5.00	2.50	7.00	5.88	8.00	20.87	34.50	30.75	23.72	6.50	10.00	0.35	0.28
LS24	59.12	86.25	95.17	77.00	4.83	5.31	1.50	3.38	0.00	0.50	3.00	2.50	4.17	6.88	17.17	25.87	27.86	19.23	1.00	9.50	0.11	0.42
LS25	59.62	83.62	101.00	90.00	0.00	6.92	0.00	3.50	0.00	0.00	1.00	3.50	0.00	7.04	12.50	26.54	0.00	35.18	1.50	10.75	0.04	0.41
LS26	46.12	52.62	86.75	83.50	4.67	5.63	5.45	3.05	2.00	1.50	2.00	6.00	4.00	5.24	6.00	7.24	33.33	82.28	1.75	2.50	0.15	0.22
Mean	63.59	73.23	90.61	87.67	5.12	6.83	5.64	6.46	1.92	2.58	3.90	4.71	4.28	7.21	15.30	20.01	29.92	43.20	5.25	7.10	0.44	0.40
p-value	ns	**	*	ns	*	*	ns	ns	*	*	*	ns	*	ns	*	*	ns	*	ns	ns	*	ns
SED	11.73	12.84	6.29	7.49	1.73	2.13	3.19	3.77	2.41	3.25	1.85	2.17	2.29	2.95	7.63	9.81	21.16	20.58	3.10	3.29	0.37	0.17
LSD (5%)	34.24	26.44	12.96	15.97	5.05	4.39	9.32	7.77	7.05	6.69	5.41	4.48	6.68	6.08	22.26	20.21	43.67	42.38	9.04	6.77	1.09	0.35
CV (%)	18.48	17.53	6.94	8.54	34.77	31.22	55.83	58.42	65.12	56.10	47.68	46.12	56.79	40.93	50.20	49.02	69.23	47.63	58.87	46.35	63.74	42.66

PH: plant height, DTM: days to maturity, FPL: fresh pod length, DPL: dry pod length, DPW: dry pod weight, NPPP: number of pods per plant, PYPP: pod yield per plant, AGB: above-ground biomass, HI: harvest index. RW: root weight, RSR: root: shoot ratio; NS: non-stressed, DS: drought-stressed, SED: standard deviation, LSD: least significant different, CV: coefficient of variation, * at 5% level of significance, ** significant at 1% level of significance, ^{ns} non-significant.

Table 6. Pearson correlation coefficients showing the magnitude of associations of phenotypic traits among okra accessions under drought-stressed (upper diagonal) and non-stressed (lower diagonal) conditions.

Traits	PH	DTM	FPL	DPL	DPW	NPPP	PYPP	AGB	HI	RW	RSR
PH		0.01 ^{ns}	0.06 ^{ns}	0.14 ^{ns}	0.07 ^{ns}	0.10 ^{ns}	0.24 ^{ns}	0.52 **	−0.19 ^{ns}	0.81 **	0.32 ^{ns}
DTM	−0.22 ^{ns}		−0.20 ^{ns}	−0.24 ^{ns}	−0.27 ^{ns}	−0.05 ^{ns}	0.52 **	−0.06 ^{ns}	−0.29 ^{ns}	−0.05 ^{ns}	0.07 ^{ns}
FPL	0.16 ^{ns}	0.12 ^{ns}		0.21 ^{ns}	0.12 ^{ns}	0.43 *	0.81 **	0.36 ^{ns}	0.85 **	0.03 ^{ns}	−0.23 ^{ns}
DPL	0.24 ^{ns}	0.03 ^{ns}	0.26 ^{ns}		0.71 **	0.30 ^{ns}	−0.03 ^{ns}	−0.35 ^{ns}	0.25 ^{ns}	0.29 ^{ns}	0.28 ^{ns}
DPW	0.41 *	−0.29 ^{ns}	0.06 ^{ns}	0.58 **		0.27 ^{ns}	0.04 ^{ns}	−0.14 ^{ns}	0.17 ^{ns}	0.03 ^{ns}	0.05 ^{ns}
NPPP	−0.03 ^{ns}	0.15 ^{ns}	0.39 **	0.24 ^{ns}	0.38 ^{ns}		0.26 ^{ns}	0.19 ^{ns}	0.34 ^{ns}	0.14 ^{ns}	−0.15 ^{ns}
PYPP	0.46 **	0.08 ^{ns}	0.83 **	0.16 ^{ns}	0.15 ^{ns}	0.37 ^{ns}		0.69 **	0.67 **	0.13 ^{ns}	−0.27 ^{ns}
AGB	0.88 **	−0.14 ^{ns}	0.36 ^{ns}	0.27 ^{ns}	0.30 ^{ns}	−0.06 ^{ns}	0.60 **		0.13 ^{ns}	0.34 ^{ns}	0.27 ^{ns}
HI	−0.47 *	0.22 ^{ns}	0.50 **	−0.07 ^{ns}	−0.11 ^{ns}	0.61 **	0.31 ^{ns}	−0.48 *		−0.09 ^{ns}	−0.13 ^{ns}
RW	0.70 **	−0.02 ^{ns}	−0.24 ^{ns}	−0.12 ^{ns}	−0.05 ^{ns}	−0.32 ^{ns}	−0.08 ^{ns}	0.62 **	−0.69 **		0.58 **
RSR	−0.18 ^{ns}	0.20 ^{ns}	−0.65 **	−0.29 ^{ns}	−0.33 ^{ns}	−0.35 ^{ns}	−0.57 **	−0.31 ^{ns}	−0.37 ^{ns}	0.49 *	

PH: plant height, DTM: number of days to maturity, FPL: fresh pod length, DPL: dry pod length, DPW: dry pod weight, NPPP: number of pods per plant, PYPP: yield per plant, AGB: above-ground biomass, HI: harvest index, RW: root weight, RSR: root: shoot weight, * significant at 5% level of significance, ** significant at 1% level of significance, ^{ns} non-significant.

3.5.2. Associations among Phenotypic Traits under Drought-Stressed and Non-Stressed Conditions

Understanding the associations between phenotypic traits provides a useful guide for the selection and improvement of desired traits. The levels of associations of the assessed phenotypic traits among accessions under DS and NS conditions in a GH environment are presented in Table 6. Correlation analysis provides a measure of associations among traits for effective selection. The poor associations recorded between DTM with DPL and NPPP suggest small trade-offs in pod yield. The high correlations between PYPP and DTM under drought stress suggest large trade-offs in yield responses. Under DS conditions, significant and positive associations were observed between PYPP with FPL (r = 0.81, $p \leq 0.001$). AGB had higher significant association with PYPP (r = 0.69, $p \leq 0.001$). HI positively and significantly correlated with FPL (r = 0.85, $p \leq 0.001$) and PYPP (r = 0.67, $p \leq 0.001$) under DS conditions, indicating that harvest index has a direct influence on pod yield. According to Kyriakopoulou et al. [10], crops under water-limited conditions show a significantly reduced harvest index. This is attributable to reduced photosynthesis, which could change pod yield. Positive and significant correlations were observed between RW and PH (r = 0.81, $p \leq 0.001$). RSR significantly and positively correlated with DPW (r = 0.58, $p \leq 0.001$) under DS conditions.

FPL positively and significantly correlated with PYPP (r = 0.83, $p \leq 0.001$) and HI (r = 0.85, $p \leq 0.001$) under NS condition. A positive and significant correlation between FPL and PYPP has been reported in okra [19,37], indicating that fresh pod length is vital for direct selection to improve the fresh pod yield in okra. Positive and highly significant correlations were recorded between AGB with PH (r = 0.88, $p \leq 0.001$) and PYPP (r = 0.60, $p \leq 0.001$) under NS conditions. There was a high positive association between above-ground biomass and plant height under drought stress conditions. This indicates that drought stress has a maximum impact on plant height due to the declined cell enlargement and cell growth due to low turgor pressure and more leaf senescence. Hence, more leaf senescence and reduced photosynthesis result in low biomass production in crops grown under water-limited conditions [7,12]. A suppression in dry biomass production in response to abiotic stress has been reported by Kaur et al. [12]. RW was positively and significantly correlated with PH (r = 0.70, $p \leq 0.001$) and AGB (r = 0.62, $p \leq 0.001$) and negatively correlated with HI (r = −0.69, $p \leq 0.001$). RSR was negatively and significantly correlated with FPL (r = −0.56, $p \leq 0.001$) and PYPP (r = −0.57, $p \leq 0.001$) but positively and significantly correlated with RW (r = 0.49, $p \leq 0.05$). The strong associations between the assessed phenotypic traits in the present study allow effective genotype selection and genetic advancement.

3.5.3. Principal Component Analysis (PCA)

PCA showing the loading scores and cumulative variations for phenotypic traits under DS and NS conditions are presented in Table 7. PCA is the most frequently used multivariate statistical analysis [21]. Three and four principal components (PCs) were identified for assessed traits under DS and NS conditions, accounting for a cumulative variance of 70.07% and 85.34%, respectively. Under DS conditions, PC1 was positively correlated with FPL, NPPP, PYPP, and HI, which accounted for 29.69% of the total variation. The results indicated that the tested okra accessions were genetically diverse. PH, RW, and RSR were positively correlated with PC2, accounting for 21.64% of the total variation under DS conditions. DPL and DPW were negatively correlated while AGB was positively associated with PC3, which accounted for 19.37% of the total variation. Under NS conditions, PC1 was positively associated with FPL, DPL, NPPP, and PYPP and negatively correlated with RSR, which accounted for 32.24% of the total variation. PC2 was positively associated with PH, AGB, and RW and negatively correlated with HI, which accounted for 28.99% of the total variation among the test accessions. PC3 was negatively correlated with DPW, while PC4 was positively associated with DTM, accounting for 14.22% and 9.89% of the total variation, respectively. The current PCA results successfully identified variables that contribute most to the response of okra accessions against drought stress.

Table 7. Principal component loading scores explained and cumulative variances of phenotypic traits among 26 okra accessions under drought-stressed and non-stressed conditions.

Traits	Drought-Stressed			Non-Stressed			
	PC1	PC2	PC3	PC1	PC2	PC3	PC4
PH	0.32	**0.74**	0.46	0.55	**0.76**	0.05	0.01
DTM	−0.33	−0.03	0.32	−0.09	−0.27	0.48	**0.74**
FPL	**0.89**	−0.23	−0.07	**0.79**	−0.31	0.37	−0.09
DPL	0.29	0.50	**−0.73**	**0.51**	0.05	−0.53	0.41
DPW	0.28	0.30	**−0.69**	0.53	0.17	**−0.71**	0.17
NPPP	**0.53**	0.07	−0.22	**0.53**	−0.49	−0.15	0.37
PYPP	**0.87**	−0.20	0.30	**0.85**	−0.01	0.46	−0.05
AGB	0.58	0.01	**0.70**	0.64	**0.71**	0.19	−0.02
HI	**0.78**	−0.29	−0.27	0.24	**−0.90**	0.16	−0.01
RW	0.26	**0.87**	0.30	−0.06	**0.88**	0.31	0.21
RSR	−0.20	**0.74**	−0.07	**−0.76**	0.30	0.10	0.39
Explained variance (eigenvalue)	3.27	2.38	2.13	3.55	3.19	1.56	1.09
Proportion of total variance (%)	29.69	21.64	19.37	32.24	28.99	14.22	9.89
Cumulative variance (%)	29.69	51.33	70.70	32.24	61.23	75.45	85.34

PH: plant height, DTM: number of days to maturity, FPL: fresh pod length, DPL: dry pod length, DPW: dry pod weight, NPPP: number of pods per plant, PYPP: pod yield per plant, AGB: above-ground biomass, HI: harvest index, RW: root weight, RSR: root: shoot ratio, PCs with ≥0.5 loading scores are boldfaced.

The relationship between the accessions and the studied phenotypic traits is illustrated using principal component biplots (Figure 4). Angles less than 45 °C between the dimensions of two variables indicate high trait associations, whereas longer vectors show the discriminating ability of a particular trait. As a result, accessions excelling in a particular trait were plotted to the vector line. Accessions LS17 and LS02 were grouped together based on the high values for RSR under DS conditions. Under NS conditions, accessions LS14, LS16, LS18, LS23, and LS01 were grouped together based on high values of PH, FPL, DPW, DPL, NPPP, and PYPP. Accessions LS21, LS01, LS12, LS14, and LS09 were grouped together based on high values of PH, FPL, DPW, DPL, NPPP, and PYPP under DS conditions. Accessions LS17, LS12, LS09, LS19, and LS09 were grouped together based on the high values of PH, FPL, DPW, DPL, NPPP, and PYPP under NS conditions.

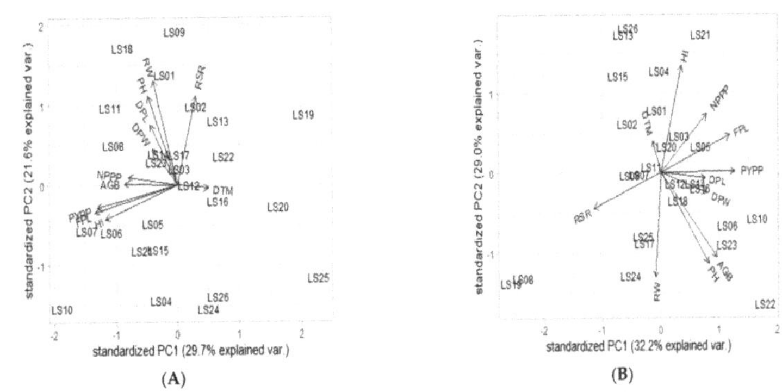

Figure 4. Principal component biplot of PC1 vs. PC2 showing groupings of 26 okra accessions based on phenotypic traits under drought-stressed (**A**) and non-stressed (**B**) conditions. PH: plant height, DTM: number of days to maturity, FPL: fresh pod length, DPL: dry pod length, DPW: dry pod weight, NPPP: number of pods per plant, PYPP: pod yield per plant, AGB: above-ground biomass, HI: harvest index, RW: root weight, RSR: root: shoot ratio.

3.5.4. Phenotypic Hierarchical Clustering

Hierarchical cluster analysis using phenotypic data allocated the okra accessions into three groups under drought-stressed conditions (Figure 5). The largest cluster (cluster I) consisted of 11 accessions, followed by cluster III with 8 accessions and cluster I with 7 accessions. High-yielding accessions (e.g., LS05 and LS07) were grouped in cluster II. Cluster I consisted of accessions which were characterized by taller plant height and with the highest number of pods per plant. Cluster III contained accessions with low pod yield. Accessions LS10 and LS18, which were grouped in cluster I under drought-stressed conditions, can be selected to develop breeding populations for enhanced pod yield. This cluster also contained taller accessions, which usually have higher biomass than shorter plants and contribute to carbon sequestration for better soil health [37]. The test accessions were also grouped into three clusters under non-stressed conditions (Figure 6). The largest cluster (cluster I) contained 12 accessions, while the second largest cluster (cluster II) contained 8 accessions, and the smallest cluster III consisted of only 6 accessions. Cluster I comprised accessions with a higher number of pods, whereas cluster III had accessions with higher harvest index and early maturity, critical attributes for drought escape due to accelerated growth and development.

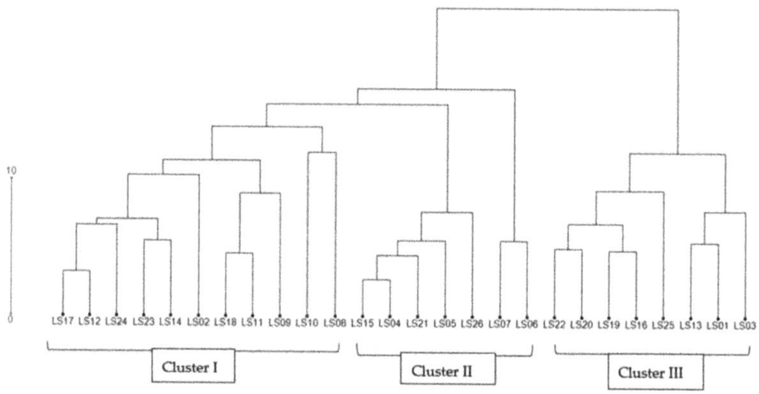

Figure 5. Hierarchical clustering of 26 okra accessions based on phenotypic traits evaluated under drought-stressed conditions.

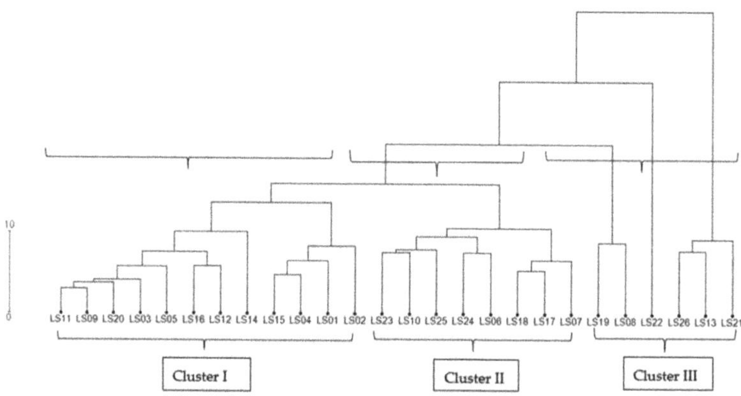

Figure 6. Hierarchical clustering of 26 okra accessions based on phenotypic traits under non-stressed conditions.

3.6. Comparison of Phenotypic and Genotypic Hierarchical Clusters

Genetic markers have proven to be a powerful tool for assessing genetic variation and elucidating genetic relationships within and among okra species, while phenotypic traits are essential indicators of genotypes in a given environment. A comparison of phenotypic and genotypic clusters was conducted to establish genotype compatibility among different dendrograms. None of the accessions maintained their positions when phenotypic hierarchical clusters were compared to genotypic hierarchical clustering under drought-stressed conditions (Figure 7). Similarly, under non-stressed conditions (Figure 8), the phenotypic clustering was opposite to the phenotypic cluster. The tanglegram comparison indicated that 42% of the accessions under drought-stressed conditions maintained their cluster membership in the phenotypic and genotypic hierarchical clustering (Figure 7). Under non-stressed conditions, 69% of the accessions maintained their membership in the phenotypic and genotypic hierarchical clustering (Figure 8). The phenotype and genotype clusters under drought and non-stressed conditions were inconsistent due to the genotype-by-environment interactions, resulting in variation in the phenotypic expression of the phenotypic traits [36]. Lower consistency in the phenotypic and genotypic clustering under drought-stressed conditions compared to non-stressed conditions is attributable to the selection pressure exerted by the drought treatment.

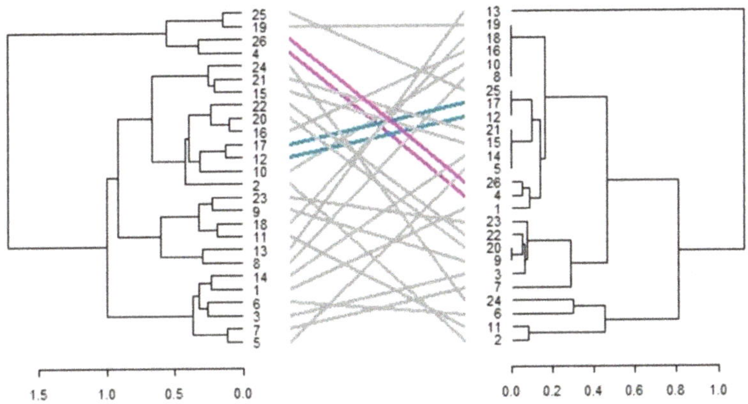

Figure 7. Tanglegram comparison of phenotypic and genotypic hierarchical clusters of 26 okra accessions based on 9 SSR markers and phenotypic data measured under drought-stressed conditions.

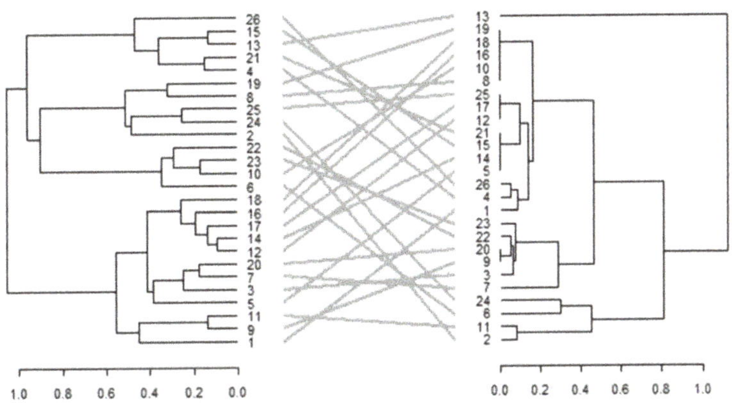

Figure 8. Tanglegram comparison of phenotypic and genotypic hierarchical clusters of 26 okra accessions based on 9 SSR markers and phenotypic data measured under non-stressed conditions.

4. Conclusions

The present study evaluated the genetic and phenotypic diversity and relationships among selected okra accessions as a guide for selecting parental accessions for breeding. SSR-assisted phenotypic and genotype evaluation and classification in the present study suggest sufficient genetic diversity in okra accessions to initiate a trait-based pre-breeding program. Genetically unrelated accessions such as LS04, LS05, LS06, LS07, LS08, LS10, LS11, LS15, LS18, LS23, LS24, and LS26 were selected based on their high yield potential and related yield-improving traits under drought stress conditions. The identified accessions are recommended as suitable breeding parents for hybridization and selection programs to improve the yield potential of okra under drought-stressed environments.

Author Contributions: Conceptualization, S.S.M., H.S., A.S.G. and J.M.; Formal analysis, S.S.M. and A.S. Funding acquisition, H.S. and A.S.G.; Investigation, S.S.M.; Methodology, S.S.M., H.S. and J.M.; Project administration, S.S.M., H.S. and A.S.G.; Resources, H.S. and A.S.G.; Supervision, H.S. and A.S.G.; Writing—original draft, S.S.M.; Writing—review and editing, S.S.M., H.S., A.S.G. and J.M. All authors have read and agreed to the published version of the manuscript.

Funding: This research was funded by the National Research Foundation, South Africa, grant number MND200618533646, the Agricultural Research Council through the PDP program, grant number 10013464, and the Moses Kotane Institute, grant number 8804030907080.

Institutional Review Board Statement: Not applicable.

Data Availability Statement: Not applicable.

Acknowledgments: The University of KwaZulu-Natal is acknowledged for the support of the project, as well as the Agricultural Research Council through the Professional Development Programme for providing plant material and support. Anthon Wanga is acknowledged for assisting in data analysis.

Conflicts of Interest: The authors declare no conflict of interest.

References

1. Düzyaman, E. Phenotypic Diversity within a Collection of Distinct Okra (*Abelmoschus esculentus*) Cultivars Derived from Turkish Land Races. *Genet. Resour. Crop Evol.* **2005**, *52*, 1019–1030. [CrossRef]
2. Kumar, M.; Sharma, V.R.; Kumar, N.; Sirohi, U.; Naresh, R.; Chaudhary, V. Screening of Microsatellite Markers for Genetic Diversity Assessment and Conservation of Germplasm in Okra (*Abelmoschus esculentus* L. Moench). *Int. J. Curr. Microbiol. Appl. Sci.* **2017**, *6*, 509–520. [CrossRef]
3. Komolafe, R.J.; Ariyo, O.J.; Alake, C.O. Correlation and Path Coefficient Analysis of Fruit Yield Attributes in Forty Genotypes of Okra (*Abelmoschus esculentus*). *Agric. Res.* **2021**, *11*, 15–23. [CrossRef]
4. Abed, M.Y.; Ibrahim, E.A.; El-Shoura, A.M. Development of Okra (*Abelmoschus esculentus* L. Moench) Hybrids Derived from Selected Inbreds under Drought Stress. *J. Plant Prod.* **2020**, *11*, 61–69. [CrossRef]
5. Saifullah, M.; Rabbani, M.G. Evaluation, and characterization of okra (*Abelmoschus esculentus* L. Moench) genotypes. *SAARC J. Agric.* **2009**, *7*, 92–99.
6. Mihretu, Y.; Wayessa, G.; Adugna, D. Multivariate analysis among okra (*Abelmoschus esculentus* (L.) Moench) collection in South-Western Ethiopia. *J. Plant Sci.* **2014**, *9*, 43–50.
7. Reddy, M.T.; Haribabu, K.; Ganesh, M.; Redyy, K.C.; Begum, H. Genetic divergence analysis of indigenous and exotic collections of okra (*Abelmoschus esculentus* L. Moench). *J. Agric. Technol.* **2012**, *8*, 611–623.
8. Alake, C.O. Genetic variability and diversity in okra landraces using agromorphological traits and seed elemental minerals. *Int. J. Veg. Sci.* **2020**, *26*, 127–149. [CrossRef]
9. Chaturvedi, A.K.; Surendran, U.; Gopinath, G.; Chandran, K.M.; Anjali, N.K.; Ct, M.F. Elucidation of stage specific physiological sensitivity of okra to drought stress through leaf gas exchange, spectral indices, growth and yield parameters. *Agric. Water Manag.* **2019**, *222*, 92–104. [CrossRef]
10. Kyriakopoulou, O.G.; Arens, P.; Pelgrom, K.T.; Karapanos, I.; Bebeli, P.; Passam, H.C. Genetic and morphological diversity of okra (*Abelmoschus esculentus* [L.] Moench.) genotypes and their possible relationships, with particular reference to Greek landraces. *Sci. Hortic.* **2014**, *171*, 58–70. [CrossRef]
11. Kaur, K.; Pathak, M.; Kaur, S.; Pathak, D.; Chawla, N. Assessment of morphological and molecular diversity among okra (*Abelmoschus esculentus* (L.) Moench.) germplasm. *Afr. J. Biotechnol.* **2013**, *12*, 3160–3170.
12. Massucato, L.R.; Nakamura, K.K.; Ruas, P.M.; Zefa, D.M.; Da Silva, D.J.H.; Gonçalves, L.S.A. Genetic diversity among Brazilian okra landraces detected by morphoagronomic and molecular descriptors. *Acta Sci. Agron.* **2020**, *42*, e43426. [CrossRef]

13. Aladele, S.E.; Ariyo, O.J.; Lapena, R.D.E. Genetic relationships among West African okra (*Abelmoschus caillei*) and Asian genotypes (*Abelmoschus esculentus*) using RAPD. *Afr. J. Biotechnol.* **2008**, *7*, 1426–1431.
14. Yuan, C.; Zhang, C.; Wang, P.; Hu, S.; Chang, H.; Xiao, W.; Lu, X.; Jiang, S.; Ye, J.; Guo, X. Genetic diversity analysis of okra (*Abelmoschus esculentus* L.) by inter-simple sequence repeat (ISSR) markers. *Genet. Mol. Res.* **2014**, *13*, 3165–3175. [CrossRef] [PubMed]
15. Kpodo, F.; Agbenorhevi, J.; Alba, K.; Bingham, R.; Oduro, I.; Morris, G.; Kontogiorgos, V. Pectin isolation and characterization from six okra genotypes. *Food Hydrocoll.* **2017**, *72*, 323–330. [CrossRef]
16. Salameh, N.M. Genetic diversity of okra (*Abelmoschus esculentus* L.) genotypes from different agroecological regions revealed by amplified fragment length polymorphism analysis. *Am. J. Appl. Sci.* **2014**, *11*, 1157–1163. [CrossRef]
17. Fougat, R.S.; Purohit, A.R.; Kumar, S.; Parekh, M.J.; Kumar, M. SSR based genetic diversity in Abelmoschus species. *Indian J. Agric. Sci.* **2015**, *85*, 1223–1228.
18. Mohammed, W.; Amelework, B.; Shimelis, H. Simple sequence repeat markers revealed genetic divergence and population structure of okra [*Abelmoschus esculentus*] collections of diverse geographic origin. *Aust. J. Crop Sci.* **2020**, *14*, 1032–1041. [CrossRef]
19. El-Fattah, B.E.S.A.; Haridy, A.G.; Abbas, H.S. Response to planting date, stress tolerance and genetic diversity analysis among okra (*Abelmoschus esculentus* (L.) Moench.) varieties. *Genet. Resour. Crop Evol.* **2020**, *67*, 831–851. [CrossRef]
20. Ahmed, Z.G.; El-Sayed, M.A. Influence of drought stress on physiological traits of crossed okra varieties. *Jordan J. Biol. Sci.* **2021**, *14*, 253–260.
21. Schafleitner, R.; Kumar, S.; Lin, C.-Y.; Hegde, S.G.; Ebert, A. The okra (*Abelmoschus esculentus*) transcriptome as a source for gene sequence information and molecular markers for diversity analysis. *Gene* **2013**, *517*, 27–36. [CrossRef] [PubMed]
22. Ouedraogo, M.H.; Sawadogo, N.; Batieno, T.B.J.; Zida, W.M.S.F.; Bougma, A.L.; Barro, A.; Kiebre, Z.; Sawadogo, M. Evaluation of genetic diversity of okra accessions [*Abelmoschus esculentus* (L. Moench)] cultivated in Burkina Faso using microsatellite markers. *Afr. J. Biotechnol.* **2018**, *17*, 126–132. [CrossRef]
23. Peakall, R.; Smouse, P.E. GenAlex 6.5: Genetic analysis in Excel. Population software for teaching and research-an update. *Bioinformatics* **2012**, *28*, 2537–2539. [CrossRef] [PubMed]
24. Nei, M.; Li, W.H. Mathematical model for studying genetic variation in terms of restriction endonucleases. *Proc. Natl. Acad. Sci. USA* **1979**, *76*, 5269–5273. [CrossRef]
25. Nagy, S.; Poczai, P.; Cernák, I.; Gorji, A.M.; Hegedűs, G.; Taller, J. PICcalc: An Online Program to Calculate Polymorphic Information Content for Molecular Genetic Studies. *Biochem. Genet.* **2012**, *50*, 670–672. [CrossRef]
26. Metsalu, T.; Vilo, J. ClustVis: A web tool for visualizing clustering of multivariate data using Principal Component Analysis and heatmap. *Nucleic Acids Res.* **2015**, *43*, W566–W570. [CrossRef]
27. Perrier, X.; Jacquemoud-Collet, J.P. DARwin Software. Dissimilarity Analysis and Representation for Windows. Available online: http://www.darwin.cirad.fr/darwin.html (accessed on 6 September 2022).
28. Maechler, M.; Rousseeuw, P.; Struyf, A.; Hubert, M.; Hornik, K.; Studer, M. Package 'Cluster'. Available online: https://cran.microsoft.com/snapshot/2014-10-10/web/packages/cluster/cluster2013.pdf (accessed on 12 November 2021).
29. Gower, J.C. A General Coefficient of Similarity and Some of Its Properties. *Biometrics* **1971**, *27*, 857–871. [CrossRef]
30. Kumar, S.; Parekh, M.J.; Fougat, R.S.; Patel, S.K.; Patel, C.B.; Kumar, M.; Patel, B.R. Assessment of genetic diversity among okra genotypes using SSR markers. *J. Plant Biochem. Biotechnol.* **2017**, *26*, 172–178. [CrossRef]
31. Petropoulos, S.; Fernandes, Â.; Barros, L.; Ferreira, I.C. Chemical composition, nutritional value and antioxidant properties of Mediterranean okra genotypes in relation to harvest stage. *Food Chem.* **2018**, *242*, 466–474. [CrossRef]
32. Pradip, K.; De Akotkar, D.K.; Pal, A.K. Genetic variability and diversity in okra [*Abelmoschus esculentus* (L) Moench]. *Electron. J. Plant Breed.* **2010**, *1*, 393–398.
33. Ravishankar, K.V.; Muthaiah, G.; Mottaiyan, P.; Gundale, S.-K. Identification of novel microsatellite markers in okra (*Abelmoschus esculentus* (L.) Moench) through next-generation sequencing and their utilization in analysis of genetic relatedness studies and cross-species transferability. *J. Genet.* **2018**, *97*, 39–47. [CrossRef]
34. Eshiet, A.J.; Brisibe, E.A. Morphological Characterization and Yield Traits Analysis in Some Selected Varieties of Okra (*Abelmoschus esculentus* L. Moench). *Adv. Crop Sci. Technol.* **2015**, *3*, 197. [CrossRef]
35. Adejumo, S.A.; Ezeh, O.S.; Mur, L.A. Okra growth and drought tolerance when exposed to water regimes at different growth stages. *Int. J. Veg. Sci.* **2018**, *25*, 226–258. [CrossRef]
36. Mkhabela, S.S.; Shimelis, H.; Gerrano, A.S.; Mashilo, J. Phenotypic and genotypic divergence in Okra [*Abelmoschus esculentus* (L.) Moench] and implications for drought tolerance breeding: A review. *South Afr. J. Bot.* **2021**, *145*, 56–64. [CrossRef]
37. Rad, S.K.; Madani, H.; Sharifabadi, H.H.; Mahmoudi, M.; Nourmohamadi, G. Effects of different irrigation intervals and sowing time on yield attributing traits of okra (*Abelmoschus esculentus* L.). *Arab. J. Geosci.* **2022**, *15*, 740. [CrossRef]

diversity

MDPI

Article

Spatial Patterns and Determinants of Endemic Taxa Richness in the Genus *Viburnum* (Adoxaceae) in China

Wenjun Lyu [1,2,†], Shenglan Du [3,†], Jiali Ying [1,2,4], Veronicah Mutele Ngumbau [5,6], Sheng Huang [3,7], Shengwei Wang [1,2,5,*] and Hongtao Liu [1,2,4,*]

1 National Germplasm Repository of Viburnum, Wuhan Botanical Garden, Chinese Academy of Sciences, Wuhan 430074, China
2 Center of Conservation Biology, Core Botanical Gardens, Chinese Academy of Sciences, Wuhan 430074, China
3 College of Forestry and Horticulture, Hubei Minzu University, Enshi 445000, China
4 University of Chinese Academy of Sciences, Beijing 100049, China
5 Sino-Africa Joint Research Center, Chinese Academy of Sciences, Wuhan 430074, China
6 East African Herbarium, National Museums of Kenya, P.O. Box 45166, Nairobi 00100, Kenya
7 Enshi Dongsheng Plant Development Co., Ltd., Enshi 445000, China
* Correspondence: wangshengwei@wbgcas.cn (S.W.); lhongtao@wbgcas.cn (H.L.)
† These authors contributed equally to this work.

Citation: Lyu, W.; Du, S.; Ying, J.; Ngumbau, V.M.; Huang, S.; Wang, S.; Liu, H. Spatial Patterns and Determinants of Endemic Taxa Richness in the Genus *Viburnum* (Adoxaceae) in China. *Diversity* **2022**, *14*, 744. https://doi.org/10.3390/d14090744

Academic Editors: Hong-Hu Meng and Yi-Gang Song

Received: 23 August 2022
Accepted: 6 September 2022
Published: 10 September 2022

Abstract: Understanding the distribution patterns and formation mechanisms of endemic taxa is essential for effective biodiversity conservation. China is an important distribution and endemic center for genus *Viburnum* in Asia. However, the distribution pattern and formation mechanism of endemic taxa of *Viburnum* remains unclear in China. In this study, we determined the distribution information of 61 endemic taxa of *Viburnum* through specimens' review and field surveys. Species distribution models were used to clarify the distribution patterns of the endemic taxa of *Viburnum*. The findings shows that the hotspot for overall endemic taxa of *Viburnum* in China is mainly distributed in temperate and subtropical mountainous areas, and the highest richness in the mountainous regions were around the Yunnan-Guizhou Plateau and the Sichuan Basin. About one-third of the endemic taxa of *Viburnum* were rare species, whose distribution area was scattered and lacked protection. The distribution pattern of the endemic taxa of genus *Viburnum* can be well explained within the three hypotheses of environmental energy, water availability and climate seasonality. This study provides additional understanding and explanation of endemic species richness distribution and their formation mechanisms. In addition, it provides conservation measures for endemic taxa of genus *Viburnum* to guide conservationists and policy makers in China.

Keywords: endemic taxa; *Viburnum*; environmental factors; geographical pattern; conservation

1. Introduction

The heterogeneous distribution patterns of biodiversity and its formation mechanisms have long been a research topic among ecologists and conservationists [1,2]. Species richness is an important part of biodiversity and understanding its geographic patterns can also provide important information for spatial planning conservation, and sustainable use of biodiversity resources [3,4]. Meanwhile, species richness of endemic, rare or threatened plants play a central role in identifying the priority sites for plant conservation [5].

Over the past few decades, hundreds of macroecological and biogeographical hypotheses have been used to explain the patterns of species richness and the formation mechanisms [6–8]. In particularly, the synergistic effects between climate and habitat heterogeneity drive species richness at large scales [9–11]. The water-energy hypothesis is one of the most widely used of these hypotheses, suggesting that areas with high energy and water availability will promote high species diversity and richness [12,13]. The habitat heterogeneity hypothesis suggests that areas of high topographic complexity can promote species differentiation through geographical isolation [4,14]. Furthermore, climatic seasonality (severe winters) and climate change since the Quaternary can also influence species richness patterns [4,15]. Recent studies suggest that species richness patterns, particularly those of endemic and rare species with low dispersal ability and narrow ranges, may be affected by the combined effects of interactions between contemporary climate and long-term climate stability [16,17]. Furthermore, the interpretation of species richness patterns and the relative importance of explanatory factors may vary depending on the study extents, research scale, spatial resolution, and species groups [18–21]. However, the richness patterns and explanatory hypotheses for woody plants genus with high proportion of endemism have received less attention and are not well understood.

Viburnum L (Adoxaceae), is a highly ornamental woody flowering genus, and popular for its horticultural values which are characterized by its showy flowers, brightly colored fruit, and attractive foliage [22–24]. The genus *Viburnum* has over 200 species worldwide, mainly distributed in the subtropical and temperate regions of Asia, South America and Europe [22,25]. It was first placed in the traditional Caprifoliaceae family and later placed in the Adoxaceae based on morphological and molecular evidence [26,27], and can be divided into 10 sections in traditional taxonomy [28]. Among those sections, the Sect. *Megalotinus*, Sect. *Solenotinus*, Sect. *Tinus*, Sect. *Tomentosa*, Sect. *Viburnum* and Sect. *Pseudotinus* were mainly distributed in Asia, the Sect. *Lentago* and Sect. *Oreinotinus* were mainly distributed in America, the Sect. *Odontotinus* was mainly distributed in both Asia and America, with the Sect. *Opulus* could be found on all continents in the circumpolar Arctic region [28]. According to the recent molecular phylogenetic findings, the global genus *Viburnum* could be broadly divided into 18 sections [29,30]. The genus *Viburnum* probably originated and initially diversified in Old World montane tropical forests, and several lineages gradually migrated to the colder forests in the higher latitudes of Asia. Subsequently, some lineages extended into Europe, and some lineages may have entered the New World through Belincia in the Eocene [30,31]. From the Oligocene onward, there was less species formation and more extinction in the lowland tropical lineages, and increased diversification in temperate lineages. After the Miocene, it spread again to tropical forests and boreal cold forests [30]. Today, the center of phylogenetic diversity is in Southeast Asia, but the greatest species diversity is found in central and southern China [32–35].

China has the largest species number of the genus *Viburnum* in the world and the distribution center of this genus in Asia [25]. According to the Flora of China and the latest research of new distribution records and new species, there are 73 species (98 taxa) of the genus *Viburnum* in China [25,36], of which 45 are endemic to China [25]. The Flora Reipublicae Popularis Sinicae describes the genus *Viburnum* is widespread in all provinces of China, with the largest number of species in the southwest [37]. However, the geographical distribution pattern and endemic centers of endemic taxa of the genus *Viburnum* in China and their driving factors are unclear, which has hampered research on the evolution and conservation of endemics of the genus *Viburnum*.

In this study, we aimed at solving the following problems: (1) what is the geographical distribution pattern of endemic taxa of *Viburnum* in China; (2) What are the possible causes of the pattern of endemic taxa of *Viburnum* in China? This study will provide a basis for research on the evolution and conservation of endemic taxa of *Viburnum* in China.

2. Methods

2.1. Plant Taxa and Distribution Data

The list of plant taxa and taxonomic information in this study was based on the Flora of China [25]. Information on new species and new distribution of *Viburnum* in China has also been recorded through literature review. The distribution data of endemic taxa of *Viburnum* in China was mainly obtained from digitized specimens and field surveys. Specimens data were obtained from the Chinese Virtual Herbarium (www.cvh.ac.cn, accessed on 10 September 2021), the National Specimen Information Infrastructure (www.nsii.org.cn, accessed on 10 September 2021), and Herbariums from other research institutes, which include Shanghai Chenshan Botanical Garden, Hangzhou Botanical Garden, Xishuangbanna Tropical Botanical Garden of the Chinese Academy of Sciences, South China Botanical Garden of the Chinese Academy of Sciences, and Wuhan Botanical Garden of the Chinese Academy of Sciences. From the digitized specimen information, we obtained latitude and longitude data. For the specimens which just had collection location name, geographical coordinates were obtained from the Chinese National Geographical Names Database (www.nfgis.nsdi.gov.cn/, accessed on 10 September 2021) and Google Earth. The field survey data was obtained from the Wild Flora Introduction Department of Wuhan Botanical Garden of the Chinese Academy of Sciences from 1980 to 2021, and from Enshi Dongsheng Plant Development Co., Ltd, Enshi, China. from 2008 to 2021. Finally, a total of 7119 valid distribution information was obtained after removing doubtful, duplicate records, ambiguous information, and misidentified specimens.

2.2. *Species Distribution Model*

In this study, Maximum entropy model (MaxEnt) [38] was chosen for predicting the potential range of endemic species of genus *Viburnum*. Maxent was specifically developed to model species distributions with presence-only data, and previous studies have shown that it performs well with few distributed data and has been less affected when positional errors occur [39–41]. In this study, Maxent was run according to the following modelling rules: linear features and quadratic features for $\geq$5 and <15 collection records, and adding hinge features for $\geq$15 records [40,42]. The random test percentage was 25% with 100 replicates. The prediction accuracy of the model was evaluated by the ROC curve. The ROC curve is a methodological technique, and the AUC value can qualitatively assess the prediction accuracy [43,44].

In this case, 19 bioclimatic variables (the average for the years 1970–2000) from the WorldClim (https://www.worldclim.org/, accessed on 10 September 2021) dataset [45], and elevation (www.diva-gis.org, accessed on 10 September 2021) with a spatial resolution of 2.5 min (~5 km) were selected as environmental predictors for the species distribution model in this study. Since multicollinearity of variables can result overfit species distribution modeling [40,42,46], in this study we performed Spearman's rank correlation tests on 19 environmental factors, and eight factors with a Spearman's correlation of less than 0.8 were selected to participate in the modeling (Table S2).

2.3. *Environmental Variables*

Four categories of environmental variables were selected for the analysis of the drivers of species richness patterns. To test the effect of energy on species richness, we selected mean annual temperature (MAT), annual range of temperature (ART) and mean temperature of the coldest quarter (MTC). To test the effect of water on species richness, we selected mean annual precipitation (MAP) and mean precipitation of the driest month (PDM). To test the effect of climatic seasonality on species richness, we selected precipitation seasonality (PS) and temperature seasonality (TS). To test the effect of heterogeneity on species richness, we choose the range in elevation of each grid cell (maximum minus minimum elevation, REL, www.earthenv.org, accessed on 10 September 2021), terrain ruggedness index (TRI; http://www.earthenv.org/, accessed on 10 September 2021) and normalized difference vegetation index (NDVI, www.geodata.cn, accessed on 10 September 2021). We selected the climate change since the Last Glacial Maximum to track the historic climate change (MATano and MAPano), which is the absolute value of the difference in MAT and MAP between the LGM and the present [47]. Climate data were obtained from the WorldClim (http://www.worldclim.org, accessed on 10 September 2021) with the spatial resolution of 1 arc-min.

2.4. *Data Analysis*

For this study, the species richness and environmental variables were statistically analyzed in grid cells at a resolution of 50 $\times$ 50 km in ArcGIS 10.2. The potential distribution area of each species was rasterized at a resolution of 50 × 50 km and the number of species in each grid cells was counted as species richness. Rare species were marked as species listed in the IUCN list and species with small distribution ranges, i.e., species with raster counts less than 100 (which also almost satisfied with the sample collection data points of less than 25). The environmental variables were extracted by their respective average values in each grid cells. All environmental variables and richness data were log-transformed because of their highly skewed distributions. To analyze the predictive power of these hypotheses for the endemic richness of *Viburnum*, we chose standard ordinary least squares regression (OLS) models that ignored spatial autocorrelation, and regression models that take spatial autocorrelation into account to assess the variables in these models. The two most commonly used spatial autoregressive models, the conditional autoregressive (CAR) and simultaneous autoregressive (SAR) models were chosen for this study. The three models were calculated in SAM (version 4.0, https://www.ecoevol.ufg.br/sam/, accessed on 10 September 2021) [48], along with the Akaike Information Criterion (AIC) used to rank the environmental models [8]. To assess the pattern of spatial autocorrelation in the residuals of all regression models, we plotted the correlation of Moran's I. To further explore the relative importance of environmental factors for endemic species richness in each section of *Viburnum*, we used the "calc.relimp" function in the R package "relaimpo" [49] in R 4.0.5 [50].

3. Results

3.1. Endemic Richness of Genus Viburnum in China

Through the examination of specimens and field data, we assessed 61 endemic taxa (including 38 species, 15 varieties, 5 subspecies and 3 forma) of *Viburnum* (Table S1; Figure 1), accounting for 62% of the taxa (including inner species taxa) of *Viburnum* in China. The endemic taxa belong to seven sections of genus *Viburnum* in China. Of these, Sect. *Odontotinus* is the most diverse with 21 taxa (14 species, 6 varieties and 1 subspecies), Sect. *Solenotinus* has 16 taxa (10 species, 4 varieties and 2 subspecies), Sect. Viburnum 10 taxa (6 species, 2 subspecies and 2 forma), Sect. *Megalotinus* 6 taxa (3 species and 3 varieties), Sect. *Tinus* 5 species (3 species, 1 variety and 1 forma), Sect. *Tomentosa* 2 taxa (1 species and 1 variety), and Sect. *Pseudotinus* is represented by only 1 species.

3.2. Geographic Patterns of Endemic Richness of Genus Viburnum in China

According to statistics of the results of the species distribution model and its distribution in 50 × 50 km grid cells. A total of 4162 grid cells were sampled, but only 1764 grid cells had recorded species occurrence of genus *Viburnum* hence they were used for analysis. 57.62% of the total grid cells sampled had recorded 0 endemic species. The endemic taxa richness of this genus varied considerably in China with taxa richness values ranging from 0 to 26 species per grid cell (Figure S1).

The geographic patterns of specimen collection of endemic species of the genus *Viburnum* shows a heterogeneous pattern, with its specimen collection mostly from the southern and southwestern of China (Figure 2A,B). The endemic taxa richness of genus *Viburnum* is distributed in 28 provincial administrations in China, of which Yunnan Province has the highest richness, followed by Guangxi and Sichuan Province. The northernmost distribution area is in Urumqi city (Xinjiang Uygur Autonomous Region), the southernmost area is in Ledong County (Hainan Province), the westernmost area in Shigatse City (Tibet Autonomous Region), and the easternmost area in Taipei City, Taiwan Province (Figure 2C). Based on the raster statistics of the potential distribution areas of endemic taxa, we found that the highest richness of endemic taxa of the genus *Viburnum* in the mountainous areas around the Sichuan Basin and Yunnan-Guizhou Plateau, such as the Qionglai-Wumeng Mountains, the Qinling-Daba Mountains and the Wuling Mountains regions. At the same time, the regions of Wuyi Mountains, the Nanling Mountains and the Central Mountains of Taiwan in the south-east of China also showed high richness (Figure 2D).

Figure 1. Some endemic taxa of the genus *Viburnum* in China. 1 *V. buddleifolium*, 2 *V. chinshanense*, 3 *V. glomeratum* subsp. *magnificum*, 4 *V. macrocephalum* f. *macrocephalum*, 5 *V. macrocephalum* f. *keteleeri*, 6 *V. rhytidophyllum*, 7 *V. schensianum*, 8 *V. utile*, 9 *V. congestum*, 10 *V. brachybotryum*, 11 *V. brevitubum*, 12 *V. chingii* var. *chingii*, 13 *V. corymbiflorum* subsp. *corymbiflorum*, 14 *V. corymbiflorum* subsp. *malifolium*, 15 *V. farreri*, 16 *V. henryi*, 17 *V. oliganthum*, 18 *V. taitoense*, 19 *V. tengyuehense* var. *tengyuehense*, 20 *V. trabeculosum*, 21 *V. amplifolium*, 22 *V. leiocarpum* var. *leiocarpum*, 24 *V. punctatum* var. *lepidotulum*, 24 *V. ternatum*, 25 *V. hanceanum*, 26 *V. sympodiale*, 27 *V. cinnamomifolium*, 28 *V. davidii*, 29 *V. atrocyaneum* f. *harryanum*, 30 *V. propinquum* var. *mairei*, 31 *V. triplinerve*, 32 *V. betulifolium*, 33 *V. chunii*, 34 *V. dalzielii*, 35 *V. foetidum* var. *rectangulatum*, 36 *V. foetidum* var. *ceanothoides*, 37 *V. fordiae*, 38 *V. formosanum* subsp. *leiogynum*, 39 *V. formosanum* var. *pubigeru*, 40 *V. kansuense*, 41 *V. lancifolium*, 42 *V. melanocarpum*, 43 *V. sempervirens* var. *sempervirens*, 44 *V. sempervirens* var. *trichophorum*, 45 *V. setigerum*, 46 *V. squamulosum*.

The distribution pattern of the endemic richness of the sections of genus *Viburnum* varied considerably. Of the seven sections, the Sect. *Viburnum* had a larger distribution range, with its distribution center in the area surrounding the Sichuan Basin (Figure 3A). The Sect. *Pseudotinus* (only 1 species) occurred in the areas between Sichuan and Shanghai (Figure 3B). The Sect. *Tinus* mainly occurred in south-western China, with its center of distribution in the Qionglai-Wumeng Mountains and its extensions into Guangxi Province (Figure 3C). The distribution area of Sect. *Solenotinus* was found to be widely spread, and the distribution hotspots were found to be in the southwest of China, especially in Yunnan Province and Wuling and Nanling Mountains (Figure 3D). The Sect. *Tomentosa* occurs only in the southeastern region of China, with Taiwan Province hosting all the species of this section (Figure 3E). The Sect. *Megalotinus* is mainly distributed in the southwestern and southern coastal areas, and the highest richness areas are Hainan, Taiwan, and southern Yunnan Provinces (Figure 3F). The Sect. *Odontotinus* had a wide distribution range, but the distribution center is in the south-eastern region, especially in the Wuyi and Nanling Mountains (Figure 3G). The rare species were mainly distributed in the southwest and southeast regions, with higher richness in Yunnan and Sichuan, Hainan and Taiwan Province (Figure 3H).

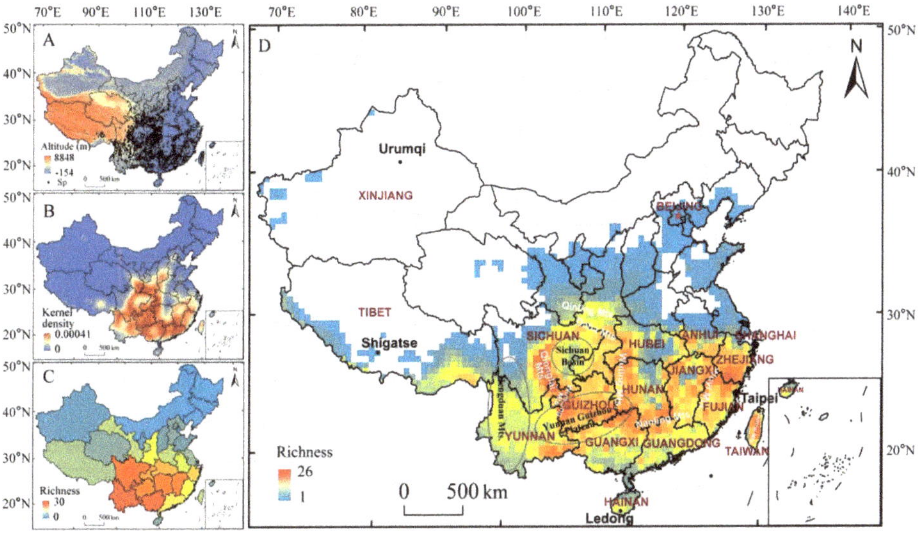

Figure 2. Geographical pattern of hotspots for endemic taxa of *Viburnum* in China. (**A**) Specimen Collection Patterns; (**B**) Kernel density of specimen collection; (**C**) Species richness at the provincial level; (**D**) Species richness at the 50 × 50 km grids.

3.3. Assessment of Variables on Species Richness

Overall, the endemic richness of genus *Viburnum* in China tends to increase with increasing longitude and decrease with increasing latitude, and areas of high richness are concentrated between 103° E and 115° E and between 25° N and 33° N (Figure S3). The endemic taxa of the genus *Viburnum* in China are mainly distributed at lower altitudes, with 50.82% of the taxa distributed at altitudes below 2000 m. Only 36.07% of the taxa are distributed at the highest altitudes in the region between 2000 m and 3000 m, while only 13.11% of the taxa have a maximum altitude above 3000 m but below 4000 m (Table S1).

The 12 explanatory variables were significant and entered the model in OLS regression analysis (Figure S3). The ART, MTC, MAP and TS were the most important variables in the model. MATano was the least important independent variable, followed by MAPano, REL, NDVI and TRI (Table 1). The OLS model explained 75.4% ($p < 0.0001$) of the variance in the transformed data (Figure 4). The Climatic seasonality hypothesis (52.9%) was the most influential factor for species richness, followed by Water availability (50.5%) and Environmental energy (49.3%) hypotheses. The least important hypothesis was Historic climate changes (16.9%). The OLS residuals showed positive spatial autocorrelation within a distance of approximately 1000 km (Figure 5). In addition, the OLS residuals were spatially clumped, showing areas of clustering with highly spatially correlated values (Figure S4).

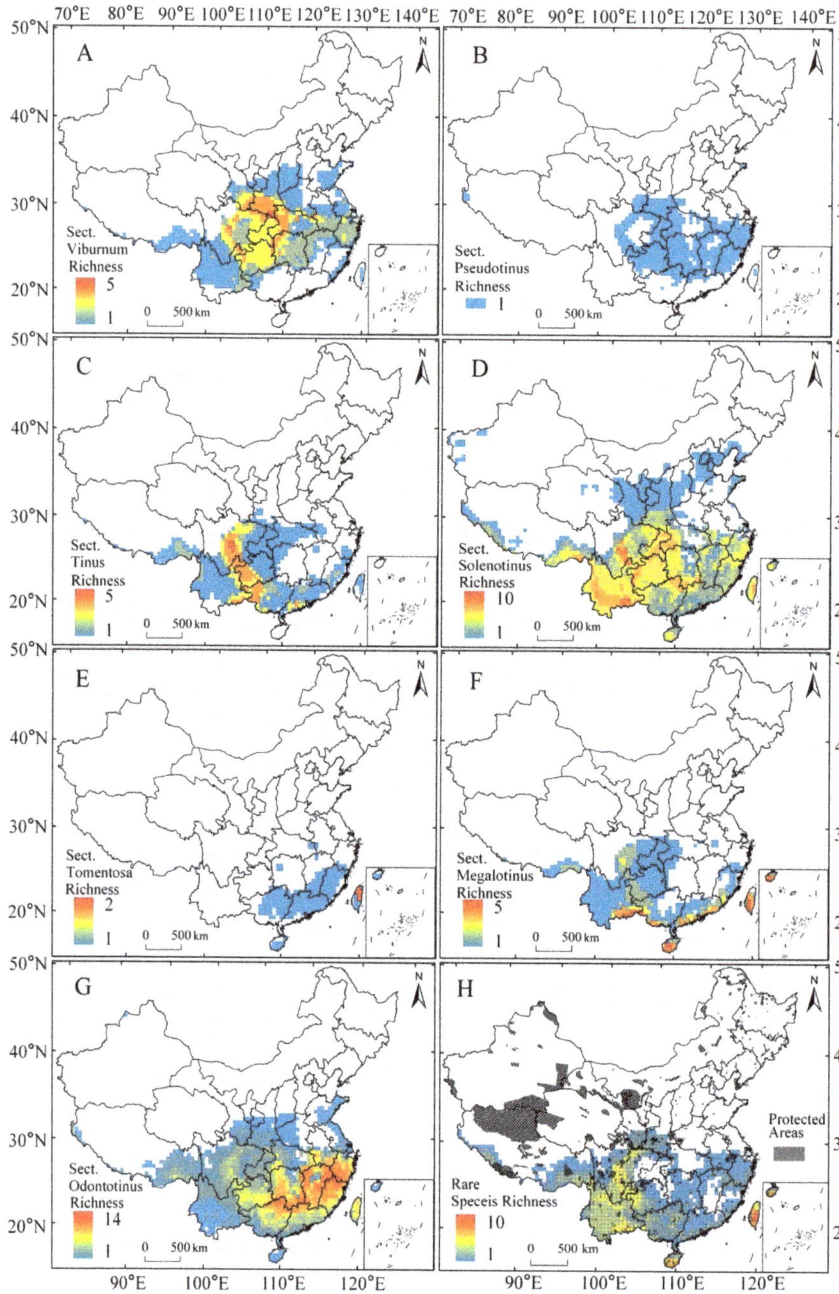

Figure 3. Geographical pattern of hotspots for the endemic richness of seven Sections (**A–G**) and rare taxa (**H**) of genus *Viburnum* in China.

Table 1. Relationships between the richness of endemic taxa of *Viburnum* and explained variables using simple ordinary least squares (OLS) model, simple conditional autoregressive (CAR) model and the simultaneous autoregressive in (SAR) model.

	OLS Models			CAR Models			SAR Models		
	t	R^2	AIC	t	R^2	AIC	t	R^2	AIC
				Environmental energy					
MAT	26.498	0.285	913.599	21.825	0.300	875.970	12.282	0.496	295.932
ART	−37.480	0.444	467.920	−31.713	0.458	424.184	−28.419	0.679	−497.299
MTC	35.452	0.416	555.496	28.107	0.416	556.969	19.114	0.571	13.626
				Water availability					
MAP	41.050	0.489	321.442	35.986	0.504	268.409	29.583	0.673	−467.821
PDM	24.220	0.250	998.358	21.927	0.269	953.337	15.860	0.531	168.554
				Climatic seasonality					
TS	−26.890	0.291	898.774	−22.896	0.316	836.727	−24.031	0.639	−289.754
PS	−24.840	0.260	975.488	−23.118	0.289	903.153	−13.334	0.501	280.869
				Heterogeneity					
REL	10.030	0.054	1407.314	13.530	0.109	1300.729	19.866	0.538	143.528
TRI	13.680	0.096	1327.069	16.096	0.146	1226.710	19.768	0.540	135.663
NDVI	−10.880	0.063	1390.441	−10.389	0.091	1337.344	−5.611	0.444	468.735
				Historic climate change					
MATano	9.466	0.048	1417.672	6.257	0.064	1388.166	4.388	0.447	462.120
MAPano	19.650	0.180	1155.614	15.310	0.168	1179.985	14.132	0.519	214.806

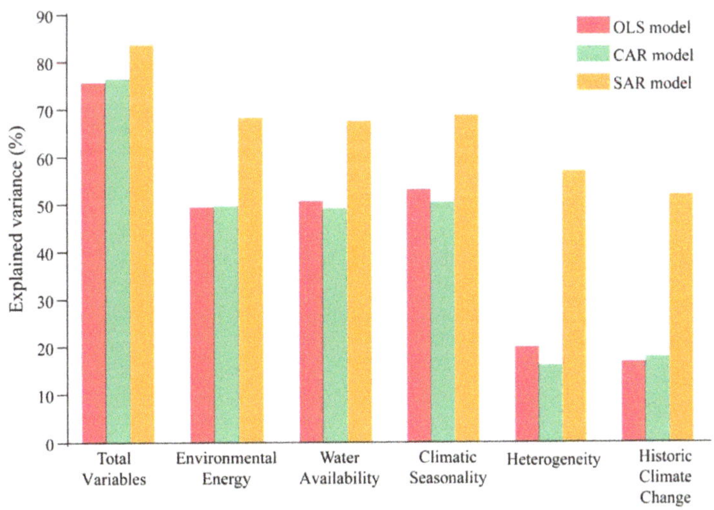

Figure 4. The explanation of endemic richness of genus *Viburnum* in China by different hypothetical models.

The most important explanatory factor in the CAR model was MAP, followed by ART, MTC, and TS. The explanatory rates for factors related to habitat heterogeneity and historical climate change were both low (<20%). In total, the CAR model accounted for 76.2% of the species richness pattern (Figure 3), with climate seasonality hypothesis alone explaining the highest rate (50.3%), followed by the environmental energy (49.6%) and the water availability (49.1%) hypothesis. With only 16.2 percent and 18.0 percent, respectively, habitat heterogeneity and the historical climate change hypotheses were inadequately explained (Table 1). The most important explanatory factor in the SAR model was ART, followed by MAP, MTC, and TS. The least explained factors were NDVI and MATano (<50%). SAR models explained 83.3% of the species richness pattern (Figure 4), with the higher individual explanations being climate seasonality (68.5%) and the environmental energy hypothesis (68%), followed by the water availability hypothesis (67.3%), habitat heterogeneity (56.8%) and the historical climate change hypothesis (51.9%) (Table 1).

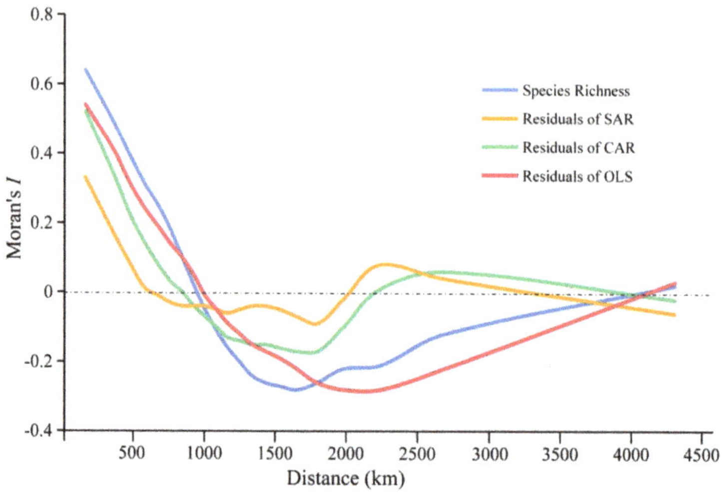

Figure 5. Moran's I correlogram for the residuals of OLS, CAR, and SAR models.

The residual examination of the three models shows that the OLS model has a strong spatial autocorrelation, with a maximum positive Moran's I index of 0.54 and a maximum negative Moran's I index of −0.28. The residuals of the CAR model had a high similarity to OLS in terms of positive spatial autocorrelation, but the negative spatial autocorrelation was much weaker than OLS. Compared to the OLS and CAR models. The residuals of the SAR model show lower spatial autocorrelation, and although it shows some positive correlation up to 600 km (the highest positive correlation Moran's I index is 0.33), the remaining distance shows less spatial autocorrelation, and the SAR model residuals show the greatest spatial heterogeneity (Figures 5 and S4).

The relative importance of ART, MTC and MAP in explaining the endemic richness pattern of genus *Viburnum* was high. The explanatory variables varied in the interpretation of endemic richness patterns in the seven sections of the genus *Viburnum*. The PS and TS were the best interpreters of the richness patterns in Sect. *Viburnum*; PS and PDM were the best interpreters of the richness patterns of Sect. *Pseudotinus*; ART and MTC being the best interpreters of the richness patterns of Sect. *Tinus*. TS and PS were the best interpretations of the richness pattern of Sect. *Solenotinus*; MAP and PDM were the best interpretations of the richness pattern of Sect. *Tomentosa*; ART and MT were the best interpretations of the richness pattern of Sect. *Megalotinus*; PDM and PS were the best interpretations of the richness pattern of Sect. *Odontotinus* (Figures 6 and S5).

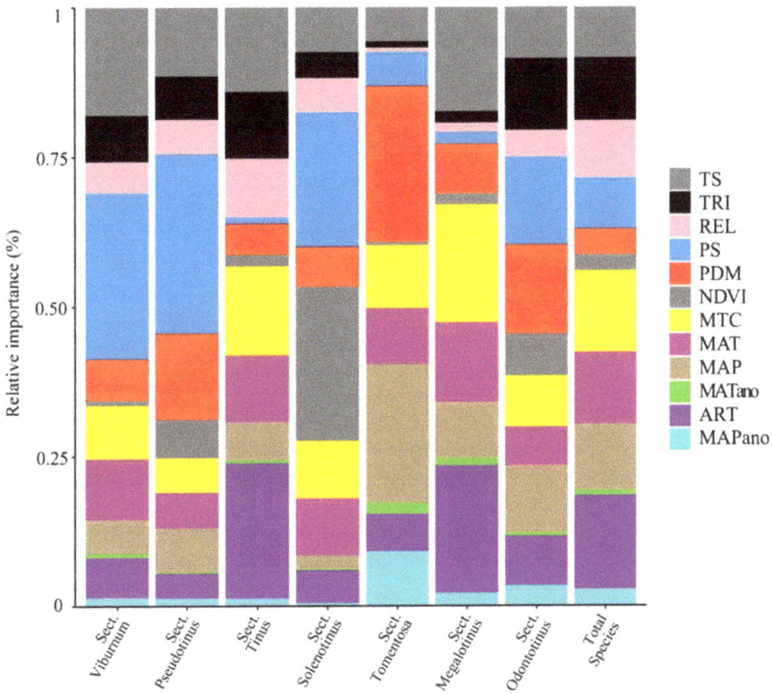

Figure 6. The relative importance of each explanatory factors for the richness of seven Sections of endemic taxa of genus *Viburnum* in China.

4. Discussion

Endemic members of genus *Viburnum* in China exhibit a very high richness, accounting for about 62% of all species (including inner species taxa) of the genus in China. They are widely distributed in China and are absent only in a few northern provinces. According to our results, China is undoubtedly an important distribution and endemic center for the genus *Viburnum* in Asia. However, the centers show inconsistent results at different scales. When assessed at the provincial scale, the endemic centers are in Yunnan and Sichuan provinces. When analyzed by raster (50 km × 50 km) processing, endemism centers were found to be mainly in the mountainous regions of southern China, such as the Qionglai-Wumeng Mts., the Qinling-Daba Mt, the Wuling Mts., the Nanling Mts., the Wuyi Mts. and the Central Mts. of Taiwan. Compared to administrative units, the rasterized method provides a clearer description of species distribution patterns and is more conducive to species conservation studies.

Our study also supports the climate seasonality hypothesis, with climate seasonality being the strongest explanation for the richness pattern of endemic taxa of *Viburnum* in China. The same findings have been reported in explanatory studies of the plant richness patterns in China for family Gesneriaceae, Moraceae, genus Rhododendron among others [4,51,52]. There is large seasonal differences in climate between the north and south of China. For example, the extreme temperature in winter, when the fluctuations are most pronounced, are higher in the northern regions than in the southern regions [53]. Such extremes temperatures may be an important reason for the restricted distribution of genus *Viburnum* in northern China, while the moderate climatic seasonality of southern and southwestern China creates stable climatic conditions for the survival and spread of taxa of this genus endemics [54], which supports the tropical niche conservatism theory (most lineages originated in tropical climates, and that they colonized extratropical and temperate areas more recently) [55,56]. Meanwhile, since many taxa of *Viburnum* seeds have dormant properties, such as *V. plicatum* var. *formosanum*; *V. sargentii*; *V. betulifolium* and *V. parvifolium* [18,57,58]. The seasonality of the climate (especially the low temperature and its duration) may promote the de-dormancy of the seeds, which will also affect the distribution pattern of endemic species of *Viburnum*.

Habitat heterogeneity and historical climatic changes have a relatively weak influence on the geographic pattern of endemic taxa richness of *Viburnum* compared to the influence of water-energy synergistic effects. However, it is undeniable that mountainous areas with high habitat heterogeneity are the regions with the highest endemic taxa richness of *Viburnum* spp. These areas of high habitat heterogeneity are mainly due to altitudinal differentiation and have a wide range of habitats with diverse hydrothermal conditions [53,59]. Many studies have demonstrated that habitats with high heterogeneity can promote the coexistence and differentiation of multiple species, and also limit the dispersal of some species which are more dependent on specific habitats [9,35,60]. These mountains also act as refuges for species in harsh climatic conditions, for example, the last glaciation affected species richness through habitat isolation and limited migrations [61]. Moreover, the highly heterogeneous tropical and subtropical mountain ranges in China have been proved to be the cradle of the Chinese flora, thus dominating the species richness and concentrations of narrow endemic species [53,61–63].

The geographic patterns and distribution centers of different sections of endemic taxa of the genus *Viburnum* also show inconsistencies. Such divergent patterns can facilitate the understanding of species evolution and biogeography of genus *Viburnum* in China. According to the distribution pattern of the seven sections obtained in this study, and combined with the phylogenetic tree of *Viburnum* constructed by Ran [23], we speculate that the temperate forest in Southwest China may be an important differentiation center of *Viburnum*, which gradually spread and evolved to the eastern and southeastern regions of China. This may be the reason for the frequent niche shifts (especially the multiple and multi-directional migration between the colder deciduous temperate forest and the warmer evergreen temperate forest) and the multiple evolutionary shifts of the genus *Viburnum* since the Oligocene have been responsible for the species diversification [31,35], while the southwest mountainous area with diverse terrain provides a natural place for species preservation and differentiation. In the process of its eastward and southeastern migration, differences in the geographical pattern of the current differentiation centers of the seven sections were gradually formed due to the geographical constraints of the mountains as well as climatic differences. However, more evidence is needed to prove the geographical distribution and evolutionary history of *Viburnum* in China.

To date, nine endemic taxa of *Viburnum* have been listed on the IUCN categories in China. These species have been categorized as: Vulnerable (2 species), Near Threatened (3 species), Endangered (2 species) and one species as Critically Endangered (Appendix). Narrow distribution, small populations and over-harvesting for their high ornamental value are the main reasons for the endangerment of these endemic species. Actually, in addition to these species already listed on the IUCN categories, there are many other endemic species of the genus *Viburnum* in China that are also threatened by such negative factors. In particular, the 24 endemic and rare taxa of the genus *Viburnum* that we have collated in this study. For example, *V. omeiense* and *V. tengyuehense* var. *polyneurum* have only type specimen records and no further wild populations have been reported in past recent decades, and are most likely extinct in the wild. The *V. squamulosum*, *V. fengyangshanense* and *V. corymbiflorum* subsp. *malifolium* have an extremely narrow distribution range and extremely sparse populations in the wild. The *V. leiocarpum* var. *punctatum* and *V. triplinerve* are less common in the wild due to the severe destruction of their habitat. Despite the fact that many mountains in China have been classified as protected areas, which greatly contributing to the conservation of plant diversity. However, the coverage of protected areas is still incomplete and discontinuous in the distribution hotspots of endemic and rare species of *Viburnum* obtained in this study (Figure 3H). What's more, human activities have significantly altered the landscape and will also inhibit the spread of many species. Thus, endemic taxa of *Viburnum* face serious challenges in the future of global changes. Here, we hereby call on conservationist to focus on the following issues in the future to promote the conservation of *Viburnum* spp. in China: (a) strengthening field distribution and population size surveys and reassessing the IUCN categories of all its taxa; (b) strengthen research on the ecological adaptation mechanisms and reproductive biology of those species with narrow distributions; (c) strengthen research into the causes of the endangerment and rarity of some taxa; (d) increase the artificial propagation of endangered and highly ornamental species to promote the conservation of wild populations; (e) the impact of climate change on the distribution patterns of endemic species and how to cope with it.

Supplementary Materials: The following supporting information can be downloaded at: https://www.mdpi.com/article/10.3390/d14090744/s1, Figure S1: Frequency distribution of endemic richness of genus *Viburnum* in China within the 50 × 50 km grids; Figure S2. Correlation between the endemic richness of genus *Viburnum* and longitude (A) and latitude (B) in China; Figure S3. Linear relationships between explanatory variables and endemic richness of genus *Viburnum;* Figure S4. Geographic distribution of residuals for the three model tests for endemic richness of genus *Viburnum* in China; Figure S5. Magnitude of variation in each of the explanatory factors for the richness of seven Sections of endemic of genus *Viburnum* in China; Table S1. Checklist of endemic species of the genus *Viburnum* in China, with their endangered status, and AUC values for species distribution model results; Table S2. Spearman's rank correlation coefficients for 8 bio-climate predictors.

Author Contributions: Conceptualization, S.W. and W.L.; data curation, W.L. and J.Y.; formal analysis, S.W., S.D., W.L. and J.Y.; funding acquisition, S.W. and H.L.; investigation, W.L., S.H. and J.Y.; methodology, S.W.; resources, W.L.; software, S.D.; supervision, H.L.; writing—original draft, S.W.; Writing—review and editing, S.W., V.M.N. and W.L. All authors have read and agreed to the published version of the manuscript.

Funding: This work was supported by the National Natural Science Foundation of China (32100166) and the Special Research Assistant Program of Chinese Academy of Sciences.

Institutional Review Board Statement: Not applicable.

Informed Consent Statement: Not applicable.

Data Availability Statement: Not applicable.

Acknowledgments: We thank Ming Tang from Jiangxi Agricultural University for helping to provide distribution information for *V. aquamulosum*. Thanks to Qiang Fu from Huazhong Agricultural University, Boshun Xia from Wuhan Botanical Garden, Chinese Academy of Sciense, Xiaoling Chen from Lijiang Aline Botanic Garden, Renkun Li from Enshi Dongsheng Plant Development Co., Ltd. for helping to Provide photo 6, 28, 40, 46 of Figure 1.

Conflicts of Interest: The authors declare no conflict of interest.

References

1. Gaston, K.J. Global patterns in biodiversity. *Nature* **2000**, *405*, 220–227. [CrossRef] [PubMed]
2. Liang, J.; Ding, Z.; Lie, G.; Zhou, Z.; Singh, P.B.; Zhang, Z.; Hu, H. Species richness patterns of vascular plants and their drivers along an elevational gradient in the central Himalayas. *Glob. Ecol. Conserv.* **2020**, *24*, e01279. [CrossRef]
3. Arvid Grytnes, J.; Vetaas, O.R. Species Richness and Altitude: A Comparison between Null Models and Interpolated Plant Species Richness along the Himalayan Altitudinal Gradient, Nepal. *Am. Nat.* **2002**, *159*, 294–304. [CrossRef] [PubMed]
4. Wang, H.-F.; Xu, X.; Cheng, X.-L.; Liu, Y.; Luo, A.; Lyu, T.; Wang, W.-L.; Nizamani, M.M.; Wang, Z. Spatial patterns and determinants of Moraceae richness in China. *J. Plant Ecol.* **2022**, rtac025. [CrossRef]
5. Dakhil, M.A.; Li, J.; Pandey, B.; Pan, K.; Liao, Z.; Olatunji, O.A.; Zhang, L.; Eid, E.M.; Abdelaal, M. Richness patterns of endemic and threatened conifers in south-west China: Topographic-soil fertility explanation. *Environ. Res. Lett.* **2021**, *16*, 034017. [CrossRef]
6. Rahbek, C.; Graves, G.R. Multiscale assessment of patterns of avian species richness. *Proc. Natl. Acad. Sci. USA* **2001**, *98*, 4534–4539. [CrossRef]
7. Xu, H.; Cao, M.; Wu, J.; Cai, L.; Ding, H.; Lei, J.; Wu, Y.; Cui, P.; Chen, L.; Le, Z.; et al. Determinants of Mammal and Bird Species Richness in China Based on Habitat Groups. *PLoS ONE* **2015**, *10*, e0143996.
8. Chen, S.-B.; Jiang, G.-M.; Ouyang, Z.-Y.; Xu, W.-H.; Xiao, Y. Relative importance of water, energy, and heterogeneity in determining regional pteridophyte and seed plant richness in China. *J. Syst. Evol.* **2011**, *49*, 95–107. [CrossRef]

9. Kerr, J.T.; Packer, L. Habitat heterogeneity as a determinant of mammal species richness in high-energy regions. *Nature* **1997**, *385*, 252–254. [CrossRef]
10. Qian, H.; Kissling, W.D. Spatial scale and cross-taxon congruence of terrestrial vertebrate and vascular plant species richness in China. *Ecology* **2010**, *91*, 1172–1183. [CrossRef]
11. Barajas-Barbosa, M.P.; Weigelt, P.; Borregaard, M.K.; Keppel, G.; Kreft, H. Environmental heterogeneity dynamics drive plant diversity on oceanic islands. *J. Biogeogr.* **2020**, *47*, 2248–2260. [CrossRef]
12. O'Brien, E. Water-energy dynamics, climate, and prediction of woody plant species richness: An interim general model. *J. Biogeogr.* **1998**, *25*, 379–398. [CrossRef]
13. Fang, W.-J.; Cai, Q.; Zhao, Q.; Ji, C.-J.; Zhu, J.-L.; Tang, Z.-Y.; Fang, J.-Y. Species richness patterns and the determinants of larch forests in China. *Plant Divers.* **2022**. [CrossRef]
14. Stein, A.; Gerstner, K.; Kreft, H. Environmental heterogeneity as a universal driver of species richness across taxa, biomes and spatial scales. *Ecol. Lett.* **2014**, *17*, 866–880. [CrossRef] [PubMed]
15. Jetz, W.; Rahbek, C. Geographic Range Size and Determinants of Avian Species Richness. *Science* **2002**, *297*, 1548–1551. [CrossRef]
16. Harrison, S.; Noss, R. Endemism hotspots are linked to stable climatic refugia. *Ann. Bot.* **2017**, *119*, 207–214. [CrossRef] [PubMed]
17. Svenning, J.-C.; Skov, F. Ice age legacies in the geographical distribution of tree species richness in Europe. *Glob. Ecol. Biogeogr.* **2007**, *16*, 234–245. [CrossRef]
18. Chen, S.-Y.; Liu, C.-P.; Baskin, C.C.; Chien, C.-T. Deep complex morphophysiological dormancy in seeds of *Viburnum plicatum* var. *formosanum* (Adoxaceae) from subtropical mountains. *Seed Sci. Res.* **2021**, *31*, 236–242. [CrossRef]
19. Giladi, I.; Ziv, Y.; May, F.; Jeltsch, F. Scale-dependent determinants of plant species richness in a semi-arid fragmented agro-ecosystem. *J. Veg. Sci.* **2011**, *22*, 983–996. [CrossRef]
20. Leo, M.; Calleja, J.A.; Lara, F.; Garilleti, R.; Medina, N.G. Drivers of plant richness patterns of Mediterranean riparian forests at local and regional scales have bottom-up and top-down effects. *J. Veg. Sci.* **2019**, *30*, 485–497. [CrossRef]
21. Huang, E.; Chen, Y.; Fang, M.; Zheng, Y.; Yu, S. Environmental drivers of plant distributions at global and regional scales. *Glob. Ecol. Biogeogr.* **2021**, *30*, 697–709. [CrossRef]
22. Xie, W.-J.; Leus, L.; Wang, J.-H.; Van Laere, K. Fertility barriers in interspecific crosses within *Viburnum*. *Euphytica* **2017**, *213*, 34. [CrossRef]
23. Ran, H.; Liu, Y.; Wu, C.; Cao, Y. Phylogenetic and Comparative Analyses of Complete Chloroplast Genomes of Chinese *Viburnum* and *Sambucus* (Adoxaceae). *Plants* **2020**, *9*, 1143. [CrossRef] [PubMed]
24. Sharifi-Rad, J.; Quispe, C.; Vergara, C.V.; Kitic, D.; Kostic, M.; Armstrong, L.; Shinwari, Z.K.; Khalil, A.T.; Brdar-Jokanovic, M.; Ljevnaic-Masic, B.; et al. Genus *Viburnum*: Therapeutic Potentialities and Agro-Food-Pharma Applications. *Oxid. Med. Cell. Longev.* **2021**, *2021*, 26. [CrossRef]
25. Yang, Q.; Valéry, M. *Flora of China. Volume 19: Viburnum L. through Adoxaceae*; Science Press: Beijing, China, 2011; pp. 570–611.
26. Backlund, A.; Bremer, B. Phylogeny of the *Asteridae* s.str. based on *rbc*L sequences with particular reference to the *Dipsacales*. *Plant Syst. Evol.* **1997**, *207*, 225–254. [CrossRef]
27. Angiosperm Phylogeny Group. An update of the Angiosperm Phylogeny Group classification for the orders and families of flowering plants: APG IV. *Bot. J. Linn. Soc.* **2016**, *181*, 1–20. [CrossRef]
28. Hara, H. *A Revision of the Caprifoliaceae of Japan with Reference to Allied Plants in Other Districts and the Adoxaceae*; Academia Scientific Books Inc.: Toyko, Japan, 1983.
29. Winkworth, R.C.; Donoghue, M.J. *Viburnum* phylogeny based on combined molecular data: Implications for taxonomy and biogeography. *Mol. Phylogenet. Evol.* **2005**, *92*, 653–666. [CrossRef]
30. Landis, M.J.; Eaton, D.A.R.; Clement, W.L.; Park, B.; Spriggs, E.L.; Sweeney, P.W.; Edwards, E.J.; Donoghue, M.J. Joint estimation of geographic movements and biome shifts during the global diversification of *Viburnum*. *Syst. Biol.* **2020**, *70*, 67–85. [CrossRef]
31. Schmerler, S.B.; Clement, W.L.; Beaulieu, J.M.; Chatelet, D.S.; Sack, L.; Donoghue, M.J.; Edwards, E.J. Evolution of leaf form correlates with tropical-temperate transitions in *Viburnum* (Adoxaceae). *Proc. Biol. Sci.* **2012**, *279*, 3905–3913. [CrossRef]
32. Lens, F.; Vos, R.; Charrier, G., van der Niet, T.; Merckx, V.; Baas, P.; Aguirre, G.J.; Jacobs, B.; Chacon, D.L.; Smets, E.; et al. Scalariform-to-simple transition in vessel perforation plates triggered by differences in climate during the evolution of Adoxaceae. *Ann. Bot.* **2016**, *118*, 1043–1056. [CrossRef]
33. Clement, W.L.; Donoghue, M.J. Dissolution of *Viburnum* Section *Megalotinus* (Adoxaceae) of Southeast Asia and Its Implications for Morphological Evolution and Biogeography. *Int. J. Plant Sci.* **2011**, *172*, 559–573. [CrossRef]
34. Clement, W.; Arakaki, M.; Sweeney, P.; Edwards, E.J.; Donoghue, M.J. A chloroplast genome tree for *Viburnum* (Adoxaceae): Implications for character evolution and phylogenetic classification. *Am. J. Bot.* **2014**, *101*, 1029–1049. [CrossRef] [PubMed]
35. Spriggs, E.L.; Clement, W.L.; Sweeney, P.W.; Madriñán, S.; Edwards, E.J.; Donoghue, M.J. Temperate radiations and dying embers of a tropical past: The diversification of *Viburnum*. *New Phytol.* **2015**, *207*, 340–354. [CrossRef] [PubMed]
36. Qiu, B.-L.; Ye, L.-X.; Chen, F.; Xie, W.-Y.; Chen, Z.-H. An update of the data of the genus *Viburnum* in Zhejiang. *J. Hangzhou Norm. Univ. Nat. Sci. Ed.* **2020**, *19*, 261–266.
37. Xu, B. *Flora Reipublicae Popularis Sinicae*; Science Press: Beijing, China, 1988; Volume 72.
38. Phillips, S.J.; Anderson, R.P.; Schapire, R.E. Maximum entropy modeling of species geographic distributions. *Ecol. Model.* **2006**, *190*, 231–259. [CrossRef]

39. Graham, C.H.; Elith, J.; Hijmans, R.J.; Guisan, A.; Townsend Peterson, A.; Loiselle, B.A. The influence of spatial errors in species occurrence data used in distribution models. *J. Appl. Ecol.* **2008**, *45*, 239–247. [CrossRef]
40. Zhang, M.-G.; Slik, J.W.F.; Ma, K.-P. Using species distribution modeling to delineate the botanical richness patterns and phytogeographical regions of China. *Sci. Rep.* **2016**, *6*, 22400. [CrossRef]
41. Wan, J.-N.; Mbari, N.J.; Wang, S.-W.; Liu, B.; Mwangi, B.N.; Rasoarahona, J.R.E.; Xin, H.-P.; Zhou, Y.-D.; Wang, Q.-F. Modeling impacts of climate change on the potential distribution of six endemic baobab species in Madagascar. *Plant Divers.* **2021**, *43*, 117–124. [CrossRef]
42. Wang, S.; Zhou, Y.; Musili, P.M.; Mwachala, G.; Hu, G.; Wang, Q. Inventory incompleteness and collecting priority on the plant diversity in tropical East Africa. *Biol. Conserv.* **2020**, *241*, 108313. [CrossRef]
43. Hosmer, D.W. *Wiley Series in Probability and Statistics*; Chap. 2; Multiple logistic regression. Applied logistic regression; John Wiley & Sons, Inc.: New York, NY, USA, 2000; pp. 31–46.
44. Javidan, N.; Kavian, A.; Pourghasemi, H.R.; Conoscenti, C.; Jafarian, Z.; Rodrigo-Comino, J. Evaluation of multi-hazard map produced using MaxEnt machine learning technique. *Sci. Rep.* **2021**, *11*, 6496. [CrossRef]
45. Hijmans, R.J.; Cameron, S.E.; Parra, J.L.; Jones, P.G.; Jarvis, A. Very high resolution interpolated climate surfaces for global land areas. *Int. J. Climatol.* **2005**, *25*, 1965–1978. [CrossRef]
46. Pearson, R.G.; Raxworthy, C.J.; Nakamura, M.; Peterson, A.T. Predicting species distributions from small numbers of occurrence records: A test case using cryptic geckos in Madagascar. *J. Biogeogr.* **2007**, *34*, 102–117. [CrossRef]
47. Sandel, B.; Arge, L.; Dalsgaard, B.; Davies, R.G.; Gaston, K.J.; Sutherland, W.J.; Svenning, J.C. The Influence of Late Quaternary Climate-Change Velocity on Species Endemism. *Science* **2011**, *334*, 660–664. [CrossRef]
48. Rangel, T.F.; Diniz-Filho, J.A.F.; Bini, L.M. SAM: A comprehensive application for Spatial Analysis in Macroecology. *Ecography* **2010**, *33*, 46–50. [CrossRef]
49. Groemping, U. Relative importance for linear regression in R: The package relaimpo. *J. Stat. Softw.* **2006**, *17*, 1–27.
50. R Core Team. *R: A Language and Environment for Statistical Computing*; R Foundation for Statistical Computing: Vienna, Austria, 2021; Available online: http://www.R-project.org/ (accessed on 10 August 2021).
51. Liu, Y.; Shen, Z.; Wang, Q.; Su, X.; Zhang, W.; Shrestha, N.; Xu, X.; Wang, Z. Determinants of richness patterns differ between rare and common species: Implications for *Gesneriaceae* conservation in China. *Divers. Distrib.* **2017**, *23*, 235–246. [CrossRef]
52. Shrestha, N.; Su, X.; Xu, X.; Wang, Z. The drivers of high Rhododendron diversity in south-west China: Does seasonality matter? *J. Biogeogr.* **2018**, *45*, 438–447. [CrossRef]
53. Pandey, B.; Khatiwada, J.R.; Zhang, L.; Pan, K.; Dakhil, M.A.; Xiong, Q.; Yadav, R.K.P.; Siwakoti, M.; Tariq, A.; Olatunji, O.A.; et al. Energy-water and seasonal variations in climate underlie the spatial distribution patterns of gymnosperm species richness in China. *Ecol. Evol.* **2020**, *10*, 9474–9485. [CrossRef]
54. Panda, R.M.; Behera, M.D.; Roy, P.S.; Biradar, C. Energy determines broad pattern of plant distribution in Western Himalaya. *Ecol. Evol.* **2017**, *7*, 10850–10860. [CrossRef]
55. Romdal, T.S.; Araujo, M.B.; Rahbek, C. Life on a tropical planet: Niche conservatism and the global diversity gradient. *Glob. Ecol. Biogeogr.* **2013**, *22*, 344–350. [CrossRef]
56. Löwenberg-Neto, P.; de Carvalho, C.J.; Hawkins, B.A. Tropical niche conservatism as a historical narrative hypothesis for the Neotropics: A case study using the fly family Muscidae. *J. Biogeogr.* **2011**, *38*, 1936–1947. [CrossRef]
57. Chien, C.-T.; Chen, S.-Y.; Tsai, C.-C.; Baskin, J.M.; Baskin, C.C.; Kuo-Huang, L.-L. Deep simple epicotyl morphophysiological dormancy in seeds of two *Viburnum* species, with special reference to shoot growth and development inside the seed. *Ann. Bot.* **2011**, *108*, 13–22. [CrossRef] [PubMed]
58. Zhang, Q.; Li, Y.; Sun, X.; Xing, S.; Cong, W.; Liu, X. Study on Dormancy Mechanism and Breaking Dormancy Method of *Viburnum sargentii* Seeds. *Am. J. Plant Sci.* **2019**, *10*, 65–78. [CrossRef]
59. Currie, D.J.; Paquin, V. Large-scale biogeographical patterns of species richness of trees. *Nature* **1987**, *329*, 326–327. [CrossRef]
60. Jimenez-Alfaro, B.; Girardello, M.; Chytry, M.; Svenning, J.-C.; Willner, W.; Gegout, J.-C.; Agrillo, E.; Antonio Campos, J.; Jandt, U.; Kacki, Z.; et al. History and environment shape species pools and community diversity in European beech forests. *Nat. Ecol. Evo.* **2018**, *2*, 483–490. [CrossRef]
61. Liao, Z.; Zhang, L.; Nobis, M.P.; Wu, X.; Pan, K.; Wang, K.; Dakhil, M.A.; Du, M.; Xiong, Q.; Pandey, B.; et al. Climate change jointly with migration ability affect future range shifts of dominant fir species in Southwest China. *Divers. Distrib.* **2020**, *26*, 352–367. [CrossRef]
62. Lu, L.-M.; Mao, L.-F.; Yang, T.; Ye, J.-F.; Liu, B.; Li, H.-L.; Sun, M.; Miller, J.T.; Mathews, S.; Hu, H.-H.; et al. Evolutionary history of the angiosperm flora of China. *Nature* **2018**, *554*, 234–238. [CrossRef]
63. Ding, W.-N.; Ree, R.H.; Spicer, R.A.; Xing, Y.-W. Ancient orogenic and monsoon-driven assembly of the world's richest temperate alpine flora. *Science* **2020**, *369*, 578–581. [CrossRef]

diversity

MDPI

Article

Conservation Significance of the Rare and Endangered Tree Species, *Trigonobalanus doichangensis* (Fagaceae)

Ling Hu [1,2], Xin-Gui Le [3], Shi-Shun Zhou [4], Can-Yu Zhang [5], Yun-Hong Tan [4], Qiang Ren [1], Hong-Hu Meng [1,4,*], Yupeng Cun [6,7,*] and Jie Li [1,8,*]

1 Plant Phylogenetics and Conservation Group, Center for Integrative Conservation, Xishuangbanna Tropical Botanical Garden, Chinese Academy of Sciences, Kunming 650223, China
2 University of Chinese Academy of Sciences, Beijing 100049, China
3 Yangjifeng National Nature Reserve of Jiangxi Province, Guixi 335400, China
4 Southeast Asia Biodiversity Research Institute, Chinese Academy of Sciences, Naypyidaw 05282, Myanmar
5 Yunnan Normal University, Kunming 650500, China
6 Pediatric Research Institute, Ministry of Education Key Laboratory of Child Development and Disorders, National Clinical Research Center for Child Health and Disorders, China International Science and Technology Cooperation Base of Child Development and Critical Disorders, Chongqing Key Laboratory of Translational Medical Research in Cognitive Development and Learning and Memory Disorders, Children's Hospital of Chongqing Medical University, Chongqing 400014, China
7 Germplasm Bank of Wild Species, Kunming Institute of Botany, Chinese Academy of Sciences, Kunming 650201, China
8 Center of Conservation Biology, Core Botanical Gardens, Chinese Academy of Sciences, Mengla 666303, China
* Correspondence: menghonghu@xtbg.ac.cn (H.-H.M.); cunyp@cqmu.edu.cn (Y.C.); jieli@xtbg.ac.cn (J.L.)

Abstract: *Trigonobalanus doichangensis* is a rare and endangered species with important evolutionary value and extremely small populations. We investigated the genetic diversity of *T. doichangensis* to provide information on its effective preservation. We used genotyping-by-sequencing (GBS) technology to assess the genetic diversity, genetic structure and gene flow of the six populations of *T. doichangensis*. Analysis of SNPs indicated that there was high genetic diversity in the ML and XSBN populations of *T. doichangensis*. F_{ST} values showed moderate genetic differentiation among the populations of *T. doichangensis*. Meanwhile, admixture, principal components and gene flow analyses indicated that the populations of *T. doichangensis* are not genetically separated in accordance with their geographical distributions. Habitat destruction and excessive exploitation may have led to a low gene flow, which has in turn resulted in the differences in seed and seedling morphological traits among populations. Based on these findings, we recommend that *T. doichangensis* be conserved through in situ approaches and artificial seedlings, including preservation of each extant population. Particularly, the ML and XSBN populations have high diversity and more ancestral information, so these two populations should be considered as conservation priorities, and seeds should be collected to obtain germplasm and increase the genetic diversity.

Keywords: biodiversity; conservation; *T. doichangensis*; genetic structure; gene flow; conservation priority

Citation: Hu, L.; Le, X.-G.; Zhou, S.-S.; Zhang, C.-Y.; Tan, Y.-H.; Ren, Q.; Meng, H.-H.; Cun, Y.; Li, J. Conservation Significance of the Rare and Endangered Tree Species, *Trigonobalanus doichangensis* (Fagaceae). *Diversity* **2022**, *14*, 666. https://doi.org/10.3390/d14080666

Academic Editor: Maurizio Rossetto

Received: 20 July 2022
Accepted: 14 August 2022
Published: 17 August 2022

1. Introduction

The arrival of the Anthropocene is bringing an unprecedented challenge to the biodiversity that is essential to humans and enhancing the value of many of the benefits of nature to human beings [1]. Biodiversity is vital to human well-being and an important pillar of ecosystem balance [2]. Since the adoption of the Convention on Biological Diversity (CBD), various countries have made several achievements and contributions in different areas with the aim of maintaining biodiversity and safeguarding the Earth's resources, but biodiversity loss and ecosystem degradation remain important problems [3]. China is rich in biological resources and contains several global biodiversity hotspots. Its unique geographical distribution and climatic zones have shaped a wide range of diverse biotypes and it is considered a major center of origins and diversity [4]. However, with rapid economic growth and dramatic population increases, conflicts between economic development and biological conservation have come to the fore, and China is becoming one of the countries with the most serious threats to biodiversity [5]. An assessment of over 35,000 species of wild higher plants in China found that approximately 10.84% of species are in a threatened state [6]. Continued loss of biodiversity and chronic environmental conflicts suggest that conservation communities, including authorities, institutes and scholars, need to re-examine the assumptions and practices upon which the conservation endeavor has been founded, particularly in tropical China, which harbors hyper-diverse plant species [7,8]. In recent years, the National Implementation Plan for Rescuing and Conserving China's Plant Species with Extremely Small Populations (PSESP) was formulated following the presentation of the concept of PSESP [9]. In China, it is these species that are the focus of the PSESP program, species with extremely small populations are precisely the ones most in need of urgent conservation action [10]. The goal of the PSESP program, developed and implemented in 2005 in Yunnan Province, is to try to secure a long-term future for these plants in peril [11]. From then on, some taxa of PSESP have been brought into focus; e.g., *Aristolochia delavayi* [12], *Camellia huana* [13], *Cycas panzhihuaensis* [14], *Horsfieldia tetratepala* [15], *Rhododendron meddianum* [16], *Rhododendron pubicostatum* [17], etc. The introduction of the PSESP program provides a target for the conservation of endangered plants in China; however, the endangerment mechanism and threat status of species are important indicators for species conservation [18]. Only by clarifying the actual threat status, endangerment mechanism and genetic diversity of species can the direction of conservation be targeted and conservation measures clarified [18].

Genetic diversity is one of the key pillars of biodiversity, and high genetic diversity increases wild plants ability to survive and reduces the risk of extinction for species, thus allowing the prediction of the fitness of species through the study of genetic diversity [19,20]. In addition, genetic differentiation and gene flow are important elements in understanding the evolutionary and adaptive potential of populations. Previous studies have shown that many endangered plants tend to have low genetic diversity and high genetic differentiation due to overexploitation and habitat destruction [15,21,22]. Inbreeding can occur between populations as a result of fewer individuals existing, leading to the gradual accumulation of deleterious mutations between populations, producing severe geographical isolation and placing the survival of populations at risk [23]. The loss of endangered species habitats also disrupts gene flow between populations and leads to population isolation, a negative effect that seriously threatens the development and diversity structure of species, while high gene flow reduces the incidence of inbreeding and population differentiation by increasing the exchange of genetic material between populations [22]. However, there are some endangered species that exhibit a genetic structure with high genetic diversity and low genetic differentiation [24,25]. In general, plant genetic diversity is influenced by seed dispersal, reproductive systems, life history, geographic range and evolutionary history. Thus, species with long life histories, extensive seed dispersal routes and outcrossing will likely retain more genetic variation, allowing them to resist environmental change, but over-exploitation and habitat destruction, as well as limitations within the species themselves, will lead to population contraction and a dramatic reduction in the number of individuals. Then, over time, inbreeding and genetic drift will increase population differentiation and reduce the gene flow between populations, leading again to a high degree of differentiation between populations [15,20,26]. Therefore, the combination of various aspects in the exploration of species diversity levels can help us to propose more accurate conservation strategies [22].

The fields of ecological and conservation genetics have developed significantly in recent decades thanks to the use of molecular markers to investigate organisms in their natural habitat and to evaluate the effect of anthropogenic disturbances [27]. However, previous studies mainly focused on genetic diversity, population structure, genetic characterization based on chloroplast DNA sequence, single-copy nuclear genes and microsatellite markers. Few studies have used next-generation sequencing (NGS) to investigate single nucleotide polymorphisms (SNPs), which have become the marker of choice for determining population structure as they are abundant, stable in the genome and can be accurately scored [28]. Among the NGS methods used in population genetics, the genotyping-by-sequencing (GBS) approach is suitable for population studies, germplasm characterization, breeding and trait mapping in diverse organisms [29]. GBS methods offer major advantages for population genomics thanks to their capacity to screen thousands of polymorphisms throughout the genome and highlight the full range of evolutionary histories (variation in drift, selection, recombination, mutation) and consequences for genetic variation [27]. This method is based on genome-wide sequencing of loci adjacent to conservative restriction sites, which makes it possible to obtain thousands of homologous loci for a set of species with no prior genome data [30].

Trigonobalanus shows many ancestral features typical of the Fagus-Quercus taxa [31,32]. *Trigonobalanus doichangensis* belongs to *Trigonobalanus*, which includes *T. doichangensis*, *T. verticillata* and *T. excelsa*. *T. doichangensis* is restricted to south Yunnan, China—i.e., Ximeng, Menglian, Puer, Lancang and Cangyuan—and Chiang Mai in northern Thailand (Figure 1) [32,33]. *T. doichangensis* and *T. verticillata* are distributed across tropical Asia, while *T. excelsa* is found in Columbia, Central America, which is important to understand the phylogeny and biogeography of Fagaceae in relation to continental drifts, climatic shifts and the past global environment [8].

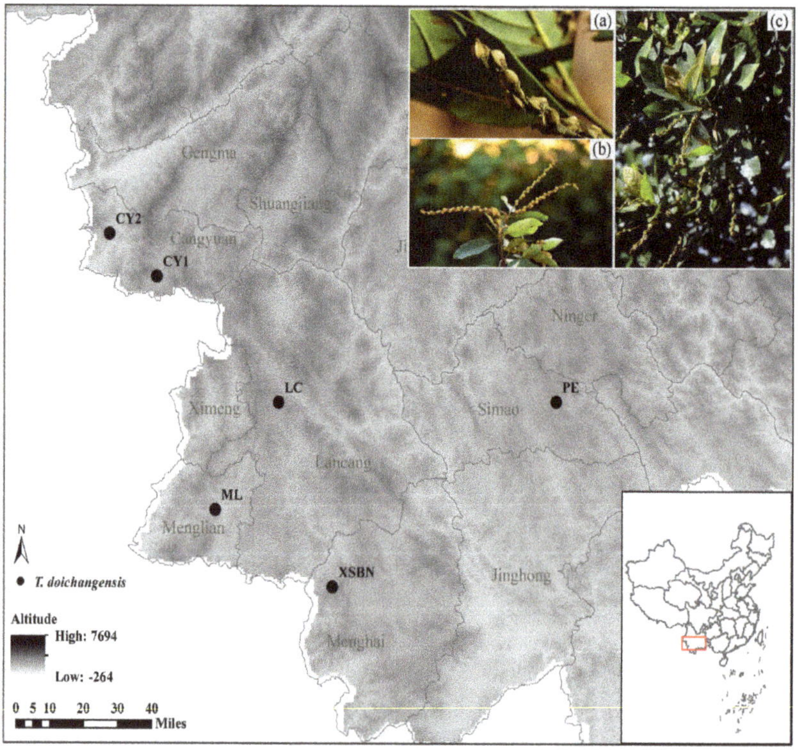

Figure 1. Sampling locations of *T. doichangensis*. (**a**,**b**) pistillate inflorescences of *T. doichangensis*; (**c**) leaf of *T. doichangensis*.

T. doichangensis has been felled for firewood, house building and agricultural tools [34]. Due to this heavy exploitation, along with habitat degradation caused by clearing areas for agriculture, *T. doichangensis* has been pushed to the verge of extinction [35]. Despite being a scientifically important tree with endangered status, the species is rare in China and, thus, poorly represented in herbaria; the species has been proposed as the second ranked endangered plant for national protection in the China Species Red List [36].

For the conservation of the rare and endangered *T. doichangensis* tree in southwest China, the extent of the genetic variation in seed and seedling traits suggested seed collection of this species for ex situ conservation and species restoration [34]. Historically, most studies on ecological and conservation genetics have relied on a small number of putatively neutral molecular markers [27]. Similarly, the population genetics of *T. doichangensis* have long been assessed using random amplified polymorphic DNA (RAPD), which suggested that habitat degradation, overexploitation and reproductive barriers are most likely to be the factors threatening the species [35]. To date, there has been little research on the population structure and genetic diversity of *T. doichangensis* based on genome-wide SNP information, although this information is important for the development of meaningful conservation management strategies for this species.

In this study, we used GBS to investigate relationships between populations of *T. doichangensis* based on 39 samples and determine the genetic diversity, population structure and genetic distance and gene flow of *T. doichangensis* in Yunnan, China. Our aim in this study was to develop meaningful conservation strategies and suggestions for *T. doichangensis*.

2. Materials and Methods

2.1. Sample Collection and Genotyping-by-Sequencing

We collected 39 individuals from six populations of *T. doichangensis* from its distribution areas in China (Figure 1; Table 1). The total genomic DNA of *T. doichangensis* was extracted using a Plant Genomic DNA Kit (Tiangen Biotech, Beijing, China). A GBS library was prepared by Shanghai Majorbio Bio-pharm Technology Co., Ltd. Double digestion was performed using *Mse*I and *Taqα*I then sequenced on an Illumina Hi-seq sequence platform with paired-end (PE) 150 sequencing.

Table 1. Sample information for *T. doichangensis*.

No.	Populations	No.	Locations	Longitude	Latitude	Elevation (m)
1	XSBN	9	Padianliangzi, Xishuangbanna, Yunnan	100°04′58″	22°02′32″	2038
2	LC	6	Donghuixiang, Lancang, Yunnan	99°48′50″	22°43′20″	1400
3	CY1	5	Chengbian, Cangyuan, Yunnan	99°14′28″	23°10′18″	1482
4	CY2	8	Nanbancun, Cangyuan, Yunnan	99°01′04″	23°19′27″	1482
5	PE	5	Dazhaishuiku, Puer, Yunnan	101°03′43″	22°45′44″	1564
6	ML	6	Dengzhanzhai, Menglian, Yunnan	99°32′30″	22°18′47″	1153

2.2. SNP Calling and Quality Filter

We used FastQC (https://www.bioinformatics.babraham.ac.uk/projects/fastqc, accessed on 14 Qctober 2021) and Fastp (http://github.com/OpenGene/fastp, accessed on 14 Qctober 2021) for quality control of the raw data, which was undertaken by Shanghai Majorbio Bio-pharm Technology Co., Ltd. skrTools (https://github.com/sharkLoc/skrTools, accessed on 7 November 2021) was used to calculate the number of reads, the GC content and the average quality of the data (Q30) for the GBS of *T. doichangensis*.

We employed the *process_radtags* module in STACKS v 2.55 [37] to filter low-quality reads and ambiguous bases. Due to STACKS need for consistent sequence lengths, the sequence lengths were trimmed to 135 bp to ensure subsequent analysis: -r -c -q -t 135. Clean reads were aligned to the reference genome of *Castanea mollissima* [38] with default parameters in BWA-MEM v 0.7.17 [39] and sorted via SAMtools v 1.8 [40], which provided the final BAM file for each sample. We built loci from the aligned paired-end data with the *gstacks* module in STACKS. Finally, we used the *populations* module in STACKS for SNP calling, specifying a genotyping rate of at least 75% of individuals within populations and a minimum number of three populations per locus, outputting the first SNP per locus.

To ensure downstream analysis, we conducted sample quality control on the SNPs, using VCFtools v 0.1.16 [41] to view the missing rates of samples and loci, randomly identifying samples with a missing rate of more than 70% according to the population and merging the BAM file to increase the signal of variation sites using SAMtools-merge. We reran the *gstacks* module and *populations* module in STACKS for the SNP calling; the parameters were the same as before, and randomly sampling was conducted in R v 4.1 [42].

The vcf file was further filtered in VCFtools using the following parameters: SNPs with missing rates greater than 20% were removed (–max-missing 0.8); SNPs with minor allele counts lower than 3 (–mac 3) were retained; SNPs with average depths greater than or equal to 3 were reserved (–minDP 3); minor allele frequencies lower than 0.05 were also removed (–maf 0.05).

2.3. Calculation of Genetic Diversity Parameter and AMOVA

The parameters of population genetic diversity included private alleles (PAs), expected heterozygosity (*He*), observed heterozygosity (*Ho*), pairwise genetic distance (F_{ST}), inbreeding coefficient (F_{IS}) and nucleotide polymorphism (*Pi*), which were calculated using the *populations* module in STACKS; the F_{ST} value was calibrated using the *p*-value.

In order to determine whether genetic variation was present within the populations, we used Arlequin v3.5.2 [43] for analysis of molecular variance (AMOVA), where groups were defined based on the results of the structural clustering. The corresponding file formats were converted using PGDspider v 2.1.1.5 [44].

2.4. Isolation by Distance

F_{ST} was used as the genetic distance between populations. Geographic distances between populations were calculated according to their geographic coordinate information using the *geosphere* package [45]. The relationship and significant differences between genetic distance and geographic distance were assessed with the Mantel test using the *ade4* package [46] in R, and 9999 permutations were performed to determine significance.

2.5. Population Structure and Gene Flow

Population structure was analyzed in terms of the different numbers of ancestors (*K*) in Admixture v 1.3 [47], where *K* was chosen between 1 and 7 (assumed populations + 1) [48]. The most likely number of genetic clusters was computed with a fivefold cross-validation (CV) error. Finally, population structure was visualized via the *pophelper* package [49] in R. Principal components analysis (PCA) clearly showed hierarchical clustering for each population. Based on the previous results, PLINK v 1.9 [50] was used to generate files for PCA analysis, and then the *ggplot2* package [51] in R software were used for the PCA display. Discriminant analysis of principal components (DAPC) was used to infer the population structure by determining the number of clusters observed without prior knowledge, and the principal components from the PCA obtained from the data dimensionality reduction were used to perform linear discriminant analysis (LDA) [52]. Since DAPC is more robust than PCA [53,54], we added DAPC to verify the results from Admixture and the PCA. We used the *adegenet* package [55] in R to plot DAPC results. TreeMix v 1.12 [56] was used to infer splitting and mixture patterns among populations of *T. doichangensis*. Migration events (m) from 0 to 6 were specified between populations, and 10 iterations per event were tested. For this analysis, "-global" and "-se" options were used to calculated the standard errors, the "-noss" parameter prevented overcorrection of the sample size and the "-bootstrap -k 1000" parameter was used to build a maximum likelihood tree (ML) by resampling blocks of 1000 SNPs [57]. The *OptM* package [58] in R was used to determine the optimal number of migration edges using the output file from the TreeMix, and the R script *plotting_funcs.R* (https://github.com/joepickrell/pophistory-tutorial/blob/master/example2/plotting_funcs.R, accessed on 6 March 2022) was used to visualize the migration results. Then, we ran the *threepop* module and *fourpop* module in TreeMix to analyze the f3-statistics and f4-statistics with -k 1000. In the f3-statistics (C; A, B), a significantly negative statistic for the f3 score (Z-score < −3) indicated that C was a mixture of A and B. In the f4-statistics (A, B; C, D), a significantly negative statistic for the f4 score (Z-score < −3) indicated gene flow between populations related to A and D or B and C, and a significantly positive statistic for the f4 score (Z-score > 3) indicated gene flow between populations related to A and C or B and D [59].

3. Results

3.1. Genotyping-by-Sequencing and Quality Control

All 39 samples of *T. doichangensis* were sequenced using GBS, which produced 412,410,372 clean reads and the average value for the quality scores of the reads (Q30) was 88.52%, while the GC content was 41.74%. After running the *process_radtags* module, a total 377,488,916 clean reads were obtained, in which the GC content was 41.66% and the Q30 was 89.06%.

SNP calling was conducted according to the method described in Section 2.2. Since the missing rates in the sampling locations XSBN and ML were relatively high (Figure 2), the XSBN and ML populations were randomly sampled and merged as new BAM files for SNP calling. A total of 38,182 variant sites were obtained, and 7591 filtered SNPs were used for downstream analysis.

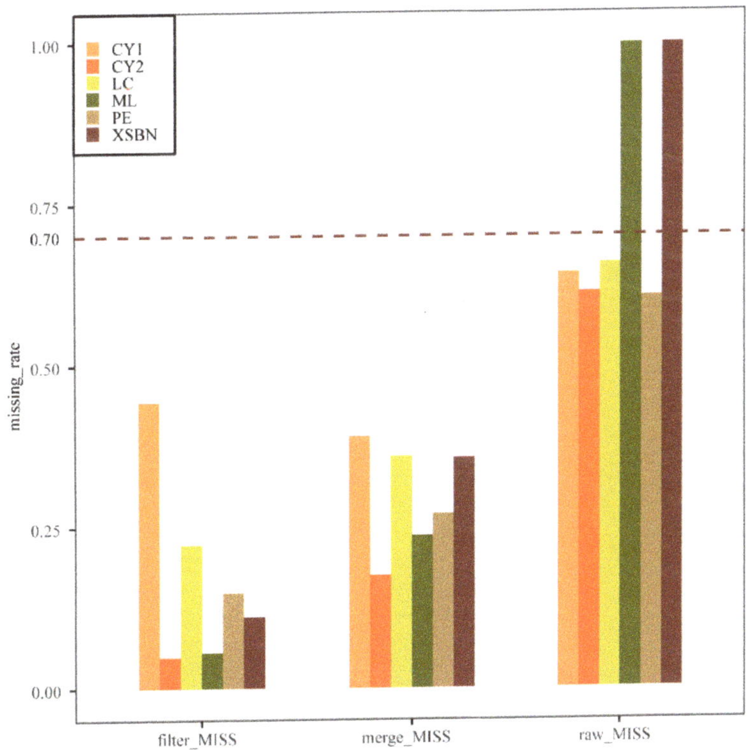

Figure 2. The missing SNP rate for each population without merging (raw_MISS); The missing SNP rate for each population after merging (merge_MISS); and the missing SNP rate for each population after filtering (filter_MISS).

3.2. Genetic Diversity and Genetic Differentiation

The results for the genetic diversity parameters of *T. doichangensis* are summarized in Table 2 and Figure 3. The observed heterozygosity (*Ho*) ranged from 0.3129 to 0.5571; the expected heterozygosity (*He*) ranged from 0.2232 to 0.3059; nucleotide diversity (*Pi*) ranged from 0.2504 to 0.3343; the inbreeding coefficient (F_{IS}) ranged from −0.4213 to −0.0879 (Table 2). The ML population showed the highest genetic diversity (*Ho* = 0.5571; *He* = 0.3059; *Pi* = 0.3343; Table 2) followed by the XSBN population, and the PE population showed the lowest genetic diversity (*Ho* = 0.3129; *He* = 0.2232; *Pi* = 0.2504; Table 2). Furthermore, the XSBN population had the most private alleles, and the CY1 population had the fewest private alleles, while the F_{IS} values of the six populations were negative, indicating that there was no inbreeding within the six populations.

Table 2. Genetic diversity parameters for *T. doichangensis* populations based on GBS.

Populations	PA	*Ho*	*He*	*Pi*	F_{IS}
XSBN	642	0.5410	0.3044	0.3235	−0.4213
LC	10	0.3169	0.2383	0.2621	−0.1022
CY1	3	0.3129	0.2335	0.2637	−0.0879
CY2	48	0.3150	0.2430	0.2598	−0.1033
PE	32	0.3129	0.2232	0.2504	−0.1151
ML	284	0.5571	0.3059	0.3343	−0.4166

Notes: PA, private allele; *Ho*, observed heterozygosity; *He*, expected heterozygosity; *Pi*, nucleotide diversity; F_{IS}, inbreeding coefficient.

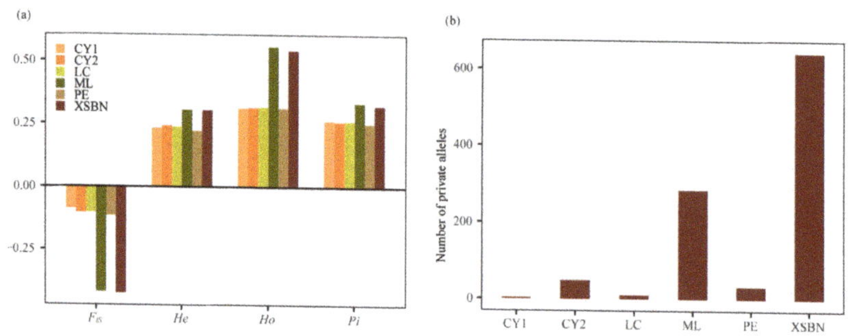

Figure 3. The results for the diversity parameters of *T. doichangensis* populations. (**a**) Expected heterozygosity (*He*), observed heterozygosity (*Ho*), inbreeding coefficient (F_{IS}) and nucleotide polymorphism (*Pi*); (**b**) number of private alleles (PAs).

The values of F_{ST} ranged from 0.0462 to 0.1402 (Table 3, Figure 4a) with an average of 0.0921, and the average $Nm = (1 - F_{ST})/4 * F_{ST} = 2.7194$. The results showed that there was genetic differentiation and gene flow between populations. In addition, the CY1 and CY2 populations showed the lowest F_{ST} (F_{ST} = 0.0462) and the ML and XSBN populations showed the largest F_{ST} (F_{ST} = 0.1402), indicating that only 4.62% of the variation occurred between the CY1 and CY2 populations and most (95.38%) occurred within populations. CY1 and CY2 were geographically closest and in the same region (Figure 1; Table 3), which may also have been the reason for the low genetic differentiation between CY1 and CY2. However, after we analyzed the correlation between genetic distance and geographic distance, we found that there was no significant correlation between genetic distance and geographic distance (Figure 4b). The AMOVA results also indicated that variation occurred within populations rather than among populations (Table 4).

Table 3. Genetic distance and geographic distance between populations of *T. doichangensis*. Below diagonal: genetic distance among populations (F_{ST}). Above diagonal: geographic distance between populations (km).

Populations	XSBN	LC	CY1	CY2	PE	ML
XSBN	-	80.5984	152.6067	179.8030	128.7604	63.4062
LC	0.0980	-	77.1407	105.5585	128.2083	53.4627
CY1	0.1014	0.0573	-	28.4602	192.1101	100.4373
CY2	0.1121	0.0561	0.0462	-	218.5402	124.7333
PE	0.1210	0.0779	0.0829	0.0779	-	164.1128
ML	0.1402	0.0957	0.1004	0.1013	0.1133	-

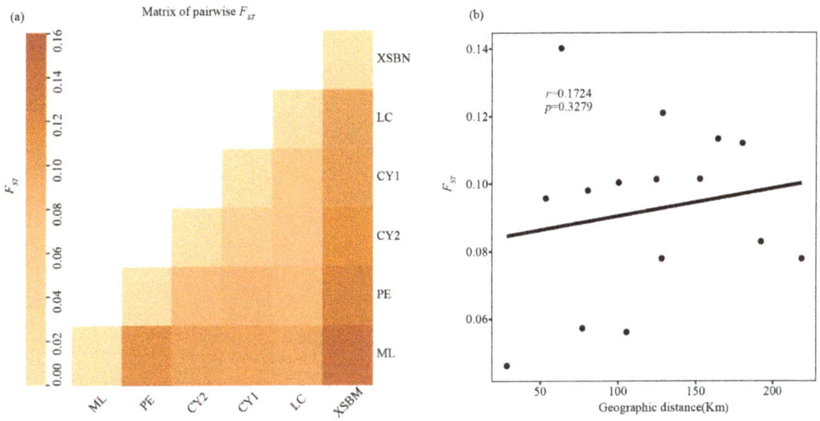

Figure 4. Pairwise F_{ST} for *T. doichangensis* populations (**a**) and the correlation between genetic distance and geographic distance (**b**).

Table 4. Analysis of molecular variance (AMOVA) of the genetic variation in *T. doichangensis*.

Source of Variation	d.f.	Sum of Squares	Variance Components	Percentage of Variation
Among groups	2	3210.467	49.1535	9.97
Among populations within groups	3	1617.129	8.8010	1.79
Within populations	72	31312.635	434.8977	88.24
Total	77	36140.231	492.8522	-

Notes: Groups are optimal results of genetic structure. d.f., degree of freedom.

3.3. Genetic Structure of *T. doichangensis*

We performed PCA, DAPC and genetic structure analysis to understand the genetic structure of each population. Fivefold cross-validation (CV) was chosen for the genetic structure analysis, and the results showed that the optimal number of ancestral components for the *T. doichangensis* populations was 3 (Figure S1). When K = 3, XSBN and ML populations formed cluster 1 (red) and cluster 3 (purple), respectively, except for a few admixed individuals. The LC, CY1, CY2 and PE populations formed a mixed-component cluster (cluster 2, green) (Figure 5c). Moreover, the results of the PCA analysis also confirmed the results from Admixture (Figure 5a). The PCA results based on 7591 SNPs showed that principal components (PC1, PC2) represented 45.3% and 29.4% of all variation, respectively (Figure 5a). The XSBN and ML populations from Xishuangbanna and Menglian each formed a single cluster, while the LC, CY1, CY2 and PE populations formed close clusters (Figure 5a). In addition, we increased the number of principal components and further analyzed the first three principal components. We found that, with three principal components, the PE population was further separated (Figure S2). The PE population is located in Puer city, Yunnan, China, and is separated from the XSBN, LC, CY1 and CY2 populations on both sides of the Lancang River. Therefore, we assumed that the geographical barrier was the reason for the separation of the PE population. We used the first three PC axes and two discriminant axes to perform DAPC, using the Bayesian information criterion (BIC) to define the best cluster (Figure S3). The first eigenvalue was 369.2 and the second eigenvalue was 151.3. The results are the same as in the genetic structure analysis and the first two principal components of the PCA (Figure 5b).

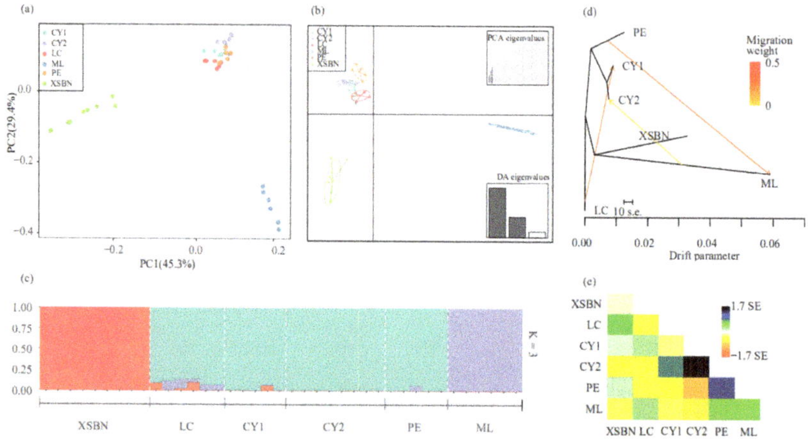

Figure 5. (**a**) Principal components analysis (PCA) plot generated for *T. doichangensis*. (**b**) Discriminant analysis of principal components (DAPC) plot generated for *T. doichangensis*. (**c**) Genetic structure bar plots at K = 3 from clustering analysis among six populations using Admixture. (**d**) Unrooted maximum likelihood phylogenies from ten iterations of three migration events. The variance of relatedness explained by this model was estimated to be 99.98%. The line represents gene flow; the direction of the arrow is the direction of gene flow; the color bar shows the migration weight, with red representing strong gene flow and yellow representing weak gene flow. The direction of gene flow was PE into ML, LC into CY1 and ML into CY2. (**e**) Residuals matrix for migration model.

3.4. Gene Flow Analysis

In order to infer the split pattern and mixtures among populations of *T. doichangensis*, we used allele frequencies from six populations to infer admixture events. Supplementary Figure S4 shows the maximum likelihood value and variance explained for each event based on *OptM* package analysis in TreeMix, which was used to choose migration events varying from 0 to 6. The model without migration events explained 99.71% of the variance, with the highest score for the variance (99.98%) being achieved when m = 3; adding more migration events did not explain more variance. Figure 5d shows the optimal "migrate" result for the last iterations. The result indicated that the direction of gene flow was PE into ML, LC into CY1 and ML into CY2. Although TreeMix detected no strong signal of migration between CY1 and CY2, the residues showed moderate gene flow between CY1 and CY2 (Figure 5e). The ML into CY2 inferred migration event did not match geographical patterns, indicating a special flow channel between the populations, which affected the introgression. We then used the three-population and four-population test (f3-statistic/f4-statistic) to evaluate the statistical significance of the mixtures (Tables S1 and S2). The three-population test indicated that a mixture existed between populations, in which CY1 was the mixing result of multiple populations (Table S1). The four-population test confirmed that there was greater gene flow between the two populations CY1 and CY2 in Cangyuan county with a relatively high mixing degree (Table S2), which also suggested low genetic differentiation between CY1 and CY2. Meanwhile, the TreeMix results were corroborated by the four-population tests. Furthermore, the f4-statistic results showed that there was lower gene flow between most populations, and it has been suggested that the existence of gene flow can maintain the stability of small populations [60]. Although we did not add outgroups to the analysis, the results can still be used as a reference for relations of populations.

4. Discussion

4.1. Genetic Diversity

The genetic variation of species is the premise of local adaptation and evolution, and it is also considered an important parameter to determine the priority of population conservation in the protection of endangered plants [12,61]. According to the CBD, genetic diversity is one of the three basic elements of biodiversity, and it is the focus of many conservation genetics studies [62]. Therefore, genetic diversity is one of the most important values in assessing biodiversity for conservation.

When studying the genetics of wild populations, it is desirable to sample tens, hundreds or even thousands of individuals. However, there are some rare and endangered taxa with narrow distributions. The genetic diversity of narrowly distributed species is likely to be lower than widespread plant species because of inbreeding depression and genetic drift [16,63]. A previous study on the genetic diversity of *T. doichangensis* based on random amplified polymorphic DNA indicated high genetic differentiation, a low level of genetic diversity and a poor gene flow [35]. Ramanatha and Hodgkin [64] stated that different molecular markers show different levels of genetic diversity and, unexpectedly, GBS indicated that the nucleotide genetic diversity of *T. doichangensis* is high (Figure 3; Table 2). In all populations, the results for the diversity parameters for *T. doichangensis* populations utilizing the expected heterozygosity (*He*), observed heterozygosity (*Ho*) and nucleotide polymorphism (*Pi*) indicated that ML and XSBN had higher genetic diversity (Figure 3). Moreover, the number of private alleles was the highest in XSBN (Table 2). The genetic diversity of species is generally influenced by their distribution range, life history, breeding system, seed dispersal mechanism and evolutionary history [16]. According to previous studies, *T. doichangensis* has a low germination rate and low fruiting rate in nature, thus limiting the development of the population [65]. Furthermore, *T. doichangensis* is distributed in evergreen broad-leaved forests where sunlight cannot reach the ground through the tree layer, and the dense trees increase the mortality of young *T. doichangensis* seedlings competing for sunlight and space to survive [66]. This is one of the reasons for the lower genetic diversity (*He* = 0.2232–0.3059; Table 2) compared to other genera in the family; e.g., *Castanopsis* (EST-SSR, *He* = 0.644; [67]) or *Quercus castanea* (nSSR, *He* = 0.762; [68]). *T. doichangensis* is also facing the dual impact of biological invasion and human activities, and its habitat has also suffered serious damage. Shrinking habitats, lower population sizes, and fewer individuals in species seriously affect genetic diversity [35]. In this study, the populations of *T. doichangensis* still retained a certain degree of genetic diversity (Table 2; Figure 3), which showed that populations of *T. doichangensis* still have a certain ability to resist external risks. However, in species protection, only relying on the species itself to resist the adverse environment is not desirable. Accelerated habitat fragmentation and population size reduction can exacerbate genetic drift and inbreeding, resulting in rapid loss of genetic diversity.

In addition, genetic diversity is usually higher in outcrossing species than in selfing species [69,70]. Notably, inbreeding coefficients (F_{IS}) of all populations were negative (Table 2; Figure 3a), implying that there is no inbreeding in the populations of *T. doichangensis*, thus increasing the genetic diversity. We assume that, in the past few decades, with the rapid reduction of the population size, the genetic drift effects have not yet accumulated, and the species are not strongly affected by genetic drift or inbreeding and retain relatively high heterozygosity and genetic variation in limited populations.

4.2. Genetic Differentiation, Genetic Structure and Gene Flow

Genetic differentiation is often strongly influenced by selection pressure, gene flow and life history [71]. For endangered populations, isolated populations and broken habitats are highly susceptible to producing high genetic differentiation [72]. The F_{ST} values of *T. doichangensis* ranged from 0.0462 to 0.1402, with an average F_{ST} value of 0.0921 (Table 3), revealing that moderate genetic differentiation among the populations occurred according to Wright's theory (Wright, 1965). However, it is generally accepted that gene flow can block genetic differentiation due to genetic drift when $Nm > 1$ [73]. We calculated that $Nm = (1 - F_{ST})/4 * F_{ST} = 2.7194$ in the *T. doichangensis* populations, which implied that there was gene flow among the populations counteracting genetic differentiation [73,74]. Furthermore, AMOVA showed that 88.24% of the genetic variation occurred within *T. doichangensis* populations but not among populations (Table 4), while the F_{IS} results showed that the inbreeding coefficients of all the populations of *T. doichangensis* were negative (Table 2), indicating that there was no inbreeding within populations and, thus, genetic differentiation was lower [23]. For rare and endangered plants, population genetic differentiation is usually affected by large geographical distances between sampled populations [69,70]. However, the genetic distance and geographic distance between populations of *T. doichangensis*, the Admixture analyses, the PCA and the gene flow demonstrated that the populations of *T. doichangensis* are not genetically separated in accordance with their geographical distributions (Figures 4b and 5). The previous finding that pairwise genetic distances between populations were not correlated with geographical distances was supported by the observation that one of the Chinese populations is most similar to the Thai population [75].

The distance among the populations in this study was about 50 km, which hardly allows gene flow (pollen or seeds) between populations. However, there were three gene flow events among the six populations in TreeMix; i.e., PE to ML, LC to CY1 and ML to CY2 (Figure 5d). Three-population and four-population tests also indicated partial gene flow between populations (Tables S1 and S2). Admixture, PCA and DAPC showed that the optimal cluster of populations was 3, and the genetic structure of some populations was mixed (Figure 5). All the above results indicated that there was some gene flow among the populations of *T. doichangensis* and that their genetic structure is not strongly geographically heterogeneous. At the same time, we found that Admixture, PCA and DAPC all pointed to the possibility that the more independent genetic components for XSBN and ML populations may retain more ancestral information, while the PE populations may have gradually separated due to river barriers after dispersal.

Previous studies found that most of the seed and seedling traits of *T. doichangensis* showed significant differences among populations [75], confirming the hypothesis that *T. doichangensis* possesses very strong genetic differentiation within its populations. In addition, in line with our observations, seedlings grow sporadically around the parent tree. Meanwhile, Sun [35] noted the presence of small beetles in the male flowers of *T. doichangensis*, so it can be presumed that pollinator beetles may be active. However, the flowers of *T. doichangensis* are small and dull in color, so they cannot attract more pollinating insects, and pollination by beetles will cause *T. doichangensis* pollen to spread over a short distance and make the gene flow between populations difficult, leading to differences between populations. Hence, the differentiation among the populations could be attributed to distance-limited pollen flow and short-distance seed dispersal. We suggest that habitat destruction and excessive exploitation may have led to low gene flow, which in turn resulted in the differences in seed and seedling morphological traits among populations.

4.3. Conservation Significance

Trigonobalanus doichangensis is a rare and endangered fagaceous plant with evolutionary significance for the understanding of the phylogeny and biogeography of Fagaceae and even Chinese flora more generally. It is currently restricted to a few sites in Yunnan province in southwestern China, as well as one in northern Thailand [35,75]. *T. doichangensis* was placed on the national Rare and Endangered Species List of China in 1984 because of its limited distribution and the destruction of its habitat within China [66,76]. *T. doichangensis* is China's second-ranked taxon for priority of national protection [36] because of its endangered status and its scientific value. However, the rare and endangered *T. doichangensis* is exposed to the risk of extinction due to the destruction and degradation of forest habitats, agriculture, silviculture and the harvesting of wood for fuel and tool making [75].

The disappearance of rare plants prevents the tracking of the evolutionary and biogeographic history of plants, which not only has consequences for medicinal and other economic uses but also limits the potential resources about past climate change and future implications [77]. In addition, although only a small number of plant species have been exploited by humans, many others play important roles in natural ecosystems, and rare species may also have novel traits that could be useful in the future. Thus, the impacts of economic development on rare plants and their habitats need to be recognized and addressed before these potential resources are lost forever.

The genetic diversity of *T. doichangensis* obtained through GBS has important conservation significance for narrowly distributed species. The most effective approach to conserving endangered species is in situ conservation [25]. A previous study based on RAPD suggested that conservation of this species should include preservation of each extant population [75]. Seedlings and saplings are rarely found in wild populations of *T. doichangensis*, and its habitat has been disturbed and destroyed as a result of various factors, including insects and low seed-germination percentages [34,35]. Therefore, it is imperative to establish conservation plots to protect its natural habitat. In accordance with its genetic diversity and structure, we suggested that ML and XSBN should be protected as in situ conservation priorities as soon as possible so they can be used as germplasms with high diversity. Moreover, because moderate genetic differentiation and certain genetic structure were revealed among populations—particularly, the gene flow in the ML and XSBN populations—germplasm resources from each population should be not mixed and instead should be used separately to avoid the risk of outbreeding depression. Considering that the habitats of *T. doichangensis* have been devastated and fragmented by human activity, and that the seed setting percentage is low and seedling growth is limited, artificial seedlings should also be raised and transplanted to areas with similar habitats in order to expand the population size and protect the gene pool of the species. In addition, conservation campaigns should be conducted in towns and villages around the distribution area of *T. doichangensis* to protect the original habitat and restore the function of the surrounding plant community.

5. Conclusions

In summary, we used the high density of SNP loci generated by GBS for population genomic analysis of *T. doichangensis* with 39 individuals from 6 populations. Our data indicated that there was high genetic diversity and moderate genetic differentiation in the six populations of *T. doichangensis*. At the same time, most of the genetic variation in *T. doichangensis* occurred among populations, and there was some gene flow among populations to counteract the genetic differentiation caused by genetic drift. Based on these results, we propose a strategy for in situ conservation of the ML and XSBN populations of *T. doichangensis* with high genetic diversity and highlight the importance of germplasm collection, artificial seedlings and conservation promotion, providing an important reference and guidance for the conservation of *T. doichangensis* populations and similarly endangered species.

Supplementary Materials: The following supporting information can be downloaded at: https://www.mdpi.com/article/10.3390/d14080666/s1, Figure S1: The results of cross-validation error (a) and ancestry composition when K = 2 to K = 6 in Admixture analysis (b); Figure S2: The results of PCA based principal component 1 to principal component 3 (PC1 and PC2 and PC3); Figure S3: Variance explained by PCA eigenvalues (a) and optimal K-value results based on BIC (b) in DAPC analysis; Figure S4: *OptM* results for TreeMix when run from 0 - 6 migration event(s); Table S1: Three-population test for *T. doichangensis*; Table S2: Four-population test for *T. doichangensis*.

Author Contributions: Conceptualization, L.H., H.-H.M., Y.C. and J.L.; Funding acquisition, H.-H.M. and Y.C.; Project administration, H.-H.M. and J.L.; Visualization, L.H. and H.-H.M.; Formal analysis, L.H.; Writing—original draft, L.H. and H.-H.M.; Writing—review and editing, L.H., X.-G.L., S.-S.Z., C.-Y.Z., Y.-H.T., Q.R., H.-H.M., Y.C. and J.L. All authors have read and agreed to the published version of the manuscript.

Funding: This research was supported by funding from the Southeast Asia Biodiversity Research Institute, Chinese Academy of Sciences (No. Y4ZK111B01), to H.-H.M.; from the CAS "Pioneer Hundred Talents" Program awarded to Y.C.; from the Youth Innovation Promotion Association, Chinese Academy of Sciences (No. 2018432), to H.-H.M.; and from the CAS "Light of West China" Program to H.-H.M.

Institutional Review Board Statement: Not applicable.

Data Availability Statement: The Data and Code supporting the current study is available from the first author (Ling Hu; huling20@outlook.com or huling@xtbg.ac.cn) on requeste.

Acknowledgments: We thank Jian-Hua Xiao and Chao-Nan Cai for help in the GBS and AMOVA. We also thank Xiao-Yan Zhang for kind help in using ArcGIS. Thanks to Chu-Meng Zhu, Huan-Xing Ren and Liu-Qing Yang for help in using R and Linux.

Conflicts of Interest: The authors declare no conflict of interest.

References

1. Meng, H.H.; Gao, X.Y.; Song, Y.G.; Cao, G.L.; Li, J. Biodiversity arks in the Anthropocene. *Reg. Sustain.* **2021**, *2*, 109–115. [CrossRef]
2. Folke, C.; Polasky, S.; Rockström, J.; Galaz, V.; Westley, F.; Lamont, M.; Scheffer, M.; Österblom, H.; Carpenter, S.R.; Chapin, F.S.; et al. Our future in the Anthropocene biosphere. *Ambio* **2021**, *50*, 834–869. [CrossRef] [PubMed]
3. Hu, Y.; Wang, M.; Ma, T.; Huang, M.; Huang, G.; Zhou, W.; Ping, X.; Lu, Y.; Wei, F. Integrated index-based assessment reveals long-term conservation progress in implementation of Convention on Biological Diversity. *Sci. Adv.* **2022**, *8*, eabj8093. [CrossRef]
4. López-Pujol, J.; Zhang, F.M.; Ge, S. Plant biodiversity in China: Richly varied, endangered, and in need of conservation. *Biodivers. Conserv.* **2006**, *15*, 3983–4026. [CrossRef]
5. Mi, X.; Feng, G.; Hu, Y.; Zhang, J.; Chen, L.; Corlett, R.T.; Hughes, A.C.; Pimm, S.; Schmid, B.; Shi, S.; et al. The global significance of biodiversity science in China: An overview. *Natl. Sci. Rev.* **2021**, *8*, nwab032. [CrossRef] [PubMed]
6. Qin, H.N.; Zhao, L.N. Evaluating the threat status of higher plants in China. *Biodivers. Sci.* **2017**, *25*, 689–695. [CrossRef]
7. López-Bao, J.V.; Chapron, G.; Treves, A. The Achilles heel of participatory conservation. *Biol. Conserv.* **2017**, *212*, 139–143. [CrossRef]
8. Meng, H.H.; Zhou, S.S.; Li, L.; Tan, Y.H.; Li, J.W.; Li, J. Conflict between biodiversity conservation and economic growth: Insight into rare plants in tropical China. *Biodivers. Conserv.* **2019**, *28*, 523–537. [CrossRef]
9. Yang, W.Z.; Xiang, Z.Y.; Zhang, S.S.; Kang, H.M.; Shi, F.Q. Plant species with extremely small populations (PSESP) and their significance in China's national plant conservation strategy. *Biodivers. Sci.* **2015**, *23*, 419–425. [CrossRef]
10. Crane, P. Conserving our global botanical heritage: The PSESP plant conservation program. *Plant Divers.* **2020**, *42*, 319–322. [CrossRef]
11. Sun, W.B.; Ma, Y.P.; Blackmore, S. How a new conservation action concept has accelerated plant conservation in China. *Trends Plant Sci.* **2019**, *24*, 4–6. [CrossRef] [PubMed]
12. Yu, Y.L.; Wang, H.C.; Yu, Z.X.; Schinnerl, J.; Tang, R.; Geng, Y.P.; Chen, G. Genetic diversity and structure of the endemic and endangered species *Aristolochia delavayi* growing along the Jinsha River. *Plant Divers.* **2021**, *43*, 225–233. [CrossRef] [PubMed]
13. Li, S.; Liu, S.L.; Pei, S.Y.; Ning, M.M.; Tang, S.Q. Genetic diversity and population structure of *Camellia huana* (Theaceae), a limestone species with narrow geographic range, based on chloroplast DNA sequence and microsatellite markers. *Plant Divers.* **2020**, *42*, 343–350. [CrossRef] [PubMed]
14. Xiao, S.Y.; Ji, Y.H.; Liu, J.; Gong, X. Genetic characterization of the entire range of *Cycas panzhihuaensis* (Cycadaceae). *Plant Divers.* **2020**, *42*, 7–18. [CrossRef] [PubMed]
15. Cai, C.N.; Xiao, J.H.; Ci, X.Q.; Conran, J.G.; Li, J. Genetic diversity of *Horsfieldia tetratepala* (Myristicaceae), an endangered Plant Species with Extremely Small Populations to China: Implications for its conservation. *Plant Syst. Evol.* **2021**, *307*, 50. [CrossRef]
16. Zhang, X.J.; Liu, X.F.; Liu, D.T.; Cao, Y.R.; Li, Z.H.; Ma, Y.P.; Ma, H. Genetic diversity and structure of *Rhododendron meddianum*, a plant species with extremely small populations. *Plant Divers.* **2021**, *43*, 472–479. [CrossRef]
17. Zhang, X.M.; Qin, H.T.; Xie, W.J.; Ma, Y.P.; Sun, W.B. Comparative population genetic analyses suggest hybrid origin of *Rhododendron pubicostatum*, an endangered plant species with extremely small populations endemic to Yunnan, China. *Plant Divers.* **2020**, *42*, 312–318. [CrossRef]

18. Sun, W.B.; Han, C.Y. Researches and conservation for plant species with extremely small populations (PSESP). *Biodivers. Sci.* **2015**, *23*, 426–429. [CrossRef]
19. Phillips, R.D.; Dixon, K.W.; Peakall, R. Low population genetic differentiation in the Orchidaceae: Implications for the diversification of the family. *Mol. Ecol.* **2012**, *21*, 5208–5220. [CrossRef]
20. Zhang, Y.H.; Jia, H.X.; Wang, Z.B.; Sun, P.; Cao, D.M.; Hu, J.J. Genetic diversity and population structure of *Populus yunnanensis*. *Biodivers. Sci.* **2019**, *27*, 355–365.
21. Xiao, J.H.; Ding, X.; Li, L.; Ma, H.; Ci, X.Q.; van der Merwe, M.; Conran, J.G.; Li, J. Miocene diversification of a golden-thread nanmu tree species (*Phoebe zhennan*, Lauraceae) around the Sichuan Basin shaped by the East Asian monsoon. *Ecol. Evol.* **2020**, *10*, 10543–10557. [CrossRef] [PubMed]
22. Waqar, Z.; Moraes, R.C.S.; Benchimol, M.; Morante-Filho, J.C.; Mariano-Neto, E.; Gaiotto, F.A. Gene flow and genetic structure reveal reduced diversity between generations of a tropical tree, *Manilkara multifida* Penn., in atlantic forest fragments. *Genes* **2021**, *12*, 2025. [CrossRef] [PubMed]
23. Pekkala, N.; Knott, K.E.; Kotiaho, J.S.; Nissinen, K.; Puurtinen, M. The effect of inbreeding rate on fitness, inbreeding depression and heterosis over a range of inbreeding coefficients. *Evol. Appl.* **2014**, *7*, 1107–1119. [CrossRef] [PubMed]
24. Gordon, S.P.; Sloop, C.M.; Davis, H.G.; Cushman, J. Population genetic diversity and structure of two rare vernal pool grasses in central California. *Conserv. Genet.* **2012**, *13*, 117–130. [CrossRef]
25. Wu, F.Q.; Shen, S.K.; Zhang, X.J.; Wang, Y.H.; Sun, W.B. Genetic diversity and population structure of an extremely endangered species: The world's largest Rhododendron. *AoB Plants* **2015**, *7*, plu082. [CrossRef]
26. Ci, X.Q.; Chen, J.Q.; Li, Q.M.; Li, J. AFLP and ISSR analysis reveals high genetic variation and inter-population differentiation in fragmented populations of the endangered *Litsea szemaois* (Lauraceae) from Southwest China. *Plant Syst. Evol.* **2008**, *273*, 237–246. [CrossRef]
27. Narum, S.R.; Buerkle, C.A.; Davey, J.W.; Miller, M.R.; Hohenlohe, P.A. Genotyping-by-sequencing in ecological and conservation Genomics. *Mol. Ecol.* **2013**, *22*, 2841–2847. [CrossRef]
28. Deschamps, S.; Llaca, V.; May, G.D. Genotyping-by-sequencing in plants. *Biology* **2012**, *1*, 460–483. [CrossRef]
29. Elshire, R.J.; Glaubitz, J.C.; Sun, Q.; Poland, J.A.; Kawamoto, K.; Buckler, E.S.; Mitchell, S.E. A Robust, Simple Genotyping-by-Sequencing (GBS) approach for high diversity species. *PLoS ONE* **2011**, *6*, e19379. [CrossRef]
30. Shekhovtsov, S.V.; Shekhovtsova, I.N.; Kosterin, O.E. Genotyping-by-Sequencing Analysis Shows That Siberian Lindens Are Nested within *Tilia cordata* Mill. *Diversity* **2022**, *14*, 256. [CrossRef]
31. Forman, L. *Trigonobalanus*, a new genus of Fagaceae, with notes on the classification of the family. *Kew Bull.* **1964**, *17*, 381–396. [CrossRef]
32. Nixon, K.C.; Crepet, W.L. *Trigonobalanus* (Fagaceae): Taxonomic status and phylogenetic relationships. *Am. J. Bot.* **1989**, *76*, 828–841. [CrossRef]
33. Hsu, Y.C.; Wang, C.J.; Wu, C.Y.; Li, H.W. *Trigonobalanus* Forman-A new recorded genus of Fagaceae in China. *Acta Bot. Yunnan* **1981**, *3*, 213–215.
34. Zheng, Y.L.; Sun, W.B.; Zhou, Y.; Coombs, D. Variation in seed and seedling traits among natural populations of *Trigonobalanus doichangensis* (A. Camus) Forman (Fagaceae), a rare and endangered plant in southwest China. *New For.* **2009**, *37*, 285–294. [CrossRef]
35. Sun, W.B.; Zhou, Y.; Han, C.Y.; Zeng, C.X.; Shi, X.D.; Xiang, Q.B.; Coombes, A. Status and conservation of *Trigonobalanus doichangensis* (Fagaceae). *Biodivers. Conserv.* **2006**, *15*, 1303–1318. [CrossRef]
36. Wang, S.; Xie, Y. *China Species Red List*; High Education Press: Beijing, China, 2004.
37. Catchen, J.; Hohenlohe, P.A.; Bassham, S.; Amores, A.; Cresko, W.A. Stacks: An analysis tool set for population genomics. *Mol. Ecol.* **2013**, *22*, 3124–3140. [CrossRef]
38. Staton, M.; Addo-Quaye, C.; Cannon, N.; Yu, J.; Zhebentyayeva, T.; Huff, M.; Islam-Faridi, N.; Fan, S.; Georgi, L.L.; Nelson, C.D.; et al. A reference genome assembly and adaptive trait analysis of *Castanea mollissima* 'Vanuxem,'a source of resistance to chestnut blight in restoration breeding. *Tree Genet. Genomes* **2020**, *16*, 1392. [CrossRef]
39. Li, H. Aligning sequence reads, clone sequences and assembly contigs with BWA-MEM. *arXiv* **2013**, arXiv:1303.3997v2.
40. Danecek, P.; Bonfield, J.K.; Liddle, J.; Marshall, J.; Ohan, V.; Pollard, M.O.; Whitwham, A.; Keane, T.; McCarthy, S.A.; Davies, R.M.; et al. Twelve years of SAMtools and BCFtools. *GigaScience* **2021**, *10*, giab008. [CrossRef]
41. Danecek, P.; Auton, A.; Abecasis, G.; Albers, C.A.; Banks, E.; DePristo, M.A.; Handsaker, R.E.; Lunter, G.; Marth, G.T.; Sherry, S.T.; et al. The variant call format and VCFtools. *Bioinformatics* **2011**, *27*, 2156–2158. [CrossRef]
42. R Core Team. *R: A Language and Environment for Statistical Computing*; R Foundation for Statistical Computing: Vienna, Austria, 2020.
43. Excoffier, L.; Lischer, H.E. Arlequin suite ver 3.5: A new series of programs to perform population genetics analyses under Linux and Windows. *Mol. Ecol. Resour.* **2010**, *10*, 564–567. [CrossRef] [PubMed]
44. Lischer, H.E.; Excoffier, L. PGDSpider: An automated data conversion tool for connecting population genetics and genomics programs. *Bioinformatics* **2012**, *28*, 298–299. [CrossRef] [PubMed]
45. Hijmans, R.J.; Karney, C.; Williams, E.; Vennes, C. Package "geosphere": Spherical Trigonometry. 2021. Available online: https://cran.r-project.org/web/packages/geosphere/geosphere.pdf (accessed on 10 March 2022).
46. Dray, S.; Dufour, A.B.; Chessel, D. The ade4 package-II: Two-table and K-table methods. *R News* **2007**, *7*, 47–52.

47. Alexander, D.H.; Novembre, J.; Lange, K. Fast model-based estimation of ancestry in unrelated individuals. *Genome Res.* **2009**, *19*, 1655–1664. [CrossRef] [PubMed]
48. Rodríguez-Ezpeleta, N.; Bradbury, I.R.; Mendibil, I.; Álvarez, P.; Cotano, U.; Irigoien, X. Population structure of Atlantic mackerel inferred from RAD-seq-derived SNP markers: Effects of sequence clustering parameters and hierarchical SNP selection. *Mol. Ecol. Resour.* **2016**, *16*, 991–1001. [CrossRef] [PubMed]
49. Francis, R.M. pophelper: An R package and web app to analyse and visualize population structure. *Mol. Ecol. Resour.* **2017**, *17*, 27–32. [CrossRef] [PubMed]
50. Purcell, S.; Neale, B.; Todd-Brown, K.; Thomas, L.; Ferreira, M.A.R.; Bender, D.; Maller, J.; Sklar, P.; De Bakker, P.I.; Daly, M.J.; et al. PLINK: A tool set for whole-genome association and population-based linkage analyses. *Am. J. Hum. Genet.* **2007**, *81*, 559–575. [CrossRef]
51. Wickham, H. *ggplot2: Elegant Graphics for Data Analysis*; Springer: New York, NY, USA, 2016.
52. Jombart, T.; Devillard, S.; Balloux, F. Discriminant analysis of principal components: A new method for the analysis of genetically structured populations. *BMC Genet.* **2010**, *11*, 94. [CrossRef]
53. Liu, Y.; Chen, L.; Duan, X.Z.; Zhao, D.S.; Sun, J.T.; Hong, X.Y. Genome-wide single nucleotide polymorphisms are robust in resolving fine-scale population genetic structure of the small brown planthopper, *Laodelphax striatellus* (Fallén) (Hemiptera: Delphacidae). *J. Econ. Entomol.* **2019**, *112*, 2362–2368. [CrossRef]
54. Dierickx, E.G.; Shultz, A.J.; Sato, F.; Hiraoka, T.; Edwards, S.V. Morphological and genomic comparisons of Hawaiian and Japanese Black-footed Albatrosses (*Phoebastria nigripes*) using double digest RADseq: Implications for conservation. *Evol. Appl.* **2015**, *8*, 662–678. [CrossRef]
55. Jombart, T. adegenet: A R package for the multivariate analysis of genetic markers. *Bioinformatics* **2008**, *24*, 1403–1405. [CrossRef]
56. Pickrell, J.K.; Pritchard, J.K. Inference of population splits and mixtures from genome-wide allele frequency data. *PLoS Genet.* **2012**, *8*, e1002967. [CrossRef] [PubMed]
57. Ma, Y.; Liu, D.; Wariss, H.M.; Zhang, R.; Tao, L.; Milne, R.I.; Sun, W. Demographic history and identification of threats revealed by population genomic analysis provide insights into conservation for an endangered maple. *Mol. Ecol.* **2022**, *31*, 767–779. [CrossRef] [PubMed]
58. Fitak, R.R. OptM: Estimating the optimal number of migration edges on population trees using Treemix. *Biol. Methods Protoc.* **2021**, *6*, bpab017. [CrossRef]
59. Pickrell, J.K.; Patterson, N.; Barbieri, C.; Berthold, F.; Gerlach, L.; Güldemann, T.; Kure, B.; Mpoloka, S.W.; Nakagawa, H.; Naumann, C.; et al. The genetic prehistory of southern Africa. *Nat. Commun.* **2012**, *3*, 1143. [CrossRef] [PubMed]
60. Liu, Y.F.; Huang, H.W. Gene flow dynamics and related adaptive evolution in plant populations. *Chin. Bull. Bot.* **2009**, *44*, 351.
61. Laikre, L.; Allendorf, F.W.; Aroner, L.C.; Baker, C.S.; Gregovich, D.P.; Hansen, M.M.; Jackson, J.A.; Kendall, K.C.; McKELVEY, K.; Neel, M.C.; et al. Neglect of genetic diversity in implementation of the convention on biological diversity. *Conserv. Biol.* **2010**, *24*, 86–88. [CrossRef] [PubMed]
62. Zhang, Y.; Song, Y.G.; Zhang, C.Y.; Wang, T.R.; Su, T.H.; Huang, P.H.; Meng, H.H.; Li, J. Latitudinal diversity gradient in the changing world: Retrospectives and perspectives. *Diversity* **2022**, *14*, 334. [CrossRef]
63. Setoguchi, H.; Mitsui, Y.; Ikeda, H.; Nomura, N.; Tamura, A. Genetic structure of the critically endangered plant *Tricyrtis ishiiana* (Convallariaceae) in relict populations of Japan. *Conserv. Genet.* **2011**, *12*, 491–501. [CrossRef]
64. Ramanatha, R.V.; Hodgkin, T. Genetic diversity and conservation and utilization of plant genetic resources. *Plant Cell Tissue Organ Cult.* **2002**, *68*, 1–19. [CrossRef]
65. Zhou, Y. Preliminary study on geographic populations of *Trigonobalanus doichangensis* and its effect on seed germination and seedling. *Seed* **2003**, *4*, 3–4.
66. Sun, W.B.; Zhou, Y.; Zhao, J.C.; Chen, G. Current distribution, population attributes and biological characters of *Trigonobalanus doichangensis* in relation to its conservation. *Acta Ecol. Sin.* **2004**, *24*, 352–358.
67. Aoki, K.; Ueno, S.; Kamijo, T.; Setoguchi, H.; Murakami, N.; Kato, M.; Tsumura, Y. Genetic differentiation and genetic diversity of *Castanopsis* (Fagaceae), the dominant tree species in Japanese broadleaved evergreen forests, revealed by analysis of EST-associated microsatellites. *PLoS ONE* **2014**, *9*, e87429. [CrossRef] [PubMed]
68. Valencia, C.L.; Piñero, D.; Mussali, G.P.; Valencia-Ávalos, S.; Tovar-Sánchez, E. Effect of a red oak species gradient on genetic structure and diversity of *Quercus castanea* (Fagaceae) in Mexico. *Tree Genet. Genomes* **2014**, *10*, 641–652. [CrossRef]
69. Hamrick, J.L.; Godt, M.J.W. *Plant Population Genetics, Breeding, and Genetic Resources*; Sinauer: Sunderland, MA, USA, 1990.
70. Nybom, H. Comparison of different nuclear DNA markers for estimating intraspecific genetic diversity in plants. *Mol. Ecol.* **2004**, *13*, 1143–1155. [CrossRef]
71. Gamba, D.; Muchhala, N. Global patterns of population genetic differentiation in seed plants. *Mol. Ecol.* **2020**, *29*, 3413–3428. [CrossRef]
72. Li, S.; Gan, X.H.; Han, H.Y.; Zhang, X.M.; Tian, Z.Q. Low within-population genetic diversity and high genetic differentiation among populations of the endangered plant *Tetracentron sinense* Oliver revealed by inter-simple sequence repeat analysis. *Ann. For. Sci.* **2018**, *75*, 74. [CrossRef]
73. Souza, F.L.; Cunha, A.F.; Oliveira, M.A.; Pereira, G.A.; Reis, S.F.d. Estimating dispersal and gene flow in the neotropical freshwater turtle *Hydromedusa maximiliani* (Chelidae) by combining ecological and genetic methods. *Genet. Mol. Biol.* **2002**, *25*, 151–155. [CrossRef]

74. Wright, S. The interpretation of population structure by F-statistics with special regard to systems of mating. *Evolution* **1965**, *19*, 395–420. [CrossRef]
75. Sun, W.B.; Han, C.Y.; Gao, L.M.; Wilson, C.A. Genetic diversity of the rare Asian plant, *Trigonobalanus doichangensis* (Fagaceae). *Aust. J. Bot.* **2007**, *55*, 10–17. [CrossRef]
76. Fu, L.G. *China Plant Red Data Book*; Science Press: Beijing, China, 1992; Volume 1, pp. 302–303.
77. Meng, H.H.; Zhou, S.S.; Jiang, X.L.; Gugger, P.F.; Li, L.; Tan, Y.H.; Li, J. Are mountaintops climate refugia for plants under global warming? A lesson from high-mountain oaks in tropical rainforest. *Alp. Bot.* **2019**, *129*, 175–183. [CrossRef]

diversity

MDPI

Article

Development of EST-SSR Markers Related to Polyphyllin Biosynthesis Reveals Genetic Diversity and Population Structure in *Paris polyphylla*

Xiaoyang Gao [1,†], Qixuan Su [1,2,†], Baolin Yao [1,2], Wenjing Yang [1,3], Weisi Ma [4], Bin Yang [4] and Changning Liu [1,5,6,*]

1 CAS Key Laboratory of Tropical Plant Resources and Sustainable Use, Xishuangbanna Tropical Botanical Garden, Chinese Academy of Sciences, Kunming 650223, China; gaoxiaoyang@xtbg.ac.cn (X.G.); suqixuan@mail.ustc.edu.cn (Q.S.); yblbio@mail.ustc.edu.cn (B.Y.); yangwenjing@xtbg.ac.cn (W.Y.)
2 School of Life Sciences, University of Science and Technology of China, Hefei 230026, China
3 College of Life Sciences, University of Chinese Academy of Sciences, Beijing 100049, China
4 Institute of Medicinal Plants, Yunnan Academy of Agricultural Sciences, Kunming 650205, China; masilaoq@aliyun.com (W.M.); yb@yaas.org.cn (B.Y.)
5 Center of Economic Botany, Core Botanical Gardens, Chinese Academy of Sciences, Xishuangbanna 666303, China
6 The Innovative Academy of Seed Design, Chinese Academy of Sciences, Kunming 650223, China
* Correspondence: liuchangning@xtbg.ac.cn; Tel.: +86-0691-8715071
† These authors contributed equally to this work.

Citation: Gao, X.; Su, Q.; Yao, B.; Yang, W.; Ma, W.; Yang, B.; Liu, C. Development of EST-SSR Markers Related to Polyphyllin Biosynthesis Reveals Genetic Diversity and Population Structure in *Paris polyphylla*. *Diversity* **2022**, *14*, 589. https://doi.org/10.3390/d14080589

Academic Editor: Mario A. Pagnotta

Received: 3 July 2022
Accepted: 20 July 2022
Published: 23 July 2022

Abstract: *Paris polyphylla* is an important medicinal plant that can biosynthesize polyphyllins with multiple effective therapies, ranging from anti-inflammation to antitumor; however, the genetic diversity of *Paris polyphylla* is still unclear. To explore the genetic characteristics of cultivation populations in primary planting areas, we developed 10 expressed sequence tag simple sequence repeat (EST-SSR) markers related to polyphyllin backbone biosynthesis and utilized them in 136 individuals from 10 cultivated populations of *P. polyphylla* var. *yunnanensis*. The genetic diversity index showed that ten loci had relatively high genetic polymorphism levels. Shannon information of loci suggested that more information occurred within population and less information occurred among population. In addition, the overall populations exhibited a low degree of differentiation among populations, but maintained a high degree of genetic diversity among individuals, resulting in high gene flow and general hybridization. The genetic structure analysis revealed that 10 populations possibly derived from two ancestral groups and all individuals were found with different levels of admixture. The two groups were different from the cultivation groups at population level, suggesting the cross-pollination among cultivars. These findings will provide insights into the genetic diversity of the germplasm resources and facilitate marker-assisted breeding for this medicinal herb.

Keywords: *Paris polyphylla*; transcriptome; polyphyllin; EST-SSR markers; genetic diversity

1. Introduction

Paris polyphylla Smith is an important medicinal perennial herb, mainly distributed from Southwest China to the pan-Himalayan region [1]. *P. polyphylla* is the most in demand of the genus, and has dominated the industrialization and utilization of medicinal plants in Southwest China. Due to the remarkable effects on hemostasis, anti-inflammation, and anti-cancer, its dried rhizome becomes the key raw material for about 80 kinds of famous patented medicines [2]. According to previous phytochemical studies, polyphyllins are regarded as the chief active ingredients in this plant, and 174 different polyphyllins have been identified so far, which account for 54% (323) of the total number of known bioactive compounds [3]. Notably, there has been a 300-fold increase in the market price paid for the rhizomes during the past nearly forty years and approximately 1000 t of the rhizomes are sold annually [4]. However, the scarcity of *P. polyphylla* becomes the bottleneck of the related pharmaceutical industry in recent years, mainly because of its long growth cycles (ca. 7–10 years), long-term seed dormancy, and the sharp increase in the demand for rhizomes by the pharmaceutical industry. In addition, wild resources are threatened by long-term over-exploration and habitat fragmentation, and natural resources of *P. polyphylla* become increasingly endangered. Therefore, it is necessary to investigate the germplasm resources and understand the differentiation for genetic resource conservation and sustainable utilization.

The genetic information of plants plays an essential part in formulating conservation strategies. Molecular markers become useful tools to study the genetic diversity and population structure of germplasm resources in non-model plants with no reference genomes [5]. A variety of molecular markers, including amplified restriction fragment length polymorphism (ARFLP), random amplified of polymorphic DNA (RAPD), restriction site amplified polymorphism (RSAP), start codon-targeted polymorphism (SCoT), sequence-related amplified polymorphism (SRAP), inter simple sequence repeat (ISSR), and simple sequence repeat (SSR) have extensively been used in plant source conservation and genetic breeding [6]. Among the different classes of molecular markers, SSR markers have become a particularly important tool because they are co-dominant, polymorphic, low-cost, and high efficiency; the putative function of SSR markers can often be deduced by a homology search [7]. Compared to SSR, expressed sequence tag simple sequences repeat (EST-SSR) has the advantage of offering more transferability among plant species and is widely used in plant genetic mapping [8,9]. To date, there are only a few reports about the genetic diversity of *P. polyphylla* by the aforementioned molecular markers. Previously, genetic diversity of the three cultivated populations of *P. polyphylla* var. *yunnanensis* was slightly higher than that of the three wild populations (0.153 vs. 0.151) through ISSR markers, suggesting the introduction and artificial selection of cultivars from comparatively wide areas of origin and subsequent gene flow among populations in cultivation [10]. Only 1.35% of genetic variations existed between 15 wild and 17 cultivated populations of *P. polyphylla* var. *yunnanensis* using AFLP markers, which indicates that there is no obvious genetic differentiation between wild and cultivated populations as result of the relatively short history of the domestication of cultivated populations [11]. SCoTs and SRAPs were developed to investigate genetic diversity using 33 *P. polyphylla* samples in the Dabie Mountains, which found that the polymorphisms and marker efficiency of SCoTs were higher than those of SRAPs [12]. Nine random EST-SSRs were detected based on a root transcriptome of *P. polyphylla* var. *yunnanensis*. Maker efficiency was then validated in 55 samples [13]. However, the lack of marker information for molecular phylogeny and genetic structure limited *P. polyphylla* collection, conservation, and utilization. *P. polyphylla* and other *Paris* species possess giant genomes, and none of their complete genomes have been sequenced so far [14]. There are, as of yet, few available EST sequences of *Paris* L. in the GenBank database. In addition, few studies have explored SSRs related to polyphyllin biosynthesis based high quality transcriptome.

In the present study, we identified a large number of EST-SSRs based on the transcriptome sequencing data of 36 tissue samples from our previous studies. The aims of this study were to: (1) develop SSR markers related to polyphyllin biosynthesis and validate their polymorphism levels; (2) explore the genetic background between germplasms from the primary planting areas of *P. polyphylla*. This study will provide novel insights into the genetic diversity of the germplasms from the major planting areas of *P. polyphylla* and aid in the conservation and utilization of this important medicinal plant.

2. Materials and Methods

2.1. Plant Material and DNA Extraction

In this study, healthy whole plants of 7-year-old *P. polyphylla* var. *yunnanensis* during the fruiting stage were widely sampled from the main grow areas in Yunnan Province, Southwest China. A total of 136 individuals from 10 populations of *P. polyphylla* var. *yunnanensis* were collected from major production areas, and all individuals were transplanted in the green house of Xishuangbanna Tropical Botanical Garden, Chinese Academy of Sciences (Kunming, China). The samples collected covered the central, northwest, southwest, and west of Yunnan (Figure 1 and Table 1), and the samples collected included two kinds of widely grown varieties, namely short-stalked variety and long-stalked variety. The two varieties have obvious differences in stalk length, size of rhizome and leaf, fruit yield, etc. The characteristics of stalk are usually used to distinguish the two varieties. The detailed information for each population is shown in Figure 1 and Table 1. The fresh leaves were sampled and stored at −80 °C. The genomic DNA was extracted using CTAB method [15]. The DNA concentration was estimated with a NanoDrop-1000 spectrophotometer (Nano Drop Technologies, Wilmington, DE, USA) and normalized to 30 ng/µL for polymerase chain reaction (PCR).

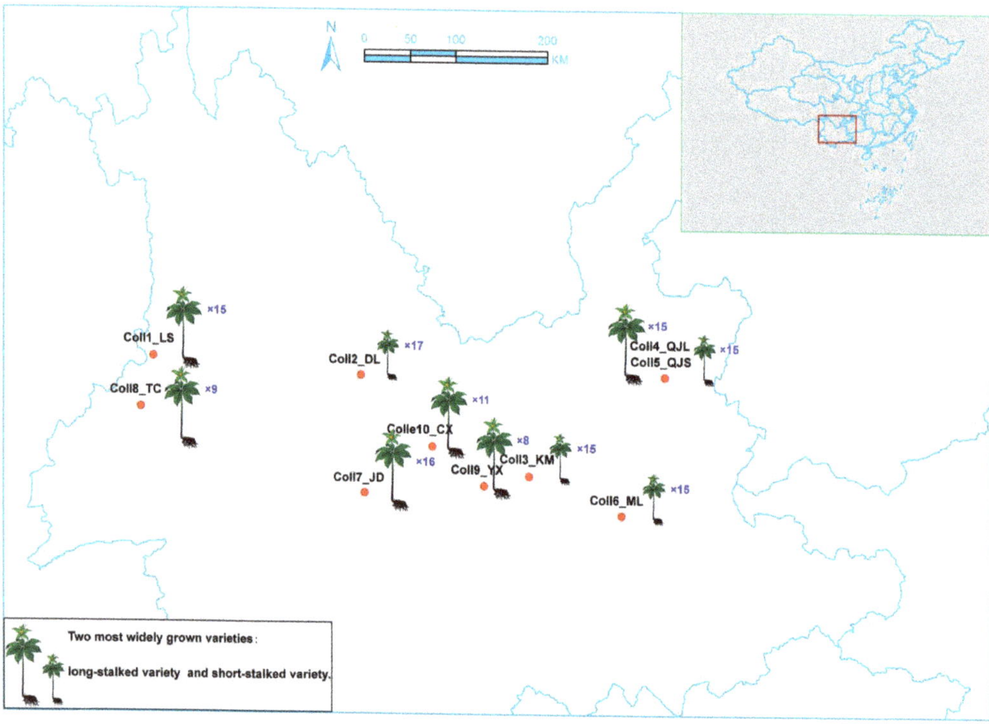

Figure 1. Sampling locations of 10 populations of *P. polyphylla* var. *yunnanensis* in the present study. The varieties with different stem heights represent the long-stalked variety and the short-stalked variety, respectively. The population size is also denoted.

Table 1. Sampling locations and number of samples analyzed in the present study.

Population	Sampling Location	City	Population Size	Varieties	Longitude (°)	Latitude (°)	Altitude (m)
Coll1_LS	Lushui	Nujiang	15	long stalk	98.78 (E)	25.91 (N)	1446
Coll2_DL	Xiangyun	Dali	17	short stalk	100.84 (E)	25.74 (N)	1775
Coll3_KM	Xishan	Kunming	15	short stalk	102.50 (E)	24.78 (N)	2008
Coll4_QJL	Zhanyi	Qujing	15	long stalk	103.83 (E)	25.73 (N)	2204
Coll5_QJS	Zhanyi	Qujing	15	short stalk	103.83 (E)	25.73 (N)	2204
Coll6_ML	Mile	Honghe	15	short stalk	103.41(E)	24.41 (N)	1711
Coll7_JD	Jingdong	Puer	16	long stalk	100.88 (E)	24.62 (N)	2237
Coll8_TC	Tengchong	Baoshan	9	short stalk	98.65(E)	25.42 (N)	1850
Coll9_YX	Yimen	Yuxi	8	long stalk	102.06 (E)	24.69 (N)	1873
Coll10_CX	Chuxiong	Chuxiong	11	long stalk	101.49 (E)	24.95 (N)	1857

2.2. EST-SSR Identification and Marker Development

All types of SSRs were identified through transcriptome analysis of *P. polyphylla* var. *yunnanensis*. A total of 36 tissues transcriptome sequencing data were from our previous study (12 tissue samples) [16] and subsequent transcriptome sequencing data (24 tissues samples, unpublished). The data are available in NCBI (PRJNA682903 and PRJNA630028). SSRs were ascertained using the microsatellite identification tool MISA 1.0 (Thiel Thomas, Seeland, Germany) [17]. The SSRs were considered to contain mono-, di-, tri-, tetra-, penta-, and hexa-nucleotides with minimum repeat numbers of 10, 5, 4, 3, 3, and 3, respectively. The distance between adjacent SSRs $\leq$ 100 bp was defined as compound SSR. The functional annotation of the gene contained SSRs were obtained through homology by searching against the public database GO, KEGG, Swiss-Prot, and Pfam using BLAST with an E-value cutoff of 10^{-5}. The non-redundancy gene sequences associated with polyphyllin biosynthesis contained SSRs were filtered for primer design by Primer 3.0 (Andreas Untergasser, Heidelberg, Germany) [18]. Flank sequence length of SSR < 20 bp and sequence contained mononucleotide repeats were removed according to SSR locus. The primer length ranged 16–26 bp, production of PCR was 100–450 bp, optimum Tm was 55–57 °C, GC content was 40–60%, and oligonucleotides were synthesized at Shanghai Sangon Biological Engineering Technology (Shanghai, China).

2.3. Marker Validation

SSR-PCR amplification for all designed markers was initially carried out using 30 random individuals from 10 populations. The PCR reaction system: 2.5 μL 10× Tag buffer, 2 μL of 2.5 mmol·L^{-1} dNTPs, 1 μL of 10 μmol·L^{-1} for each of forward and reverse primer of DNA, 0.25 μL 2.5 Taq Plus DNA polymerase, 1 μL DNA template. The PCR reaction conditions and procedures were performed as follows: initial denaturation step of 95 °C for 3 min, followed by denaturation at 94 °C for 30 s, annealing temperature at 60 °C for 30 s, extension at 72 °C for 30 s, 10 cycles; denaturation at 94 °C for 30 s, annealing temperature at 55 °C for 30 s, extension at 72 °C for 30 s, 10 cycles. The final extension was performed at 72 °C for 7 min. the amplified PCR products were detected by 8% non-denaturing polyacrylamide gels and stained by nucleic acid dye. The selected PCR products labelled with TAMRA, FAM, and HEX were pooled before separation in ABI 3730XL (Applied Biosystems). The PCR products were separated using a 96-capillary 3730XL DNA Analyzer (Thermo Fisher Scientific, Waltham, MA, USA) and the peak patterns were sized by Genemapper 4.0 (Thermo Fisher Scientific, Waltham, MA, USA). The primer pairs and marker were evaluated and determined, which yielded clear, reproducible, and polymorphic bands with an expected size and clear fluorescence signal that were selected for subsequent allele identification of all individuals.

2.4. Data Analysis

The artificial proofreading for raw data was implemented by checking the capillary electrophoresis (CE) peak diagram. The bands with the same base size were represented by a similar peak at the same locus. The EST-SSRs were tested for selective neutrality by means of an F_{ST} outlier method using LOSITAN [19,20]. After the preliminary runs to estimate the mean neutral F_{ST}, 20,000 simulations with the infinite allele model (IAM) were performed, according to parameter settings set by Ohtani et al. [21]. Outlier loci under positive or balancing selection were determined based on 99.5% confidence intervals. The clustering pattern of individuals and populations were revealed by neutral loci. The loci with F_{ST} outliers were excluded from the following analyses. The EST-SSR loci data was formatted for subsequent analyses using GenAlEx 6.5b2 (Peakall Rod, Canberra, Australia) [22]. The number of observed alleles (*Na*), the number of efficient alleles (*Ne*), the observed heterozygosity (*Ho*), the expected heterozygosity (*He*), *Nei*'s genetic diversity index (*h*), Shannon diversity index (*I*), polymorphic information content (PIC), major allele frequency (MP), etc., were calculated using POPGENE 1.32 (Naoko Takezaki, Kagawa, Japan) and Powermaker 3.25 (Kejun Liu, Raleigh, USA) [23,24]. The genetic differentiation coefficient among populations (F_{ST}), intraspecific inbreeding coefficient (F_{IS}), population inbreeding coefficient (F_{IT}), and gene flow (*Nm*) were calculated using Arlequin 3.5 (Laurent Excoffier, Lausanne, Switzerland) [25]. *Nm* was calculated followed $Nm = 0.25(1 - F_{ST})/F_{ST}$ [26]. The Hardy–Weinberg equilibrium (HWE) with the chi-squared test for each population and loci was analyzed using POPGENE 1.32 (Naoko Takezaki, Kagawa, Japan), which was adjusted using Bonferroni for multiple tests [27]. The principal coordinate analysis (PCoA) via covariance matrix with data standardization was conducted using GenAlEx 6.50b (Rod Peakall, Canberra, Australia) [22].

The *Nei*'s (1983) standard genetic distance among populations, individuals, and clustering trees based on the unweighted pair group method with arithmetic means (UPGMA) algorithm (bootstrap: 1000) were calculated and analyzed using PowerMarker 3.25 (Kejun Liu, Raleigh, NC, USA) [24]. The consensus tree was generated, edited, and visualized using Phylip 3.68 (Jacques D. Retief, Totowa, NJ, USA), MEGA 5.10 (Sudhir Kumar, Tokyo, Japan), and FigTree 1.4.2 (A. Rambaut, Edinburgh, UK), respectively [28–30]. The population genetic structure was determined by utilizing a Bayesian clustering analysis using SRUCTURE 2.3.4 (K Pritchard Jonathan, Oxford, UK) [31]. A total of ten independent simulations for each *K* ranging from 1 to 10 were performed with a burn-in period of 100,000 steps followed by 100,000 Markov Chain Monte Carlo (MCMC) iterations using the Admixture Model. The most probable number of population groups (*K*) was determined with delta $K(\Delta K)$ through web-based STRUCTURE HARVESTER [32,33]. Repeated sampling analysis and genetic structural plots were analyzed by CLUMPP 1.1.2 (Mattias Jakobsson, Ann Arbor, MI, USA) and visualized by DISTRUCT 1.1 (Noah A. Rosenberg, Los Angeles, CA, USA) [34,35].

3. Results

3.1. EST-SSR Identification

Novel EST-SSR markers were developed based on the transcriptome assembled from different tissues during developmental stages [16]. In total, 102,472 different EST-SSRs were identified from 341,191 unigenes, distributed in 73,770 sequences, with an average of 0.22 SSR per unigene and a distribution density of one SSR per 2.61 kb. The repeated motifs of SSRs were diverse; mononucleotide (34.32%) and dinucleotide (37.91%) were the most common repeated motif types. Among these, A/T repeat motif was the most abundant type (91.74% of total mononucleotide repeats), followed by AG(24.66%)/CT(15.45%)/TC(18.17%) (Supplementary Material Figure S1). The compound SSRs were also identified, and the number of this type of SSR was approximately 13,630. The copy number of repeat motifs was unevenly distributed in different unit types. The most frequent copy number of mononucleotide repeats was 91, whereas the most frequent copy number of hexanucleotide was 16. The copy number of repeat motifs significantly decreased with the increasing length of repeat unit, particularly from dinucleotide to hexanucleotide. The genes that contained SSRs were functionally annotated by the public database GO, SwissProt, Pfam, and KEGG. The annotation result showed that 12,723 unigenes containing SSRs have similarities to the homologs of GO terms. 4568 were by KEGG pathways, 8365 were by Swiss-Prot, and 7950 were by Pfam domains. A total of 244 unigenes related to polyphyllin biosynthesis were identified after length filtering and redundancy removing. Among these, 64 unigenes contained different SSRs. Similar to the distribution of SSR motif types in the transcriptome, mononucleotide and dinucleotide accounted for the larger proportion (55%), which was followed by compound SSRs, accounting for 15% (Figure 2).

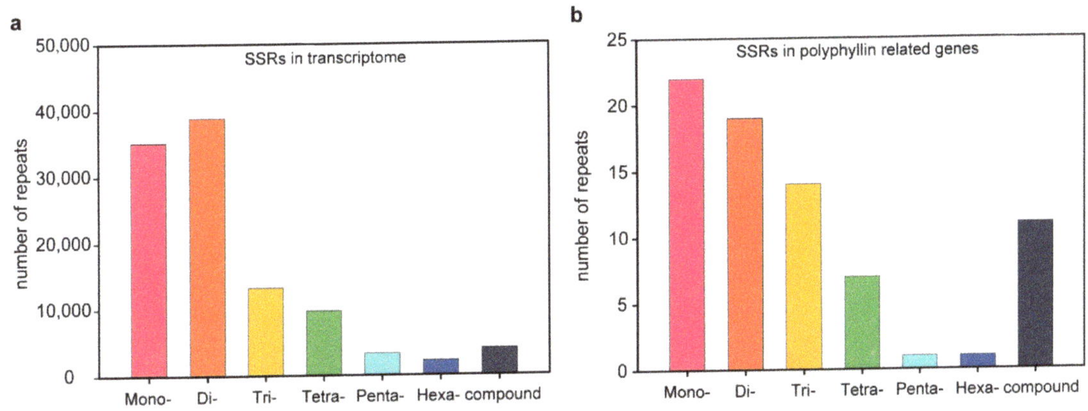

Figure 2. Statistics of EST-SSR based on the transcriptomic data. (**a**) number of different SSR types identified in the transcriptome: Mono-, Di-, Tri-, Tetra-, Penta-, Hexa-, and compound represent mononucleotide, dinucleotide, trinucleotide, pentanucleotide, hexanucleotide, and compound nucleotide, respectively; (**b**) number of different SSR related to polyphyllin biosynthesis.

3.2. Development of EST-SSR Makers

A total of 34 SSR primers related to polyphyllin biosynthetic genes were designed based on the conserved sequences at the 5′ or 3′ ends of SSR, taking into account of Tm values, hairpin structure, length of PCR product, etc. 34 pairs of SSR-PCR primers were firstly validated in 30 samples in the preliminary experiment, and their PCR products were evaluated by the results of agarose gel electrophoresis and capillary electrophoresis (CE). Finally, 12 primer pairs (35%) were determined and were applied in SSR-PCR and the genotype analysis of 136 samples (Table 2). All of the products successfully yielded clear, reproducible, and polymorphic bands with an expected size and clear fluorescence signal. As shown in Table 2, products from 6 primer pairs contained dinucleotide repeats, products from 3 primer pairs contained trinucleotide repeats, products from one primer pair contained tetranucleotide repeats, products from 2 primer pairs contained compound nucleotide repeats.

Table 2. Primers associated with polyphyllin biosynthetic gene designed in this study.

Primers	Sequence (5′ to 3′)	Tm (°C)	SSR Type	Expected Product Size (bp)	5′Modification	Gene Candidate	Polyphyllin Backbone Biosynthesis
STR1035-9F STR1035-9R	CTATCGGAGAGTCTGACCCTAC GTAACCATTGATTTCCAGCTG	55	(GT)6	130	5′HEX	*STE24*	downstream
STR1035-11F STR1035-11R	CAGAATAAAGACGGTGAATTAAAAT CCCATGCATATGATCCTCTG	56	(CGC)4	115	5′HEX	*SMT2*	downstream
STR1035-13F STR1035-13R	AAGCTGGAATCAACCATAAACT AGAGCAGGAGAAACCCTAGAA	55	(AG)5	124	5′HEX	*SQLE*	downstream
STR1035-14F STR1035-14R	TGCTAAAAAGGCTGGTGATATC CGGCTTTCACTGTTTCACATA	57	(AG)11*(A)10	111	5′HEX	*DXS*	upstream
STR1035-15F STR1035-15R	CAAATAATATGATCCCTACAGAAGA TAATAATAGCAGTTCCACATTCAGT	56	(TTA)4	191	5′6-FAM	*HMGS*	upstream
STR1035-18F STR1035-18R	GCAGAAACTGTACCATGAGGAG CGTCTTGCTTGATTAACTAGGATT	57	(CAAA)3	268	5′6-FAM	*FNTA*	downstream
STR1035-22F STR1035-22R	CGATCCGAATCCTCTGTTAAA GTCACCATTAGGATCCATTTCT	56	(CT)5	191	5′6-FAM	*MVD*	upstream
STR1150-1F STR1150-1R	CAAGCTATTCGCCGTCCT CTGCCCCAGAATCGAGC	56	(CGC)4*(ACG)4	427	5′6-FAM	*HMGR*	upstream
STR1150-3F STR1150-3R	ATCTCCACGCCTTCCCTT CTCTGCTTCTCTTTTCGCAAT	57	(CCA)4	170	5′6-FAM	*ispD*	upstream
STR1150-4F STR1150-4R	AGGATAACTAACAAAAGAGAGGATG TCTTCCTATAGAGGTTGAGTGCT	56	(TC)5	190	5′6-FAM	*ispE*	upstream
STR1150-7F STR1150-7R	TGCCCCCCCTCATCTC GGAAATTCTTGAGCTTGCAGT	56	(TC)5	140	5′6-FAM	*TGL4*	downstream
STR1150-9F STR1150-9R	GTGCCCGTTCCATTCAAG TGCTCGCCGGAGAGTATG	57	(GA)10	119	5′6-FAM	*MVK*	upstream

3.3. Polymorphism Analysis of SSR Loci

The EST-SSR markers in this study were developed based on expressed sequence tags derived from the transcriptome data. Firstly, the SSR loci were tested for selective neutrality, and the SSRs with F_{ST} outliers were filtered from the following analyses. The LOSITAN analysis detected two F_{ST} outliers, indicating that the loci 1035P11 and 1035P14 were probably under positive selection (Figure 3); these two outlier loci were excluded from all subsequent analyses; the remaining loci under neutral selection were reserved for the subsequent genetic variation analyses.

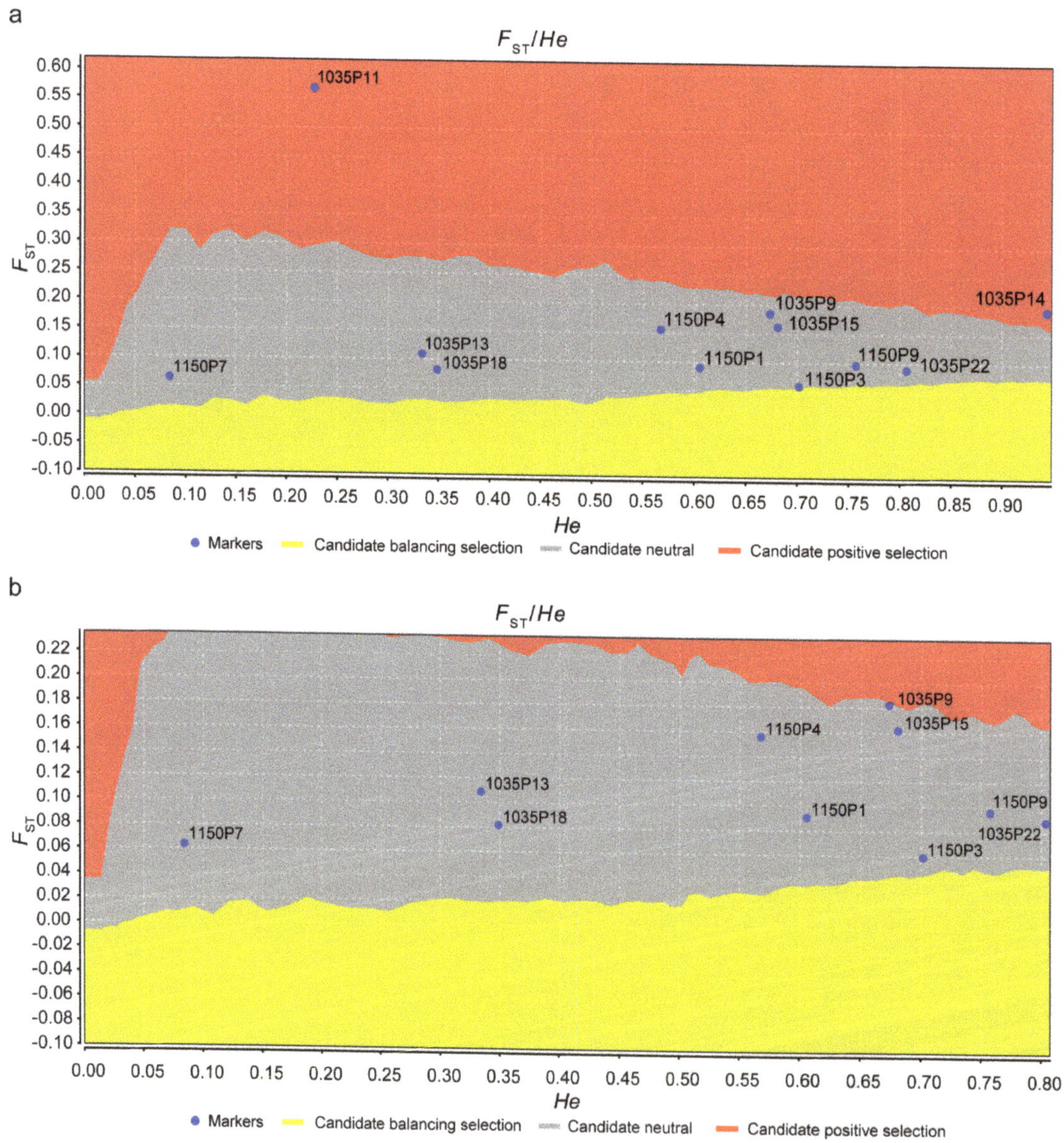

Figure 3. Assessment of F_{ST} outlier EST−SSR loci and neutrality tests. (**a**) analysis of F_{ST} outliers of 12 SSR loci; (**b**) neutrality tests for 10 SSR loci.

The genetic variation analysis of these loci was implemented based on 136 individuals from 10 populations. As shown in Table 3, *Na* was 99 alleles in total, and it ranged from 5 to 13, with an average of 9.90 alleles per locus. *Ne* was 27.92 in total, ranging from 1.1018 to 4.9541. *Ho* ranged from 0.3088 to 0.9599. *He* ranged from 0.0927 to 0.8011. *h* ranged from 0.0924 to 0.7981. PIC can detect and reflect the genetic variation level [36]. PIC value is grouped into highly polymorphic (PIC > 0.5), moderate polymorphic (0.5 > PIC > 0.25), and low polymorphic (PIC < 0.25) categories. PIC of 10 SSR loci ranged from 0.0842 to 0.7684. According to the Bostein theory, PIC values of 9 loci were over 0.25 and they were relatively high polymorphic loci, except locus 1150P7. Locus 1035P22 has the highest PIC value (0.7684), and the Shannon's information index of this locus is 1.8423. PIC values of the loci were consistent with *He* values of the corresponding loci. In general, the average *Ne*, *He*, and PIC were 2.7920, 0.5600, 0.5225, indicating that 10 screened EST-SSR loci had relatively high genetic polymorphism levels. Among the 10 loci, 1150P7 had the lowest level of genetic diversity and 1035P22 had the highest level of genetic diversity.

Table 3. The genetic diversity and genetic variation of EST-SSR loci. ** represent deviating from Hardy-Weinberg equilibrium (HWE) at the level of $p < 0.01$.

Loci	*Na*	*Ne*	*I*	PIC	*Ho*	*He*	*Nei (h)*	HWE	F_{IS}	F_{ST}	F_{IT}	*Nm*
1035P9	9	3.0711	1.3396	0.6202	0.5515	0.6769	0.6744	0.11	−0.0311	0.1983	0.1734	1.0105
1035P13	5	1.5449	0.5833	0.3102	0.2132	0.354	0.3527	**	0.3055	0.1281	0.3945	1.7010
1035P15	7	3.1617	1.3559	0.6505	0.1544	0.6862	0.6837	**	0.7419	0.1781	0.7879	1.1533
1035P18	11	1.4961	0.821	0.3207	0.1103	0.3328	0.3316	**	0.5952	0.1091	0.6393	2.0421
1035P22	13	4.9541	1.8423	0.7684	0.3235	0.8011	0.7981	**	0.5045	0.1131	0.5605	1.9598
1150P1	12	2.5313	1.4029	0.5785	0.1838	0.6072	0.6049	**	0.6646	0.1145	0.7030	1.9332
1150P3	12	3.3544	1.4452	0.6559	0.5368	0.7045	0.7019	**	0.1581	0.0866	0.2310	2.6380
1150P4	5	2.3126	0.9829	0.4781	0.1691	0.5697	0.5676	**	0.6610	0.1720	0.7193	1.2035
1150P7	5	1.1018	0.2441	0.0842	0.0441	0.0927	0.0924	**	0.4801	0.0855	0.5246	2.6727
1150P9	20	4.3902	2.1041	0.7585	0.6912	0.7751	0.7722	0.02	−0.0132	0.1182	0.1065	1.8653

F-statistic estimates for 10 SSR loci of 10 populations showed the genetic differentiation and inbreeding coefficients. F_{ST} values for these loci were estimated 0.0855–0.1983, with an average of 0.1304, which indicated that there was little genetic variation among the populations and 86.96% of the genetic variation was within populations. The low F_{ST} implied a low level of differentiation among the populations of *P. polyphylla* var. *yunnanensis*. F_{IS} values of 5 loci were over 0.50, and the high F_{IS} implied a considerable degree of inbreeding. Whereas, F_{IS} values of locus 1035P9 and 1150P9 showed the excess heterozygosity with negative F_{IS} values (−0.0311, −0.0312), suggesting outbreeding. F_{IT} values ranged from 0.1065 to 0.7879, with an average value of 0.4840. In this study, *Nm* values were estimated to be ranged from 1.0105 to 2.6727. The average *Nm* value of the loci was 1.82 (*Nm* > 1). Two loci (locus 1035P9 and locus 1150P9) were in accordance with the Hardy–Weinberg equilibrium (HWE), but eight loci showed significant departures from HWE after Bonferroni correction, apparently due to heterozygote deficiency. According to the Shannon informational diversity statistics portioned by population and total for codominant data averaged across loci, 25% of total information occurred among populations, 75% of total information occurred within population (Figure S2a).

3.4. Genetic Diversity and Genetic Variation of Populations

The genetic diversity parameters of 10 populations were also estimated and the populations displayed abundant genetic diversity (Table 4). The *Na* ranged from 2.4 to 4.9 with an average value of 3.79. *Ne* ranged from 1.69 to 2.78 with an average value of 2.32. *Ho* value ranged from 0.2182 to 0.3800 with average value of 0.29. Population QJL, QJS, and LS had high *Ho* values. *He* values ranged from 0.35 to 0.57 with average value of 0.49. Population QJL and QJS had high *He* values. Shannon diversity index (*I*) ranged from 0.53 to 1.09, and the maximum and the minimum of *I* were from population QJS and

TC, respectively. The values of *h* can reflect the population variations, which ranged from 0.33 to 0.56 with average value of 0.47. The *h* values and the ranks were in accordance with results of *He* values. Populations QJL, QJS, and LS had higher genetic diversity ($h > 0.50$, $Ne > 2.3$), the population from TC had lowest genetic diversity ($h = 0.33$, $Ne = 1.69$). Based on the above, population QJS, LS had higher genetic diversity than other populations, whereas population TC had the lowest genetic diversity. AMOVA was carried out to assess the overall distribution of diversity within and among populations (Figure S2b). Of the total genetic diversity, 1% of the variation occurred among populations, 69% occurred among individuals, and 30% occurred within individual; thus, AMOVA supported the results of *Nei*'s genetic statistics and the Shannon diversity estimation that there is a low degree of population differentiation.

Table 4. The genetic diversity and genetic differentiation of cultivated populations.

Pop	Major Allele Frequency	Genotype Number	*Na*	*Ne*	Gene Diversity	PIC	*I*	*Ho*	*He*	*Nei* (*h*)
LS	0.58	4.9	4.4	2.7809	0.5531	0.5078	1.0689	0.3000	0.5685	0.5496
DL	0.61	4.1	3.2	2.2729	0.4635	0.4124	0.8135	0.2353	0.4718	0.4580
KM	0.66	5.3	4.7	2.4983	0.4593	0.4315	0.9585	0.3200	0.4759	0.4600
QJL	0.56	5.5	4.9	2.7692	0.5553	0.5073	1.0883	0.3733	0.5740	0.5549
QJS	0.59	6.0	5.0	2.4917	0.5489	0.5092	1.0906	0.3800	0.5678	0.5489
ML	0.58	4.3	3.4	2.3365	0.5087	0.4419	0.8852	0.3133	0.5262	0.5087
JD	0.63	4.9	4.4	2.2769	0.4801	0.4343	0.9317	0.2687	0.5016	0.4859
TC	0.75	2.3	2.2	1.6859	0.3278	0.2782	0.5346	0.2889	0.3471	0.3278
YX	0.69	2.6	2.4	1.9034	0.4094	0.3509	0.671	0.2375	0.4367	0.4094
CX	0.67	4.0	3.3	2.1356	0.4343	0.3905	0.7938	0.2182	0.4550	0.4343

3.5. Genetic Structure and Population Clustering

To infer the genetic structure, the coancestry relationships of the populations were analyzed based on a Bayesian module using the STRUCTURE program. The results showed that when $K = 2$, ΔK reached the maximum value, which indicates that 10 populations mainly came from two ancestral groups (Figure 4a,b). The individuals in the groups were composed with admixed populations. In addition, UPGMA clustering of populations was constructed based on *Nei*'s (1983) genetic distance among populations. The unrooted tree of populations revealed that the populations were clustered into two groups, which was broadly consistent with the genetic population structure results (Figure 4c). Following the same analysis procedure, the tree of individuals was conducted subsequently (Figure 5). The clustering results showed that individuals of different populations clustered together and the populations with higher genetic diversity index were inclined to have individuals admixing with other populations. The PCoA results showed that the first and second principal components accounted for 36.8% and 33.0% of the total genetic variation, respectively (Figure 4d); it also showed that populations from two cultivated groups got close to each other, such as population JD and QJS. Most individuals were roughly clustered according to their corresponding populations (Figure S3).

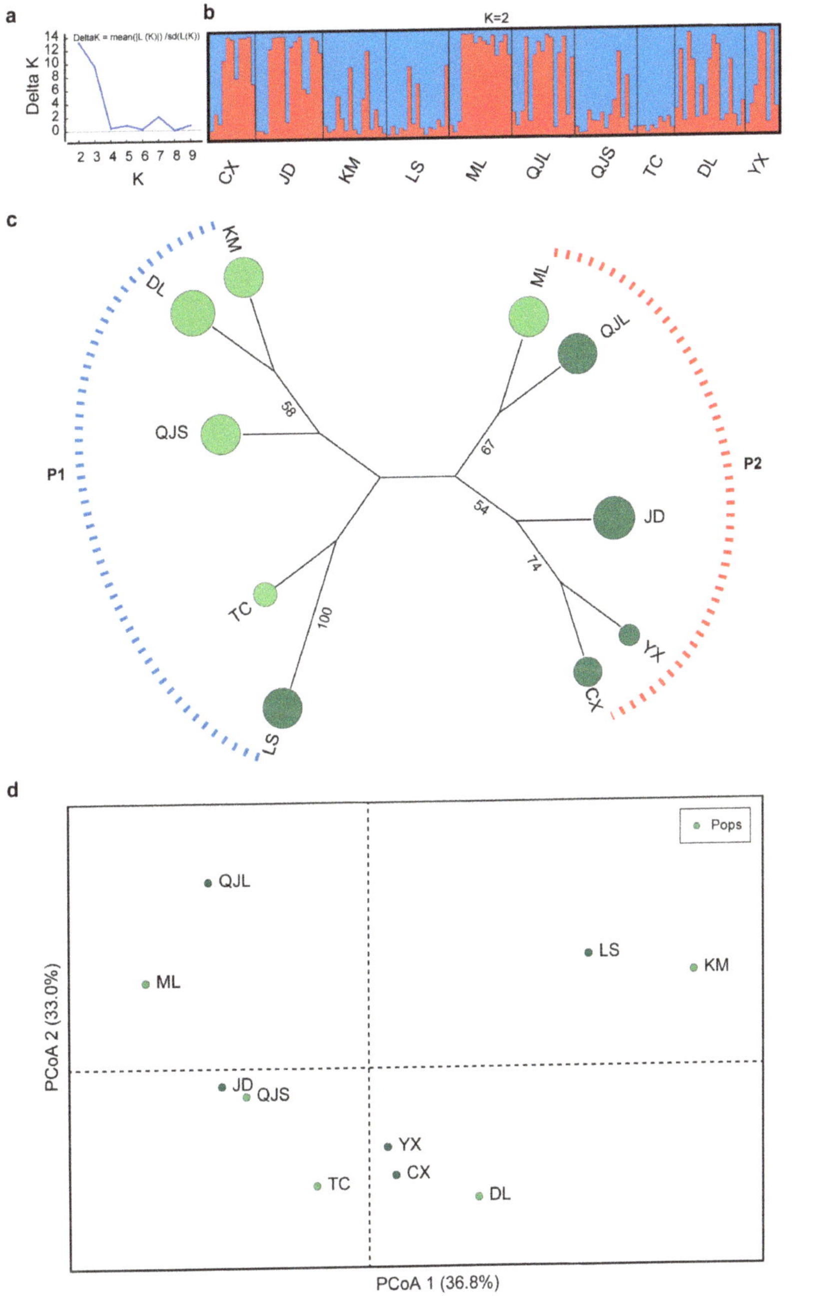

Figure 4. The population genetic structure of *P. polyphylla* var. *yunnanensis*. (**a**) relationships between the number of clusters (K) and the corresponding log probability of the data L(ΔK); (**b**) assignment of individuals to K = 2 genetically distinguishable group. (**c**) genetic divergence of 10 populations based on UPGMA cluster analysis. The two cultivated populations are denoted with dark green and light green. (**d**) principal coordinate analysis (PCoA) of 10 populations.

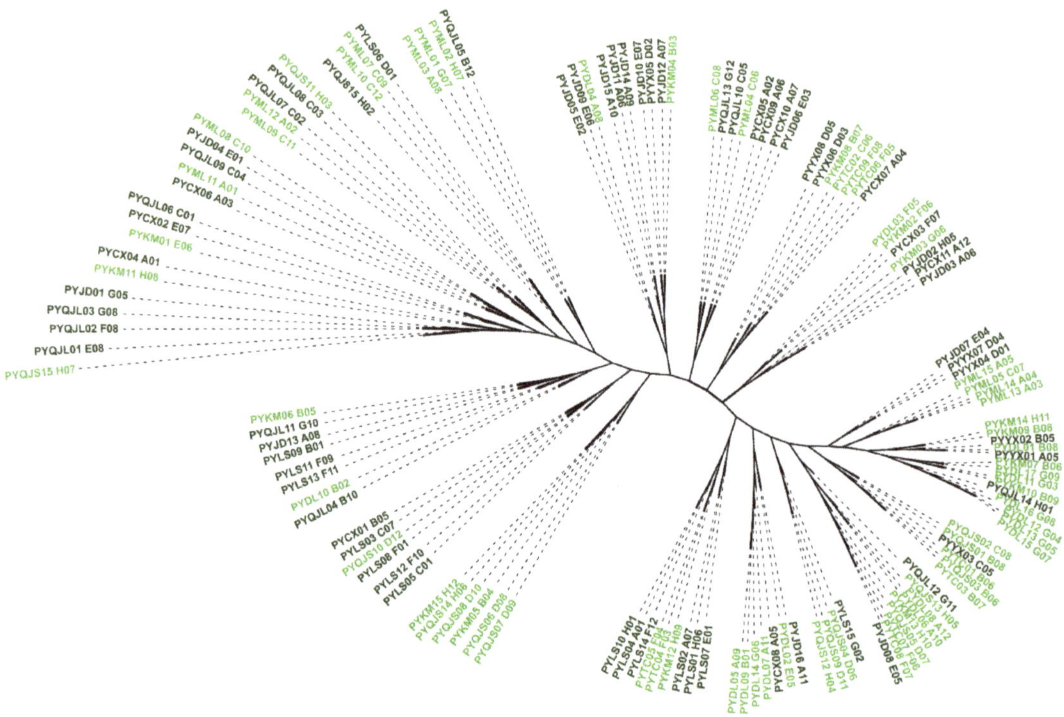

Figure 5. Genetic divergence of 136 individuals based on UPGMA cluster analysis. The long-stalked variety and the short-stalked variety are denoted with dark green and light green.

4. Discussion

4.1. SSR Frequency and Distribution

SSR holds a great promise for exploiting genetic diversity, characterizing accumulated phenotypic variation, and associating markers with traits in plant germplasm [37]. Unigenes derived microsatellite markers overcome the problem of redundancy in the EST database and have the advantage of assaying variation in the transcribed regions with their unique identities and positions [38]. The majority of SSRs in this study were dinucleotide repeats (38,848, 37.91%) based on the transcriptomic data of *P. polyphylla* var. *yunnanensis*. Another previous study of *P. polyphylla* var. *yunnanensis* transcriptome showed different conclusions that suggested that the monucleotide repeat type (56.3%) was the most abundant [39]. Nevertheless, distribution results in the present study are consistent with the frequencies of microsatellites among the gene indices of 24 plants in the previous study, which indicates that dinucleotide repeats are the majority of SSRs [40]. In addition, A/T repeat motif accounted for the largest proportion (91.74% of total mononucleotide repeats) in this study, which confirmed that the plant is rich in AT repeats [37]. The distribution density was estimated as one SSR per every 2.61 kb in this study. SSR loci in *Glycine* was proved to be three times more likely to occur in translated regions when derived from transcriptomic data than genomic data [41]. The density appears to vary significantly across plants through SSR density analyses based on transcriptomic data, i.e., *Populus wulianensis* (2.64 kb) [42], *Fagopyrum esculentum* (8.21 kb) [43], and arrowhead (9.13 kb) [44]. In contrast to microsatellite markers developed from genomic library, EST–SSRs can contribute to direct allele selection because they have known or putative functions and may be associated with the targeted trait [8]. Previous research has shown that SSRs have many important functions in terms of development, gene regulation, evolution, etc. [45]. The locations of SSRs appear to determine the types of functional role SSRs might play, and changes in SSRs in different genetic locations can lead to changes in the phenotypes of an organism [46]. As polyphyllins are the main bioactive ingredients of the *Paris* species, this study developed EST-SSR markers based on the unigenes related to polyphyllin biosynthesis for the first time. SSRs scattered in the gene candidates were involved in the upstream and downstream of polyphyllin backbone biosynthesis. SSRs in coding regions can determine whether or not a gene gets activated or whether the protein product is truncated [46]. The most common SSR motif types related to polyphyllin (mononucleotide and dinucleotide) were in the accordance with those of the transcriptome.

4.2. Marker Polymorphism

A total of 10 SSR markers developed here are the first set of microsatellites related to the bioactive ingredient biosynthesis for *Paris* species. For all loci analyzed, the average *Ho* was obviously lower than the average *He*, indicating that self-pollination may be more common than is usually assumed in *P. polyphylla*; however, an excess of homozygotes may also result from sub-population (Wahlund effect). The average *He* is similar to that of the makers (0.5251) from isolating microsatellites from a (CT)n-enriched genomic library of *P. poyphylla* var. *chinensis* [47] and the average PIC is similar to that of random SSR makers (0.5355) from the root transcriptome of *P. polyphylla* [13]. As a whole, the average *Ne* (2.792), *He* (0.5600), and PIC (0.5225) showed the relatively high genetic polymorphism levels of these 10 loci [34]. Among the loci, locus 1150P7 had the lowest level of genetic diversity, whereas locus 1035P22 had the highest level of genetic diversity. The Shannon information of the loci showed that less information occurred among population, and more information occurred within population. In addition, locus 1035P9 and locus 1150P9 were in accordance with HWE, but the rest showed significant departures from HWE after Bonferroni correction, apparently due to heterozygote deficiency. HWE departures can be caused by intrinsic factors in the studied sample and by specific marker characteristics like mutation rates [48]. To clarify, the high number of loci deviating from HWE could partly be a result of sampling, the presence of null alleles, or might arise from selective pressure on the coding regions [49]. Null alleles can occur due to mutations in primer binding sites and lead to the overestimation of homozygosity [50]. Low levels of *Ho* here partly support the latter hypothesis. According to previous studies, there are approximately 30 microsatellites developed with small samples (10–60 samples) using a (CT)n-enriched genomic library, a magnetic bead enrichment strategy, or the transcriptome of a root [13,47,51]. However, all of these random markers without functional data were subsequently applied in 115 samples for a genetic diversity study of *P. polyphylla* var. *yunnanensis* and only 7 of them were validated to be efficient [52]. In this study, 10 markers of EST-SSR were derived from SSR related to the polyphyllin backbone biosynthesis; they were screened from 34 candidate markers after conducting experimental evaluation multiple times. Although the study was based on a limited number of markers, the results should be considered in future germplasm utilization and molecular-assisted breeding for *P. polyphylla*. The steady progress in microsatellite markers will benefit the genetic diversity and molecular breeding of *P. polyphylla* and ultimately help increase yields for this medicinal herb and other *Paris* plants.

4.3. Relationships in the Germplasm Diversity

To preserve the natural population and ensure a steady and renewable source of *P. polyphylla* for ethnomedical purposes, thriving cultivation of seedlings and planting has become essential in recent years [53]. The wild *P. polyphylla* and its varieties are rather rare; thus, populations collected in this study are cultivars of *P. polyphylla* var. *yunnanensis* from the representative growing areas, including the short-stalked variety and the long-stalked variety widely planted. The genetic diversity analysis of populations showed that populations QJS and LS had the higher genetic diversity than other populations, whereas population TC had the lowest genetic diversity. Population has a high level of genetic diversity, suggesting that it has strong capability to adapt to stressful environmental conditions [54]. The overall populations exhibited a low degree of differentiation among populations, but maintained a high degree of genetic diversity among individuals, which were revealed by the results of AMOVA and genetic diversity estimation. A considerable degree of differentiation among individuals can be explained by cross-pollination and hybridization, since *P. polyphylla* is an insect-pollinated plant [55]. Although the high F_{IS} of 5 loci suggested a high degree of inbreeding, the self-pollination rate is found to be low in the agricultural cultivation [1,36]. This situation is even more striking when the different cultivars are introduced and grown simultaneously in the plant base. The low F_{ST} also implied a low level of differentiation among the populations. As the high average *Nm* (>1.82) was detected, gene flow played a role in homogenization for populations and effectively suppressed the genetic differentiation that resulted from gene drift [56]. The considerably high gene flow might be indicative of an earlier period of more pronounced gene flow when the species had a more continuous distribution [57]. The genetic structure revealed that 10 populations probably derived from two ancestral groups and all germplasms were found to be have different levels of admixture. However, the two groups did not quite tally with the two cultivation groups at population level and the samples from the two cultivation groups from different populations were mixed with one another at the individual level. Moreover, cultivated populations also showed high genetic variation, consisting of the genetic diversity investigation of wild and cultivated populations of *P. polyphylla* using ISSR [10]. It can be speculated that they have originated from mixed provenances; thus, screening for superior provenances should be carried out as soon as possible [11]. Most populations of endangered species are commonly subdivided into different breeding groups, such as different breeds in the case of domestic plants, which are, in turn, subdivided into smaller reproductive units more or less interconnected [58]. Hence, the populations with higher genetic diversity have more utilization potential for resource conservation and selection of breeding materials.

5. Conclusions

In this study, a total of 10 EST-SSR makers related to polyphyllin backbone biosynthesis were developed based on the transcriptome of *P. polyphylla* var. *yunnanensis*. The novel SSR loci showed relatively high genetic polymorphism levels. The overall populations exhibited a low degree of differentiation among populations, but maintained abundant genetic diversity among individuals. The clustering groups of populations were different from cultivated groups, resulting in interspecific and intervarietal hybridization. The ten novel markers of EST-SSR provide an important tool for exploring the genetic diversity of *P. polyphylla*, and they will assist in developing efficient strategies for the germplasm resource management and conservation of this medicinal plant. The findings of this study may facilitate maker-assisted breeding and genetic engineering schemes involving this species, and other medicinal plants of the genus *Paris*.

Supplementary Materials: The following supporting information can be downloaded at: https://www.mdpi.com/article/10.3390/d14080589/s1, Figure S1: The distribution of six SSR motifs identified in the transcriptome. (a) number of mononucleotide repeats; (b) number of dinucleotide repeats; (c) number of trinucleotide repeats (display the top 50% of total number); (d) number of tetranucleotide repeats (display repeat number > 23); (e) number of pentanucleotide repeats (display repeat number > 10); (f) number of hexanucleotide repeats (display repeat number > 5). (c–f) only display the repeats with high occurrence frequency; Figure S2: The diversity and variation analysis within populations and among populations. (a) Shannon informational diversity statistics partitioned by population and total for codominant data (b) analysis of molecular variance using allelic distance matrix for F-statistics. Figure S3: PCoA of 136 individuals. The ten populations are denoted with different color.

Author Contributions: C.L. and X.G. brought the idea and managed the research. X.G., W.M. and B.Y. (Bin Yang) collected the samples. Q.S., B.Y. (Baolin Yao) and W.Y. implemented biological. X.G. conducted the data analysis. X.G. prepared the manuscript. All authors have read and agreed to the published version of the manuscript.

Funding: This research was funded by National Natural Science Foundation of China, grant number 31800273, 31970609; Yunnan Fundamental Research Projects, grant number 202001AT070114; Crop Varietal Improvement and Insect Pests Control by Nuclear Radiation; Startup Fund from Xishuangbanna Tropical Botanical Garden; 'Top Talents Program in Science and Technology' from Yunnan Province. Publication costs are funded by National Natural Science Foundation of China (Grant No. 31800273), Yunnan Fundamental Research Projects (Grant No. 202001AT070114), and Crop Varietal Improvement and Insect Pests Control by Nuclear Radiation.

Institutional Review Board Statement: Not applicable.

Informed Consent Statement: Not applicable.

Data Availability Statement: The data presented in this study are available in NCBI database with accession number PRJNA682903 and PRJNA630028.

Acknowledgments: We thank the following people for kind help in this study: Guoqiang Zhang (Shandong University) and Hanrui Bai (University of Science and Technology of China). We also thank the Central Laboratory of Public Technology Service Center at Xishuangbanna Tropical Botanical Garden, Chinese Academy of Sciences for providing the computer resources and technical support.

Conflicts of Interest: The authors declare no conflict of interest.

References

1. He, L. *The Genus Paris Plants*; Science Press: Beijing, China, 2008; p. 33.
2. Pei, Y.F.; Zhang, Q.Z.; Wang, Y.Z. Application of Authentication Evaluation Techniques of Ethnobotanical Medicinal Plant Genus *Paris*: A Review. *Crit. Rev. Anal. Chem.* **2020**, *50*, 405–423. [CrossRef] [PubMed]
3. Ding, Y.G.; Zhao, Y.L.; Zhang, J.; Zuo, Z.T.; Zhang, Q.Z.; Wang, Y.Z. The traditional uses, phytochemistry, and pharmacological properties of *Paris* L. (Liliaceae): A review. *J. Ethnopharmacol.* **2021**, *278*, 114293. [CrossRef] [PubMed]
4. Cunningham, A.B.; Brinckmann, J.A.; Bi, Y.F.; Pei, S.J.; Schippmann, U.; Luo, P. *Paris* in the spring: A review of the trade, conservation and opportunities in the shift from wild harvest to cultivation of *Paris polyphylla* (Trilliaceae). *J. Ethnopharmacol.* **2018**, *222*, 208–216. [CrossRef] [PubMed]
5. Parida, S.K.; Kalia, S.K.; Kaul, S.; Dalal, V.; Hemaprabha, G.; Selvi, A.; Pandit, A.; Singh, A.; Gaikwad, K.; Sharma, T.R.; et al. Informative genomic microsatellite markers for efficient genotyping applications in sugarcane. *Theor. Appl. Genet.* **2009**, *118*, 327–338. [CrossRef]
6. Grover, A.; Sharma, P.C. Development and use of molecular markers: Past and present. *Crit. Rev. Biotechnol.* **2016**, *36*, 290–302. [CrossRef]
7. Powell, W.; Machray, G.C.; Provan, J. Polymorphism revealed by simple sequence repeats. *Trends Plant Sci.* **1996**, *1*, 215–222. [CrossRef]
8. Varshney, R.K.; Graner, A.; Sorrells, M.E. Genic microsatellite markers in plants: Features and applications. *Trends Biotechnol.* **2005**, *23*, 48–55. [CrossRef]
9. Ellis, J.R.; Burke, J.M. EST-SSRs as a resource for population genetic analyses. *Heredity* **2007**, *99*, 125–132. [CrossRef]
10. He, J.; Wang, H.; Li, D.Z.; Chen, S.F. Genetic diversity of *Paris polyphylla* var. *yunnanensis*, a traditional Chinese medicinal herb, detected by ISSR markers. *Planta Med.* **2007**, *73*, 1316–1321.
11. Huang, Y.; Zhou, N.; Yang, M.; Shen, Y.X.; Zhang, D.Q. A comparative study of the population genetics of wild and cultivated populations of *Paris polyphylla* var. *yunnanensis* based on amplified fragment length polymorphism markers. *Ecol. Evol.* **2019**, *9*, 10707–10722.
12. Zhao, X.P.; Zou, G.F.; Zhao, J.; Hu, L.Y.; Lan, Y.F.; He, J.L. Genetic relationships and diversity among populations of *Paris polyphylla* assessed using SCoT and SRAP markers. *Physiol. Mol. Biol. Plants.* **2020**, *26*, 1281–1293. [CrossRef] [PubMed]
13. Wang, L.; Yang, Y.; Zhao, Y.; Yang, S.; Udikeri, S.; Liu, T. De Novo Characterization of the Root Transcriptome and Development of EST-SSR Markers in *Paris polyphylla* Smith var. *yunnanensis*, an Endangered Medical Plant. *J. Agric. Sci. Technol.* **2016**, *18*, 437–452.
14. Pellicer, J.; Kelly, L.J.; Leitch, I.J.; Zomlefer, W.B.; Fay, M.F. A universe of dwarfs and giants: Genome size and chromosome evolution in the monocot family Melanthiaceae. *New Phytol.* **2014**, *201*, 1484–1497. [CrossRef]
15. Cota-Sanchez, J.H.; Remarchuk, K.; Ubayasena, K. Ready-to-use DNA extracted with a CTAB method adapted for herbarium specimens and mucilaginous plant tissue. *Plant Mol. Biol. Rep.* **2006**, *24*, 161–167. [CrossRef]
16. Gao, X.Y.; Zhang, X.; Chen, W.; Li, J.; Yang, W.J.; Zhang, X.W.; Li, S.Y.; Liu, C.N. Transcriptome analysis of *Paris polyphylla* var. *yunnanensis* illuminates the biosynthesis and accumulation of steroidal saponins in rhizomes and leaves. *Phytochemistry* **2020**, *178*, 12460.
17. Thiel, T.; Michalek, W.; Varshney, R.K.; Graner, A. Exploiting EST databases for the development and characterization of gene-derived SSR-markers in barley (*Hordeum vulgare* L.). *Theor. Appl. Genet.* **2003**, *106*, 411–422. [CrossRef] [PubMed]
18. Untergasser, A.; Cutcutache, I.; Koressaar, T.; Ye, J.; Faircloth, B.C.; Remm, M.; Rozen, S.G. Primer3-new capabilities and interfaces. *Nucleic Acids Res.* **2012**, *40*, e115. [CrossRef]

19. Antao, T.; Lopes, A.; Lopes, R.J.; Beja-Pereira, A.; Luikart, G. LOSITAN: A workbench to detect molecular adaptation based on a F(st)-outlier method. *BMC Bioinform.* **2008**, *9*, 323. [CrossRef]
20. Beaumont, M.A.; Nichols, R.A. Evaluating loci for use in the genetic analysis of population structure. *Proc. R. Soc. B-Biol. Sci.* **1996**, *263*, 1619–1626.
21. Ohtani, M.; Kondo, T.; Tani, N.; Ueno, S.; Lee, L.S.; Ng, K.K.S.; Muhammad, N.; Finkeldey, R.; Na'iem, M.; Indrioko, S.; et al. Nuclear and chloroplast DNA phylogeography reveals Pleistocene divergence and subsequent secondary contact of two genetic lineages of the tropical rainforest tree species *Shorea leprosula* (Dipterocarpaceae) in South-East Asia. *Mol. Ecol.* **2013**, *22*, 2264–2279. [CrossRef]
22. Peakall, R.; Smouse, P.E. GenAlEx 6.5: Genetic analysis in Excel. Population genetic software for teaching and research—An update. *Bioinformatics* **2012**, *28*, 2537–2539. [CrossRef] [PubMed]
23. Takezaki, N.; Nei, M.; Tamura, K. POPTREE2: Software for Constructing Population Trees from Allele Frequency Data and Computing Other Population Statistics with Windows Interface. *Mol. Biol. Evol.* **2010**, *27*, 747–752. [CrossRef] [PubMed]
24. Liu, K.J.; Muse, S.V. PowerMarker: An integrated analysis environment for genetic marker analysis. *Bioinformatics* **2005**, *21*, 2128–2129. [CrossRef] [PubMed]
25. Excoffier, L.; Lischer, H.E.L. Arlequin suite ver 3.5: A new series of programs to perform population genetics analyses under Linux and Windows. *Mol. Ecol. Resour.* **2010**, *10*, 564–567. [CrossRef]
26. Wright, S. Evolution in Mendelian populations. *Genetics* **1931**, *16*, 97–159. [CrossRef]
27. Bland, J.M.; Altman, D.G. Multiple Significance Tests—The Bonferroni Method. *Br. Med. J.* **1995**, *310*, 170. [CrossRef]
28. Retief, J.D. Phylogenetic analysis using PHYLIP. *Methods Mol. Biol.* **2000**, *132*, 243–258.
29. Kumar, S.; Tamura, K.; Nei, M. MEGA: Molecular Evolutionary Genetics Analysis software for microcomputers. *Comput. Appl. Biosci.* **1994**, *10*, 189–191. [CrossRef]
30. Rambaut, A. FigTree 1.4. 2 Software. Institute of Evolutionary Biology, Univ. Edinburgh. 2014. Available online: http://tree.bio.ed.ac.uk/software/figtree (accessed on 1 July 2022).
31. Pritchard, J.K.; Stephens, M.; Donnelly, P. Inference of Population Structure Using Multilocus Genotype Data. *Genetics* **2000**, *155*, 945–959. [CrossRef]
32. Evanno, G.; Regnaut, S.; Goudet, J. Detecting the number of clusters of individuals using the software STRUCTURE: A simulation study. *Mol. Ecol.* **2005**, *14*, 2611–2620. [CrossRef]
33. Earl, D.A.; Vonholdt, B.M. STRUCTURE HARVESTER: A website and program for visualizing STRUCTURE output and implementing the Evanno method. *Conserv. Genet. Resour.* **2012**, *4*, 359–361. [CrossRef]
34. Jakobsson, M.; Rosenberg, N.A. CLUMPP: A cluster matching and permutation program for dealing with label switching and multimodality in analysis of population structure. *Bioinformatics* **2007**, *23*, 1801–1806. [CrossRef] [PubMed]
35. Rosenberg, N.A. DISTRUCT: A program for the graphical display of population structure. *Mol. Ecol. Notes* **2004**, *4*, 137–138. [CrossRef]
36. Botstein, D.; White, R.L.; Skolnick, M.; Davis, R.W. Construction of a Genetic-Linkage Map in Man Using Restriction Fragment Length Polymorphisms. *Am. J. Hum. Genet.* **1980**, *32*, 314–331. [PubMed]
37. Kalia, R.K.; Rai, M.K.; Kalia, S.; Singh, R.; Dhawan, A.K. Microsatellite markers: An overview of the recent progress in plants. *Euphytica* **2011**, *177*, 309–334. [CrossRef]
38. Parida, S.K.; Kumar, K.A.R.; Dalal, V.; Singh, N.K.; Mohapatra, T. Unigene derived microsatellite markers for the cereal genomes. *Theor. Appl. Genet.* **2006**, *112*, 808–817. [CrossRef]
39. Li, B.; Peng, L.; Sun, X.C.; Huang, W.J.; Wang, N.; He, Y.H.; Shi, X.B.; Liu, Y.R.; Zhang, P.; Yang, X.J.; et al. Organ-specific transcriptome sequencing and mining of genes involved in polyphyllin biosynthesis in *Paris polyphylla*. *Ind. Crop. Prod.* **2020**, *156*, 112775. [CrossRef]
40. Von Stackelberg, M.; Rensing, S.A.; Reski, R. Identification of genic moss SSR markers and a comparative analysis of twenty-four algal and plant gene indices reveal species-specific rather than group-specific characteristics of microsatellites. *BMC Plant Biol.* **2006**, *6*, 9. [CrossRef]
41. Hodel, R.G.J.; Gitzendanner, M.A.; Germain-Aubrey, C.C.; Liu, X.X.; Crowl, A.A.; Sun, M.; Landis, J.B.; Segovia-Salcedo, M.C.; Douglas, N.A.; Chen, S.C.; et al. A New Resource for the Development of SSR Markers: Millions of Loci from a Thousand Plant Transcriptomes. *Appl. Plant Sci.* **2016**, *4*, 1600024. [CrossRef]
42. Wu, Q.C.; Zang, F.Q.; Xie, X.M.; Ma, Y.; Zheng, Y.Q.; Zang, D.K. Full-length transcriptome sequencing analysis and development of EST-SSR markers for the endangered species *Populus wulianensis*. *Sci. Rep.* **2020**, *10*, 16249. [CrossRef]
43. Liu, Y.; Fang, X.M.; Tang, T.; Wang, Y.D.; Wu, Y.H.; Luo, J.Y.; Wu, H.T.; Wang, Y.Q.; Zhang, J.; Ruan, R.W.; et al. Inflorescence Transcriptome Sequencing and Development of New EST-SSR Markers in Common Buckwheat (*Fagopyrum esculentum*). *Plants* **2022**, *11*, 742. [CrossRef] [PubMed]
44. You, Y.N.; Huang, X.F.; Liu, H.B.; Cheng, T.; Zheng, X.F.; Diao, Y.; Bao, Z.Z.; Dong, C.; Ke, W.D.; Hu, Z.L. Leaf Transcriptome Analysis and Development of EST-SSR Markers in Arrowhead (*Sagittaria trifolia* L. var. *Sinensis*). *Trop. Plant Biol.* **2020**, *13*, 189–200. [CrossRef]
45. Lawson, M.J.; Zhang, L.Q. Distinct patterns of SSR distribution in the *Arabidopsis thaliana* and rice genomes. *Genome Biol.* **2006**, *7*, R14. [CrossRef] [PubMed]

46. Li, Y.C.; Korol, A.B.; Fahima, T.; Nevo, E. Microsatellites within genes: Structure, function, and evolution. *Mol. Biol. Evol.* **2004**, *21*, 991–1007. [CrossRef] [PubMed]
47. Zheng, J.Y.; Wang, H.; Chen, X.X.; Wang, P.; Gao, P.; Li, X.N.; Zhu, G.P. Microsatellite markers for assessing genetic diversity of the medicinal plant *Paris polyphylla* var. *chinensis* (Trilliaceae) . *Genet. Mol. Res.* **2012**, *11*, 1975–1980. [CrossRef]
48. Aranzana, M.J.; Illa, E.; Howad, W.; Arus, P. A first insight into peach [*Prunus persica* (L.) Batsch] SNP variability. *Tree Genet. Genomes* **2012**, *8*, 1359–1369. [CrossRef]
49. Menken, S.B.J.; Smit, E.; DenNijs, H.C.M. Genetical population structure in plants: Gene flow between diploid sexual and triploid asexual dandelions (*Taraxacum* section *Ruderalia*). *Evolution* **1995**, *49*, 1108–1118. [CrossRef]
50. Callen, D.F.; Thompson, A.D.; Shen, Y.; Phillips, H.A.; Richards, R.I.; Mulley, J.C.; Sutherland, G.R. Incidence and Origin of Null Alleles in the (Ac)N Microsatellite Markers. *Am. J. Hum. Genet.* **1993**, *52*, 922–927.
51. Song, Y.; Li, M.F.; Xu, J.; Zhao, Z.; Chen, N.Z. Polymorphic microsatellite markers in the traditional Chinese medicinal plant *Paris polyphylla* var. *yunnanensis*. *Genet. Mol. Res.* **2015**, *14*, 9939–9942. [CrossRef]
52. Chen, Z.S.Z. Genetic diversity of *Paris polyphylla* var. *yunnanensis* by SSR marker. *Chin. Tradit. Herb. Drugs* **2017**, *48*, 1834–1838.
53. Qi, J.J.; Zheng, N.; Zhang, B.; Sun, P.; Hu, S.N.; Xu, W.J.; Ma, Q.; Zhao, T.Z.; Zhou, L.L.; Qin, M.J.; et al. Mining genes involved in the stratification of *Paris Polyphylla* seeds using high-throughput embryo transcriptome sequencing. *BMC Genomics* **2013**, *14*, 358. [CrossRef] [PubMed]
54. Sgro, C.M.; Lowe, A.J.; Hoffmann, A.A. Building evolutionary resilience for conserving biodiversity under climate change. *Evol. Appl.* **2011**, *4*, 326–337. [CrossRef]
55. Ren, Z.X.; Wang, H.; Bernhardt, P.; Li, D.Z. Insect Pollination and Self-Incompatibility in Edible and/or Medicinal Crops in Southwestern China, a Global Hotspot of Biodiversity. *Am. J. Bot.* **2014**, *101*, 1700–1710. [CrossRef] [PubMed]
56. Hutchison, D.W.; Templeton, A.R. Correlation of pairwise genetic and geographic distance measures: Inferring the relative influences of gene flow and drift on the distribution of genetic variability. *Evolution* **1999**, *53*, 1898–1914. [CrossRef] [PubMed]
57. Zhang, D.Q.; Zhou, N. Genetic diversity and population structure of the endangered conifer *Taxus wallichiana* var. *mairei* (Taxaceae) revealed by Simple Sequence Repeat (SSR) markers. *Biochem. Syst. Ecol.* **2013**, *49*, 107–114.
58. Toro, M.A.; Caballero, A. Characterization and conservation of genetic diversity in subdivided populations. *Philos. Trans. R. Soc. Lond. B-Biol. Sci.* **2005**, *360*, 1367–1378. [CrossRef]

diversity

MDPI

Article

The Chromosome-Level Genome of *Elaeagnus moorcroftii* Wall., an Economically and Ecologically Important Tree Species in Drylands

Xinxing Fu [1], Jingjing Wu [2], Xiaohui Ma [2], Kunpeng Li [3,4], Hui Zhang [1], Shengdan Wu [2,*] and Kun Sun [1,*]

1 College of Life Sciences, Northwest Normal University, Lanzhou 730070, China; fuxinxing@nwnu.edu.cn (X.F.); zhanghui@nwnu.edu.cn (H.Z.)
2 State Key Laboratory of Grassland Agro-Ecosystems, College of Ecology, Lanzhou University, Lanzhou 730000, China; wujj21@lzu.edu.cn (J.W.); maxh20@lzu.edu.cn (X.M.)
3 State Key Laboratory of Systematic and Evolutionary Botany, Institute of Botany, Chinese Academy of Sciences, Beijing 100093, China; likunpeng@ibcas.ac.cn
4 University of Chinese Academy of Sciences, Beijing 100049, China
* Correspondence: wusd@lzu.edu.cn (S.W.); kunsun@nwnu.edu.cn (K.S.)

Abstract: *Elaeagnus moorcroftii* Wall. (Elaeagnaceae) is an important tree species naturally growing in arid Northwest China that has great economic and ecological values in drylands. In this study, we de novo assembled a chromosome-level genome for *E. moorcroftii* by using PacBio's high-fidelity (HiFi) sequencing and Hi-C-assisted assembly technology. The assembled genome size was 529.56 Mb, of which 94.56% was anchored to 14 pseudochromosomes with a contig N50 up to 28.21 Mb. In total, 29,243 protein-coding genes were annotated, and 98.5% of the Benchmarking Universal Single-Copy Orthologs (BUSCOs) were captured in the genome. Evolutionary genomic analysis showed that *E. moorcroftii* split with *Elaeagnus mollis* 9.38 million years ago (Ma), and contrasted evolutionary trajectories of gene family expansion and contraction were observed for these two closely related species. Furthermore, we identified two successive whole genome duplication (WGD) events occurred in the genome of *E. moorcroftii*, in addition to the ancient *gamma* hexaploidization event shared by core eudicots. Together, the chromosome-level genome assembly for *E. moorcroftii* decoded here provides valuable genomic information for the further genetic improvement and molecular breeding of this indigenous species in drylands.

Keywords: *Elaeagnus moorcroftii* Wall.; PacBio's high-fidelity sequencing; Hi-C-assisted assembly; whole genome duplication; xerophyte; drylands

Citation: Fu, X.; Wu, J.; Ma, X.; Li, K.; Zhang, H.; Wu, S.; Sun, K. The Chromosome-Level Genome of *Elaeagnus moorcroftii* Wall., an Economically and Ecologically Important Tree Species in Drylands. *Diversity* **2022**, *14*, 468. https://doi.org/10.3390/d14060468

Academic Editors: Hong-Hu Meng and Yi-Gang Song

Received: 30 May 2022
Accepted: 10 June 2022
Published: 11 June 2022

1. Introduction

Land degradation and desertification constitute one of the most serious environmental problems facing the world. Drylands cover about 41% of the global land area and are home to more than 38% of the world's population [1,2]. Ecosystems in drylands are fragile and vulnerable to climate change and human activities [3]. The degree of desertification in these drylands is likely to increase rapidly, the areas of drylands will continue to expand, and the risk of ecological degradation will be further exacerbated [2]. Desert plants play an important role in maintaining the stability of dryland ecosystems and provide ecological services for the production and life of people in drylands. Therefore, it is very meaningful to conduct scientific research on desert plants. With the rapid development of sequencing technologies, the genomes of some important desert plants have been deciphered recently, such as the sea buckthorn (*Hippophae rhamnoides*) with medicinal and edible value [4,5], wild and perennial legume forage *Medicago ruthenica* [6], and xerophytic plant *Haloxylon ammodendron* [7].

E. moorcroftii is a kind of deciduous tree of the Elaeagnaceae family, up to 10 m tall, mainly distributed in the desert areas of Northwest China, including the Xinjiang, Gansu, and Inner Mongolia provinces. It has excellent characteristics of drought resistance and salt tolerance and is an important tree species for windbreaks, sand fixation, and soil and water conservation in Northwest China [8]. This species is a non-leguminous nitrogen-fixing plant, its roots are symbiotic with *Frankia* actinomycetes, which can play the role of biological nitrogen fixation and soil improvement [4,9], and its rhizosphere arbuscular mycorrhizal fungi can improve their resistance to salt stress [10]. Therefore, *E. moorcroftii* can be introduced to barren desert and saline-alkali land for soil improvement and afforestation, which has important ecological value. In addition, this species has high economic value because of its edible fruits, medicinal whole plants, and ornamental flowers. The species exhibits a high fruit yield; the fruit can be eaten directly and used in jam, vinegar, wine, pastries, livestock feed, etc. [11]. The branches, leaves, and flowers of the species have remarkable observed biological activities and are widely used to treat many health issues like aging, burns, dyspepsia, diarrheal, pain, bronchitis, and neurasthenia [11,12]. The flowers are attractive and aromatic and are also used for extracting aromatic oil [11]. However, the molecular-level study of *E. moorcroftii* is limited to taxonomic relationships [8]. The lack of genomic information hinders a comprehensive understanding of the evolutionary history and unique biology of *E. moorcroftii*.

In this study, we first report a chromosomal-level genome assembly of *E. moorcroftii* (2n = 2x = 28) with PacBio's long-read single molecule high-fidelity (HiFi) reads and high-throughput chromosome conformation capture (Hi-C) data, and then we used it to explore the evolutionary trajectories of *E. moorcroftii* and other Elaeagnaceae species, including *H. rhamnoides* [4,5] and *E. mollis* [13] as recently published, by comparative genomic analysis. The genome sequence of *E. moorcroftii* presented here will provide valuable genomic resources for further in-depth study and utilization of this indigenous species in drylands.

2. Materials and Methods

2.1. Plant Materials and Sequencing

The fresh young leaves used for genomic DNA sequencing were collected from an adult plant of *E. moorcroftii* growing in Minqin Desert Botanical Garden, Gansu Province, China. The total genomic DNA was extracted using a modified trimethylammonium bromide (CTAB) method [14]. For PacBio's HiFi sequencing, SMRTbell libraries were constructed using the protocol of Pacific Biosciences with 20 kb inserts and sequenced using circular consensus sequencing (CCS) mode on a PacBio Sequel II platform, generating a total of more than 28 Gb high-quality CCS reads. For Hi-C sequencing, cross-linked chromatin was first digested with *Dpn* II, end-labeled with biotin-14-dATP, and then used for in situ DNA ligation. The ligated DNA was sheared into 300–600 bp fragments, blunt-end repaired, purified through biotin–streptavidin-mediated pull-down, and then sequenced on the Illumina HiSeq 2500 platform, generating a total of more than 52 Gb raw sequencing data.

2.2. Genome Assembly and Assessment

The high-quality HiFi reads were assembled into contigs using hifiasm v0.14 [15] with default parameters. The Hi-C data were aligned to the contig assembly using JUICER v.1.5 [16]. The contigs of the *E. moorcroftii* assembly were further clustered, ordered, and oriented onto chromosomes based on the contact frequency calculated from the mapped Hi-C read pairs by 3d-DNA pipeline [17] with parameters of '-m haploid –r 2'. Orientation and placement errors were manually corrected via Juicebox Assembly Tools (https://github.com/aidenlab/Juicebox (accessed on 17 November 2021)).

The completeness of the genome assembly of *E. moorcroftii* was assessed by transcript alignment and Benchmarking Universal Single-Copy Orthologs (BUSCO) analysis [18]. RNA-seq reads were assembled and mapped to the genome by HISAT2 v2.1.0 [19]. BUSCO analysis of the final assembly and annotation was performed using BUSCO v4 [18] with the Embryophyta obd10 database to evaluate the completeness of the reference genome of *E. moorcroftii*.

2.3. Genome Annotation

Tandem repeats in the *E. moorcroftii* genome were annotated using GMATA v2.2 [20] and Tandem Repeats Finder v4.09 [21]. The transposable elements (TEs) in the genome were predicted using a combination of homology-based and de novo approaches. In the homology-based approach, we used RepeatMasker v.4.1.0 [22] with the known repeat database Repbase v.21.01 [23] to search for the TEs in the *E. moorcroftii* genome. For the de novo approach, we used MITE-Hunter [24] and RepeatModeler v.2.0 [25] to construct a de novo repeat sequence database for *E. moorcroftii* and then used RepeatMasker v.4.1.0 [22] to search for repeats in the genome. After removing overlapping repeats, the repeats identified by different methods were combined into the final repeat annotation.

Protein-coding genes were predicted based on the repeat masked genome using three approaches, including homology search, de novo prediction, and estimation from transcriptome evidence. Homology-based prediction was performed using GeMoMa v1.3.1 [26] with the protein sequences of four closely related species in eudicots, *E. mollis* [13], *Ziziphus jujuba* [27], *Arabidopsis thaliana* [28], and *Vitis vinifera* [29]. The de novo prediction was conducted using Augustus v3.2.3 [30] with default parameters. RNA-seq based gene prediction was performed using STAR v2.7.3a [31], Stringtie v2.0.1 [32], and PASA v2.0.2 [33], and the public available RNA-seq data for mixed tissue samples including the leaf, root, stem, and fruit were downloaded from the National Center for Biotechnology Information (NCBI) under the accession numbers of SRR12569922, SRR12569923, and SRR12569924. Finally, the outputs from the above three approaches were integrated into a final gene set by EvidenceModeler (EVM) v1.1.1 [34].

For gene function annotation, the predicted gene models were blasted against the SwissProt v5.3 (https://www.expasy.org/resources/uniprotkb-swiss-prot (accessed on 29 July 2021)), NCBI non-redundant (NR) (https://www.ncbi.nlm.nih.gov/refseq/about/nonredundantproteins/ (accessed on 29 July 2021)), and Clusters of Orthologous Groups for Eukaryotic complete Genomes (KOG) v5.3 (http://genome.jgi-psf.org/help/kogbrowser.jsf (accessed on 29 July 2021)) databases for the best matches using BLASTP with an *E*-value cut-off of 1e-5. The protein motifs and domains were annotated using InterProScan v5.31 [35]. The Gene Ontology (GO) entries were searched using Blast2GO v2.5 [36]. Pathway information for each gene was assigned using the Kyoto Encyclopedia of Genes and Genomes (KEGG) v5.24 database [37].

In addition, the non-coding RNA genes were annotated. tRNAs were predicted by tRNAscan-SE v1.3.1 [38] with eukaryote parameters. MicroRNA, small nuclear RNA, and small nucleolar RNA were predicted using INFERNAL v1.1.2 [39] based on the Rfam [40] and miRbase databases [41]. The rRNAs and their subunits were predicted using RNAmmer v1.2 [42].

2.4. Comparative and Evolutionary Genomic Analysis

The protein sequences of *E. moorcroftii* and ten other sequenced plant species, *H. rhamnoides*, *E. mollis*, *Malus domestica*, *Prunus persica*, *Z. jujuba*, *Morus notabilis*, *Cannabis sativa*, *Populus trichocarpa*, *A. thaliana*, *V. vinifera*, were used for the phylogenetic analysis (see Table S1 for the summary of genomic information of these species). Single-copy orthogroups among the 11 species were identified using OrthoFinder v2.3.11 [43]. The amino acid alignments of each single-copy orthogroup were aligned by MAFFT v.7 [44], and nucleotide alignments were generated according to the corresponding amino acid alignments using PAL2NAL [45]. A maximum likelihood phylogeny was constructed based on the concatenated alignments of all single-copy genes using IQ-TREE v.1.6.12 [46]. The species divergence time was estimated using the program MCMCTREE in the PAMLv.4.9 package [47]. We selected three fossil calibration points from the TimeTree database (http://www.timetree.org (accessed on 1 March 2022)) for the split of: (1) *Malus-Prunus* at 30–61 million years ago (Ma); (2) *Morus-Cannabis* at 53–97 Ma; (3) *Populus-Arabidopsis* at 98–117 Ma. In addition, the time calibration of family Rhamnaceae (that is the split of *Ziziphus* with three Elaeagnaceae species) was set to >99 Ma based on an old fossil in the extant genus *Phylica* of Rhamnaceae [48]. The gene family expansion and contraction of 11 species were analyzed using CAFE v3.1 [49], and the expanded and contracted gene families in *E. moorcroftii* were subjected to GO enrichment to analyze their functions. The different modes of gene duplications were identified by DupGen_finder software [50].

2.5. Whole Genome Duplication Analysis

The protein-coding sequences of *E. moorcroftii*, *H. rhamnoides*, and *E. mollis* were self-aligned and aligned with each other using BLASTP with an *E*-value cut-off of 1e-5. Syntenic blocks within a genome or between genomes were identified using MCScanX [51] with default parameters based on the above protein-sequence alignments. For each syntenic gene pair, the synonymous substitution rate (*Ks*) was calculated using the Nei-Gojobori method [52] by yn00 program of the PAML package [47]. The macro-syntenic relationships among the three species were visualized by the python version of MCScan software (https://github.com/tanghaibao/jcvi/wiki/MCscan-(Python-version) (accessed on 29 July 2021)).

3. Results

3.1. Genome Assembly of E. Moorcroftii

To obtain a high-quality reference genome of *E. moorcroftii*, we used a combination of HiFi long reads and Hi-C data to construct a chromosome-level assembly. We generated a total of 28 Gb high-quality HiFi long reads, with a maximum and average read length of 64.20 and 14.78 kb, respectively (Table S2 and Figure S1, distribution of PacBio's HiFi sequencing data). De novo assembly of these HiFi reads with hifiasm v0.14 [16] yielded a preliminary assembly comprising 168 contigs with contig N50 size of 28.21 Mb, and the total length of the assembled genome was 529.56 Mb (Table 1). A total of 52 Gb Hi-C reads were generated to refine the assembly, resulting in 500.73 Mb (94.56%) of the contig sequences anchored onto 14 chromosomes (Figure 1, Table 1).

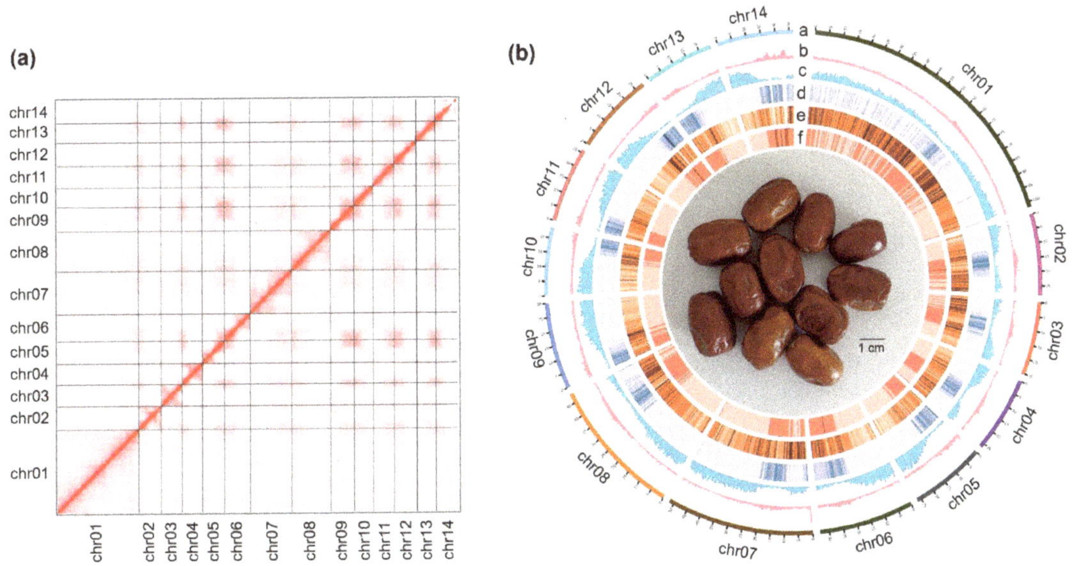

Figure 1. The genome features of *E. moorcroftii*. (**a**) Heatmap of Hi-C interactions for 14 pseudochromosomes. (**b**) Circos plot showing the genomic landscape of *E. moorcroftii*. The tracks from outer to inner circles indicate the following: a: 4 pseudochromosomes in megabases; b: GC content; c: gene density; d: density of *Gypsy* LTR retrotransposons; e: density of *Copia* LTR retrotransposons; f: LTR retrotransposons density. The center of the circos plot shows the fruits of *E. moorcroftii* (photo by X.F.).

Table 1. Assembly and annotation statistics of *E. moorcroftii* genome.

Assembly	
Length of genome assembly (Mb)	529.56
Anchored to chromosome (Mb)	500.73
Contig N50 (Mb)	28.21
Longest contig (Mb)	101.76
BUSCO score of assembly (%)	96.7%
Annotation	
GC content	30.39%
Percentage of repeat sequences (%)	60.95%
Number of protein-coding gene (%)	29243
Average gene length (bp)	4318.92
Average exon length (bp)	220.21
BUSCO score of annotation (%)	98.5%

To assess the quality and completeness of this assembled genome, we first mapped RNA-seq reads to the assembled genome, more than 93% of which were properly mapped (Table S3). Furthermore, BUSCO analysis for the genome assembly showed that 96.7% of the 1614 core plant genes was captured, including 95.8% complete BUSCOs and 0.9% fragmented BUSCOs (Table S4). These evidences together indicated that the *E. moorcroftii* assembly has high quality and completeness.

3.2. Annotation of the E. Moorcroftii Genome

We identified 322.78 Mb of non-redundant repetitive sequences in the *E. moorcroftii* genome, representing 60.95% of the genome assembly (Table S5). Long terminal repeat (LTR) retrotransposons were the most abundant type, accounting for 31.95% of the whole genome, of which *Copia* and *Gypsy* were the two most frequent LTR types, accounting for about 16.02% and 13.30% of the genome, respectively (Table S5).

Using genomic and transcriptomic data, we predicted 29,243 protein-coding gene models in the *E. moorcroftii* genome, with an average gene length of 4318.92 bp and an average exon length of 220.21 bp (Table 1). Among the predicted protein-coding genes, 95.10% were annotated through at least one of the following protein-related databases: the NCBI NR database (94.75%), the SwissProt protein database (81.38%), the KOG database (91.24), the InterProScan database (86.10), the GO database (64.91%), and the KEGG database (3.87%) (Table S6). In total, 95.10% of the protein-coding genes were functionally annotated by various databases. In addition, our annotation captured 98.5% of BUSCOs, including 97.4% complete gene models plus 1.1% fragmented gene models (Table S7).

We also predicted 183 micro RNAs (miRNAs), 2461 small-nuclear RNAs (snRNAs), 1636 transfer RNAs (tRNAs), and 20,568 ribosomal RNAs (rRNAs), with calculated average lengths of 124.86, 107.52, 75.21, and 238.47 bp, respectively (Table S8).

3.3. Evolutionary History of E. Moorcroftii

To investigate the evolutionary history of *E. moorcroftii*, we performed a gene family clustering using *E. moorcroftii* and ten other representative angiosperm species, including two related species of the Elaeagnaceae family (*E. mollis*, *H. rhamnoides*), five species of the same order Rosales (Rosaceae: *M. domestica*, *P. persica*, Rhamnaceae: *Z. jujuba*; Moraceae: *M. notabilis*; Cannabaceae: *C. sativa*), and three outgroup species of eudicot clade (*A. thaliana*, *P. trichocarpa*, and *V. vinifera*) (Table S1). We identified 175 single-copy orthogroups and used them to reconstruct the phylogenetic tree of *E. moorcroftii* and ten other plant species (Figure 2). The results showed that *E. moorcroftii* clustered a monophyletic group with its related species *E. mollis* and *H. rhamnoides*, which in turn formed sister to *Z. jujuba* of Rhamnaceae. The divergence time between *E. moorcroftii* and *E. mollis* was estimated to be around 9.38 million years ago (Ma), and the most recent common ancestor of two *Elaeagnus* species split with *H. rhamnoides* occurred at about 30.52 Ma (Figure 2).

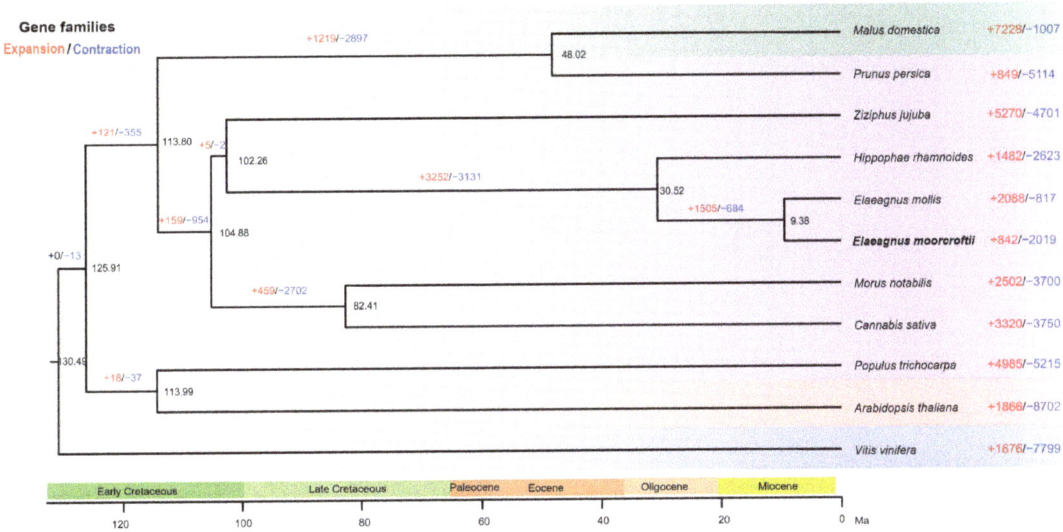

Figure 2. Comparative and evolutionary genomic analysis of *E. moorcroftii* and 10 other plant species. A phylogenetic tree among 11 species was reconstructed based on 175 single-copy genes, and their divergence times were also estimated. The numbers of expansion (blue) and contraction (red) gene families are shown above the branches.

Furthermore, to explore lineage-specific dynamic changes in gene families, the expansion and contraction of gene families based on the birth-and-death model were identified by CAFE v3.1 [49]. We detected 842 expansion and 2019 contraction gene families in *E. moorcroftii* genome, relative to the most recent common ancestor of *E. moorcroftii* and *E. mollis* (Figure 2), whereas an opposite trend was observed in its close relative *E. mollis*, which included more expanded (2088) than contracted (817) gene families. This difference might be partly resulted from the different levels of contribution from various modes of gene duplications. For example, dispersed duplications contributed to more in *E. mollis* than in *E. moorcroftii* with relation to the expanded gene families (16.45% versus 7.64%). In contrast, tandem duplications contributed to a higher percentage of expanded genes in *E. moorcroftii* than in *E. mollis* (14.34% versus 6.40%) (Figure S2). GO enrichment analysis revealed that the expanded gene families in *E. moorcroftii* were mainly involved in multiple biosynthetic processes (e.g., diterpenoid biosynthetic process and ATP biosynthetic process) and various metabolic process (e.g., ATP metabolic process, diterpenoid metabolic process, and purine-containing compound metabolic process), as well as immune responses (innate immune response and immune system process) (Figure S3), while contracted gene families were mainly related to protein depolymerization biological process and their enriched molecular function including ADP binding, ion channel activity, hormone binding, etc. (Figure S4).

A detailed comparative analysis among *E. moorcroftii*, *E. mollis*, *H. rhamnoides*, and *Z. jujuba* identified 12,118 common gene families shared by these four species, and 278 gene families uniquely appeared in *E. moorcroftii* (Figure S5). This number is comparable with the unique gene families in *E. mollis* (388) and *H. rhamnoides* (396), but it is much fewer than the unique gene families in *Z. jujuba* (2083) (Figure S5). The unique gene families are mainly involved in cellular respiration and oxidative phosphorylation biological processes (Figure S6).

3.4. Whole-Genome Duplication Events in E. Moorcroftii

The distributions of synonymous substitutions per synonymous site (K_S) of paralogous gene pairs in *E. moorcroftii* genome showed two recent clear peaks around 0.38 and

0.45 (Figure 3a). Similar K_S peaks also were identified in the genomes of its two closely related species, *E. mollis* and *H. rhamnoides* (Figure 3a). Moreover, the large-scale gene duplications in three species occurred earlier than the time of their divergence (Figure 3a). Therefore, these two closely occurring peaks might reflect two successive whole-genome duplication (WGD) events shared by three Elaeagnaceae species. To further confirm this, an intragenomic synteny analysis of *E. moorcroftii* identified one syntenic block corresponding to three homologous syntenic blocks (Figure S7), again supporting two relatively recent WGD events that occurred in *E. moorcroftii*. In addition, syntenic analysis among the genomes of *E. moorcroftii* and the most closely related species (*E. mollis* and *H. rhamnoides*) was performed to explore WGD history that occurred in the Elaeagnaceae family. By using a *Vitis vinifera*, a basal core eudicot lineage lacking any further WGD event after the ancient *gamma* event shared by core eudicots [53], genome as a reference, the intergenomic synteny comparisons between *V. vinifera* and *E. moorcroftii* revealed a clearly syntenic depth ratios of 1:4, indicating two WGD events occurred after the split of the two species (Figure 3b). Furthermore, the genomes of *E. moorcroftii* and *E. mollis* present highly conservative synteny, and consistent syntenic depth ratios were found for *E. mollis* and *H. rhamnoides* (Figure 3b). Thus, the integrated evidence showed that two successive WGD events might have occurred in the common ancestor of *E. moorcroftii*, *E. mollis*, and *H. rhamnoides*.

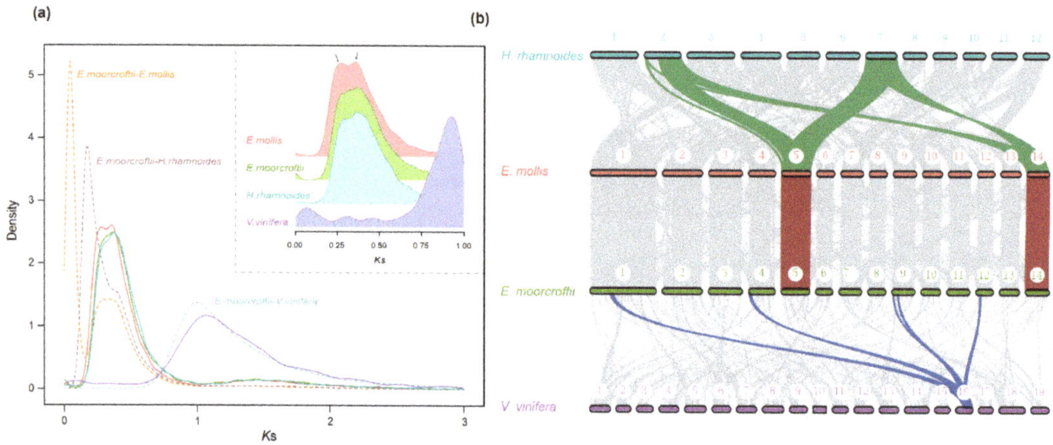

Figure 3. Integrating Ks and synteny analyses reveal two successive WGD events in *E. moorcroftii*. (**a**) The distributions of K_S of paralogous gene pairs of *E. mollis* (pink line), *E. moorcroftii* (green line), *H. rhamnoides* (blue line), and *V. vinifera* (purple line). The dashed lines represent K_S distributions related to the species divergence of the corresponding species pairs. (**b**) Macro-syntenic comparisons among three Elaeagnaceae species and *V. vinifera*.

4. Discussion

4.1. A High-Quality Dryland Tree Species Genome

Land degradation and desertification is an ongoing environmental problem that threatens the sustainable development of human beings, and this is particularly serious for lives in drylands [1–3]. A native or indigenous species naturally growing in drylands, to some extent, might hold the key to solve this problem. Recently, a growing number of studies have focused on decoding the genomic information of typical arid plants, for example, *Populus euphratica* [54], *H. rhamnoides* [4,5], *H. ammodendron* [7], etc. To our knowledge, besides the well-known desert poplar *P.euphratica*, our genome assembly for *E. moorcroftii* represents the second genome for high-tree species naturally growing in arid northwest China. Meanwhile, the chromosome-level genome of *E. moorcroftii* is of high quality, with contig N50 up to approximately 30 Mb (Table 1), which is much higher than that of its two closest relatives (*H. rhamnoides* with N50 of 3.56 Mb in [4] and 2.15 Mb in [5], and *E. mollis* with N50 of 18.40 Mb) of Elaeagnaceae with published genomes.

4.2. Differential Evolutionary Dynamics of Gene Families between E. Moorcroftii and E. Mollis

Our phylogenomic and molecular dating analyses showed that the divergence time between *E. moorcroftii* and its close relative *E. mollis* was about 9.38 Ma. After the split from their most recent common ancestor (MRCA), these two species experienced independent gene family evolutionary dynamics, which can be observed by the contrasted expansion and contraction trend (Figure 2). The number of contracted gene families is more than twice the expanded gene families in *E. moorcroftii*, while this trend is opposite in *E. mollis*, which have more expanded than contracted gene families (Figure 2). A detailed analysis suggested that these differential evolutionary dynamics of gene families *between E. moorcroftii* and *E. mollis* might partly result from contribution levels of different modes of gene duplications (Figure S2). Nevertheless, the opposite evolutionary trend of gene families might reflect their unique biology of these two species.

4.3. The Potential Evolutionary Significance of two Successive WGD Events in E. Moorcroftii

For Elaeagnaceae plants, a recent genome study of *H. rhamnoides* revealed it experienced a recent whole-genome duplication after the divergence between *H. rhamnoides* and the most closely related species jujube, with a signature *Ks* peak at c. 0.38 [4]. However, almost simultaneously, another study of *H. rhamnoides* genome indicated that it experienced two rounds of WGDs, one recent WGD (*Ks* peak at ~0.38) and one older WGD (*Ks* peak at ~0.45) [5]. In this study, our integrated *Ks* and synteny analyses provide strong evidence for the occurrence of two successive WGD events in *E. moorcroftii* genome (similar with the report by [5]), in addition, these two events possibly being shared by *E. moorcroftii* and its two close relatives (*E. mollis* and *H. rhamnoides*). It has been well-acknowledged that the frequent occurrence of WGD events has played an important role on the plant's adaptive evolution and diversification [55–59]. For example, the co-retained duplicated genes, after multiple independent WGDs in different angiosperms lineages were selected by environmental stresses during the Cretaceous–Paleocene (K-Pg) mass extinction, possibly enhanced the plant's survival and adaptation [59]. Furthermore, it was demonstrated that not only did the MRCA of angiosperms experience an ancient WGD event, but the MRCA of all extant Gymnosperms also shared an ancient WGD event, as revealed by a recent report of the *Cycas* genome, which may have contributed to seed-related morphology innovation [60]. Considering the evolutionary significance of the recognized WGD, it should be postulated that these two successive WGD events might have contributed to the evolution and adaptation of *E. moorcroftii*, although it needs to be further investigated.

5. Conclusions

As an indigenous species naturally growing in arid Northwest China, *E. moorcroftii* presents ideal tree species with great economic and ecological values that can be widely cultured in drylands for economic development and ecological restoration. In this study, the high-quality genome sequence of *E. moorcroftii* decoded here represents a small but essential step forward to understanding the evolution and adaptation of this species in drylands. Our results not only provided new insights into the genomic evolution of Elaeagnaceae species but also revealed clear evidence for two successive WGD events that occurred in the common ancestor of *E. moorcroftii*, *E. mollis*, and *H. rhamnoides*. Further, by using the *E. moorcroftii* genome as reference, more studies should be encouraged to investigate the unique biology of this species. For example, how could this species respond to long-term drought and salt stresses? Moreover, whole-genome resequencing analysis of samples with certain properties (such as resistance to abiotic factors) is helpful to screen for better germplasm resources and breed new varieties. In total, the genomic resource of *E. moorcroftii* paves ways for the further genetic improvement and widespread cultivation of this indigenous species in drylands.

Supplementary Materials: The following supporting information can be downloaded at: https://www.mdpi.com/article/10.3390/d14060468/s1, Figure S1: The distribution frequency of PacBio's HiFi sequencing data; Figure S2: The different modes of gene duplications related to expanded and contracted gene families in *E. moorcroftii* and *E. mollis*; Figure S3: The enriched GO terms for the expanded gene families of *E. moorcroftii*; Figure S4: The enriched GO terms for the contracted gene families of *E. moorcroftii*; Figure S5: Venn diagram showing the shared and specific gene families among *E. moorcroftii*, *E. mollis*, *H. rhamnoides*, and *Z. jujube*; Figure S6: The enriched GO terms for the unique gene families of *E. moorcroftii*; Figure S7: Syntenic dot plot of the self-comparison of *E. moorcroftii*; Table S1: Information of plant genomes used in this study; Table S2: Basic statistics of PacBio's HiFi sequencing data; Table S3: Mapping rates of RNA-seq reads onto the assembly genome; Table S4: Genome assembly completeness evaluated based on BUSCO; Table S5: Information of different classes of repetitive sequences in *E. moorcroftii* genome; Table S6: Functional annotation of the predicted genes in *E. moorcroftii*; Table S7: *E. moorcroftii* genome annotation completeness evaluated based on BUSCO; Table S8: List of non-coding RNA genes in *E. moorcroftii* genome.

Author Contributions: Conceptualization, X.F., S.W. and K.S.; methodology, J.W., X.M. and K.L.; software, J.W. and X.M.; validation, X.F., S.W. and K.S.; formal analysis, J.W. and X.F.; investigation, X.F.; resources, S.W.; data curation, J.W.; writing—original draft preparation, X.F.; writing—review and editing, S.W., K.S. and H.Z.; visualization, X.F. and J.W.; supervision, K.S.; project administration, X.F.; funding acquisition, X.F. and K.S. All authors have read and agreed to the published version of the manuscript.

Funding: This research was partly funded by the Shenzhen Key Laboratory of Southern Subtropical Plant Diversity, Fairy Lake Botanical Garden (Grant No. SSTLAB-2022-01 to X.F.), 2022 Key Talent Project of Gansu Province (to K.S.), and the Fundamental Research Funds for the Central Universities (lzujbky-2021-49) (to S.W.).

Institutional Review Board Statement: Not applicable.

Data Availability Statement: All raw sequencing data and the assembled genome for this project were deposited in the National Genomics Data Center (NGDC; https://ngdc.cncb.ac.cn/ (accessed on 29 July 2021)) under accession number CRA007134.

Acknowledgments: We would like to thank the staff in Minqin Desert Botanical Garden for their help with the plant materials' collection.

Conflicts of Interest: The authors declare no conflict of interest.

References

1. Reynolds, J.F.; Smith, D.M.S.; Lambin, E.F.; Turner, B.L., II; Mortimore, M.; Batterbury, S.P.J.; Downing, T.E.; Dowlatabadi, H.; Fernández, R.J.; Herrick, J.E.; et al. Global desertification: Building a science for dryland development. *Science* **2007**, *316*, 847–851. [CrossRef] [PubMed]
2. Huang, J.; Yu, H.; Guan, X.; Wang, G.; Guo, R. Accelerated dryland expansion under climate change. *Nat. Clim. Change* **2016**, *6*, 166–172. [CrossRef]
3. Huang, J.; Zhang, G.; Zhang, Y.; Guan, X.; Wei, Y.; Guo, R. Global desertification vulnerability to climate change and human activities. *Land Degrad. Dev.* **2020**, *31*, 1380–1391. [CrossRef]
4. Wu, Z.; Chen, H.; Pan, Y.; Feng, H.; Fang, D.; Yang, J.; Wang, Y.; Yang, J.; Sahu, S.K.; Liu, J.; et al. The genome of *Hippophae rhamnoides* provides insights into a conserved molecular mechanism in actinorhizal and rhizobial symbiosis. *New Phytol.* **2022**, *235*, 276–291. [CrossRef]
5. Yu, L.; Diao, S.; Zhang, G.; Yu, J.; Zhang, T.; Luo, H.; Duan, A.; Wang, J.; He, C.; Zhang, J. Genome sequence and population genomics provide insights into chromosomal evolution and phytochemical innovation of *Hippophae rhamnoides*. *Plant Biotechnol. J.* **2022**. [CrossRef]
6. Wang, T.; Ren, L.; Li, C.; Zhang, D.; Zhang, X.; Zhou, G.; Gao, D.; Chen, R.; Chen, Y.; Wang, Z.; et al. The genome of a wild *Medicago* species provides insights into the tolerant mechanisms of legume forage to environmental stress. *BMC Biol.* **2021**, *19*, 96. [CrossRef]
7. Wang, M.; Zhang, L.; Tong, S.; Jiang, D.; Fu, Z. Chromosome-level genome assembly of a xerophytic plant, *Haloxylon ammodendron*. *DNA Res.* **2022**, *29*, dsac006. [CrossRef]
8. Ma, W.; Wang, P.; Huang, J.; Zhang, D.; Yang, W.; Pan, B.; Shi, W. The taxonomic relationship between *Elaeagnus moorcroftii* and *E. angustifolia* (Elaeagnaceae) based on morphological similarities and simple-sequence repeat markers. *Nord. J. Bot.* **2021**, *39*, e03011. [CrossRef]

9. Kucho, K.; Yamanaka, T.; Sasakawa, H.; Mansour, S.R.; Uchiumi, T. Different dynamics of genome content shuffling among host-specificity groups of the symbiotic actinobacterium Frankia. *BMC Genom.* **2014**, *15*, 609. [CrossRef]
10. Liang, B.B.; Wang, W.J.; Fan, X.X.; Kurakov, A.V.; Liu, Y.F.; Song, F.Q.; Chang, W. Arbuscular mycorrhizal fungi can ameliorate salt stress in *Elaeagnus angustifolia* by improving leaf photosynthetic function and ultrastructure. *Plant Biol.* **2021**, *23*, 232–241. [CrossRef]
11. Nazir, N.; Zahoor, M.; Nisar, M. A review on traditional uses and pharmacological importance of genus *Elaeagnus* species. *Bot. Rev.* **2020**, *86*, 247–280. [CrossRef]
12. Wang, Y.; Zhao, P.; Wang, Y.L.; Zhang, Y. Nutritional composition of wild *Elaeagnus angustifolia* fruits. *J. Gansu Agric. Univ.* **2006**, *41*, 130–132.
13. Ren, B.; Ru, D.; Chen, L.; Duan, N.; Li, Y.; Shi, J.; Cao, J.; Liu, B. Genome Sequence of *Elaeagnus mollis*, the first chromosome-level genome of the family Elaeagnaceae. *Genome Biol. Evol.* **2021**, *13*, evab266. [CrossRef] [PubMed]
14. Allen, G.C.; Flores-Vergara, M.A.; Krasynanski, S.; Kumar, S.; Thompson, W.F. A modified protocol for rapid DNA isolation from plant tissues using cetyltrimethylammonium bromide. *Nat. Protoc.* **2006**, *1*, 2320–2325. [CrossRef] [PubMed]
15. Cheng, H.; Concepcion, G.T.; Feng, X.; Zhang, H.; Li, H. Haplotype-resolved de novo assembly using phased assembly graphs with hifiasm. *Nat. Methods* **2021**, *18*, 170–175. [CrossRef]
16. Durand, N.C.; Shamim, M.S.; Machol, I.; Rao, S.S.; Huntley, M.H.; Lander, E.S.; Aiden, E.L. Juicer provides a one-click system for analyzing loop-resolution Hi-C experiments. *Cell Syst.* **2016**, *3*, 95–98. [CrossRef]
17. Dudchenko, O.; Batra, S.S.; Omer, A.D.; Nyquist, S.K.; Hoeger, M.; Durand, N.C.; Shamim, M.S.; Machol, I.; Lander, E.S.; Aiden, A.P.; et al. De novo assembly of the *Aedes aegypti* genome using Hi-C yields chromosome-length scaffolds. *Science* **2017**, *356*, 92–95. [CrossRef]
18. Simao, F.A.; Waterhouse, R.M.; Ioannidis, P.; Kriventseva, E.V.; Zdobnov, E.M. BUSCO: Assessing genome assembly and annotation completeness with singlecopy orthologs. *Bioinformatics* **2015**, *31*, 3210–3212. [CrossRef]
19. Kim, D.; Langmead, B.; Salzberg, S.L. HISAT: A fast spliced aligner with low memory requirements. *Nat. Methods* **2015**, *12*, 357–360. [CrossRef]
20. Wang, X.; Wang, L. GMATA: An integrated software package for genome-scale SSR mining, marker development and viewing. *Front Plant Sci.* **2016**, *7*, 1350. [CrossRef]
21. Gary, B. Tandem repeats finder: A program to analyze DNA sequences. *Nucleic Acids Res.* **1999**, *2*, 573–580.
22. Tarailo-Graovac, M.; Chen, N. Using RepeatMasker to identify repetitive elements in genomic sequences. *Curr. Protoc. Bioinformatics* **2009**, *25*, 4.10.1–4.10.14. [CrossRef] [PubMed]
23. Bao, W.; Kojima, K.K.; Kohany, O. Repbase update, a database of repetitive elements in eukaryotic genomes. *Mob. DNA* **2015**, *6*, 11. [CrossRef] [PubMed]
24. Han, Y.; Wessler, S.R. MITE-Hunter: A program for discovering miniature inverted-repeat transposable elements from genomic sequences. *Nucleic Acids Res.* **2010**, *38*, e199. [CrossRef] [PubMed]
25. Flynn, J.M.; Hubley, R.; Goubert, C.; Rosen, J.; Clark, A.G.; Feschotte, C.; Smit, A.F. REPEATMODELER2 for automated genomic discovery of transposable element families. *Proc. Natl. Acad. Sci. USA* **2020**, *117*, 9451–9457. [CrossRef]
26. Keilwagen, J.; Wenk, M.; Erickson, J.L.; Schattat, M.H.; Grau, J.; Hartung, F. Using intron position conservation for homology-based gene prediction. *Nucleic Acids Res.* **2016**, *44*, e89. [CrossRef]
27. Huang, J.; Zhang, C.; Zhao, X.; Fei, Z.; Wan, K.; Zhang, Z.; Pang, X.; Yin, X.; Bai, Y.; Sun, X. The Jujube genome provides insights into genome evolution and the domestication of sweetness/acidity taste in fruit trees. *PLoS Genet.* **2016**, *12*, e1006433. [CrossRef]
28. Kaul, S.; Koo, H.L.; Jenkins, J.; Rizzo, M.; Rooney, T.; Tallon, L.J.; Feldblyum, T.; Nierman, W.; Benito, M.-I.; Lin, X.; et al. Analysis of the genome sequence of the flowering plant *Arabidopsis thaliana*. *Nature* **2000**, *408*, 796–815.
29. Jaillon, O.; Aury, J.M.; Noel, B.; Policriti, A.; Clepet, C.; Casagrande, A.; Choisne, N.; Aubourg, S.; Vitulo, N.; Jubin, C.; et al. The grapevine genome sequence suggests ancestral hexaploidization in major angiosperm phyla. *Nature* **2007**, *449*, 463–467.
30. Stanke, M.; Keller, O.; Gunduz, I.; Hayes, A.; Waack, S.; Morgenstern, B. AUGUSTUS: Ab *initio* prediction of alternative transcripts. *Nucleic Acids Res.* **2006**, *34*, W435–W439. [CrossRef]
31. Dobin, A.; Davis, C.A.; Schlesinger, F.; Drenkow, J.; Zaleski, C.; Jha, S.; Batut, P.; Chaisson, M.; Gingeras, T.R. STAR: Ultrafast universal RNA-seq aligner. *Bioinformatics* **2013**, *29*, 15–21. [CrossRef] [PubMed]
32. Kovaka, S.; Zimin, A.V.; Pertea, G.M.; Razaghi, R.; Salzberg, S.L.; Pertea, M. Transcriptome assembly from long-read RNA-seq alignments with StringTie2. *Genome Biol.* **2019**, *20*, 1–13. [CrossRef] [PubMed]
33. Campbell, M.A.; Haas, B.J.; Hamilton, J.P.; Mount, S.M.; Buell, C.R. Comprehensive analysis of alternative splicing in rice and comparative analyses with *Arabidopsis*. *BMC Genom.* **2006**, *7*, 327. [CrossRef] [PubMed]
34. Haas, B.J.; Salzberg, S.L.; Zhu, W.; Pertea, M.; Allen, J.E.; Orvis, J.; White, O.; Buell, C.R.; Wortman, J.R. Automated eukaryotic gene structure annotation using EVidenceModeler and the program to assemble spliced alignments. *Genome Biol.* **2008**, *9*, R7. [CrossRef]
35. Jones, P.; Binns, D.; Chang, H.Y.; Fraser, M.; Li, W.; McAnulla, C.; McWilliam, H.; Maslen, J.; Mitchell, A.; Nuka, G.; et al. InterProScan 5: Genome-scale protein function classification. *Bioinformatics* **2014**, *30*, 1236–1240. [CrossRef]
36. Conesa, A.; Götz, S.; García-Gómez, J.M.; Terol, J.; Talón, M.; Robles, M. Blast2GO: A universal tool for annotation, visualization and analysis in functional genomics research. *Bioinformatics* **2005**, *21*, 3674–3676. [CrossRef]

37. Kanehisa, M.; Sato, Y.; Kawashima, M.; Furumichi, M.; Tanabe, M. KEGG as a reference resource for gene and protein annotation. *Nucleic Acids Res.* **2016**, *44*, D457–D462. [CrossRef]
38. Lowe, T.M.; Eddy, S.R. tRNAscan-SE: A program for improved detection of transfer RNA genes in genomic sequence. *Nucleic Acids Res.* **1997**, *25*, 955–964. [CrossRef]
39. Nawrocki, E.P.; Eddy, S.R. Infernal 1.1: 100-fold faster RNA homology searches. *Bioinformatics* **2013**, *29*, 2933–2935. [CrossRef]
40. Griffiths-Jones, S.; Bateman, A.; Marshall, M.; Khanna, A.; Eddy, S.R. RFAM: An RNA family database. *Nucleic Acids Res.* **2003**, *31*, 439–441. [CrossRef]
41. Kozomara, A.; Griffiths-Jones, S. miRBase: Annotating high confidence microRNAs using deep sequencing data. *Nucleic Acids Res.* **2014**, *42*, D68–D73. [CrossRef] [PubMed]
42. Lagesen, K.; Hallin, P.; Rødland, E.A.; Staerfeldt, H.H.; Rognes, T.; Ussery, D.W. RNAmmer: Consistent and rapid annotation of ribosomal RNA genes. *Nucleic Acids Res.* **2007**, *35*, 3100–3108. [CrossRef] [PubMed]
43. Emms, D.M.; Kelly, S. OrthoFinder: Phylogenetic orthology inference for comparative genomics. *Genome Biol.* **2019**, *20*, 1–14. [CrossRef]
44. Katoh, K.; Standley, D.M. MAFFT multiple sequence alignment software v.7: Improvements in performance and usability. *Mol. Biol. Evol.* **2013**, *30*, 772–780. [CrossRef] [PubMed]
45. Suyama, M.; Torrents, D.; Bork, P. PAL2NAL: Robust conversion of protein sequence alignments into the corresponding codon alignments. *Nucleic Acids Res.* **2006**, *34*, W609–W612. [CrossRef] [PubMed]
46. Nguyen, L.T.; Schmidt, H.A.; von Haeseler, A.; Minh, B.Q. IQ-TREE: A fast and effective stochastic algorithm for estimating maximum-likelihood phylogenies. *Mol. Biol. Evol.* **2015**, *32*, 268–274. [CrossRef]
47. Yang, Z. PAML 4: Phylogenetic analysis by maximum likelihood. *Mol. Biol. Evol.* **2007**, *24*, 1586–1591. [CrossRef]
48. Shi, C.; Wang, S.; Cai, H.H.; Zhang, H.R.; Long, X.X.; Tihelka, E.; Song, W.; Feng, Q.; Jiang, R.; Cai, C.; et al. Fire-prone Rhamnaceae with South African affinities in Cretaceous Myanmar amber. *Nat. Plants* **2022**, *8*, 125–135. [CrossRef]
49. De Bie, T.; Cristianini, N.; Demuth, J.P.; Hahn, M.W. CAFE: A computational tool for the study of gene family evolution. *Bioinformatics* **2006**, *22*, 1269–1271. [CrossRef]
50. Qiao, X.; Li, Q.; Yin, H.; Qi, K.; Li, L.; Wang, R.; Zhang, S.; Paterson, A.H. Gene duplication and evolution in recurring polyploidization-diploidization cycles in plants. *Genome Biol.* **2019**, *20*, 38. [CrossRef]
51. Wang, Y.; Tang, H.; Debarry, J.D.; Tan, X.; Li, J.; Wang, X.; Lee, T.H.; Jin, H.; Marler, B.; Guo, H.; et al. MCScanX: A toolkit for detection and evolutionary analysis of gene synteny and collinearity. *Nucleic Acids Res.* **2012**, *40*, e49. [CrossRef] [PubMed]
52. Nei, M.; Gojobori, T. Simple methods for estimating the numbers of synonymous and nonsynonymous nucleotide substitutions. *Mol. Biol. Evol.* **1986**, *3*, 418.
53. Jiao, Y.; Leebens-Mack, J.; Ayyampalayam, S.; Bowers, J.E.; McKain, M.R.; McNeal, J.; Rolf, M.; Ruzicka, D.R.; Wafula, E.; Wickett, N.J.; et al. A genome triplication associated with early diversification of the core eudicots. *Genome Biol.* **2012**, *13*, R3. [CrossRef] [PubMed]
54. Ma, T.; Zhou, G.; Yue, Z.; Hu, Q.; Chen, Y.; Liu, B.; Qiu, Q.; Wang, Z.; Zhang, J.; Wang, K.; et al. Genomic insights into salt adaptation in a desert poplar. *Nat. Commun.* **2014**, *4*, 2797. [CrossRef]
55. Jiao, Y.; Wickett, N.J.; Ayyampalayam, S.; Chanderbali, A.S.; Landherr, L.; Ralph, P.E.; Tomsho, L.P.; Hu, Y.; Liang, H.; Soltis, P.S.; et al. Ancestral polyploidy in seed plants and angiosperms. *Nature* **2011**, *473*, 97–100. [CrossRef]
56. Wang, J.; Sun, P.; Li, Y.; Liu, Y.; Yang, N.; Yu, J.; Ma, X.; Sun, S.; Xia, R.; Liu, X.; et al. An overlooked paleotetraploidization in Cucurbitaceae. *Mol. Biol. Evol.* **2018**, *35*, 16–26. [CrossRef]
57. Soltis, P.S.; Soltis, D.E. Ancient WGD events as drivers of key innovations in angiosperms. *Curr. Opin. Plant Biol.* **2016**, *30*, 159–165. [CrossRef]
58. Van de Peer, Y.; Mizrachi, E.; Marchal, K. The evolutionary significance of polyploidy. *Nat. Rev. Genet.* **2017**, *18*, 411–424. [CrossRef]
59. Wu, S.; Han, B.; Jiao, Y. Genetic Contribution of Paleopolyploidy to Adaptive Evolution in Angiosperms. *Mol. Plant* **2020**, *13*, 59–71. [CrossRef]
60. Liu, Y.; Wang, S.; Li, L.; Yang, T.; Dong, S.; Wei, T.; Wu, S.; Liu, Y.; Gong, Y.; Feng, X.; et al. The *Cycas* genome and the early evolution of seed plants. *Nat. Plants* **2022**, *8*, 389–401. [CrossRef]

diversity

MDPI

Article

Phylogeography and Population History of *Eleutharrhena macrocarpa* (Tiliacoreae, Menispermaceae) in Southeast Asia's Most Northerly Rainforests

Shijie Song [1,2], Jianyong Shen [1], Shishun Zhou [1], Xianming Guo [3], Jinchao Zhao [4], Xinghui Shi [5], Zhiyong Yu [6], Qiangbang Gong [7], Shaohua You [8] and Sven Landrein [1,*]

1 Xishuangbanna Tropical Botanical Garden, Chinese Academy of Sciences, Xishuangbanna 666303, China; songshijie@xtbg.ac.cn (S.S.); shenjianyong@xtbg.ac.cn (J.S.); zss@xtbg.org.cn (S.Z.)
2 College of life sciences, University of Chinese Academy of Sciences, Beijing 101408, China
3 Research Institute of Xishuangbanna National Nature Reserve, Xishuangbanna 666100, China; xianmingguo56@gmail.com
4 Management Bureau of Nangunhe National Nature Reserve, Lincang 677400, China; jinchaozhao91@gmail.com
5 Pingbian Branch of Management and Protection Bureau of Daweishan National Nature Reserve, Honghe Hani and Yi Autonomous Prefecture 661299, China; xinghuishi26@gmail.com
6 Management Bureau of Fenshuiling National Nature Reserve, Honghe Hani and Yi Autonomous Prefecture 661500, China; zhiyongyu41@gmail.com
7 Tongbiguan Provincial Nature Reserve, Dehong Dai and Jingpo Autonomous Prefecture 678400, China; qiangbanggong@gmail.com
8 Xishuangbanana National Nature Reserve, Xishuangbanna 666100, China; youshaohua266@gmail.com
* Correspondence: sven@xtbg.ac.cn

Abstract: The diversification of Tiliacoreae and the speciation of *Eleutharrhena* are closely linked to Southeast Asia's most northerly rainforests which originate from the Himalayan uplift. Migration routes across biogeographical zones within the Asian clade, including those of *Eleutharrhena*, *Pycnarrhena*, and *Macrococculus*, and their population structures are still unexplored. We combine endocarp morphology, phylogenetic analyses, divergence time estimation, ancestral area reconstruction, as well as SCoT method to reconstruct the past diversification of *Eleutharrhena macrocarpa* and to understand their current distribution, rarity, and evolutionary distinctiveness. The disjunct, monospecific, and geographically restricted genera *Eleutharrhena* and *Macrococculus* both have a dry aril, a unique feature in Menispermaceae endocarps that further confirms their close relationship. *Pycnarrhena* and *Eleutharrhena* appeared during the end of the Oligocene c. 23.10 million years ago (Mya) in Indochina. *Eleutharrhena* speciation may be linked to climate change during this time, when humid forests became restricted to the northern range due to the Himalayan uplift. Differentiation across the Thai–Burmese range could have contributed to the isolation of the Dehong populations during the Miocene c. 15.88 Mya, when exchange between India and continental Asia ceased. Dispersal to the Lanping–Simao block and further differentiation in southeastern and southern Yunnan occurred during the Miocene, c. 6.82 Mya. The specific habitat requirements that led to the biogeographic patterns observed in *E. macrocarpa* contributed to a low genetic diversity overall. Population 1 from Dehong, 16 from Pu'er, and 20 from Honghe on the East of the Hua line have a higher genetic diversity and differentiation; therefore, we suggest that their conservation be prioritized.

Keywords: aril; conservation genetics; *Eleutharrhena macrocarpa*; evolutionary distinctiveness; Himalayan uplift; Menispermaceae; southern Yunnan rainforests; Tiliacoreae

Citation: Song, S.; Shen, J.; Zhou, S.; Guo, X.; Zhao, J.; Shi, X.; Yu, Z.; Gong, Q.; You, S.; Landrein, S. Phylogeography and Population History of *Eleutharrhena macrocarpa* (Tiliacoreae, Menispermaceae) in Southeast Asia's Most Northerly Rainforests. *Diversity* **2022**, *14*, 437. https://doi.org/10.3390/d14060437

Academic Editor: Juan J. Morrone

Received: 3 February 2022
Accepted: 24 May 2022
Published: 30 May 2022

1. Introduction

Eleutharrhena macrocarpa (Diels) Forman is a woody liana belonging to the Tiliacoreae tribe in the Menispermaceae or moonseed family. It is primarily distributed across China (southern Yunnan), Laos, Myanmar, and India [1] (Figure 1). Like many species of the moon-seed family, members of the Tiliacoreae are typically dioecious lianas inhabiting tropical rainforests and monsoon forests in South America, Africa, Asia, and Oceania [2]. Tiliacoreae Miers sensu Ortiz [2] includes 16 genera and 111 species, among which 6 genera are monospecific, including *Eleutharrhena* [2–4]. Although the sister relationships within the Tiliacoreae are weakly supported, the monophyly of the tribe is well established (24 species, 10 genera, and 3 plastid regions were used in analyses by Ortiz, Wang, Jacques and Chen [2], whereas 26 species, 11 genera, and 7 plastid plus 2 nuclear regions were used in the analyses by Lian, et al. [5]). The tribe is morphologically characterized by male flowers with more than four whorls of sepals, longitudinally grooved seed endocarps that are ribbed or rugose abaxially, hippocrepiform seeds without endosperm, and subcylindrical embryos [2]. Seeds play an important part in the classification of Menispermaceae and are highly diagnostic [6]. Fruits of Menispermaceae have woody endocarps and the seeds are often curved around an invagination of the lateral sides, termed the condyle [7]. The origin of the condyle is described by Ortiz [7] as resulting from the development of the ovary wall. Two main types of condyle were identified. Only the *Menispermum* condyle type occurs in Tiliacoreae, which have deeply curved seeds resulting from a bilaterally compressed, laminiform condyle.

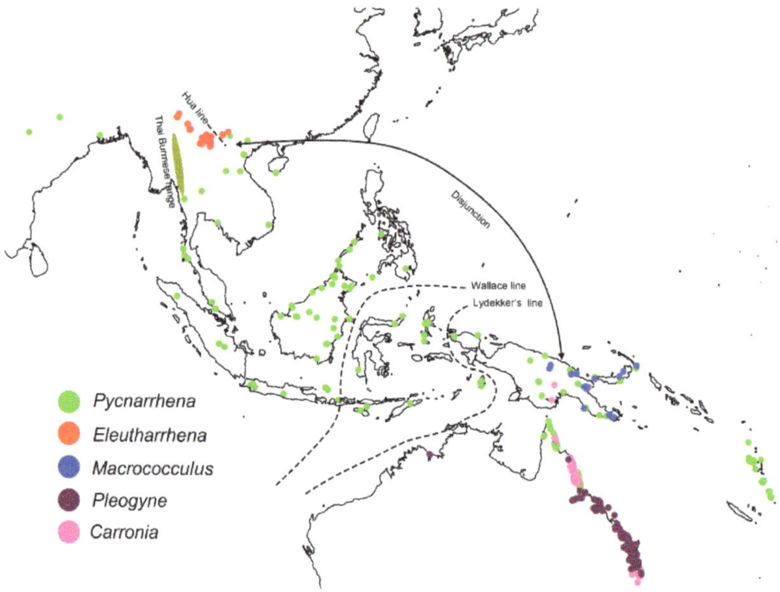

Figure 1. Distribution map of *Eleutharrhena* and sister genera *Macrococculus*, *Pycnarrhena*, *Pleogyne*, and *Carronia*. Distribution records downloaded from GBIF except for *Eleutharrhena macrocarpa* records.

Eleutharrhena is a sister to *Pycnarrhena* [5] and is morphologically similar to *Macrococculus* [8]. Their ranges overlap, but *E. macrocarpa* morphologically differs by its free stamens, stipitate drupelets, and grouped stomata on the abaxial leaf surface. It can become a large woody vine reaching the canopy and produces very large drupelets, although these are not as large as those of *Macrococculus*, which are said to be dispersed by flightless cassowary birds (*Casuarius*) [9]. *Eleutharrhena macrocarpa* was assessed to be a critically endangered species [10] and a plant species with extremely small populations (PSESP) in China [11]. Hou et al. [12] reported approximately 40 plants surviving in Yunnan, China, and Lang et al. [1] estimated that 60 individuals occurred in Yunnan and we sampled 48 individuals in our study (Figure 2).

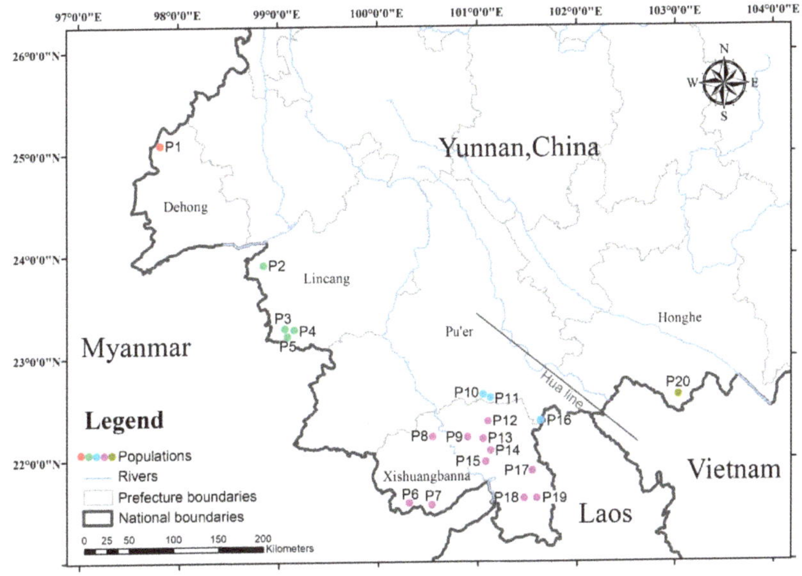

Figure 2. Distribution map of *Eleutharrhena macrocarpa* populations used for genetic diversity and structure analysis.

The origin of Menispermaceae has been evaluated to be c. 109 Mya in the Indo-Malayan region and it was suggested that the radiation of the family is linked to the simultaneous appearance of modern neotropical and paleotropical forests around the world following the mass extinction at the Cretaceous–Paleogene boundary [13]. The tropical rainforests in southern China may have appeared during the late Tertiary accompanied by the uplift of the Himalayan mountains and the development of a monsoon climate [14]. The structure of these forests is similar to that of SE Asian lowland rainforests, but they have a deciduous tree layer, fewer megaphanerophytes and epiphytes, and more abundant lianas [14].

Several distinct floristic regions and biogeographical areas can be recognized in tropical southern China: Taiwan, Hainan, part of Guangxi, southeastern Yunnan, and southern Yunnan [14]. The movement of the Indochina geoblock and the Lanping–Simao block together with the Himalayan uplift has profoundly influenced the flora of southern Yunnan [15]. The floras of southern and southeastern Yunnan have a high proportion of tropical Asian elements, but the southern part is more closely related to the Indo-Malaysian flora, suggesting that a boundary exists between these two parts [15]. *Eleutharrhena* is a typical Indo-Malaysian element and its distribution may closely correspond to the biogeographical events that led to the formation of southern Yunnan flora. Several lines have been proposed to explain floral boundaries in Yunnan; the Hua line [15] between southern and southeastern Yunnan floras; the Tanaka line [16] which was used to describe the difference between the Sino Himalayan forests to the west and the Sino Japanese forests to the east; and the Salween–Mekong divide, separating the east Himalayan from the Hengduan mountains during glacial periods [17]. However, the influence of these different lines and the period of occurrence may not all apply to the tropical flora of southern Yunnan or may only be valid for a part of it.

Although Menispermaceae are closely linked to the expansion of tropical rainforests and their speciation could be linked to interglacial periods as well as distinct biogeographic events, few studies have reported genetic structures at the species level. Examples include the distribution of *Sinomenium acutum* (Thumb.) Rehd. et Wils. in subtropical China using chloroplast marker haplotypes [18], the genetic structure of *Stephania yunnanensis* H.S.Lo using DALP [19], and the link between *Chasmanthera dependens* Hochst. climatic refugia and genetic diversity in West Africa using AFLP and plastid markers [20].

In this study, we combine phylogenetics, time-measured phylogenies, ancestral area reconstruction, and population genetics based on the Start Codon Targeted (SCoT) polymorphism methods to answer questions about the phylogeography and the population history of *Eleutharrhena macrocarpa*. Specifically, we investigate how the genetic diversity is linked to the diversification of the southern Yunnan flora, the Himalayan mountain uplift, interglacial periods, and the occurrence of distinct population partitions. Furthermore, we focus on the seed morphology and its evolution within Tiliacoreae. Finally, we provide data on the populations' genetic structure that can be used to prioritize populations for conservation.

2. Materials and Methods

2.1. Morphological Observations and Analysis

One fruiting specimen of *Eleutharrhena macrocarpa* was selected and fruits were boiled for three hours and then left to soak overnight. Endocarps were then dissected and observed under a stereomicroscope (Nikon corporation). For comparison with other seeds having a strongly curved embryo, a conspicuous condyle and intrusive tissue such as the raphe, we also dissected the fresh fruits of *Haematocarpus validus* (Miers) Bakh.f. ex Forman (Pachygoneae). *Haematocarpus* is often misidentified as *E. macrocarpa* in the field but markedly differs in its seed morphology [5]. The specimens were collected in Menglun forest reserve in Xishuangbanna, Yunnan. For description and terminology, we followed Wefferling, Hoot and Neves [6]. Endocarp characteristics were mapped on the Bayesian phylogenetic tree (see 2.3) using Mesquite v. 3.70 (https://www.mesquiteproject.org/ accessed on 5 May 2022) to observe the changes in character states. A maximum likelihood (ML) analysis was performed to determine the ancestral character states. Trees were visualized using the ball and stick method, with pie charts at each node indicating the proportional likelihood of ancestral characteristics. A heuristic search with Maxtrees set to 100 and a majority rule consensus tree based on 17 morphological characters [3] (Table A1) was also performed for all the genera in Tiliacoreae using Mesquite v. 3.70.

2.2. Sampling, DNA Extraction, PCR, and Electrophoresis

Thirty-three genetic sequences and one outgroup [21] from Tiliacoreae were downloaded from the National Center for Biotechnology Information (NCBI) for the phylogenetic study. Seven plastid regions (*matK, ndhF, rbcL, atpB, trnL-F, trnH-psbA*, and *rps16*) and two nuclear regions (*26S rDNA* and *ITS*) were selected according to several recently published Menispermaceae phylogenetic studies [2,5,22] (see Table A2). One additional DNA sample collected in Xishuangbanna was extracted and its whole chloroplast was sequenced using NGS, aligned, assembled, and annotated by us (GenBank: MZ502223, detailed methods not shown here). The regions of interest were then extracted using Geneious v. 8.1.3 and added to our matrix. The sequences were aligned using the Muscle Algorithm [23] and the resulting sequences were edited with Geneious v. 8.1.3.

Silica-dried leaves of 48 individuals were collected in China for the SCoT analysis. Each population (from Dehong to Honghe) was more than 4 km apart. DNA was extracted using the Geneon Plant Genomic DNA Kit (Geneon, Inc. Changchun, China) and five geographically distant samples were selected to test 36 universal SCoT primer pairs [24]. PCR amplifications were conducted in a 50 μL volume, including 45 μL Tsingke master mix, 4 μL primers (10 μM), and 1 μL DNA extract. A C1000 Touch thermocycler (Bio-Rad Laboratories, Inc. China) was used to perform amplifications with the following program: 95 °C for 5 min; 35 cycles at 98 °C for 30 s, 52 °C for 30 s, 72 °C for 1 min, and 72 °C for 5 min, and finally held at 4 °C. An ABI-3730xl Sequencer (Applied Biosystems, Ltd. UK) was used to perform the capillary electrophoresis using 5′-labeled fluorescent primers. Electrophoresis results were transferred to a 0–1 matrix for further analysis.

2.3. Phylogenetic Analysis, Dating, and Ancestral Area Reconstruction

Phylogenetic trees were inferred with maximum likelihood (ML) and Bayesian inference (BI). ML analyses were performed using RAxML-HPC v. 8 on XSEDE [25] using the CIPRES Science Gateway v. 3.3 portal [26] based on GTR-CAT model with 1000 replications. The Bayesian analysis was implemented with MrBayes v. 3.2.3 [27] in PhyloSuite v. 1.2.2 platform [28] specifying the DNA substitution model selected by partitionFinder2 [29]. We performed four runs using 10 million generations with four chains, sampling trees every 1000 generations. The first 20% of trees were discarded and a 50% majority rule consensus tree was reconstructed from the remaining post-burn-in trees. The average standard deviation of the split frequency was used to verify that all runs reached stationarity and converged on the same distribution. To estimate the divergence time in Menispermaceae and Tiliacoreae, we ran a Bayesian relaxed molecular clock analysis with the plastid and nuclear combined dataset in BEAST v. 2.6.6 [30]. Our partitions were set in BEAUTi part of the BEAST package, and one fossil calibration point was inserted following Lian et al. [21] (*Triclisia inflata* 17.7 Mya can be assigned to extant *Triclisia* [31]). The parameters were the relaxed clock with a lognormal distribution and the Yule model. The maximum age of the Tiliacoreae root was set to 49.3 Mya [13] with a normal distribution and standard deviation of three. One run with 100 million generations and sampling every 5000 generations was conducted in BEAST. The first 10% of trees were discarded as burn-in. We used Tracer v. 1.7 [32] to assess the convergence and effective sample size. Tree Annotator v. 1.8.0 (part of the BEAST package) was used to summarize the set of post-burn-in trees and their parameters. FigTree v. 1.4.2 (http://tree.bio.ed.ac.uk/software/figtree/ accessed on 1 February 2021) was used to visualize the tree and divergence times. Tiliacoreae clade (2), including *Eleutharrhena* and its sister genera, was extracted from the maximum clade credibility tree inferred with BEAST, and input in BioGeoBEARS [33] implemented in RStudio v. 1.4.1717 [34]. Four major geographical ranges were designed: Indochina, southern Yunnan, Sundaland including Wallacea and Australasia including New Guinea. Six models were tested, namely DEC, DEC + J, DIVALIKE, DIVALIKE + J, BAYAREALIKE, and BAYAREALIKE + J [33], and the most fitting model was selected by calculating the best value for the Likelihood Ratio Test (LRT). The resulting probabilities of the ancestral states were drawn as pie charts at the node on the provided tree.

2.4. Population Genetic Analysis

POPGENE v. 3.2 [35] was used to analyze the observed number of alleles (Na), the effective number of alleles (Ne), Shannon's information index (I), the percentage of polymorphic loci (PPL), and Nei's genetic diversity index (H). Genetic distance was visualized and interpreted using a PCoA analysis in GenAIEx v. 6.1 [36]. STRUCTURE v. 2.3.4 [37] was used to infer population structure following the Falush, et al. [38] input method, with a burn-in period of 100,000 iterations, 1,000,000 Markov Chain Monte Carlo (MCMC) repetitions, 1–5 K ranges, and 10 independent runs. Evanno's method [39] was used to determine the best number of subpopulations in the Structure Harvester v. 0.6.94 [40]. CLUMPP v. 1.1.2 [41] was used to align clusters and Distruct v. 1.1 [42] was used to visualize the results. Populations including more than one individual were used to infer genetic differentiation using phiPT (Φpt), a measure that allows intra-individual variation to be suppressed and is therefore ideal for comparing codominant binary data [43,44] and gene flow (Nm = $[(1/\Phi pt) - 1]$). An analysis of molecular variance (AMOVA) was performed among populations and within populations. Mantel's test was performed to determine the relationship between genetic and geographic distances in GenAIEx v. 6.51.

3. Results

3.1. Endocarps

The druplets of both *Eleutharrhena macrocarpa* and *Haematocarpus validus* are very similar in terms of size, number, and the stipitate base. The exocarps of *H. validus* (Figure 3) differ by their deep red and fleshy cells, as well as the position of the style scar, which is basal and not subapical as in *E. macrocarpa* (Figure 3). Endocarps have a deep dorsal longitudinal groove as well as several lateral grooves in *H. validus*, whereas only a ventral ridge and faint transversal ridge can be seen on the lateral side of *E. macrocarpa* endocarps. Two large extruded funicle apertures can be observed at the base of the ventral part in *E. macrocarpa*. The endocarp of *H. validus* is clearly deeply curved with a pronounced bilaterally compressed condyle, which is absent in *E. macrocarpa*, and the raphe is clearly visible and intrusive inside the condyle in *H. validus*. Both genera have seeds without endosperm and fleshy cotyledons, although *E. macrocarpa* cotyledons are unequal. *E. macrocarpa*'s endocarps are more similar to *Pycnarrhena*'s but markedly differ by the presence of two basal and ventral funicle apertures and the presence of a well-developed bilobed aril surrounding the cotyledons (Figure 3). We could not obtain the fruit of *Macrococculus*, but Forman [8] observed that the seed is covered with a reticulate membrane as well as two basal and ventral funicle apertures and is thus very similar to *Eleutharrhena*.

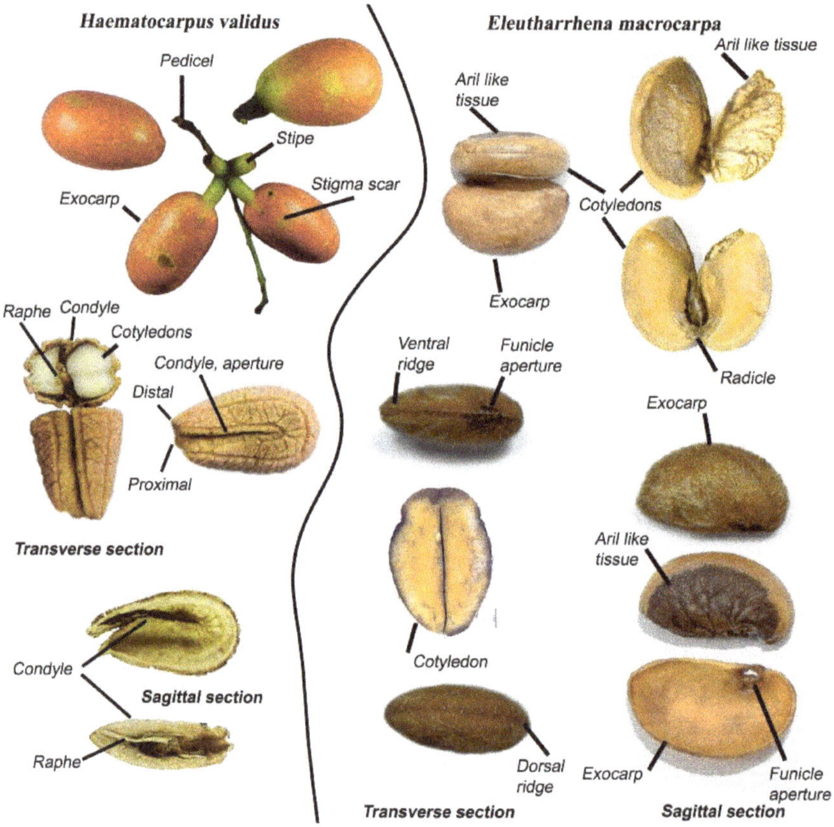

Figure 3. Fruits morphology of *Haematocarpus validus* and *Eleutharrhena macrocarpa* (sagittal and transverse sections).

The node at the base of the Tiliacoreae indicates a 99% maximum likelihood (ML) that the common ancestor had endocarps with deeply curved embryo and a conspicuous condyle. The node at the base of clade (2) had a 95% ML for a deeply curved embryo with intrusive condyle. Additionally, the node at the base of *Pycnarrhena* and *Eleutharrhena* had a 45% ML for weakly curved embryo and weak condyle and a 43% ML for a straight embryo with an intrusive aril (Figure 4). Based on 17 morphological characters (Table A1), the heuristic search produced an incongruent tree with the molecular data (Figure A1). However, *Macrococculus* and *Eleutharrhena* had a 98% support for their sister relationship.

Figure 4. Endocarp characteristics mapped on the Bayesian phylogenetic tree of Tiliacoreae using mesquite. All nodes have more than 0.9 posterior probabilities (PP) support except when noted with *, meaning less than 0.8 PP. Endocarp illustrations: A. *Pycnarrhena pleniflora*; B. *Pycnarrhena ozantha*; C. *Eleutharrhena macrocarpa*; D. *Pleogyne cunninghamii*; E. *Carronia thyrsiflora*; F. *Tiliacora triandra*; G. *Albertisia scandens*; H. *Anisocycla jollyana*; I. *Beirnaertia cabindensis*; J. *Syrrheonema fasciculatum*; K. *Curarea* sp.; L. *Chondrodendron platyphyllum*; M. *Sciadotenia pubistaminea*. D and J were not sampled and added to the analysis because only a few DNA regions were available and their branches had no support.

3.2. Phylogenetic Analyses

The matrix was 10657 bp-long, including 469 parsimony-informative characters. The ML and Bayesian analyses of the plastid and nuclear-combined dataset were mostly congruent (see Figure A2). Three groups within Tiliacoreae can be identified: (1) the neotropical genera *Curarea, Sciadotenia,* and *Chondrodendron* (BS = 100%, PP = 1); (2) the Asian and Australasian genera *Carronia, Eleutharrhena,* and *Pycnarrhena* (BS = 95%, PP = 0.99); and (3) the remaining paleotropical and Australasian genera *Tiliacora, Albertisia, Anisocycla, Triclisia,* and *Beirnaertia* (BS = 80%, PP = 0.99).

3.3. Dating Analyses

The maximum clade credibility tree of the plastid and nuclear combined alignment using the Yule model is displayed in Figure 5. The neotropical clade (1) crown speciated during the Oligocene c. 31.19 Mya (95% highest posterior density interval (HPD): 23.13–39.73 Mya). The mostly African clade (3) diverged from clade (2) during the Eocene c. 40.06 Mya (95% HPD: 33.28–46.67 Mya). The differentiation between *Pycnarrhena* and *Eleutharrhena* occurred during the Oligocene c. 23.10 Mya (95% HPD: 17.45–29.57 Mya), whereas one individual of *Eleutharrhena macrocarpa* from Dehong diverged much earlier than the other individuals during the Miocene c. 15.88 Mya (95% HPD: 10.43–22.01 Mya). *Carronia* diverged from *Eleutharrhena* and *Pycnarrhena* during the Eocene c. 35.87 Mya (95% HPD: 27.93–43.57 Mya).

3.4. Ancestral Area Reconstruction of *Eleutharrhena* and *Pycnarrhena*

BAYAREALIKE + J model (Bayesian approach with a large number of areas under ML) had higher statistical support (see Table A3). The ancestral range of *Eleutharrhena* and *Pycnarrhena* was Indochina c. 23.10 Mya. The ancestral range of *Eleutharrhena* was Indochina c. 15.88 Mya and it dispersed to southern Yunnan c. 6.82 Mya (Figure 5). The ancestral area of *Pycnarrhena* was Indochina with a dispersal to Australasia c. 15.51 Mya and finally Sundaland c. 2.16 Mya.

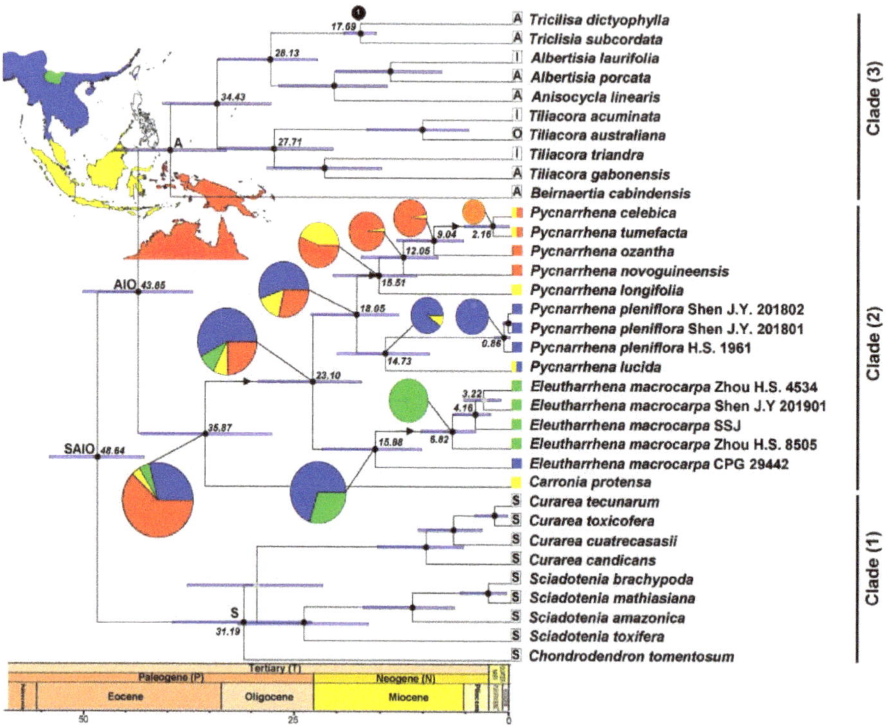

Figure 5. Chronogram of tribe Tiliacoreae resulting from the analysis of the plastid and nuclear combined dataset using the Yule model in BEAST 2. Nodes with black circles have posterior probabilities (PPs) between 0.9 and 1; grey circles have a PP between 0.6 and 0.8. Blue bars correspond to the 95% highest posterior density intervals (HPD) for the corresponding node. Fossils: 1. *Triclisia inflata* 17.7 Mya. Ancestral range reconstruction of tribe Tiliacoreae using the package BioGeoBEARS in R. The best model for clade (2) was BAYAREALIKE + J with 4 areas max, d = 0.003, e = 0, j = 0.074 and LnL = −21.97. Arrows indicate a dispersal event. Areas for *Eleutharrhena* and *Pycnarrhena*—red: Australasia; blue: Indochina; yellow: Sundaland; green: southern Yunnan; orange: Australasia + Sundaland. Areas for the remaining genera in Tiliacoreae: S: South America; A: Australasia; I: Indochina; O: Australasia.

3.5. Population Genetic Diversity

SCoT primers 7, 8, and 23 were the most polymorphic, producing unambiguous clear bands and were subsequently selected for amplification and analysis. Forty-eight individuals generated 536 bands in total. Genetic diversity indices for populations including more than one individual are presented in Table 1. Nei's genetic diversity index (H) ranged from 0.228 to 0.3120, and Shannon's information index (I) ranged from 0.335 to 0.4715. Populations 20, 1 and 16, located in Honghe, Dehong and Pu'er, respectively, had the highest genetic diversity (H = 0.3120, I = 0.4715; H = 0.3062, I = 0.4567 and H = 0.297, I = 0.4506, respectively). The mean genetic diversity value (H = 0.2705, I = 0.4106) is low, the gene flow (N_m = 0.6983) is medium according to the standard used in [45] and the coefficient of gene differentiation G_{ST} (among populations) = 0.4173. All these values show a high population differentiation [46]. The genetic diversity was H = 0.169 and I = 0.2873 for all 48 individuals.

Table 1. Genetic diversity of *Eleutharrhena macrocarpa* populations.

Populations	Individuals	Na	Ne	H	I	PPL
P1	3	1.8217	1.5187	0.3062	0.4567	82.2%
P4	2	1.5539	1.3917	0.2294	0.335	55.4%
P5	2	1.6201	1.4385	0.2568	0.375	62.0%
P7	3	1.8267	1.4623	0.2866	0.4363	82.7%
P9	7	1.9349	1.4148	0.2612	0.4115	93.5%
P13	6	1.9487	1.4140	0.2618	0.4144	94.9%
P16	3	1.8468	1.483	0.297	0.4506	84.7%
P18	7	1.8968	1.3546	0.228	0.3663	89.7%
P19	2	1.6425	1.4543	0.2661	0.3885	64.3%
P20	3	1.8779	1.512	0.312	0.4715	87.8%

Na = observed number of alleles; Ne = effective number of alleles; H = Nei's gene diversity; I = Shannon's information index; PPL = the percentage of polymorphic loci.

3.6. Cluster Analysis and Population Structure

The principal coordinate analysis (PCoA) including all samples shows that three subpopulations can be observed (Figure 6a). Subpopulation 1 has two individuals from Dehong. Individuals from Lincang are all in subpopulation 2. Subpopulations 2 and 3 are mixed between Xishuangbanna, Dehong, Honghe, Lincang, and Pu'er individuals. PCoA results indicate that subpopulations 2 and 3 are distinct, although some individuals from Xishuangbanna belong to subpopulation 2 or 3. Most individuals from southeastern Yunnan, eastern Xishuangbanna and Pu'er are grouped in subpopulation 3 and most individuals from western southern Yunnan and western Xishuangbanna are in subpopulation 2. No natural geographical barriers exist between subpopulations 2 and 3 except for the Lancang river. Structure analyses and Evanno's methods also indicated the best K = 3, and Honghe individuals have a similar admixture to that of the Xishuangbanna, Pu'er, and Lincang populations, whereas two individuals from subpopulation 1 in Dehong have more private alleles, as also reflected in the plastid phylogenetic analysis (Figure 6b).

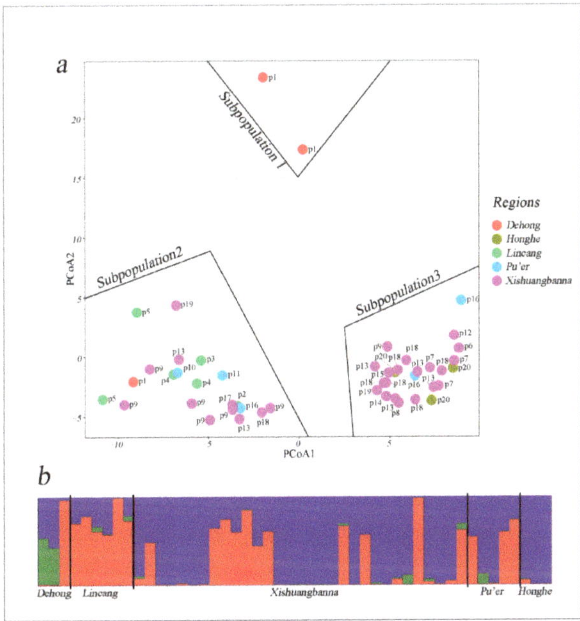

Figure 6. (**a**) Principal coordinates analysis of *Eleutharrhena macrocarpa* based on the genetic similarity matrix of SCoT markers; and (**b**) STRUCTURE analysis revealed that the best K is 3.

3.7. *Analysis of Molecular Variance and Mantel's Test*

The AMOVA results show that most of the genetic variation occurred within populations (83%) whereas only 17% of the variance occurred among populations (see Table A4). The Mantel test shows no correlation between the geographical and genetic distance (R^2 = 0.2299), (see Figure A3).

4. Discussion

4.1. *Endocarp Morphology*

Tiliacoreae is characterized by male flowers with more than four whorls of sepals, longitudinally grooved seed endocarp that are ribbed or abaxially rugose, hippocrepiform seeds without endosperm and subcylindrical embryos [2]. A lack of morphological data, especially for the flowers of both sexes and the fruit has resulted in the patchy morphological data analyses of Menispermaceae tribes [6]. Using the morphological characteristics listed in the synopsis of the family [3], very few morphological characters could be used to produce a tree congruent with the phylogenetic analyses based on seven plastid and two nuclear regions (Figure A1). However, hippocrepiform seeds without endosperm and a conspicuous condyle seem to be a shared primitive characteristic of the tribe, except within *Eleutharrhena*, *Pycnarrhena*, and *Albertisia* (Figure 4). In *Eleutharrhena* (as well as *Macrococculus*, not sampled in our phylogenetic analysis), the condyle has completely disappeared, and instead, an aril-like tissue is present (Figure 3). We hypothesize that the bilobed aril is derived from the aborted ovule and could represent the dry cotyledons that now enclose the remaining fertile ovule [47]. This character may also be linked to the intrusion of organs within the condyle, as shown by the raphe in *Haematocarpus validus* (Figure 3) and does not seem to have an ecological function. Further apomorphies include the perforate condyle in *Syrrheonema*, the straight endocarp and reduced condyle in *Albertisia* as well as the ruminate endosperm in *Tiliacora*. The later ruminate endosperm is said to be of secondary origin, as some *Tiliacora* species have no endosperm.

4.2. *Tiliacoreae and Eleutharrhena Biogeography*

Our phylogenetic results based on seven plastid and two nuclear regions are similar to previously obtained phylogenies [2,5] and confirm three clades within the tribe. The distribution of the tribe Tiliacoreae includes South America, Africa, Asia, and Australasia and resembles groups that have experienced western Gondwana vicariance, c. 105 Mya [48].

Pycnarrhena and *Eleutharrhena* appeared during the Oligocene c. 23.10 Mya in Indochina. During the late Eocene and Oligocene, the climate for much of Southeast Asia was seasonal and humid forests became isolated [49]. We can infer that this climatic change may have favored the speciation of *Eleutharrhena* in its northern range due to the Himalayan uplift and the apparition of a monsoon climate. *Pycnarrhena* dispersal to Australasia and Sundaland followed a distinctive pattern that can also be observed in *Tiliacora*—firstly that of a dispersal from Indochina to Australasia during the Miocene c. 15.51 Mya and then back to Sundaland c. 2.16 Mya. The main change at the beginning of the Miocene is the change in Sundaland from a seasonally wet to humid climate. This corresponds to the closure of the Indonesian throughflow, and the Australian plate collision with Southeast Asia [49]. Dispersal from Asia to Australia may have occurred when the sea level is known to have been lower, allowing for some migrations to happen across the Wallace line [50,51]. The Wallace line is a well-known biogeographical barrier between New Guinea/Australia and Sundaland [52], a deep trench that separated both floras and which could have separated earlier lineages. The disjunction of *Eleutharrhena* + *Macrococculus* (based on morphological data only) across the Wallace line may be explained by the narrow habitat requirements of these genera and is linked to the distribution of humid rainforests.

Eleutharrhena macrocarpa samples from Xishuangbanna diverged very early from the individuals from Dehong during the Miocene c. 15.88 Mya (Figure 5). Differentiation across the Thai–Burmese border range (Tenasserim Hills, Dawna Range, and Karen Hills) was proposed to explain the diversity in several bird and mammal species [53]. Molecular data [54] suggest that exchanges between the Indian subcontinent and mainland Asia peaked from 44 Mya in the Eocene to the mid Miocene, and then decreased after 14 Mya due to the drier conditions developing in northern India. It is therefore possible that the Dehong population may represent a separate refugium across the Thai–Burmese range that was previously connected to northern India and later became separated.

4.3. *Eleutharrhena Macrocarpa Population Structure*

A relatively high number of bands was obtained from the SCoT analysis, although between 53712 and 78921 unigenes were detected in *Menispermum canadense* and *M. dauricum* [55]. Our total bands therefore represent between 0.7% and 1% of the total number of unigenes, which may be due to the use of fluorescent primers and capillary electrophoresis.

Menispermaceae species are known to be outbreeding species which should result in reduced population differentiation. *Sinomenium acutum* populations were shown to have a high gene flow and low population differentiation, although refugia populations were identified by their higher genetic diversity (H_T = 0.828, H_S = 0.710, $N_{ST} < G_{ST}$, AMOVA within populations 83.62%, among populations 16.38% N_m = 2.552) [18]. *Tinospora cordifolia* had relatively low genetic diversity but this study was only based on five individuals (H = 0.2114) [56]. *Menispermum canadense* and *M. dauricum* were found to have a relatively high inbreeding coefficient and low genetic diversity (inbreeding coefficient of 0.198, H_o = 0.377, and H_e = 0.342 and no significant deviation from the Hardy–Weinberg equilibrium) [55]. The genetic diversity of *Chasmanthera dependens* showed opposite results from the cpDNA haplotypes and AFLP results (high population differentiation and weak geographical correlation for cpDNA, with F_{ST} = 0.797 and low population differentiation and strong geographical correlation for AFLP, with F_{ST} = 0.064) [20]. *Stephania yunnanensis* showed high population differentiation and relatively low gene flow (N_a = 1.6192; N_e = 1.4001; H = 0.2298; I = 0.43401) [19]. The view that most Menispermaceae species are outbreeding species with weak population differentiation contrasts with these previous and recent studies and could also show that the differentiation may be due to refugia, the founder effect, geographical isolation, and selection pressure [57].

Eleutharrhena macrocarpa shows low genetic diversity but high population differentiation and medium gene flow. The low level of genetic diversity and medium gene flow among populations but high population differentiation might be due to mixed forces involving (1) small population sizes and specific habitat requirements along with (2) sympatric barriers such as mountain ranges or rivers and (3) pollinator specificity [58]. Fragmentation and small population sizes could explain the relatively low differentiation of the populations from Honghe at the eastern limit of the range across the Hua line. The Hua line [15] was used to describe the different flora composition between southern and southeastern Yunnan. It has been suggested that this could correspond to the movement of the Indochina geoblock as well as the Lanping–Simao block together with the Himalayan uplift [15]. However, this line does not seem to have strongly affected the dispersal of *E. macrocarpa* to southeastern China.

Finally, *Eleutharrhena* has highly specific habitat requirements [59] which may be reflected by the higher population differentiation despite its low genetic diversity.

4.4. *Evolutionary Distinctiveness*

Our results suggest low genetic diversity and high population differentiation in *E. macrocarpa* and indicate that populations p1, p16, and p20 should be prioritized for conservation because of their higher genetic diversity and differentiation. The conservation value can be assessed using several methods and priorities such as threats, economic benefits, environment quality, and phylogenetic distinctiveness [60]. Our results show that *Eleutharrhena* has a high phylogenetic distinctiveness value because it is morphologically related to *Macrococculus* both having an unusual aril-like structure surrounding the seed. It has a high biogeographic value, as shown by the disjunction between these two genera as well as the isolation of the Dehong population. This is the first time that we have documented differentiation across the Thai–Burmese border range for any plant species in southern Yunnan. *Eleutharrhena macrocarpa* is a relict plant of the mighty rainforests that has colonized and persisted in the northern range of the Southeast Asian biodiversity hotspot.

5. Conclusions

Eleutharrhena and *Macrococculus* are morphologically similar and *Eleutharrhena* is a sister to *Pycnarrhena*. *Eleutharrhena* may have evolved during the Himalayan uplift during the Miocene c. 15.88 Mya and its speciation may be linked to changes in climate during this time, when humid forests became restricted to the northern range. Dispersal to the Lanping–Simao block and further differentiation between southeastern and southern Yunnan occurred during the Miocene c. 6.82 Mya. The conservation of species with rich evolutionary histories and that are linked to major biogeographic events such as the Himalayan uplift should be prioritized.

Further studies involving the sampling of *Macrococculus*, the remaining genera within tribe Tiliacoreae, and the use of codominant genetic markers would improve our understanding of such complex biogeographical patterns.

Author Contributions: Conceptualization, S.S. and S.L.; methodology, S.S. and S.L.; software, S.S. and S.L.; validation, S.S. and S.L.; formal analysis, S.S. and S.L.; investigation, S.S., J.S., S.Z., X.G., J.Z., X.S., Z.Y., Q.G. and S.Y.; data curation, S.S. and S.L.; writing—original draft preparation, S.S.; writing—review and editing, S.L.; visualization, S.S. and S.L.; supervision, S.L.; project administration, S.L.; funding acquisition, S.L. All authors have read and agreed to the published version of the manuscript.

Funding: This research was funded by the Anthony Hitchcock Species Recovery Fund, grant number GBGF 2020/44717 and Center for Gardening and Horticulture, Xishuangbanna Tropical Botanical Garden, Chinese Academy of Sciences, grant number PPKP014B06.

Institutional Review Board Statement: Not applicable for studies not involving humans or animals.

Informed Consent Statement: Not applicable.

Data Availability Statement: Not applicable.

Acknowledgments: The authors thank Jo Osborne for manuscript review and editing. Thanks to Bulong Nature Reserve, Yiwu Nature Reserve and Nanpenghe Nature Reserve as well as the Forest Resources Management Station of Jinuo Mountain for providing help in conducting the surveys.

Conflicts of Interest: The authors declare no conflict of interest.

Appendix A

Table A1. Morphological characters and character states used in this study.

No.	Morphological Characters	State 1	State 2	State 3	State 4	State 5
1	Number of male sepals	6	9	12	>15	-
2	Number of male petals	0	3	6	-	-
3	Number of stamens	3	6	9	>12	-
4	Stamen fusion	Free	Basally fused	Fused	-	-
5	Anthers dehiscence	Transverse/oblique	Longitudinal	-	-	-
6	Connective	Absent	Present	-	-	-
7	Number of staminodes	0	3	6	-	-
8	Number of carpels	3	6	9	>12	-
9	Condyle/seed	Absent/embryo straight or slightly curved	Absent/aril present, embryo straight	Weak/embryo slightly curved	double lameliform/embryo strongly curved	Perforate/embryo strongly curved
10	Endosperm	Absent	Present	-	-	-
11	Cauliflory	Yes	No	-	-	-

Table A1. *Cont.*

No.	Morphological Characters	State 1	State 2	State 3	State 4	State 5
12	Pseudocorolla	Yes	No	-	-	-
13	Clustered stomata	Yes	No	-	-	-
14	Petals in female flower	Yes	No	-	-	-
15	Inflorescence	Cymose/fasciculate	Capituliform	Flowers single	-	-
16	Glands on petals	Yes	No	-	-	-
17	Coroniform sepals	Yes	No	-	-	-

Table A2. DNA sequences used for maximum likelihood and Bayesian phylogenetic analyses (new sequences are labeled with an asterisk *.

Taxon, voucher, locality, *rbcL*, *atpB*, *matK*, *ndhF*, *trnL-F*, *trnH-psbA*, *rps16*, *26S rDNA*, and *ITS*.
Albertisia laurifolia, Hong Y.P. 99371 (PE), Hainan, FJ626590, HQ260813, EF143849, JN051700, MW633412, MW633442, MW621979, EF143841; ***Albertisia porcata***, McPherson 16678 (MO), Gabon, HQ260758, HQ260814, KX384047, EF624261, KX384110, MW633413, MW633443, MW621980, -; ***Anisocycla linearis***, Hong-Wa et al. 466 (MO), Madagascar, HQ260759, HQ260816, JN051805, EF624263, JN051739, MW633414, MW633444, MW621981, MW621214; ***Beirnaertia cabindensis***, Walters and Niangadouma 1267 (MO), Gabon, HQ260766, HQ260822, JN051811, EF624270, JN051745, MW633415, MW633445, MW621982, MW621215; ***Carronia protensa***, van der Werff and Gray 17049 (MO), Australia, HQ260769, HQ260825, JN051815, EF624274, JN051749, MW633416, MW633446, MW621983, MW621216; ***Chondrodendron tomentosum***, Ortiz et Vásquez 217 (AMAZ, MO), Peru, HQ260771, HQ260828, JN051818, EF624278, JN051752, MW633417, MW633447, KM364844, -; ***Curarea candicans***, Torke 310 (MO), Guyana, HQ260776, HQ260832, JN051824, EF624288, JN051758, MW633418, MW633448, -, MW621217; ***Curarea cuatrecasasii***, Ortiz and Aguilar 324 (INB, MO), Costa Rica, MW633353, MW633367, KX384061, EF624289, KX384124, MW633419, MW633449, -, MW621218; ***Curarea tecunarum***, Ortiz and Vásquez 214 (AMAZ, MO), Peru, MW633354, MW633368, KX384062, EF624290, KX384125, MW633420, MW633450, -, MW621219; **Curarea toxicofera**, Ortiz 184 (AMAZ, MO), Peru, FJ026480, FJ026420, KX384063, EF624291, KX384126, -, -, -, -; ***Eleutharrhena macrocarpa* CPG29442**, CPG29442 (PE), Yunnan, MW633358, MW633371, MW633383, MW633394, MW633404, MW633424, MW633454, MW621986, MW621223; ***Eleutharrhena macrocarpa* Shen J.Y. 201901**, Shen J.Y. 201901 (HITBC), Yunnan, MW633357, MW633370, MW633382, MW633393, MW633403, MW633423, MW633453, MW621985, MW621222; ***Eleutharrhena macrocarpa* SSJ**, SSJ (HITBC), Yunnan, MZ502223 *, MZ502223 *, MZ502223 *, MZ502223 *, MZ502223 *, MZ502223 *, MZ502223 *, -, -; ***Eleutharrhena macrocarpa* Zhou H.S. 4534**, Zhou H.S. 4534 (HITBC, PE), Yunnan, MW633355, MW633369, MW633380, MW633391, MW633401, MW633421, MW633451, -, MW621220; ***Eleutharrhena macrocarpa* Zhou H.S. 8505**, Zhou H.S. 8505 (HITBC, PE), Yunnan, MW633356, -, MW633381, MW633392, MW633402, MW633422, MW633452, MW621984, MW621221; ***Limacia blumei***, F. Jacques 07 (P), cult. West Java, JN051683, JN051879, JN051836, EF624309, JN051770, -, -, -, -; ***Pycnarrhena celebica***, F. Jacques 014 (P), cult. Bogor, FJ026503, FJ026443, JN051846, -, -, -, -, -, -; ***Pycnarrhena longiflora***, F. Jacques 015 (P), cult. Bogor, EU526993, EU526965, JN051845, JN051716, JN051780, MW633425, MW633455, MW621987, MW621224; ***Pycnarrhena lucida***, A.F.G. Kerr 17887 (P), Thailand, MW633359, MW633372, MW633384, MW633395, MW633405, -, MW633456, MW621988, MW621225; ***Pycnarrhena novoguineensis***, Gray 8794 (MO), Australia, HQ260795, HQ260851, JN051847, EF624326, JN051782, MW633426, MW633457, MW621989, MW621226; ***Pycnarrhena ozantha***, T. Haevermans 418 (P), New Caledonia, MW633360, MW633373, MW633385, MW633396, MW633406, -, MW633458, MW621990, MW621227; ***Pycnarrhena pleniflora* H.S. 1961**, Zhou H.S. 1961 (PE), cult. Xishuangbanna, MW633361, MW633374, MW633386, MW633397, MW633407, MW633427, MW633459, MW621991, MW621228; ***Pycnarrhena pleniflora* Shen J.Y. 201801**, Shen J.Y. 201801(HITBC), Yunnan, MW633362, MW633375, MW633387, MW633398, MW633408, MW633428, MW633460, MW621992, MW621229;

Table A2. *Cont.*

Pycnarrhena pleniflora* Shen J.Y. 201802**, Shen J.Y. 201802 (HITBC), Yunnan, MW633363, MW633376, MW633388, MW633399, MW633409, MW633429, MW633461, MW621993, MW621230; ***Pycnarrhena tumefacta, CPG33074 (PE), Bogor, Indonesia, MW633365, MW633378, MW633390, MW633400, MW633411, MW633432, MW633464, MW621995, MW621232; ***Sciadotenia amazonica***, Ortiz and Zárate 264 (AMAZ, MO), Peru, HQ260799, HQ260855, JN051851, EF624330, JN051786, MW633433, MW633465, MW621996, MW621233; ***Sciadotenia brachypoda***, Ortiz and Farroñay 222 (AMAZ, MO), Peru, -, -, -, EF624331, -, -, -, -, -; ***Sciadotenia mathiasiana***, Ortiz and al. 259 (AMAZ, MO), Peru, MW633366, MW633379, KX384078, EF624332, KX384142, MW633434, MW633466, -, MW621234; ***Sciadotenia toxifera***, Ortiz and al. 231 (AMAZ, MO), Peru, HQ260800, HQ260856, KX384079, EF624333, KX384143, MW633435, MW633467, -, MW621235; ***Tiliacora acuminata***, F. Jacques 11 (P), cult. West Java, JN051696, JN051895, JN051861, JN051730, JN051796, MW633436, MW633468, MW621997, MW621236; ***Tiliacora australiana***, Gray 9132 (QRS), Australia, JN051697, JN051896, JN051862, JN051731, JN051797, MW633437, MW633469, MW621998, MW621237; ***Tiliacora funifera***, D. Kenfack 2100 (MO), Ghana, FJ026512, FJ026452, JN051863, EF624340, JN051798, MW633438, MW633470, KM364880, MK288774; ***Tiliacora gabonensis***, Walters and Niangadouma 1159 (MO), Gabon, JN051698, -, JN051864, EF624341, JN051799, -, MW633471, MW621999, MW621238; ***Triclisia dictyophylla***, Kenfack and Zapfack 2038 (MO), Cameroon, HQ260810, HQ260866, JN051866, EF624344, JN051801, MW633439, MW633472, MW622000, MW621239; ***Triclisia subcordata***, Kenfack2101 (MO), Ghana, HQ260811, HQ260867, JN051867, EF624345, JN051802, MW633440, MW633473, MW622001, MW621240.

Table A3. Results from BioGeoBEARS models testing.

Model	Number of Parameters	LnL	d	e	j	AIC	AICc	AIC Weight
DEC	2	−25.14	0.007	0	0	54.29	55.29	0.074
DEC + J	3	−23.91	0.005	0	0.042	53.83	56.01	0.052
DIVALIKE	2	−24.29	0.010	0	0	52.59	53.59	0.174
DIVALIKE + J	3	−24.18	0.008	0	0.02	54.36	56.54	0.040
BAYAREALIKE	2	−23.76	0.004	0.04	0	51.53	52.53	0.296
BAYAREALIKE + J	3	−21.97	0.003	0	0.074	49.95	52.13	0.362

AIC and AICc comparisons of different models of biogeographical range evolution and estimates. AIC, Akaike Information Criterion; d, dispersal; e, extinction; j, weight of jump dispersal/founder speciation; LnL, log-likelihood of the model.

Table A4. AMOVA results among populations and within populations.

Source	df	SS	MS	Est. Var.	%
Among populations	9	852.278	94.698	10.873	17%
Within populations	28	1528.143	54.577	54.577	83%
Total	37	2380.421		65.449	100%

Df = degree of freedom; SS = sum of squares; MS = mean squares; Est. Var. = estimate of variance; % = percentage of total variation.

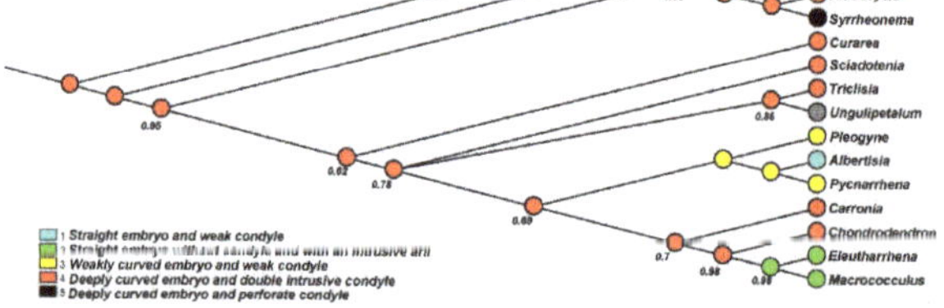

Figure A1. Heuristic search and majority rule consensus of Tiliacoreae based on 17 morphological characters using Mesquite. Endocarp ancestral character states mapped and visualized using the ball and stick method. Consensus tree supports 1 except when stated.

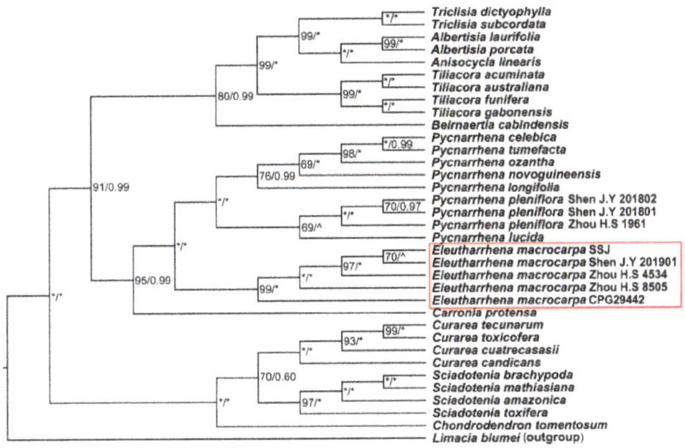

Figure A2. Maximum likelihood and Bayesian phylogenetic tree of the tribe Tiliacoreae based on the combination of *rbcL, atpB, matK, ndhF, trnL-F, trnH-psbA, rps16, 26S rDNA,* and *ITS* regions. * indicates BS = 100% or PP = 1.0; ^ indicates incongruency between ML and Bayesian phylogenetic trees.

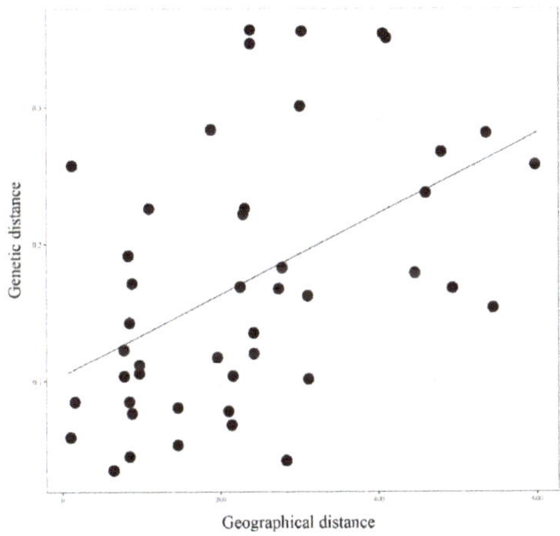

Figure A3. Mantel's test between genetic distance and geographical distance correlation. Equation: $Y = 0.0003\,x + 0.1029$, $R^2 = 0.2299$, $p = 0.01$.

References

1. Lang, X.D.; Li, Z.H.; Liu, W.D.; Xu, C.H.; Su, J.R.; Li, S.F.; Wang, F.Z.; Huang, X.B. A new study report on the endangered plant *Eleutharrhena macrocarpa* (Diels) Forman, with extremely small populations. *Acta Bot. Boreali-Occident. Sin.* **2017**, *37*, 203–210.
2. Ortiz, R.D.; Wang, W.; Jacques, F.M.B.; Chen, Z.D. Phylogeny and a revised tribal classification of *Menispermaceae* (moonseed family) based on molecular and morphological data. *Taxon* **2016**, *65*, 1288–1312. [CrossRef]
3. Kessler, P. Menispermaceae. In *The Families and Genera of Vascular Plants II. Flowering Plants: Dicotyledons*; Kubitzki, K., Rohwer, J.G., Bittrich, V., Eds.; Springer: Berlin, Germany, 1993; pp. 402–418.
4. Stevens, P. *The families and Genera of Vascular Plants. Vol. 2. Flowering Plants· Dicotyledons: Magnoliid, Hamamelid and Caryophyllid Families*; Springer: New York, NY, USA, 1994; pp. 1–663.
5. Lian, L.; Shen, J.Y.; Ortiz, R.D.; Yu, S.X.; Chen, Z.D.; Wang, W. Phylogenetic analyses and modelling distributions guide conservation of a critically endangered liana species, *Eleutharrhena macrocarpa* (Menispermaceae). *Taxon* **2021**, *70*, 931–945. [CrossRef]
6. Wefferling, K.M.; Hoot, S.B.; Neves, S.S. Phylogeny and fruit evolution in Menispermaceae. *Am. J. Bot.* **2013**, *100*, 883–905. [CrossRef]
7. Ortiz, R.D. Seed diversity in Menispermaceae: Developmental anatomy and insights into the origin of the condyle. *Int. J. Plant Sci.* **2012**, *173*, 344–364. [CrossRef]

8. Forman, L. The Menispermaceae of Malesia and adjacent areas: VI: *Pycnarrhena, Macrococculus & Haematocarpus*. *Kew Bull.* **1972**, *26*, 405–422.
9. Forman, L.L. Menispermaceae. In *Flora Malesiana*; Van Steenis, C.G.G.J., de Wilde, W.J.J.O., Eds.; Martinus Nijhoff Publishers: London, United Kingdom, 1986; Volume 10, pp. 157–253.
10. Wang, S.; Xie, Y. *China Species Red List*; Higher Education Press: Beijing, China, 2004; pp. 1–224.
11. Yang, J.; Cai, L.; Liu, D.T.; Chen, G.; Gratzfeld, J.; Sun, W.B. China's conservation program on Plant Species with Extremely Small Populations (PSESP): Progress and perspectives. *Biol. Conserv.* **2020**, *244*, 108535. [CrossRef]
12. Hou, Z.Q.; Zhou, D.; Hou, S.N.; Ma, C.C.; Dai, J.; Xiao, Z.Q.; Du, F. Present situation of *Eleutharrhena macrocarpa* in China. *Plant Divers.* **2015**, *37*, 640–646.
13. Wang, W.; Ortiz, R.D.C.; Jacques, F.M.B.; Xiang, X.G.; Li, H.L.; Lin, L.; Li, R.Q.; Liu, Y.; Soltis, P.S.; Soltis, D.E.; et al. Menispermaceae and the diversification of tropical rainforests near the Cretaceous-Paleogene boundary. *New Phytol.* **2012**, *195*, 470–478. [CrossRef]
14. Hua, Z. Advances in biogeography of the tropical rain forest in southern Yunnan, southwestern China. *Trop. Conserv. Sci.* **2008**, *1*, 34–42. [CrossRef]
15. Hua, Z. Geographical patterns of Yunnan seed plants may be influenced by the clockwise rotation of the Simao-Indochina geoblock. *Front. Earth Sci.* **2015**, *3*, 1–9.
16. Zhao, Y.J.; Gong, X. Genetic divergence and phylogeographic history of two closely related species (*Leucomeris decora* and *Nouelia insignis*) across the'Tanaka Line'in Southwest China. *BMC Evol. Biol.* **2015**, *15*, 1–13. [CrossRef] [PubMed]
17. Li, Y.; Zhai, S.N.; Qiu, Y.X.; Guo, Y.P.; Ge, X.J.; Comes, H.P. Glacial survival east and west of the 'Mekong–Salween Divide'in the Himalaya–Hengduan Mountains region as revealed by AFLPs and cpDNA sequence variation in *Sinopodophyllum hexandrum* (Berberidaceae). *Mol. Phylogenet. Evol.* **2011**, *59*, 412–424. [CrossRef] [PubMed]
18. He, Y.; Guo, C.; Zeng, X.; Yang, H.; Xiong, X.; Qiu, P. Genetic population variation and phylogeny of *Sinomenium acutum* (Menispermaceae) in subtropical China through chloroplast marker. *bioRxiv* **2018**, 449900.
19. Ma, Y.S.; Yu, H.; Li, Y.Y.; Yan, H.; Cheng, X. A study of genetic structure of *Stephania yunnanensis* (Menispermaceae) by DALP. *Biochem. Genet.* **2008**, *46*, 227–240. [CrossRef]
20. Iloh, A.C.; Schmidt, M.; Muellner-Riehl, A.N.; Ogundipe, O.T.; Paule, J. Pleistocene refugia and genetic diversity patterns in West Africa: Insights from the liana *Chasmanthera dependens* (Menispermaceae). *PLoS ONE* **2017**, *12*, e0170511. [CrossRef]
21. Lian, L.; Ortiz, R.D.; Jabbour, F.; Zhang, C.F.; Xiang, X.G.; Erst, A.S.; Gao, T.G.; Chen, Z.D.; Wang, W. Phylogeny and biogeography of Pachygoneae (Menispermaceae), with consideration of the boreotropical flora hypothesis and resurrection of the genera *Cebatha* and *Nephroia*. *Mol. Phylogenet. Evol.* **2020**, *148*, 106825. [CrossRef]
22. Wang, W.; Ortiz, R.D.C.; Jacques, F.M.B.; Chung, S.W.; Liu, Y.; Xiang, X.G.; Chen, Z.D. New insights into the phylogeny of *Burasaieae* (Menispermaceae) with the recognition of a new genus and emphasis on the southern Taiwanese and mainland Chinese disjunction. *Mol. Phylogenet. Evol.* **2017**, *109*, 11–20. [CrossRef]
23. Edgar, R.C. MUSCLE: Multiple sequence alignment with high accuracy and high throughput. *Nucleic Acids Res.* **2004**, *32*, 1792–1797. [CrossRef]
24. Collard, B.C.Y.; Mackill, D.J. Start Codon Targeted (SCoT) polymorphism: A simple, novel DNA marker technique for generating gene-targeted markers in plants. *Plant Mol. Biol. Rep.* **2008**, *27*, 86–93. [CrossRef]
25. Stamatakis, A. RAxML version 8: A tool for phylogenetic analysis and post-analysis of large phylogenies. *Bioinformatics* **2014**, *30*, 1312–1313. [CrossRef] [PubMed]
26. Miller, M.A.; Pfeiffer, W.; Schwartz, T. The CIPRES science gateway: A community resource for phylogenetic analyses. In Proceedings of the the 2011 TeraGrid Conference: Extreme Digital Discovery, Salt Lake City, UT, USA, 18–21 July 2011; pp. 1–8.
27. Ronquist, F.; Teslenko, M.; van der Mark, P.; Ayres, D.L.; Darling, A.; Hohna, S.; Larget, B.; Liu, L.; Suchard, M.A.; Huelsenbeck, J.P. MrBayes 3.2: Efficient Bayesian phylogenetic inference and model choice across a large model space. *Syst. Biol.* **2012**, *61*, 539–542. [CrossRef] [PubMed]
28. Zhang, D.; Gao, F.L.; Jakovlic, I.; Zou, H.; Zhang, J.; Li, W.X.; Wang, G.T. PhyloSuite: An integrated and scalable desktop platform for streamlined molecular sequence data management and evolutionary phylogenetics studies. *Mol. Ecol. Resour.* **2020**, *20*, 348–355. [CrossRef]
29. Lanfear, R.; Frandsen, P.B.; Wright, A.M.; Senfeld, T.; Calcott, B. PartitionFinder 2: New methods for selecting partitioned models of evolution for molecular and morphological phylogenetic analyses. *Mol. Biol. Evol.* **2017**, *34*, 772–773. [CrossRef] [PubMed]
30. Bouckaert, R.; Heled, J.; Kühnert, D.; Vaughan, T.; Wu, C.-H.; Xie, D.; Suchard, M.A.; Rambaut, A.; Drummond, A.J. BEAST 2: A software platform for Bayesian evolutionary analysis. *PLoS Comput. Biol.* **2014**, *10*, e1003537. [CrossRef] [PubMed]
31. Chesters, K.I. The miocene flora of Rusinga Island, Lake Victoria, Kenya. *Palaeontogr. Abt. B.* **1957**, 30–71.
32. Rambaut, A.; Drummond, A.J.; Xie, D.; Baele, G.; Suchard, M.A. Posterior summarization in Bayesian phylogenetics using Tracer 1.7. *Syst. Biol.* **2018**, *67*, 901–904. [CrossRef]
33. Matzke, N.J. *BioGeoBEARS: Biogeography with bayesian (and likelihood) evolutionary analysis in R scripts*; University of California: Berkeley, SC, USA, 2013.
34. Team, R. RStudio: Integrated development environment for R. Available online: http://www.rstudio.com/ (accessed on 10 September 2021).

35. Yeh, F.C.; Yang, R.C.; Boyle, T.B.; Ye, Z.H.; Mao, J.X. POPGENE, the user-friendly shareware for population genetic analysis. *J. Hered.* **1997**, *10*, 295–301.
36. Peakall, R.; Smouse, P.E. GenAlEx 6.5: Genetic analysis in Excel. Population genetic software for teaching and research-an update. *Bioinformatics* **2012**, *28*, 2537–2539. [CrossRef]
37. Pritchard, J.K.; Stephens, M.; Donnelly, P. Inference of population structure using multilocus genotype data. *Genetics* **2000**, *155*, 945–959. [CrossRef]
38. Falush, D.; Stephens, M.; Pritchard, J.K. Inference of population structure using multilocus genotype data: Dominant markers and null alleles. *Mol. Ecol. Notes* **2007**, *7*, 574–578. [CrossRef] [PubMed]
39. Evanno, G.; Regnaut, S.; Goudet, J. Detecting the number of clusters of individuals using the software STRUCTURE: A simulation study. *Mol. Ecol.* **2005**, *14*, 2611–2620. [CrossRef] [PubMed]
40. Earl, D.A.; vonHoldt, B.M. STRUCTURE HARVESTER: A website and program for visualizing STRUCTURE output and implementing the Evanno method. *Conserv. Genet. Resour.* **2011**, *4*, 359–361. [CrossRef]
41. Jakobsson, M.; Rosenberg, N.A. CLUMPP: A cluster matching and permutation program for dealing with label switching and multimodality in analysis of population structure. *Bioinformatics* **2007**, *23*, 1801–1806. [CrossRef]
42. Rosenberg, N.A. DISTRUCT: A program for the graphical display of population structure. *Mol. Ecol. Notes* **2004**, *4*, 137–138. [CrossRef]
43. Assoumane, A.; Zoubeirou, A.M.; Rodier-Goud, M.; Favreau, B.; Bezançon, G.; Verhaegen, D. Highlighting the occurrence of tetraploidy in *Acacia senegal* (L.) Willd. and genetic variation patterns in its natural range revealed by DNA microsatellite markers. *Tree Genet. Genomes* **2013**, *9*, 93–106. [CrossRef]
44. Teixeira, H.; Rodriguez-Echeverria, S.; Nabais, C. Genetic diversity and differentiation of *Juniperus thurifera* in Spain and Morocco as Determined by SSR. *PloS ONE* **2014**, *9*, e88996.
45. Govindaraju, D.R. Relationship between dispersal ability and levels of gene flow in plants. *Oikos* **1988**, *52*, 31–35. [CrossRef]
46. Nei, M. Analysis of gene diversity in subdivided populations. *Proc. Natl. Acad. Sci. USA* **1973**, *70*, 3321–3323. [CrossRef]
47. Joshi, A. Morphology of *Tinospora cordifolia*, with some observations on the origin of the single integument, nature of synergidae, and affinities of the *Menispermaceae*. *Am. J. Bot.* **1939**, *26*, 433–439. [CrossRef]
48. Davis, C.C.; Bell, C.D.; Mathews, S.; Donoghue, M.J. Laurasian migration explains Gondwanan disjunctions:Evidence from Malpighiaceae. *Proc. Natl. Acad. Sci. USA* **2002**, *99*, 6833–6837. [CrossRef] [PubMed]
49. Morley, R.J. Assembly and division of the South and South-East Asian flora in relation to tectonics and climate change. *J. Trop. Ecol.* **2018**, *34*, 209–234. [CrossRef]
50. Tsai, C.C.; Liao, P.C.; Ko, Y.Z.; Chen, C.H.; Chiang, Y.C. Phylogeny and historical biogeography of *Paphiopedilum Pfitzer* (Orchidaceae) based on nuclear and plastid DNA. *Front. Plant Sci.* **2020**, *11*, 126. [CrossRef] [PubMed]
51. Wu, X.; Liu, X.; Kodrul, T.; Quan, C.; Jin, J. *Dacrycarpus* pattern shedding new light on the early floristic exchange between Asia and Australia. *Natl. Sci. Rev.* **2019**, *6*, 1086–1090. [CrossRef] [PubMed]
52. Van Oosterzee, P. *Where worlds collide: The Wallace Line*; Cornell University Press: New York, NY, USA, 1997.
53. Baveja, P.; Garg, K.M.; Chattopadhyay, B.; Sadanandan, K.R.; Prawiradilaga, D.M.; Yuda, P.; Lee, J.G.; Rheindt, F.E. Using historical genome-wide DNA to unravel the confused taxonomy in a songbird lineage that is extinct in the wild. *Evol. Appl.* **2021**, *14*, 698–709. [CrossRef]
54. Klaus, S.; Morley, R.J.; Plath, M.; Zhang, Y.-P.; Li, J.-T. Biotic interchange between the Indian subcontinent and mainland Asia through time. *Nat. Commun.* **2016**, *7*, 12132. [CrossRef]
55. Hina, F.; Yisilam, G.; Wang, S.; Li, P.; Fu, C. De novo transcriptome assembly, gene annotation and SSR marker development in the moon seed genus *Menispermum* (Menispermaceae). *Front. Genet.* **2020**, *11*, 380. [CrossRef]
56. De Britto, A.J.; Raj, T.L.S.; Petchimuthu, K.; Kumar, P.B.J.R.; Sujin, R.M.; Dharmar, K. Inter population genetic variability in *Tinospora cordifolia* (Wild.) Miers ex Hook. F. & Thoms. (Menispermaceae), through RAPD marker. *Sci. Acta Xaver.* **2010**, *1*, 57–64.
57. Godt, M.J.W.; Hamrick, J. Genetic structure of two endangered pitcher plants, *Sarracenia jonesii* and *Sarracenia oreophila* (Sarraceniaceae). *Am. J. Bot.* **1996**, *83*, 1016–1023. [CrossRef]
58. Huang, Y.; Zhang, C.Q.; Li, D.Z. Low genetic diversity and high genetic differentiation in the critically endangered *Omphalogramma souliei* (Primulaceae): Implications for its conservation. *J. Syst. Evol.* **2009**, *47*, 103–109. [CrossRef]
59. Song, S.; Shen, J.; Zhou, S.; Guo, X.; Zhao, J.; Shi, X.; Yu, Z.; Gong, Q.; You, S.; Landrein, S. Highlighting the plight of threatened lianas: Distribution, dynamics, and potential habitat areas of a critically endangered species. *Glob. Ecol. Conserv.* **2022**, *36*, e02130. [CrossRef]
60. Capmourteres, V.; Anand, M. 'Conservation value': A review of the concept and its quantification. *Ecosphere* **2016**, *7*, e01476. [CrossRef]

diversity

MDPI

Article

Taxonomic Uncertainty and Its Conservation Implications in Management, a Case from *Pyrus hopeiensis* (Rosaceae)

Xian-Yun Mu [1,*], Jiang Wu [2] and Jun Wu [3]

1 College of Ecology and Nature Conservation, Beijing Forestry University, Beijing 100083, China
2 College of Biological Sciences and Technology, Beijing Forestry University, Beijing 100083, China; wujiang9212@163.com
3 Centre of Pear Engineering Technology Research, State Key Laboratory of Crop Genetics and Germplasm Enhancement, Nanjing Agricultural University, Nanjing 210095, China; wujun@njau.edu.cn
* Correspondence: xymu85@bjfu.edu.cn

Abstract: Improved taxonomies and phylogenies are essential for understanding the evolution of organisms, the development of conservation plans, and the allocation of funds and resources, especially for threatened species with uncertain identities. Pears are an economically and nutritionally important fruit, and wild pear species are highly valued and protected because of their utility for the development of cultivars. *Pyrus hopeiensis* is an endangered species endemic to North China, which is sympatric with and difficult to distinguish from the widely distributed and morphologically similar species *P. ussuriensis*. To clarify its taxonomic identity, principal coordinate analysis was performed using 14 quantitative and qualitative characters from *P. hopeiensis*, *P. ussuriensis*, and *P. phaeocarpa*, and phylogenomic analysis was performed based on whole-genome resequencing and whole plastome data. *Pyrus hopeiensis* was synonymized with *P. ussuriensis* based on morphological and phylogenetic evidence, as well as our long-term field studies. *Pyrus hopeiensis* is proposed to be excluded from the list of local key protected wild plants. Given that the holotype of *P. ussuriensis* was not designated, a lectotype was designated in this work. Integrative evidence-based taxonomic study including museomics is suggested for organisms with uncertain identities, which will contribute to biodiversity conservation.

Keywords: endangered species; integrative evidence; *Pyrus hopeiensis*; *Pyrus ussuriensis*; taxonomic uncertainty; biodiversity conservation

Citation: Mu, X.-Y.; Wu, J.; Wu, J. Taxonomic Uncertainty and Its Conservation Implications in Management, a Case from *Pyrus hopeiensis* (Rosaceae). *Diversity* **2022**, *14*, 417. https://doi.org/10.3390/d14060417

Academic Editor: Mario A. Pagnotta

Received: 6 May 2022
Accepted: 23 May 2022
Published: 24 May 2022

1. Introduction

The intensification of human activities, especially since the Industrial Revolution, has negatively affected global climate and biodiversity and is responsible for the Sixth Mass Extinction [1]. Rare and endangered plants are important components of biodiversity, and many of them are facing high extinction pressure because of inadequate conservation management [2], as well as taxonomic uncertainty [3,4]. Accurately identifying plants is a huge challenge, and taxonomic stability is an elusive goal [5,6]. Although studies of the evolution and taxonomic identity of species can greatly contribute to our knowledge of the organisms that we are interested in protecting, taxonomic problems might occur because of improvements in our knowledge of the phylogeny and evolution of organisms, as well as the recognition of previously made nomenclatural errors [7]. More than half of economically important tropical African ginger specimens from 40 herbaria in 21 countries are likely to be wrongly named [8], which can affect the conservation status assessment of related species. *Haematocarpus subpeltatus* Merr. (Menispermaceae), a misidentified critically endangered species that has received much conservation focus, has been identified as a new record in China, but the real entity that should be given priority for conservation, *Eleutharrhena macrocarpa* (Diels) Forman, has been clarified by molecular phylogenetic analyses [9]. These findings highlight the significance of taxonomy and molecular phylogeny in biodiversity conservation, especially for species with taxonomic uncertainty.

Pears are an economically and nutritionally important temperate fruit that have been cultivated for more than 3000 years [10]. The genus *Pyrus* L. (Rosaceae) is geographically divided into oriental and occidental pears [11]. Considerable morphological variation and extensive hybridization have been documented in the genus [12,13]. The number of species recognized in *Pyrus* varies from 21 to more than 80 [11,14], and 73 of them are abundant in Eurasia [15]. Although remarkable advances have been made in the genetics and breeding of *Pyrus* in recent years, and whole genomes of several species have been published, e.g., [10,13,16–18], the taxonomy, diversification and phylogeny of *Pyrus* remain unclear, several currently accepted species such as *P. caucasica* Fed., *P. pyraster* (L.) Burgsd. and *P. spinosa* Forssk. are not monophyletic in the phylogenetic trees [12,19], suggesting the necessity of further taxonomic and phylogenetic studies. One reason for the lack of robustness of the molecular trees of *Pyrus* in previous studies is the few numbers of molecular markers employed. Whole nuclear genome sequences can provide vital information for reconstruction species' phylogenies [20], and complete plastomes can also provide key information for phylogenetic studies as well as for detecting hybridization events [21,22]. A hybrid usually neighbors one of its parents in the nuclear tree and group with another parent in the plastome tree, and phylogenetic discordance (cytonuclear discordance) is usually detected for hybrid species. Hence, extensive phylogenomic analyses are necessary for advancing our understanding of the phylogeny of *Pyrus*. This will shed light on the taxonomy and evolution of pear species and enhance pear production.

China is a diversity center for oriental pears, as 14 native species, including 5 primary wild species and over 2000 cultivars, have been reported to date [13,23,24]. Among them, *Pyrus hopeiensis* Yü, a taxonomically controversial species endemic to North China, is thought to be a hybrid of *P. ussuriensis* Maxim. and *P. phaeocarpa* Rehd. [25]. It is classified as an endangered species and is listed as a key protected wild plant of Hebei Province; it has thus received much conservation attention, e.g., [26–29]. *Pyrus hopeiensis* can be distinguished from *P. phaeocarpa* from several morphological characteristics, such as its persistent calyx and spiny serrate leaf margin (vs. caducous calyx and serrate leaf margin in *P. phaeocarpa*). However, differentiating *P. hopeiensis* and *P. ussuriensis* is more challenging. *Pyrus ussuriensis* is an important germplasm resource that is widely distributed in northern China, the Russian Far East, North Korea, and Japan, and more than 150 cultivars have been obtained from this species in China [30]. *Pyrus hopeiensis* has been documented in Beijing, Hebei, and Shandong provinces, and its type locality is Jieshi Mountain, a small hill with an altitude of 695 m in Changli County, Hebei Province, which is in the eastside of Yan Mountain in North China [23,25,31]. Based on the original description [25], *P. hopeiensis* can be distinguished from *P. ussuriensis* by its brown fruit and obvious spots on its surface (vs. yellow fruit with fewer spots). However, numerous individuals with intermediate morphological characteristics have been identified in the field and herbaria. Taxonomic ambiguity should be clarified, as this would aid conservation efforts for endangered and regional protected species. Hence, further morphological and phylogenomic studies should be performed to unravel the identity of *P. hopeiensis*.

In the present study, we aim to: (1) assess morphological differences among *P. hopeiensis*, *P. phaeocarpa*, and *P. ussuriensis*, (2) reconstruct the molecular phylogenetic tree and test the hybrid identity of *P. hopeiensis* based on whole genome resequencing data, and (3) clarify the taxonomic identity of *P. hopeiensis*. This study provides novel insight for the conservation management of *Pyrus* species and highlight the importance of plant taxonomy and phylogenomic analysis for biodiversity conservation.

2. Materials and Methods

2.1. Morphological Study

All specimens of *P. hopeiensis* (including type specimens) deposited in the herbarium of the Institute of Botany, Chinese Academy of Science (PE) and hundreds of specimens of

P. ussuriensis and *P. phaeocarpa* in PE were examined (Appendix A). Fourteen qualitative and quantitative characters were selected (Table 1), which included 1 binary and 13 continuous characters. Given differences in leaf morphology during the flowering and fruiting period in *Pyrus*, morphological analyses were performed for both periods. Aside from one specimen collected from an individual cultivated at PE (*Ren 2*, 21 August 1962), only specimens of *P. hopeiensis* identified by Yü and Ku were used to avoid misidentification. Two data matrices were made for 45 flower specimens and 61 fruit specimens, and each was treated as an operational taxonomic unit. Following Wang [32], the data matrix was standardized using a zero-mean normalization method. The formula $X^* = (X - m)/s$ was used, where "X" is the sample, "m" is the arithmetic mean, and "s" is the standard deviation. A principal coordinate analysis was performed based on the Gower general similarity coefficient analysis for mixed data sets using MVSP-Version 3.13b software.

Table 1. Morphological characteristics used in the principal coordinate analysis and their coding.

	Specimens of the Flowering Period		Specimens of the Fruiting Period
1	Leaf blade length [cm]	1	Leaf blade length [cm]
2	Leaf blade width [cm]	2	Leaf blade width [cm]
3	Ratio of leaf blade width to length	3	Ratio of leaf blade width to length
4	Leaf serration: (1) long; (0) short	4	Leaf serration: (1) long; (0) short
5	Length of flower stalk [cm]	5	Length of fruit [cm]
6	Indumentum density on stalk: (0) sparse; (1) dense	6	Persisting calyx: (0) no; (1) yes
		7	Length of pedicel [cm]
		8	Ratio of fruit to pedicel Length

2.2. Phylogenomic Inference

In this study, whole-genome resequencing data of 27 samples of *Pyrus* representing 13 species from [33,34] were used for phylogenomic inference. Specifically, a total of 22 samples from Wu et al. [33] were employed, which included three individuals of *P. phaeocarpa*, five *P. ussuriensis*, and two *P. hopeiensis*. Five samples of *P. hopeiensis* (PWH8, PWH11, PWH18, PWH19, and PWH20) were used from Li et al. [34]. All samples were numbered following [33,34]. *Malus* × *domestica* L. was used as the outgroup, and its whole-genome sequence [35] was downloaded from NCBI.

Raw reads were filtered by removing several types of low-quality paired reads, including reads with adapters, paired reads with N content greater than 10%, and low-quality ($Q < 10$) paired reads contained in single-end sequencing reads that exceeded 50% of the length of the read. The accessions were then mapped to the apple genome using the MEM algorithm of Burrows–Wheeler Aligner in BWA software [36], and single nucleotide polymorphisms (SNPs) were called using the HaplotypeCaller module in GATK [37] and filtered using the following parameters: QD < 2.0||MQ < 40.0||FS > 60.0||QUAL < −12.5||ReadPosRankSum < −8.0—clusterSize 2—clusterWindowSize 5. The obtained SNPs were further filtered to construct a high-quality data matrix for phylogenetic inference with the minor allele frequency greater than 0.05, and less than 0.8 missing rates of the confirmed credible genotype from all accessions for each site and biallelic SNPs. Finally, a maximum likelihood (ML) phylogenetic tree was generated using IQTREE [38], and ultrafast bootstrap support values (BS_{ML}) were estimated with 1000 replicates [39].

A total of sixteen complete plastomes of *Pyrus* species were downloaded from NCBI, including one for each of the three species. *Malus kansuensis* (Batal.) Schneid. was used as the outgroup. Four samples of *P. hopeiensis* (PWH8, PWH11, PWH18, and PWH20) were selected from [34], one *P. ussuriensis* (1) was selected from [33]. The whole chloroplast genomes of these five samples were generated from the whole-genome resequencing data using the software GetOrganelle [40]. Plastomes of all samples were aligned by MAFFT [41]. After cutting both sides of the aligned sequences, a final data matrix with a length of 162,337 bp was obtained and used for phylogenetic analysis in IQTREE.

3. Results

3.1. Morphological Study

The results of principal coordinate analysis based on multiple characters showed that the four flower specimens of *P. hopeiensis* clustered with *P. ussuriensis* but were clearly separated from the *P. phaeocarpa* cluster (Figure 1A). Likewise, the six fruit specimens of *P. hopeiensis* were nested within the *P. ussuriensis* cluster, and this mixed group was distinct from that of *P. phaeocarpa* (Figure 1B).

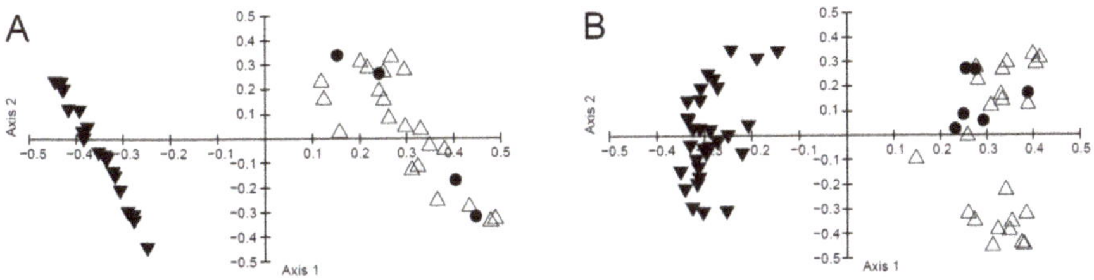

Figure 1. Principal coordinate analysis of *Pyrus hopeiensis*, *P. ussuriensis* and *P. phaeocarpa* based on morphological characters. (**A**) specimens of flowering period, (**B**) specimens of fruiting period. Black triangles: *P. phaeocarpa*, white triangles: *P. ussuriensis*, black circles: *P. hopeiensis*.

3.2. Phylogenomic Inference

After filtration, a total of 204.87 G clean data were obtained from whole-genome resequencing data, and a high-quality SNP data matrix with a length of 1,340,483 bp was generated. The ML tree was well resolved, and the seven samples of *P. hopeiensis* were divided into two groups. Two of them (PWH11 and PWH20) were grouped with five samples of *P. ussuriensis*, forming a monophyletic clade with full support (BS_{ML} = 100, Figure 2A); this clade was sister to another fully supported monophyletic clade that comprised the rest of the *P. hopeiensis* samples (PWH8, PWH18, PWH19, 1 and 2). The three samples of *P. phaeocarpa* comprised a fully supported monophyletic clade (BS_{ML} = 100, Figure 2A), and this clade was distinct from the *P. ussuriensis–P. hopeiensis* clade. Similar results were obtained in the plastome tree, with samples of *P. hopeiensis* and *P. ussuriensis* combined into a fully supported monophyletic clade (BS_{ML} = 100, Figure 2B). A sample of *P. hopeiensis* from Shandong Province [28] was embedded within the subclade of *P. ussuriensis*. The clade of *P. phaeocarpa* was distinct from the *P. ussuriensis–P. hopeiensis* clade, and no phylogenetic discordance was detected among these three species.

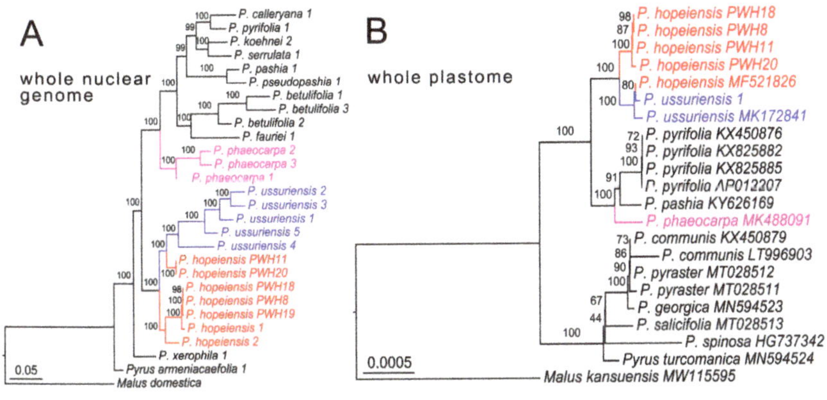

Figure 2. Maximum likelihood phylogenetic tree inferred from the SNP data matrix based on the whole-genome resequencing data (**A**) and whole plastid genome sequences (**B**). The numbers above

the nodes indicate bootstrap values generated from maximum likelihood analysis, scale bar indicates substitutions per site. Whole plastome sequences of species downloaded from NCBI are given their GenBank numbers following species' name in the plastome tree (**B**). Samples of *P. hopeiensis* are presented with red, *P. ussuriensis* with blue, and *P. phaeocarpa* with pink.

4. Discussion

4.1. Integrative Evidence for Reappraisal of the Identity of Pyrus hopeiensis

Morphological characters are important for species identification, and some morphological characteristics vary greatly among habitats and at different growth stages; this can result in inconsistencies in the designation of the taxonomic identity of species. The color of fruit and the spots on the surface of fruit are key morphological characteristics for discriminating between *P. hopeiensis* and *P. ussuriensis* [25]. The holotype (PE00020671, Figure 3A), isotype (PE00020670), and paratypes (*FS Chang 238, 269, 276*) of *P. hopeiensis* were collected in September 1953; this species was described 10 years later [25]. Although the fruit was described as brown in the original paper (Pomum . . . , fuscum, . . .) [25] (p. 232), we found no fruit in these type specimens. However, the fruit is described as "mud-like yellow" (in Chinese) in the collection record of paratype *FS Chang 269* (Figure S1). Several brown or black fruits are present on two specimens collected in Shandong Province (*Zhou* et al. *1178, 1590*) on 6 and 7 June 1959 and identified as *P. hopeiensis* by Yü. Although "green fruit" is written clearly in Chinese on the record of *Zhou* et al. *1178*, these immature pomes turned black in herbaria (Figure S2). Hence, the fruit color of *P. hopeiensis* recorded in previous studies (brown) was not consistent with that of individuals observed in the field (yellow).

We conducted a long-term field investigation in North China since 2010. The fruit color of *P. ussuriensis* recorded in previous studies (yellow) is unstable. A dried fruit preserved in the lectotype of *P. ussuriensis* (Figure 3B) is pale brown. The color of mature fruit varies from greenish yellow to reddish yellow in the field (Figure 3C–E). One specimen collected by P. Zhang (*742*) on 13 September 1956 from Changli County, Hebei Province (Figure S3A) has a mature fruit that is reddish in color. Another specimen (*P. Zhang 766*, Figure S3B) presents three densely spotted fruits, and the fruit color (brown) is clearly indicated on the collection record (in Chinese). These two specimens were both identified by Yü as *P. hopeiensis* on 29 April 1960. The density of spots on the fruits both varies greatly within and among individuals; for example, there are several spots in Figure 3C (Dongling Mountain, in the north end of Taihang Mountain) and 3D (Yudu Mountain, in the west of Yan Mountain), but almost none in Figure 3E (Song Mountain, neighboring Yudu Mountain). Furthermore, persistent and caducous calyxes associated with mature fruits were observed in the same trees in the field (e.g., Figure 3B). Carpel number (4 vs. 5) and the length of the fruit pedicel (long vs. short) are sometimes used to differentiate *P. hopeiensis* from *P. ussuriensis*. However, our morphological analyses of specimens (Figure 1) and our field observations (Figure 3C–I) indicate that this trait has no taxonomic relevance.

We conducted two field investigations at the type locality—Jieshi Mountain, Xingshuyuan Village, Changli County, Hebei Province—at the end of August in 2015 and 2017. This is a small hill like the others located in the Yan Mountains: its altitude is 695 m, and it is 15 km from the Bohai Sea. Besides *P. hopeiensis* and *P. ussuriensis* documented by Li et al. [34], we found several individuals of *P. phaeocarpa* and *P. betulifolia*. The calyx was absent in mature fruit of these two species; fruits of the former are approximately 2 cm in diameter (vs. <1 cm in the latter), and the carpel number is 3–4 (vs. 2–3) (Figure 4). In 2015, we met an employee of the local forestry bureau who identified an individual as the endemic species, *P. hopeiensis* (Figure 4A). In Figure 4, fruits Aa, Bb, and C were collected from this individual. Key morphological characters (such as calyx on fruit, carpel number and leaf margin) indicate that this individual is instead *P. phaeocarpa*. Aside from a large number of wild *Pyrus* individuals, the Jieshi Mountain type locality includes a large orchard that contains *Pyrus* cultivars. One of the *P. betulifolia* trees growing here has fruits with an unusual green color (Figure 4B(c),D). This special germplasm resource has also been discovered in Shanxi Province [42]. Unfortunately, this individual in the Jieshi Mountain died because of the growth of *Pueraria montana* (Loureiro) Merrill (Fabaceae) during our visit in 2017.

Figure 3. Type specimens of *Pyrus hopeiensis* and *P. ussuriensis* and fruit morphological characteristics of *P. ussuriensis* populations in North China. (**A**) Holotype of *P. hopeiensis*, (**B**) lectotype of *P. ussuriensis*, (**C–E**) fruit color, spots, and calyx, (**F**,**G**) fruit size, length of pedicel and calyx, (**H**,**I**) carpel number (4 vs. 5).

Figure 4. Fruits and leaves of *Pyrus phaeocarpa* and *P. betulifolia* from the type locality of *P. hopeiensis*, the Jieshi Mountain, Changli County, Hebei Province. (**A**,**B**) Fruits and leaves of *P. phaeocarpa* and *P. betulifolia*, (**C**,**D**) cross-cut of fruits of *P. phaeocarpa* and *P. betulifolia*, respectively, indicating 3–4 carpels per fruit. (**A**(**a**),**B**(**b**),**C**) were collected from the same tree of *P. phaeocarpa*, (**B**(**c**),**D**) were collected from the same tree of *P. betulifolia*.

Given the uncertain morphological differences between *P. hopeiensis* and *P. ussuriensis*, a phylogenetic study provides a powerful tool for clarifying the relationships between these taxa. However, the high level of genetic exchange and hybridization in Rosaceae at both the species and genus level [34,43–45] precludes the resolution of phylogenetic relationships among *Pyrus* species when traditional molecular markers are used, e.g., [12,19,46]. With whole-genome resequencing data, a robust molecular tree of *Pyrus* was constructed based on the SNP data matrix [33]. Two individuals of *P. hopeiensis* collected from the Zhengzhou Fruit Research Institute, Chinese Academy of Agricultural Sciences (CAAS), and Research Institute of Pomology, CAAS, were sampled. Both were embedded within a fully supported monophyletic clade of *P. ussuriensis*, and they were distinct from the *P. phaeocarpa* clade. We performed phylogenomic inference in this study based on the whole-genome resequencing data of Wu et al. [33] and Li et al. [34]. Our results demonstrate that all the *P. hopeiensis* samples were mixed with *P. ussuriensis*, and they together comprised one fully supported monophyletic clade (Figure 2). However, no cytonuclear discordance was detected for these three pear species between the nuclear and plastid genome data. Hence, a hybrid origin of *P. hopeiensis* from *P. ussuriensis* and *P. phaeocarpa* was not supported in this study. Both the co-existence of *P. ussuriensis* and *P. hopeiensis* at the same locations and their synchronous flowering and fruiting phenology suggest that there is no niche differentiation between them. Hence, evidence from morphological comparisons, phylogenomic inference, and our extensive field studies indicate that the locally endemic and protected *P. hopeiensis* is a synonym of the widely distributed *P. ussuriensis*. Consequently, 'the chosen one' of wild pear species (*P. hopeiensis*), which has received multiple conservation resources, should be excluded from the list of local key protected wild plants (Figure 5).

Figure 5. Integrative evidence indicates that the locally endemic *Pyrus hopeiensis* is a synonym of *P. ussuriensis* and should be excluded from the list of key protected wild plants.

4.2. Integrative Evidence-Based Taxonomy for Biodiversity Conservation

Climate change and biodiversity loss are two critical events of the Anthropocene, and the existence and distribution of biodiversity need to be better documented prior to proposing conservation strategies [47,48]. Although changes in the names of plants can be frustrating for conservationists, taxonomy is a science that involves classifying continuous variation in discrete categories and clarifying the phylogenetic relationships among taxa; taxonomists also have the responsibility of removing unnatural taxa that conflict with evolutionary and phylogenetic theory [49]. Species' names should change when a more rational and testable hypothesis of the species is produced based on more evidence [50]. For example, *Juglans hopeiensis* Hu (Juglandaceae), a species endemic to North China and listed as a key protected wild plant of Hebei Province, is thought to be a hybrid of *J. regia* L. and *J. mandshurica* Maxim. [51]. Whole plastome data suggested that it was closely related to *J. mandshurica*, whereas the reduced-representation genomic data indicated a close relationship with *J. regia* [21]. Based on dense sampling whole-genome data, Zhang et al. [52] confirmed its hybrid identity and demonstrated that only the first generation of *J. hopeiensis* is viable, which precludes further speciation. On the one hand, the accurate identification of endangered species is a prerequisite before conservation action can be taken. On the other hand, species incorrectly listed as endangered species can result in a waste of valuable resources, and the prime of this are *P. hopeiensis* and *E. macrocarpa* [9]. Thus, integrative evidence is needed to assess the identity of controversial species, and these approaches should figure prominently in conservation management.

Because of limited resources, setting priorities for the conservation of threatened species is critically important. Species with unresolved taxonomies can pose problems for endangerment status assessments and impede conservation management; genetic data in the genomic era can provide information that can help clarify the taxonomic status of species [53]. In addition to living organisms, specimens preserved in museums and herbaria can provide valuable morphological and ecological data as well as genomic resources that can be used to resolve taxonomic uncertainties, reconstruct accurate phylogenies, and assess the mechanisms of ecological adaptation [54]. In some cases, type specimen may be the only credible remnant available for species identification. Hence, genetic resources from voucher specimens and even type specimens are vital for accurate identification, taxonomic designations, and conservation [55]. Although extracting DNA from museum specimens can be a major challenge and potentially be destructive [56–58], the growth of museomics, the study of DNA sequences obtained from museum specimens, combined with phenetic data collected from key specimens, will provide new taxonomic, evolutionary, and conservation insights.

4.3. Taxonomic Treatment

Based on the morphological comparison of specimens, phylogenomic inference, and our field studies, *P. hopeiensis* is a synonym of *P. ussuriensis*. The holotype of *P. ussuriensis* was not previously designated. Although several specimens are deposited in herbaria (e.g., GH, K and P), one of them preserved in LE that fit the original description of *P. ussuriensis* was designated as the lectotype.

Pyrus ussuriensis Maxim. ex. Rupr., Bull. Acad. Imp. Sci. Saint-Pétersbourg, sér. 2. 15: 132. 1857.—**Lectotype (designated here)**: Russia, *CJ Maximowicz s.n.* (LE, barcode LE01009639, [photo!]; isolectotype: LE barcode LE01009638, [photo!]). = *Pyrus hopeiensis* Yü, Acta Phytotax. Sin. 8: 232. 1963. ***syn. nov.***—Type: China, Hebei Province, Changli County, Xingshuyuan village, 21 September 1953, *FS Chang 268* (PE barcode 00020671!; isolectotype: PE barcode 00020670!).

Other Specimens Examined

Pyrus ussuriensis Maxim. ex. Rupr.: **RUSSIA. Ussuri:** *CJ Maximowicz 96* (K barcode K000758073 [photo]), *CJ Maximowicz s.n.* (GH barcode GH00032508 [photo], LE barcode LE01009633 [photo], LE01009634 [photo]). **CHINA. Beijing:** Changping, 4 May 2001, *Changping Investigation Team 173* (PE barcode 01274862!), Haidian, June 1960, *Anonymous s.n.* (PE barcode 01274861!), 21 August 1962, *G Ren 2* (PE barcode 01608659!), Jiangou, 25 April 1959, *Hebei Investigation First Team 116* (PE barcode 00548756!), Mentougou, Xiaolongmen, July, 1978, *Beijing Normal University Team s.n.* (PE barcode 01274865!), Shangfangshan, 21 May 1959, *Hebei Investigation First Team 341* (PE barcode 00548750!); **Hebei Province**: Changli County, Xingshuyuan village, 1 Sept. 1953, *FS Chang 238* (PE barcode 00548753!), 21 Sept. 1953, *FS Chang 269* (PE barcode 00548751!), 24 Sept. 1953, *FS Chang 276* (PE barcode 00548755!), 13 September 1956, *P Zhang 742* (PE barcode 01448598!), Chicheng County, Zhenanpu town, 7 September 1959, *Anonymous 4100* (PE barcode 00548752!, 01274860!), Funing County, 24 April 1959, *Anonymous 58* (PE barcode 00548760!), Laiyuan County, 26 July 1959, *Anonymous 4-3743* (PE barcode 00548754!), Yi County, 24 July 1959, *Anonymous 3246* (PE barcode 01274859!), 25 July 1959, *Anonymous 3287* (PE barcode 01274858!),

Yu County, Xiaowutai Mountain, 27 August 1950, *YW Cui 2706* (PE barcode 01274864!), 28 August 1950, *YW Cui 2864* (PE barcode 01274863!); **Inner Mongolia Autonomous Region:** Zhaomengkaqi, Xiaoxigoumen, 4 June 1962, *Meng-Ning Comprehensive investigation Team 252* (PE barcode PE00548749!), 4 Aug. 1962, *Meng-Ning Comprehensive investigation Team* 1292 (PE barcode PE00548748!); **Ningxia Autonomous Region:** Helan Mountain, 25 May 1959, *YQ He 02424* (PE barcode 00548759!); **Shandong Province:** Qingdao City, Laoshan, 7 June 1959, *TY Zhou* et al. *1590* (PE barcode 00548758!), 8 June 1959, *TY Zhou* et al. *1178* (PE barcode 00548757!); **Shanxi Province:** Xing County, *Yellow River Investigation Second Team 2695* (PE barcode 00548761!).

5. Conclusions

Plant names may change because of the development of taxonomic studies, which can aid biodiversity conservation. Because of frequent gene exchange and hybridization among species, the phylogenetic relationships among *Pyrus* species remain unclear, and the taxonomic identities of several species still require examination, such as *P. hopeiensis*, a regional key protected and potential hybrid species endemic to North China. Comprehensive morphological studies were performed based on flowering and fruiting specimens of *P. hopeiensis* and its potential parent, *P. ussuriensis*, and *P. phaeocarpa*. Extensive phylogenomic inferences were performed based on a high-quality SNP data matrix that was generated from whole-genome resequencing and the whole plastome data. A long-term field investigation of *Pyrus* species in North China was also conducted. Both morphological and phylogenomic studies indicate a close relationship between samples of *P. hopeiensis* and *P. ussuriensis*, and this mixed cluster consists of a fully supported monophyletic clade and is distinct from other species. Hence, *P. hopeiensis* is a synonym of *P. ussuriensis*. The lectotype of *P. ussuriensis* was designated here. Our work provides information for both the taxonomic study and conservation management of *Pyrus* and highlights the significance of integrative evidence-based taxonomy in biodiversity conservation.

Supplementary Materials: The following supporting information can be downloaded at: https://www.mdpi.com/article/10.3390/d14060417/s1, Figure S1. Paratype of *Pyrus hopeiensis*. *FS Chang 269*, PE barcode: 00548751. Figure S2. Specimen of *Pyrus hopeiensis*. *TY Zhou* et al., *1178*, PE barcode: 00548757. Figure S3. Specimens of *Pyrus hopeiensis*. A: *P Zhang 742* (PE barcode: 01448598), B: *P Zhang 766* (PE barcode: 01449638).

Author Contributions: X.-Y.M. conceived the work, performed field investigation, and prepared, wrote, and revised the manuscript; J.W. (Jiang Wu) carried out specimens' morphological statistics and the principal coordinate analysis; J.W. (Jun Wu) provided the original resequencing data of *Pyrus* species described in the manuscript. All authors have read and agreed to the published version of the manuscript.

Funding: This research was funded by the Natural Science Foundation of China (grant number 32070235), the Project of the Second Investigation of the National Key Protected Wild Plants of Beijing, and the Project of Agricultural Wild Plant Resources Investigation of Beijing (grant number 13200346).

Institutional Review Board Statement: Not applicable.

Data Availability Statement: All data are available in this article.

Acknowledgments: We thank Irina Illarionova (LE) for providing photos of the type specimen of *Pyrus ussuriensis*, and the curator and staff at PE for their help. We also thank Qin-Wen Lin, Yi-Xuan Zhu, Xue-Li Shen, Ling Tong, Wan-Jie Jiang, and De-Huai Zhang for their help during our field investigation.

Conflicts of Interest: The authors declare no conflict of interest.

Appendix A

The information of specimens used for principal coordinate analysis given in the Appendix is taken from specimens deposited in PE.

Pyrus hopeiensis Yü: **CHINA. Beijing:** Changping District, 4 May 2001, *Changping Investigation Team 173* (PE barcode 01274862), Haidian, June 1960, *Anonymous s.n.* (PE barcode 01274861), 21 August 1962, *G Ren 2* (PE barcode 01608659), Jiangou, 25 April 1959, *Hebei Investigation First Team 116* (PE barcode 00548756), Mentougou, Xiaolongmen, July, 1978, *Beijing Normal University Team s.n.* (PE barcode 01274865); **Hebei Province:** Changli County, Xingshuyuan village, 13 September 1956, *P Zhang 742* (PE barcode 01448598), 19 September 1956, *P Zhang 766* (PE barcode 01449638), 22 September 1956, *P Zhang 782* (PE barcode 01449639, 01449640), Funing County, 24 April 1959, *Anonymous 58* (PE barcode 00548760).

Pyrus phaeocarpa Rehd.: **CHINA. Beijing:** Haidian District, Cheying village, 22 August 1972, *Anonymous Hai-676* (PE barcode 01204095), Mentougou District, Dongyangtuo village, 19 April 1954, *ZL Yan 167* (PE barcode 01448580), Mengwu village, 20 April 1954, *ZL Yan 171* (PE barcode 01448530); 20 August 1954, *ZL Yan 168* (PE barcode 01448527); **Hebei Province:** Changli County, Fenghuangshan, 21 April 1953, *TT Yü 35* (PE barcode 01448549, 01448550), near Railway Station Office, 27 April 1953, *TT Yü 092* (PE barcode 01448535), Xishan, *Anonymous s.n.* (PE barcode 01448554), Xingshuyuan, 17 April 1953, *TT Yü 13* (PE barcode 01448538), 20 April 1953, *TT Yü 30* (PE barcode 01448555), *TT Yü 227* (PE barcode 01448609, 01448611), West Zhanggezhuang, *Anonymous 14* (PE barcode 01448534), 16 September 1953, *Anonymous 6* (PE barcode 01448532), *Anonymous 9* (PE barcode 01448540), 24 April 1956, *P Zhang 647* (PE barcode 01448561, 01448562), 13 September 1956, *P Zhang 738* (PE barcode 01448567), *P Zhang 739* (PE barcode 01448560), *P Zhang 739* (PE barcode 01448569), *P Zhang 740* (PE barcode 01448569), *P Zhang 741* (PE barcode 01449634), Liugezhuang, 6 October 1953, *FS Chang 284* (PE barcode 00549049), Dongling, 22 April 1936, *DF Jin K-901* (PE barcode 00549045, 00549062), Funing County, Jiejia village, 22 September 1956, *FS Chang 776* (PE barcode 01448557, 01448558), *FS Chang 777* (PE barcode 01448559), *FS Chang 780* (PE barcode 01448575), Liugezhuang, 6 October 1953, *FS Chang 283* (PE barcode 00549046), 19 September 1956, *P Zhang 766* (PE barcode 01449638), Shibeigou, 24 April 1953, *TT Yü 58* (PE barcode 00549044), Yuanjiagou, 23 April 1953, *TT Yü 53* (PE barcode 01448543, 01448544), Qinghuangdao City, chitushan, 4 May 1953, *TT Yü 141* (PE barcode 01448545); **Henan Province:** Song County, 15 September 1956, *Bureau of Henan Forestry 986* (PE barcode 00549057), *Bureau of Henan Forestry 1259* (PE barcode 00549055, 00549058), 26 August 1959, *Anonymous 35149* (PE barcode 00549054), 1 October 1959, *Bureau of Henan Forestry 1259* (PE barcode 00549056); **Gansu Province:** Wushan County, 4 June 1956, *Yellow River Investigation Team 4480* (PE barcode 00549059); **Jiangxi Province:** Guangchang County, 7 October 1958, *QM Hu 5398* (PE barcode 00549061), Suichuan County, 22 September 1963, *JS Yue* et al. *4168* (PE barcode 00549060); **Shaanxi Province:** Muhuguan, 22 June 1960, *Anonymous 0496* (PE barcode 01274917, 01274918), Shangzhou City, 4 July 1960, *Anonymous 1074* (PE barcode 01274921), Shanyang County, 12 May 1963, *ZY Zhang 15906* (PE barcode 01274919), 12 May 1964, *JX Yang 2656* (PE barcode 01274923), 22 May 1964, *JX Yang 2733* (PE barcode 01274920), 10 June 1964, *JX Yang 2929* (PE barcode 01274916), Banmiao, 6 May 1964, *JX Yang 2558* (PE barcode 01274922). **USA.** 10 May 1918, *HH Chung 4255* (PE barcode 01682167), 6 May 1930, *CEK and FPM 17501* (PE barcode 01682166).

Pyrus ussuriensis Maxim. ex Rupr.: **CHINA. Beijing:** Changping District, Nankou, 29 May 1956, *Herbarium Team 1198* (PE barcode 00549475), Fangshan District, 24 August 1956, *Herbarium Team 3529* (PE barcode 00549474), Haidian District, Xishan, Biyunsi, 22 April 1955, *Herbarium Team 561* (PE 00549442, 00549478), 10 July 1955, *Wofosi Investigation Team 153* (PE barcode 01448240), 1 August 1957, *YJ Zhang 327* (PE barcode 01448238), Western Park, 15 April 1951, *FZ Wang 11* (PE barcode 01449781), Mentougou District, Baihuashan, 22 July 1956, *CJ Liu and DY Xing 170* (PE barcode 00549483, 00549498), Dajuesi, April 1936, *DF Jin 12008* (PE barcode 01449776, 01449777), Zhoujiaxiang, April 1936, *Y Liu 12013* (PE barcode 01449775), 2 August 1936, *DF Jin 181* (PE barcode 01449774), *DF Jin 188* (PE 01449773), Miaofengshan, 23 May 1930, *HF Chow 40274* (PE barcode 00549481), 6 May 1953, *F Zhao 71* (PE barcode 00549506), Miyun County, Wulingshan, 4 May 1951, *Y Liu and J Zhang 15067* (PE barcode 00549448, 00549451, 00549487), Pinggu District, Beiyang Bridge,

23 May 1972, *Anonymous Ping-188* (PE barcode 01204098); **Hebei Province:** 1930, *HF Chow 40336* (PE barcode 00549496), 21 April 1936, *Anonymous 379* (PE barcode 00549493), Changli County, West Zhanggezhuang, 26 April 1956, *P Zhang 651* (PE barcode 01449433), Dingzhou City, Xinxingzhuang, 15 April 1956, *JY Zhong 529* (PE barcode 01449727), *JY Zhong 530* (PE barcode 01449735), Zhaozhuang, 22 May 1956, *JY Zhong 551* (PE barcode 01449737), Dongling, 23 April 1930, *HF Chow 40325* (PE barcode 00549479), 1 May 1932, *HF Chow 41931* (PE barcode 00549480), 22 April 1936, *DF Jin K-900* (PE barcode 00549445), 20 June 1956, *JX Duan 27* (PE barcode 00549441, 00549446), 25 June 1956, *JX Duan 180* (PE barcode 00549450), Funing County, 30 April 1956, *P Zhang 668* (PE barcode 01448400), Jiejiagou, 22 September 1956, *FS Chang 778* (PE barcode 01448651), Neiqiu County, Fuanmugou, *XY Liu and F Zhao 431* (PE barcode 01449778), Yu County, Xiaowutaishan, 13 September 1956, *Herbarium Team 2417* (PE barcode 00549476), Zhangjiakou City, 25 April 1956, *JX Duan 180* (PE barcode 00549447, 00549484, 00549485), Xiling, Huangtupo, 8 June 1953, *F Zhao 222* (PE barcode 00549488); **Tianjin:** Ji County, Panshan, 4 July 1956, *Herbarium Team 1941* (PE barcode 00549444); Precise locality unknown, *Anonymous 12049* (PE barcode 01449783), *Anonymous K-919* (PE barcode 01449782).

References

1. Cowie, R.H.; Bouchet, P.; Fontaine, B. The Sixth Mass Extinction: Fact, fiction or speculation? *Biol. Rev.* **2022**, *97*, 640–663. [CrossRef] [PubMed]
2. Ma, Y.; Chen, G.; Edward Grumbine, R.; Dao, Z.; Sun, W.; Guo, H. Conserving plant species with extremely small populations (PSESP) in China. *Biodivers. Conserv.* **2013**, *22*, 803–809. [CrossRef]
3. Lu, Z.; Qin, H.; Jin, X.; Zhang, Z.; Yang, Q.; Hong, D.; Li, D.; Li, K.; Yuan, L.; Zhou, Z. On the necessity, principle, and process of updating the List of National Key Protected Wild Plants. *Biodivers. Sci.* **2021**, *29*, 1577–1582. [CrossRef]
4. Yang, Y.; Tan, C.; Yang, Z. Conservation of gymnosperms in China: Perspectives from the List of National Key Protected Wild Plants. *Biodivers. Sci.* **2021**, *29*, 1591–1598. [CrossRef]
5. Heywood, V.H.; Davis, P.H. *Principles of Angiosperm Taxonomy*; Oliver and Boyd: Edinburgh, UK; London, UK, 1963.
6. Dubois, A. Phylogeny, taxonomy, and nomenclature: The problem of taxonomic categories and nomenclatural ranks. *Zootaxa* **2007**, *1519*, 27–58. [CrossRef]
7. Morrison, W.R.; Lohr, J.L.; Duchen, P.; Wilches, R.; Trujillo, D.; Mair, M.; Renner, S.S. The impact of taxonomic change on conservation: Does it kill, can it save, or is it just irrelevant? *Biol. Conserv.* **2009**, *142*, 3201–3206. [CrossRef]
8. Goodwin, Z.; Harris, D.J.; Filer, D.; Wood, J.R.I.; Scotland, R.W. Widespread mistaken identity in tropical plant collections. *Curr. Biol.* **2015**, *25*, R1066–R1067. [CrossRef]
9. Lian, L.; Shen, J.Y.; Ortiz, R.D.C.; Yu, S.X.; Chen, Z.D.; Wang, W. Phylogenetic analyses and modelling distributions guide conservation of a critically endangered liana species, *Eleutharrhena macrocarpa* (Menispermaceae). *Taxon* **2021**, *70*, 931–945. [CrossRef]
10. Zhang, M.; Xue, C.; Hu, H.; Li, J.; Xue, Y.; Wang, R.; Fan, J.; Zou, C.; Tao, S.; Qin, M.; et al. Genome-wide association studies provide insights into the genetic determination of fruit traits of pear. *Nat. Commun.* **2021**, *12*, 1144. [CrossRef]
11. Rubtsov, G.A. Geographical distribution of the genus *Pyrus* and trends and factors in its evolution. *Am. Nat.* **1944**, *78*, 358–366. [CrossRef]
12. Korotkova, N.; Parolly, G.; Khachatryan, A.; Ghulikyan, L.; Sargsyan, H.; Akopian, J.; Borsch, T.; Gruenstaeudl, M. Towards resolving the evolutionary history of Caucasian pears (*Pyrus*, Rosaceae)-Phylogenetic relationships, divergence times and leaf trait evolution. *J. Syst. Evol.* **2018**, *56*, 35–47. [CrossRef]
13. Dong, X.; Wang, Z.; Tian, L.; Zhang, Y.; Qi, D.; Huo, H.; Xu, J.; Li, Z.; Liao, R.; Shi, M.; et al. De novo assembly of a wild pear (*Pyrus betuleafolia*) genome. *Plant Biotechnol. J.* **2020**, *18*, 581–595. [CrossRef] [PubMed]
14. Browicz, K. Concept and chorology of the genus *Pyrus* L. *Arboret. Kórnick.* **1993**, *38*, 17–33.
15. Robertson, K.R.; Rohrer, J.R.; Rohrer, J.B.; Smith, P.G. A synopsis of genera in subfamily Maloideae (Rosaceae). *Syst. Bot.* **1991**, *16*, 376–394. [CrossRef]
16. Wu, J.; Wang, Z.; Shi, Z.; Zhang, S.; Ming, R.; Zhu, S.; Khan, M.A.; Tao, S.; Korban, S.S.; Wang, H.; et al. The genome of the pear (*Pyrus bretschneideri* Rehd.). *Genome Res.* **2013**, *23*, 396–408. [CrossRef] [PubMed]
17. Li, X.; Singh, J.; Qin, M.; Li, S.; Zhang, X.; Zhang, M.; Khan, A.; Zhang, S.; Wu, J. Development of an integrated 200K SNP genotyping array and application for genetic mapping, genome assembly improvement and genome wide association studies in pear (*Pyrus*). *Plant Biotechnol. J.* **2019**, *17*, 1582–1594. [CrossRef] [PubMed]
18. Shirasawa, K.; Itai, A.; Isobe, S. Chromosome-scale genome assembly of Japanese pear (*Pyrus pyrifolia*) variety 'Nijisseiki'. *DNA Res.* **2021**, *28*, b1. [CrossRef]
19. Zheng, X.; Cai, D.; Potter, D.; Postman, J.; Liu, J.; Teng, Y. Phylogeny and evolutionary histories of *Pyrus* L. revealed by phylogenetic trees and networks based on data from multiple DNA sequences. *Mol. Phylogenet. Evol.* **2014**, *80*, 54–65. [CrossRef]

20. Feng, S.; Bai, M.; Rivas-González, I.; Li, C.; Liu, S.; Tong, Y.; Yang, H.; Chen, G.; Xie, D.; Sears, K.E.; et al. Incomplete lineage sorting and phenotypic evolution in marsupials. *Cell* **2022**, *185*, 1646–1660.e18. [CrossRef]
21. Mu, X.Y.; Sun, M.; Yang, P.F.; Lin, Q.W. Unveiling the identity of wenwan walnuts and phylogenetic relationships of Asian *Juglans* species using restriction site-associated DNA-sequencing. *Front. Plant Sci.* **2017**, *8*, 1708. [CrossRef]
22. Liu, B.B.; Liu, G.N.; Hong, D.Y.; Wen, J. *Eriobotrya* belongs to *Rhaphiolepis* (Maleae, Rosaceae): Evidence from chloroplast genome and nuclear ribosomal DNA data. *Front. Plant Sci.* **2020**, *10*, 1731. [CrossRef] [PubMed]
23. Yü, T.T.; Ku, T.C. Pyrus. In *Flora Reipublicae Popularis Sinicae*; Yü, T.T., Ed.; Science Press: Beijing, China, 1974; Volume 36, pp. 354–372.
24. Wuyun, T.; Ma, T.; Uematsu, C.; Katayama, H. A phylogenetic network of wild Ussurian pears (*Pyrus ussuriensis* Maxim.) in China revealed by hypervariable regions of chloroplast DNA. *Tree Genet. Genomes* **2013**, *9*, 167–177. [CrossRef]
25. Yü, T.T.; Kuan, K.C. Taxa nova Rosacearum Sinicarum (I). *Acta Phytotaxon. Sin.* **1963**, *8*, 202–236.
26. Liang, T.T.; Ma, Y.; Guo, J.; Zang, D.K. Transcriptome sequencing and analysis of wild pear (*Pyrus hopeiensis*) using the Illumina Platform. *Arab. J. Sci. Eng.* **2016**, *41*, 45–53. [CrossRef]
27. Zang, R.G.; Dong, M.; Li, J.Q.; Chen, X.Y.; Zeng, S.J.; Jiang, M.X.; Li, Z.Q.; Huang, J.H. Conservation and restoration for typical critically endangered wild plants with extremely small population. *Acta Ecol. Sin.* **2016**, *36*, 7130–7135. [CrossRef]
28. Li, W.; Lu, Y.; Xie, X.; Li, B.; Han, Y.; Sun, T.; Xian, Y.; Yang, H.; Liu, K. Development of chloroplast genomic resources for *Pyrus hopeiensis* (Rosaceae). *Conserv. Genet. Resour.* **2018**, *10*, 511–513. [CrossRef]
29. Li, Y.; Zhang, J.; Li, L.; Gao, L.; Xu, J.; Yang, M. Structural and comparative analysis of the complete chloroplast genome of *Pyrus hopeiensis* "Wild Plants with a Tiny Population" and three other *Pyrus* species. *Int. J. Mol. Sci.* **2018**, *19*, 3262. [CrossRef]
30. Cao, Y.; Tian, L.; Gao, Y.; Liu, F. Genetic diversity of cultivated and wild Ussurian Pear (*Pyrus ussuriensis* Maxim.) in China evaluated with M13-tailed SSR markers. *Genet. Resour. Crop Evol.* **2012**, *59*, 9–17. [CrossRef]
31. Liu, Q.; Kang, M.; Jiang, Y. Some new recorded plants from Beijing and Hebei (II). *J. Beijing Norm. Univ.* **2003**, *39*, 674–676.
32. Wang, Q. Identity of *Microtoena coreana* (Lamiaceae), a doubtful species from Korea. *Phytotaxa* **2015**, *195*, 79–85. [CrossRef]
33. Wu, J.; Wang, Y.; Xu, J.; Korban, S.S.; Fei, Z.; Tao, S.; Ming, R.; Tai, S.; Khan, A.M.; Postman, J.D.; et al. Diversification and independent domestication of Asian and European pears. *Genome Biol.* **2018**, *19*, 77. [CrossRef] [PubMed]
34. Li, Y.; Zhang, J.; Wang, S.; Zhang, Y.; Yang, M. The distribution and origins of *Pyrus hopeiensis*-"Wild Plant With Tiny Population" using whole genome resequencing. *Front. Plant Sci.* **2021**, *12*, 668796. [CrossRef] [PubMed]
35. Velasco, R.; Zharkikh, A.; Affourtit, J.; Dhingra, A.; Cestaro, A.; Kalyanaraman, A.; Fontana, P.; Bhatnagar, S.K.; Troggio, M.; Pruss, D.; et al. The genome of the domesticated apple (*Malus* × *domestica* Borkh.). *Nat. Genet.* **2010**, *42*, 833–839. [CrossRef] [PubMed]
36. Li, H.; Durbin, R. Fast and accurate short read alignment with Burrows-Wheeler transform. *Bioinformatics* **2009**, *25*, 1754–1760. [CrossRef] [PubMed]
37. McKenna, A.; Hanna, M.; Banks, E.; Sivachenko, A.; Cibulskis, K.; Kernytsky, A.; Garimella, K.; Altshuler, D.; Gabriel, S.; Daly, M.; et al. The genome analysis toolkit: A MapReduce framework for analyzing next-generation DNA sequencing data. *Genome Res.* **2010**, *20*, 1297–1303. [CrossRef]
38. Nguyen, L.; Schmidt, H.A.; von Haeseler, A.; Minh, B.Q. IQ-TREE: A fast and effective stochastic algorithm for estimating maximum-likelihood phylogenies. *Mol. Biol. Evol.* **2015**, *32*, 268–274. [CrossRef] [PubMed]
39. Hoang, D.T.; Chernomor, O.; von Haeseler, A.; Minh, B.Q.; Vinh, L.S. UFBoot2: Improving the ultrafast bootstrap approximation. *Mol. Biol. Evol.* **2018**, *35*, 518–522. [CrossRef]
40. Jin, J.J.; Yu, W.B.; Yang, J.B.; Song, Y.; DePamphilis, C.W.; Yi, T.S.; Li, D.Z. GetOrganelle: A fast and versatile toolkit for accurate de novo assembly of organelle genomes. *Genome Biol.* **2020**, *21*, 241. [CrossRef]
41. Katoh, K.; Rozewicki, J.; Yamada, K.D. MAFFT online service: Multiple sequence alignment, interactive sequence choice and visualization. *Brief. Bioinform.* **2019**, *20*, 1160–1166. [CrossRef]
42. Qu, Y.; Wang, R.; Ma, C.; Li, D.; Song, J. Genetic diversity analysis of pear genus in Shanxi Province by SSR markers. *North. Hort.* **2015**, *39*, 11–17.
43. Xiang, Y.; Huang, C.; Hu, Y.; Wen, J.; Li, S.; Yi, T.; Chen, H.; Xiang, J.; Ma, H. Evolution of Rosaceae fruit types based on nuclear phylogeny in the context of geological times and genome duplication. *Mol. Biol. Evol.* **2017**, *34*, 262–281. [CrossRef] [PubMed]
44. Zhang, S.D.; Jin, J.J.; Chen, S.Y.; Chase, M.W.; Soltis, D.E.; Li, H.T.; Yang, J.B.; Li, D.Z.; Yi, T.S. Diversification of Rosaceae since the Late Cretaceous based on plastid phylogenomics. *New Phytol.* **2017**, *214*, 1355–1367. [CrossRef] [PubMed]
45. Liu, B.B.; Ren, C.; Kwak, M.; Hodel, R.G.J.; Xu, C.; He, J.; Zhou, W.B.; Huang, C.H.; Ma, H.; Qian, G.Z.; et al. Phylogenomic conflict analyses in the apple genus *Malus* s.l. reveal widespread hybridization and allopolyploidy driving diversification, with insights into the complex biogeographic history in the Northern Hemisphere. *J. Integr. Plant Biol.* **2022**, *64*, 1020–1043. [CrossRef] [PubMed]
46. Zheng, X.; Hu, C.; Spooner, D.; Liu, J.; Cao, J.; Teng, Y. Molecular evolution of Adh and LEAFY and the phylogenetic utility of their introns in *Pyrus* (Rosaceae). *BMC Evol. Biol.* **2011**, *11*, 255. [CrossRef]
47. Corlett, R.T. Safeguarding our future by protecting biodiversity. *Plant Divers.* **2020**, *42*, 221–228. [CrossRef]
48. Raven, P.; Wackernagel, M. Maintaining biodiversity will define our long-term success. *Plant Divers.* **2020**, *42*, 211–220. [CrossRef]
49. Bremer, K.; Bremer, B.; Karis, P.; Kallersjo, M. Time for change in taxonomy. *Nature* **1990**, *343*, 202. [CrossRef]

50. Raposo, M.A.; Stopiglia, R.; Brito, G.R.R.; Bockmann, F.A.; Kirwan, G.M.; Gayon, J.; Dubois, A. What really hampers taxonomy and conservation? A riposte to Garnett and Christidis (2017). *Zootaxa* **2017**, *4317*, 179–184. [CrossRef]
51. Hu, H.H. Notulae systematicae ad Florem Sinensium V. *Bull. Fan Mem. Inst. Biol.* **1934**, *5*, 305–306.
52. Zhang, W.P.; Cao, L.; Lin, X.R.; Ding, Y.M.; Liang, Y.; Zhang, D.Y.; Pang, E.L.; Renner, S.S.; Bai, W.N. Dead-end hybridization in walnut trees revealed by large-scale genomic sequence data. *Mol. Biol. Evol.* **2021**, *39*, msab308. [CrossRef]
53. Krofel, M.; Hatlauf, J.; Bogdanowicz, W.; Campbell, L.A.D.; Godinho, R.; Jhala, Y.V.; Kitchener, A.C.; Koepfli, K.P.; Moehlman, P.; Senn, H.; et al. Towards resolving taxonomic uncertainties in wolf, dog and jackal lineages of Africa, Eurasia and Australasia. *J. Zool.* **2022**, *316*, 155–168. [CrossRef]
54. Albani Rocchetti, G.; Armstrong, C.G.; Abeli, T.; Orsenigo, S.; Jasper, C.; Joly, S.; Bruneau, A.; Zytaruk, M.; Vamosi, J.C. Reversing extinction trends: New uses of (old) herbarium specimens to accelerate conservation action on threatened species. *New Phytol.* **2021**, *230*, 433–450. [CrossRef] [PubMed]
55. Sánchez-C., D.; Richardson, J.E.; Hart, M.; Serrano, J.; Cárdenas, D.; Gonzalez, M.A.; Cortés-B., R. A plea to DNA barcode type specimens: An example from *Micropholis* (Sapotaceae). *Taxon* **2022**, *71*, 154–167. [CrossRef]
56. Hart, M.L.; Forrest, L.L.; Nicholls, J.A.; Kidner, C.A. Retrieval of hundreds of nuclear loci from herbarium specimens. *Taxon* **2016**, *65*, 1081–1092. [CrossRef]
57. Silva, C.; Besnard, G.; Piot, A.; Razanatsoa, J.; Oliveira, R.P.; Vorontsova, M.S. Museomics resolve the systematics of an endangered grass lineage endemic to north-western Madagascar. *Ann. Bot.* **2017**, *119*, 339–351. [CrossRef] [PubMed]
58. Raxworthy, C.J.; Smith, B.T. Mining museums for historical DNA: Advances and challenges in museomics. *Trends Ecol. Evol.* **2021**, *36*, 1049–1060. [CrossRef]

diversity

MDPI

Article

Biogeography and Diversification of the Tropical and Subtropical Asian Genus *Gastrochilus* (Orchidaceae, Aeridinae)

Yang Li [1], Weitao Jin [1], Liguo Zhang [1], Peng Zhou [1], Yan Luo [2], Ziwei Zhu [3] and Xiaoguo Xiang [1,*]

[1] Jiangxi Province Key Laboratory of Watershed Ecosystem Change and Biodiversity, Institute of Life Science, School of Life Sciences, Nanchang University, Nanchang 330031, China; liyang192116@163.com (Y.L.); jin1234933@126.com (W.J.); zhangliguo1997@163.com (L.Z.); zhoupengrjcck@163.com (P.Z.)
[2] Xishuangbanna Tropical Botanical Garden, Chinese Academy of Sciences, Mengla 650223, China; luoyan@xtbg.org.cn
[3] Jiangxi Academy of Forest, Nanchang 330013, China; ziweiangel@163.com
* Correspondence: xiangxg2010@163.com

Abstract: Tropical and subtropical Asia are major orchid diversity and endemism centers. However, the evolutionary dynamics of orchids in these areas remain poorly studied. *Gastrochilus* D. Don, a species-rich orchid genus from tropical and subtropical Asian forests, was employed to investigate the issue. We firstly used eight DNA regions to reconstruct the phylogeny and estimate the divergence times within *Gastrochilus*. We inferred the ancestral ranges and conducted a diversification analysis based on empirical and simulated data. Subsequently, we assessed the ancestral niche state and tested for phylogenetic signals in the evolution of niche conditions. Our results suggested that the most recent common ancestor of *Gastrochilus* occurred in the subtropical area of the East Asiatic region in the late Miocene (8.13 Ma). At least eight dispersal events and four vicariant events were inferred to explain the current distribution of *Gastrochilus*, associated with the global cooling from the Plio-Pleistocene. The genus experienced a slowly decreasing diversification rate since its origin, and no significant correlation between current niches and phylogenetic relatedness was observed. The diversification of *Gastrochilus* was attributed to accumulation through time, integrated with the intensification of the Asian Monsoon system during the Plio-Pleistocene, pollination, and epiphytism.

Keywords: *Gastrochilus*; biogeography; diversification; niche; Asian Monsoon

Citation: Li, Y.; Jin, W.; Zhang, L.; Zhou, P.; Luo, Y.; Zhu, Z.; Xiang, X. Biogeography and Diversification of the Tropical and Subtropical Asian Genus *Gastrochilus* (Orchidaceae, Aeridinae). *Diversity* **2022**, *14*, 396. https://doi.org/10.3390/d14050396

Academic Editors: Jorge Capelo and Mario A. Pagnotta

Received: 7 March 2022
Accepted: 14 May 2022
Published: 17 May 2022

1. Introduction

Exploring the patterns of plant diversity today is a basic issue for biogeographers and evolutionists. To better understand the dynamics of plant diversity, it is vital to integrate the historical biogeography and the niche requirements of species [1]. Over the past two decades, the phylogenetic niche conservatism (PNC) and niche evolution (NE) hypotheses were proposed to account for species diversity [2,3]. PNC deems that most species tend to maintain their ancestral niches, survive in similar climatic environments, and differentiate in-situ [4]. Several studies have shown that the PNC hypothesis may explain the diversity of groups with different evolutionary histories [5–8]. For example, the diversity patterns of Zygophyllaceae at the global scale can be attributed to the strong phylogenetic conservatism in their precipitation-related niches [7]. In turn, NE posits that species may expand their niche breadth or occupy new conditions and can diversify in new habitats and climatic regimes [2]. For example, the diversity of *Hakea* (Proteaceae) in different geographic regions was explained by the frequency of evolutionary biome shifts [9]. Although there are many studies strongly supporting the role of PNC or NE in explaining clade diversification [4,7,9–11], several researchers suggested that there was no significant correlation between current niches and phylogenetic relatedness [12,13], especially the organisms in isolated habitats such as the birds in high-altitude regions [14], which implied that the species diversity was attributed to accumulation through time.

Tropical and subtropical Asia are major orchid diversity centers [15,16]. These regions are characterized by their high plant diversity and endemism [17,18] and have also been considered both a "Cradle" and a "Museum" for vascular plants since the Cretaceous [17,19,20]. During the Cenozoic, the Asian mainland experienced a series of complex geological and climate changes, such as the uplift of the Himalaya-Tibetan Plateau [21] and the establishment and intensification of the Asian monsoon [22]. The multistage uplifts of the Himalayas resulted in significant climate changes, new geophysical environments, novel ecological niches, and the formation of physical and physiological isolation barriers across the faunal and floral elements of Asia [23,24]. On the one hand, the uplift of the Himalayas provided many new niches and is attributed to organisms' diversification [25]. On the other hand, the four periods of intensification of the East Asia Summer Monsoon (EASM) during the Cenozoic have possibly brought abundant rainfall [26] and are positively correlated with plant richness [27–30]. Therefore, the uplift of the Himalayas and the EASM produced more new niches and conditions for organisms and are proposed to explain biological diversification in East Asia [29–31].

Gastrochilus D. Don (1825) (Aeridinae, Vandeae, Epidendroideae, Orchidaceae) is an epiphytic orchid genus widely distributed in tropical and subtropical Asia [15,32]. Thanks to its high morphological diversity and brightly colored flowers, it has potential horticultural value [15]. Since the latest preliminary revision of *Gastrochilus* [32], nearly 20 new species have been found in south China (Chongqing, Yunnan, Taiwan), Vietnam, Myanmar, Nepal and India [33–45]. Additionally, *Haraella* Kudo and *Luisiopsis* C.S.Kumar and P.C.S.Kumar have been transferred to *Gastrochilus* [46,47]. Therefore, the genus *Gastrochilus* consists now of 69 species, of which many are narrow endemics, with a species diversity center in the South-East Asian archipelago [36,48,49]. Recently, Liu et al. [49] revealed that *Gastrochilus* is monophyletic and divided into five clades based on five DNA regions (ITS, *matK*, *psbA-trnH*, *psbM-trnD*, *trnL-F*) and inferred that pollination system shifts in *Gastrochilus* have occurred independently at least three times. Liu et al. [50] reconstructed the phylogenetic relationships within the *Cleisostoma–Gastrochilus* clades (Aeridinae) based on the complete chloroplast genome, strongly supporting the monophyly of *Gastrochilus*. However, the spatio-temporal evolution of the genus is still unclear.

In this study, our objectives are (1) to estimate divergence times within *Gastrochilus* using eight plastid and nuclear DNA regions, (2) to investigate the historical biogeography of *Gastrochilus*, and (3) to explore the factors that have led to its diversification.

2. Materials and Methods

2.1. Taxon Sampling and Molecular Data

In this study, we sampled 34 species of *Gastrochilus*, comprehensively covering the distribution range of this genus. Based on Pridgeon et al. [15] and Farminhão et al. [51], 18 species closely related to *Gastrochilus* from Aeridinae and four species from Angraecinae were used as outgroups. All sequence data were downloaded from the GenBank (https://www.ncbi.nlm.nih.gov/ (accessed on 19 December 2021)). The phylogenetic analysis of Epidendroideae showed that a proportion of potentially parsimony informative sites of the internal transcribed spacer (ITS) *trnL-F*, *matK* and *rbcL* were 64%, 28%, 28%, and 11%, respectively, and they showed their strong ability to resolve species relationships [52]. It has been suggested that the addition of non-coding chloroplast regions could provide higher relative variability in resolving species relationships [53,54]. A total of eight DNA markers were employed in this study, including one nuclear marker (ITS) and seven chloroplast DNA markers (*atpH-I*, *matK*, *psbA-trnH*, *psbM-trnD*, *rbcL*, *trnL-F*, and *rps19-rpl22*). Taxon information and GenBank accession numbers are listed in Table S1.

2.2. Phylogenetic Analysis

DNA sequences were aligned and subsequently manually adjusted in BioEdit [55]. Topological congruence between the chloroplast and nuclear data was evaluated using the incongruence length difference (ILD) test [56]. The partition homogeneity test for plastid

DNA and ITS shows character incongruence ($p = 0.01$). Visual inspection indicates that there are very few "hard" conflicts between the plastid vs. ITS trees, and such conditions have been interpreted by Wendel and Doyle [57] as a soft incongruence, which might disappear with additional data. Gatesy et al. [58] demonstrated that concatenating truly incongruent data sets could still increase resolution and branch support. Therefore, we combined the datasets for subsequent analyses. All characters were unordered and had equal weight. Gaps were treated as missing data.

Three phylogenetic reconstruction methods were performed, including maximum parsimony (MP), maximum likelihood (ML), and Bayesian inference (BI). MP analyses were performed in PAUP* 4.0b10 [59]. Heuristic searches were conducted with 1000 replicates of random addition, in combination with tree-bisection-reconnection (TBR) branch-swapping, MulTrees in effect, and steepest descent off. Bootstrap support values were conducted with 1000 replicates with 10 random taxon additions and heuristic search options.

Based on the Akaike information criterion (AIC), the best-fit nucleotide substitution model of DNA regions was chosen using ModelTest v.3.7 [60]. ML analyses were conducted in RAxML v.8.4 [61]. We conducted a rapid bootstrap analysis (1000 replicates) and searched for the best-scoring ML tree simultaneously. BI analyses were performed in MrBayes v.3.2 [62]. Four Markov Chain Monte Carlo tests were run, sampling one tree every 1000 generations for 3,000,000 generations. Tracer v.1.5 was used to assess chain convergence and ensure that the effective sample sizes (ESS) are above 200 for all parameters [63]. Majority rule (>50%) consensus trees were constructed after removing the "burn-in" samples (the first 20% of the sampled trees).

2.3. Time Estimation

Firstly, the likelihood ratio test (LRT) [64] was conducted to determine whether the data evolved in a clock-like fashion. Log-likelihood ratios of the clock and non-clock model were compared. The degree of freedom is equivalent to the number of terminal taxa minus two, and significance was assessed by comparing two times the log-likelihood difference to a chi-square distribution [65]. The LRT test rejected a clock-like evolution ($\delta = 1363.8014$, df = 54, $p < 0.001$), and therefore. we used a relaxed lognormal clock model to estimate the divergence in BEAST v.2.6.0 [66]. There is no *Gastrochilus* fossil nor any fossils of one of its close relatives in Aeridinae and Angraecinae; thus, two calibration points were set based on Givnish et al. [16,67]: (1) the split age of Aeridinae and Angraecinae (21.21 Ma) was used for the tree root age, and a prior normal distribution (SD = 3.05) for the calibration point was assigned following the suggestion of Ho [68]; (2) the crown age of Aeridinae was set to 16 Ma with a normal distribution (SD = 1.0). The speciation prior was set as YULE, and the substitution model of DNA regions was selected as the GTR+I+Γ model. Markov Chain Monte Carlo (MCMC) searches were run for 100,000,000 generations and sampled every 1000 generations. Convergence was assessed by Tracer v.1.5 [63], and the effective sampling size for all parameters was >200. The maximum clade credibility (MCC) tree was computed by TreeAnnotator v.1.7.4 [69].

2.4. Biogeographical Analyses

Based on the extant distribution of *Gastrochilus* and outgroups, four main regions were categorized based on Takhajan [70]: East Asiatic region (A), Indian region (B), Indo-Chinese region (C) and Malesian region (D). The ancestral range reconstruction was inferred using the Statistical Dispersal–Extinction–Cladogenesis (S-DEC) model, as implemented in RASP [71]. In S-DEC, it summarizes biogeographic reconstructions across all user-supplied trees. The DEC model is applied to each ultrametric tree within a posterior distribution resulting from a Bayesian phylogenetic analysis. Subsequently, we calculated the probability of an ancestral range x at node n on a summary tree [71]. The MCC tree obtained from BEAST was chosen as the summary tree. The random 1000 trees from BEAST trees after burn-in were input to estimate probabilities of ancestral range at each node.

2.5. Ancestral State Reconstruction and Correlates of Diversification

Ancestral state reconstruction was performed using the maximum likelihood method implemented in BayesTrait v.4.0 [72]. Information about species habitats was compiled from online databases (www.gbif.org (accessed on 23 January 2022); www.orchidspecies.com (accessed on 23 January 2022); www.africanorchids.dk (accessed on 6 March 2022); www.iplant.cn (accessed on 10 January 2022)), and the taxonomic literatures [32–45], and we defined two states: (1) tropical (state 0); (2) subtropical (state 1) (Table S2).

The binary state speciation and extinction model (BiSSE) was used to examine whether the climatic zone is directly correlated with differential rates of diversification implemented in DIVERSITREE 0.9-6 [73]. To correct for non-random, incomplete sampling, we specified sampling fractions, i.e., the proportion of species in tropical Asia and in subtropical Asia that are included in the tree.

2.6. Diversification Analysis

Birth–death likelihood (BDL) models were used to test the significance of heterogeneity or the consistency of the temporal diversification rate [74]. The model selection was based on the difference in the AIC scores between the best-fitting rate-constant and rate-variable models (ΔAIC_{RC}). The calculations were performed using laser 2.3 [74].

To better understand diversification rates in *Gastrochilus*, we employed two methods to analyze rates. First, semi-logarithmic lineage-through-time (LTT) plots were constructed using the R package ape 2.5-1 [75]. The MCC tree was used to generate the tempo of diversification, and 1000 trees were sampled randomly from the converged BEAST trees to calculate a 95% credibility interval. Second, we used CLaDS (cladogenetic diversification rate shift model), a model-based approach to estimate speciation rates [76]. CLaDS applies a Bayesian approach to infer speciation rates along a phylogeny and assumes that rates change after every speciation event.

To evaluate the effect of the missing species, we add all 35 missing species randomly in the MCC tree in the R package 'phytools' 0.4-60 [77]. Then we carried out a diversification analysis in LTT and CLaDS.

2.7. Collection of Species Distribution Data and Environmental Variables

Distribution data of *Gastrochilus* were collected from online databases (the global biodiversity information facility, https://www.gbif.org/ (accessed on 23 January 2022)), herbaria (Herbarium, Institute of Botany, Academia Sinica (PE), and Herbarium of Jiangxi University (JXU)), and our fieldwork. These datasets were carefully assessed, and some erroneous records (i.e., occurrences in the oceans, ice sheets, and deserts), duplicates, and cultivation records were removed. Finally, a total of 262 unique distribution records from 33 species were used in this study (Table S3). We also collected 20 environmental variables including 19 bioclimatic variables and one topographical layer (elevation) (https://www.worldclim.org/data/worldclim21.html (accessed on 22 January 2022)) [78]. All environmental variables are at a resolution of 30 arc seconds. Mean values of the variables for each species were used in the further analysis.

2.8. Estimation of Evolutionary Rate in Niche Traits

To estimate the evolutionary rate of niche in *Gastrochilus*, we firstly ordinated all environmental variables and niche data using phylogenetic principal component analysis (PCA) implemented in R package phytools 0.7-70 [77] with the "phyl.pca" function. Then, we conducted complementary runs using the BAMM trait model on the first axis of the phylogenetic PCA of niche traits in BAMM 2.5.0 [79]. For niche rate, the MCMC was run for 10 million generations and sampled every 5000 generations. Prior values were selected using the "setBAMMpriors" function. Postrun analysis and visualization used the R package BAMMtools 2.17 [79]. The initial 25% of samples of the MCMC run were discarded as burn-in, and the remaining data were assessed for convergence using the CODA package [80] to ensure that the ESS values were above 200.

2.9. Detection of Phylogenetic Signals of Niche Traits

Phylogenetic signals were measured using Blomberg's K [81] and Pagel's λ [82]. We estimated the phylogenetic signals based on the time-calibrated tree and the first axis of the phylogenetic PCA of niche traits of *Gastrochilus* using the "phylosig" function in the R package phytools 0.7-70 [77].

3. Results

3.1. Phylogenetic Relationships and Divergence Time Estimates within Gastrochilus

The total length of combined DNA sequences was 10,065 bp, of which 1351 characters were variable, and 551 characters were parsimony informative. The monophyly of *Gastrochilus* was strongly supported (BI-PP = 1.00, ML-BP = 97, MP-BS = 95; Figure 1). The inter-species relationships within *Gastrochilus* were supported by moderate to high supporting values, but the relationships among the main clades have not been resolved (Figure 1).

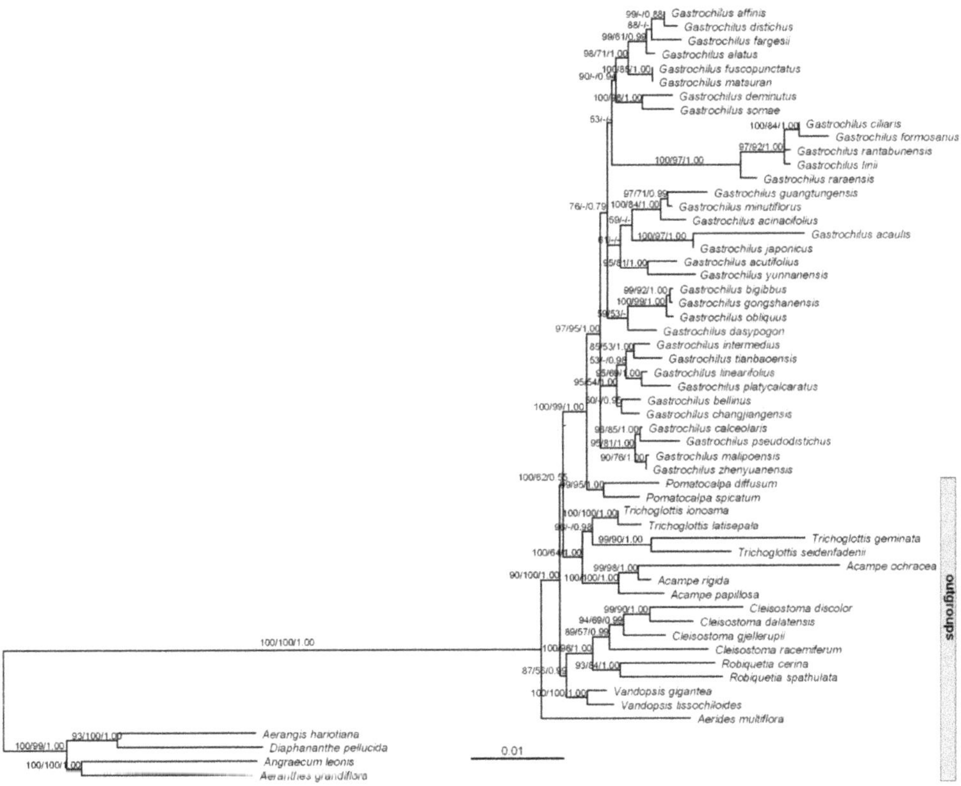

Figure 1. Phylogenetic tree obtained by the maximum likelihood method of the combination of nuclear and plastid regions. Numbers above the branches indicate supported values (>50%) from maximum likelihood, maximum parsimony, and Bayesian Inference methods, respectively. Numbers at the nodes are bootstrap percentages and Bayesian posterior probabilities, respectively. A dash (-) indicates that a node is not supported in the analysis.

Our results suggest a stem age of *Gastrochilus* at 9.49 Ma (95% highest probability density (HPD): 6.61–12.55; Figure 2a, node 1), and a crown age of *Gastrochilus* at 8.13 Ma (95% HPD: 5.51–10.83; Figure 2, node 2). Most of the species originated during the Pliocene and early Pleistocene (Figure 2a).

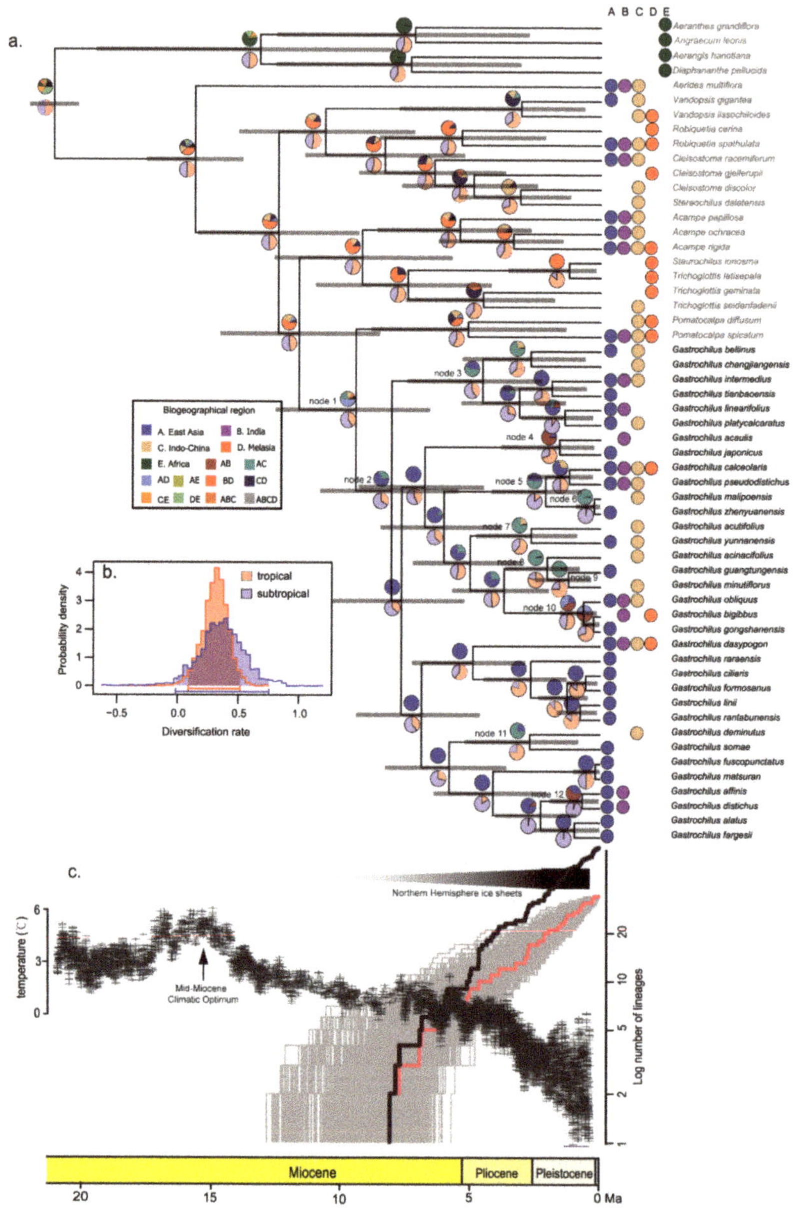

Figure 2. (**a**) Ancestral range reconstruction of *Gastrochilus* based on the chronogram. The chronogram was generated in BEAST analysis. Grey bars show 95% highest posterior density intervals. Nodes of interest were numbered from 1 to 12. The pie charts above the branches represent the results of ancestral range reconstruction, and those under the branches represent the results of habitat reconstruction. (**b**) Habitat-dependent posterior probability distribution of net diversification rates from BiSSE analyses. (**c**) LTT plots of *Gastrochilus* for empirical data (pink line) and simulated data (black line), respectively. The depiction of temperature changes is modified from Zachos et al. [83].

3.2. Ancestral Range Reconstruction

The ancestral area reconstruction of *Gastrochilus* based on S-DEC is shown in Figure 2a. The ancestral area of *Gastrochilus* is uncertain, although it probably originated in East Asia or Malesia (node 1). The range of the most recent common ancestor (MRCA) of the genus is inferred in East Asia (node 2). The current distribution of *Gastrochilus* is inferred to be the result of eight dispersal events and four vicariant events. There are three dispersal events from the East Asiatic region to the Indian region at 1.79 Ma (95% HPD: 0.25–3.86; node 4), 0.74 Ma (95% HPD: 0.13–1.56; node 10), and 0.80 Ma (95% HPD: 0.03–1.92; node 12), respectively. The remaining five dispersal events from the East Asiatic region to Indo-Chinese region happened at 4.70 Ma (95% HPD: 2.27–7.50; node 3), 2.29 Ma (95% HPD: 0.73–4.23; node 5), and 2.79 Ma (95% HPD: 0.89–4.81; node 7), 2.15 Ma (95% HPD: 0.79–3.68; node 8), and 2.86 Ma (95% HPD: 0.82–5.23; node 11), respectively. Additionally, there are three vicariant events that happened between East Asiatic and Indo-Chinese regions at 0.33 Ma (95% HPD: 0–0.95; node 6), 1.33 Ma (95% HPD: 0.35–2.52; node 9), and 2.86 Ma (95% HPD: 0.82–5.23; node 11), respectively. Only one diverged event occurred between East Asiatic and Indian regions (1.79 Ma, 95% HPD: 0.25–3.86; node 4).

3.3. Diversification of Gastrochilus

A positive ΔAIC_{RC} value suggests that the data are best approximated by a rate-variable model of diversification [74], so the BDL analysis rejected the null hypothesis of temporally homogeneous diversification rates within *Gastrochilus* ($\Delta AIC_{RC} = 2.07$). The BiSSE analysis indicated that the tropical Asian lineages and the subtropical Asian lineages presented the nearly same diversification rate (Figure 2b). The LTT plot showed that *Gastrochilus* exhibited a high rate of lineage accumulation since its divergence and then decreased slowly through time (Figure 2c, red line). Furthermore, the ClaDS showed there was no significant mean speciation rate shift during its evolutionary history, and the mean speciation rate decelerated very slowly from 8.13 Ma (95% HPD: 5.51–10.83) to the present (Figure 3c,d (green line)). The simulated analysis of LTT plots and CLaDS analysis also showed that the diversification rate and speciation rate increased at the early evolutionary stage and then decreased since the latest Miocene (Figure 2c (black line), Figure 3d (black line)), respectively.

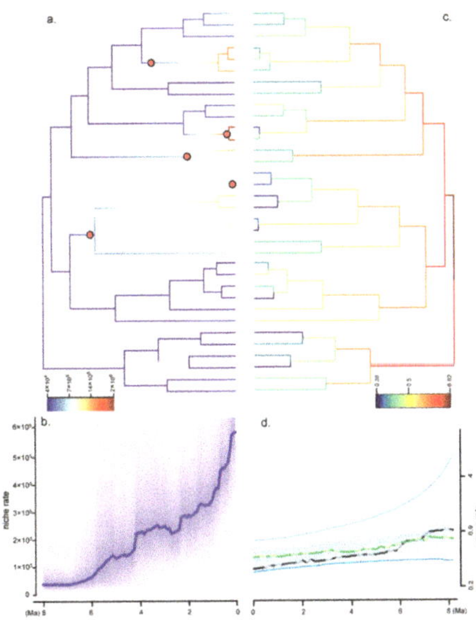

Figure 3. Niche analysis and diversification analysis of *Gastrochilus*. (**a**) Niche evolution and shift of

Gastrochilus. (**b**) Niche rate during the evolutionary history of *Gastrochilus*. (**c**) Inferred lineage-specific speciation rates for *Gastrochilus* phylogeny. (**d**) Inferred mean speciation rate of *Gastrochilus* through time, with individual MCMC iterations (thin blue line), the 95% credibility interval for each time point (thick blue line), the mean rate for each time point (dotted green) of empirical data, and the mean rate for each time point (dotted black) of simulated data. The unit of the diversification rate is speciation events per million years; niche rates are unitless.

3.4. Niche Evolution and Phylogenetic Signals

PC1 had a higher contribution from Bio12 (Annual precipitation), Bio16 (Precipitation of wettest quarter), elevation, Bio2 (Mean diurnal range), Bio18 (Precipitation of warmest quarter) and Bio13 (Precipitation of wettest month) (Table S4). Annual precipitation is the main influencing factor with a high PC1 loading value of about 0.88 (Table S4). Nine major shifts of the evolutionary rate of environmental factors were found in the genus *Gastrochilus*, six of which occurred in the last 2 Ma (Figure 3a). However, evolutionary rates of niche experienced a strong increase toward the present, beginning around the early Pliocene (Figure 3b). Notably, the annual precipitation seemed to play a key role in the rate shift of niche evolution (Table S4). The values of Blomberg's K and Pagel's λ are 0.175 and 0.023 with $p > 0.05$, respectively. This suggests that no significant phylogenetic signals were detected in the niche traits of *Gastrochilus*.

4. Discussion

4.1. Temporal and Spatial Mode of Gastrochilus

In the present study, our result strongly supported that *Gastrochilus* is monophyletic, which is consistent with the previous studies [49,50]. The S-DEC result inferred that the MRCA of *Gastrochilus* lived in the East Asiatic region (Figure 2a, node 2). The ancestor of *Gastrochilus* has migrated from the East Asiatic region to its adjacent regions (Indian region and Indo-Chinese region) since the early Pliocene at least eight times (Figure 2a). Following the middle Miocene climatic optimum at approximately 15 Mya, a period of global cooling began from ~11 Mya followed by the drastic temperature fluctuations during the Pliocene and Pleistocene [83]. The cooling climate caused many species to migrate southward or to lower altitudes [84–86], which also meant *Gastrochilus* dispersed from the East Asian region into the Indian and Indo-Chinese regions. However, it is well-known that the monsoon system (South Asian Summer Monsoon (SASM) and East Asian Summer Monsoon (EASM)) strengthened in the Asian mainland during the Miocene and Pliocene, especially between c. 15–4 Ma [87–89]. Recent studies reported that both South and East Asian Summer monsoons played a decisive role in the landscape evolution of the Himalayas and the adjoining areas in the Indo-Malayan Realm [90]. The intensifications of the EASM during the Late Cenozoic brought abundant rainfall and, therefore, significantly promoted the survival and differentiation of plants in tropical and subtropical Asian mainland [19,28–30,91]. During the dynamic evolutionary processes of *Gastrochilus*, five of eight migration events occurred from the East Asiatic region to Indo-Chinese region in the Pliocene to the early Pleistocene (4.70–2.15 Ma; Figure 2a, node 3, 5, 7, 8, 11), in agreement with the timing of the intensification of the EASM (3.6–2.6 Ma) [26] and SASM (3.57–2.78 Ma) [92]. Additionally, the other three events from the East Asiatic region to the Indian region happened during the late Pleistocene (1.79–0.74 Ma; Figure 2a, node 4, 10, 12), which is consistent with major changes in the monsoon cyclicity that occurred through the Mid-Pleistocene Transition between c. 1.25 and 0.7 Ma [89]. In addition, due to the drastic decline of temperature since the late Pliocene and the frequent temperature fluctuations during the Quaternary, the lineages of *Gastrochilus* experienced four vicariant events immediately (Figure 2a, node 4, 6, 9) or after dispersal events (Figure 2a, node 11). In general, the current distribution pattern of *Gastrochilus* does not appear to have occurred via long-distance dispersal. Rather, range expansions associated with a few vicariances are suggested here to explain this pattern.

4.2. Diversification and Niche Evolution of *Gastrochilus*

Our LTT plots indicate that the lineages of *Gastrochilus* had accumulated over its evolutionary time since it diverged from the sister groups (8.13 Ma, 95% HPD: 5.51–10.83; Figure 2c). The rate-through-time plots by CLaDS suggested that the mean speciation rates of *Gastrochilus* decreased slowly through its evolutionary history (Figure 3c,d). The diversification analyses of simulated data also indicated the same tendency of its evolutionary dynamics (Figures 2c and 3d). The same diversification pattern has been detected in *Cirrhopetalum* alliance (*Bulbophyllum*, Orchidaceae) [93]. Since the late Pliocene, the global cooling has intensified [83,88], and it might have brought about the slow speciation rates of *Gastrochilus*.

Although there are nine significant niche shifts in the evolutionary history of *Gastrochilus* (Figure 3a), both the values of Blomberg's K (0.75, $p > 0.05$) and Pagel's λ (0.023, $p > 0.05$) inferred that there are no significant phylogenetic signals in the niche traits of *Gastrochilus*. Furthermore, both the lineages in tropical Asia and subtropical Asia demonstrated a similar diversification rate (Figure 2b), although more than 65% of currently recognized species are restricted to tropical Asia. Our results imply that the species diversity of *Gastrochilus* is explained by accumulation through time. This result is similar to the diversification pattern of *Bulbophyllum* in tropical and subtropical Asia, in which species richness is most likely the result of a time-for-speciation effect since the late Miocene [94]. Our environmental niche analyses demonstrated that annual precipitation is an important environmental variable determining the distribution of *Gastrochilus* (Table S4). Statistically, more than 70% of the extant diversity within this genus was generated in the late Pliocene and the Early Pleistocene (Figure 2a). The intensifications of the EASM during the Pliocene brought abundant rainfall to the tropical and subtropical Asian mainland and probably facilitated the diversification of *Gastrochilus* with numerous dust-like seeds. Moreover, Givnish et al. [16] proposed that the remarkable diversity of orchids is apparently driven in part by the acquisition of pollinia, epiphytism, tropical distributions, CAM photosynthesis, pollination syndromes, and life on extensive tropical cordilleras. They also pointed out that shifts in net diversification are scale-dependent, and multiple factors—several of them interconnected—have contributed to orchid diversification at the genus level. Liu et al. [49] showed that the presence of epichile hairs has switched many times in *Gastrochilus*, representing a character state evolving as an adaptation to bee pollination [95], and thus speculated that pollination system shifts occurred independently at least three times in *Gastrochilus* [49]. The pollination shifts in promoting speciation are recorded in *Holcoglossum* [96]. Furthermore, except for a few species growing on rocks (e.g., *G. gongshanensis*), the genus is mainly found on the tree trunks in rainforests, broadleaved forests, or coniferous forests (Table S2) [37,48,50]. In a word, the diversification of *Gastrochilus* is not only a result of the intensification of monsoons in the last c. 10 Ma but is also attributed to the integration of pollination syndromes and epiphytism.

Supplementary Materials: The following are available online at https://www.mdpi.com/article/10.3390/d14050396/s1, Table S1: The samples and GenBank accession numbers used in this study, Table S2: The distribution, climatic zone, phenology, and habitat of species in this study, Table S3: The species distribution data used in this study, Table S4: Niche PCA loadings.

Author Contributions: Y.L. (Yang Li), Z.Z. and X.X. designed the research. P.Z., Y.L. (Yan Luo), L.Z., W.J. and X.X. collected and performed analyses. W.J., X.X. and Z.Z. drafted the manuscript, W.J., X.X., Y.L. (Yang Li), Y.L. (Yan Luo), P.Z., L.Z. and Z.Z. revised the manuscript. All authors have read and agreed to the published version of the manuscript.

Funding: This research was funded by the National Natural Science Foundation of China–Yunnan Joint Fund Project (grant number: U1802242), the National Natural Science Foundation of China (grant numbers: 31670212, 31300181, and 32060056), and Guangxi Key Laboratory Construction Project (grant number: 19-185-7).

Institutional Review Board Statement: Not applicable.

Data Availability Statement: Not applicable.

Conflicts of Interest: The authors declare no conflict of interest.

References

1. Smith, S.A.; Donoghue, M.J. Combining historical biogeography with niche modeling in the *Caprifolium* clade of *Lonicera* (Caprifoliaceae, Dipsacales). *Syst. Biol.* **2010**, *59*, 322–341. [CrossRef] [PubMed]
2. Wiens, J.; Donoghue, M.J. Historical biogeography, ecology and species richness. *Trends Ecol. Evol.* **2004**, *19*, 639–644. [CrossRef] [PubMed]
3. Donoghue, M.J.; Edwards, E.J. Biome shifts and niche evolution in plants. *Annu. Rev. Ecol. Evol. Syst.* **2014**, *45*, 547–572. [CrossRef]
4. Wiens, J.; Ackerly, D.D.; Allen, A.P.; Anacker, B.L.; Buckley, L.B.; Cornell, H.V.; Damschen, E.I.; Davies, J.T.; Grytnes, A.; Harrison, S.P.; et al. Niche conservatism as an emerging principle in ecology and conservation biology. *Ecol. Lett.* **2010**, *13*, 1310–1324. [CrossRef]
5. Couvreur, T.L.P.; Forest, F.; Baker, W.J. Origin and global diversification patterns of tropical rain forests: Inferences from a complete genus-level phylogeny of palms. *BMC Biol.* **2011**, *9*, 44. [CrossRef]
6. Pyron, R.A.; Costa, G.C.; Patten, M.A.; Burbrink, F.T. Phylogenetic niche conservatism and the evolutionary basis of ecological speciation. *Biol. Rev.* **2015**, *90*, 1248–1262. [CrossRef]
7. Wang, Q.; Wu, S.D.; Su, X.Y.; Zhang, L.J.; Xu, X.T.; Lyu, L.S.; Cai, H.Y.; Shrestha, N.; Liu, Y.P.; Wang, W.; et al. Niche conservatism and elevated diversification shape species diversity in drylands: Evidence from Zygophyllaceae. *Proc. R. Soc. B* **2018**, *285*, 20181742. [CrossRef]
8. Gagnon, E.; Ringlberg, J.J.; Bruneau, A.; Bruneau, A.; Lewis, G.P.; Hughes, C. Global Succulent Biome phylogenetic conservatism across the pantropical *Caesalpinia* Group (Leguminosae). *New Phytol.* **2019**, *222*, 1994–2008. [CrossRef]
9. Cardillo, M.; Weston, P.H.; Reynolds, Z.K.M.; Olde, P.M.; Mast, A.R.; Lemmon, E.M.; Lemmon, A.R.; Bromham, L. The phylogeny and biogeography of *Hakea* (Proteaceae) revealed the role of biome shifts in a continental plant radiation. *Evolution* **2017**, *71*, 1928–1943. [CrossRef]
10. Crisp, M.D.; Arroyo, M.T.K.; Cook, L.G.; Gandolfo, M.A.; Jordan, G.J.; McGlone, M.S.; Weston, P.H.; Westoby, M.; Wilf, P.; Linder, H.P. Phylogenetic biome conservation on a global scale. *Nature* **2009**, *458*, 754–756. [CrossRef]
11. Rice, A.; Smarda, P.; Novosolov, M. The global biogeography of polyploid plants. *Nat. Ecol. Evol.* **2019**, *3*, 265–273. [CrossRef]
12. Losos, J.B. Phylogenetic niche conservatism, phylogenetic signal and the relationship between phylogenetic related and ecological similarity among species. *Ecol. Lett.* **2018**, *11*, 995–1107. [CrossRef]
13. Liu, H.C.; Jacquemyn, H.; Chen, W.; Janssens, S.B.; He, X.Y.; Yu, S.; Huang, Y.Q. Niche evolution and historical biogeography of lady slipper orchids in North America and Eurasia. *J. Biogeogr.* **2020**, *48*, 2727–2741. [CrossRef]
14. Rice, N.H.; Martınez-Meyer, E.; Peterson, T. Ecological niche differentiation in the *Aphelocoma* jays: A phylogenetic perspective. *Biol. J. Linn. Soc.* **2003**, *80*, 369–383. [CrossRef]
15. Pridgeon, A.; Cribb, P.; Chase, M.; Rasmussen, F.N. *Genera Orchidacearum: Epidendroideae (Part Three)*; Oxford University Press: New York, NY, USA, 2014.
16. Givnish, T.J.; Spalink, D.; Ames, M.; Lyon, S.P.; Hunter, S.J.; Zuluaga, A.; Iles, W.J.D.; Clements, M.A.; Arroyo, M.T.K.; Leebens, M.J.; et al. Orchid phylogenomics and multiple drivers of their extraordinary diversification. *Proc. R. Soc. B* **2015**, *282*, 20151553. [CrossRef]
17. Wen, J. Evolution of eastern Asian and eastern North American disjunct distributions in flowering plants. *Annu. Rev. Ecol. Syst.* **1999**, *30*, 421–455. [CrossRef]
18. Wu, Z.Y.; Wu, S. A proposal for a new floristic kingdom (realm)—The East Asiatic Kingdom, its delineation and characteristics. In *Floristic Characteristics and Diversity of East Asian Plants*; Zhang, A.L., Wu, S.G., Eds.; Higher Education Press: Beijing, China; Springer Verlag Press: Berlin/Heidelberg, Germany, 1998; pp. 1–42.
19. Lu, L.M.; Mao, L.F.; Yang, T.; Ye, J.F.; Liu, B.; Li, H.L.; Sun, M.; Miller, J.T.; Mathews, S.; Hu, H.H.; et al. Evolutionary history of the angiosperm flora of China. *Nature* **2018**, *554*, 234–238. [CrossRef]
20. Mittermeier, R.A.; Fonseca, P.; Gil, R.; Hoffmann, M.; Pilgrim, J.; Brooks, T.; Mittermeier, C.G.; Lamoreux, J.; Fonseca, A.B.G. *Hotspots Revisited: Earth's Biologically Richest and Most Endangered Terrestrial Ecoregions*; Cermex: Mexico City, Mexico, 2005.
21. An, Z.S.; Kutzbach, J.E.; Prell, W.L.; Porter, W.L. Evolution of Asian monsoons and phased uplift of the Himalaya-Tibetan plateau since late Miocene times. *Nature* **2001**, *411*, 62–66.
22. Sun, X.J.; Wang, P.X. How old is the Asian monsoon system? Palaeobotanical records from China. *Palaeogeogr. Palaeoclimat. Palaeoecol.* **2005**, *222*, 181–222. [CrossRef]
23. Manish, K.; Pandit, M.K. Geophysical upheavals and evolutionary diversification of plant species in the Himalaya. *Peer J.* **2018**, *6*, e5919. [CrossRef]
24. Pusok, A.E.; Stegman, D.R. The convergence history of India-Eurasia records multiple subduction dynamics processes. *Sci. Adv.* **2020**, *6*, eaaz8681. [CrossRef] [PubMed]
25. Xing, Y.; Ree, R.H. Uplift-driven diversification in the Hengduan Mountains, a temperate biodiversity hotspot. *Proc. Natl. Acad. Sci. USA* **2017**, *114*, E3444–E3451. [CrossRef] [PubMed]

26. An, Z.S. *Late Cenozoic Climate Change in Asia: Loess, Monsoon and Monsoon-Arid Environment Evolution*; Springer: Dordrecht, The Netherlands, 2014.
27. Ashokan, A.; Xavier, A.; Suksathan, P.; Ardiyani, M.; Leong-Škorničková, J.; Newman, M.; Kress, W.J.; Gowda, V. Himalayan orogeny and monsoon intensification explain species diversification in an endemic ginger (*Hedychium*: Zingiberaceae) from the Indo-Malayan Realm. *Mol. Phylogenet. Evol.* **2022**, *170*, 107440. [CrossRef] [PubMed]
28. Xiang, X.G.; Mi, X.C.; Zhou, H.L.; Li, J.W.; Chung, S.W.; Li, D.Z.; Huang, W.C.; Jin, W.T.; Li, Z.Y.; Huang, L.Q.; et al. Biogeogrpahical diversification of mainland Asian *Dendrobium* (Orchidaceae) and its implications for the historical dynamics of evergreen broad-leaved forests. *J. Biogeogr.* **2016**, *43*, 1310–1323. [CrossRef]
29. Yu, X.Q.; Gao, L.M.; Soltis, D.E.; Soltis, P.S.; Yang, J.B.; Fang, L.; Yang, S.X.; Li, D.Z. Insights into the historical assembly of east Asian subtropical evergreen broadleaved forests revealed by the temporal history of the tea family. *New Phytol.* **2017**, *215*, 1235–1248. [CrossRef]
30. Chen, X.H.; Xiang, K.L.; Lian, L.; Peng, H.W.; Erst, S.A.; Xiang, X.G.; Chen, Z.D.; Wang, W. Biogeographic diversification of *Mahonia* (Berberidaceae): Implications for the origin and evolution of East Asian subtropical evergreen broadleaved forests. *Mol. Phylogenet. Evol.* **2020**, *151*, 106910. [CrossRef]
31. Li, X.Q.; Xiang, X.G.; Zhang, Q.; Jabbour, F.; Ortiz, R.D.C.; Erst, A.S.; Li, Z.Y.; Wang, W. Immigration dynamics of tropical and subtropical Southeast Asian limestone karst floras. *Proc. R. Soc. B* **2022**, *289*, 20211308. [CrossRef]
32. Tsi, Z.H. A preliminary revision of *Gastrochilus* (Orchidaceae). *Guihaia* **1996**, *16*, 123–154.
33. Yi, S.R.; Huang, Y.; Xiao, B.; Shen, M.L.; Li, J. *Gastrochilus brevifimbriatus* (Orchidaceae, Aeridinae), a New Species from Chongqing, China. *Novon* **2010**, *20*, 113–116.
34. Kumar, P.; Gale, S.W.; Kocyan, A.; Fischer, G.A.; Averyanov, L.V.; Borosova, R.; Bhattacharjee, A.; Li, J.H.; Pang, K.S. *Gastrochilus kadooriei* (Orchidaceae), a new species from Hong Kong, with notes on allied taxa in section *Microphyllae* found in the region. *Phytotaxa* **2014**, *164*, 91–103. [CrossRef]
35. Raskoti, B.B. A new species of *Gastrochilus* and new records for the orchids of Nepal. *Phytotaxa* **2015**, *233*, 179. [CrossRef]
36. Averyanov, L.V.; Van, C.N.; Truong, V.B.; Maisak, T.; Luu, H.T.; Nguyen, K.S.; Diep, D.Q.; Hoang, T.; Tuan, N.H.; Chu, X.C.; et al. New orchids (Orchidaceae: Cymbidieae and Vandeae) in the Flora of Vietnam. *Taiwania* **2018**, *63*, 119–138.
37. Hsu, T.C.; Hsieh, S.I.; Wu, J.H.; Hung, H.C. *Gastrochilus deltoglossus* (Orchidaceae: Epidendroideae: Vandeae: Aeridinae), a new species from Taiwan. *Taiwania* **2018**, *63*, 360–365.
38. Rao, W.H.; Liu, Z.J.; Zhang, G.Q.; Chen, X.H.; Huang, J.; Chen, G.Z.; Chen, L.J. A new epiphytic species of *Gastrochilus* (Orchidaceae: Epidendroideae) from Yunnan, China. *Phytotaxa* **2019**, *413*, 296–300.
39. Wu, X.F.; Ye, D.P.; Pan, B.; Lin, X.Q.; Jiang, H.; Liu, Q. Validation of *Gastrochilus prionophyllus* (Vandeae, Orchidaceae), a new species from Yunnan province, China. *PhytoKeys* **2019**, *130*, 161–169. [CrossRef]
40. Liu, Q.; Tan, Y.H.; Gao, J.Y. A new species of *Gastrochilus* (Aeridinae, Vandeae, Orchidaceae) and a new record species from Yunnan, China. *Phytotaxa* **2016**, *282*, 66–70. [CrossRef]
41. Liu, Q.; Gao, J.Y. *Gastrochilus dulongjiangensis* (Aeridinae, Vandeae, Epidendroideae, Orchidaceae), a new species from Yunnan province, China. *Phytotaxa* **2018**, *340*, 293–296. [CrossRef]
42. Liu, Q.; Kumar, P.; Gao, J.Y. Notes on *Gastrochilus gongshanensis* (Orchidaceae). *Kew Bull.* **2019**, *74*, 21. [CrossRef]
43. Liu, Q.; Zhou, S.S.; Li, R.; Tan, Y.H.; Zyaw, M.; Xing, X.K.; Gao, J.Y. Notes on the genus *Gastrochilus* (Orchidaceae) in Myanmar. *PhytoKeys* **2020**, *138*, 113–123. [CrossRef]
44. Ormerod, P.; Kumar, C.S. A new Bhutanese *Gastrochilus* (Orchidaceae), and another transfer in Indian *Cylindrolobus*. *Harv. Pap. Bot.* **2020**, *25*, 125–127. [CrossRef]
45. Li, J.W.; Ya, J.D.; Ye, D.P.; Liu, C.; Liu, Q.; Pan, R.; He, Z.X.; Pan, B.; Cai, J.; Lin, D.L.; et al. Taxonomy notes on Vandeae (Orchidaceae) from China: Five new species and two new records. *Plant Diver.* **2021**, *43*, 379–389. [CrossRef] [PubMed]
46. Smith, J.J. *Gastrochilus odoratus* (Kudô) J.J. Smith. *Bull. Jard. Bot. Buitenzorg* **1937**, *14*, 168.
47. Chase, M.W.; Cameron, K.M.; Freudenstein, J.V.; Pridgeon, A.M.; Salazar, G.; Berg, C.V.D.; Schuiteman, A. An updated classification of Orchidaceae. *Bot. J. Linn. Soc.* **2015**, *177*, 151–174. [CrossRef]
48. Chen, S.C.; Tsi, Z.H.; Wood, J.J. Gastrochilus D. Don. In *Flora of China Vol. 25 (Orchidaceae)*; Wu, Z.Y., Raven, P.H., Hong, D.Y., Eds.; Science Press: Beijing, China; Missouri Botanical Garden Press: St. Louis, MI, USA, 2009; pp. 491–498.
49. Liu, Q.; Song, Y.; Jin, X.H.; Gao, J.Y. Phylogenetic relationships of *Gastrochilus* (Orchidaceae) based on nuclear and plastid DNA data. *Bot. J. Linn. Soc.* **2019**, *189*, 228–243. [CrossRef]
50. Liu, D.K.; Tu, X.D.; Zhao, Z.; Zeng, M.Y.; Zhang, S.; Ma, L.; Zhang, G.Q.; Wang, M.M.; Liu, Z.J.; Lan, S.R.; et al. Plastid phylogenomic data yield new and robust insights into the phylogeny of *Cleisostoma-Gastrochilus* clades (Orchidaceae, Aeridinae). *Mol. Phylogenet. Evol.* **2020**, *145*, 10729. [CrossRef] [PubMed]
51. Van den Berg, C.; Goldman, D.H.; Freudenstein, J.V.; Pridgeon, A.M.; Cameron, K.M.; Chase, M.W. An overview of the phylogenetic relationships within Epidendroideae inferred from multiple DNA regions and circumscription of Epidendreae and Arethuseae (Orchidaceae). *Am. J. Bot.* **2005**, *92*, 613–645. [CrossRef]
52. Farminhão, J.N.M.; Verlynde, S.; Kaymak, E.; Droissart, V.; Simo-Droissart, M.; Collobert, G.; Martos, F.; Stévart, T. Rapid radiation of angraecoids (Orchidaceae, Angraecinae) in tropical Africa characterised by multiple karyotypic shifts under major environmental instability. *Mol. Phylogenet. Evol.* **2021**, *159*, 107105. [CrossRef]

53. Shaw, J.; Lickey, E.B.; Beck, J.T.; Farmer, S.B.; Liu, W.S.; Miller, J.; Siripun, K.C.; Winder, C.T.; Schilling, E.E.; Small, R.L. The tortoise and the hare II: Relative utility of 21 noncoding chloroplast DNA sequences for phylogenetic analysis. *Am. J. Bot.* **2005**, *92*, 142–166. [CrossRef]
54. Shaw, J.; Lickey, E.B.; Schilling, E.E.; Small, R.L. Comparison of whole chloroplast genome sequences to choose noncoding regions for phylogenetic studies in angiosperms: The tortoise and the hare III. *Am. J. Bot.* **2007**, *94*, 275–288. [CrossRef]
55. Hall, T.A. BioEdit: A user-friendly biological sequence alignment editor and analysis program for Windows 95/98/NT. *Nucl. Acids Symp. Ser.* **1999**, *41*, 95–98.
56. Farris, J.S.; Källersjö, M.; Kluge, A.G.; Bult, C. Testing significance of incongruence. *Cladistics* **1994**, *10*, 315–319. [CrossRef]
57. Wendel, J.; Doyle, J. Phylogenetic incongruence, window into genome history and evolution. In *Molecular Systematics of Plants II: DNA Sequencing*; Soltis, D.E., Soltis, P.S., Doyle, J.J., Eds.; Kluwer Academic Publishing: Dordrecht, The Netherlands, 1998; pp. 265–296.
58. Gatesy, J.; O'Grady, P.; Baker, R.H. Corroboration among data sets in simultaneous analysis, hidden support for phylogenetic relationships among higher level Artiodactyl taxa. *Cladistics* **1999**, *15*, 271–313. [CrossRef]
59. Swofford, D.L. *PAUP 4.0b10: Phylogenetic analysis using parsimony (and other methods)*; Sinauer: Sunderland, UK, 2002.
60. Posada, D.; Crandall, K.A. Modeltest: Testing the model of DNA substitution. *Bioinformatics* **1998**, *14*, 817–818. [CrossRef]
61. Stamatakis, A. RAxML version 8: A tool for phylogenetic analysis and post-analysis of large phylogenies. *Bioinformatics* **2014**, *30*, 1312–1313. [CrossRef]
62. Ronquist, F.; Teslenko, M.; van der Mark, P.; Ayres, D.L.; Darling, A.; Höhna, S.; Larget, B.; Liu, L.; Suchard, M.A.; Huelsenbeck, J.P. MrBayes 3.2: Efficient Bayesian phylogenetic inference and model choice across a large model space. *Syst. Biol.* **2012**, *61*, 539–542. [CrossRef]
63. Rambaut, A.; Drummond, A.J. Tracer v1.5 [online]. Available online: http://beast.bio.ed.ac.uk/Tracer (accessed on 30 November 2009).
64. Felsenstein, J. Phylogenies from molecular sequences: Inference and reliability. *Ann. Rev. Genet.* **1988**, *22*, 521–565. [CrossRef]
65. Sorhannus, U.; Bell, C.V. Testing for equality of molecular evolutionary rates: A comparison between a relative-rate test and a likelihood ratio test. *Mol. Biol. Evol.* **1999**, *16*, 849–855. [CrossRef]
66. Bouckaert, R.; Heled, J.; Kühnert, D.; Vaughan, T.; Wu, C.H.; Xie, D.; Suchard, M.A.; Rambaut, A.; Drummond, A.J. BEAST 2: A Software platform for Bayesian evolutionary analysis. *PLoS Comput. Biol.* **2014**, *10*, e1003537. [CrossRef]
67. Givnish, T.J.; Spalink, D.; Ames, M.; Lyon, A.P.; Hunter, S.J.; Zuluaga, A.; Doucette, A.A.; Caro, G.G.; McDaniel, J.; Clements, M.A.; et al. Orchid historical biogeography, diversification, Antarctica and the paradox of orchid dispersal. *J. Biogeogr.* **2016**, *43*, 1905–1916. [CrossRef]
68. Ho, S.Y.W. Calibrating molecular estimates of substitution rates and divergence times in bird. *J. Avian Biol.* **2007**, *38*, 409–414. [CrossRef]
69. Drummond, A.J.; Rambaut, A. BEAST: Bayesian evolutionary analysis by sampling trees. *BMC Evol. Biol.* **2007**, *7*, 214. [CrossRef]
70. Takhtajan, A. *Floristic Regions of the World*; University of California Press: Berkeley, CA, USA; Los Angeles, CA, USA, 1986.
71. Yu, Y.; Harris, A.J.; Blair, C.; He, X.J. RASP (Reconstruct Ancestral State in Phylogenies): A tool for historical biogeography. *Mol. Phylogenet. Evol.* **2015**, *87*, 46–49. [CrossRef]
72. Pagel, M.; Meade, A. BayesTraits: Computer Program and Documentation Version 4.0. 2022. Available online: http://www.evolution.reading.ac.uk/BayesTraitsV4.0.0.html (accessed on 1 February 2022).
73. FitzJohn, R.G. Diversitree: Comparative phylogenetic analyses of diversification in R. *Mol. Phylogenet. Evol.* **2012**, 3, 1084–1092. [CrossRef]
74. Rabosky, D.L. LASER: A maximum likelihood toolkit for detecting temporal shits in diversification rates from molecular phylogenies. *Evol. Bioinform. Online* **2006**, *2*, 257–260. [CrossRef]
75. Paradis, E.; Claude, J.; Strimmer, K. APE: Analyses of phylogenetics and evolution in R language. *Bioinformatics* **2004**, *20*, 289–290. [CrossRef] [PubMed]
76. Maliet, O.; Morlon, H. Fast and accurate estimation of species-specific diversification rates using data augmentation. *Syst. Biol.* **2022**, *71*, 353–366. [CrossRef] [PubMed]
77. Revell, L.J. Phytools: An R package for phylogenetic comparative biology (and other things). *Mol. Phylogenet. Evol.* **2012**, *3*, 217–223. [CrossRef]
78. Fick, S.E.; Hijmans, R.J. WorldClim 2: New 1-km spatial resolution climate surfaces for global land areas. *Int. J. Climatol.* **2017**, *37*, 4302–4315. [CrossRef]
79. Rabosky, D.L.; Grundler, M.; Anderson, C.; Title, P.; Shi, J.J.; Brown, J.W.; Huang, H.; Larson, J.G.; Kembel, S. BAMMtools: An R package for the analysis of evolutionary dynamics on phylogenetic trees. *Mol. Phylogenet. Evol.* **2014**, *5*, 701–707. [CrossRef]
80. Plummer, M.; Best, N.; Cowles, K.; Vines, K. CODA: Convergence diagnosis and output analysis for MCMC. *R News* **2006**, *6*, 7–11.
81. Blomberg, S.P.; Garland, T.; Ives, A.R. Testing for phylogenetic signal in comparative data: Behavioral traits are more labile. *Evolution* **2003**, *57*, 717–745. [CrossRef]
82. Pagel, M. Inferring the historical patterns of biological evolution. *Nature* **1999**, *401*, 877–884. [CrossRef]
83. Zachos, J.C.; Dickens, G.R.; Zeebe, R.E. An early Cenozoic perspective on greenhouse warming and carbon-cycle dynamics. *Nature* **2008**, *451*, 279–283. [CrossRef]

84. Zhang, J.B.; Li, R.Q.; Xiang, X.G.; Manchester, S.R.; Lin, L.; Wang, W.; Wen, J.; Chen, Z.D. Integrated fossil and molecular data reveal the biogeographic diversification of the eastern Asian-eastern North American disjunct hickory genus (*Carya* Nutt.). *PLoS ONE* **2013**, *8*, e70449. [CrossRef]
85. Kou, Y.X.; Cheng, S.M.; Tian, S.; Li, B.; Fan, D.M.; Chen, Y.J.; Soltis, D.E.; Soltis, P.S.; Zhang, Z.Y. The antiquity of *Cyclocarya paliurus* (Juglandaceae) provides new insights into the evolution of relict plants in subtropical China since the late Tertiary Miocene. *J. Biogeogr.* **2016**, *43*, 351–360. [CrossRef]
86. Ye, X.Y.; Ma, P.F.; Yang, G.Q.; Guo, C.; Zhang, Y.X.; Chen, Y.M.; Guo, Z.H.; Li, D.Z. Rapid diversification of alpine bamboos associated with the uplift of the Hengduan Mountains. *J. Biogeogr.* **2019**, *46*, 2678–2689. [CrossRef]
87. Filipelli, G.M. Intensification of the Asian monsoon and a chemical weathering event in the late Miocene-early Pliocene: Implications for late Neogene climate change. *Geology* **1997**, *25*, 27–30. [CrossRef]
88. Osborne, C.P. Atmosphere, ecology and evolution: What drove the Miocene expansion of C4 grasslands? *J. Ecol.* **2008**, *96*, 35–45. [CrossRef]
89. Ding, W.N.; Ree, R.H.; Spicer, R.A.; Xing, Y.W. Ancient orogenic and monsoon-driven assembly of the world's richest temperate alpine flora. *Science* **2020**, *369*, 578–581. [CrossRef]
90. Spicer, R.A.; Yang, J.; Herman, A.B.; Kodrul, T.; Aleksandrova, G.; Maslova, N.; Spicer, T.E.V.; Ding, L.; Xu, Q.; Shukla, A.; et al. Paleogene monsoons across India and South China: Drivers of biotic change. *Gondwana Res.* **2017**, *49*, 350–363. [CrossRef]
91. Chang, Z.; Xiao, J.; Lu, L.; Yao, H. Abrupt shifts in the Indian monsoon during the Pliocene marked by high-resolution terrestrial records from the Yuanmou Basin in southwest China. *J. Asian Earth Sci.* **2010**, *37*, 166–175. [CrossRef]
92. Willeit, M.; Ganopolski, A.; Calov, R.; Brovkin, V. Mid-Pleistocene transition in glacial cycles explained by declining CO_2 and regolith removal. *Sci. Adv.* **2019**, *5*, eaav7337. [CrossRef]
93. Hu, A.Q.; Gale, S.W.; Liu, Z.J.; Fischer, G.A.; Saunders, R.M.K. Diversification slowdown in the *Cirrhopetalum* alliance (*Bulbophyllum*, Orchidaceae): Insights from the evolutionary dynamics of crassulacean acid metabolism. *Front. Plant Sci.* **2022**, *13*, 794171. [CrossRef]
94. Gamisch, A.; Comes, H.P. Clade-age-dependent diversification under high species turnover shapes species richness disparities among tropical rainforest lineages of *Bulbophyllum* (Orchidaceae). *BMC Evol. Biol.* **2019**, *19*, 93. [CrossRef]
95. Takamiya, T.; Wongsawad, P.; Sathapattayanon, A.; Tajima, N.; Suzuki, S.; Kitamura, S.; Shioda, N.; Handa, T.; Kitanaka, S.; Lijima, H.; et al. Molecular phylogenetics and character evolution of morphologically diverse groups, *Dendrobium* section *Dendrobium* and allies. *AoB Plants* **2014**, *6*, plu045. [CrossRef]
96. Xiang, X.G.; Li, D.Z.; Jin, X.H.; Hu, H.; Zhou, H.L.; Jin, W.T.; Lai, Y.J. Monophyly or paraphyly– the taxonomy of *Holcoglossum* (Aeridinae: Orchidaceae). *PLoS ONE* **2012**, *7*, e52050. [CrossRef] [PubMed]

diversity

Article

Effects of Bamboo Forest Type and Density on the Growth of *Bletilla striata* and Root Endophytic Fungi

Hao Fu [1,2], Zhilin Song [2], Shanmin Li [1], Siren Lan [2], Xinhua Zeng [1,*] and Weichang Huang [1,2,*]

[1] Eastern China Conservation Centre for Wild Endangered Plant Resources, Shanghai Chenshan Botanical Garden, Shanghai 201602, China; fuhaoxsxz@163.com (H.F.); lishanmin2022@163.com (S.L.)
[2] College of Art and Landscape Architecture, Fujian Agriculture and Forestry University, Fuzhou 350002, China; songzlinlin@163.com (Z.S.); lsr9636@163.com (S.L.)
* Correspondence: zengxinhua@csnbgsh.cn (X.Z.); huangweichang@csnbgsh.cn (W.H.)

Abstract: *Bletilla striata* is a terrestrial orchid with high ornamental and medicinal values that is widely interplanted in bamboo forests. However, little is known about the effects of bamboo forest type and density on the growth of *B. striata* and its symbiotic relationship with root endophytic fungi. In this study, the growth state of *B. striata*, the community composition and diversity of its root endophytic fungal, and the fungal nutritional function were investigated in *Phyllostachys edulis*, *P. iridescens* and *P. glauca* forests with three densities. We found that the type and density of the bamboo forest had significant effects on the growth of *B. striata*, with the biomass, leaf width, root quantity and width being the highest in the low-density of the *P. edulis* forest. The community composition and abundance of root endophytic fungi in *B. striata* varied among different bamboo forests and densities, with *P. edulis* and *P. iridescens* forests dominated by Basidiomycota and *Serendipita*, while *P. glauca* prevailed by Ascomycota and *Dactylonectria*. The trophic modes of root endophytic fungi were also affected by forest types and densities. The abundance of symbiotroph fungi was the highest in *P. edulis* and *P. iridescens* forests and greatly varied with density gradient, and saprotrophic fungi comprised the highest proportion in the *Ph. glauca* forest. These results provide basic data for further research and the configuration between bamboo species and terrestrial orchids.

Keywords: *Bletilla striata*; compound system of bamboo and orchid; endophytic fungi; fungal diversity; symbiotic relationship

Citation: Fu, H.; Song, Z.; Li, S.; Lan, S.; Zeng, X.; Huang, W. Effects of Bamboo Forest Type and Density on the Growth of *Bletilla striata* and Root Endophytic Fungi. *Diversity* **2022**, *14*, 391. https://doi.org/10.3390/d14050391

Academic Editor: Kevin Cianfaglione

Received: 8 April 2022
Accepted: 12 May 2022
Published: 14 May 2022

1. Introduction

As an artificial ecosystem that considers ecological, social and economic benefits, the agroforestry system is of great significance for ecological environmental protection and sustainable development of agroforestry [1]. Agroforestry systems are conducive to optimizing the rational use of resources and space by plants, improving system biodiversity and the internal ecological environment of the system (e.g., light, heat, water, soil, air, fertilizer, microorganism) [2,3]. Because of their advantages of diverse output and income balance, agroforestry systems are widely adopted in many parts of the world, especially in developing countries, and play an important role in the national economy and in people's livelihood [4].

Agroforestry systems involve a wide range of woodland types, with bamboo forests being one of the most important types. Bamboo is one of the most important species in terrestrial forest ecosystems that are widely distributed in tropical and subtropical regions, being the second largest forest in the world [5,6]. Bamboo forests have important economic, ecological and social values [7]. Compared with single crop systems, an agroforestry system based on bamboo forests has higher net primary productivity, which has attracted wide international attention especially in tropical regions [8,9]. A bamboo forest composite system has many modes, mainly including bamboo-grass, bamboo-medicinal plants, bamboo-fungus, bamboo cultivation and forest tourism [7,10], of which bamboo-medicinal herbs are one of the most popular modes at this stage. Researchers have found that there are a variety of medicinal plants suitable for planting in bamboo forests, such as *Polygonatum sibiricum*, *Lophatherum gracile*, *Sarcandra glabra*, *Bletilla striata* and *Polygonatum odoratum* [11], and the impact of bamboo forest composite systems on medicinal plants varies among different stand types, bamboo forest density and habitats [12,13]. Huang et al., reported that canopy density of 0.4~0.6 is the most effective stand density for polysaccharide accumulation of *Polygonatum cyrtonema* [14]. Zhou found that the understory habitat had a significant impact on the growth traits of *Polygonatum cyrtonema*, and the Moso bamboo forest was the best interplanting site [13]. Feng found that different medicinal plants have different density requirements for bamboo forests, with *Polygonatum cyrtonema* adapting to all density ranges, while *Curculigo orchioides* prefers low density and *Paris polyphylla* medium densities [15]. However, at present, research on composite systems of bamboo-medicinal plants have mainly focused on their economic and social value, and thus, little is known about their ecological value, especially soil physical and chemical properties, plant landscape and the interaction between plant species in bamboo forests.

Bletilla striata (Thunb. ex A. Murray) Rchb. f. is a perennial herb mainly distributed across China, the Korean Peninsula and Japan, and it belongs to the genus *Bletilla* in the tribe Arethuseae, Orchidaceae [16]. Owing to its high ornamental and medicinal value, *B. striata* is widely used as an ornamental plant and as traditional medicinal material in Asia [17]. *Bletilla striata* is a typical mycorrhizal plant which relies on root endophytic fungi (both mycorrhizal and non-mycorrhizal fungi) throughout its life cycle, especially during seed germination and seedling recruitment periods [18,19]. Researchers have found that root endophytic fungi can not only transport carbohydrates and break down cellulose in the matrix but can also directly provide nutrients and hormones (e.g., amino acids, gibberellins and jasmonate) for plant growth [20,21]. In addition, root endophytic fungi were found to promote the absorption of macronutrient elements and micronutrient elements by plants [22,23] and facilitate the production of metabolites, including antibiotics, phenolic compounds, peroxidase and hydrolase, thus enhancing disease resistance and stress tolerance in orchids [24,25]. Mycorrhizal partners can also influence orchid distributions and determine which habitats allow orchid growth and what environmental factors are critical for orchid recruitment [26].

As a typical plant with a high potential for compound cultivation, *B. striata* has been widely interplanted with bamboo forests in East China. The growth of *B. striata* and its relationship with root endophytic fungi are greatly influenced by many factors such as temperature, light intensity, nutrients, humidity and soil pH, which are strongly influenced by bamboo forest type and density. Zeng et al. found that there were significant temporal variations in the diversity and community composition of root endophytic fungi of *B. striata* [27]. Ma et al. found that nitrogen had a strong influence on the growth and polysaccharides accumulation of *B. striata*, and ammonium-nitrate mixed nitrogen at the concentration of 15 mmol$\cdot$L^{-1} had the best results [28]. By adjusting the density of bamboo forests in time, the understory can meet various demands of light radiation [12]. Meanwhile, the diurnal and seasonal changes of temperature, moisture and light conditions varied among different bamboo forests and stand densities [13]. Currently, there have been some studies on the cultivation of *B. striata* under bamboo forests. However, most of these studies focused on the water and soil conservation function, land use capability and production efficiency of bamboo forests [29,30]. Few studies have emphasized the landscape effect and root endophytic microbial community of *B. striata* under different bamboo species and its densities.

In this study, we selected three common types of bamboo (*Phyllostachys edulis* (Carr.) J.Houz., *Phyllostachys iridescens* C. Y. Yao et S. Y. Chen and *Phyllostachys glauca* McClure) in Shanghai, China, intercropped with *B. striata* to study the effects of different types of bamboo and bamboo density on the growth of *B. striata* and the community composition of root endophytic fungi. Our main objectives are to answer the following questions: (1) What are the mycorrhizal partners of *B. striata* in a bamboo forest? (2) Does the community composition of root endophytic fungi vary among different bamboo forest types or different stand densities? (3) What are the most suitable types of bamboo forest and planting density for the growth of *B. striata* and its mycorrhizal association. Our study will provide data and research ideas for the plant configuration under bamboo forests and the application of orchids under the forests.

2. Materials and Methods

2.1. Overview of the Study Area

The research site was located in Shanghai Chenshan Botanical Garden, which is in the north subtropical monsoon humid climate, with an annual average temperature of 15.6 °C, annual sunshine duration of 1817 h and precipitation of 1313 mm [31]. The bamboo garden was constructed north of the botanical garden in 2016 and covered an area of approximately 3 hm^2. There are more than 70 bamboo species, and *Phyllostachys edulis*, *P. iridescens* and *P. glauca* were chosen as our experiment materials.

2.2. The Sample Set

Bamboo forests of *Phyllostachys edulis*, *P. iridescens* and *P. glauca* were divided into high-, middle- and low-density bamboo forests. The densities of the *P. edulis* forest were 14,000, 8500 and 4500 plants/hm^2, respectively; the densities of the *P. iridescens* forest were 30,000, 22,000 and 15,000 plants/hm^2, respectively, and the densities of the *P. glauca* forest were 60,000, 45,000 and 30,000 plants/hm^2, respectively. In February 2021, three 4 × 4 m plots were established in each forest density. Each plot was repeated three times, which resulted in a total of 27 plots. One-year tissue culture plantlets of *B. striata* which derived from the same batch and suffered the same hardening–seedling process were used in this study. *B. striata* was planted at 30 × 30 cm spacing in each plot in early March 2021.

2.3. Sample Collection and Processing

Root and soil samples of *B. striata* were collected at the end of July 2021. Five healthy *B. striata* plants were randomly selected from each plot. Two vegetative roots were selected from each plant, and the root samples from five plants were mixed as one root sample. The soil on the surface of *B. striata* root samples was washed first with clean water and then with sterile water for 30 s in an ultra-clean workbench. The samples were then washed in 75% alcohol for 30 s, soaked in 3.5% NaClO for 4 min, and washed three times with sterile water. The treated *B. striata* roots were placed in a clean plastic bag and stored at −80 °C for high-throughput sequencing. Simultaneously, the soil samples around the root were collected, and the fresh soil samples were separated as soon as possible and preserved at −20 °C. The rest of the soil was ground after natural air drying and screened through a 60-mesh sieve to determine soil physical and chemical properties together with fresh soil samples.

In mid-October 2021, 10 healthy *B. striata* plants were randomly selected from each plot and taken back to the laboratory for washing. After washing, the plant height, leaf length, leaf width and root length were measured using a measuring tape. Vernier calipers were used to measure the stem thickness, root width and other indicators. The fresh weight of *B. striata* was measured on an electronic balance. Finally, the *B. striata* plants were heated at 105 °C for 20 min and dried to a constant weight at 65 °C. The dry weight was then measured.

2.4. Environmental Factor Index Measurement

Soil pH was measured using the potentiometric method. Electrical conductivity (EC) was measured using the extraction and drying method. Total nitrogen (TN), total phosphorus (TP) and total potassium (TK) were determined using the Kjeldahl method, the molybdenum–antimony resistance colorimetric method and NaOH melting flame spectrophotometry, respectively. Ammonium nitrogen (AN) and nitrate nitrogen (NN) were determined using potassium chloride-indophenol blue colorimetry and ultraviolet spectrophotometry, respectively. Available phosphorus (AP) and available potassium (AK) were determined by sodium bicarbonateHCl extraction and ammonium acetate extraction—flame spectrophotometry, respectively. Organic matter (OM) and microbial carbon (MBC) were determined using potassium dichromate external heating and chloroform fumigation methods, respectively. The moisture content (MC) and soil bulk density (SBD) were measured using the drying and ring knife methods, respectively [32]. Light intensity (LI) in the bamboo forests was measured using a Xima-AS823 illuminance meter (Shenzhen, China) in the growing season of *B. striata* (March to August). Five points were measured in each plot, and the average value was determined six times a month.

2.5. High-Throughput Sequencing

An E.Z.N.A.® soil DNA kit (Omega Bio-Tek, Norcross, GA, USA) was used to extract the total community DNA, and 1% agarose gel electrophoresis was used to assess the quality of the DNA. The nuclear ribosomal internal transcribed spacer-1 (ITS-1) region was amplified with the primers ITS1F (5′-CTTGGTCATTTAGagGaAGTAA-3′) and ITS2R (5′-gCTGCGTTCTTCATCGATG-3). The amplification process included: pre-denaturation at 95 °C for 3 min, denaturation at 95 °C for 30 s, annealing at 55 °C for 30 s and extension at 72 °C for 45 s, 35 cycles. It was stably extended at 72 °C for 10 min and finally stored at 10 °C. The PCR reaction system was as follows: 5× TransStart FastPfu Buffer 2 μL, 2.5 mmol/L dNTPs 2 μL, 5 μmol/L upstream primer 0.8 μL, 5 μmol/L downstream primer 0.8 μL, TransStart FastPfu DNA polymerase 0.2 μL and template DNA10 ng, fill to 20 μL. The PCR product was extracted from 2% agarose gel and purified using the AxyPrep DNA Gel Extraction Kit (Axygen Biosciences, Union City, CA, USA) according to manufacturer's instructions and quantified using a Quantus™ Fluorometer (Promega, Madison, WI, USA).

Purified amplicons were pooled in equimolar and paired-end sequenced (2 × 300) on an Illumina MiSeq platform (Illumina, San Diego, CA, USA) according to the standard protocols from Majorbio Bio-Pharm Technology Co. Ltd. (Shanghai, China). Raw fastq files were demultiplexed and quality filtered with Trimmomatic according to the following criteria: (i) The 300 bp reads were truncated at any site receiving an average quality score of <20 over a 50 bp sliding window, and the truncated reads shorter than 50 bp were discarded; reads containing ambiguous characters were also discarded. (ii) Only overlapping sequences longer than 10 bp were assembled according to their overlapped sequence. The maximum mismatch ratio of overlap region is 0.2. Reads that could not be assembled were discarded.

2.6. Data Processing and Analysis

The cloud platform of Shanghai Majorbio Technology Co., Ltd. (Shanghai, China) was used for interactive cloud analysis of the biological information (http://www.i-sanger.com, accessed on 3 May 2021). UPARSE software (version 7.1 http://drive5.com/uparse/, accessed on 3 May 2021) was used to perform operational taxonomic unit (OTU) clustering and eliminate chimeras based on 97% similarity. Each sequence was annotated for species classification using the RDP classifier (http://rdPh.cme.msu.edu/, accessed on 3 May 2021) and compared with the SILVA and UNITE databases. The OTUs were used to analyze the composition of the microbial community structure, its α-diversity, microbial correlations and microbial function prediction, and then the graphs were created. SPSS 22.0 (IBM, Inc., Armonk, NY, USA) was used for statistical analyses. One-way ANOVA was used to compare the differences of the growth indices of *B. striata* and the diversity of root endophytic fungi under different bamboo forests and different stand densities for each bamboo forest, and a least significant difference (LSD) multiple comparison was used to analyze the significance ($p < 0.05$). All the graphs were plotted in Origin 2018 (OriginLab, Northampton, MA, USA) and Adobe Illustrator CC 2018 (San Jose, CA, USA).

3. Results

3.1. Effects of Bamboo Forest Type and Density on the Growth of *B. striata*

The leaf width and dry weight of the aboveground and underground biomass, root length, root width and root number of *B. striata* were significantly higher in the *Phyllostachys edulis* forest than in the *P. iridescens* forest, while those in the *P. iridescens* forest were significantly higher than those in the *P. glauca* forest (Figure 1C–E,G–I). The dry weight ratio of the aboveground to the underground biomass of *B. striata* in the *P. iridescens* forest was significantly the highest, and plant height and leaf length were the largest in the three bamboo forests (Figure 1A,B,F).

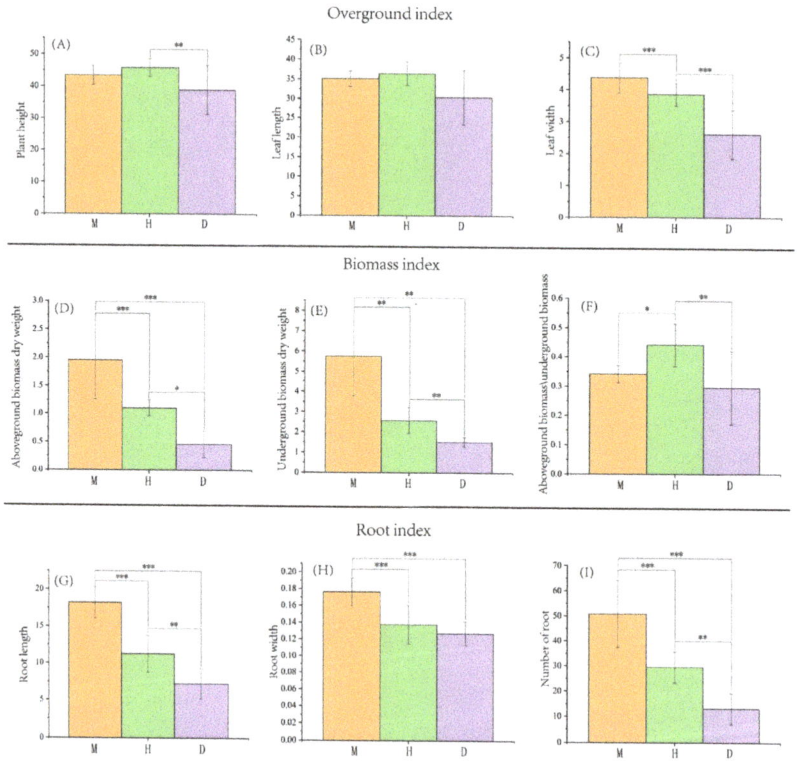

Figure 1. Growth indices of *Bletilla striata* under different bamboo forests. (**A**) Plant height; (**B**) Leaf length; (**C**) Leaf width; (**D**) Aboveground biomass dry weight; (**E**) Underground biomass dry weight; (**F**) Aboveground biomass/underground biomass; (**G**) Root length; (**H**) Root width; (**I**) Number of root; M, H, D: *Phyllostachys edulis, P. iridescens, P. glauca*, respectively. * $p < 0.05$. ** $p < 0.01$. *** $p < 0.001$.

There were also differences in the growth of *B. striata* under different densities of bamboo forests. The leaf width, root width, dry weight of the aboveground and belowground biomass of *B. striata* in *P. edulis* forest were significantly higher in the low-density than in the high-density sections of the forest. The leaf length in the low-density group in *P. iridescens* was significantly higher than that in the high-density group, and the root number and belowground biomass dry weight in the medium density group were significantly higher than those in the high-density group; the dry weight ratio of the aboveground and belowground biomass in the high-density group was significantly higher than that in the low-density group (Table 1).

Table 1. Growth indices of *Bletilla striata* under different planting densities (mean ± SD).

		Plant Height (cm)	Leaf Length (cm)	Leaf Width (cm)	Root Length (cm)	Root Width (cm)	Number of Roots (num.)	Aboveground Biomass (g)	Belowground Biomass (g)	Aboveground/Belowground Biomass
P. edulis	high	44.24 ± 4.97 [a]	35.51 ± 0.78 [a]	4.01 ± 0.22 [b]	17.60 ± 2.38 [a]	0.16 ± 0.00 [b]	41.3 ± 4.49 [a]	1.32 ± 0.18 [b]	3.6 ± 0.28 [b]	0.36 ± 0.02 [a]
	medium	42.38 ± 0.08 [a]	34.39 ± 1.19 [a]	4.28 ± 0.48 [a b]	18.44 ± 2.15 [a]	0.17 ± 0.01 [a b]	48.83 ± 14.37 [a]	1.92 ± 0.57 [a b]	5.9 ± 1.17 [a]	0.32 ± 0.03 [a]
	low	43.64 ± 2.70 [a]	35.39 ± 3.49 [a]	4.84 ± 0.28 [a]	18.50 ± 2.70 [a]	0.19 ± 0.02 [a]	62.27 ± 12.56 [a]	2.63 ± 0.52 [a]	7.75 ± 1.14 [a]	0.34 ± 0.02 [a]
P. iridescens	high	45.45 ± 2.93 [a]	36.12 ± 3.29 [a b]	3.97 ± 0.44 [a]	10.14 ± 2.14 [a]	0.14 ± 0.03 [a]	22.77 ± 1.00 [b]	0.97 ± 0.05 [a]	1.89 ± 0.200 [b]	0.52 ± 0.07 [a]
	medium	43.89 ± 1.93 [a]	34.14 ± 1.82 [b]	3.92 ± 0.48 [a]	10.01 ± 2.80 [a]	0.14 ± 0.02 [a]	33.97 ± 6.69 [a]	1.18 ± 0.19 [a]	2.77 ± 0.62 [a]	0.43 ± 0.03 [a b]
	low	48.18 ± 2.10 [a]	39.03 ± 1.88 [a]	3.69 ± 0.16 [a]	13.62 ± 0.38 [a]	0.12 ± 0.02 [a]	31.73 ± 2.29 [a]	1.15 ± 0.05 [a]	3.06 ± 0.26 [a]	0.38 ± 0.02 [b]
P. glauca	high	40.53 ± 3.17 [a]	31.39 ± 2.76 [a]	2.64 ± 0.38 [a]	7.08 ± 2.20 [a]	0.12 ± 0.02 [a]	13.05 ± 3.27 [a]	0.5 ± 0.17 [a]	1.65 ± 0.13 [a]	0.30 ± 0.08 [a]
	medium	33.68 ± 6.35 [a]	25.67 ± 5.82 [a]	2.17 ± 0.60 [a]	6.61 ± 1.79 [a]	0.12 ± 0.01 [a]	9.72 ± 3.28 [a]	0.31 ± 0.20 [a]	1.29 ± 0.27 [b]	0.24 ± 0.12 [a]
	low	42.44 ± 11.30 [a]	33.85 ± 10.17 [a]	3.03 ± 1.10 [a]	7.83 ± 2.65 [a]	0.13 ± 0.02 [a]	16.55 ± 9.58 [a]	0.57 ± 0.31 [a]	1.61 ± 0.05 [a]	0.35 ± 0.18 [a]

Same lowercase letters within a column of the same bamboo forest indicate no significant difference among different densities ($p < 0.05$).

3.2. Composition of Endophytic Fungal OTUs in *B. striata* under Different Bamboo Forest Types and Densities

The classification diagram of endophytic fungal OTUs in *B. striata* roots shows the number of common and unique OTUs in the root samples from three bamboo forests and the OTU comparison among different densities in the same bamboo forest. As shown in Figure 2, the number of endophytic fungal OTUs in *B. striata* under different bamboo species was as high as 421 in the *P. iridescens* forest. There were 80 fungal OTUs shared among the three bamboo forests, while there were 213, 247 and 108 fungal OTUs specific to a single forest of *P. edulis*, *P. iridescens* and *P. glauca*, respectively. There also existed some fungal OTUs shared by two bamboo forests (Figure 2A).

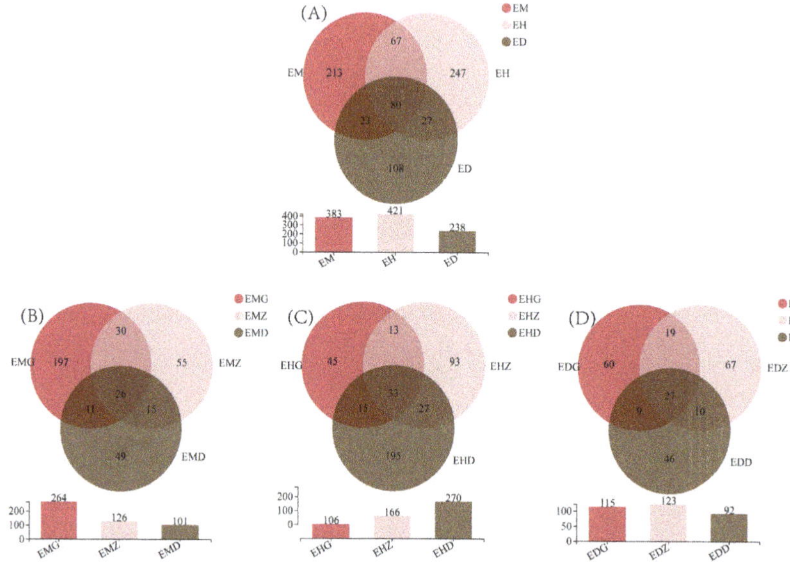

Figure 2. Venn diagram of the OTU classification of endophytic fungi in the *Bletilla striata* roots under different planting types and densities: (**A**) three bamboo forest models; (**B**) *Phyllostachys edulis* model; (**C**) *P. iridescens* model; (**D**) *P. glauca* model; EM, EH, ED: *P. edulis, P. iridescens, P. glauca*, respectively; EMG, EMZ, EMD: *P. edulis* at high-, medium- and low-density, respectively; EHG, EHZ, EHD: *P. iridescens* at high-, medium- and low-density, respectively; EDG, EDZ, EDD: *P. glauca* at high-, medium- and low-density, respectively. OTU, operational taxonomic unit.

The number of OTUs of *B. striata* under different densities of bamboo also differed. The number of OTUs in the high-density group of *P. edulis* was as high as 264, and the number of endemic species was 197, which was significantly higher than that in the medium-density group (all: 126, endemic species: 55) and the-low density group (all: 101, endemic species: 49) (Figure 2B). However, the trend was opposite in the *P. iridescens* forest. The highest number of OTUs was 270 in the low-density group, and 195 were endemic species, which were higher than those in the medium-density group (all: 166, endemic species: 93) and high-density group (all: 106, endemic species: 45) (Figure 2C). The difference in the numbers of OTUs was relatively small under different densities in the *P. glauca* forest. The highest number of OTUs was 123 in the medium-density group (Figure 2D).

3.3. Diversity Analysis of the Endophytic Fungal Community in *B. striata*

At the level of OTUs, the results of the sample coverage index showed that the coverage rate in the collected samples was 99.977 to 99.984% (Figure 3C). These results indicated that the microbial information in each sample was fully measured, and the results represented the actual situation of endophytic fungi in the *B. striata* roots. The α-diversity index of the endophytic fungi in the *B. striata* roots from three bamboo forests did not differ significantly, but all showed that *P. iridescens* > *P. edulis* > *P. glauca* (Figure 3A,B).

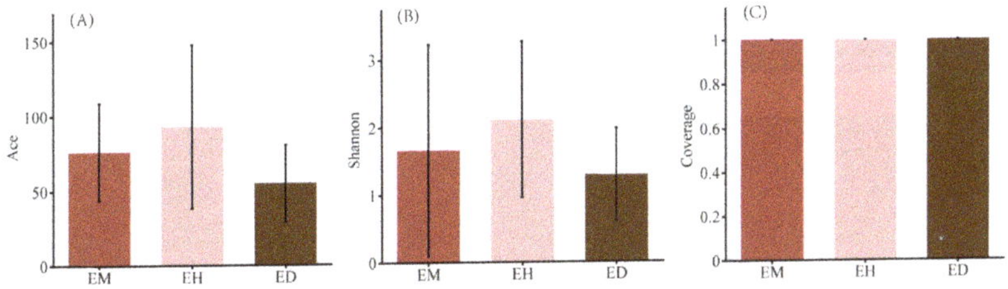

Figure 3. Analysis of the α-diversity of endophytic fungi in the *Bletilla striata* roots at OTU level under different bamboo forests: EM, *Phyllostachys edulis* model; EH, *P. iridescens* model; ED, *P. glauca* model. (**A**) Ace; (**B**) Shannon; (**C**) Coverage.

Compared with the bamboo forest type, the stand density had a stronger influence on the diversity and richness of endophytic fungi in the *B. striata* roots. The diversity index in the *P. edulis* high-density group was 3.32, which was significantly higher than that in the low-density group (0.39), and the richness index was the largest in the high-density group. However, the difference was not significant (Figure 4A,D). Among the three density groups of the *P. iridescens* forest, the richness index of the low-density group was significantly the highest (159.7), followed by the medium-density group (69.9) and the high-density group (48.16) (Figure 4B). There was no significant difference in the richness and diversity index of the *P. glauca* forest, which were both relatively low in the low-density group (Figure 4C,F).

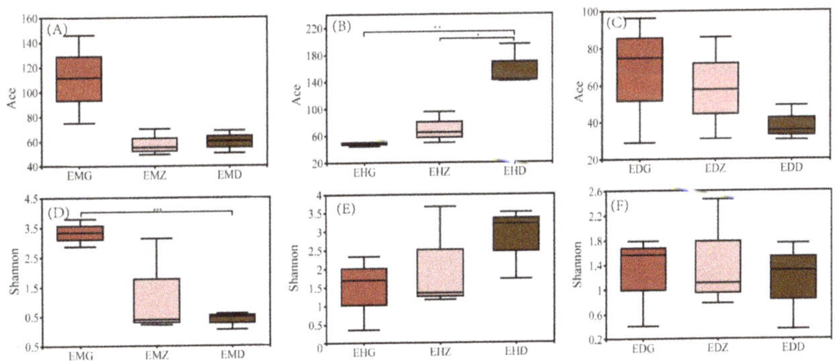

Figure 4. Analysis of the α-diversity of the endophytic fungi in *Bletilla striata* roots at OTU level at different densities: EMG, EMZ, EMD, *Phyllostachys edulis* at high-, medium- and low-density, respectively; EHG, EHZ, EHD, *P. iridescens* at high-, medium- and low-density, respectively; EDG, EDZ, EDD, *P. glauca* at high-, medium- and low-density, respectively. (**A–C**), Ace for *P. edulis*, *P. iridescens* and *P. glauca*, respectively; (**D–F**), Shannon for *P. edulis*, *P. iridescens* and *P. glauca*, respectively. * $p < 0.05$. ** $p < 0.01$. *** $p < 0.001$.

3.4. Analysis of the Community Composition of Endophytic Fungi in B. striata Roots

The community composition of endophytic fungi in *B. striata* roots under different bamboo forest types and densities was studied by high-throughput sequencing technology. As indicated in Figure 5, the endophytic fungi (relative abundance > 0.01) in the roots of *B. striata* in the bamboo forests included five phyla. Those in which the phyla were dominated by Basidiomycota and Ascomycota comprised the highest proportion (Figure 5A).

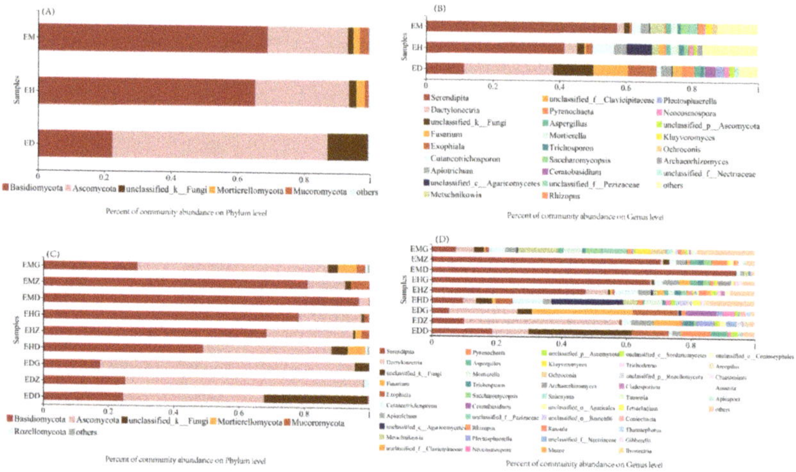

Figure 5. Abundance of the endophytic fungi from *Bletilla striata* roots in different grouping patterns. Community composition of the endophytic fungi from *B. striata* roots of three bamboo forests at the phylum (**A**) and genus (**B**) levels. Community composition of endophytic fungi of the *B. striata* roots at the phylum (**C**) and genus (**D**) levels at different densities: EM, EH, ED, *Phyllostachys edulis, P. iridescens, P. glauca*, respectively; EMG, EMZ, EMD, *P. edulis* at high-, medium- and low-density, respectively; EHG, EHZ, EHD, *P. iridescens* at the high-, medium- and low-density, respectively; EDG, EDZ, EDD, *P. glauca* at the high-, medium- and low-density, respectively.

The relative abundance of *B. striata* fungi differed significantly among different types of bamboo forests. Basidiomycota was the dominant phylum of endophytic fungi in the *P. edulis* and *P. iridescens* forests, comprising 68.87% and 65.18%, respectively, while Ascomycota comprised 24.31% and 28.43%, respectively. The opposite was true in the *P. glauca* forest. Ascomycota was the dominant phylum that comprised 64.72%, while Basidiomycota comprised 22.47% (Figure 5A). The abundance of endophytic fungi in the *B. striata* roots also clearly changed under different densities. The abundance of Basidiomycota increased with the decrease in density of *P. edulis* forests, comprising 28.95% in the high-density group and 96.74% in the low-density group. The opposite was true in the *P. iridescens* forest. The abundance of Basidiomycota decreased with the decrease in bamboo density, comprising 78.18% in the high-density group and 48.96% in the low-density group. The overall abundance of Basidiomycota was low in the *P. glauca* forest, and the variation in the abundance was small among the three density groups. In addition, the abundance of Ascomycota at different densities changed in response to changes in Basidiomycota (Figure 5C).

At the genus level, *Serendipita* was the dominant genus of endophytic fungi in *B. striata* in the *P. edulis* and *P. iridescens* forests, comprising 57.30% and 41.46%, respectively. The community composition of endophytic fungi in *B. striata* in the *P. glauca* forest tended to be complex. The proportion of *Dactylonectria* increased to 26.70%, while that of *Serendipita* was only 11.38% (Figure 5B). The dominant genera of endophytic fungi in *B. striata* in the *P. edulis* and *P. iridescens* forests varied substantially under different densities of bamboo forests. The abundance of *Serendipita* increased as the density in *P. edulis* forest decreased, comprising 7.48% in the high-density group and 93.91% in the low-density group. The opposite was true in the *P. iridescens* forest. The abundance of *Serendipita* decreased with decreasing density, comprising 67.30% in the high-density group and 9.71% in the low-density group. *Dactylonectria* was the dominant species in the *P. glauca* forest, comprising the highest proportion (47.92%) in the medium-density group. The abundance of *Serendipita* changed similarly in the *P. edulis* forest, comprising 5.40% in high-density and 18.69% in low-density conditions (Figure 5D).

3.5. LefSe Difference Analysis of Endophytic Fungi Community in the B. striata Roots

A linear discriminant analysis effect size (LefSe) was used to study the groups with significant effects on the diversity of abundance under different bamboo forests and densities. As shown in Figure 6, the endophytic fungi in the *B. striata* roots of three bamboo forests differed significantly. There were 11 different indicator species in the *P. edulis* forest, including Basidiomycota and Chytridiomycota. The rest included g_*Serendipita* (f_Serendipitaceae, o_Sebacinales order to genus), c_Agaricomycetes, g_Metschnikowia (f_Metschnikowiacea, o_Saccharomycetales, c_Saccharomycetes class to genus) and f_Dipodascaceae (o_Saccharo mycetales, c_Saccharomycetes class to family). There were nine different indicator species in the *P. iridescens* forest, including g_*Cutaneotrichosporon*, g_*Candida* (f_Saccharomycetales_fam _Incertae_sedis family to genus), g_*Acremonium*, g_*Penicillium*, g_*Cyberlindnera* (f_Phaffomy cetaceae family to genus), g_*Chaetomium* and g_*Trichosporon*. There were three different indicator species in the *P. glauca* forest, including o_Hypocreales (c_Sordariomycetes, p_Ascomycota phylum to order) (Figure 6A).

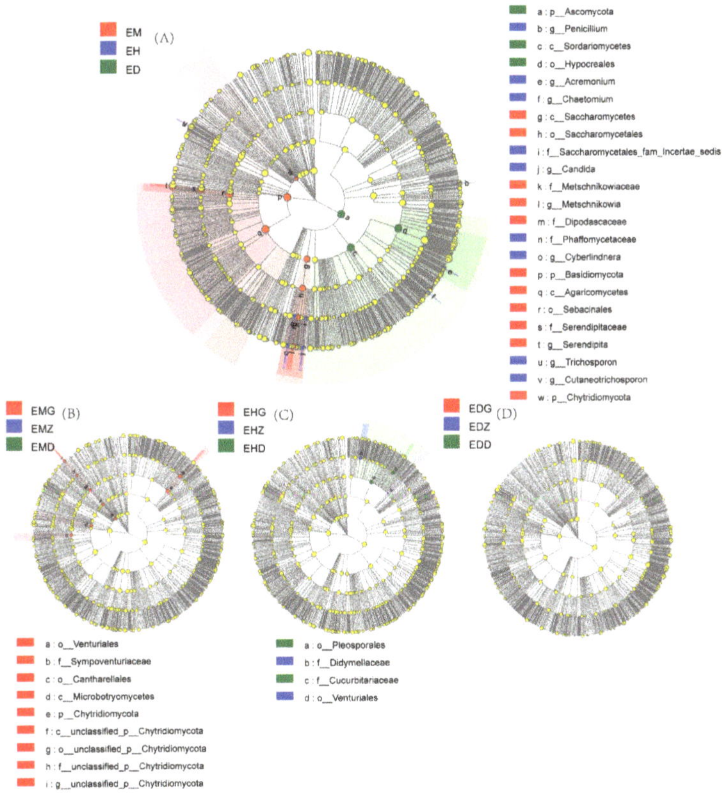

Figure 6. LefSe discriminant diagram of the endophytic fungi communities of *Bletilla striata* under different bamboo forest types and densities: (**A**) three bamboo forest models; (**B**) *Phyllostachys edulis* model; (**C**) *P. iridescens* model; (**D**) *P. glauca* model. Different color nodes indicate the groups that are enriched in the corresponding groups and have a significant influence on the differences between groups. The yellow nodes represent the microbial groups with no significant difference among the different groups, and the node diameter represents the species abundance: EM, EH, ED, *P. edulis*, *P. iridescens*, *P. glauca*, respectively; EMG, EMZ, EMD, *P. edulis* at high-, medium- and low-density, respectively; EHG, EHZ, EHD, *P. iridescens* at high-, medium- and low-density, respectively; EDG, EDZ, EDD, *P. glauca* at high-, medium- and low-density, respectively; LEfSe, linear discriminant analysis effect size.

Under the three densities, the different species in the high-density of the *P. edulis* forest were f_Sympoventuriaceae (o_Venturiales order to family), o_Cantharellales and c_Microbotryomycetes (Figure 6B). F_Didymellaceae and o_Venturiale were two distinct species in the medium-density forest of *P. iridescens*, F_Cucurbitariaceae (o_Pleosporales order) was found in the low-density forest (Figure 6C). There were no different species in the three densities of *P. glauca* (Figure 6D).

3.6. Correlation Analysis of Environmental Factors and the Endophytic Fungi Community in B. striata Roots

Collinearity analysis was performed on 18 environmental factors, including the pH, EC, OM, TN, AN, NN, TP, AP, TK, AK, MBC, MC, SBD, SOC, C:N, C:P, N:P and LI, and 5 soil indices, including TN, NN, TP, SOC and C:P. Indices that had variance influence factor (VIF) values > 10 were removed, and then the redundancy analysis (RDA) was started. The results showed that environmental factors explained the changes in endophytic fungi community in 71.87% of *B. striata* at the phylum level, with TK, AP, LI, SBD, pH and MC positively correlated with Basidiomycota and negatively correlated with Ascomycota. C:N, EC, OM, N:P, AK, AN and MBC positively correlated with Ascomycota and negatively correlated with Basidiomycota (Figure 7A). The SBD ($p = 0.001$, $R^2 = 0.57$) and LI ($p = 0.002$, $R^2 = 0.33$) had significant effects on the community composition of endophytic fungi in *B. striata*. N:P ($p = 0.05$, $R^2 = 0.24$), MBC ($p = 0.08$, $R^2 = 0.19$) and OM ($p = 0.07$, $R^2 = 0.22$) also had some influence on the community composition of endophytic fungi in *B. striata* (Table 2).

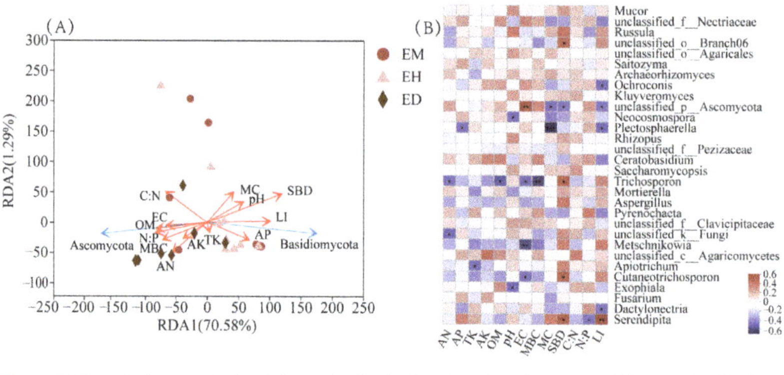

Figure 7. Correlation analysis of the endophytic fungi and environmental factors in *Bletilla striata* roots of different bamboo forests: (**A**) RDA chart at the phylum level; (**B**) heatmap analysis chart at the genus level: EM, *Phyllostachys edulis*; EH, *P. iridescens*; ED, *P. glauca*; RDA, redundancy analysis. The blue and red arrows represent endophytic fungi and environmental factors, respectively.

Table 2. Correlation data table of RDA analysis between environmental factors and endophytic fungi of *Bletilla striata* roots.

		AN	AP	TK	AK	OM	Ph	EC	MBC	MC	SBD	C:N	N:P	LI
Endophytic fungi	*p*	0.13	0.18	0.96	0.48	0.07	0.17	0.19	0.08	0.25	0.00	0.12	0.05	0.02
	R^2	0.17	0.14	0.00	0.06	0.22	0.14	0.14	0.19	0.12	0.57	0.17	0.24	0.33

$p < 0.05$ represents a significant influence; R^2, higher values indicate a closer association.

The effects of different environmental factors on the endophytic fungal community of *B. striata* roots could be observed from the heatmap. In the *P. iridescens* and *P. edulis* forests, *Serendipita* significantly positively correlated with the SBD and LI and negatively correlated with N:P. The dominant genus *Dactylonectria* significantly negatively correlated with LI in the *P. glauca* forest (Figure 7B).

3.7. FUNGuild Function Prediction Analysis of the Endophytic Fungi in *B. striata* Roots

Based on the FUNGuild function prediction, endophytic fungi in the roots of *B. striata* were divided into nine categories, including Undefined, Pathogen–Saprotroph–Symbiotroph, Pathotroph, Pathotroph–Saprotroph, Pathotroph–Saprotroph–Symbiotroph, Pathotroph–Symbiotroph, Saprotroph, Saprotroph–Symbiotroph and Symbiotroph. Symbiotrophs comprised 59% of the *B. striata* roots in *P. edulis*, 43% in *P. iridescens*, and 12% in *P. glauca*. Correspondingly, the proportion of saprotrophic fungi in the *P. glauca* forest increased significantly, comprising 42%, followed by the *P. iridescens* forest (24%), and the *P. edulis* forest (23%) (Figure 8A).

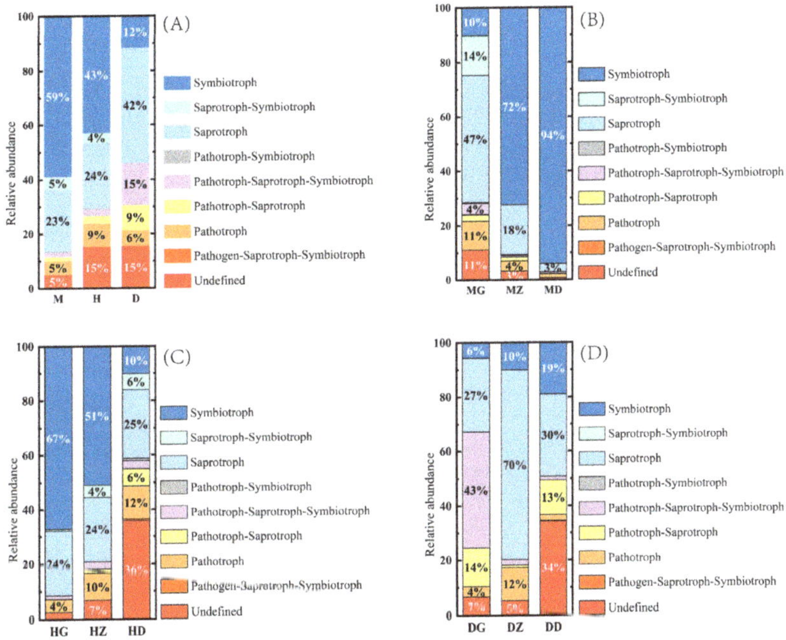

Figure 8. FUNGuild function prediction of the endophytic fungi of *Bletilla striata* in different grouping models: (**A**) three bamboo forest models; (**B**) *Phyllostachys edulis* model; (**C**) *P. iridescens* model; (**D**) *P. glauca* model. M, H, D, *P. edulis, P. iridescens, P. glauca*, respectively; MG, MZ, MD, *P. edulis* at high-, medium- and low-density, respectively; HG, HZ, HD, *P. iridescens* at high-, medium- and low-density, respectively; DG, DZ, DD, *P. glauca* at high-, medium- and low-density, respectively.

In addition, the density of bamboo forest had a significant effect on the nutrient type of endophytic fungi in the B. striata roots. In the *P. edulis* forest, saprotrophs (47%) comprised the highest proportion in the high-density group, while symbiotrophs comprised only 10%. However, symbiotrophs comprised 72% and 94% in the medium- and high-density groups, respectively, which were absolutely dominant (Figure 8B). Saprotrophs and pathotrophs were the primary nutrient types in the *P. glauca* forest. Saprotrophs comprised 70% of the medium-density group, and Pathotroph–Saprotroph–Symbiotroph comprised 43% of the high-density group (Figure 8D). In the *P. iridescens* forest, symbiotrophs and saprotrophs were the primary types of nutrient consumers, but their changing trends differ from those of the other two bamboo forests. The symbiotrophic fungi were 67% at high-density and 10% at low-density. Saprotrophs comprised 24–25% at all three densities (Figure 8C).

4. Discussion

4.1. Effects of Different Bamboo Forest Types and Densities on the Growth of B. striata

In this study, bamboo forest type and stand density had significant influences on the growth of *B. striata*. The growth of the aboveground and belowground parts of *B. striata* in the *P. edulis* stand was clearly superior among the three bamboo stands, and the dry matter accumulation of roots, tubers, stems and leaves was significantly higher ($p < 0.05$), indicating that the environment under *P. edulis* forest is the most suitable for the growth of *B. striata*. The plant height and leaf lengths of *B. striata* in the *P. iridescens* forest have some advantages, and the dry weight ratio of the aboveground and belowground biomass was significantly higher than that of the other bamboo forests, indicating that the growth of *B. striata* in the *P. iridescens* forest is more focused on the aboveground than the belowground parts, which is similar to the results of a study that found that the biomass distribution of plants under low light conditions is more directed to the aboveground part [33,34]. The soil nutrient indices, such as N, K, OM and MBC, were all superior in the *P. glauca* forest, but the growth of *B. striata* was significantly inhibited compared with the other bamboo forests. This could be owing to the extreme forest canopy density and the complexity of the root microbial composition in *B. striata*. Saprophytic and pathogenic fungi, such as *Dactylonectria*, increased. In addition, the leaf width of *B. striata* in the *P. iridescens* and *P. glauca* forests decreased significantly, because under high canopy density, the plants would sacrifice part of their photosynthetic processes to form narrower, thinner or heavier leaves, thus, increasing the density of leaf tissue [35]. Observing the growth of *B. striata* under different densities of the same bamboo forest enabled the observation that the lower density of bamboo forest has clear advantages to the growth of *B. striata*. Studies have shown that *B. striata* can grow normally under a canopy density of 0.2~0.9, but the optimal shade is 45~55% [36,37]. Bamboo forests usually have a high canopy density. In this study, the average LI in the *P. edulis* forest decreased by 76.53% compared with the LI outside of the company, and the LI in the low-density group decreased the least (52.72%). Therefore, low-density *P. edulis* forests affect the growth of *B. striata* by best meeting its demand for light.

4.2. Effects of Different Bamboo Forest Types and Densities on the Endophytic Fungal Community in B. striata Roots

Plant growth is closely related to the colonization of fungi, and the colonization of endophytic fungi in plant roots is also affected by its environment. This study found that the change in bamboo species and density affected the abundance and diversity of endophytic fungi in the *B. striata* roots. In comparison, the abundance and diversity of endophytic fungi in the *B. striata* roots were the highest in *P. iridescens* forests, indicating that the environment under the *P. iridescens* forest may be favorable to the growth of endophytic fungi in the roots of *B. striata*. However, the effect on *B. striata* is double-sided, which could be caused by the decrease in mycorrhizal fungi and the increase in pathogenic fungi [38]. This is consistent with the results of FUNGuild functional prediction analysis in this study.

In this study, Basidiomycota and Ascomycota were the dominant fungi in the roots of *B. striata*, and the composition of their abundance changed significantly under different bamboo forest types and densities, indicating that the external environment would significantly affect the endophytic fungal community composition of *B. striata*, which was similar to the results of previous studies [39,40]. Simultaneously, these changes had a clear correlation with the nutrient types of endophytic fungi in *B. striata* roots. There were 45 genera of endophytic fungi with relative abundance greater than 1% in the *Bletilla striata* roots, including nine nutrient types. Symbiotrophic, saprotrophic and pathotrophic fungi differed in their abundance in the three bamboo forests. This difference effectively reflected the different selection of endophytic fungi in the *B. striata* roots when grown under different bamboo forests. In addition, 10 genera of Orchidaceae mycorrhizal fungi have been reported in three bamboo forests, including *Serendipita, Fusarium, Pyrenochaeta, Aspergillus, Ceratobasidium, Neocosmospora, Russula, Trichoderma, Thanatephorus,* and *Chaetomium* [41,42].

The LEfSe method was used to analyze the different indicator species of the composition of the endophytic fungi community in the *B. striata* roots in different groups, which could reflect the selection results of endophytic fungi in bamboo forests to some extent. Symbiotic fungi were the primary endophytic fungi in *B. striata* in the *P. edulis* forests. *Serendipita* ranked the first, and it is a mycorrhizal fungus of the orchid family. Pathogenic and saprophytic fungi increased in the *P. iridescens* forest, but there were still symbiotic fungi and orchid mycorrhizal fungi such as *Penicillium* and *Chaetomium*. The proportion of saprophytic fungi increased substantially in the *P. glauca* forest, and almost no symbiotic fungi were found. Thus, the *P. edulis* forest was the most suitable environment for the growth of symbiotic fungi in *B. striata*. In addition, there were more saprophytic and pathogenic differential species in groups in bamboo forests with higher densities. The results showed that the low-density environment was suitable for symbiotic fungal colonization, and thus, the growth of *B. striata* under the bamboo forest was affected.

4.3. Potential Effects of Serendipita *on the Growth of* B. striata

Serendipita was the most abundant endophytic fungus in the *B. striata* roots in the bamboo forest. It is a member of the Serendipitaceae family in Basidiomycota, is widely distributed in nature and can form highly diversified root symbionts with many plants. Studies have shown that the root system of host plants was often significantly increased by artificial inoculation [43,44], which was consistent with the results of this study that *Serendipita* was the most abundant in *P. edulis* forests where the root length, root width and root numbers of *B. striata* were significantly the highest. This result was the most pronounced in *P. edulis* low-density forests. In addition, the abundance of *Serendipita* positively correlated with most of the growth indices of *B. striata*. This is probably because the Serendipitaceae have distinctive molecular mechanisms of action on host plants. It can promote plant growth and enhance plant resistance to abiotic and pathogen stresses [45].

The abundance of *Serendipita* displayed obvious gradient changes in different bamboo forest groups, suggesting that the bamboo forest environment could significantly change the abundance of *Serendipita*. A correlation analysis showed that the abundance of *Serendipita* on *B. striata* grown under bamboo was significantly affected by the SBD, LI and N:P, which was consistent with previous studies on the correlation between the abundance of *Sebacinales* fungi and elements such as N and pH. Simultaneously, *Sebacinales* mycorrhizae often exist in barren soils, such as those in southwestern Australia, or in plants that thrive in severe soils and exhibit some drought resistance [43]. Therefore, *Serendipita* can be used to inoculate orchids in unfavorable soils to produce green applications to improve the competitiveness and stress resistance of plants, expand the planting range of plants, and provide more effective guidance for the introduction, cultivation, production and sustainable development of orchids.

5. Conclusions

The growth and development of *B. striata* were significantly affected by the types and densities of bamboo forests. The environment in the *Phyllostachys edulis* forest was the most favorable for the whole growth of *B. striata* and the best under low density, and the primary growth of *B. striata* in the *P. iridescens* forest was in the aboveground parts. The community composition and diversity of endophytic fungi in the *B. striata* roots differed under different bamboo forest types and densities. Basidiomycota was the main endophytic fungi in the *P. edulis* and *P. iridescens* forests, and symbiotrophic fungi were the most dominant type based on the analysis of nutrient usage. Ascomycota was the dominant phylum in the *P. glauca* forest, and saprotrophic fungi were the main nutrient types. The soil bulk density and light intensity were the primary factors that affected the abundance of these two phyla. The dominant species of endophytic fungi in the *B. striata* roots also differed in different types of bamboo forests. *Serendipita* was the dominant genus in the *P. edulis* and *P. iridescens* forests, and *Dactylonectria* was the dominant species in the *P. glauca* forest. The abundances of Basidiomycota and *Serendipita* increased with the increase in bamboo density in the

P. edulis forest but showed an opposite trend in *P. iridescens* forests. In this study, *Serendipita* was the primary mycorrhizal fungi of *B. striata* in three bamboo forests, and most growth indices positively correlated with *Serendipita*, which was greatly affected by the density of bamboo. Therefore, in the application and production of afforestation, we can select suitable strains based on site conditions and prepare fungal agents to inoculate plants to improve their adaptability and stress resistance. This study provides new ideas to deepen the study of plant configuration under bamboo forests and the application of growing orchids under these forests.

Author Contributions: Conceptualization, W.H. and X.Z.; methodology, H.F. and X.Z.; software, H.F. and X.Z.; validation, W.H. and S.L. (Siren Lan); formal analysis, H.F.; investigation, H.F and Z.S.; resources, Z.S. and S.L. (Shanmin Li); data curation, X.Z.; writing—original draft preparation, H.F.; writing—review and editing, H.F. and X.Z.; visualization, H.F.; supervision, W.H., S.L. (Siren Lan) and X.Z.; project administration, W.H.; funding acquisition, W.H. and X.Z. All authors have read and agreed to the published version of the manuscript.

Funding: This research was funded by Shanghai Agricultural Development Project of Science and Technology (2021-02-08-00-12-F00778), National Natural Science Foundation of China (31902108), and the Special Fund for Scientific Research of Shanghai Landscaping & City Appearance Administrative Bureau (G222406).

Institutional Review Board Statement: Not applicable.

Data Availability Statement: Data are available upon request from the corresponding author.

Acknowledgments: We acknowledge and thank all the teachers and classmates from Shanghai Chenshan Botanical Garden and Fujian Agriculture and Forestry University who offer help in the surveys.

Conflicts of Interest: The authors declare no conflict of interest. The funders had no role in the design of the study; in the collection, analyses, or interpretation of data; in the writing of the manuscript, or in the decision to publish the results.

References

1. Fan, W.; Meng, P.; Zhang, J.S. *Chinese Agroforestry Research*; China Forestry Press: Beijing, China, 2003.
2. Mortimer, R.; Saj, S.; David, C. Supporting and regulating ecosystem services in cacao agroforestry systems. *Agroforest. Syst.* **2018**, *92*, 1639–1657. [CrossRef]
3. Wolz, K.J.; Branham, B.E.; de Lucia, E.H. Reduced nitrogen losses after conversion of row crop agriculture to alley cropping with mixed fruit and nut trees. *Agr. Ecosyst. Environ.* **2018**, *258*, 172–181. [CrossRef]
4. Li, W.H.; Lai, S.D. *Agroforestry in China*; Science Press: Beijing, China, 1994.
5. Fan, L.J.; Guo, X.Y.; Ma, N.X. Comparative study on phylogenetics and sequences composition of bamboos and cereals. *Forest Res.* **2006**, *2*, 165–169.
6. Yang, C.B. *Characteristics Stabilization Mechanisms Soil Organic Carbon Moso Bamboo Plantations*; Chinese Academy of Forestry: Beijing, China, 2019.
7. Peng, L.Y.; Peng, C.; Ai, W.S.; Yang, M.; Meng, Y.; Tu, J.; Li, M.Q. Research advance and forecast on the tridimensional compound management of bamboo forest. *Hunan Forest. Sci. Technol.* **2019**, *46*, 72–76.
8. Lobovikov, M.; Schoene, D.; Lou, Y. Bamboo in climate change and rural livelihoods. *Mitig. Adapt. Strat. Gl.* **2012**, *17*, 261–276. [CrossRef]
9. Bhardwaj, G.D.R.; Kaushal, R.; Devi, M. Performance of ginger crop in response to integrated nutrient management under bamboo-based agroforestry system in mid-hill sub-humid conditions of Himachal Pradesh. *Int. J. Farm Sci.* **2016**, *6*, 24–29.
10. Fan, Y.R.; Chen, S.L.; Yang, Q.P.; Li, Y.C.; Guo, Z.W.; Chen, S. Population growth and biomass allocation of *Polygonatum cyrtonema* within a *Phyllostachys edulis* forest utilizing bamboo density treatments. *J. Zhejiang Agrofor. Univ.* **2013**, *30*, 199–205.
11. Cai, C.J.; Fan, S.H.; Liu, G.L.; Wang, S.M.; Feng, Y. Research and development advance of compound management of bamboo forests. *World Bamboo Ratt.* **2018**, *16*, 47–52.
12. Chen, S.L.; Chen, C.Y.; Yang, Q.P.; Wang, W.H. A study on the optimized mode of forest structure of *Dendrocalamus latiflorus* for Bamboo shoot. *Acta Agri. Univ. Jiangxiensis* **2005**, *17*, 191–194.
13. Zhou, X.H.; Xiao, Z.Y.; Zeng, P.S.; Yao, J.B.; Wu, X.Y.; Huang, Y.T. Effects of different forest habitats and growth years on the growth of *Polygonatum cyrtonema* and the content of medicinal active components. *J. Southwest Forest. Univ.* **2019**, *39*, 155–160.

14. Huang, Y.P.; Wang, B.F.; Fan, F.R.; Shen, Q.T.; Chen, S.H.; Huang, S.M.; Su, S.J. Effect of forest types and canopy density on polysaccharide content of *Polygonatum cyrtonema*. *Chin. Agr. Sci. Bull.* **2016**, *32*, 102–105.
15. Feng, J.Y. *Effects of Bamboo Density and Drought Stress on Photosynthetic Characteristics of Five Medicinal Plants*; Zhejiang A&F University: Hangzhou, China, 2021.
16. Flora of China Editorial Committee. *Flora of China*; Science Press: Beijing, China, 1999.
17. Qian, C.D.; Jiang, F.S.; Yu, H.S.; Fu, Y.H.; Cheng, D.Q.; Gan, L.S.; Ding, Z.S. Antibacterial biphenanthrenes from the fibrous roots of *Bletilla striata*. *J. Nat. Prod.* **2015**, *78*, 939–943. [CrossRef] [PubMed]
18. Herrera, H.; Valadares, R.; Oliveira, G.; Fuentes, A.; Almonacis, L.; Bashan, Y.; Arriagada, C. Adaptation and tolerance mechanisms developed by mycorrhizal *Bipinnula fimbriata* plantlets (Orchidaceae) in a heavy metal-polluted ecosystem. *Mycorrhiza* **2018**, *28*, 651–663. [CrossRef] [PubMed]
19. Smith, S.E.; Read, D.J. *Mycorrhizal Symbiosis*, 3rd ed.; Academic Press: New York, NY, USA, 2008.
20. Schiebold, J.M.I.; Bidartondo, M.I.; Karasch, P.; Gravendeel, B.; Gebauer, G. You are what you get from your fungi: Nitrogen stable isotope patterns in *Epipactis* species. *Ann. Bot.* **2017**, *119*, 1085–1095. [CrossRef]
21. Liu, S.S.; Chen, J.; Li, S.C.; Zeng, X.; Meng, Z.X.; Guo, S.X. Comparative transcriptome analysis of genes involved in GA-GID1-DELLA regulatory module in symbiotic and asymbiotic seed germination of *Anoectochilus roxburghii* (Wall.) Lindl. (Orchidaceae). *Int. J. Mol. Sci.* **2015**, *16*, 30190–30203. [CrossRef]
22. Gonneau, C.; Jersakova, J.; de Tredern, E.; Till-Bottraud, I.; Saarinen, K.; Sauve, M.; Roy, M.; Hajek, T.; Selosse, M.A. Photosynthesis in perennial mixotrophic *Epipactis* spp. (Orchidaceae) contributes more to shoot and fruit biomass than to hypogeous survival. *J. Ecol.* **2014**, *102*, 1183–1194. [CrossRef]
23. Liebel, H.T.; Bidartondo, M.I.; Gebauer, G. Are carbon and nitrogen exchange between fungi and the orchid *Goodyera repens* affected by irradiance? *Ann. Bot.* **2015**, *115*, 251–261. [CrossRef]
24. Burke, R.M.; Cairney, J.W.G. Carbohydrate oxidases in ericoid and ectomycorrhizal fungi: A possible source of Fenton radicals during the degradation of lignocelluloses. *New Photol.* **1998**, *139*, 637–645. [CrossRef]
25. Slezack, S.; Dumas-Gaudot, E.; Rosendahl, S.; Kjoller, R. Endoproteolytic activities in pea roots inoculated with the arbuscular mycorrhizal fungus *Glomus mosseae* and/or *Aphanomyces euteiches* in relation to bioprotection. *New Photol.* **1999**, *142*, 517–529. [CrossRef]
26. McCormick, M.K.; Taylor, D.L.; Juhaszova, K.; Burnett, J.R.; Whigham, D.F.; Oneill, J.P. Limitations on orchid recruitment: Not a simple picture. *Mol. Ecol.* **2012**, *21*, 1511–1523. [CrossRef]
27. Zeng, X.H.; Diao, H.X.; Ni, Z.Y.; Shao, L.; Jiang, K.; Hu, C.; Huang, Q.J.; Huang, W.C. Temporal variation in community composition of root associated endophytic fungi and carbon and nitrogen stable isotope abundance in two *Bletilla species* (*Orchidaceae*). *Plants* **2021**, *10*, 18. [CrossRef] [PubMed]
28. Ma, J.H.; Qi, Y.Q.; Ren, Q.F.; Liu, F.; Ou, M.Z.; Chen, Y.F. Effects of nitrogen on the growth and polysaccharides accumulation of *Bletilla striata*. *Plant Physiol. J.* **2022**, *58*, 331–338.
29. Tao, J.X.; Lv, H.B.; Zhou, W.H.; Luo, Z.Y. Exploration of natural Rubber+ *Bletilla striata* (Thunb.) Reichb. f. planting mode. *Mod. Agric. Sci. Technol.* **2018**, *17*, 138–144.
30. Huang, X.J.; Zhang, W.B.; Lai, G.W.; Zhang, F.G.; Jiang, Z.H.; Qiu, W.Q. Effects of different shading degrees on the growth of *Bletilla striata* and rhizoma huangjing under compound management. *Mod. Agric. Sci. Technol.* **2019**, *4*, 49–51.
31. Liu, X.Q. *The Botanical Garden in Shanghai Mountain Bamboo Landscaping Art of W Transform the Design Application*; Shanghai Normal University: Shanghai, China, 2016.
32. Lu, R.K. *Methods of Soil Agrochemical Analysis*; China Agricultural Science and Technology Press: Beijing, China, 2000; pp. 12–264.
33. Xu, W.D. *Effects of Light, Planting Density and Harvesting Time on Yield and Quality of Lettuce in Plant Factory*; Nanjing Agricultural University: Nanjing, China, 2015.
34. Wang, X.W.; Wang, Y.G.; Fu, Q.S.; Zhao, B.; Guo, Y.D. Effects of low light stress on morphological trait, physiological characters and leaf ultrastructure of tomato (*Lycopersicon esculentum*) seedlings. *North China Agri. J.* **2009**, *24*, 144–149.
35. Bai, H.N.; Niu, X.; Wang, B.; Song, Q.F.; Tao, Y.Z. Effects of expansion of *Phyllostachys edulis* on community diversity of endophytic fungi in leaves of *Rhododendron latoucheae*. *Chin. J. Ecol.* **2021**, *12*, 1–13.
36. Wang, W.S. Seedling and understory cultivation techniques of *Bletilla striata*. *Anhui Forest. Sci. Technol.* **2018**, *44*, 61–63.
37. Zhou, Y.R. *Effects of Different Light Intensities on Photosynthetic and Physiological Characteristics of Bletilla striata*; Zhengzhou University: Zhengzhou, China, 2020.
38. Tuominen, A. Defensive strategies in *Geranium sylvaticum*, Part 2: Roles of water-soluble tannins, flavonoids and phenolic acids against natural enemies. *Phytochemistry* **2013**, *95*, 408–420. [CrossRef]
39. Xi, G.J.; Li, J.B.; Shi, J.; Han, Z.M. Diversity of endophytic fungi in *Bletilla striata*. *Acta Agric. Zhejiangensis* **2017**, *29*, 2077–2083.
40. Den, W.X.; Zhao, M.L.; Wang, Z.L.; Li, Y.M. Analysis of species composition of endophytic fungi in roots of *Bletilla striata* based on high throughput sequencing. *J. Southwest Forest. Univ.* **2021**, *41*, 76–84.
41. Reiter, N.; Phillips, R.D.; Swarts, N.D.; Wright, M.; Holmes, G.; Sussmilch, F.C.; Davis, B.J.; Whitehead, M.R.; Linde, C.C. Specific mycorrhizal associations involving the same fungal taxa in common and threatened *Caladenia* (*Orchidaceae*): Implications for conservation. *Ann. Bot.* **2020**, *5*, 1–51. [CrossRef] [PubMed]

42. Oktalira, F.T.; May, T.W.; Dearnaley, J.D.W.; Linde, C.C. Seven new *Serendipita* species associated with Australian terrestrial orchids. *Mycologia* **2021**, *113*, 968–987. [CrossRef] [PubMed]
43. Bokati, D.; Craven, K.D. The cryptic Sebacinales: An obscure but ubiquitous group of root symbionts comes to light. *Fungal Ecol.* **2016**, *22*, 115–119. [CrossRef]
44. Qin, Y.; Druzhinina, I.S.; Pan, X.Y.; Yuan, Z.L. Microbially mediated plant salt tolerance and microbiome-based solutions for saline agriculture. *Biotechnol. Adv.* **2016**, *34*, 1245–1259. [CrossRef] [PubMed]
45. Weiß, M.; Waller, F.; Zuccaro, A.; Selosse, M.A. Sebacinales-one thousand and one interactions with land plants. *New Phytol.* **2016**, *211*, 20–40. [CrossRef]

diversity

MDPI

Article

Evolution of Reproductive Traits and Implications for Adaptation and Diversification in the Yam Genus *Dioscorea* L.

Min Chen [1], Xiaoqin Sun [1], Jia-Yu Xue [2], Yifeng Zhou [1] and Yueyu Hang [1,*]

[1] Institute of Botany, Jiangsu Province and Chinese Academy of Sciences, Nanjing 210014, China; chenmin@cnbg.net (M.C.); xiaoqinsun@cnbg.net (X.S.); njgzhou@163.com (Y.Z.)
[2] College of Horticulture, Academy for Advanced Interdisciplinary Studies, Nanjing Agricultural University, Nanjing 210014, China; xuejy@njau.edu.cn
* Correspondence: hangyueyu@cnbg.net; Tel.: +86-025-84347052

Abstract: *Dioscorea* is a pantropical monocotyledonous genus encompassing several well-known tuber crops and medicinal plants. It possesses remarkable morphological diversity, especially in reproductive characteristics, which are suggested to play important roles in species adaptation and diversification. Yet there have been few studies that consider the evolutionary pattern followed by these characters in this genus. In this study, the phylogenetic relationships among Chinese yams were reconstructed from five chloroplast and two mitochondrial DNA sequences. The evolutionary histories of bulbil possession, inflorescence architecture, the color of the male flowers and the degree of male flower opening were reconstructed. The results suggested that yam bulbils evolved after the divergence between *D.* sect. *Testudinaria* and other species of *Dioscorea* except for in *D.* sect. *Stenophora* and *D.* sect. *Apodostemon*. The evolutionary trend in the degree of male flower opening ranged from fully open to nearly closed. Male flowers with dark colors and panicles were shown to be derived in *Dioscorea*. These characteristics were found to be closely associated with the reproductive patterns and pollinating mechanisms of the *Dioscorea* species. The findings also shed light on the systematic relationships within this genus.

Keywords: phylogeny; character evolution; bulbil; inflorescence architecture; floral color; perianth opening degree

Citation: Chen, M.; Sun, X.; Xue, J.-Y.; Zhou, Y.; Hang, Y. Evolution of Reproductive Traits and Implications for Adaptation and Diversification in the Yam Genus *Dioscorea* L.. *Diversity* **2022**, *14*, 349. https://doi.org/10.3390/d14050349

Academic Editors: Hong-Hu Meng and Yi-Gang Song

Received: 19 April 2022
Accepted: 27 April 2022
Published: 29 April 2022

1. Introduction

Dioscorea L. is the largest genus of the Dioscoreaceae, comprising about 630 species [1], mainly distributed in tropical and subtropical areas, sometimes expanding to temperate regions [2]. Species of this genus are mostly dioecious vines with underground storage organs [3]. The edible starchy tuber makes it the third most important tropical tuberous crop globally [4], and the metabolite-rich rhizomes of some species are used as a source of pharmaceutical compounds [5]. Apart from its economic importance, *Dioscorea* is one of the most critical taxa in monocot systematics since it is laid near the basal position in the phylogenetic tree of monocotyledonous plants and has a number of similarities with dicots [6,7]. However, there are challenges in the taxonomic and systematic investigation of *Dioscorea* due to its great morphological diversity, dioecy and small flowers [2,8]. There are about 1600 taxonomic names attributed to *Dioscorea*, but most of them are considered synonyms [1], indicating the controversy on the circumscription of species boundaries. Moreover, the classification system of *Dioscorea* differs greatly among authors, and many of the proposed infrageneric taxa do not quite represent natural lineages [2]. Based on morphological and anatomic characteristics, *Dioscorea* species have been assigned to between 24 and 58 sections in traditional taxonomic studies [3,9–11]. Recent phylogenetic analyses based on molecular data sets, including cpDNA regions and nuclear genes, have led to the recognition of 10–11 main clades in this genus [2,12–15]. The results greatly simplified previous classification systems and gave rise to a fundamental phylogenetic framework for understanding the infrageneric relationships and evolutionary history of *Dioscorea*. Despite extensive study of phylogeny and biogeography, research focusing on character evolution and its relation to species diversification and ecological success are extremely limited.

During angiosperm evolution, advances in vegetative and reproductive organs generated remarkable morphological diversity and were pivotal to niche adaptation and speciation [16]. Recent phylogenetic research has indicated some lineages of *Dioscorea*, such as the Madagascar group and *D.* sect. *Enantiophyllum*, have very short phylogenetic branches, resulting from radiation events [2,12,13,17]. It has been proposed that fast evolution over a short period may reflect responses to tremendous climate change or the rise of innovative characters [18]. *Dioscorea* species exhibit extreme morphological variations; for example, underground storage organs include branching rhizomes, perennial or annual tubers, cylindrical or globose tubers, right or left stem twining and seed wings extending from the apex, the base or around the whole seed. Previous studies revealed the distribution patterns of some of these traits, including tuber morphology, stem twining direction, seed wing shape [2], stem anatomy [8], leaf venation [19], pollen characters and chromosome numbers [15]. However, these studies mainly focused on identifying synapomorphies for infrageneric lineages rather than elucidating the evolutionary history of traits and their links with the wide range of environmental conditions in which the plants lived.

Burkill [11] suggested that adaptation to limiting rainfall was the most important factor underpinning species diversification in *Dioscorea*. The restricted distribution and fewer species in *D.* sect. *Stenophora*, compared with great diversity and pantropical distribution of other clades of *Dioscorea*, led Wilkin et al. [2] to propose that tubers play an important role in the origins of diversity in this genus. Investigations focusing on African *Dioscorea* species showed that the shift from forest to open grassland was associated with changes in tuber size and orientation that protect the plants from fires. Their stem habit shifted from twining to erect due to the lack of supporting vegetation, and seed wing morphology adapted to release at low height, requiring higher wind speeds for efficient dispersal [20]. In addition, changes in flower and fruit morphology were suggested to play key roles in the exposure to radiation of the Madagascan clade [2]. These morphological traits can greatly influence the pollination process and seed dispersal, further altering the fitness of different species in given conditions. However, the evolutionary pattern of reproductive characters in *Dioscorea*, especially floral traits, remain poorly understood.

The Himalayan-Hengduan Mountains are purported to be the center of origin and diversification for the yam genus, with a large proportion of endemic species [21]. There are 52 species, 1 subspecies and 9 varieties of *Dioscorea* in China, which show enormous variability in traits related to sexual and vegetative reproduction [22]. For instance, some species produce bulbils at the leaf axils, while others do not; the inflorescence may be spikes, racemes or panicles; the color of male flowers varies between white, yellow, green to orange, purple and so on; and the perianths of male flowers are completely open, half-closed or fully closed during blooming in different species (shown in Figure 1). These features are widely used to divide plants into infrageneric groups and define species. This group of *Dioscorea* species is, therefore, an excellent candidate for exploring character evolution and implications for species diversification and ecological adaptation. In this study, we reconstructed the phylogenetic relationships between 48 *Dioscorea* species using 5 chloroplast and 2 mitochondrial DNA markers. Based on the phylogenetic framework produced, we explored the evolutionary patterns of four reproductive characters to clarify the driving force for the diversification and adaptive evolution of this important angiosperm lineage.

Figure 1. Characters related to reproduction showing morphological diversity of *Dioscorea*. (**A–D**) Bulbils. (**E–H**) Floral color. (**I–K**) Perianth opening degree. (**L–O**) Inflorescence architecture. (**A**) *D. kamoonensis*. (**B**) *D. melanophyma*. (**C**,**N**) *D. delavayi*. (**D**,**K**) *D. polystachya*. (**E**,**O**) *D. futschauensis*. (**F**) *D. zingiberensis*. (**G**) *D. yunnanensis*. (**H**) *D. subcalva*. (**I**) *D. tokoro*. (**J**) *D. bulbifera*. (**L**) *D.exalata*. (**M**) *D. panthaica*.

2. Materials and Methods

2.1. Sampling

A total of 48 *Dioscorea* taxa were sampled in this study, covering all sections distributed in China, except the monotypic section *D.* sect. *Stenocorea*. The samples consisted of 43 species, 1 subspecies and 4 varieties. Among them, 47 taxa were collected in the field and the vouchers were deposited in the Herbarium of the Institute of Botany, Jiangsu Province and the Chinese Academy of Sciences, China (NAS). The sequences of the remaining species, *D. wallichii* J. D. Hooker, were downloaded from GenBank. We selected *Tacca chantrieri* Andre (Dioscoreaceae), a representative of the genus closely related to *Dioscorea* according to the recently built phylogeny of the Dioscoreaceae [14], as the outgroup in our analyses. The geographical origin, voucher specimen information and GenBank accession numbers of all samples are listed in Table 1.

Table 1. Taxa used in this study with locality, voucher information and GenBank accession numbers.

Species	Locality	Voucher	GenBank Accession No.						
			*mat*K	*trn*L-F	*rbc*L	*psb*A-*trn*H	*rpl36*-*rps8*	*nad*1	*rps3*
Dioscorea nipponica Makino	Lin'an, Zhejiang, China	NAS 0648570	AY957600 *	DQ841308	AF307455 *	GQ265171	GQ265222	GQ265123	GQ265268
D. nipponica subsp. *rosthornii* (Prain & Burkill) C. T. Ting	Tianshui, Gansu, China	NAS 0648571	DQ974184	DQ841309	DQ408178	GQ265172	GQ265223	GQ265122	GQ265269
D. althaeoides R. Knuth	Dêqên, Yunnan, China	NAS 0648572	EU407548	EU301741	EU407550	GQ265182	GQ265233	GQ265135	GQ265281
D. tokoro Makino	Anhua, Hunan, China	NAS 0648573	DQ974186	DQ841312	DQ408180	GQ265174	GQ265225	GQ265125	GQ265271
D. zingiberensis C. H. Wright	Mt. Hengshan, Hunan, China	NAS 0646476	AY973831 *	DQ841318	AY939889 *	GQ265154	GQ265206	GQ265105	GQ265251
D. sinoparviflora C. T. Ting, M. G. Gilbert & N. J. Turland	Lijiang, Yunnan, China	NAS 0648574	DQ974179	DQ841326	DQ408171	GQ265163	GQ265212	GQ265112	GQ265258
D. deltoidea Wall. ex Griseb.	Kunming, Yunnan, China	NAS 0648575	EF614207	DQ841305	EF614218	GQ265169	GQ265220	GQ265120	GQ265266
D. panthaica Prain & Burkill	Lijiang, Yunnan, China	Y. F. Zhou & B. C. Wu 200308015	GQ265088	GQ265291	GQ265187	GQ265184	GQ265235	GQ265136	GQ265283
D. biformifolia C. Pei & C. T. Ting	Mt. Eshan, Yunnan, China	NAS 0648576	EU407549	EU301742	EU301740	GQ265147	GQ265196	GQ265097	GQ265243
D. gracillima Miq.	Mt. Lushan, Jiangxi, China	NAS 0648577	DQ974190	DQ841315	DQ408164	GQ265150	GQ265201	GQ265101	GQ265247
D. collettii Hook. f. var. *collettii*	Jinghong, Yunnan, China	NAS 0648578	DQ974178	DQ841300	DQ408173	GQ265141	GQ265192	GQ265093	GQ265239
D. collettii var. *hypoglauca* (Palibin) C. Pei & C. T. Ting	Mt. Hengshan, Hunan, China	NAS 0648579	DQ974176	DQ841319	EF614220	GQ265155	GQ265205	GQ265106	GQ265252
D. futschauensis Uline ex R. Knuth	Yongtai, Fujian, China	NAS 0648580	DQ974175	DQ841316	DQ408166	GQ265151	GQ265202	GQ265102	GQ265248
D. spongiosa J. Q. Xi, M. Mizuno & W. L. Zhao	Mt. Hengshan, Hunan, China	NAS 0648581	DQ974191	DQ841317	DQ974194	GQ265153	GQ265204	GQ265104	GQ265250
D. banzhuana C. Pei & C. T. Ting	Mengzi, Yunnan, China	NAS 0648582	DQ974182	DQ841301	DQ408174	GQ265167	GQ265218	GQ265118	GQ265264
D. simulans Prain & Burkill	Guilin, Guangxi, China	NAS 0648583	EF614206	DQ841320	EF614217	GQ265138	GQ265189	GQ265090	GQ265236
D. esculenta (Lour.) Burkill	Lingshui, Hainan, China	NAS 0648585	AY956497 *	DQ841298	AY904794 *	GQ265180	GQ265231	GQ265131	GQ265277
D. esculenta var. *spinosa* (Roxburgh ex Prain & Burkill) R. Knuth	Lingshui, Hainan, China	B. C. Wu200804023	GQ265087	GQ265290	GQ265186	GQ265181	GQ265232	GQ265134	GQ265280
D. tentaculigera Prain & Burkill	Lincang, Yunnan, China	NAS 0648465	GQ265089	GQ265292	GQ265188	GQ265183	GQ265234	GQ265137	GQ265282
D. yunnanensis Prain & Burkill	Lijiang, Yunnan, China	NAS 0648459	EF614209	GQ265288	EF614221	GQ265161	GQ265213	GQ265113	GQ265259
D. subcalva Prain & Burkill	Tianlin, Guangxi, China	NAS 0648544	EF614208	EF614222	EF614214	GQ265160	GQ265211	GQ265111	GQ265257
D. subcalva var. *submollis* (R. Knuth) C. T. Ting & P. P. Ling	Mt. Jinfo, Chongqing, China	NAS 0648545	EF614204	GQ265287	EF614216	GQ265152	GQ265203	GQ265103	GQ265249
D. nitens Prain & Burkill	Lijiang, Yunnan, China	NAS 0648586	EF614205	EF614223	EF614215	GQ265162	GQ265214	GQ265114	GQ265260
D. bulbifera L.	Jinghong, Yunnan, China	NAS 0648587	AY956488 *	EF619352	AY904791 *	GQ265178	GQ265229	GQ265132	GQ265278
D. melanophyma Prain & Burkill	Mengzi, Yunnan, China	NAS 0648548	EF614210	DQ841303	DQ408176	GQ265143	GQ265194	GQ265095	GQ265241
D. kamoonensis Kunth	Mengzi, Yunnan, China	NAS 0648549	EF028332	DQ841302	DQ408175	GQ265142	GQ265193	GQ265094	GQ265240
D. delavayi Franchet	Kunming, Yunnan, China	NAS 0648550	GQ265085	GQ265284	DQ974196	GQ265148	GQ265199	GQ265100	GQ265246
D. menglaensis H. Li	Jinghong, Yunnan, China	NAS 0648461	GQ265086	GQ265285	GQ265185	GQ265146	GQ265198	GQ265099	GQ265245
D. pentaphylla Linnaeus	Guilin, Guangxi, China	NAS 0648551	AY972483 *	DQ841327	AF307470 *	GQ265140	GQ265191	GQ265092	GQ265237
D. esquirolii Prain & Burkill	Longzhou, Guangxi, China	NAS 0648552	DQ974177	DQ841322	DQ408168	GQ265139	GQ265190	GQ265091	GQ265238
D. hispida Dennstedt.	Longzhou, Guangxi, China	NAS 0648553	AY957589 *	DQ841323	AF307463 *	GQ265145	GQ265197	GQ265098	GQ265244

Table 1. *Cont.*

Species	Locality	Voucher	GenBank Accession No.						
			*mat*K	*trn*L-F	*rbc*L	*psb*A-*trn*H	*rpl36-rps8*	*nad1*	*rps3*
D. aspersa Prain & Burkill	Mengzi, Yunnan, China	NAS 0648588	EF614211	DQ841304	EF614213	GQ265168	GQ265219	GQ265119	GQ265265
D. polystachya Turczaninow	Jurong, Jiangsu, China	NAS 0648589	EF028331	DQ841313	DQ408181	GQ265144	GQ265195	GQ265096	GQ265242
D. japonica Thunberg	Lin'an, Zhejiang, China	NAS 0648590	DQ974183	DQ841307	AF307457 *	GQ265170	GQ265221	GQ265121	GQ265267
D. cirrhosa Loureiro	Longzhou, Guangxi, China	NAS 0648591	EF028329	DQ841324	AY904792 *	GQ265158	GQ265209	GQ265109	GQ265255
D. cirrhosa var. *cylindrica* C. T. Ting & M. C. Chang	Mt. Diaoluo, Hainan, China	NAS 0648592	DQ974189	DQ841314	DQ408184	GQ265179	GQ265230	GQ265133	GQ265279
D. wallichii J. D. Hooker	-	-	AY973830 *	-	AY939888 *	-	-	-	-
D. glabra Roxburgh	Longzhou, Guangxi, China	NAS 0648593	AY956501 *	DQ841321	AF307456*	GQ265157	GQ265208	GQ265108	GQ265254
D. fordii Prain & Burkill	Guilin, Guangxi, China	NAS 0648594	EF028333	DQ841299	DQ974195	GQ265156	GQ265207	GQ265107	GQ265253
D. persimilis Prain & Burkill	Mingxi, Fujian, China	NAS 0648595	DQ974193	DQ841328	DQ408165	GQ265175	GQ265226	GQ265127	GQ265273
D. exalata C. T. Ting & M. C. Chang	Tianlin, Guangxi, China	NAS 0648596	EF028330	DQ841325	DQ408170	GQ265159	GQ265210	GQ265110	GQ265256
D. alata Linnaeus	Jinghong, Yunnan, China	NAS 0648597	AB040208 *	DQ841331	AY667098 *	GQ265165	GQ265216	GQ265116	GQ265262
D. decipiens J. D. Hooker	Jinghong, Yunnan, China	NAS 0648598	DQ974181	DQ841329	AF307454 *	GQ265166	GQ265217	GQ265117	GQ265263
D. composita Hemsl.	Jinghong, Yunnan, China	NAS 0648405	DQ974180	DQ841330	DQ408172	GQ265164	GQ265215	GQ265115	GQ265261
D. sansibarensis Pax	Botanical Garden Regen Germany	Y. F. Zhou200403004	DQ974187	DQ841296	AY939883 *	GQ265177	GQ265228	GQ265129	GQ265275
D. caucasica Lipsky	Lyon, France	NAS 0648584	DQ974188	DQ841297	DQ408182	-	-	GQ265130	GQ265276
D. elephantipes Engl.	South Africa	N. Sheng200511014	AY956496 *	DQ841306	AF307461 *	GQ265176	GQ265227	GQ265128	GQ265274
D. villosa L.	America	NAS 0648463	-	GQ265286	DQ006092 *	GQ265149	GQ265200	GQ265126	GQ265272
Tacca chantieri André	-	-	AY973837 *	FJ194472 *	AJ235810 *	EF590744 *	-	DQ786152 *	-

* Sequences obtained from Genbank.

2.2. DNA Extraction, Amplification and Sequencing

The total genomic DNA was extracted from fresh or silica-gel dried leaves following a modified CTAB method [23] and stored at −20 °C before amplification. Five chloroplast markers, *mat*K, *rbc*L, *trn*L-F, *psb*A-*trn*H and *rpl36-rps8*, plus two mitochondrial markers, *nad1* and *rps3*, were chosen, taking evolutionary rate and ease of sequencing into account. The primers used for *mat*K and *rbc*L amplification followed Gao et al.'s recommendations [24]. The amplification of *trn*L-F used primers described by Taberlet et al. [25]. The primers for *psb*A-*trn*H and *rpl36-rps8* were newly designed based on homologous sequences of *D. elephantipes* (L'Hér.) Engl. on GenBank. The primers for the *nad1* gene were designed based on the sequence of *Oryza sativa* Linn. The *rps3* gene was amplified according to Laroche and Bousquet [26]. The sequences of all primers used in this study are detailed in Table 2.

Table 2. Markers and primers used in this study.

Marker	Primer Name	Direction	Primer Sequence (5′ to 3′)
*mat*K	*mat*K MF	forward	ATT TGC GAT CTA TTC ATT CAA T
	*mat*K MR	reverse	TGA GAT TCC GCA GGT CAT T
*rbc*L	*rbc*L m3	forward	TAT CTT AGC GCC ATT CCG AGT A
	*rbc*L m4	reverse	CGC GGA TAA TTT CAT TAC CTT C
*trn*L-F	*trn*L-F c	forward	CGA AAT CGG TAG ACG CTA CG
	*trn*L-F f	reverse	ATT TGA ACT GGT GAC ACG AG
*psb*A- *trn*H	*psb*A F1	forward	AAT GCT CAC AAC TTY CCT CTA
	*trn*H R1	reverse	CCA CTG CCT TGA TCC ACT TG
*rpl*36- *rps*8	*rpl*36 F1	forward	TTA CCC YTG TCT YTG TTT ATG
	*rps*8 R1	reverse	CTA CGA GAR GGT TTT ATT GAA
*nad*1	*nad*1 F1	forward	CCT TGT GAG CAC GTT TGG AT
	*nad*1 R1	reverse	GAC AAT CTC ACT CGA ATT ACA G
*rps*3	*rps*3 F1	forward	GTT CGA TAC GTC CAC CTA C
	*rps*3 R1	reverse	GTA CGT TTC GGA TAT RGC AC

The PCR reaction mixture contained 40 ng of genomic DNA template, 2.5 mmol/L $MgCl_2$, 1 × Mg-free DNA polymerase buffer, 0.12 mmol/L dNTPs, 0.3 mmol/L of each primer, 1 U Taq DNA polymerase and water added accordingly to a final volume of 50 µL. The PCR program for *mat*K, *rbc*L, *trn*L-F, *nad1* and *rps3* was as follows: a 3 min premelt at 94 °C, followed by 35 cycles of 45 s denaturation at 94 °C, 30 s annealing at 58 °C and a 1.5 min extension at 72 °C, plus a final extension of 5 min at 72 °C. For *psb*A-*trn*H and *rpl36-rps8*, the PCR reaction included an initial denaturation at 94 °C for 3 min, followed by 35 cycles of denaturation at 94 °C for 30 s, annealing at 54 °C for 30 s, extension at 72 °C for 80 s and a final extension at 72 °C for 5 min. The PCR products were examined electrophoretically using 0.8–1.2% agarose gels and purified using a TIAN gel Midi Purification Kit (TIANGEN Biotech, Beijing, China). The purified products were sequenced using the same primer pairs with PCR. All of the sequences obtained were subjected to a BLAST (https://blast.ncbi.nlm.nih.gov/Blast.cgi (accessed on 15 April 2020)) search to detect contamination and nonspecific amplification. After confirmation, the newly generated sequences were submitted to GenBank.

2.3. Phylogenetic Analyses

The assembled sequences were aligned using MAFFT v.7 [27] with default settings and then adjusted manually for accuracy in Geneious R9 9.1.8 (https://www.geneious.com (accessed on 4 July 2020)). Character gaps were treated as missing data. Phylogenetic relationships were reconstructed using maximum likelihood (ML), maximum parsimony (MP) and Bayesian inference (BI) approaches. ML analysis was firstly performed to build a single-gene tree in online CIPRES Science Gateway v.3.3 (http://www.phylo.org/ (accessed on 18 Octobor 2020)) [28], using RAxML-HPC v8.2.10 [29,30] under the GTR + G DNA substitution model [31]. One hundred rapid bootstrap replicates were generated for each single gene matrix [32]. Since there was no strong conflict among individual markers, we conducted further ML analysis using a concatenated matrix of all seven loci under the GTR + G model as recommended by jModelTest v2.1 [33]. Bootstrap analyses were used to evaluate the support for each clade with 1000 bootstrap replicates. MP analysis was conducted in PAUP* version 4.0b10 [34]. All characters were equally weighted. Trees were inferred using the heuristic search option with tree bisection probabilities (TBR) swapping and 1000 replicates of random addition. Ten trees were held in each step during stepwise addition. The maximum number of trees was set to 10,000, and all parsimonious trees were saved. The tree length (TL), consistency index (CI), retention index (RI) and rescaled consistency index (RC) were calculated for each maximum parsimony tree. The bootstrap results were summarized in a 50% majority-rule consensus cladogram. BI analysis was implemented using MrBayes version 3.2.6. [35]. A Markov Chain Monte Carlo (MCMC) analysis was run with two independent chains with a random starting tree for 100 million generations, sampling one tree every 1000 generations. The first 25% generations were discarded as burn-in, and the remaining trees were used to construct a consensus tree with a 50% majority rule and obtain the posterior probabilities.

2.4. Character Coding and Ancestral State Reconstruction

Four characters closely related to vegetative or sexual reproduction were selected for ancestral character state reconstruction, namely bulbil formation, inflorescence structure, floral color and opening degree of the perianth. Morphological data were obtained from direct observations and the literature. The evolutionary history of each of the four characters was traced over the Bayesian 50% majority-rule tree using MP approaches available in Mesquite 3.5.1 [36]. The character states were treated as unordered and equally weighted. Morphological characters and their states were coded as follows: a. bulbil absence (0), presence (1); b. inflorescence umbel (0), spike (1), raceme (2) or panicle (3); c. floral color dark (e.g., violet, purplish red, orange) (0) or light (e.g., white, pale yellow, green) (1); d. perianth opening degree open (0), half-closed (1) or closed (2). For species in which the floral color changed during blooming, we used the color at the mature stage as the character state. The resulting codes for the sampled species are summarized in Table 3.

Table 3. Data matrix of morphological characters used in this study.

Species		Character			
		a	b	c	d
Dioscorea sect. *Stenophora*	*D. nipponica*	0	1	1	0
	D. nipponica subsp. *rosthornii*	0	2	1	0
	D. althaeoides	0	3	1	0
	D. tokoro	0	3	1	0
	D. zingiberensis	0	1	0	0
	D. sinoparviflora	0	1	0	0
	D. deltoidea	0	1	1	0
	D. panthaica	0	3	1	0
	D. biformifolia	0	3	1	0
	D. gracillima	0	1	1	0
	D. collettii	0	1	1	0
	D. collettii var. *hypoglauca*	0	1	1	0
	D. futschauensis	0	3	0	0
	D. spongiosa	0	3	1	0
	D. banzhuana	0	3	1	0
	D. simulans	0	2	0	0
	D. caucasica	0	1	1	0
	D. villosa	0	1	1	1

Table 3. *Cont.*

Species		Character			
		a	b	c	d
D. sect. *Combilium*	*D. esculenta*	0	1	1	0
	D. esculenta var. *spinosa*	0	1	1	0
D. sect. *Shannicorea*	*D. tentaculigera*	1	1	1	0
	D. yunnanensis	0	1	1	0
	D. subcalva	0	1	1	0
	D. subcalva var. *submollis*	0	1	1	0
	D. nitens	0	1	1	0
D. sect. *Opsophyton*	*D. bulbifera*	1	1	0	1
D. sect. *Botryosicyos*	*D. melanophyma*	1	2	1	2
	D. kamoonensis	1	2	1	2
	D. delavayi	1	2	1	2
	D. menglaensis	1	3	1	2
	D. pentaphylla	1	3	1	1
	D. esquirolii	1	2	1	1
D. sect. *Lasiophyton*	*D. hispida*	0	3	1	2
D. sect. *Enantiophyllum*	*D. aspersa*	0	1	1	2
	D. polystachya	1	1	1	2
	D. japonica	1	1	1	2
	D. cirrhosa	1	1	1	2
	D. cirrhosa var. *cylindrica*	1	3	1	2
	D. wallichii	0	3	1	2
	D. glabra	1	3	1	2
	D. fordii	1	3	1	2
	D. persimilis	1	3	1	2
	D. exalata	1	3	1	2
	D. alata	1	3	1	2
	D. decipiens	1	3	1	2
D. sect. *Apodostemon*	*D. composita*	0	1	1	1
D. sect. *Macroura*	*D. sansibarensis*	1	1	1	2
D. sect. *Testudinaria*	*D. elephantipes*	0	2	0	0
outgroup	*Tacca chantieri*	0	0	0	0

3. Results

3.1. Phylogenetic Analyses

According to our analyses, the chloroplast segment *trn*L-F exhibited the highest variability among the seven molecular markers, followed by *mat*K and *nad1* (see Table 4). The final concatenated matrix of all seven markers consisted of 7218 base pairs containing 1157 variable sites and 614 parsimony-informative sites. The molecular reconstruction based on the combined matrix gave rise to a highly resolved phylogenetic tree with moderate to strong support. There was no significant incongruence in topology or support values among trees generated from different methods. Thus, only the ML tree is presented in Figure 2.

Table 4. Informative parameters for the seven molecular markers.

DNA Region	Aligned Length (bp)	Variable Site (bp/%)	Informative Site (bp/%)	Tree Length	*CI*	*RI*
*mat*K	1032	237/23.0	163/15.8	317	0.852	0.959
*rbc*L	1142	134/11.7	84/7.4	220	0.664	0.867
*trn*L-F	939	280/29.8	80/8.5	385	0.816	0.905
*psb*A-*trn*H	344	113/32.8	82/23.8	160	0.850	0.955
*rpl*36-*rps*8	857	111/13.0	64/7.5	142	0.817	0.949
*nad*1	1478	207/14.0	125/8.5	246	0.902	0.949
*rps*3	1426	75/5.3	19/1.3	89	0.876	0.929
7 DNA	7218	1157/16.0	617/8.5	1640	0.782	0.913

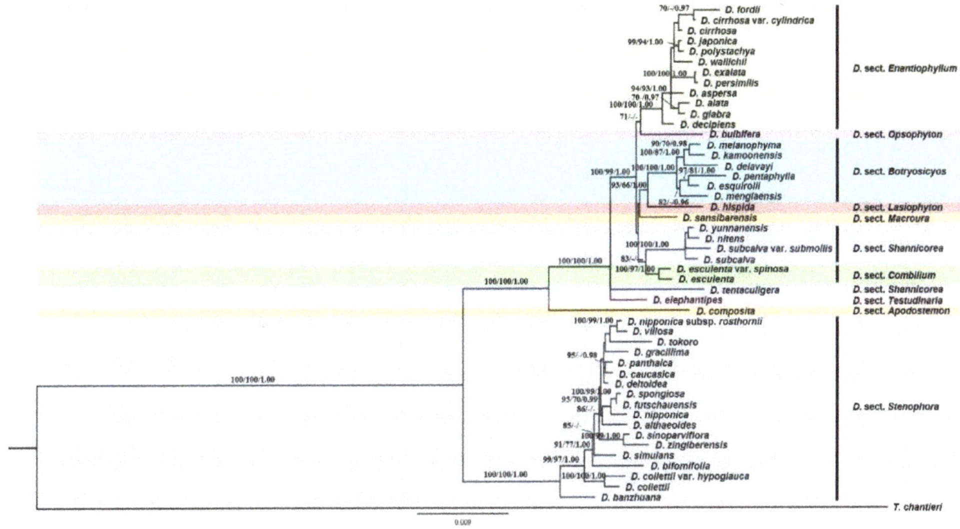

Figure 2. The maximum likelihood tree using the combined data from chloroplast *mat*K, *rbc*L, *trn*L-F, *psb*A-*trn*H, *rpl36-rps8*, as well as mitochondria *nad1* and *rps3*. Branches are labeled with maximum likelihood bootstraps higher than 70%, parsimony bootstrap proportions higher than 50% and Bayesian posterior probabilities more than 0.95.

All sampled *Dioscorea* species clustered in a monophyletic group consisting of two major clades with strong support. In accordance with previous studies, *D.* sect. *Stenophora* was the first diverging clade, sister to the other clade that included all the remaining species (BS = 100). The species of *D.* sect. *Botryosicyos* was grouped into a monophyletic clade (BS = 100). This clade was further divided into two subclades, one consisting of *D. pentaphylla* L., *D. esquirolii* Prain et Burkill and *D. menglaensis* H. Li and the other including *D. melanophyma* Prain et Burkill, *D. kamoonensis* Kunth and *D. delavayi* Franch. The only species of *D.* sect. *Lasiophyton*, *D. hispida* Dennst, was sister to *D.* sect. *Botryosicyos* (BS = 93). The position of *D. tentaculigera* Prain et Burkill remained unresolved, while other species of *D.* sect. *Shannicorea* clustered together in a well-supported clade (BS = 100). The two samples of *D.* sect. *Combilium* grouped together, and they were revealed as sisters to the Shannicorea clade with moderate support (BS = 83). All accessions of *D.* sect. *Enantiophyllum* formed a well-supported monophyletic clade (BS = 100). The species of *D.* sect. *Opsophyton* (*D. bulbifera*) was positioned as a sister to *D.* sect. *Enantiophyllum* with moderate support (BS = 71).

Some taxonomic treatments at the species level were supported by the phylogenetic tree. *Dioscorea collettii* Hook. f. and *D. collettii* var. *hypoglauca* (Palibin) C. T. Ting grouped together, while *D. esculenta* (Lour.) Burkill and *D. esculenta* var. *spinosa* (Roxburgh ex Prain & Burkill) R. Knuth were resolved as the closest relatives to each other. However, some varieties or subspecies of one species did not form a monophyletic group. For instance, *D. nipponica* Makino and *D. nipponica* subsp. *rosthornii* (Prain & Burkill) C. T. Ting, and *D. subcalva* Prain et Burkill and *D. subcalva* var. *submollis* (R. Knuth) C. T. Ting & P. P. Ling, did not group together.

3.2. Ancestral Character State Analyses

The ancestral state of four reproductive features of the *Dioscorea* species was reconstructed based on MP methods. As for character "a" the absence or presence of bulbils at the axil), state "0" (absent) was suggested to be the plesiomorphic state for *Dioscorea* (Figure 3). No species of the outgroup *Tacca* and *D.* sect. *Stenophora*, the earliest diverging group in *Dioscorea*, produces bulbils. The trait distribution in the phylogenetic tree suggests that bulbils appeared after the divergence of *D.* sect. *Stenophora,* but were subsequently lost several times independently, in *D.* sect. *Combilium*, *D.* sect. *Lasiophyton* and several species in *D.* sect. *Enantiophyllum* (*D. wallichii* Hook. f., *D. aspersa* Prain et Burkill). Bulbils persisted in most species in *D.* sect. *Botryosicyos*, *D.* sect. *Opsophyton* and *D.* sect. *Enantiophyllum*.

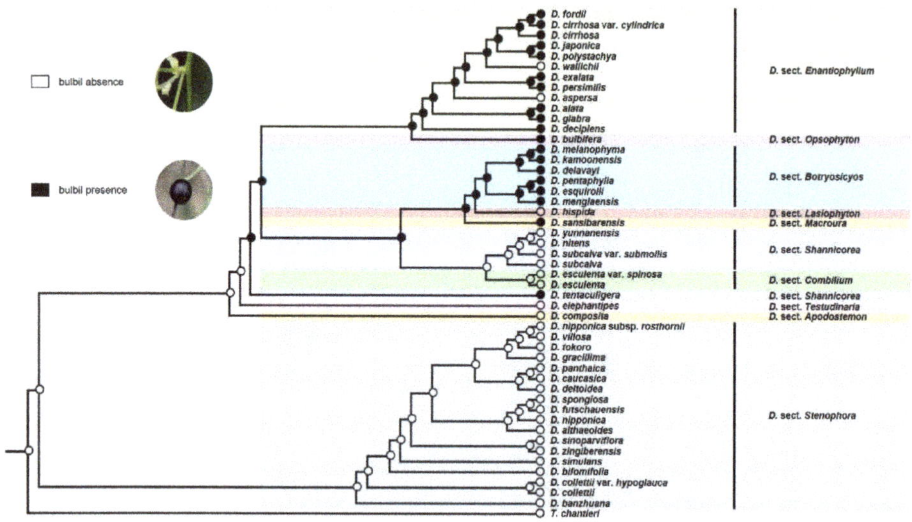

Figure 3. Ancestral character state reconstruction for bulbils using the parsimony method.

The ancestral state of the inflorescence architecture was inferred to be a spike (Figure 4). Racemes and panicles were inferred to be derived states, with the former occurring sporadically in different clades and the latter occurring mainly in *D.* sect. *Botryosicyos* and *D.* sect. *Enantiophyllum*. Spike occurred as a homoplasy in *D.* sect. *Shannicorea,* while other clades did not show a consistent type of male inflorescence.

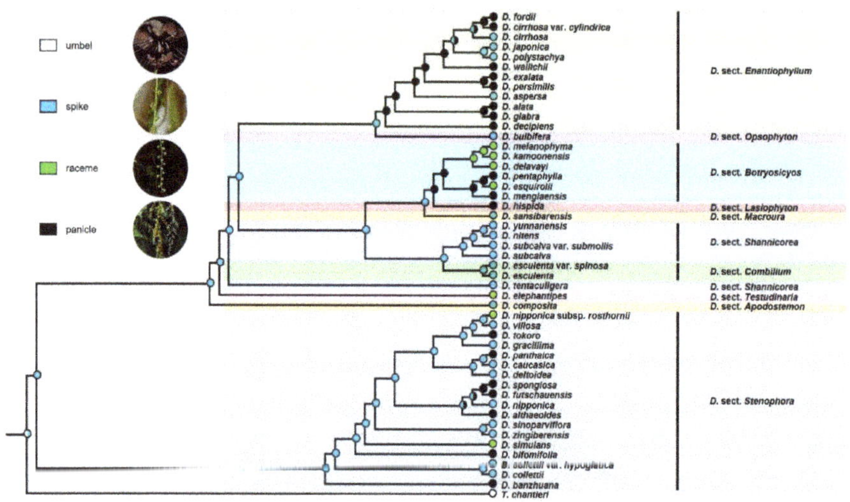

Figure 4. Ancestral character state reconstruction for the inflorescence architecture using the parsimony method.

The light color of the male flowers was uncovered as the ancestral state of *Dioscorea* (Figure 5). The dark color evolved independently several times in *D.* sect. *Stenophora* and *D.* sect. *Opsophyton*. Most species with dark colored male flowers occurred in *D.* sect. *Stenophora*. There were also some species in *D.* sect. *Shannicorea* and *D.* sect. *Enantiophyllum* that possessed male flowers that were purple in color or had a brown stripe on them. This distribution probably reflects multiple convergence events in response to similar pollination strategies.

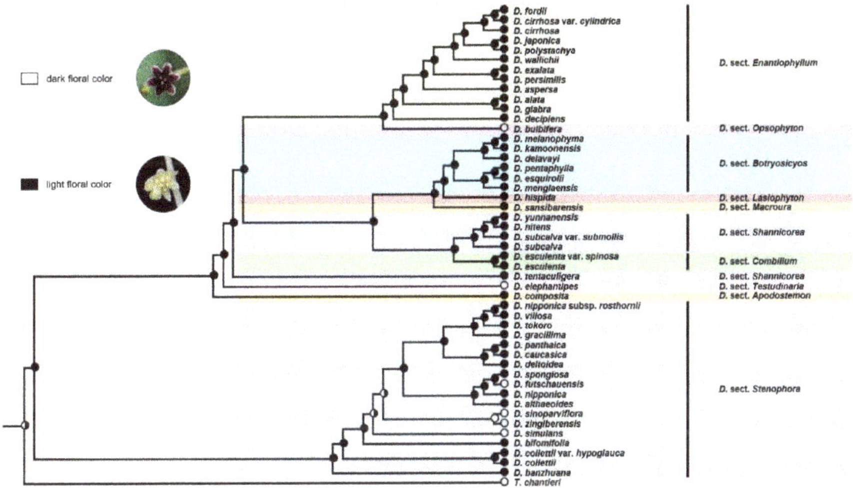

Figure 5. Ancestral character state reconstruction for the floral color using the parsimony method.

The male flowers of *Dioscorea* ancestors were suggested to open fully, and this state was retained in *D.* sect. *Stenophora* and *D.* sect. *Shannicorea* (Figure 6). The perianth of male flowers was closed to some degree in the ancestor of the clade encompassing *D.* sect. *Enantiophyllum*, *D.* sect. *Opsophyton*, *D.* sect. *Botryosicyos*, *D.* sect. *Lasiophyton* and *D.* sect. *Combilium*. Male flowers of *D.* sect. *Botryosicyos* and *D.* sect. *Enantiophyllum* were generally closed.

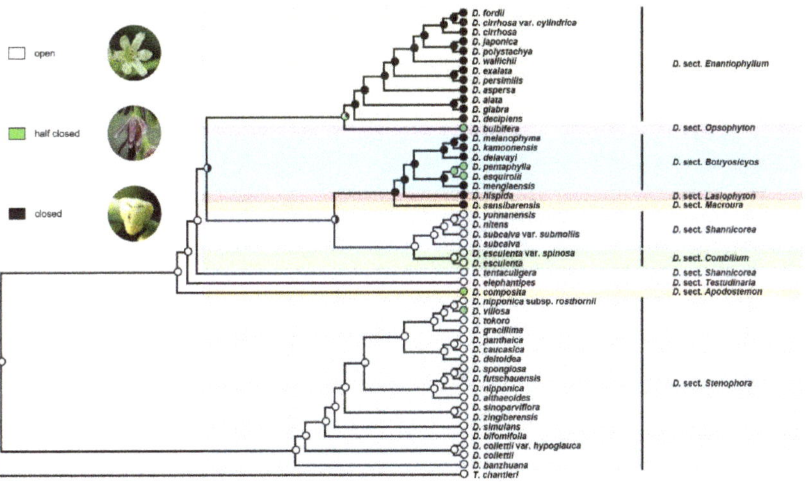

Figure 6. Ancestral character state reconstruction for the perianth opening degree using the parsimony method.

4. Discussion

4.1. Evolution of Organs for Vegetative Propagation and Diversity in the Reproductive Strategy

Vegetative propagation is a crucial mechanism that ensures plants maintain and spread their population, allowing plants to survive unfavorable conditions for sexual reproduction [37]. *Dioscorea* species possess multiple organs involved in vegetative propagation, including rhizomes, horizontal or vertical tubers and aerial bulbils, as a result of adaptation to various environmental conditions. In the early diverged *D.* sect. *Stenophora*, the underground parts consisted of perennial branched horizontal rhizomes. Perennial to annual starchy tubers took the place of rhizomes in subsequently evolved lineages to undertake the function of asexual reproduction. Rhizomatous species are supposed to be more adaptive to long growing seasons in shaded habitats [38], whereas tubers can evade destruction by animals and allow adaptation to warm and humid climates [39]. The possession of tubers was suggested to play an important role in species diversification of *Dioscorea* [2]. However, tubers are mainly storage organs for plants to live through winter and germinate in for the next growing season; the reproductive efficiency and dispersal ability of tubers are limited.

The emergence of aerial bulbils in some *Dioscorea* species greatly changed the propagation patterns of these species. The bulbil of *Dioscorea* is a minor storage organ compared to the underground tuber, arising from the axils of leaves or inflorescences and varying in form and size in different species [40]. On the basis of morphological and anatomical studies, bulbils are interpreted as modified axillary branches and miniature reproductions of the rhizome [11,41]. Functioning as a means of vegetative propagation, bulbils are released from the plants after maturation and then dispersed by gravity or by water in streams, sometimes over long distances, like seeds. Bulbils can easily germinate and grow independently, giving rise to large numbers of progeny. Because of these advantageous characteristics, species that produce bulbils are usually more vigorous and more competitive than species that reproduce only by seeds, as observed in *Allium* (Alliaceae) [42] and *Butomus* (Butomaceae) [43].

As mentioned above, the vegetative propagation strategies vary among the different lineages of *Dioscorea*, and the production of bulbils is a derived character in most clades of this genus. This may have greatly influenced the expansion of the *Dioscorea* species. In *Poa alpina* L. (Poaceae), bulbil-producing plants are better adapted to higher elevations compared with seed-producing plants. Therefore, this species is able to occupy a range of ecological niches by means of different reproductive modes [44]. Bulbils in *Fritillaria* (Liliaceae) are suspected to be an adaptation to underground dispersal [38]. In *Dioscorea*, the rhizomatous taxa are mainly distributed in South and East Asia, and many of them are endemic species only found in restricted areas, while the species that produce bulbils occupy a wide range of habitats worldwide [2]. *D. bulbifera*, the only wild species exhibiting worldwide distribution in *Dioscorea*, produces huge numbers of bulbils compared to other species. Bulbils of *D. sansibarensis* Pax are buoyant and spread through water flow, achieving wide distribution in African valleys [11]. Here, it is postulated that the production of bulbils improved the adaptive ability of *Dioscorea* species and promoted their expansion.

4.2. Evolution of Floral Characters and Their Relation to Sexual Propagation

Flower evolution is strictly linked to pollination strategy [45,46]. *Dioscorea* is predominantly dioecious and allogamous; thus sexual propagation in this genus relies on the outcrossing process. Moreover, the sticky nature of *Dioscorea* pollen grains prevents transport by wind, and, consequently, the flowers of *Dioscorea* species are pollinated mainly by insects [47]. The diverse flower exhibition characteristics of this genus were most likely shaped by pollinator-mediated selection.

According to our analyses, spikes are reconstructed as the plesiomorphy of *Dioscorea*, while panicles appear mainly in *D.* sect. *Enantiophyllum* as a derived inflorescence type. This contrasts with the opinion presented by Burkill [11], who suggested that spikes and racemes evolved from panicles. The distribution patterns of inflorescence types among species do not mirror phylogenetic relationships, with closely related lineages displaying different types, except *D.* sect. *Shannicorea*. Such a pronounced polymorphism might have been driven by pollinators. Inflorescence architecture is related to the mode and efficiency of pollination. For instance, changes in inflorescence architecture were associated with the transition from biotic (insect) to abiotic (wind) pollination in *Schiedea salicaria* Hbd. (Caryophyllaceae) [48], and affected pollinator behavior and mating success in *Spiranthes sinensis* (Pers.) Ames (Orchidaceae) [49]. Panicles are made of multiple spikes or racemes to enlarge the volume of the flower and enhance the attraction of pollinators. They possess two to three times the number of flowers and amount of pollen as spikes or racemes of the same length. The function of panicles in insect attraction compared with other types of inflorescences is worthy of investigation in *Dioscorea*.

In our results, most species of *Dioscorea* possess male flowers of light color, and dark color originated several times independently in different lineages. It is well known that floral color is among the most important visual signals in pollinator attraction. For example, hawkmoth and hummingbird pollinators prefer yellow and red morphs of *Mimulus aurantiacus* Curtis, respectively [50]. The reflectance spectrum of the dark perianth generated by UV-light is recognizable to insects [51]. Dark color and light color perianth could be visually discriminated by insects and thus will affect the visiting choice of different pollinators [52]. An alternative explanation for floral color polymorphisms among closely related species is that color divergence evolves in response to interspecific competition for pollinators as a means to decrease interspecific pollinator movements [53]. Indeed, the color of perianth in *Dioscorea* includes white, yellow, green, orange, red and purple, sometimes changing from one to another during flower development. How these variations of floral colors influence the category and behavior of the pollinators will be an interesting subject to investigate.

The evolutionary trend of the perianth opening degree in *Dioscorea* is from wholly opened to nearly closed. This character may also influence the category of pollinators. As discovered in the study of *Merianieae* (Melastomataceae), corollas of most buzz-bee syndrome species are widely open, forming bowl-shaped flowers, whereas they are more closed and form urceolate to pseudo-campanulate flowers in vertebrate-pollinated species [54]. The same was also observed in *Erica* (Ericaceae) [55] and *Ruellia* (Acanthaceae) [56]. Zhao et al. [57] reported that *D. nipponica* subsp. *rosthornii* is pollinated by halictids. In *D.* sect. *Enantiophyllum* species, male panicles were found usually to be made up of closed flowers that restrict visitations only to small insects. The pollinators of species such as *D. rotundata* Poir., *D. japonica* Thunb. and *D. polystachya* Turcz. were purported to be thrips [47,58–60]. Therefore, the evolution of male flower opening degree in *Dioscorea* may be a result of pollinator specialization.

Actually, as reported in previous studies, the pollinators of *Dioscorea* species include *Coleoptera, Diptera, Hymenoptera, Hemiptera, Thripidae, Thysanoptera, Halictus* and *Andrena*. [47,60]. The diversification of floral morphological characters in this genus is thought to be promoted by pollination-associated adaptations. Intensive investigation of the pollination strategies in other lineages is needed to better understand the adaptive evolution of reproduction in *Dioscorea*.

4.3. Infrageneric Relationships and Taxonomic States of Several Species

The phylogenetic framework reconstructed in this study was consistent with those obtained in previous studies [2,12,14,15]. The relationships indicated by molecular information were generally consistent with infrageneric division by traditional systematics.

The monophyly of *Dioscorea* sect. *Stenophora* has been certified in a number of molecular studies using various datasets. This is also supported by morphological evidence since species of *D.* sect. *Stenophora* show underground organ type, chromosome number and pollen type that set them apart from other lineages of *Dioscorea*. As indicated in this study, no species of *D.* sect. *Stenophora* generated bulbils, and they all had male flowers with fully open perianths, thus giving further support to the isolated position of this section within the *Dioscorea* genus.

Most species of *Dioscorea* sect. *Lasiophyton* have been transferred into *D.* sect. *Botryosicyos* [22]. However, there are clear morphological differences between these two taxa. Species of *D.* sect. *Lasiophyton* have compound leaves with three palmate leaflets that are also palmately veined, and all six stamens of the male flower are fertile; in contrast, in *D.* sect. *Botryosicyos*, leaves usually have more than three leaflets, are pinnately veined, and have three stamens alternating with three staminodes. In this study, the only Chinese species of *D.* sect. *Lasiophyton*, *D. hispida*, was resolved as the sister lineage to the *Botryosicyos* clade, supporting the exclusion of this species from *D.* sect. *Botryosicyos* by Ding and Gilbert [22].

Dioscorea tentaculigera was assigned to *D.* sect. *Shannicorea* by Burkill [11]. Its male flowers are light in color and fully open. However, *D. tentaculigera* produces bulbils, and its male flowers are sessile, which is an obvious departure from other species in *D.* sect. *Shannicorea*. The phylogenetic position of *D. tentaculigera* was not resolved in our analyses, as has been the case in previous studies (e.g., [2,12–14]). Maurin et al. [20] placed this species outside of the *Enantiophyllum* clade based on six cpDNAs; Soto Gomez et al. [61] positioned it as a sister to the Mediterranean clade based on 260 nuclear genes; and Noda et al. [15] suggested that it should be treated as a distinct section according to four cpDNAs. More evidence from molecular and morphological data is needed to draw a conclusion about the correct classification of this species.

Accessions of *Dioscorea* sect. *Enantiophyllum* were found to cluster into a highly supported monophyletic clade. There are several synapomorphies of this group: (i) stem twining to the right, (ii) leaves generally opposite, (iii) seeds winged all round, (iv) bulbils normally present, (v) male inflorescence panicles and (vi) perianth white to yellow and nearly closed when blooming. The limits of this section were found to be reasonable for a natural lineage [2,12,14,15,17].

The phylogenetic analyses also shed light on the taxonomy of certain species. *Dioscorea nipponica* subsp. *rosthornii* was separated from *D. nipponica* subsp. *nipponica* by a cork layer of rhizomes and male flowers that were pedicellate [62], as well as a difference in chromosome number [24,39]. However, they have not been resolved as sister groups in any molecular phylogenetic trees generated to date [12,14,24]. A similar situation was observed with *D. subcalva* var. *submolis*, which showed differences from *D. subcalva* var. *subcalva* that included sparse leaf blades and longer infructescences. We noticed that these characters varied in a continuous range between individuals among populations. The relationships between these two varieties and other species of *D.* sect. *Shannicorea* remain unresolved by molecular analyses [15,17]. Further investigation with sufficient samples may help us understand the internal relationships within this section and the circumscription of the species.

5. Conclusions

This study provides the first phylogenetic analyses focusing on the evolution of four main reproductive traits in *Dioscorea*. The development of bulbils in late-diverged lineages of *Dioscorea* has diversified the mode of vegetative propagation, which is supposed to have greatly improved reproductive efficiency and heightened plants' ability to occupy new habitats. Spikes, male flowers in light color and wholly open perianth are reconstructed as plesiomorphic in *Dioscorea*, whereas panicles, dark color flowers and nearly closed perianth are suggested derived states. The extraordinary variation of floral characters in this genus appears to be a consequence of adaptive evolution driven by pollinators. A broader sampling of *Dioscorea*, especially new-world species, together with their character information and extensive pollination studies among species with different floral, types are necessary to achieve a better understanding of the diversification mechanism of this genus.

On the other hand, this work provides a suitable phylogenetic framework for the revision of infrageneric classification and species delimitation in *Dioscorea*. Some of the morphological characters show consistency in a certain clade. Species in *D.* sect. *Stenophora* do not produce bulbils, and their male flowers are always wholly open; *D.* sect. *Shanni-*

corea is characterized by no bulbils, spikes, light color and wholly open flowers; *D.* sect. *Enantiophyllum* is distinguished by producing bulbils, with flowers in light color and nearly closed. According to our results, bulbils and perianth opening degree are informative taxonomic features at the section level in *Dioscorea*. The monophyletic status of several sections proposed in classical taxonomy, such as *D.* sect. *Stenophora* and *D.* sect. *Enantiophyllum*, were supported by morphological synapomorphies besides molecular data. Morphological characters analyzed in the present work can also explain the unexpected position of *D. tentaculigera* in phylogenetic trees. In addition, our results suggest the necessity of reconsideration of some infraspecific taxa as well.

Author Contributions: Methodology, M.C. and J.-Y.X.; software, J.-Y.X.; validation, X.S., Y.H. and Y.Z.; formal analysis, M.C. and X.S.; investigation, M.C.; resources, Y.Z. and Y.H.; data curation, M.C.; writing—original draft preparation, M.C.; writing—review and editing, X.S. and Y.H.; supervision, Y.H.; project administration, Y.H.; funding acquisition, M.C. and X.S. All authors have read and agreed to the published version of the manuscript.

Funding: This study was funded by the Youth Foundation of Jiangsu Province to M.C. (Grant No. BK20180316) and the Independent Research Project of Jiangsu Provincial Public Welfare Research Institute to X.Q.S. (Grant No. BM2018021-2).

Institutional Review Board Statement: Not applicable.

Data Availability Statement: Data generated in this study have been uploaded to NCBI (https://www.ncbi.nlm.nih.gov/) and GenBank accession numbers are listed in Table 1.

Conflicts of Interest: The authors declare no conflict of interest.

References

1. Govaerts, R.; Wilkin, P.; Raz, L.; Téllez-Valdés, O. World checklist of Dioscoreaceae. Facilitated by the Royal Botanic Gardens, Kew. 2018. Available online: https://wcsp.science.kew.org/ (accessed on 25 July 2021).
2. Wilkin, P.; Schols, P.; Chase, M.W.; Chayamarit, K.; Furness, C.A.; Huysmans, S.; Rakotonasolo, F.; Smets, E.; Thapyai, C. A plastid gene phylogeny of the yam genus, *Dioscorea*: Roots, fruits and Madagascar. *Syst. Bot.* **2005**, *30*, 736–749. [CrossRef]
3. Huber, H. Dioscoreaceae. Flowering plants. Monocotyledons. Lilianae (except Orchidaceae). In *The Families and Genera of Vascular Plants*; Kubitzk, K., Ed.; Springer: Berlin/Heidelberg, Germany, 1998; Volume 3, pp. 216–235.
4. Asiedu, R.; Sartie, A. Crops that feed the world. Yams. *Food Secur.* **2010**, *2*, 305–315. [CrossRef]
5. Price, E.J.; Wilkin, P.; Sarasan, V.; Fraser, P.D. Metabolite profiling of *Dioscorea* (yam) species reveals underutilized biodiversity and renewable sources for high-value compounds. *Sci. Rep.* **2016**, *6*, 29136. [CrossRef] [PubMed]
6. Dahlgren, G. An undated angiosperm classification. *Bot. J. Linn. Soc.* **1989**, *100*, 197–203. [CrossRef]
7. Mignouna, H.; Abang, M.; Asiedu, R.; Geeta, R. True yams (*Dioscorea*): A biological and evolutionary link between eudicots and grasses. *Cold Spring Harb. Protoc.* **2009**, *11*, pdb.emo136. [CrossRef]
8. Sheikh, N.; Kumar, Y. Foliar epidermal, stem and petiole anatomy of Meghalayan *Dioscorea* L. (Dioscoreaceae) and its systematic implication. *Bangladesh J. Plant Taxon* **2017**, *24*, 53–63. [CrossRef]
9. Uline, E.B. Eine monographie de Dioscoreaceen. *Bot. Jahrb.* **1898**, *25*, 126–165.
10. Knuth, R. Dioscoreaceae. In *Das Pflanzenreich, 87 (IV 43)*; Engler, H.G.A., Ed.; Verlag von Wilhelm Engelmann (Druck von Breitkopf & Härtel in Leipzig): Leipzig, Germany, 1924; pp. 1–387.
11. Burkill, I.H. The organography and the evolution of Dioscoreaceae, the family of the yams. *Bot. J. Linn. Soc.* **1960**, *56*, 319–412. [CrossRef]
12. Viruel, J.; Segarra-Moragues, J.G.; Raz, L.; Forest, F.; Wilkin, P.; Sanmartín, I.; Catalán, P. Late cretaceous-early eocene origin of yams (*Dioscorea*, Dioscoreaceae) in the Laurasian Palaearctic and their subsequent oligocenemiocene diversification. *J. Biogeogr.* **2016**, *43*, 750–762. [CrossRef]
13. Viruel, J.; Forest, F.; Paun, O.; Chase, M.W.; Devey, D.; Couto, R.S.; Gabriel, J.; Segarra-Moragues Catalán, P.; Wilkin, P. A nuclear *Xdh* phylogenetic analysis of yams (*Dioscorea*: Dioscoreaceae) congruent with plastid trees reveals a new Neotropical lineage. *Bot. J. Linn. Soc.* **2018**, *187*, 232–246. [CrossRef]
14. Couto, R.S.; Martins, A.C.; Bolson, M.; Lopez, R.C.; Smidt, E.C.; Braga, J.M.A. Time calibrated tree of *Dioscorea* (Dioscoreaceae) indicates four origins of yams in the Neotropics since the Eocene. *Bot. J. Lin. Soc.* **2018**, *188*, 144–160. [CrossRef]
15. Noda, H.; Yamashita, J.; Fuse, S.; Pooma, R.; Poopath, M.; Tobe, H.; Tamura, M.N. A large-scale phylogenetic analysis of *Dioscorea* (Dioscoreaceae), with reference to character evolution and subgeneric recognition. *Acta Phytotax. Geobot.* **2020**, *71*, 103–128.
16. Endress, P.K. *Diversity and Evolution Biology of Tropical Flowers*; Cambridge University Press: Cambridge, UK, 1994; p. 511.
17. Hsu, K.M.; Tsai, J.L.; Chen, M.Y.; Ku, H.M.; Liu, S.C. Molecular phylogeny of *Dioscorea* (Dioscoreaceae) in East and Southeast Asia. *Blumea J. Plant Taxon. Plant Geogr.* **2013**, *58*, 21–27. [CrossRef]

18. Biffin, E.; Brodribb, T.J.; Hill, R.S.; Thomas, P.; Lowe, A.J. Leaf evolution in Southern Hemisphere conifers tracks the angiosperm ecological radiation. *Proc. Biol Sci.* **2012**, *279*, 341–348. [CrossRef]
19. Sun, X.Q.; Xue, J.Y.; Lei, Z.; Li, M.M.; Zhang, Y.M.; Zhou, G.C.; Hang, Y.Y. Taxonomic and phylogenetic significance of leaf venation characteristics in *Dioscorea* plants. *Arch. Biol. Sci.* **2018**, *70*, 397–407. [CrossRef]
20. Maurin, O.; Muasya, A.M.; Catalán, P.; Shongwe, E.Z.; Viruel, J.; Wilkin, P.; Van der Bank, M. Diversification into novel habitats in the Africa clade of *Dioscorea* (Dioscoreaceae): Erect habit and elephant's foot tubers. *BMC Evol. Biol.* **2016**, *16*, 238. [CrossRef] [PubMed]
21. Wan, J.R.; Ding, Z.Z.; Qin, H.Z. A phytogeographical study on the family Dioscoreaceae. *Acta Bot. Boreale-Occident. Sin.* **1994**, *14*, 128–135.
22. Ding, Z.Z.; Gilbert, M.G. Dioscoreaceae. In *Flora of China*; Wu, Z.Y., Raven, P.H., Eds.; Science Press: Beijing, China; Botanical Garden Press: St. Louis, MI, USA, 2000; Volume 24, p. 284.
23. Doyle, J.J.; Doyle, J.L. A rapid DNA isolation procedure for small amounts of fresh leaf tissue. *Phytochem. Bull.* **1987**, *19*, 11–15.
24. Gao, X.; Zhu, Y.P.; Wu, B.C.; Zhao, Y.M.; Chen, J.Q.; Hang, Y.Y. Phylogeny of *Dioscorea* sect. *Stenophora based on chloroplast matK, rbcL and trnL-F sequences. J. Syst. Evol.* **2008**, *46*, 315–321.
25. Taberlet, P.T.; Gielly, L.; Patou, G.; Bouvet, J. Universal primers for amplification of three non-coding regions of chloroplast DNA. *Plant Mol. Biol.* **1991**, *17*, 1105–1109. [CrossRef]
26. Laroche, J.; Bousquet, J. Evolution of the mitochondrial *rps*3 intron in perennial and annual angiosperms and homology to *nad*5 intron 1. *Mol. Biol. Evol.* **1999**, *16*, 441–452. [CrossRef]
27. Katoh, K.; Standley, D.M. MAFFT multiple sequence alignment software version 7, improvements in performance and usability. *Mol. Biol. Evol.* **2013**, *30*, 772–780. [CrossRef]
28. Miller, M.A.; Pfeiffer, W.; Schwartz, T. Creating the CIPRES Science Gateway for inference of large phylogenetic trees. In Proceedings of the Gateway Computing Environments Workshop (GCE), New Orleans, LA, USA, 14 November 2010; pp. 1–8.
29. Stamatakis, A. RAxML-VI-HPC: Maximum likelihood-based phylogenetic analyses with thousands of taxa and mixed models. *Bioinformatics* **2006**, *22*, 2688–2690. [CrossRef]
30. Stamatakis, A. RAxML version 8: A tool for phylogenetic analysis and post-analysis of large phylogenies. *Bioinformatics* **2014**, *30*, 1312–1313. [CrossRef]
31. Yang, Z.H. Maximum-likelihood estimation of phylogeny from DNA sequences when substitution rates differ over sites. *Mol. Biol. Evol.* **1993**, *10*, 1396–1401.
32. Mirarab, S.; Bayzid, M.S.; Boussau, B.; Warnow, T. Statistical binning enables an accurate coalescent-based estimation of the avian tree. *Science* **2014**, *346*, 1250463. [CrossRef]
33. Darriba, D.; Taboada, G.L.; Doallo, R.; Posada, D. jModelTest 2: More models, new heuristics and parallel computing. *Nat. Methods* **2012**, *9*, 772. [CrossRef]
34. Swofford, D.L. *PAUP*: Phylogenetic Analysis Using Parsimony (*and Other Methods), Version 4.0b10*; Sinauer Associates: Sunderland, MA, USA, 2002.
35. Ronquist, F.; Teslenko, M. Van der Mark, P.; Ayres, D.L.; Darling, A.; Hohna, S.; Larget, B.; Liu, L.; Suchard, M.A.; Huelsenbeck, J.P. MrBayes 3.2: Efficient Bayesian phylogenetic inference and model choice across a large model space. *Syst. Biol.* **2012**, *61*, 539–542. [CrossRef]
36. Maddison, M.P.; Maddison, D.R. Mesquite: A Modular System for Evolutionary Analysis. Version 3.5.1. 2018. Available online: http://www.mesquiteproject.org (accessed on 2 November 2020).
37. Cook, R.E. Growth and development in clonal plant populations. In *Population Biology and Evolution of Clonal Organisms*; Jackson, J.B.C., Buss, L.W., Cook, R.E., Eds.; Yale University Press: New Haven, CT, USA; London, UK, 1985; pp. 259–296.
38. Rønsted, N.; Law, S.; Thornton, H.; Fay, M.F.; Chase, M.W. Molecular phylogenetic evidence for the monophyly of *Fritillaria* and *Lilium* (Liliaceae; Liliales) and the infrageneric classification of *Fritillaria*. *Mol. Phylogenet. Evol.* **2005**, *35*, 509–527. [CrossRef]
39. Chin, H.C.; Chang, M.C.; Ling, P.P.; Ting, C.T.; Dou, F.P. A cytotaxonomic study on Chinese *Dioscorea* L.—The chromosome numbers and their relation to the origin and evolution of the genus. *Acta Phytotaxon. Sin.* **1985**, *23*, 11–18.
40. Wickham, L.D.; Wilson, L.A.; Passam, H.C. The origin, development and germination of bulbils in two *Dioscorea* species. *Ann. Bot.* **1982**, *50*, 621–627. [CrossRef]
41. Ayensu, E.S. *Anatomy of the Monocotyledons, VI. Dioscoreales*; Clarendon Press: Oxforod, UK, 1972; p. 182.
42. Ronsheim, M.L.; Bever, J.D. Genetic variation and evolutionary trade-offs for sexual and asexual reproductive modes in *Allium vineale* (Liliaceae). *Am. J. Bot.* **2000**, *87*, 1769–1777. [CrossRef]
43. Eckert, C.G.; Lui, K.; Bronson, K.; Corradini, P.; Bruineau, A. Population genetic consequences of extreme variation in sexual and clonal reproduction in an aquatic plant. *Mol. Ecol.* **2003**, *12*, 331–344. [CrossRef]
44. Steiner, B.L.; Armbruster, G.F.J.; Scheepens, J.F.; Stöcklin, J. Distribution of bulbil- and seed-producing plants of *Poa alpina* (Poaceae) and their growth and reproduction in common gardens suggest adaptation to different elevations. *Am. J. Bot.* **2012**, *99*, 2035–2044. [CrossRef]
45. Willmer, P. *Pollination and Floral Ecology*; Priceton University Press: Priceson, NJ, USA, 2011; pp. 1–792.
46. Patiny, S. *Evolution of Plant-Pollinator Relationships*; Cambrige University Press: Cambridge, UK, 2012; pp. 1–477.
47. Mondo, J.M.; Agre, P.A.; Edemodu, A.; Adebola, P.; Asiedu, R.; Akoroda, M.O.; Asfaw, A. Floral biology and pollination efficiency in yam (*Dioscorea* spp.). *Agriculture* **2020**, *10*, 560. [CrossRef]

48. Weller, S.G.; Sakai, A.K.; Culley, T.M.; Campbell, D.R.; Dunbar-Wallis, A.K. Predicting the pathway to wind pollination: Heritabilities and genetic correlations of inflorescence traits associated with wind pollination in *Schiedea salicaria* (Caryophyllaceae). *J. Evol. Biol.* **2006**, *19*, 331–342. [CrossRef]
49. Iwata, T.; Nagasaki, O.; Ishii, H.S.; Ushimaru, A. Inflorescence architecture affects pollinator behaviour and mating success in *Spiranthes sinensis* (Orchidaceae). *New Phytol.* **2012**, *193*, 196–203. [CrossRef]
50. Weiss, M.R. Floral color changes as cues for pollinators. *Nature* **1991**, *354*, 227–229. [CrossRef]
51. Streisfeld, M.A.; Kohn, J.R. Environment and pollinator-mediated selection on parapatric floral races of *Mimulus aurantiacus*. *J. Evol. Biol.* **2007**, *20*, 122–132. [CrossRef]
52. Wang, H.; Conchou, L.; Bessière, J.M.; Cazals, G.; Schatz, B.; Imbert, E. Flower color polymorphism in *Iris lutescens* (Iridaceae): Biochemical analyses in light of plant-insect interactions. *Phytochemistry* **2013**, *94*, 123–134. [CrossRef]
53. Muchhala, N.; Johnsen, S.; Smith, S.D. Competition for hummingbird pollination shapes flower color variation in Andean Solanaceae. *Evolution* **2014**, *68*, 2275–2286. [CrossRef] [PubMed]
54. Dellinger, A.S.; Artuso, S.; Pamperl, S.; Michelangeli, F.A.; Penneys, D.S.; Fernández-Fernández, D.M.; Alvear, M.; Almeda, F.; Armbruster, W.S.; Staedler, Y.; et al. Modularity increases rate of floral evolution and adaptive success for functionally specialized pollination systems. *Commun. Biol.* **2019**, *2*, 453. [CrossRef] [PubMed]
55. Reich, D.; Berger, A.; von Balthazar, M.; Chartier, M.; Sherafati, M.; Schönenberger, J.; Manafzadeh, S.; Staedler, Y.M. Modularity and evolution of flower shape: The role of function, development, and spandrels in *Erica*. *New Phytol.* **2020**, *226*, 267–280. [CrossRef] [PubMed]
56. Tripp, E.A.; Dexter, K.G.; Stone, H.B. Reproductive character displacement and potential underlying drivers in a species-rich and florally diverse lineage of tropical angiosperms (*Ruellia*; Acanthaceae). *Ecol. Evol.* **2021**, *11*, 4719–4730. [CrossRef]
57. Zhao, Y.M.; Hang, Y.Y.; Zhou, Y.F. The pollination of *Dioscorea nipponica* ssp. *rosthornii*. *Acta Phytotaxon. Sin.* **2008**, *17*, 15–21.
58. Sadik, S.; Okereke, O.U. Flowering Pollen Grain Germination, Fruiting, Seed Germination and Seedling Development of White Yam, *Dioscorea rotundata* Poir. *Anm. Bot.* **1975**, *39*, 597–604. [CrossRef]
59. Mizuki, I.; Ishida, K.; Kikuzawa, K. Sexual and vegetative reproduction in the aboveground part of a dioecious clonal plant, *Dioscorea japonica* (Dioscoreaceae). *Ecol. Res.* **2005**, *20*, 387–393. [CrossRef]
60. Li, M.M.; Yan, Q.Q.; Sun, X.Q.; Zhao, Y.M.; Zhou, Y.F.; Hang, Y.Y. A preliminary study on pollination biology of three species in *Dioscorea* (Dioscoreaceae). *Life Sci. J.* **2014**, *11*, 436–444.
61. Soto Gomez, M.; Pokorny, M.B.; Kantar, F.; Forest, I.J.; Leitch, B.; Gravendeel, P.; Wilkin, S.; Graham, W.; Viruel, J. A customized nuclear target enrichment approach for developing a phylogenomic baseline for *Dioscorea* yams (Dioscoreaceae). *Appl. Plant Sci.* **2019**, *7*, e11254. [CrossRef]
62. Pei, J.; Ting, C.T.; Chin, H.C.; Su, P.; Tang, S.Y.; Chang, H.C. A preliminary systematic study of *Dioscorea* L. sect. *Stenophora* Uline. *Acta Phytotaxon. Sin.* **1979**, *17*, 61–72.

diversity

MDPI

Article

Chloroplast Genome of *Salvia* Sect. *Drymosphace*: Comparative and Phylogenetic Analysis

Ting Su [1,2], Yan-Fei Geng [3], Chun-Lei Xiang [4], Fei Zhao [4], Mei Wang [1,2], Li Gu [1,2] and Guo-Xiong Hu [1,2,*]

1 College of Life Sciences, Guizhou University, Guiyang 550025, China; sutingyo@163.com (T.S.); wmgzu20@163.com (M.W.); guligz@163.com (L.G.)
2 The Key Laboratory of Plant Resources Conservation and Germplasm Innovation in Mountainous Region Ministry of Education, Guizhou University, Guiyang 550025, China
3 College of Tea Science, Guizhou University, Guiyang 550025, China; yfgeng@gzu.edu.cn
4 CAS Key Laboratory for Plant Diversity and Biogeography of East Asia, Kunming Institute of Botany, Chinese Academy of Sciences, Kunming 650201, China; xiangchunlei@mail.kib.ac.cn (C.-L.X.); zhaofei@mail.kib.ac.cn (F.Z.)
* Correspondence: gxhu@gzu.edu.cn

Abstract: Sect. *Drymosphace* is one of eight sections of *Salvia* subg. *Glutinaria* and includes 13 species and one dubious species that hold great economic value. Although the section is well supported, interspecific relationships remain unresolved. Moreover, most of this section's plastome information remains unknown. In this study, we sequenced and assembled eight sect. *Drymosphace* plastomes and conducted comparative analyses within this section. The length of plastid genome sequences ranged from 151,330 bp to 151,614 bp, with 80 protein-coding, 30 tRNA, and four rRNA genes being annotated. The plastomes were found to be as conservative as other Lamiaceae species, showing high consistency and similarity in terms of gene content, order, and structure. Within the sect. *Drymosphace*, single-copy regions were more variable than IR regions, and the intergenic regions were more variable than the coding regions; nine hypervariable regions were detected, and some of them may be useful for the phylogenetic analysis of *Salvia*. The topologies inferred from all of the data sets indicated that sect. *Drymosphace* was monophyletic and that *S. honania* was sister to *S. meiliensis*. Compared to previous studies involving more sect. *Drymosphace* species, phylogenomic analyses can improve the phylogenetic resolution considerably.

Keywords: *Salvia miltiorrhiza*; subg. *Glutinaria*; plastid genome; phylogenomics

Citation: Su, T.; Geng, Y.-F.; Xiang, C.-L.; Zhao, F.; Wang, M.; Gu, L.; Hu, G.-X. Chloroplast Genome of *Salvia* Sect. *Drymosphace*: Comparative and Phylogenetic Analysis. *Diversity* **2022**, *14*, 324. https://doi.org/10.3390/d14050324

Academic Editors: Hong-Hu Meng, Yi-Gang Song and Mario A. Pagnotta

Received: 30 January 2022
Accepted: 15 April 2022
Published: 22 April 2022

1. Introduction

The non-monophyletic nature of traditionally defined *Salvia* led to the establishment of a broad definition of *Salvia* that reduces the five small embedded genera (*Rosmarinus*, *Perovskia*, *Dorystaechas*, *Meriandra* and *Zhumeria*) to subgenera, resulting in a total of 11 subgenera being recognized within *Salvia* [1–3]. *Salvia* is the largest genus in Lamiaceae and includes approximately 1000 species. The genus has a subcosmopolitan distribution but is mainly concentrated in South America, Southwest Asia and the Mediterranean region, and East Asia [3]. As one of biodiversity centers of *Salvia*, approximately 100 species have been recorded in East Asia, 85 of which are native to China [4–11]. Based on recent studies, with the exception of *Salvia grandifolia* and *S. deserta* belonging to subg. *Sclarea*, the rest of East Asian *Salvia* have been placed into the newly established subg. *Glutinaria* [2,12].

Sect. *Drymosphace* was first established by George Bentham in 1832–1836 [13] and was later placed in subg. *Salvia* in 1876 [14]. Based on stamen morphology, Stibal [15] transferred sect. *Drymosphace* to subg. *Sclarea*. Following Stibal, Wu [16] also classified sect. *Drymosphace* into subg. *Sclarea* and established three series (ser. *Miltiorrhizae*, ser. *Plectranthoidites*, and ser. *Honaniae*) within this section. According to Wu's circumscription, a total of 19 species and three varieties can be recognized in sect. *Drymosphace* [4,5,9,16], and all of these taxa are endemic to China, with the exception of *S. nubicola* and *S. plectranthoides*, which extend to some Himalayan countries. Recent phylogenetic studies have demonstrated that sect. *Drymosphace* and the three series sensu Wu [16] are not monophyletic [2,3]. Integrating morphological with molecular phylogenetic evidence, the redefined sect. *Drymosphace* was transferred from subg. *Sclarea* to the newly established subg. *Glutinaria* due to its synapomorphy of robust taproots, pinnate leaves, relatively long corollas (length > 2 cm), and fused deformed posterior thecae [2]. According to the synapomorphy, five species were removed from the previous sect. *Drymosphace* sensu Wu of subg. *Glutinaria* [16]: *S. cavaleriei* and *S. prionitis* were transferred to the newly established sect. *Sobiso*, *S. petrophila* was transferred to sect. *Sonchifoliae*, *S. nubicola* was transferred to sect. *Glutinaria*, *S. trijuga* was transferred to sect. *Substoloniferae*, and *S. breviconnectivata* was regarded as a dubious species [2].

Sect. *Drymosphace* are importance of medical value, with *Salvia miltiorrhiza* being most widely used. "Danshen", a famous traditional Chinese medicine originating from the dried roots of *S. miltiorrhiza*, has been extensively applied to treat coronary heart diseases and cerebrovascular diseases [17]. Due to mass market demand, the wild resource of *S. miltiorrhiza* has reduced sharply. However, although sect. *Drymosphace* is well supported by molecular and morphological evidence, interspecific relationships remain unresolved. Moreover, *S. miltiorrhiza* and its closely related species are morphologically similar, which makes it difficult to identify authentic products, and an increasing number of substitutes and adulterants have been found on the market.

Currently, phylogenetic studies based on short DNA fragments have only been able to determine the phylogenetic backbone of *Salvia* [2,18–22], and relationships among species require further focus. The shallow rate of DNA evolution is one of the main obstacles to determining the relationships between low-level taxa. Due to there being a limited amount of information, the resolutions of phylogenetic trees that has been inferred from chloroplast gene fragments are often low [2,23]. With the birth of next-generation sequencing (NGS) technology, genomics has entered an era of low-cost, large-scale, and high-throughput sequencing, and an increasing number of genomic data are being used for phylogenetic analysis. As one of the three major genomes of green plants, the genomic structure and gene content of chloroplasts is relatively conservative and easy to sequence. At present, plastid genome data are being used to solve phylogenetic relationships at different taxonomic levels and have been demonstrated to be effective [24–26]. Although attempts to utilize NGS to analyze plastome characterization and even though the phylogeny of *Salvia* have been determined, few taxa of sect. *Drymosphace* were involved in these studies [27,28].

In this study, we sequenced eight plastomes of sect. *Drymosphace* and conducted comparative genomics analyses together with two other data sets that have been published on this section (*Salvia meiliensis* and *S. miltiorrhiza*). The aims of this study were (1) to characterize the plastome structure within sect. *Drymosphace*; (2) to screen hyper-variable regions of sect. *Drymosphace*; and (3) to infer the phylogeny of sect. *Drymosphace*.

2. Materials and Methods

2.1. Plant Materials, DNA Extraction and Genome Sequencing

In this study, 14 taxa (13 species and one variety) were sampled from *Salvia* subg. *Glutinaria*, including 10 taxa from sect. *Drymosphace* (Table 1). Eight sect. *Drymosphace* plastomes were newly sequenced, and the other sequences were downloaded from the GenBank database. Additionally, three species from other subgenera were selected as outgroups (Table 1).

The total genomic DNA was extracted from fresh or silica gel-dried leaves using the modified CTAB method [29]. In order to obtain qualified DNA for library construction, a Qubit Fluorometer was used to determine and calculate the DNA concentration and yield when they were at least c > 12.5 ng/μL and m > 1 μg, and the sample integrity was assessed by means of electrophoresis on a 1% agarose gel. Illumina Hiseq-2500 platform at

BGI-Wuhan was utilized for sequencing by paired-end (PE) library of 150 bp, and 3 to 5 GB of raw short sequence data were generated for each species.

Table 1. Voucher information and GenBank and SRA accession numbers for taxa used in this study.

Taxa	Subgenus	Voucher	Locality	GenBank Accession	SRA Accession
S. bowleyana	*Glutinaria*	GX Hu and F Zhao 131	China, Fujian	MW435404 *	SRR18650098 *
S. bulleyana	*Glutinaria*	*NA*	*NA*	MH603954	*NA*
S. chanryoenica	*Glutinaria*	s.n. (KH)	South Korea	MH261357	*NA*
S. dabieshanensis	*Glutinaria*	GX Hu and F Zhao 0165	China, Anhui	MW435405 *	SRR18587923 *
S. honania	*Glutinaria*	GX Hu and F Zhao 0168	China, Henan	MW435406 *	SRR18600490 *
S. meiliensis	*Glutinaria*	GX Hu and FZ Shangguan Hu0089	China, Anhui	MN520018	*NA*
S. miltiorrhiza	*Glutinaria*	Cultivated	China, Beijing	HF586694	*NA*
S. nanchuanensis	*Glutinaria*	JX Yang and XZ He YJX-01	China, Hunan	MW435407 *	SRR18595850 *
S. nanchuanensis var. pteridifolia	*Glutinaria*	GX Hu and B Pan 615	China, Guangxi	MW435408 *	SRR18595337 *
S. plectranthoides	*Glutinaria*	GX Hu et al. 0006	China, Yunnan	MW435409 *	SRR18650097 *
S. prattii	*Glutinaria*	*NA*	China, Yunnan	MK944407	*NA*
S. przewalskii	*Glutinaria*	*NA*	*NA*	MH603953	*NA*
S. subbipinnata	*Glutinaria*	JX Yang and XZ He YJX-04	China, Zhejiang	MW435410 *	SRR18650096 *
S. yunnanensis	*Glutinaria*	GX Hu 603	China, Guizhou	MW435411 *	SRR18595336 *
S. hispanica	*Calosphace*	Cultivated, SD118	China, Shandong	MT083896	*NA*
S. rosmarinus	*Rosmarinus*	Cultivated	China, Shaanxi	KR232566	*NA*
S. officinalis	*Salvia*	So_003	*NA*	MG772529	*NA*

NA = information is unavailable. * = newly sequenced plastomes in this study. SRA = Sequence Read Archive.

2.2. Plastome Assembly and Annotation

After filtering the raw data and discarding the low-quality reads, the remaining PE reads were assembled into whole plastomes using GetOrganelle v1.6.2a [30]. The assembly graph of the generated complete plastome was verified by Bandage v.0.8.1 [31]. Plastome sequences were annotated using software PGA [32], using the *Salvia miltiorrhiza* plastome [33] as reference. The tRNA genes were further verified using the online tRNAscan-SE [34] search servers and then manually adjusted in Geneious 10.0.5 [35]. OrganellarGenomeDRAW, an online program, was used to generate circular annotated plastid genome maps [36], and the plastomes were deposited in the GenBank database (Table 1).

2.3. Codon Usage and Repeated Sequence Analysis

Relative Synonymous Codon Usage (RSCU) is a parameter that is used to evaluate the codon usage preferences of protein-coding sequences. Here, the RSCU values for all of the protein-coding sequences were computed using the program codon W 1.4.4 [37]. The codon usage in the form of heatmap was conducted by employing R language with the RSCU value. An RSCU value > 1 indicates that codon usage is highly preferred, an RSCU value = 1 means that codon usage is not preferred, and an RSCU value < 1 indicates that the codon usage is low [38].

The simple sequence repeat (SSR) of sect. *Drymosphace* plastomes were identified using MISA [39] by setting the minimum number of repeat units to 8, 4, 4, 3, 3, and 3 for the mono-, di-, tri-, tetra-, penta-, and hexanucleotides, respectively. Four types of dispersed repeats (forward, reverse, palindrome, and complement repeats) in sect. *Drymosphace* were determined using REPuter [40] with a repeat size ≥ 30 bp and sequence identities ≥ 90 (Hamming distance of 3).

2.4. Comparative Genomic Analyses

The boundaries of single-copy (SC) and reverse repeat (IR) regions of sect. *Drymosphace* were determined in Geneious 10.0.5, and the boundary expansion and contraction sketch of IR was drawn using Adobe Illustrator CC. mVISTA [41] was used to determine the variability in the complete plastome sequences of sect. *Drymosphace* in Shuffle-LAGAN mode, and *Salvia bowleyana* was selected as the reference genome. Genome rearrangement was carried out by Mauve [42] using *S. bowleyana* as a reference genome. DnaSP v.5.0 [43] was employed to analyze nucleotide diversity (Pi) within sect. *Drymosphace* by setting the step size to 200 bp with a 600 bp window length.

2.5. Phylogenetic Analysis

To infer the phylogenetic relationships of sect. *Drymosphace*, 14 plastid genomes from subg. *Glutinaria* were selected to carry out analyses with *S. rosmarinus*, *S. hispanica*, and *S. officinalis* from three other subgenera as outgroups. Seven matrices, including complete plastome (CP) sequences with one IR region excluded, large single copy (LSC), small single copy (SSC), IR, protein-coding exons (coding regions, CR), intergenic spacer and introns (non-coding region, NCR), and nine hyper-variable region (HVR) data sets were prepared for the phylogenetic analyses. The sequences were aligned with MAFFT [44] using Gblocks v0.91 [45] to exclude the ambiguously aligned positions. The neat sequence matrices were employed to infer phylogenetic trees using both maximum likelihood (ML) and Bayesian inference (BI).

Under the GTRGAMMA substitution model, ML analyses were performed using RAxML-HPC2 on XSEDE v.8.2.12 [46] on the CIPRES Science Gateway (http://www.phylo.org/ (accessed on 20 March 2022)) [47]. With the exception of setting bootstrap iterations (–# | –N) to 1000, the other parameters used default values.

BI analysis was carried out in MrBayes 3.2.6 [48] and implemented in PhyloSuite [49]. The ModelFinder [50] was utilized to select the best model according to the Akaike information criterion (AIC) (Table 2). Four Markov chain Monte Carlo (MCMC) iterations were run simultaneously for 2,000,000 generations. Each run started with a random tree, and a random tree was sampled every 1000 generations. Stationarity was considered to be reached when the average standard deviation of the split frequencies was less than 0.01. After discarding the first 25% trees as burn-in, the remaining trees were used to calculate a majority-rule consensus tree for each matrix.

Table 2. Summary of the data set information for phylogenetic analyses.

Data Matrix	Aligned Length (bp)	Constant Sites (bp)	Variable Sites (bp)	Parsimony Informative (bp)	Best Fit Model (AIC)
CP	151,218	144,550 (95.59%)	6668	2366 (1.56%)	GTR + F + I + G4
LSC	82,556	77,911 (94.37%)	4645	1637 (1.98%)	GTR + F + I + G4
SSC	17,531	15,967 (91.08%)	1564	567 (3.23%)	GTR + F + G4
IR	25,563	25,335 (99.11%)	228	80 (0.31%)	GTR + F + I
CR	66,888	64,304 (96.14%)	2584	921 (1.38%)	GTR + F + I + G4
NCR	59,877	56,165 (93.80%)	3712	1346 (2.25%)	GTR + F + I + G4
HVR	10,298	9096 (88.33%)	1064	363 (3.52%)	GTR + F + G4

3. Results and Discussion

3.1. Plastome Features of Sect. *Drymosphace*

The length of complete plastomes of sect. *Drymosphace* ranged from 151,330 bp (*Salvia dabieshanensis*) to 151,614 bp (*S. meiliensis*). All of the plastomes displayed a typical quadripartite architecture that contained a large single copy (LSC: 82,629–82,854 bp) and a small single copy (SSC: 17,541–17,587 bp) separated by two copies of an inverted repeat (IR: 25,520–25,590 bp). The total GC content varied slightly from 38.00% to 38.03%, and the IR regions had the highest GC content (43.10–43.14%) followed by those in the LSC (36.12–36.15%) and SSC (31.97–32.05%) regions (Figure 1, Table 3).

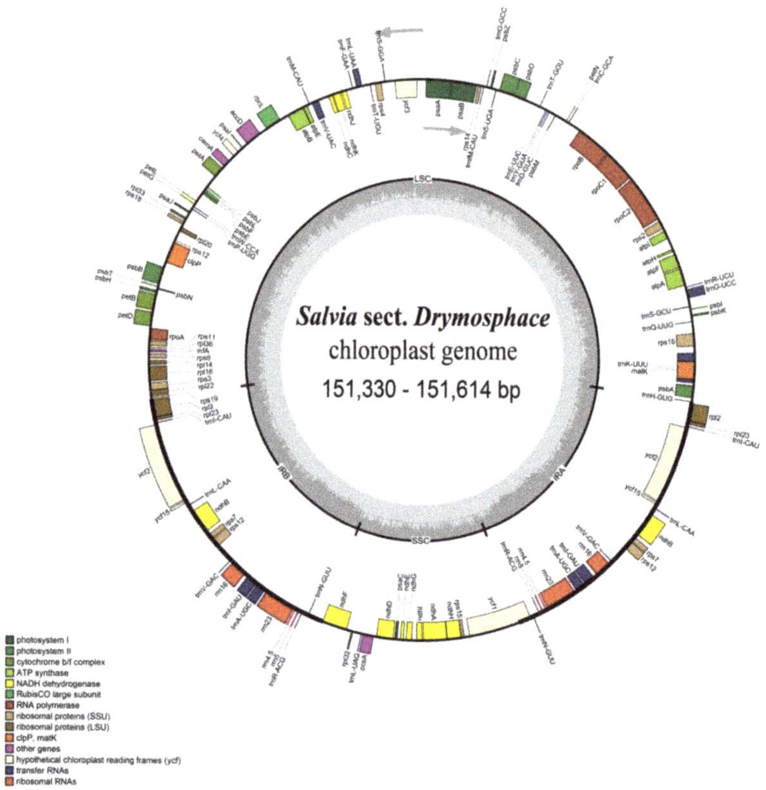

Figure 1. Complete plastome gene map of *Salvia* sect. *Drymosphace*. Genes outside of the circle are transcribed in the counterclockwise direction, and those that are inside are transcribed in the clockwise direction. LSC—large single copy; SSC—small single copy; IR—inverted repeat.

A total of 114 unique genes, including 80 protein-coding, 30 tRNA, and four rRNA genes, were detected in each species (Table 3). Four rRNA, seven tRNA, and seven protein-coding genes were duplicated in IR regions (Table 3). Of the 18 genes containing intron, 15 comprised only one intron, and three genes harbored two introns. The *rps12* gene was recognized as being a trans-spliced gene, with the 5′ end located in the LSC region and the duplicated 3′ end in the IR region (Figure 1, Table 4).

Table 3. Plastome features of 10 taxa of *Salvia* sect. *Drymosphace* presented in this study.

Taxa	Clean Reads	Reads Used for Assembly	Mean Coverage (×)	Complete		LSC		SSC		IR		Number of Genes	PCG	tRNA Genes	rRNA Genes
				Size(bp)	GC (%)	Length (bp)	GC (%)	Length (bp)	GC (%)	Length (bp)	GC (%)				
S. bowleyana	6961,322	6,160,014	526	151,508	38.01	82,809	36.14	17,585	32.01	25,557	43.12	114	80	30	4
S. dabieshanensis	11,727,732	9,644,622	504	151,330	38.02	82,629	36.14	17,587	32.01	25,557	43.12	114	80	30	4
S. honania	7,221,162	6,369,236	514	151,522	38.01	82,768	36.13	17,586	31.97	25,584	43.11	114	80	30	4
S. meiliensis	68,112,046	25,577,470	1970	151,614	38.00	82,854	36.12	17,580	32.00	25,590	43.10	114	80	30	4
S. miltiorrhiza	NA	NA	NA	151,332	38.02	82,698	36.15	17,556	32.01	25,539	43.12	114	80	30	4
S. nanchuanensis	9,188,568	7,579,704	507	151,426	38.02	82,791	36.14	17,563	31.99	25,536	43.13	114	80	30	4
S. nanchuanensis var. *pteridifolia*	15,948,806	13,152,169	514	151,455	38.01	82,831	36.12	17,584	32.05	25,520	43.14	114	80	30	4
S. plectranthoides	6,978,570	6,231,469	512	151,416	38.01	82,775	36.13	17,563	32.03	25,539	43.12	114	80	30	4
S. subbipinnata	5,368,560	4,649,062	516	151,388	38.03	82,769	36.15	17,541	32.02	25,539	43.13	114	80	30	4
S. yunnanensis	6,448,098	5,394,701	508	151,413	38.02	82,652	36.14	17,581	32.03	25,590	43.11	114	80	30	4

Table 4. Genes present in the plastomes of 10 taxa of sect. *Drymosphace*.

Gene Functions	Group of Genes	Name of Genes
Photosynthesis	Subunits of ATP synthase	*atpA*, *atpB*, *atpE*, *atpF* *, *atpH*, *atpI*
	Subunits of NADH dehydrogenase	*ndhA* *, *ndhB* * (×2), *ndhC*, *ndhD*, *ndhE*, *ndhF*, *ndhG*, *ndhH*, *ndhI*, *ndhJ*, *ndhK*
	Subunits of cytochrome	*petA*, *petB* *, *petD* *, *petG*, *petL*, *petN*
	Subunits of photosystem I	*psaA*, *psaB*, *psaC*, *psaI*, *psaJ*
	Subunits of photosystem II	*psbA*, *psbB*, *psbC*, *psbD*, *psbE*, *psbF*, *psbH*, *psbI*, *psbJ*, *psbK*, *psbL*, *psbM*, *psbN*, *psbT*, *psbZ*
	Subunit of rubisco	*rbcL*
Self-replication	Large subunit of ribosome	*rpl2* * (×2), *rpl14*, *rpl16* *, *rpl20*, *rpl22*, *rpl23* (×2), *rpl32*, *rpl33*, *rpl36*
	DNA-dependent RNA polymerase	*rpoA*, *rpoB*, *rpoC1* *, *rpoC2*
	Small subunit of ribosome	*rps2*, *rps3*, *rps4*, *rps7* (×2), *rps8*, *rps11*, *rps12* ** (×2), *rps14*, *rps15*, *rps16**, *rps18*, *rps19*
	rRNA Genes	*rrn4.5* (×2), *rrn5* (×2), *rrn16* (×2), *rrn23* (×2)
	tRNA Genes	*trnA-UGC* * (×2), *trnC-GCA*, *trnD-GUC*, *trnE-UUC*, *trnF-GAA*, *trnfM-CAU*, *trnG-GCC*, *trnG-UCC* *, *trnH-GUG*, *trnI-CAU* (×2), *trnI-GAU* * (×2), *trnK-UUU* *, *trnL-CAA* (×2), *trnL-UAA* *, *trnL-UAG*, *trnM-CAU*, *trnN-GUU* (×2), *trnP-UGG*, *trnQ-UUG*, *trnR-ACG* (×2), *trnR-UCU*, *trnS-GCU*, *trnS-GGA*, *trnS-UGA*, *trnT-GGU*, *trnT-UGU*, *trnV-GAC* (×2), *trnV-UAC* *, *trnW-CCA*, *trnY-GUA*
Other genes	Subunit of Acetyl-CoA-carboxylase	*accD*
	c-type cytochrome synthesis gene	*ccsA*
	Envelop membrane protein	*cemA*
	Protease	*clpP* **
	Translational initiation	*infA*
	Maturase	*matK*
Unknown function	Conserved open reading frames	*ycf1*, *ycf2* (×2), *ycf3* **, *ycf4*, *ycf15* (×2)

* gene with a single intron; ** gene with two introns; (×2) duplicated gene.

The plastid genomes of sect. *Drymosphace* showed high consistency and similarity in terms of their gene structure, order, and content. These results are consistent with other plastomes of Lamiaceae [26–28,33,51] and are also as conservative as most of the angiosperms that have been previously reported [52–56].

3.2. Amino Acid Abundance and Codon Usage

The number of codons in sect. *Drymosphace* ranged from 26,501 (*Salvia miltiorrhiza*) to 26,583 (*S. dabieshanensis*). Among these codons, leucine (10.59–10.64%) was the most frequently observed amino acid, and cysteine (1.10–1.12%) was the least frequently observed (Figure 2, Table S1). Most protein-coding genes had the same standard ATG sequence as the initiation codon, but the *rps19* and *ycf15* genes started with a GTG sequence. This non-ATG initiation codon has also been found in other angiosperm chloroplasts [28,57,58]. Heatmaps showed that 30 of the total 64 types of codons used in sect. *Drymosphace* were preferred codons (RSCU > 1), and with the exception of UUG (Leu), all of the preferred codons ended with an A or a U (Figure 3, Supplementary Material Table S1). This codon usage bias shows a similar trend observed in the majority of plastomes in higher plants [52,59,60].

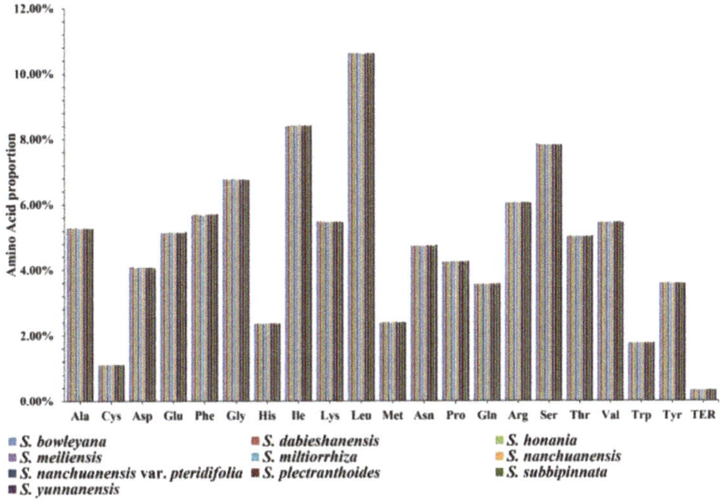

Figure 2. Amino acid frequencies in all protein-coding genes in the chloroplast genome of 10 taxa of *Salvia* sect. *Drymosphace*.

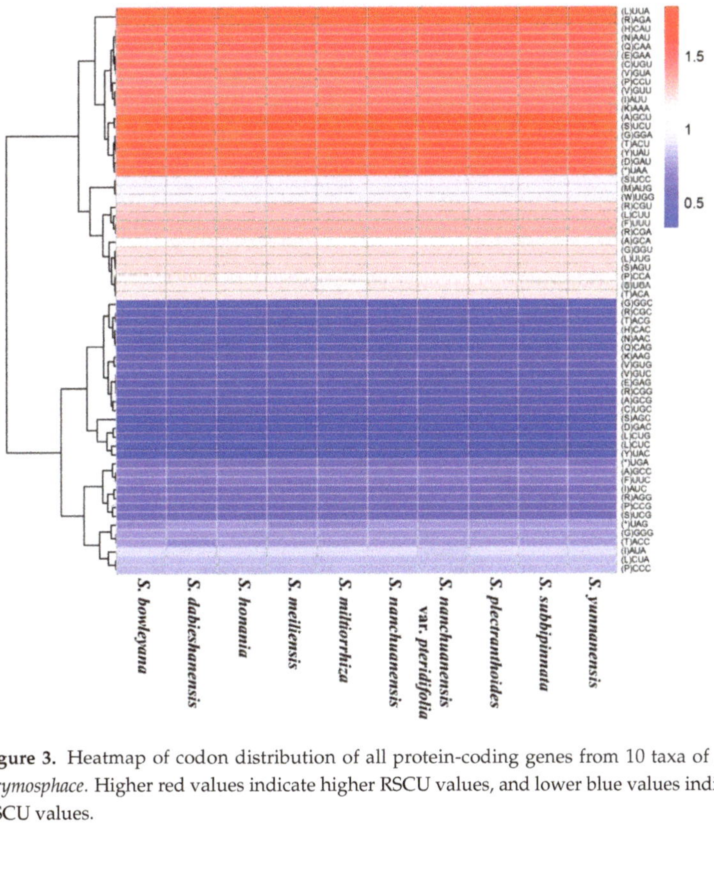

Figure 3. Heatmap of codon distribution of all protein-coding genes from 10 taxa of *Salvia* sect. *Drymosphace*. Higher red values indicate higher RSCU values, and lower blue values indicate lower RSCU values.

3.3. Simple Sequence Repeats and Long Repeat Sequences

Simple sequence repeat (SSR), also called microsatellite, is a sequence with a tandemly repeated motif that ranges in size from one to six. Due to high repeatability and polymorphism, SSR is a molecular marker that is commonly used in population genetics and evolutionary analysis [52,61–64]. In this study, a total of 1685 SSRs were identified with each taxon bearing 166–173 SSRs from 10 plastomes of sect. *Drymosphace* (Figure 4, Table S2). Five types of SSR (excluding penta-nucleotide SSR) were detected in sect. *Drymosphace,* of which the mono-nucleotide SSR was the richest (1265, 75.07%), followed by the di-nucleotide (346, 20.53%), tetra-nucleotide (69, 4.09%), hexa-nucleotide (4, 0.24%), and tri-nucleotide (1, 0.06%) SSRs. In terms of repeat units, the A/T unit in the mono-nucleotide repeat was the most abundant (1242, 73.71%), followed by the di-nucleotide AT/AT (136, 8.07%) and TA/TA (100, 5.93%). The tri-nucleotide SSR (AAT/ATT) was only detected once (1, 0.06%) in *S. nanchuanensis* var. *pteridifolia*, and the hexa-nucleotide SSRs (4, 0.24%) were detected once in *S. dabieshanensis* (ATTCAT/ATGAAT), *S. meiliensis* (AATCAA/TTGATT), *S. miltiorrhiza* (ACTTAG/CTAAGT), and *S. plectranthoides* (ACTTAG/CTAAGT). In addition, six unique repeat units (AAT/ATT, AAAT/ATTT, TTAA/AATT, ATTCAT/ATGAAT, AATCAA/TTGATT, ACTTAG/CTAAGT) were detected within sect. *Drymosphace.* The majority of the SSRs resided in LSC regions (1125, 66.77%), and the others were located in other SSC regions (320, 18.99%) and in IR regions (240, 14.24%). As reported in previous plastomes of *Salvia* of Lamiaceae [28,51] as well as in other families [55,65,66], most of the SSRs in sect. *Drymosphace* are composed of A/T mono-nucleotides, which represent a percentage of 73.71%. Similar to most *Salvia* plastomes that have been reported upon before [28,51,67], we did not detect the penta-nucleotide SSR in sect. *Drymosphace.* However, a recent study reported a penta-nucleotide SSR from *S. yunnanensis* of sect. *Drymosphace* [68]. *Salvia yunnanensis* was also included in this study as well as in another independent analysis [28], in which the samples were from Guizhou and Yunnan Province, China, respectively, but neither study identified the penta-nucleotide SSR in these species. Therefore, there is a lack of the penta-nucleotide within sect. *Drymosphace,* or at the very least, it is rare. Even though the mono-nucleotide SSRs showed significant differences among species, ranging from 123 (*S. miltiorrhiza*) to 134 (*S. yunnanensis*), no significant differences were observed in the number of the other four types of SSRs.

A total of 478 long repeat sequences with lengths greater than 30 bp, including 222 forward and 255 palindromic repeats and one complementary repeat, were detected in sect. *Drymosphace* (Figure 5, Table S3). The number of repeats for each taxa varied from 42 to 49 and had lengths that ranged from 30 to 99 bp. Most of the long repeats were located in IR (389, 81.38%) regions, followed by LSC (79, 16.53%) and SSC (10, 2.09%). In particular, most of these repeats were detected in coding regions (378, 79.08%), with a few being located in intergenic regions (50, 10.46%) and in introns (50, 10.46%). Long repeat sequences may play a pivotal t role in plant evolution and could promote variation and rearrangement in the plastid genome [69–71]. All of these repeats, together with the aforementioned SSRs, may have potential utility in population studies.

3.4. Comparative Genomic Analyses

The contraction and expansion of IR regions at the borders is a main reason for size variation in plastomes and plays an important role in the evolution of seed plants [72–75]. To explore the expansion and contraction of the IR regions, the boundaries between the SC and IR regions of sect. *Drymosphace* were analyzed. The *rps19* gene spanned the LSC/IRb boundary in all species and had a length of 234 bp or 236 bp in the LSC and 43 bp or 45 bp in the IRb, which generated a short *rps19* pseudogene (*ψrps19*) that was 43 bp or 45 bp in length at the IRa/LSC border. The *ndhF* crossed the IRb/SSC border and had an equal length of 32 bp in the IRb and of 2185 bp or 2203 bp in the SSC. The SSC/IRa boundary was located within the *ycf1* gene and had lengths ranging from 4452 bp to 4470 bp in the SSC and of 1056 bp or 1167 bp in the IRa, resulting in a pseudogene (*ψycf1*) at the IRb/SSC border with 1056 bp or 1167 bp overlapping with the *ndhF* gene. In addition, the IRa/LSC boundary resided between the *ψrps19* and *trnH*, and the distance from *trnH* was 8 bp or 10 bp (Figure 6). In *Salvia*, IR/LSC junctions are considered to be highly conserved, and the expansion/contraction of the IR regions is infrequent [28]. Although a few IR contractions have been reported in *Salvia* (e.g., *S. hispanica*, *S. mekongensis*, and *S. rosmarinus*) [28], no contractions were detected in this study, suggesting that IR contraction events may not occur in sect. *Drymosphace*.

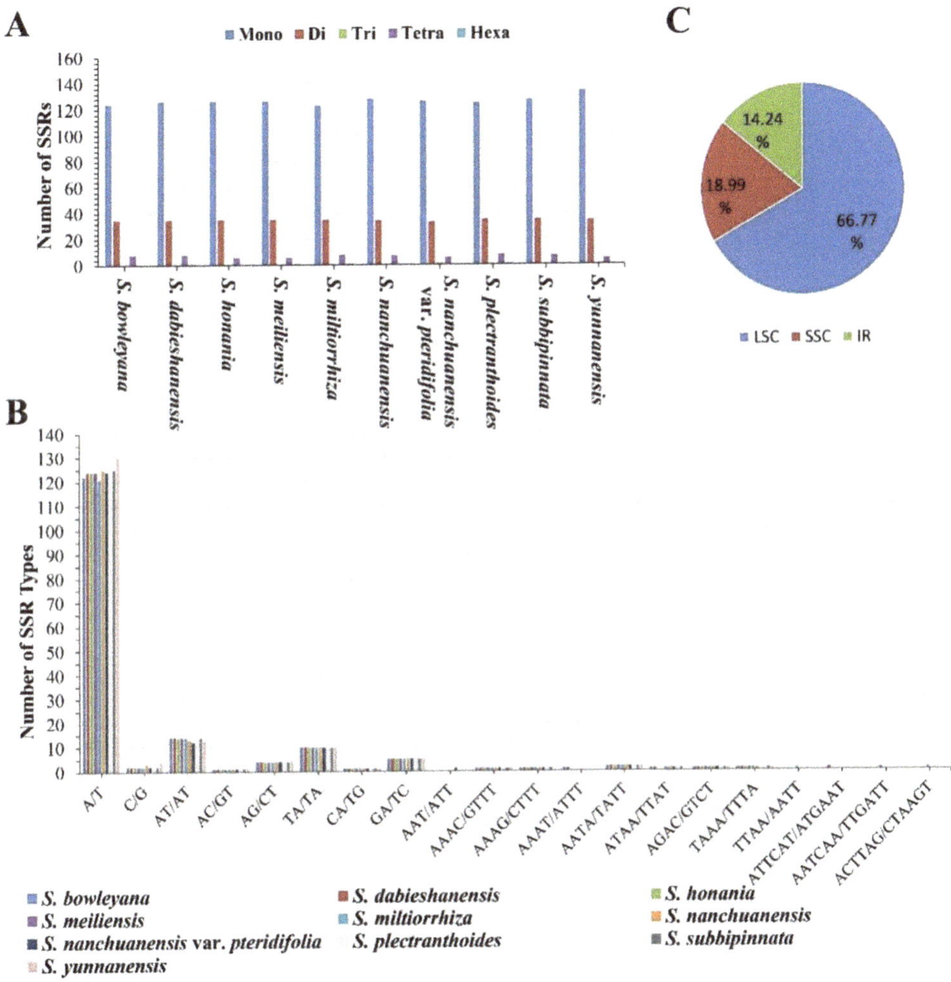

Figure 4. Comparison of simple sequence repeats (SSR) among the plastomes of 10 taxa of *Salvia* sect. *Drymosphace*. (**A**). Number of different types of SSRs; (**B**). number of different of SSR repeat units; (**C**) frequencies of SSRs in LSC, IR, and SSC regions.

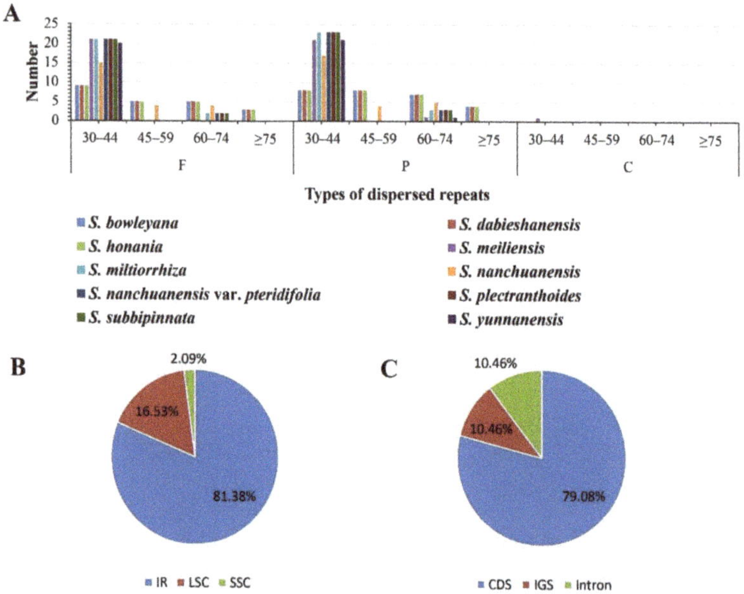

Figure 5. Analysis of repetitive sequences in the plastomes of 10 taxa of *Salvia* sect. *Drymosphace*. (**A**). Number of different types of longer repeats; (**B**). frequencies of longer repeats in the LSC, SSC, and IR regions; (**C**). frequencies of longer repeats in protein coding regions, intergenic spacers, and intron regions.

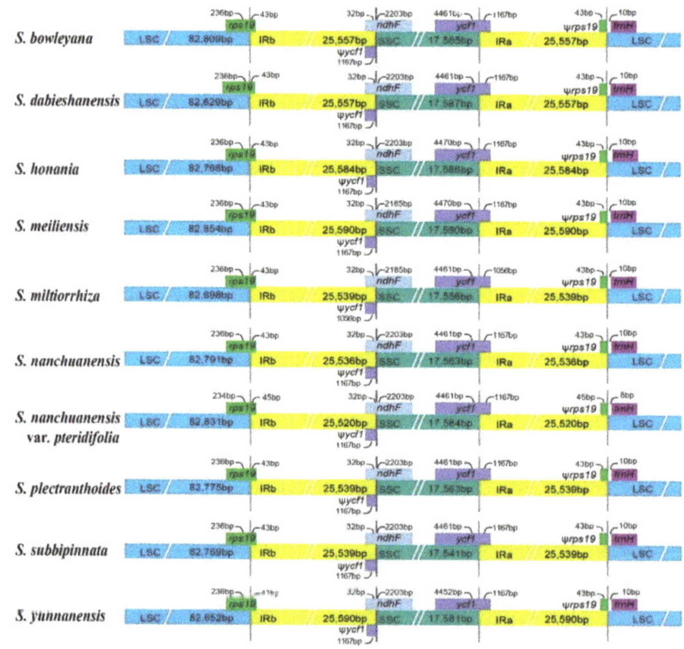

Figure 6. Comparison of junctions between the LSC, IR, and SSC regions among plastomes of 10 taxa of *Salvia* sect. *Drymosphace*. Genes are denoted by colored boxes, with *ψ* indicating a pseudogene.

Comparison of complete plastome sequences showed that the genes within sect. *Drymosphace* were arranged in the same order and that the IR regions were more conserved than the LSC and SSC regions (Figure 7), which may be related to the copy corrections caused by gene conversion between the two IR regions [76,77]. In addition, noncoding regions were found to be more divergent than coding regions, and the most varied regions occurred in intergenic regions, such as *accD-psaI*, *rps4-trnL*, *rps16-trnQ*, and *trnH-psbA*. Genome rearrangement analysis showed a collinear relationship, suggesting that there was no rearrangement or inversion events in the plastomes of sect. *Drymosphace* (Figure S1).

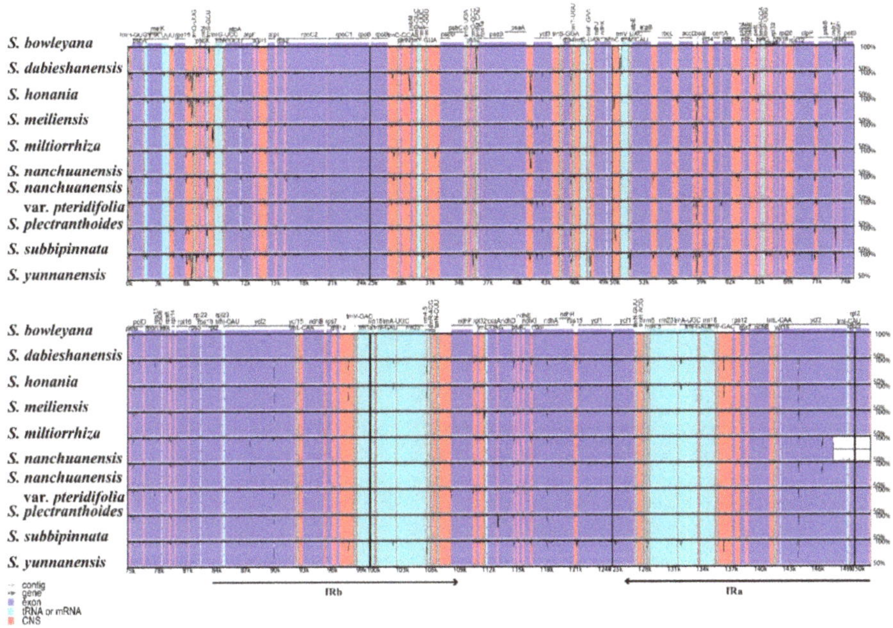

Figure 7. Sequence alignment of the whole plastomes of 10 taxa of *Salvia* sect. *Drymosphace* using mVISTA with *S. bowleyana* as a reference. The vertical scale indicates percentage of identity, ranging from 50% to 100%.

Sequence divergence analyses showed that the nucleotide variability (Pi) values of the sect. *Drymosphace* plastomes ranged from 0 to 0.01376, with an average value of 0.0014 (Figure 8). Nine highly variable regions with Pi values > 0.005 were detected, including two genes (*clpP* and *ycf1*) and seven intergenic spacers (*trnH-psbA*, *trnK-rps16*, *petN-psbM*, *rps4-trnT*, *rbcL-accD*, *rpl32-trnL-ccsA*, and *ndhD-psaC*). Among these hyper-variable regions, six loci were located in the LSC, three were located in the SSC, and none were located in the IR regions. *trnH-psbA* is the most variable region (Pi = 0.01376) among the nine hypervariable regions. Due to high rates of insertion/deletion and universal primers, *trnH-psbA* has been selected to be a plant barcode for species discrimination [78–81] and has been used for phylogenetic analyses in Lamiaceae [82–84]. However, the resolution of this DNA marker is not high enough to solve phylogenetic relationships at the section and below levels for *Salvia* [2,85]. For the two hypervariable genic regions, the *clpP* gene has not been used as a marker in phylogenetic study of *Salvia*, and *ycf1* has been demonstrated to be a promising maker for phylogenetic analyses within *Salvia* [28,86].

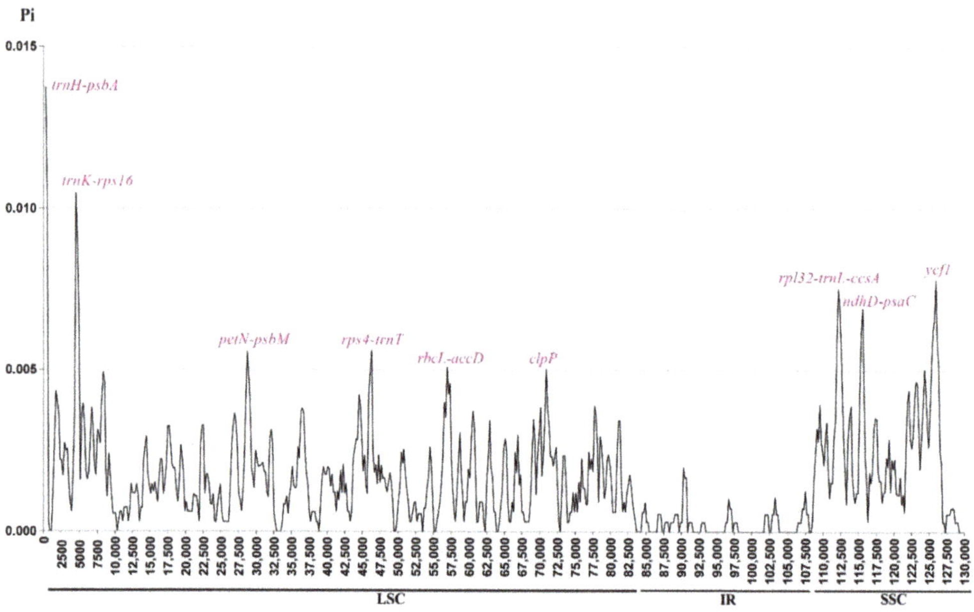

Figure 8. Sliding window analysis of plastomes of 10 taxa of *Salvia* sect. *Drymosphace*. The *X*-axis indicates the position of the midpoint of a window, and the *Y*-axis represents the nucleotide diversity of each window. The nine hyper-variable regions with Pi > 0.005 are marked.

3.5. Phylogenetic Analysis

Seven matrices including 17 taxa were employed to conduct phylogenetic analyses (Table 1). The aligned CP dataset had the longest length (145,835 bp), and the combined nine HVRs had the shortest length (10,298 bp). The highest percentage of constant sites were from the IR data set (99.11%), and the lowest were from the HVR data set (88.33%). As opposed to the constant sites, the percentage of the informative sites in the HVR data set was the highest (3.52%), and lowest percentage was in the IR regions (0.31%) (Table 2).

The topologies that were inferred by ML and BI analyses for each matrix were identical, with the exception of a few weakly supported nodes. Therefore, only an ML tree with posterior probabilities (PP) indicated above the branches and the ML bootstrap (BS) values provided below.

In all of the analyses, monophylies of subg. *Glutinaria* and sect. *Drymosphace* were recovered, and sister relationships between *Salvia honania* and *S. meiliensis*, *S. dabieshanensis*, and *S. bowleyana* within sect. *Drymosphace* were strongly supported (Figure 9 and Figures S2–S7). Within sect. *Drymosphace*, with the exception of the CR matrix, all of the analyses supported the finding that *S. honania* and *S. meiliensis* together to be sister to the rest of the section; with the exception of the IR regions, all of the topologies that were inferred from other data set showed that *S. dabieshanensis*, *S. bowleyana*, *S. miltiorrhiza*, and *S. plectranthoides* formed a strongly supported subclade. The phylogenetic positions of *S. nanchuanensis*, *S. yunnanensis*, and *S. subbipinnata* varied slightly in different data set.

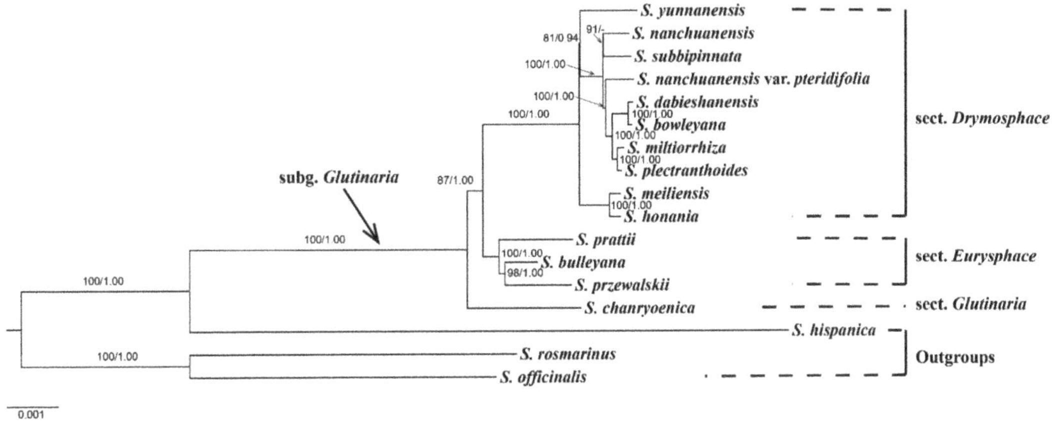

Figure 9. Phylogenetic tree inferred from maximum likelihood and Bayesian inference based on complete chloroplast genome. Posterior probabilities are provided above the branches, and the ML bootstrap values are indicated below, with the ML bootstrap values <50% and PP < 0.90 indicated by '-'.

Integrating molecular phylogenetic with morphological evidence, Hu et al. [2] established a new sect. *Drymosphace*, but the interspecific relationships within the section are confused, and the supported values for most subclades are weak. For example, *S. honania* and *S. meiliensis* are two morphologically similar species [87]. They are easily distinguished from the other species within sect. *Drymosphace*, as characterized by small and oblong upper corolla lips and clearly exserted stamens that adhere laterally to the corolla [2,87]. The two species should have been sister group. However, they did not gather together in previous analyses that were based on only a few DNA markers [87]. In this study, all of the analyses indicated that the two species formed a sister relationship, supporting the finding that *S. honania* is closely related to *S. meiliensis* [2,87]. Additionally, compared to previous studies including sect. *Drymosphace* based on short DNA fragments [2,12,85], the plastid phylogenomic analyses in this study based on all of the matrices greatly improved the phylogenetic resolution (Figure 9). Although monophyly of sect. *Drymosphace* and sister relationships of *S. honania* and *S. meiliensis* were confirmed in this study, the relationships between some of the taxa presented here are inconsistent with those inferred by morphological characteristics. Morphologically, *Salvia plectranthoides, S. nanchuanensis*, and *S. nanchuanensis* var. *pteridifolia* share long tubular corolla tubes and straight upper corolla lips and are therefore placed in ser. *Plectranthoidites* [4,16]. They should have been found to be closely related and grouped together in phylogenetic trees, especially *S. nanchuanensis* and *S. nanchuanensis* var. *pteridifolia*. However, in this study, *S. nanchuanensis* is sister to *S. subbipinnata* instead of *S. nanchuanensis* var. *pteridifolia*, and *S. plectranthoides* is sister to *S. miltiorrhiza*, indicating that *S. nanchuanensis* and ser. *Plectranthoidites* are not monophyletic. Therefore, the plastid genome is still unable to completely determine the interspecific relationships of sect. *Drymosphace*.

Of all seven data sets, the IR regions generated the poorest phylogenetic resolution, with four subclades having very low support values within sect. *Drymosphace* (PP < 0.9 and BS < 50%) (Figure S4). Wu et al. [27] also reported similar results in plastid phylogenomic results for *Salvia* [27]. The sequence length as well as informative sites may lead to this result. Previous studies indicate that more and longer DNA sequences may greatly improve the phylogenetic resolution and the support values of the branches [28,88,89]. In this study, the IR regions (25,563 bp) and HVR (10,298) were the two shortest data sets, so the phylogenetic resolutions that were inferred from them are lower than those of the other five matrices (Figures S4 and S7). On the other hand, the IR region has the lowest percentage of informative sites, which may be the result of the low resolution. Therefore, the use of IR regions separately is not suggested for phylogenetic analysis in *Salvia*.

4. Conclusions

In this study, we sequenced and assembled eight *Salvia* sect. *Drymosphace* plastomes. The reported plastid genomes are conservative, showing high levels of consistency and similarity to the chloroplast genomes of other species of Lamiaceae in terms of gene content, order, and structure. Comparative analyses revealed that there is no rearrangement or inversion events in the plastomes of *sect. Drymosphace* and that single-copy regions and intergenic regions are more variable than IR regions and coding regions. Phylogenomic analyses can recover the monophyly of the newly established sect. *Drymosphace*, considerably improve phylogenetic resolution, and determine interspecific relationships for some of the species within this section.

Supplementary Materials: The following are available at https://www.mdpi.com/article/10.3390/d14050324/s1. Table S1: Amino acid frequencies and codon distribution of 10 taxa of *Salvia* sect. *Drymosphace*, Table S2: SSRs of 10 taxa of *Salvia* sect. *Drymosphace*. Table S3: Long repeats of 10 taxa of *Salvia* sect. *Drymosphace*, Figure S1: Mauve alignment of plastomes of 10 taxa of *Salvia* sect. *Drymosphace*, Figure S2: Phylogenetic tree inferred from maximum likelihood and Bayesian inference based on LSC, Figure S3: Phylogenetic tree inferred from maximum likelihood and Bayesian inference based on SSC, Figure S4: Phylogenetic tree inferred from maximum likelihood and Bayesian inference based on IR, Figure S5: Phylogenetic tree inferred from maximum likelihood and Bayesian inference based on CR, Figure S6: Phylogenetic tree inferred from maximum likelihood and Bayesian inference based on NCR, Figure S7: Phylogenetic tree inferred from maximum likelihood and Bayesian inference based on HVR.

Author Contributions: Conceptualization, G.-X.H.; software, T.S., L.G. and F.Z.; formal analysis, T.S., M.W. and L.G.; writing—original draft preparation, T.S. and G.-X.H.; writing—review and editing, G.-X.H., C.-L.X., T.S., Y.-F.G. and F.Z.; funding acquisition, G.-X.H. and Y.-F.G. All authors have read and agreed to the published version of the manuscript.

Funding: This research was funded by the National Natural Science Foundation of China (32060048 and 31900275) and the Construction Program of Biology First-Class Discipline in Guizhou (GNYL [2017] 009).

Institutional Review Board Statement: Not applicable.

Informed Consent Statement: Not applicable.

Data Availability Statement: The plastome sequence data supporting this study are openly available in the GenBank nucleotide database and the Sequence Read Archive. Other data are contained within the article or within the Supplementary Materials.

Acknowledgments: We thank Jia-Xin Yang, Xuanze He, and Bo Pan for their help with the field work.

Conflicts of Interest: The authors declare no conflict of interest.

References

1. Drew, B.; González-Gallegos, J.G.; Xiang, C.L.; Kriebel, R.; Drummond, C.; Walker, J.; Sytsma, K. *Salvia* united: The greatest good for the greatest number. *Taxon* **2017**, *66*, 133–145. [CrossRef]
2. Hu, G.X.; Takano, A.; Drew, B.T.; Liu, E.D.; Soltis, D.E.; Soltis, P.S.; Peng, H.; Xiang, C.L. Phylogeny and staminal evolution of *Salvia* (Lamiaceae, Nepetoideae) in East Asia. *Ann. Bot.* **2018**, *122*, 649–668. [CrossRef] [PubMed]
3. Kriebel, R.; Drew, B.T.; Drummond, C.P.; Gonzalez-Gallegos, J.G.; Celep, F.; Mahdjoub, M.M.; Rose, J.P.; Xiang, C.L.; Hu, G.X.; Walker, J.B.; et al. Tracking temporal shifts in area, biomes, and pollinators in the radiation of *Salvia* (sages) across continents: Leveraging anchored hybrid enrichment and targeted sequence data. *Am. J. Bot.* **2019**, *106*, 573–597. [CrossRef]
4. Li, H.W.; Hedge, I.C. Salvia. In *Flora of China*; Wu, C.Y., Raven, P.H., Hong, D.Y., Eds.; Science Press: Beijing, China; Missouri Botanical Garden Press: St. Louis, MO, USA, 1994; Volume 17, pp. 196–224.
5. Hu, G.X.; Liu, Y.; Xu, W.B.; Liu, E.D. *Salvia petrophila* sp. nov. (Lamiaceae) from north Guangxi and south Guizhou, China. *Nord. J. Bot.* **2014**, *32*, 190–195. [CrossRef]

6. Hu, G.X.; Liu, E.D.; Zhang, T.; Cai, J.; Xiang, C.L. *Salvia luteistriata* (Lamiaceae), a new species from northeastern Sichuan, China. *Phytotaxa* **2017**, *314*, 123–128. [CrossRef]
7. Hu, G.X.; Peng, H. Identity of *Salvia weihaiensis* (Lamiaceae) from China. *Phytotaxa* **2015**, *202*, 298–300. [CrossRef]
8. Xiang, C.L.; Hu, G.X.; Peng, H. New combinations and new synonyms in Lamiaceae from China. *Biodivers. Sci.* **2016**, *24*, 719–722. [CrossRef]
9. Ding, B.Y.; Chen, Z.H.; Xu, Y.L.; Jin, X.F.; Wu, D.F.; Chen, J.B.; Wu, W.J. New species and combination of Lamiaceae from Zhejiang China. *Guihaia* **2019**, *39*, 10–15. [CrossRef]
10. Wei, Y.K.; Pendry, C.A.; Zhang, D.G.; Huang, Y.B. *Salvia daiguii* (Lamiaceae): A new species from west Hunan, China. *Edinb. J. Bot.* **2019**, *76*, 359–368. [CrossRef]
11. Wei, Y.K.; Pendry, C.A.; Huang, Y.B.; Ge, B.J.; Xiao, H.W. *Salvia subviolacea*, A new species from the Himalayas-Hengduan Mountains, China. *Edinb. J. Bot.* **2021**, *78*, 1–9. [CrossRef]
12. Hu, G.X.; Liu, E.D.; Wu, Z.K.; Sytsma, K.J.; Drew, B.T.; Xiang, C.L. Integrating DNA sequences with morphological analysis clarifies phylogenetic position of *Salvia grandifolia* (Lamiaceae): An enigmatic species endemic to Southwestern China. *Int. J. Plant Sci.* **2020**, *181*, 787–799. [CrossRef]
13. Bentham, G. *Labiatarum Genera Et Species*; Ridgeway: London, UK, 1932–1936.
14. Bentham, G. Labiatae. In *Genera Plantarum*; Bentham, G., Hooker, J.D., Eds.; Reeve and Co.: London, UK, 1876; Volume 2, pp. 1160–1196.
15. Stibal, E. A revision of Chinese and East Burmese species of *Salvia*. *Acta Horti Gotobg.* **1934**, *9*, 101–165.
16. Wu, C.Y. Salvia. In *Flora Reipublicae Popularis Sinicae*; Wu, C.Y., Ed.; Science Press: Beijing, China, 1977; Volume 66, pp. 70–196.
17. Wang, B.Q. *Salvia miltiorrhiza*: Chemical and pharmacological review of a medicinal plant. *J. Med. Plant. Res.* **2010**, *4*, 2813–2820. [CrossRef]
18. Walker, J.B.; Sytsma, K.J. Staminal evolution in the genus *Salvia* (Lamiaceae): Molecular phylogenetic evidence for multiple origins of the staminal lever. *Ann. Bot.* **2007**, *100*, 375–391. [CrossRef]
19. Jenks, A.A.; Walker, J.B.; Kim, S.C. Phylogeny of New World *Salvia* subgenus *Calosphace* (Lamiaceae) based on cpDNA (psbA-trnH) and nrDNA (ITS) sequence data. *J. Plant Res.* **2013**, *126*, 483–496. [CrossRef]
20. Will, M.; Classen-Bockhoff, R. Why Africa matters: Evolution of Old World *Salvia* (Lamiaceae) in Africa. *Ann. Bot.* **2014**, *114*, 61–83. [CrossRef]
21. Will, M.; Classen-Bockhoff, R. Time to split *Salvia* s.l. (Lamiaceae)-New insights from Old World *Salvia* phylogeny. *Mol. Phylogenet. Evol.* **2017**, *109*, 33–58. [CrossRef]
22. Fragoso-Martínez, I.; Martínez-Gordillo, M.; Salazar, G.A.; Sazatornil, F.; Jenks, A.A.; Peña, M.R.G.; Barrera-Aveleida, G.; Benitez-Vieyra, S.; Magallón, S.; Cornejo-Tenorio, G. Phylogeny of the Neotropical sages (*Salvia* subg. *Calosphace*; Lamiaceae) and insights into pollinator and area shifts. *Plant Syst. Evol.* **2018**, *304*, 43–55. [CrossRef]
23. Deng, T.; Nie, Z.L.; Drew, B.T.; Volis, S.; Kim, C.; Xiang, C.L.; Zhang, J.W.; Wang, Y.H.; Sun, H. Does the Arcto-Tertiary biogeographic hypothesis explain the disjunct distribution of Northern Hemisphere herbaceous plants? The case of *Meehania* (Lamiaceae). *PLoS ONE* **2015**, *10*, e0117171. [CrossRef]
24. Moore, M.J.; Bell, C.D.; Soltis, P.S.; Soltis, D.E. Using plastid genome-scale data to resolve enigmatic relationships among basal angiosperms. *Proc. Natl. Acad. Sci. USA* **2007**, *104*, 19363–19368. [CrossRef]
25. Parks, M.; Cronn, R.; Liston, A. Increasing phylogenetic resolution at low taxonomic levels using massively parallel sequencing of chloroplast genomes. *BMC Biol.* **2009**, *7*, 84. [CrossRef] [PubMed]
26. Zhao, F.; Li, B.; Drew, B.T.; Chen, Y.P.; Wang, Q.; Yu, W.B.; Liu, E.D.; Salmaki, Y.; Peng, H.; Xiang, C.L. Leveraging plastomes for comparative analysis and phylogenomic inference within *Scutellarioideae* (Lamiaceae). *PLoS ONE* **2020**, *15*, e0232602. [CrossRef] [PubMed]
27. Wu, H.; Ma, P.F.; Li, H.T.; Hu, G.X.; Li, D.Z. Comparative plastomic analysis and insights into the phylogeny of *Salvia* (Lamiaceae). *Plant Divers.* **2021**, *43*, 15–26. [CrossRef] [PubMed]
28. Zhao, F.; Drew, B.T.; Chen, Y.P.; Hu, G.X.; Li, B.; Xiang, C.L. The Chloroplast genome of *Salvia*: Genomic characterization and phylogenetic analysis. *Int. J. Plant Sci.* **2020**, *181*, 812–830. [CrossRef]
29. Doyle, J.J.; Doyle, J.L. A rapid DNA isolation procedure for small quantities of fresh leaf tissue. *Phytochem. Bull.* **1987**, *19*, 11–15.
30. Jin, J.J.; Yu, W.B.; Yang, J.B.; Song, Y.; De Pamphilis, C.W.; Yi, T.S.; Li, D.Z. GetOrganelle: A fast and versatile toolkit for accurate de novo assembly of organelle genomes. *Genome Biol.* **2020**, *21*, 241. [CrossRef]
31. Wick, R.R.; Schultz, M.B.; Zobel, J.; Holt, K.E. Bandage: Interactive visualization of de novo genome assemblies. *Bioinformatics* **2015**, *31*, 3350–3352. [CrossRef]
32. Qu, X.J.; Moore, M.J.; Li, D.Z.; Yi, T.S. PGA: A software package for rapid, accurate, and flexible batch annotation of plastomes. *Plant Methods* **2019**, *15*, 50. [CrossRef]
33. Chen, H.; Zhang, J.; Yuan, G.; Liu, C. Complex interplay among DNA modification, noncoding RNA expression and protein-coding RNA expression in *Salvia miltiorrhiza* chloroplast genome. *PLoS ONE* **2014**, *9*, e99314. [CrossRef]
34. Lowe, T.M.; Chan, P.P. tRNAscan-SE On-line: Integrating search and context for analysis of transfer RNA genes. *Nucleic Acids Res.* **2016**, *44*, W54–W57. [CrossRef]

35. Kearse, M.; Moir, R.; Wilson, A.; Stones-Havas, S.; Cheung, M.; Sturrock, S.; Buxton, S.; Cooper, A.; Markowitz, S.; Duran, C.; et al. Geneious Basic: An integrated and extendable desktop software platform for the organization and analysis of sequence data. *Bioinformatics* **2012**, *28*, 1647–1649. [CrossRef] [PubMed]
36. Lohse, M.; Drechsel, O.; Kahlau, S.; Bock, R. OrganellarGenomeDRAW—A suite of tools for generating physical maps of plastid and mitochondrial genomes and visualizing expression data sets. *Nucleic Acids Res.* **2013**, *41*, W575–W581. [CrossRef] [PubMed]
37. Peden, J.F. Analysis of Codon Usage. Ph.D. Thesis, University of Nottingham, Nottingham, UK, 1999.
38. Randall, T.A.; Dwyer, R.A.; Huitema, E.; Beyer, K.; Cvitanich, C.; Kelkar, H.; Fong, A.M.; Gates, K.; Roberts, S.; Yatzkan, E.; et al. Large-scale gene discovery in the Oomycete Phytophthora infestans reveals likely components of phytopathogenicity shared with true fungi. *Mol. Plant Microbe Interact.* **2005**, *18*, 229–243. [CrossRef]
39. Thiel, T.; Michalek, W.; Varshney, R.K.; Graner, A. Exploiting EST databases for the development and characterization of gene-derived SSR-markers in barley (*Hordeum vulgare* L.). *Theor. Appl. Genet.* **2003**, *106*, 411–422. [CrossRef] [PubMed]
40. Kurtz, S.; Choudhuri, J.V.; Ohlebusch, E.; Schleiermacher, C.; Stoye, J.; Giegerich, R. REPuter: The manifold applications of repeat analysis on a genomic scale. *Nucleic Acids Res.* **2001**, *29*, 4633–4642. [CrossRef] [PubMed]
41. Mayor, C.; Brudno, M.; Schwartz, J.R.; Poliakov, A.; Rubin, E.M.; Frazer, K.A.; Pachter, L.S.; Dubchak, I. VISTA: Visualizing global DNA sequence alignments of arbitrary length. *Bioinformatics* **2000**, *16*, 1046–1047. [CrossRef] [PubMed]
42. Kurtz, S.; Phillippy, A.; Delcher, A.L.; Smoot, M.; Shumway, M.; Antonescu, C.; Salzberg, S.L. Versatile and open software for comparing large genomes. *Genome Biol.* **2004**, *5*, R12. [CrossRef]
43. Librado, P.; Rozas, J. DnaSP v5: A software for comprehensive analysis of DNA polymorphism data. *Bioinformatics* **2009**, *25*, 1451–1452. [CrossRef]
44. Katoh, K.; Standley, D.M. MAFFT multiple sequence alignment software version 7: Improvements in performance and usability. *Mol. Biol. Evol.* **2013**, *30*, 772–780. [CrossRef]
45. Talavera, G.; Castresana, J. Improvement of phylogenies after removing divergent and ambiguously aligned blocks from protein sequence alignments. *Syst. Biol.* **2007**, *56*, 564–577. [CrossRef]
46. Stamatakis, A. RAxML version 8: A tool for phylogenetic analysis and post-analysis of large phylogenies. *Bioinformatics* **2014**, *30*, 1312–1313. [CrossRef] [PubMed]
47. Miller, M.A.; Pfeiffer, W.; Schwartz, T. Creating the CIPRES Science Gateway for inference of large phylogenetic trees. In Proceedings of the 2010 Gateway Computing Environments Workshop (GCE), New Orleans, LA, USA, 14 November 2010; IEEE: Manhattan, NY, USA, 2010. [CrossRef]
48. Ronquist, F.; Teslenko, M.; van der Mark, P.; Ayres, D.L.; Darling, A.; Hohna, S.; Larget, B.; Liu, L.; Suchard, M.A.; Huelsenbeck, J.P. MrBayes 3.2: Efficient Bayesian phylogenetic inference and model choice across a large model space. *Syst. Biol.* **2012**, *61*, 539–542. [CrossRef] [PubMed]
49. Zhang, D.; Gao, F.L.; Jakovlic, I.; Zou, H.; Zhang, J.; Li, W.X.; Wang, G.T. PhyloSuite: An integrated and scalable desktop platform for streamlined molecular sequence data management and evolutionary phylogenetics studies. *Mol. Ecol. Resour.* **2020**, *20*, 348–355. [CrossRef] [PubMed]
50. Kalyaanamoorthy, S.; Minh, B.Q.; Wong, T.K.F.; von Haeseler, A.; Jermiin, L.S. ModelFinder: Fast model selection for accurate phylogenetic estimates. *Nat. Methods* **2017**, *14*, 587–589. [CrossRef]
51. Qian, J.; Song, J.; Gao, H.; Zhu, Y.; Xu, J.; Pang, X.; Yao, H.; Sun, C.; Li, X.; Li, C.; et al. The complete chloroplast genome sequence of the medicinal plant *Salvia miltiorrhiza*. *PLoS ONE* **2013**, *8*, e57607. [CrossRef]
52. Gao, B.; Yuan, L.; Tang, T.; Hou, J.; Pan, K.; Wei, N. The complete chloroplast genome sequence of *Alpinia oxyphylla* Miq. and comparison analysis within the Zingiberaceae family. *PLoS ONE* **2019**, *14*, e0218817. [CrossRef] [PubMed]
53. Zong, D.; Zhou, A.; Zhang, Y.; Zou, X.; Li, D.; Duan, A.; He, C. Characterization of the complete chloroplast genomes of five *Populus* species from the western Sichuan plateau, southwest China: Comparative and phylogenetic analyses. *PeerJ* **2019**, *7*, e6386. [CrossRef] [PubMed]
54. Yang, J.X.; Hu, G.X.; Hu, G.W. Comparative genomics and phylogenetic relationships of two endemic and endangered species (*Handeliodendron bodinieri* and *Eurycorymbus cavaleriei*) of two monotypic genera within Sapindales. *BMC Genom.* **2022**, *23*, 27. [CrossRef]
55. Gu, L.; Su, T.; An, M.T.; Hu, G.X. The Complete dhloroplast genome of the vulnerable *Oreocharis esquirolii* (Gesneriaceae): Structural features, comparative and phylogenetic Analysis. *Plants* **2020**, *9*, 1692. [CrossRef]
56. Wicke, S.; Schneeweiss, G.M.; dePamphilis, C.W.; Muller, K.F.; Quandt, D. The evolution of the plastid chromosome in land plants: Gene content, gene order, gene function. *Plant Mol. Biol.* **2011**, *76*, 273–297. [CrossRef]
57. Millen, R.S.; Olmstead, R.G.; Adams, K.L.; Palmer, J.D.; Lao, N.T.; Heggie, L.; Kavanagh, T.A.; Hibberd, J.M.; Gray, J.C.; Morden, C.W.; et al. Many parallel losses of *infA* from chloroplast DNA during angiosperm evolution with multiple independent transfers to the nucleus. *Plant Cell* **2001**, *13*, 645–658. [CrossRef] [PubMed]
58. Shinozaki, K.; Ohme, M.; Tanaka, M.; Wakasugi, T.; Hayashida, N.; Matsubayashi, T.; Zaita, N.; Chunwongse, J.; Obokata, J.; Yamaguchi-Shinozaki, K.; et al. The complete nucleotide sequence of the tobacco chloroplast genome: Its gene organization and expression. *EMBO J.* **1986**, *5*, 2043–2049. [CrossRef]
59. Shimada, H.; Sugiura, M. Fine structural features of the chloroplast genome: Comparison of the sequenced chloroplast genomes. *Nucleic Acids Res.* **1991**, *19*, 983–995. [CrossRef] [PubMed]

60. Park, I.; Kim, W.J.; Yeo, S.M.; Choi, G.; Kang, Y.M.; Piao, R.; Moon, B.C. The complete chloroplast genome sequences of *Fritillaria ussuriensis* Maxim. and *Fritillaria cirrhosa* D. Don, and comparative analysis with Other *Fritillaria* species. *Molecules* **2017**, *22*, 982. [CrossRef] [PubMed]
61. Powell, W.; Morgante, M.; McDevitt, R.; Vendramin, G.G.; Rafalski, J.A. Polymorphic simple sequence repeat regions in chloroplast genomes: Applications to the population genetics of pines. *Proc. Natl. Acad. Sci. USA* **1995**, *92*, 7759–7763. [CrossRef] [PubMed]
62. Chen, C.; Zhou, P.; Choi, Y.A.; Huang, S.; Gmitter, F.G., Jr. Mining and characterizing microsatellites from *Citrus* ESTs. *Theor. Appl. Genet.* **2006**, *112*, 1248–1257. [CrossRef] [PubMed]
63. Feng, S.; He, R.; Lu, J.; Jiang, M.; Shen, X.; Jiang, Y.; Wang, Z.; Wang, H. Development of SSR markers and assessment of genetic diversity in medicinal *Chrysanthemum morifolium* cultivars. *Front. Genet.* **2016**, *7*, 113. [CrossRef] [PubMed]
64. He, Y.; Xiao, H.; Deng, C.; Xiong, L.; Yang, J.; Peng, C. The complete chloroplast genome sequences of the medicinalplant *Pogostemon cablin*. *Int. J. Mol Sci.* **2016**, *17*, 820. [CrossRef]
65. Kuang, D.Y.; Wu, H.; Wang, Y.L.; Gao, L.M.; Zhang, S.Z.; Lu, L. Complete chloroplast genome sequence of *Magnolia kwangsiensis* (Magnoliaceae): Implication for DNA barcoding and population genetics. *Genome* **2011**, *54*, 663–673. [CrossRef]
66. Du, Y.P.; Bi, Y.; Yang, F.P.; Zhang, M.F.; Chen, X.Q.; Xue, J.; Zhang, X.H. Complete chloroplast genome sequences of *Lilium*: Insights into evolutionary dynamics and phylogenetic analyses. *Sci. Rep.* **2017**, *7*, 5751. [CrossRef]
67. Liang, C.L.; Wang, L.; Lei, J.; Duan, B.Z.; Ma, W.S.; Xiao, S.M.; Qi, H.J.; Wang, Z.; Liu, Y.Q.; Shen, X.F.; et al. A comparative analysis of the chloroplast genomes of four *Salvia* medicinal plants. *Engineering* **2019**, *5*, 907–915. [CrossRef]
68. Liang, C.L.; Wang, L.; Ma, W.S.; Xu, J. A comparative study of complete chloroplast genome for the genus *Salvia*. *J. Plant Biochem. Biot.* **2021**, *30*, 117–125. [CrossRef]
69. Cavalier-Smith, T. Chloroplast evolution: Secondary symbiogenesis and multiple losses. *Curr. Biol.* **2002**, *12*, R62–R64. [CrossRef]
70. Wang, R.J.; Cheng, C.L.; Chang, C.C.; Wu, C.L.; Su, T.M.; Chaw, S.M. Dynamics and evolution of the inverted repeat-large single copy junctions in the chloroplast genomes of monocots. *BMC Evol. Biol.* **2008**, *8*, 36. [CrossRef] [PubMed]
71. Nie, X.; Lv, S.; Zhang, Y.; Du, X.; Wang, L.; Biradar, S.S.; Tan, X.; Wan, F.; Weining, S. Complete chloroplast genome sequence of a major invasive species, crofton weed (*Ageratina adenophora*). *PLoS ONE* **2012**, *7*, e36869. [CrossRef] [PubMed]
72. Davis, J.I.; Soreng, R.J. Migration of endpoints of two genes relative to boundaries between regions of the plastid genome in the *grass* family (Poaceae). *Am. J. Bot.* **2010**, *97*, 874–892. [CrossRef]
73. Li, R.; Ma, P.F.; Wen, J.; Yi, T.S. Complete sequencing of five Araliaceae chloroplast genomes and the phylogenetic implications. *PLoS ONE* **2013**, *8*, e78568. [CrossRef]
74. Menezes, A.P.A.; Resende-Moreira, L.C.; Buzatti, R.S.O.; Nazareno, A.G.; Carlsen, M.; Lobo, F.P.; Kalapothakis, E.; Lovato, M.B. Chloroplast genomes of *Byrsonima* species (Malpighiaceae): Comparative analysis and screening of high divergence sequences. *Sci. Rep.* **2018**, *8*, 2210. [CrossRef] [PubMed]
75. Shi, H.; Yang, M.; Mo, C.; Xie, W.; Liu, C.; Wu, B.; Ma, X. Complete chloroplast genomes of two *Siraitia Merrill* species: Comparative analysis, positive selection and novel molecular marker development. *PLoS ONE* **2019**, *14*, e0226865. [CrossRef]
76. Khakhlova, O.; Bock, R. Elimination of deleterious mutations in plastid genomes by gene conversion. *Plant J.* **2006**, *46*, 85–94. [CrossRef]
77. Meng, X.X.; Xian, Y.F.; Xiang, L.; Zhang, D.; Shi, Y.H.; Wu, M.L.; Dong, G.Q.; Ip, S.P.; Lin, Z.X.; Wu, L.; et al. Complete chloroplast genomes from *Sanguisorba*: Identity and variation among four species. *Molecules* **2018**, *23*, 2137. [CrossRef] [PubMed]
78. Kress, W.; Wurdack, K.J.; Zimmer, E.A.; Weigt, L.A.; Janzen, D.H. Use of DNA barcodes to identify flowering plants. *Proc. Natl. Acad. Sci. USA* **2005**, *102*, 8369–8374. [CrossRef] [PubMed]
79. Kress, W.J.; Erickson, D.L. A two-locus global DNA barcode for land plants: The coding *rbcL* gene complements the non-coding *trnH-psbA* spacer region. *PLoS ONE* **2007**, *2*, e508. [CrossRef] [PubMed]
80. Chase, M.W.; Cowan, R.S.; Hollingsworth, P.M.; Berg, C.V.D.; Wilkinson, M.J. A Proposal for a Standardised Protocol to Barcode All Land Plants. *Taxon* **2007**, *56*, 295–299. [CrossRef]
81. Group, C.P.W. A DNA barcode for land plants. *Proc. Natl. Acad. Sci. USA* **2009**, *106*, 12794–12797.
82. Meng, H.H.; Zhang, M.L. Phylogeography of *Lagochilus ilicifolius* (Lamiaceae) in relation to Quaternary climatic oscillation and aridification in northern China. *Biochem. Syst. Ecol.* **2021**, *39*, 787–796. [CrossRef]
83. Hu, G.X.; Su, T.; An, M.T.; Wang, X.Y. Rediscovery of *Pogostemon dielsianus* (Lamiaceae, Lamioideae), a rare endemic species from southwestern China, after one century. *PhytoKeys* **2021**, *171*, 61–73. [CrossRef]
84. Xiang, C.L.; Zhao, F.; Cantino, P.D.; Drew, B.T.; Li, B.; Liu, E.D.; Soltis, D.E.; Soltis, P.S.; Peng, H. Molecular systematics of *Caryopteris* (Lamiaceae) and its allies with reference to the molecular phylogeny of subfamily Ajugoideae. *Taxon* **2018**, *67*, 376–394. [CrossRef]
85. Li, Q.Q.; Li, M.H.; Yuan, Q.J.; Cui, Z.H.; Huang, L.Q.; Xiao, P.G. Phylogenetic relationships of *Salvia* (Lamiaceae) in China: Evidence from DNA sequence datasets. *J. Syst. Evol.* **2013**, *51*, 184–195. [CrossRef]
86. Drew, B.T.; Sytsma, K.J. Testing the monophyly and placement of *Lepechinia* in the tribe Mentheae (Lamiaceae). *Syst. Bot.* **2011**, *36*, 1038–1049. [CrossRef]
87. Su, S.W.; Shen, Z.A.; He, J.Q. New species of the genus *Salvia* from Anhui. *Acta Bot. Yunnanica* **1984**, *6*, 55–62.

88. Rokas, A.; Carroll, S.B. More genes or more taxa? The relative contribution of gene number and taxon number to phylogenetic accuracy. *Mol. Biol. Evol.* **2005**, *22*, 1337–1344. [CrossRef] [PubMed]
89. Philippe, H.; Brinkmann, H.; Lavrov, D.V.; Littlewood, D.T.J.; Manuel, M.; Wörheide, G.; Baurain, D. Resolving difficult phylogenetic questions: Why more sequences are not enough. *PLoS Biol.* **2011**, *9*, e1000602. [CrossRef] [PubMed]

diversity

MDPI

Article

Integrative Taxonomy Supports Two New Species of *Rhodiola* (Crassulaceae) in Xizang, China

Shiyong Meng *, Zimeng Wang and Lv Ye

School of Life Sciences, Peking University, Beijing 100871, China; xztswzm@126.com (Z.W.); yelv_123@163.com (L.Y.)
* Correspondence: msy542702@pku.edu.cn

Abstract: The Qinghai–Tibet Plateau includes the Himalayas and Hengduan Mountains and is well known for its rich biodiversity. Evolutionary radiation is one of the main ways by which plants diversify in mountains, particularly the Qinghai–Tibet Plateau. It presents a large challenge to the classification of taxa that radiate quickly. One way to overcome these challenges is to continue conducting detailed field studies while integrating morphological and molecular evidence to classify these taxa. The aim of this research was to provide a case for the systematic study of the complex taxa *Rhodiola*, which rapidly radiate. During the field study, we found two unique variants of *Rhodiola* in an alpine dry meadow and beds of pebbles on beaches, respectively. We utilized a morphological principal component analysis, scanning electron microscopy and molecular phylogenetic analysis to propose two new species: *Rhodiola wangii* S.Y. Meng and *Rhodiola namlingensis* S.Y. Meng. *R. wangii* is similar to *R. stapfii* (Hamet) S.H. Fu, but it differs in having an intensely broad rhombus and alternate leaves, a distinct petiole, stamens gathered together and reflexed purple scales. *R. namlingensis* is similar to *R. prainii* (Hamet) H. Ohba, but it differs in its exerted alternate leaves, the presence of more than four leaves on the stem, thick leaf blades, an obovate to inverted triangle, and short petioles. The conservation status of these two species was also assessed.

Keywords: Xizang; *Rhodiola namlingensis*; *Rhodiola wangii*; new species; taxonomy; radiation

Citation: Meng, S.; Wang, Z.; Ye, L. Integrative Taxonomy Supports Two New Species of *Rhodiola* (Crassulaceae) in Xizang, China. *Diversity* **2022**, *14*, 289. https://doi.org/10.3390/d14040289

Academic Editors: Hong-Hu Meng and Yi-Gang Song

Received: 23 February 2022
Accepted: 6 April 2022
Published: 12 April 2022

1. Introduction

The Qinghai–Tibet Plateau (QTP) and adjacent high-altitude areas contain more than 12,000 species of flowering plants, which renders it one of the areas of the world with the highest richness of species and a high level of endemism because of its habitats, such as high mountainous ranges, steep gorges, rocky outcrops, desert steppes and alpine meadows among others [1,2].

Evolutionary radiation is one of the main ways by which plants diversify in mountains, particularly the Qinghai–Tibet Plateau [3,4]. However, radiated taxa are often inconsistent between molecular phylogenetic and morphological classifications [5,6]. It is possible that these inconsistencies are caused by inadequate sampling or limited genomic sampling with insufficient molecular markers. In addition, some morphological characters are homogeneous [6]. Therefore, it is essential to recruit more molecular markers while conducting more field surveys to find valid characters and more types of variation. Several attempts by different types of molecular markers have been made to define the species of these genera [7–9], but the importance of field study has been neglected. However, hundreds of new plant species in the Qinghai–Tibet Plateau have been described in recent years [10], indicating that the surveys conducted so far are highly insufficient.

Rhodiola L. are members of the Crassulaceae DC., which includes more than 70 different species that are primarily distributed throughout the alpine zone of the northern hemisphere, particularly in the Qinghai–Tibet Plateau and adjacent high-altitude areas. Their major characteristics include well-developed thick rhizomes and apical parts with scaly leaves [11–13]. *Rhodiola* species occupy many habitats. For example, *R. yunnanensis* (Franch.) S.H. Fu, *R. macrocarpa* (Praeg.) S.H. Fu, and *R. liciae* (Hamet) S.H. Fu are distributed in thickets at low-to-medium altitudes; *R. sacra* (Prain ex Hamet) S.H. Fu and *R. bupleuroides* (Wall. ex Hook. f. et Thoms.) S.H. Fu grow in rocky outcrops at medium-to-high altitudes, and *R. coccinea* (Royle) Borissova and *R. crenulata* (Hook. f. et Thoms.) H. Ohba occupy mountain tops that exceed 4500 m. *Rhodiola* are highly diverse morphologically, and Fu and Fu [14] divided the genus into eight sections with seven series based on morphological studies. However, molecular phylogenetic research showed that only three of the eight sections (Sect. *Trifida* Fu, Sect. *Prainia* H. Ohba, and Sect. *Pseudorhodiola* (Diels) H. Ohba) can be considered monophyletic, while two important characters (dioecy and marcescent flowering stems) evolved multiple times within *Rhodiola* [15]. Molecular dating analysis showed that the primary diversification of *Rhodiola* occurred during two periods of the QTP uplifts, at 15–6.5 mya and 5–1.8 mya ago [15]. Rapid diversification, simplicity and the homogeneity of morphological characters lead to taxonomic difficulties in *Rhodiola*, particularly in the delimitation of species [16]. Thus, further intensive studies are still merited with more extensive sampling to clarify the systematic relationships of *Rhodiola*.

Our aim was to explore the diversity of *Rhodiola* in the Qinghai–Tibet Plateau and define species by combining morphological and molecular evidence. An extensive study of the systematic evolutionary biology research of *Rhodiola* in the wild has been conducted since 2010, which has led to the publication of some new *Rhodiola* species [17]. Here, we propose two new species of *Rhodiola* that are distributed on the platform of the Qinghai–Tibet Plateau. One of them is distributed on the shores of the plateau valley, whereas the other grows in the alpine dry meadow. These new species are very small and can easily avoid detection. The discovery of these two new species will help to understand the rapid radiation of *Rhodiola* in the Qinghai–Tibet Plateau.

2. Materials and Methods

2.1. Materials Collection and Field Investigation

Habitat plays an important role in the diversification of *Rhodiola* [15], and we took great pains to record that habitat during field investigations. Samples of two putative new species were collected in the field, including one from Lhünzê County (one population, 28°06′00″ N, 91°55′53″ E) and another from Namling County (Namling 1: 30°04′91″ N, 89°06′27″ E; Namling 2: 29°55′56″ N, 89°07′01″ E). Their close relatives were also collected from Xizang Province, China. The type specimens and other specimens of *R. prainii* and *R. stapfii* were examined, and specimens of the other related taxa were obtained from herbaria (CVH, K, PE) and examined for comparative research.

2.2. Morphological Analysis

Leaf characters play an important role in the definition of *Rhodiola* species [16], while the flowers are morphologically simple. Thus, the morphological analysis is primarily based on the leaf characters. To avoid the influence of artificiality and find key delimitation traits, we used a principal component analysis (PCA) to perform statistical analyses on the traits. We observed and measured the leaf traits and conducted the PCA using Origin 2020 (OriginLab Corporation, Northampton, MA, USA). All the data were quantitative, including the leaf length (L), leaf width (W), length of petiole (P), relative length of the petiole ($B = P/L$), the distance between the widest part of blade and the base of petiole (H), relative width of the petiole ($K = M/W$), relative length of the distance between the widest part of blade and base of petiole ($D = H/L$) and the shape of the blade ($S = W/L$) (Figure S1).

2.3. Scanning Electron Microscopy (SEM) Analysis

Although Gontcharova et al. [18] showed that the seed coats of *Rhodiola* in the Russian Far East vary considerably, they correspond to the features of gross morphology at the species level. Thus, we conducted a scanning electron microscopy (SEM) analysis using a Helios NanoLab G3 UC (Thermo Fisher, Waltham, MA, USA). First, seeds were directly fixed on the sample shuttle. Secondly, the sample was coated with gold using a vacuum sputter coater. Finally, the sample was transferred to the stage of the sample room for observation. The terminology of seed coat sculpturing and anatomy were those described by Gontcharova et al. [18]. The sizes of seeds were measured using Photoshop CC 2018 (Adobe, San Jose, CA, USA).

2.4. Phylogenetic Analysis

DNA barcoding research showed that the nuclear ribosomal internal transcribed spacer (ITS) was the best single-locus barcode, resolving 66% of the *Rhodiola* [19]. Thus, the ITS regions were used as molecular markers. The DNA of the new species was extracted and amplified by PCR as described by Zhang et al. [17]. The ITS sequences of 33 specimens of *Rhodiola* were downloaded from NCBI along with two accessions of *Phedimus*, another genus of the Crassulaceae, as outgroups. The GenBank accession numbers are shown in Table S1.

Phylogenetic analyses were performed using Bayesian inference (BI) and maximum-likelihood (ML) methods in PhyloSuite [20]. The ITS sequences were aligned with MAFFT v. 7 [21] and manually checked in PhyloSuite [20]. The evolutionary models for the ML and BI analyses were determined by ModelFinder [22] using the Akaike Information Criterion (AIC). ML trees with 1000 bootstraps (BS) were produced using an IQ-tree [23]. BS analyses were used to evaluate the support for individual clades with 5000 replicates. A BI analysis was performed using MrBayes [24]. Four chains of the Markov Chain Monte Carlo (MCMC) were run for 2,000,000 generations, sampling one tree every 100 generations, starting with a random tree. The average standard deviation of the split frequencies was used to assess the convergence of two runs. A majority rule (>50%) consensus tree was constructed after removing the burn-in period samples (the first 25% of the sampled trees) and posterior probabilities (PP) to estimate the robustness of the BI trees.

3. Results

3.1. Habitat

During the field investigation, we found that these two new species and similar ones grow on alpine meadows, but the habitats can be differentiated from each other. The Lhünzê population grows on the soil slopes of alpine meadows that are relatively dry, while *R. stapfii* grows in the moist parts of alpine meadows. The Namling populations primarily grow on the beds of pebbles on beaches or in rock crevices on the shore of the Xiangqu (Jiacuo Zangbo), which is one of the Yarlung Zangbo's tributaries. In contrast, *R. prainii* primarily grows on rocks or trees in the subtropical rain forest on the southern slope of the Himalayas and the slope rocks on the plateau of the Qinghai–Tibet Plateau.

3.2. Morphological Analysis

Morphologically, the Lhünzê population is closely related to *R. stapfii*. However, several characters differentiate them. First, rhizomes of the Lhünzê population are glossy, while those of *R. stapfii* are enfolded by the remnants of old shoots. Secondly, flowering stems with some leaves are alternate in the Lhünzê population, whereas all four to six pieces of the leaves are in whorls and on the apical part of the *R. stapfii* rhizomes. Third, the leaves of Lhünzê population have obvious petioles, while those of *R. stapfii* are very short or sessile. Finally, the stamens of the Lhünzê population are gathered together, and the scales are purple and reflexed. In contrast, the stamens of *R. stapfii* do not gather together, and the scales are white. A principal component analysis (PCA) based on the 8 qualitative leaf traits revealed considerable variability among the 44 individuals considered in this study (Figure 1A). The PC1 scores, which accounted for 48.9% of the total variation, showed very high correlation with length of the leaf (L) and petiole, including the relative width of the petiole (K = M/W) and relative length of the petiole (B = P/L). The scores of PC1 were also highly correlated with other traits, such as shape of the blade (S = W/L), relative length of the distance between the widest part of blade and base of petiole (D = H/L), and the length between the widest part of blade and base of petiole (H). The PC2 scores, which explained 31.8% of the total variation, were highly correlated with the width of the leaf (W), between the widest part of blade and the base of petiole (H) and length of the leaf (L).

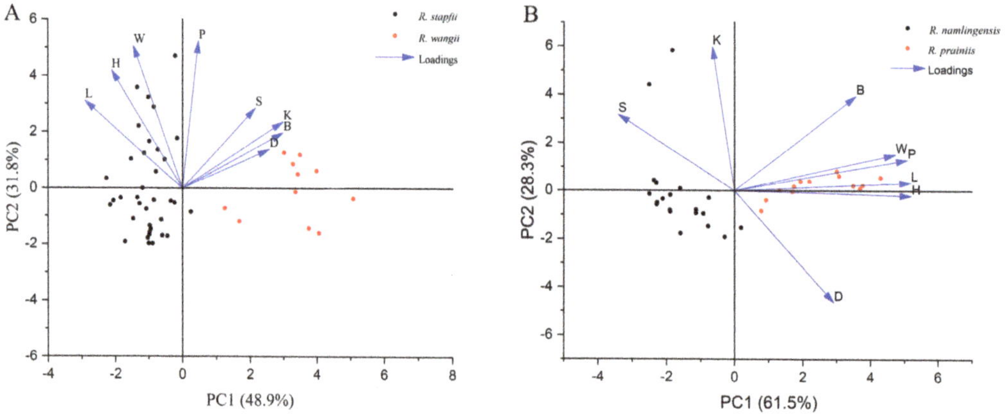

Figure 1. Principal component analysis based on the leaf traits. (**A**) Lhünzê population and *Rhodiola stapfii* (Raymond-Hamet) S.H. Fu. (**B**) *R. namlingensis* S.Y. Meng and *R. prainii* (Hamet) H. Ohba.

A morphological analysis indicated that the Namling populations are similar to those of *R. prainii* with flattened leaf blade and similar inflorescences and flower structures. However, the plants can be differentiated by several characters. First, the Namling populations contain more developed flowering stems, while *R. prainii* contains short flowering stems, which usually have only two nodes on the stem. Secondly, leaves on the stems in Namling populations are alternate, while the leaves of *R. prainii* often have four whorls at the top of stem. The lower part of the stem leaves is opposite and has a developed pseudopetiole that is 1 to 3 cm long. Third, *R. prainii* flowers from July to August, while the Namling populations flower from September to mid-October. The PCA based on the eight quantitative traits revealed a considerable variability among the 47 individuals. The first two PCs together explained 89.7% of the total variation (Figure 1B). The PC1 scores, which accounted for 61.5% of the total variation, were showed very high correlation with the length of the leaf (L), width of the leaf (W), length between the widest part of blade and base of the petiole (H) and length of the petiole (P). The scores of PC2, which explained 28.2% of the total variation, were highly correlated with the relative width of the petiole (K = M/W), relative length of the petiole (B = P/L) and relative length of the distance between the widest part of blade and base of petiole (D = H/L).

3.3. SEM Analysis

The seeds of *R. stapfii* are triangular with a wing-like projection at the chalazal end, 920 μm long, and 400 μm wide (Figure 2A). The testa cells are morphologically uniform or vary only slightly in their shape and are less regularly arranged. The outer periclinal cell walls are flat and smooth, while the anticlinal cell walls are thickened and bulging (Figure 2E). The cells are quadrangular and isodiametric, and the cell boundaries are well-defined and curved. The outer periclinal cell wall is convex. The seeds of the Lhünzê population are square and lack wing-like projections at the chalazal end, and they are 519 μm long and 324 μm wide (Figure 2B). The testa cells are not readily apparent and have dense stripes in the outer periclinal wall (Figure 2F). The seeds of *R. prainii* are similar to those of *R. stapfii*. They are triangular but small (761 μm long and 309 μm wide) (Figure 2C,G). They lack a wing-like projection at the chalazal end, and the anticlinal cell walls are thin and sunken. The outer periclinal cell wall is thickened and convex. The seeds of Namling populations are oblong and lack a wing-like projection at the chalazal end, and they are 835 μm long and 203 μm wide (Figure 2D). The testa cells have obvious longitudinal edges on the surface; the width between the longitudinal edges is 23.6 μm, and there are obvious horizontal stripes in the depressions between the edges that extend to the longitudinal edges, forming a regular pattern of orbital decoration. (Figure 2H).

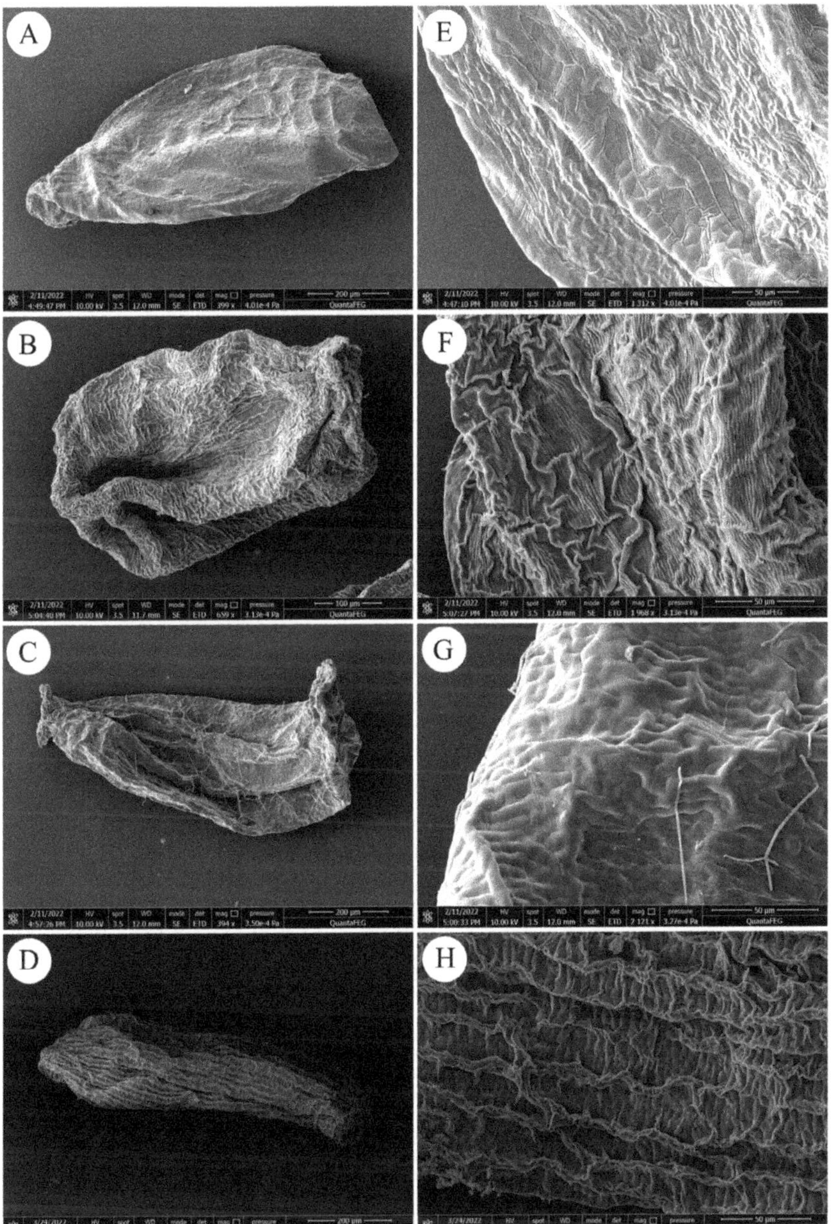

Figure 2. Scanning electron microscopy of four *Rhodiola* seeds. (**A**,**E**) *R. stapfii* (PEY0067558); (**B**,**F**) Lhünzê population (PEY0067593); (**C**,**G**) *R. prainii* (PEY0068682); (**D**,**H**) Namling population (PEY0067596); (**A**–**D**) The shape of seeds; (**E**–**H**) Local magnification of the seed micromorphology.

3.4. Molecular Analyses

Following the alignment, we obtained a matrix of 588 base pairs (bp) and selected SYM+G4 for the Bayesian and ML analyses (Figure S2). The 50% majority rule consensus tree of all the post burn-in trees is shown in Figure 3 with the Bayesian posterior probabilities (PPs).

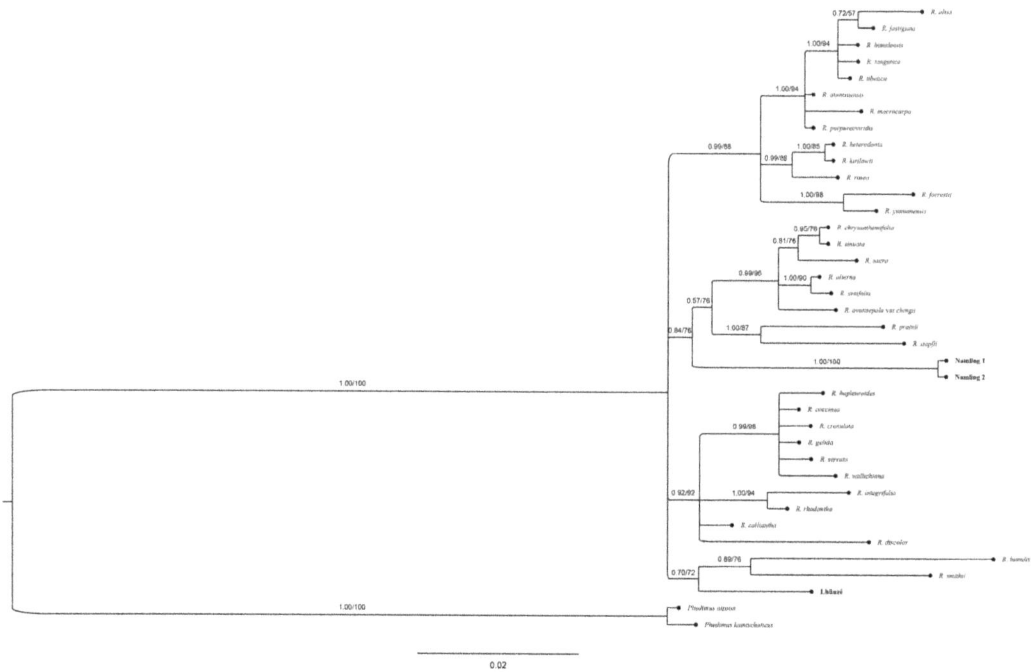

Figure 3. A Bayesian phylogenetic tree based on ITS sequences for the two new *Rhodiola* species described and their close relatives. The topology of the maximum likelihood (ML) tree was highly compatible with that of the Bayesian tree. Bayesian posterior probabilities (PPs: left) and bootstrap support (BP: right) values (>50%) are shown above the branch.

The BI tree strongly supported the monophyly of Sect. *Trifida*, Sect. *Prainia*, and Sect. *Pseudorhodiola*. The two samples of Namling populations from different sites are shown as a distinct clade (Posterior Probability [PP] = 1.0, Bootstrap value [BP] = 100%). The Namling populations formed a monophyletic clade with the species of Sect. *Trifida* and Sect. *Prainia* (PP = 0.84, BP = 64) (Figure 3), whereas the Lhünzê population formed a monophyletic clade with *R. smithii* and *R. humilis* (PP = 0.70, BP = 62).

4. Discussion

This survey found that species of *Rhodiola* have spread to the dry soil slopes of alpine meadow and beds of pebble beaches on the Qinghai–Tibet Plateau. It shows that *Rhodiola* has occupied various habitats on the Qinghai–Tibet Plateau. The habitat reflects the radiation of *Rhodiola*, which is consistent with the results of molecular systematics [15,25].

Seed micromorphology provides many characters that are potentially useful to identify species and perform phylogenetic inference in the Crassulaceae [26–28]. However, there has been little research on the seed coat micromorphology of *Rhodiola* [18]. In this study, we conducted an SEM analysis of the seed coats of *R. prainii*, *R. stapfii*, and the Namling and Lhünzê populations. The results show that the two morphological types of these four species differed from those previously reported [18]. The surface ornamentation of the seed coats of the Namling populations clearly differ from those of *R. prainii*, as do those of the Lhünzê population and *R. stapfii*. Thus, the seed coat characters of *Rhodiola* show considerable diversity and have some value in the delimitation of species. Thus, there should be more focus on the seed coat micromorphologies of *Rhodiola*, which includes more than 70 species. Recently, many molecular systematic studies have been conducted, and *Rhodiola* has been an ideal material on which conduct research on radiated taxa [15,16,25,29], but few studies have integrated the ornamentation of seed coats.

In this research, we integrated morphological studies, SEM and ITS using PCA and a phylogenetic analysis for the delimitation of *Rhodiola*. A morphological comparison indicated that the Lhünzê population is similar to that of *R. stapfii*. In contrast, a phylogenetic analysis showed that the Lhünzê population is more closely related to *R. smithii* and *R. humilis* than *R. stapfii* phylogenetically. However, scale-like leaves were identified in the rhizome apex of the Lhünzê population, whereas the leaves in the rhizome apex of *R. smithii* and *R. humilis* were developed. The Namling populations are morphologically similar to those of *R. prainii*, but the PCA results show that they can be easily differentiated by the leaf shape (the length of the leaf, width of the leaf, length of between the widest part of blade and base of petiole and length of the petiole). A phylogenetic analysis shows that the Namling populations form a monophyletic clade with the species of Sect. *Trifida* and Sect. *Prainia*. In this monophyletic clade, the Namling populations were placed at the base.

Thus, based on the phylogenetic analysis, leaf morphological analysis and habitat information, we believe that the Lhünzê population and Namling population are distinct new species of *Rhodiola* that have not yet been described, and will provide detailed information to help understand the radiation of *Rhodiola*.

5. Description of the New Species

1. *Rhodiola wangii* S.Y. Meng sp. nov. (Figures 4 and 5)
urn:lsid:ipni.org:names: 77260717-1.

Type:—China. Xizang, Lhünzê Co., 28°06′00″ N, 91°55′53″ E, 4914 m, 20 July 2014, S.Y. Meng et al. MHW0071 (holotype PEY0067593, isotype, PEY0068681).

Diagnosis:—Similar to *Rhodiola stapfii* (Raymond-Hamet) S.H. Fu but differs in having an intensely broad rhombus and alternate leaves, a distinct petiole, stamens that are gathered together and have reflexed purple scales. (vs. middle stem leaves 5- or 6-verticillate, ovate to ovate-oblong, white nectar scales).

Description:—A low and delicate herbaceous perennial, 0.5–1 cm high. Caudex nearly erect, fewer branches, slightly thicker, usually 4–8 mm across; apical part often short, branched and accrescent, crowned by the scaly radical leaves. Scaly radical leaves membranous, persistent, purple, long ovate with an entire margin, 1.4–1.7 mm long, 0.8 mm wide. Roots very slender. Flowering stems 1–3 from the apex of each caudex branch, deciduous, 0.5–1 cm long, erect, simple, terete, smooth. Leaves alternate, flattened, broad rhombus, closely arranged on the stem, ascending-spreading, distinct petiole, spurless, glabrous, 5–7 mm long, 2–5 mm wide, entire. Inflorescences terminal, 1–3 flower buds form cymes, inconspicuous bracteates, bracts leafy. Flower green, 4(5)-merous, dioecious. Female flowers (♀) white-green, sepal 4 (5), free, long triangular, apex acuminate, entire, green, slightly fleshy, 1.2–1.6 mm long, 0.7–0.8 mm wide. Petals 4 (5), triangular free, green, with a short tip at the apex, 2 mm long, 0.8 mm wide, entire along the margin. Nectar-scales purple, trapezoid, rolling outward, truncate at the apex, 0.5 mm long, 0.4 mm wide. Carpel 4–5, free, erect, triangular ovate, ca. 2 mm long, style 0.5 mm long, curved outward. Follicles erect. Ovules ca. 5–9 in each locule. Seed orbicular-ovate ca. 0.8 mm long, 0.6 mm wide. Male flowers (♂) 4(5)-merous, sepals 4(5), green, free, long triangular, with a short tip at the apex, 1.2–1.5 mm long, 0.8–1.0 mm wide. Petals 4 (5) free, pale green, 1.8–2.0 mm long, 0.8–1.0 mm wide. Stamens 8–10, shorter than petals, antepetalous ones inserted 0.5 mm from petal base, long ca. 1.8 mm, antesepalous ones ca. 2 mm long, anther yellow. Nectar scales 4(5), broadly quadrangular, purple-red, 0.5 mm long, 0.2 mm wide. Carpel 4(5), erect, undeveloped.

Distribution and Habitat:—Perennial herb on mountain slopes, 4914 m. The distribution of *R. wangii* S.Y. Meng is somewhat limited (Figure 6). To date, only one population has been found on an arid hillside in the southeast region of Xizang, China.

Phenology:—Flowers from July to August, and fruits from August to September.

Etymology:—The epithet 'wangii' is used to commemorate the famous Chinese botanist and plant science popularization pioneer, Professor Wang Jinwu, Peking University. He published many popular articles and an illustrated handbook of botanical taxonomy, which promoted public attention and love for plants.

Common name (assigned here):—Jing Wu Hong Jing Tian (劲武红景天; Chinese name).

Proposed IUCN conservation status:—The new species grows in the arid meadows of Mt. Shangala. We collected only one population near Mt. Shangala, while another population was found between the road of Lhünzê County to Comai County in Lhünzê County in 2021. Although the habitat of the Qinghai–Tibet Plateau is relatively stable, more active economic and construction activities, such as grazing, may affect the population. The species is considered to be "Vulnerable" (VUD1) according to the IUCN Red List Criteria [30].

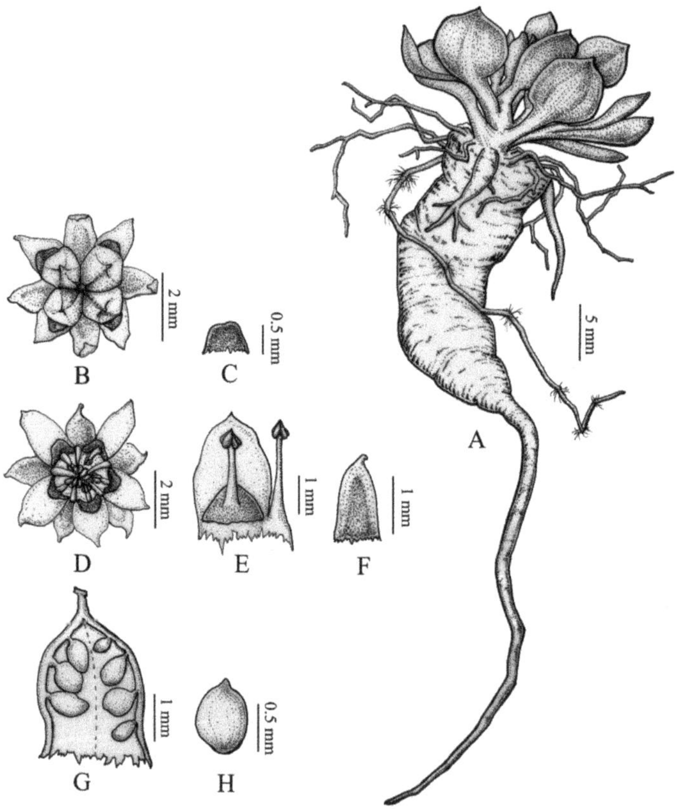

Figure 4. *Rhodiola wangii* S.Y. Meng. (**A**). Rhizome and flowering stem; (**B**). Female flower; (**C**). Nectar; (**D**). Male flower; (**E**). Petal and Stamens; (**F**). Sepal; (**G**). Follicles; (**H**). Seeds. Drawing by Ye Lv from PEY00067593.

Figure 5. *Rhodiola wangii* S.Y. Meng. (**A**). Rhizome and flowering stem; (**B**). Female plants; (**C**). Male plants.

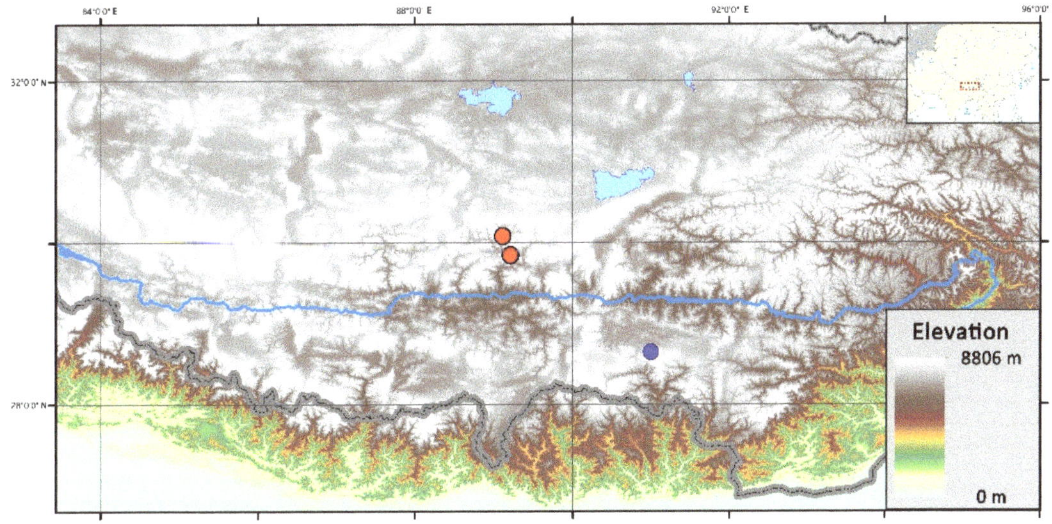

Figure 6. Geographic distribution. Blue dot shows *Rhodiola wangii* S.Y. Meng. Red dots show *R. namlingensis* S.Y. Meng.

2. *Rhodiola namlingensis* S.Y. Meng sp. nov. (Figures 7 and 8)
urn:lsid:ipni.org:names: 77260718-1.

Type:—China. Xizang: Namling Co., 30°04′91″ N, 89°06′27″ E, 4208 m, 18 September 2015, S.Y. Meng et al. 2015091802 (holotype, PEY0067596; isotype PEY0067595).

Diagnosis:—Similar to *Rhodiola prainii* (Hamet) H. Ohba but differs in its exerted alternate leaves, more than four stem leaves, thick leaf blades, obovate to inverted triangle, short petiole (vs. stem leaves 4, verticillate, oblong-elliptic leaf blades, ovate, broadly ovate, or reniform-orbicular, base abruptly narrowed to long attenuate.)

Description:—A perennial herb, 2–4 cm tall. Roots slightly thicker, branches, 4–9 cm long. Caudex cylindrical, slender, 2–7 mm thick, erect, the apical part densely covered with scaly radical-leaves. Scaly radical-leaves broadly triangular with entire margin, persistent, acuminate at the apex, 1.5–1.8 mm long, 1.5 mm wide, reddish brown. Flowering stems 1–4 from each branch apex of rhizomes, erect, simple, terete, glabrous. Leaves alternate, densely arranged on the upper part, ascending-spreading, spurless, thick herbaceous, flattish, yellowish green, obovate to inverted triangle, round at the apex; very short attenuate at the base; entire along the margin, 12–16 mm long, 6–8 mm wide, glabrous on both surfaces. The costa not obvious, with a short petiole. Inflorescences simple or few branches, corymbiform, 1–4 flowered; bracts shortly petiolate, obovate, 5–7 × 3–4 mm. Flowers bisexual, white or red, unequally 5-merous; pedicel 4–6 mm. Sepals 5, green, succulent, triangular-ovate, 5–7 mm long, 3–3.5mm wide, obtuse at the apex, entire. Petals 5, white or red, sometimes white with red plaques, oblong-ovate, 10–12 mm long, the united part 0.5 mm long, the free part 9–11 mm long, 6–8 mm wide, apex acuminate, tapering to the base. The petals do not fully open at maturity but stand upright surrounded by stamens and ovaries. Stamens 10, slightly shorter than petals, antepetalous ones inserted 2–3 mm from petal base, long ca. 5 mm, antesepalous ones 7–8 mm long, the filaments slender, sub-linear, anther tip, blue. Nectar scales trapezoid, ca. 1 mm long, ca. 0.8 mm wide, apex emarginate. Carpels 5, slightly connate at base, erect, long elliptic, 6–7 mm long, style 2–3 mm long, erect; each carpel has about 20 ovules. Seeds long elliptic, 1.5 mm long and 0.5 mm wide.

Distribution and Habitat: *R. namlingensis* was only known from southeast Xizang (Namling), China (Figure 6). Now, two populations have been found. It grows in gravel crevices or rock crevices on the beach of the Jiacuo Zangbo Valley, between 3800 and 4000 m above sea level.

Etymology:—The specific epithet is derived from Namling county, southeast of Xizang, China.

Phenology:—Flowering in August to September, fruiting in September to October.

Common name (assigned here):—Nan Mu Lin Hong Jing Tian (南木林红景天; Chinese name).

Proposed IUCN conservation status:—This new species is only known from southeast Xizang (Namling) where two populations were found in gravel crevices or rock crevices on the beach of the Jiacuo Zangbo Valley, between 3800 and 4000 m above sea level. The species is considered to be "Vulnerable" (VUD1) according to the IUCN Red List Criteria [30].

Additional specimens examined (paratype):—China. Xizang: Namling Co., 4313 m, 18 September 2015, S.Y. Meng et al. 2015091804 (PEY0068680, PEY0067594); Namling Co., 4313 m, 30 July 2015, S.Y. Meng & Z.M. Wang MWH110 (PEY0066822); Namling Co., 4212 m, 11 August 2010, S.Y. Meng msy05 (PEY0068679).

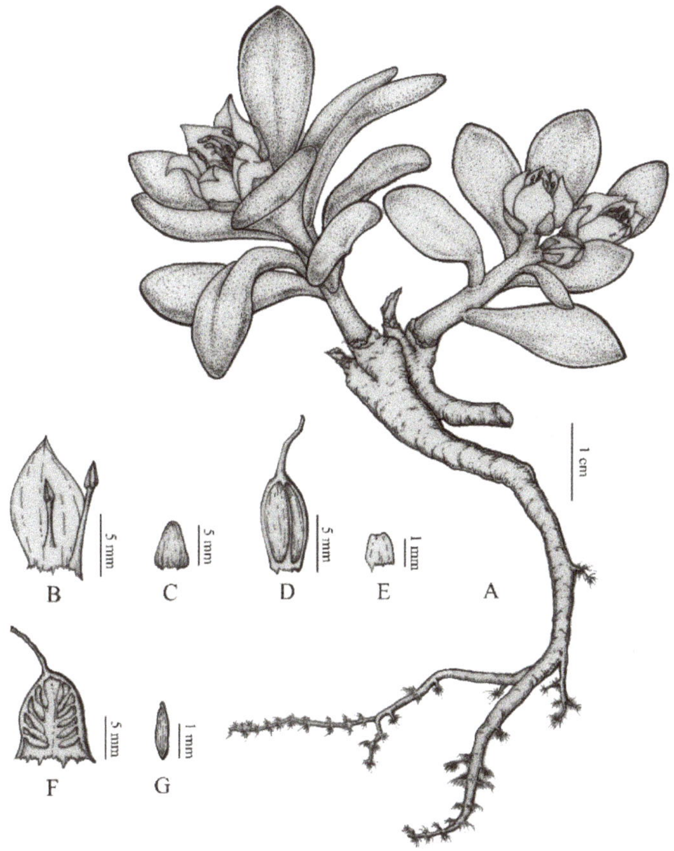

Figure 7. *Rhodiola namlingensis* S.Y. Meng. (**A**). Rhizome and flowering stem; (**B**). Petal and stamens; (**C**). Sepal; (**D**). Carpel; (**E**). Nectar; (**F**). Follicle; (**G**). Seed. Drawing by Ye Lv from PEY0067595.

Figure 8. *Rhodiola namlingensis* S.Y. Meng. (**A**). A plant and its habitat; (**B**). Side face of the plant.

Supplementary Materials: The following supporting information can be downloaded at: https://www.mdpi.com/article/10.3390/d14040289/s1, Figure S1: Leaf measurement. Figure S2: ML tree based on ITS sequences for these two new *Rhodiola* species and their close relatives. Table S1: Plant materials of 36 accessions of the Rhodiola taxa and two outgroup taxa with their collection locality, voucher information, and accession numbers of ITS sequences reported by Zhang et al. (2014).

Author Contributions: The experimental design was completed by S.M. and Z.W. Sample collection and treatment were conducted by S.M. The data were analyzed, and the figures and tables prepared with assistance from L.Y. and Z.W. The manuscript was drafted by S.M. edited the manuscript for structure, language, and scientific content. All authors have read and agreed to the published version of the manuscript.

Funding: This research was funded by the National Natural Science Foundation of China (No. 31600159).

Institutional Review Board Statement: Not applicable.

Data Availability Statement: Not applicable.

Acknowledgments: The authors thank the Core Facilities at the School of Life Sciences, Peking University for assistance with scanning electron microscope observation. We also thank Hong-ya Gu, Guang-yuan Rao and Sodmergen, Jia-chen Hao, and Yi-hao Shi for field assistance.

Conflicts of Interest: The authors declare no conflict of interest.

References

1. Wu, Z.Y. Hengduan Mountain flora and her significance. *J. Jpn. Bot.* **1988**, *63*, 297–311.
2. Wu, S.G.; Yang, Y.P.; Fei, Y. On the flora of the alpine region in the Qinghai-Xizang (Tibet) plateau. *Acta Bot. Yunnan* **1995**, *17*, 233–250.
3. Hughes, C.E.; Atchison, G.W. The ubiquity of alpine plant radiations: From the Andes to the Hengduan Mountains. *New Phytol.* **2015**, *207*, 275–282. [CrossRef]
4. Xia, X.M.; Yang, M.Q.; Li, C.L.; Huang, S.X.; Jin, W.T.; Shen, T.T.; Wang, F.; Li, X.H.; Yoichi, W.; Zhang, L.H.; et al. Spatiotemporal Evolution of the Global Species Diversity of Rhododendron. *Mol. Biol. Evol.* **2022**, *39*, 1–15. [CrossRef] [PubMed]
5. Zhang, J.Q.; Meng, S.Y.; Wen, J.; Rao, G.Y. Phylogenetic relationships and character evolution of *Rhodiola* (Crassulaceae) based on nuclear ribosomal ITS and plastid trnL-F and psbA-trnH sequences. *Syst. Bot.* **2014**, *39*, 441–451. [CrossRef]
6. Yu, W.B.; Liu, M.L.; Wang, H.; Mill, R.R.; Ree, R.H.; Yang, J.B.; Li, D.Z. Towards a comprehensive phylogeny of the large temperate genus *Pedicularis* (Orobanchaceae), with an emphasis on species from the Himalaya-Hengduan Mountains. *BMC Plant Biol.* **2015**, *15*, 1–14. [CrossRef]
7. Ouchon, C.H.P.; Ernández, A.N.F.; Assar, J.A.M.N.; Oyer, F.R.B.; Ubert, S.E.A. Phylogenomic analysis of the explosive adaptive radiation of the *Espeletia* complex (Asteraceae) in the tropical Andes. *Syst. Biol.* **2018**, *67*, 1041–1060. [CrossRef]
8. Carter, K.A.; Liston, A.; Bassil, N.V.; Alice, L.A.; Bushakra, J.M.; Sutherland, B.L.; Mockler, T.C.; Bryant, D.W.; Hummer, K.E. Target Capture Sequencing Unravels Rubus Evolution. *Front. Plant Sci.* **2019**, *10*, 1–18. [CrossRef]
9. Naciri, Y.; Christe, C.; Bétrisey, S.; Song, Y.G.; Deng, M.; Garfì, G.; Kozlowski, G. Species delimitation in the East Asian species of the relict tree genus *Zelkova* (Ulmaceae): A complex history of diversification and admixture among species. *Mol. Phylogenet. Evol.* **2019**, *134*, 172–185. [CrossRef]
10. Du, C.; Liao, S.; Boufford, D.E.; Ma, J. Twenty years of Chinese vascular plant novelties, 2000 through 2019. *Plant Divers.* **2020**, *42*, 393–398. [CrossRef]
11. Fu, K.T.; Ohba, H. Crassulaceae. In *Flora of China*; Wu, C.Y., Raven, P.H., Eds.; Science Press: Beijing, China, 2001; Volume 8, pp. 202–268.
12. Ohba, H. Crassulaceae. In *Illustrated Handbook of Succulent Plants*; Eggli, U., Ed.; Springer: Berlin/Heidelberg, Germany; New York, NY, USA, 2003; Volume 14, pp. 210–227.
13. Thiede, J.; Eggli, U. Crassulaceae. In *The Families and Genera of Vascular Plants*; Kubitzki, K., Ed.; Springer: Berlin/Heidelberg, Germany; New York, NY, USA, 2007; Volume 9, pp. 83–118.
14. Fu, S.H.; Fu, K.T. Crassulaceae. In *Flora Reipublicae Popularis Sinicae*; Chen, W.Q., Ruan, Y.Z., Eds.; Science Press: Beijing, China, 1984; Volume 34, pp. 33–216.
15. Zhang, J.Q.; Meng, S.Y.; Allen, G.A.; Wen, J.; Rao, G.Y. Rapid radiation and dispersal out of the qinghai-tibetan plateau of an alpine plant lineage *Rhodiola* (Crassulaceae). *Mol. Phylogenet. Evol.* **2014**, *77*, 147–158. [CrossRef] [PubMed]
16. Li, Y.C.; Wen, J.; Ren, Y.; Zhang, J.Q. From seven to three: Integrative species delimitation supports major reduction in species number in *Rhodiola* section *Trifida* (Crassulaceae) on the Qinghai-Tibetan Plateau. *Taxon* **2019**, *68*, 268–279. [CrossRef]
17. Zhang, J.Q.; Meng, S.Y.; Rao, G.Y. Two new species of *Rhodiola* (Crassulaceae) from the Qinghai-Tibetan Plateau. *Phytotaxa* **2015**, *224*, 159–172. [CrossRef]
18. Gontcharova, S.B.; Gontcharov, A.A.; Yakubov, V.V.; Kondo, K. Seed surface morphology in some representatives of the Genus *Rhodiola* sect. *Rhodiola (Crassulaceae) in the Russian Far East. Flora* **2009**, *204*, 17–24.
19. Zhang, J.Q.; Meng, S.Y.; Wen, J.; Rao, G.Y. DNA barcoding of *Rhodiola* (Crassulaceae): A case study on a group of recently diversified medicinal plants from the Qinghai-Tibetan Plateau. *PLoS ONE* **2015**, *10*, e0119921. [CrossRef]
20. Zhang, D.; Gao, F.; Jakovlić, I.; Zou, H.; Zhang, J.; Li, W.X.; Wang, G.T. PhyloSuite: An integrated and scalable desktop platform for streamlined molecular sequence data management and evolutionary phylogenetics studies. *Mol. Ecol. Resour.* **2020**, *20*, 348–355. [CrossRef]

21. Katoh, K.; Standley, D.M. MAFFT multiple sequence alignment software version 7: Improvements in performance and usability. *Mol. Biol. Evol.* **2013**, *30*, 772–780. [CrossRef]
22. Kalyaanamoorthy, S.; Minh, B.Q.; Wong, T.K.F.; Von Haeseler, A.; Jermiin, L.S. ModelFinder: Fast model selection for accurate phylogenetic estimates. *Nat. Methods* **2017**, *14*, 587–589. [CrossRef]
23. Nguyen, L.T.; Schmidt, H.A.; Von Haeseler, A.; Minh, B.Q. IQ-TREE: A fast and effective stochastic algorithm for estimating maximum-likelihood phylogenies. *Mol. Biol. Evol.* **2015**, *32*, 268–274. [CrossRef]
24. Ronquist, F.; Teslenko, M.; Van Der Mark, P.; Ayres, D.L.; Darling, A.; Höhna, S.; Larget, B.; Liu, L.; Suchard, M.A.; Huelsenbeck, J.P. Mrbayes 3.2: Efficient bayesian phylogenetic inference and model choice across a large model space. *Syst. Biol.* **2012**, *61*, 539–542. [CrossRef]
25. Zhao, D.N.; Ren, C.Q.; Zhang, J.Q. Can plastome data resolve recent radiations? *Rhodiola* (Crassulaceae) as a case study. *Bot. J. Linn. Soc.* **2021**, *197*, 513–526. [CrossRef]
26. Bywater, M. Observations on seeds of *Crassula* sect. *Rosulares*. *Kew Bull.* **1980**, *35*, 401–402. [CrossRef]
27. Weng, M.L.; Kuo-Huang, L.-L. Comparative Anatomy and Histochemical Study of the Seeds of *Sedum formosanum* N. E. br. And *Sedum morrisonense* Hayata. *Taiwania* **1998**, *43*, 307–315.
28. Wickens, G.E.; Bywater, M. Seed studies in *Crassula* subgen. *Disporocarpa*. *Kew Bull.* **1980**, *34*, 629–637. [CrossRef]
29. Zhu, R.W.; Li, Y.C.; Zhong, D.L.; Zhang, J.Q. Establishment of the most comprehensive ITS2 barcode database to date of the traditional medicinal plant *Rhodiola* (Crassulacaee). *Sci. Rep.* **2017**, *7*, 1–9.
30. IUCN. Guidelines for Using the IUCN Red List Categories and Criteria; Version 15. 2022. Available online: http://www.iucnredlist.org/documents/RedListGuidelines.pdf (accessed on 2 April 2022).

diversity

MDPI

Article

Effect of Hermaphrodite–Gynomonoecious Sexual System and Pollination Mode on Fitness of Early Life History Stages of Offspring in a Cold Desert Perennial Ephemeral

Jannathan Mamut [1,2], Junhui Cheng [3], Dunyan Tan [1,2,*], Carol C. Baskin [1,4,5] and Jerry M. Baskin [1,4]

1 College of Life Sciences, Xinjiang Agricultural University, Urumqi 830052, China; jinaiti@163.com (J.M.); carol.baskin@uky.edu (C.C.B.); jerry.baskin@yahoo.com (J.M.B.)
2 Ministry of Education Key Laboratory for Western Arid Region Grassland Resources and Ecology, College of Grassland Sciences, Xinjiang Agricultural University, Urumqi 830052, China
3 College of Resources and Environment, Xinjiang Agricultural University, Urumqi 830052, China; cjhgraymice@126.com
4 Department of Biology, University of Kentucky, Lexington, KY 40506, USA
5 Department of Plant and Soil Sciences, University of Kentucky, Lexington, KY 40546, USA
* Correspondence: tandunyan@163.com

Abstract: Gynomonoecy, the occurrence of both pistillate (female) and perfect (hermaphroditic) flowers on the same plant, has received little attention compared to gynodioecy and other plant sexual systems. *Eremurus anisopterus* is a perennial ephemeral in the cold desert of northwest China with a hermaphrodite–gynomonoecious sexual system in the same population. The primary aim of this study was to compare the early life history traits and inbreeding depression between progeny from pistillate and hermaphrodite flowers in hermaphrodites and gynomonoecious individuals. All of the traits of progeny from outcrossed pistillate flowers on gynomonoecious plants were significantly greater than for other pollination types. Selfing (vs. outcrossing) resulted in a decrease in all traits, indicating inbreeding depression (ID) during early life history stages of gynomonoecious and hermaphroditic plants. ID for seed mass, seed germination and seedling survivorship under water stress for pistillate flowers on gynomonoecious plants was significantly higher than it was for hermaphrodite flowers on both gynomonoecious and hermaphrodite plants. The advantage of the offspring of pistillate (vs. hermaphrodite) flowers may contribute to the maintenance of gynomonoecy in *E. anisopterus* in its cold desert sand dune habitat.

Keywords: early seedling growth; *Eremurus anisopterus*; hermaphrodite–gynomonoecious sexual system; inbreeding depression; offspring fitness; seed germination; seed mass

Citation: Mamut, J.; Cheng, J.; Tan, D.; Baskin, C.C.; Baskin, J.M. Effect of Hermaphrodite–Gynomonoecious Sexual System and Pollination Mode on Fitness of Early Life History Stages of Offspring in a Cold Desert Perennial Ephemeral. *Diversity* **2022**, *14*, 268. https://doi.org/10.3390/d14040268

Academic Editors: Michael Wink, Yi-Gang Song and Hong-Hu Meng

Received: 26 January 2022
Accepted: 30 March 2022
Published: 1 April 2022

1. Introduction

Flowering plants exhibit a diversity of sexual systems that have combined male and female functions within and among individuals in seemingly every possible combination [1–3]. These sexual systems include hermaphroditism, monoecy, andromonoecy, gynomonoecy, dioecy, androdioecy and gynodioecy [3–5]. In recent decades, evolutionary biologists have been interested in the evolution and maintenance of diverse sexual systems because they might be an important mechanism in the promotion of outcrossing [6]. In flowering plants, this process is demonstrated by the transition from bisexual (hermaphrodite) to unisexual flowers and from bisexual to unisexual individuals [7]. Plants exhibit remarkable diversity in their sexual systems, which acts as a major driver of the genetic and evolutionary dynamics of flowering plants [8–10]. The evolution of plant sexual systems has been an important research issue since Darwin's studies [1]. Theoretical models suggest that the fitness consequences of selfing and outcrossing, the optimal allocation of resources to female and male functions, the genetic control of sex expression and environmental conditions may be the result of the selective pressure of sex expression and can affect the sexual systems of flowering plants [9,11–13].

Gynomonoecy refers to the co-occurrence of female (pistillate) and perfect (hermaphrodite) flowers on the same plant [2,14,15]. Within populations, occurrence of the two types of flowers on gynomonoecious plants is due to genetic effects, including incomplete restoration of a male-sterile cytoplasm by one or many nuclear loci, and maternal effects, pollination system, resource allocation and herbivory effects [16]. Several hypotheses have been proposed to explain the evolution and maintenance of gynomonoecy. (1) As in other systems with unisexual flowers, the presence of the two flower types may permit the flexible allocation of resources to female and male reproductive functions in response to variations in environmental factors [17–19]. (2) Pistillate flowers may outcross more than perfect flowers. The outcrossing rate of pistillate ray florets of the radiate morph was significantly greater than that of perfect disc florets of either the radiate or nonradiate morph [20]. (3) The advantage of the pistillate flowers in aster heads lies in their attractiveness to pollinators [14]. (4) Pistillate flowers would be favored if perfect flowers were susceptible to biased predation by florivores [3,21,22]. In recent decades, most studies on gynomonoecy have focused on flowering phenology and female fitness [23], breeding system [24,25] floral sex ratios [14,26], sexual phenotypes [27], sex allocation and reproductive success [5], phylogeny [4,7,28] and the effects of biotic and abiotic factors of the environment on sex expression [3,29] on plants with gynomonoecy or the co-occurrence of gynomonoecious and other sexual systems. However, few studies have examined the effect of gynomonoecious sexual systems on progeny performance [30].

Theoretical and empirical work suggests that inbreeding depression (ID) will coevolve with the mating system, resulting in a positive correlation between outcrossing rates and inbreeding depression [31–33], and this is thought to play a key role in the evolution of plant breeding systems [11,34–37]. ID can vary between stages of the plant life cycle, and several studies have compared ID for seed production, seed mass, germination, survival and other plant life history traits in gynodioecious species (coexistence of females (male-sterile) and hermaphroditic individuals) [7,38,39]. However, other than studies on the eudicot genus *Silene* (*S. acaulis* [40] and *S. nutans* [41]), little is known about the effects of selfing versus outcrossing on the progeny fitness of gynomonoecious species.

Eremurus anisopterus Regel. (Xanthorrhoeaceae, formerly Liliaceae), is an early spring flowering perennial ephemeral that occurs in the cold deserts of Kazakstan and northwest China [42]. In China, *E. anisopterus* grows on fixed and semi-fixed sand dunes of the Gurbantunggut Desert of northern Xinjiang. Generally, the shoots of *E. anisopterus* begin to emerge aboveground in late March [15], and the temperature is low and fluctuates greatly during this period [43]. This species has a hermaphrodite–gynomonoecious sexual system [5]. About 30% of the pollen received by perfect (hermaphrodite) flowers of *E. anisopterus* is outcrossed pollen, while more than 60% of that received by pistillate (female) flowers is outcrossed pollen (J. Mamut, unpublished data). Furthermore, pistillate flowers produce larger seeds than hermaphroditic flowers in gynomonoecious plants, thereby compensating (in part) for the loss of pollen (male gamete) production [5]. Thus, we hypothesized that the progeny of pistillate flowers would have a greater performance in the early life history stages than those from hermaphrodite flowers in both hermaphrodites and gynomonoecious individuals. To test this hypothesis, we compared the progeny performance and inbreeding depression from different pollination types of pistillate and hermaphrodite flowers in both hermaphrodites and gynomonoecious individuals under stressful conditions.

2. Materials and Methods

2.1. Study Species and Seed Collection Site

Each flowering individual of *E. anisopterus* produces one large raceme with dozens of flowers. The racemes of gynomonoecious individuals have basal pistillate and distal perfect flowers. Flowering begins from late April to early May. Flowers do not have nectar and are mainly pollinated by halictid bees and honeybees, which usually fly upwards when visiting receme inflorescences. Thus, pistillate flowers are more likely to produce outcrossed seeds [5]. *E. anisopterus* populations have both hermaphrodite (H) and gynomonoecious (Gm) plants, and the proportion of Gm individuals varies significantly from 2 to 17% among the populations [19].

Freshly matured seeds were collected from 900 plants in a large population of *E. anisopterus* growing on a cold desert sand dune in the Mosuowan region of the Gurbantunggut Desert (44°55′19.4″ N, 85°33′20.2″ E; 313 m a.s.l) in Xinjiang Uyghur Autonomous Region, northwest China. This region has a wide range of mean monthly temperatures. The coldest month is January (−23 °C), and the hottest month is July (33 °C), with a mean annual temperature of 7 °C. The annual precipitation (including rain and snow) is 147.3 ± 8.6 mm, with ~64% occurring in spring and summer. The monthly precipitation ranges from 5 (February) to 22 mm (July) (data from Mosuowan weather station, 1991–2010). Mean annual potential evaporation is 1942 mm [44].

2.2. Pollination Type

To examine the characteristics of seeds from different flower morph types and different types of pollination, we randomly labeled 900 hermaphroditic and gynomonoecious individuals in the population on 24–28 April 2014, and performed the following pollination treatments: (1) For hermaphrodite flowers, we tagged 300 hermaphroditic and 300 gynomonoecious individuals and established three pollination types (S, Op, Ou) of 100 hermaphrodite (H) and three pollination types of gynomonoecious (Gm) plants (see left and middle group of pollination types in Figure 1). Ten hermaphroditic (H) flowers were marked at the bud stage on each plant and enclosed in a mesh bag to avoid accidental pollination events. As soon as flowers opened, their stamens were cut, and the following hand pollinations were performed once stigmas became receptive: (a) open (designated as treatment Op), natural pollination without any floral manipulation; (b) self-pollinations were performed with pollen collected from other flowers of the same plant (S); or (c) outcross pollinations were made using a mixture of pollen from three hermaphrodites from the same population (Ou) (Figure 1). (2) For pistillate flowers, of the 300 selected Gm plants, we also marked 10 flowers in a basal position, using between four and seven pistillate (P) flowers on each plant, because *E. anisopterus* individuals generally produce between one and seven pistillate flowers. The pistillate flowers in each study group received three similar pollination types (see the group of pollination types on the right in Figure 1) as in the hermaphroditic plants. Flowers receiving treatments 2 and 3 were bagged after pollination to exclude any subsequent open pollination. Nonpollinated flowers were removed to keep the number of flowers per mother plant similar and to avoid resource allocation to nontarget flowers.

To easily distinguish the results of different plant, flower and pollination types, we used three-letter acronyms to refer to specific groups of results. The three letters of the acronym refer to the plant type (hermaphrodite (H) vs. gynomonoecious (Gm), the flower type (hermaphroditic flowers of H, hermaphroditic and pistillate flowers of Gm) and the pollination type (open pollination, selfing and outcrossing) (Figure 1). For example, GmP-S refers to seeds produced by a pistillate flower on a gynomonoecious plant that was self-pollinated.

2.3. Seed and Seedling Traits Measured

For seeds produced using the above treatments, we measured seed set, seed mass, embryo growth, germination percentage and seedling survival and growth.

2.3.1. Seed Set

Seed set was measured by randomly sampling one fruit each from 20 individuals of each pollination type, and the seed set per fruit was determined for all treatments.

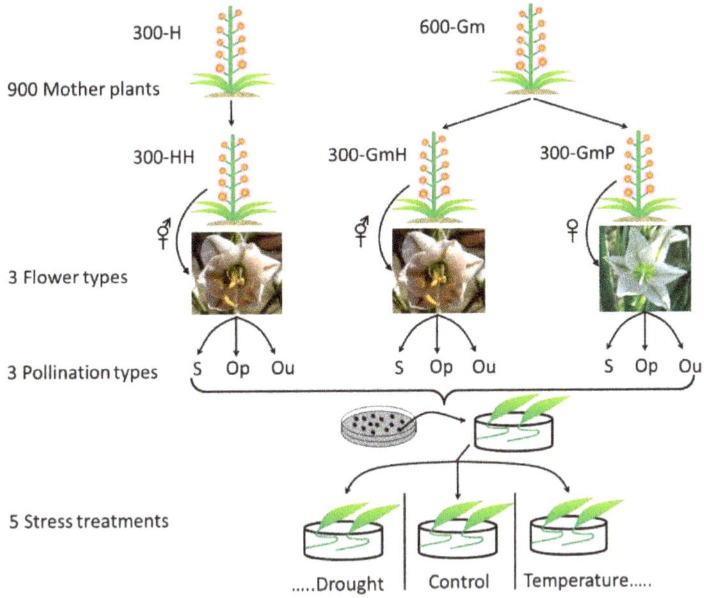

Figure 1. Schematic diagram showing the sequence of pollination types and stress treatments. H—hermaphrodite plants; Gm—gynomonoecious plants; HH—hermaphrodite flowers on hermaphroditic plants; GmH—hermaphrodite flowers on gynomonoecious plants; GmP—pistillate flowers on gynomonoecious plants. S—selfed; Op—open pollinated; Ou—crossed.

2.3.2. Seed Mass

Seed mass was measured by randomly sampling 30 seeds (one seed per fruit from each of 30 plants) from each pollination type, and the individual seeds were weighed using a Sartorius BS210S electronic balance (0.0001 g).

2.3.3. Phenology of Embryo Growth

Embryos in seeds of *E. anisopterus* are underdeveloped, and consequently they must grow inside the seed before radicle emergence (germination) [45]. To determine the initial length (size) of the embryo, 25 seeds from each pollination type were placed on moist filter paper in Petri dishes at room temperature on 27 June 2014. After 24 h, each seed was cut open with a razor blade, and embryo length and seed length (E:S ratio) were determined using a dissecting microscope equipped with a micrometer.

To monitor the phenology of embryo growth, 30 freshly mature seeds of each pollination type were placed in 10 fine-mesh nylon bags and buried in clay pots (30 cm diameter, 30 cm height) to a depth of 3 cm in sand from the natural habitat of *E. anisopterus*. The pots were buried in soil (top of pot level with soil surface) in an experimental garden on the campus of Xinjiang Agricultural University, Urumqi, on 27 June 2014. Sand in the pots was kept moist throughout the monitoring period. After burial, 25 or 30 seeds (some seeds decayed during burial) were removed from one haphazardly selected bag every 15 d until 26 October 2014 and the E:S ratio was determined as described above. Sand temperatures at a depth of 3 cm were recorded at 2 h intervals throughout the burial period using a Tiny Tag data logger (Model Micro Lite LITE5016, Fourier Technologies, Beijing, China). Daily mean, maximum and minimum temperatures were calculated from these data.

2.3.4. Seed Germination

One hundred freshly mature seeds of each of the nine levels of pollination types were sown on wet filter paper with 12 cm diameter Petri dishes and cold stratified at

5/2 °C (12/12 h) for 8 weeks in darkness [45]. After cold stratification, four replicates of 25 seeds each were transferred to 9 cm diameter Petri dishes under green light and incubated in darkness at 15/2 °C for 30 days. Seeds incubated in darkness were checked after 30 d for germination, and distilled water was added every 7 d (in green light). After the germination trials were complete, non-germinated seeds were tested for viability, and germination percentages were calculated based on number of viable seeds [45].

2.3.5. Seedling Growth and Survival

To compare the early growth of seedlings from the nine pollination types, we used seeds that had been stored dry at room conditions for 6 months. One hundred seeds from each of the nine pollination types were placed on moist sand in 12 cm diameter Petri dishes and cold stratified at 5/2 °C for 8 weeks in darkness. After treatment, four replicates of 25 seeds were transferred under green light to new 12 cm diameter Petri dishes and incubated in darkness at the 15/2 °C temperature regime for 4 weeks. These procedures have been shown to break seed dormancy, thus allowing for the measurement of subsequent seedling traits.

In addition to growing seedlings in the 15/2 °C optimal condition, we also examined seedlings' response to two types of stressful conditions often found in the natural environment of this species—specifically, high temperature and low soil water content.

Temperature

For these treatments, we transferred four replicates of 10 3-week-old seedlings from moist filter paper in Petri dishes into a hydroponic system. In this system, individual seedlings were placed into 1 cm diameter holes (1 seedling per hole) bored into round pieces of styrofoam (1 cm thick templates) that covered the Petri dishes (12 cm diameter, 2.5 cm deep) containing 80 mL of 1/2 strength Hoagland nutrient solution. Roots of the seedlings reached into the nutrient solution, which was replaced with fresh Hoagland solution 2 weeks after the initial transplantation, at the mid-point of the experiment. Using this setup, we incubated two groups of seedlings at daily (12/12 h) temperature regimes of 5/2 and 20/10 °C in light for 28 d, representing suboptimal, and superoptimal temperature, respectively. The optimum temperature regime for the seed germination of *E. anisopterus* is 15/2 °C [45]. Thus, we grew an additional group of seedlings at 15/2 °C to serve as the control. Seedlings were arranged haphazardly, and all Petri dishes were rotated several times weekly to reduce the possibility of positional effects. Seedling survival was monitored at 1-week intervals and height, root length and mass of seedlings were measured only at the end of the experiment.

Water Stress

The design used to grow seedlings under water stress was very similar to the one used to test the effect of temperature stress on seedling growth. Three replication of 10 3-week-old seedlings were transferred from the moist filter paper into Petri dishes containing 0 (control, 1/2 strength Hoagland nutrient solution only (HNS)), −0.15 and −0.51 MPa PEG (HNS+ PEG) solutions at 15/2 °C for 28 d [46]. Seedling survival was monitored at 1-week intervals and height, root length and mass of seedlings were measured only after 28 d.

2.4. *Effect of Inbreeding Depression on Seed Set, Mass, Germination and Seedling Growth and Survival*

Using the measurements from different pollination types, we calculated the inbreeding depression (δ) for individual traits (seed set, seed mass, embryo growth, seed germination and seedling growth and survival) using the equation $\delta = 1 - (W_S/W_O)$, where W_S and W_O were the measurement of individual fitness traits for selfed and out-crossed progeny, respectively. Thus, (W_S/W_O) is relative fitness [47,48]. In this study, W_S was lower than W_O in all cases; thus, our measurement of inbreeding depression was always positive.

2.5. Multiplicative Fitness of Early Life History Traits

The effects of inbreeding are cumulative (multiplicative) across the life cycle [48]. Thus, the cumulative fitness of the early life history stages of *E. anisopterus* was estimated by calculating the product of relative fitness (W_s/W_o) for seed set, seed mass, embryo growth, germination percentage and seedling survival/growth (cumulative relative fitness, CRF); 1—CRF = total ID for these three stages of *E. anisopterus*.

All calculations for inbreeding depression were conducted separately for the three types of flowers: hermaphrodite and pistillate flowers on gynomonoecious plants, and hermaphrodite flowers on hermaphrodite plants.

2.6. Data Analysis

A linear-mixed model was used to explore the effects of explanatory variable (plant type, flower type and pollination type) on response variable (seed set, seed mass, embryo growth and seed germination). In the linear-mixed model, fixed effects were detected by the main and all potential interactive effects among these explanatory variables except for the interactive effect between plant and flower types, since flower type was not fully crossed with plant type. Plant type was also considered a random factor nested with individuals, which was randomly sampled. The linear-mixed model was conducted by "lme" function in "nlme" packages in R software [49]. If fixed factors had significant main or interactive effects in the linear-mixed model, we further used the "lsmean" function in "lsmean" package to conduct multiple comparisons [50].

To express the relative importance of each explanatory variable on response variables, we firstly calculated adjusted R^2 for the whole linear-mixed model. Then, the main and interactive effects of each explanatory variable were removed from the whole model, and their adjusted R^2 was calculated as suggested by de Vries et al. [51]. The relative importance of each explanatory variable was depicted by its adjusted R^2 calculated from the removed model [51]. In particular, the adjusted R^2 was calculated by the "r. squared LR" function in "MuMIn" packages [52]. Finally, we used the linear regression to examine the relationship between seed mass and seed germination and seedling traits.

3. Results

3.1. Seed Traits

The linear-mixed model results revealed that flower type and pollination type significantly affected all measured traits except embryo growth (however, flower type significantly affected the final E:S ratio) (Table 1). There were no significant effects of plant type on seed mass or seed germination, nor was there a significant interaction between flower type and pollination type or between plant type and pollination type. However, plant type significantly affected seed set (Table 1). More specifically, flowers that were cross-pollinated produced more (Table 1; Figure 2A) and larger (Table 1; Figure 2B) seeds than self-pollinated flowers, and the open pollinated flowers exhibited intermediate trait values in all cases. The germination of cross-pollinated flowers was significantly higher (Table 1; Figure 2C) than that of self-pollinated flowers except HH.

3.2. Phenology of Embryo Growth

The initial E: S ratio (28 June) for fresh seeds of each pollination type ranged from 0.707 ± 0.009 for selfed seeds of HH to 0.735 ± 0.018 for outcrossed seeds of GmP, and differences among pollination types were not significant (Table 1, Figure 3).

Embryos grew only a little during summer 2014; however, between 11 September and 26 October 2014, during which time mean daily maximum and minimum sand temperatures were 14.8 and 4.6 °C, respectively, embryos grew rapidly (Figure 3A). The final E: S ratio (26 October) ranged from 0.868 ± 0.007 for HH-S to 0.899 ± 0.007 for GmP-Ou, and the differences among the flower types were significant (Table 1). Embryos from GmP-Ou and GmP-Op reached the critical E:S ratio required for germination on 26 October 2014, at which time 2% of the seeds in the bags had germinated.

Table 1. Results of the linear mixed model showing the effect of plant type (hermaphroditic vs. gynomonoecious), flower type (hermaphroditic flowers of H, hermaphroditic and pistillate flowers of Gm), pollination type (open pollination, selfing and outcrossing) and their interactions in terms of seed set, seed mass and seed germination. Effects with $p < 0.05$ are shown in boldface type.

Source of Variation	d.f.	*F*	*p*
Seed set			
Plant type (PT)	1, 19	20.799	**0.0002**
Flower type (FT)	1, 133	86.986	**<0.0001**
Pollination type (PT′)	2, 133	11.448	**<0.0001**
FT * PT′	2, 133	10.258	**0.0001**
PT * PT′	2, 133	0.0241	0.976
Seed mass			
Plant type (PT)	1, 29	2.215	0.147
Flower type (FT)	1, 203	44.407	**<0.0001**
Pollination type (PT′)	2, 203	32.104	**<0.0001**
FT * PT′	2, 203	0.967	0.382
PT* PT′	2, 203	1.359	0.259
Initial E:S ratio			
Plant type (PT)	1, 218	0.620	0.433
Flower type (FT)	1, 218	0.300	0.255
Pollination type (PT′)	2, 218	0.130	0.324
FT * PT′	2, 218	0.010	0.987
Final E:S ratio			
Plant type (PT)	1, 218	1.230	0.268
Flower type (FT)	1, 218	5.110	0.024
Pollination type (PT′)	2, 218	1.580	0.209
FT * PT′	2, 218	0.920	0.401
Seed germination			
Plant type (PT)	1, 3	9.320	0.055
Flower type (FT)	1, 21	25.329	**0.0001**
Pollination type (PT′)	2, 21	24.533	**<0.0001**
FT * PT′	2, 21	1.4868	0.249
PT* PT′	2, 21	0.8807	0.429

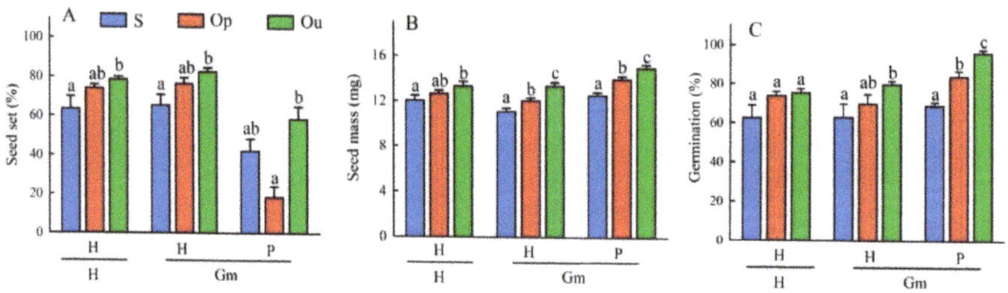

Figure 2. Mean (± SE) seed set (**A**), seed mass (**B**), and germination percentage (**C**) of seeds from selfed (S), open pollinated (Op) and crossed (Ou) flowers of *Eremurus anisopterus*. HH, hermaphrodite flowers on hermaphrodite plants; GmH, hermaphrodite flowers on gynomonoecious plants; GmP, pistillate flowers on gynomonoecious plants. Different letters above each group indicate significant differences between pollination types within each flower type ($p < 0.05$).

3.3. *Seedling Traits*

Flower type, pollination type and stress treatment had significant effects on seedling survival and biomass (Table 2). There were no significant effects of plant type on survival, but plant type significantly affected biomass. However, flower type, pollination type and stress treatment interactions ($p < 0.05$) had no significant effects on seedling survival or flower type, and pollination type and stress treatment interactions ($p < 0.05$) had no significant effects on biomass (Table 2). Seedling survival and biomass were higher for the control than for seedlings exposed to stress treatments. Selfed progeny had significantly lower survivorship and lower biomass than outcrossed progeny incubated in –0.51 MPa in three flower types except the GmH in biomass. Seedlings from the two groups of hermaphroditic flowers (from the H and Gm plants) were similar in terms of survival and biomass (Figure 4).

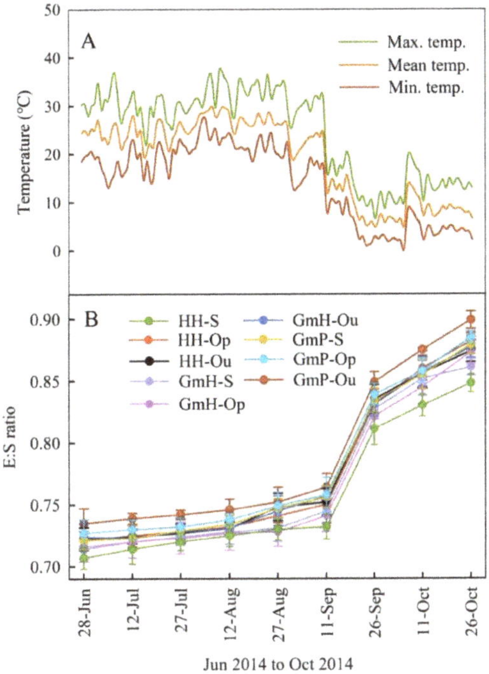

Figure 3. Daily maximum, daily minimum and daily mean sand temperatures at a depth of 3 cm (**A**) and phenology of embryo growth in seeds from each pollination type in *Eremurus anisopterus* and (**B**) in the experimental garden. Error bars in (**B**) are ±s.e.

Table 2. Results of linear mixed model showing the effect of plant type (hermaphroditic vs. gynomonoecious), flower type (hermaphroditic flowers of H, hermaphroditic and pistillate flowers of Gm), pollination type (open pollination, selfing and outcrossing), stress treatment (control, 2 temperatures, and 2 water stresses) and their interactions in terms of seedling survival and seedling biomass. Effects with $p < 0.05$ are shown in boldface type.

Source of Variation	Survival			Biomass		
	d.f.	*F*	*p*	d.f.	*F*	*p*
Plant type (PT)	1, 2	4.939	0.1563	1, 19	44.918	**<0.0001**
Flower type (FT)	1, 86	10.623	**0.0016**	1, 817	131.391	**<0.0001**
Pollination type (PT′)	2, 86	26.427	**<0.0001**	2, 817	126.087	**<0.0001**
Stress treatment (ST)	4, 86	124.124	**<0.0001**	4, 817	333.552	**<0.0001**
FT * PT′	2, 86	1.213	0.3023	2, 817	5.477	**0.0043**
FT * ST	4, 86	3.896	**0.0059**	4, 817	2.831	**0.0238**
CT * ST	8, 86	3.53	**0.0014**	8, 817	1.402	0.1918
PT * PT′	2, 86	3.049	0.0525	2, 817	6.697	0.0013
PT * ST	4, 86	3.721	**0.0077**	4, 817	0.074	0.9901
FT * PT′ * ST	8, 86	0.257	0.9778	8, 817	1.199	0.2967
PT * PT′ * ST	8, 86	0.426	0.9023	8, 817	0.058	0.9999

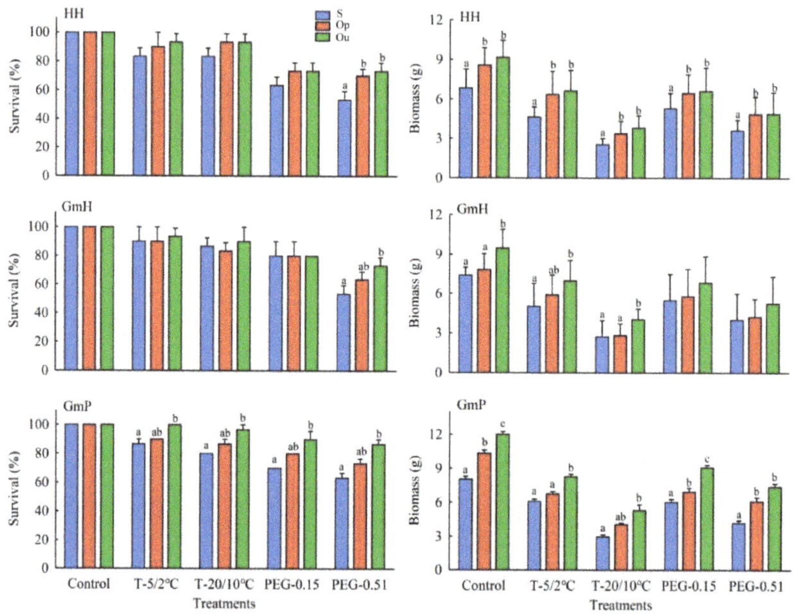

Figure 4. Mean (±SE) seedling survival (left) and biomass (right) of offspring from seeds of selfed (S), open pollinated (Op) and crossed (Ou) flowers of *Eremurus anisopterus* under different stress conditions. HH, hermaphrodite flowers on hermaphrodite plants; GmH, hermaphrodite flowers on gynomonoecious plants; GmP, pistillate flowers on gynomonoecious plants. Different letters above each group indicate significant differences between pollination types within each treatment ($p < 0.05$). Marginally significant differences are not shown here.

3.4. *Effect of Seed Mass on Germination and Seedling Traits*

Germination percentage of different pollination type seeds from H and P flowers from hermaphrodite and gynomonoecious plants was positively correlated with seed mass ($R^2 = 0.93$, $p < 0.001$) (Figure 5). Outcrossed seeds from pistillate flowers on gynomonoecious plants were significantly larger than those of selfed seeds from gynomonoecious and hermaphroditic plants. ANCOVA showed that germination percentage was independent of seed mass (Table 3).

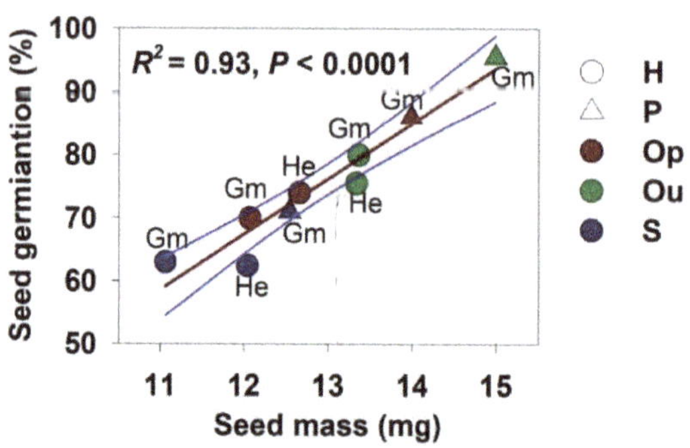

Figure 5. Relationship between seed mass and seed germination in *Eremurus anisopterus*. Red lines show linear regressions, and blue lines show ± 95% confidence intervals. He—hermaphrodite plants; Gm—gynomonoecious plants; H—hermaphrodite flowers P—pistillate flowers; S—selfed—Op—open pollinated; Ou—crossed.

Table 3. Analysis of covariance (ANCOVA) of the effect of seed size on germination.

Source of Variation	d.f.	MS	*F*	*p*
Pollination type	8	202.272	3.239	0.011
Seed mass	1	49.132	0.787	0.383
Error	26	62.456		
$R^2 = 0.588$				

Seed mass tended to affect seedling survival and biomass, and its effect varied under stressful conditions. However, there was no significant effect in the control and PEG —0.15. The survival and biomass of selfed progeny from larger outcrossed seeds from pistillate flowers on gynomonoecious plants (which were larger than selfed seeds from gynomonoecious and hermaphroditic plants) was significantly higher than that of selfed seeds from both gynomonoecious and hermaphroditic plants (which were smaller than outcrossed seeds from female flowers on gynomonoecious plants), especially for outcrossed progeny from pistillate flowers on gynomonoecious plants under PEG —0.51 MPa (Figure 6).

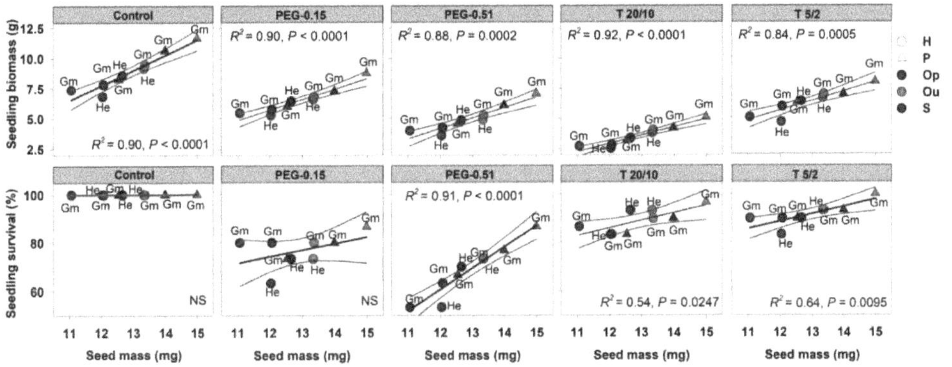

Figure 6. Relationship between seed mass and seedling survival and biomass in *Eremurus anisopterus*. Red lines show linear regressions, and blue lines show ±95% confidence intervals. NS, no significant difference. Abbreviations as in Figure 5.

3.5. *Effect of Inbreeding Depression on Seed and Seedling Traits*

Inbreeding depression for seed mass and germination was much stronger for GmP than for GmH and HH. However, seed set and embryo growth did not differ in inbreeding depression among three flower types. All of the germinated seeds survived in the control, suggesting that inbreeding did not affect survival under benign conditions. Under suboptimal and superoptimal temperatures (5/2 °C and 20/10 °C), seedling survivor exhibited weak inbreeding depression. However, under water stress (−0.51 Mpa) seedling survivorship exhibited a much stronger inbreeding depression (Table 4).

Total ID for the early growth stages for GmH, GmP and HH ranged from 0.53 to 0.65, 0.59 to 0.80 and 0.46 to 0.64, respectively (Table 5).

Table 4. Inbreeding depression for seed set, seed mass, embryo growth, germination and seedling survival of *Eremurus anisopterus* under different stress conditions. GmH—hermaphrodite flowers on gynomonoecious plants; GmP—pistillate flowers on gynomonoecious plants; HH—hermaphrodite flowers on hermaphrodite plants. S-C; seedling survivorship in control (1/2 strength Hoagland nutrient solution); S-T—seedling survivorship under temperature stress; S-W—seedling survivorship under water stress. Different letters within a row indicate significant differences (Tukey's HSD, $p < 0.05$).

Treatment	Flower Type			F	p
	GmH	GmP	HH		
Seed set	0.188 ± 0.082^{a}	0.188 ± 0.073^{a}	0.164 ± 0.211^{a}	0.010	0.990
Seed mass	0.102 ± 0.006^{a}	0.165 ± 0.007^{b}	0.097 ± 0.009^{a}	26.219	0.000
Initial E:S ratio	0.024 ± 0.006^{a}	0.026 ± 0.013^{a}	0.021 ± 0.008^{a}	0.061	0.941
Final E:S ratio	0.019 ± 0.002^{a}	0.024 ± 0.006^{a}	0.017 ± 0.004^{a}	0.502	0.607
Germination	0.205 ± 0.031^{a}	0.256 ± 0.017^{b}	0.202 ± 0.046^{a}	16.523	0.003
S-C	0.000 ± 0.000^{a}	0.000 ± 0.000^{a}	0.000 ± 0.000^{a}	—	—
S-T 5/2 °C	0.086 ± 0.016^{a}	0.088 ± 0.003^{a}	0.062 ± 0.003^{a}	2.353	0.176
S-T 20/10 °C	0.074 ± 0.016^{a}	0.077 ± 0.021^{a}	0.052 ± 0.003^{a}	0.773	0.503
S-W —0.15 Mpa	0.129 ± 0.012^{a}	0.157 ± 0.007^{b}	0.125 ± 0.010^{a}	13.251	0.001
S-W —0.51 Mpa	0.268 ± 0.016^{a}	0.294 ± 0.005^{b}	0.271 ± 0.030^{a}	12.933	0.003

Table 5. Total inbreeding depression (TID) for seed set, seed mass, final E:S ratio, seed germination and seedling survival/growth of *Eremurus anisopterus* under different environmental stresses. GmH—hermaphrodite flowers on gynomonoecious plants; GmP—pistillate flowers on gynomonoecious plants; HH—hermaphrodite flowers on hermaphrodite plants. S-C; seedling survivorship in control (1/2 strength Hoagland nutrient solution); S-T—seedling survivorship under temperature stress; S-W—seedling survivorship under water stress.

Treatment	GmH	GmP	HH
S-C	0.529	0.719	0.567
S-T 5/2 °C	0.604	0.714	0.597
S-T 20/10 °C	0.605	0.593	0.627
S-W —0.15 Mpa	0.566	0.772	0.464
S-W —0.51 Mpa	0.648	0.798	0.639

4. Discussion

All traits of outcross seeds from pistillate flowers on gynomonoecious plants were significantly higher than they were for those of the other pollination types. Thus, our hypothesis that the life history traits of pistillate flowers are more fit and that ID would be higher for the progeny of pistillate flowers than for those of hermaphrodite flowers in both hermaphrodites and gynomonoecious individuals is supported.

Seeds produced by different pollination types may affect the fitness of the post-germination traits of the progeny [47,53,54]. In natural populations of *E. anisopterus*, pistillate flowers received more outcross pollen than perfect flowers, resulting in 60% of the seeds being outcrossed, while perfect flowers received more pollinator visits than pistillate flowers but had a higher proportion of geitonogamy, resulting in 30% of the seeds being outcrossed (J. Mamut, unpublished data). Hence, the presence of pistillate flowers on gynomonoecious plants could help reduce selfing rates, in support of the outcrossing hypothesis of gynomonoecy.

Seed mass is important in the life history of plants [55–57], because it has a strong influence on seed germination, seedling emergence, survivorship, seedling size and seedling competitive ability [56,58]. Various studies have shown that seeds from female flowers of gynodioecious species are larger than those of perfect flowers [30,59]. Perfect flowers of *E. anisopterus* produced significantly more seeds than pistillate flowers [5]. However, the mass of individual outcrossed seeds from pistillate flowers on gynomonoecious plants was significantly greater than that from the other pollination types (Figure 2B, Table 1), suggesting that pollination type had a significant effect on seed mass and that the pistillate flowers on plants of this gynomonoecious species increase female function via seed quality.

Seeds of *E. anisopterus* are dispersed in late June and, at this time, the embryos of all pollination types of seeds are underdeveloped. The embryos began to elongate in summer, but they grew slowly. Between mid-September and late October, embryos grew rapidly and reached their critical length for radical emergence (germination) [5]. The critical E: S ratio of GmP-Ou was significantly greater than that of other pollination types, indicating that embryos of outcrossed seeds of pistillate flowers (GmP-Ou) elongate more than those of selfed (GmH-S, GmP-S and HH-S) embryos. Ours is the first study on the effect of flower cross-type on embryo growth.

Some species described as gynodioecious are truly gynomonoecious–gynodioecious, with three phenotypes: females, perfect-flowered hermaphrodites and gynomonoecious individuals [41,60]. In the gynomonoecious–gynodioecious species *Silene acaulis,* seeds from three flower types (pistillate flowers on females, pistillate flowers on gynomonoecious and perfect flowers on gynomonoecious plants) did not differ significantly in germination percentage in either the greenhouse or field [40]. In another gynomonoecious–gynodioecious species, *S. nutans,* Dufay et al. [41] observed that crossed seeds of hermaphrodites germinated to a higher percentage than selfed seeds of both gynomonoecious and hermaphrodite plants. In *E. anisopterus,* the germination percentage and seedling performance of larger outcrossed seeds from pistillate flowers on Gm plants were significantly higher than those of selfed seeds from hermaphrodite flowers on both Gm and H plants [56,61,62].

The relative frequency of selfing and outcrossing influences the offspring fitness and genetic diversity of flowering plants [63]. The production of pistillate flowers enhances the opportunity for outcrossing. Many studies of gynodioecious species found that the offspring of females germinates to a higher percentage and grows larger than the offspring of hermaphrodites [38,64,65]. In *E. anisopterus,* pollination type had a significant effect on early seedling growth and survival. Under stressful conditions, seedlings from cross-pollinated pistillate flowers on gynomonoecious plants grew and survived better than those of seedlings from the other pollination types (Figure 4). This suggests that seedlings from female flowers on gynomonoecious plants have more potential to grow and survive in the unpredictable environment (with regard to timing and amount of precipitation) of the cold deserts in Central Asia in early spring than those from perfect flowers of gynomonoecious or hermaphroditic plants.

Inbreeding depression is thought to be a major selective factor in the evolution of reproductive system diversity in flowering plants, particularly in maintaining outcrossing in spite of the automatic gene transmission advantage of self-fertilization [66–68]. For *E. anisopterus,* we found inbreeding depression in gynomonoecious and hermaphroditic plants for different life cycle traits. A decrease in both seed mass and germination percentage resulting from selfing in *E. anisopterus* is consistent with the results reported for gynomonoecious–gynodioecious species *S. acaulis* [40], *Dianthus sylvestris* [69] and *S. nutans* [41]. Furthermore, we found significant inbreeding depression in GmH, GmP and HH for different life cycle stages. ID for seed mass, germination and seedling survivorship under water stress for GmP was significantly higher than ID for GmH and HH. However, there was no difference between GmH and HH, suggesting that flower type had a significant effect on inbreeding depression. ID may or may not increase with stress (See Table 1 in [48] for references on the effect of competition and physical environment on inbreeding depression in plants). For *E. anisopterus,* inbreeding depression was found for the survivorship of seedlings under environmental stress, and especially for those under water stress. The outcrossed progeny of *E. anisopterus* had higher survivorship than selfed progeny, suggesting that, for this species, recessive alleles are more deleterious under stressful conditions [39].

5. Conclusions

Several hypotheses have been proposed to explain the evolution and maintenance of gynomonoecy. They suggest that this sexual system may enhance outcrossing [5,21,69], avoid pollen–pistil interference [14,26], act as a defense against herbivores [3,21,23], enhance attractiveness to pollinators [21,70] and/or permit flexible resource allocation [17]. Our study shows that total ID was higher for the progeny of pistillate flowers than for those of hermaphrodite flowers in hermaphrodites and gynomonoecious individuals across the early history stages. Seed mass, seed germination and seedling survival/growth of outcrossed seeds from pistillate flowers on gynomonoecious plants were significantly higher than they were for those of the other pollination types. Thus, the female flower has an advantage over the hermaphrodite flower in *E. anisopterus*. These results are consistent with the outcrossing-benefit hypothesis for gynomonoecy, which proposes that female flowers of gynomonoecious plants can partially avoid inbreeding depression by favoring outcrossing rates [21,29].

Author Contributions: J.M.: conceptualization (equal), investigation (equal), methodology (equal), software (equal), writing—original draft (lead); J.C.: conceptualization (supporting), methodology (supporting), software (equal); D.T.: conceptualization (equal), methodology (equal), writing—review and editing (lead); C.C.B.: conceptualization (equal), methodology (equal), writing—review and editing (lead); J.M.B.: conceptualization (equal), methodology (equal), writing—review and editing (lead). All authors have read and agreed to the published version of the manuscript.

Funding: This work was supported in part by National Natural Science Foundation of China (31960053), Tianshan Youth Program of Xinjiang Uygur Autonomous Region (2019Q019) and National Natural Science Foundation of China (U1603231).

Institutional Review Board Statement: Not applicable.

Informed Consent Statement: Not applicable.

Data Availability Statement: Not applicable.

Acknowledgments: We thank Shu-Mei Chang from University of Georgia for valuable comments that helped us to improve the manuscript.

Conflicts of Interest: The authors declare no conflict of interest.

References

1. Darwin, C. *The Different Forms of Flowers on Plants of the Same Species*; John Murray: London, UK, 1877.
2. Yampolsky, C.; Yampolsky, H. Distribution of sex forms in the phanerogamic flora. *Bibl. Genet.* **1922**, *3*, 1–62.
3. Wise, M.J.; Coffey, L.E.; Abrahamson, W.G. Nutrient stress and gall flies interact to affect floral-sex ratio in gynomonoecious *Solidago altissima* (Asteraceae). *Am. J. Bot.* **2008**, *95*, 1233–1239. [CrossRef] [PubMed]
4. Torices, R.; Anderberg, A.A. Phylogenetic analysis of sexual systems in Inuleae (Asteraceae). *Am. J. Bot.* **2009**, *96*, 1011–1019. [CrossRef] [PubMed]
5. Mamut, J.; Xiong, Y.Z.; Tan, D.Y.; Huang, S.Q. Pistillate flowers experience more pollen limitation and less geitonogamy than perfect flowers in a gynomonoecious herb. *New Phytol.* **2014**, *201*, 670–677. [CrossRef] [PubMed]
6. Charlesworth, B.; Charlesworth, D. A model for the evolution of dioecy and gynodioecy. *Am. Nat.* **1978**, *112*, 975–997. [CrossRef]
7. Chang, S.M. Gender-specific inbreeding depression in a gynodioecious plant, *Geranium maculatum* (Geraniaceae). *Am. J. Bot.* **2007**, *94*, 1193–1204. [CrossRef]
8. Méndez, M.; Munzinger, J. *Planchonella*, first record of gynomonoecy for the family Sapotaceae. *Plant Syst. Evol.* **2010**, *287*, 65–73. [CrossRef]
9. Charlesworth, D. *Gender and Sexual Dimorphism in Flowering Plants*; Geber, M.A., Dawson, T.E., Delph, L.F., Eds.; Springer: Berlin, Germany, 1999; pp. 33–60.
10. Charlesworth, D. Evolution of plant breeding systems. *Curr. Biol.* **2006**, *16*, R726–R735. [CrossRef]
11. Barrett, S.C.H. The evolution of plant sexual diversity. *Nat. Rev. Genet.* **2002**, *3*, 274–284. [CrossRef]
12. Hesse, E.; Pannell, J.R. Sexual dimorphism in a dioecious population of the wind-pollinated herb Mercurialis annua: The interactive effects of resource availability and competition. *Ann. Bot.* **2011**, *107*, 1039–1045. [CrossRef]
13. Barrett, S.C.H.; Hough, J. Sexual dimorphism in flowering plants. *J. Exp. Bot.* **2013**, *64*, 67–82. [CrossRef] [PubMed]
14. Bertin, R.I.; Kerwin, M. Floral sex ratios and gynomonoecy in *Aster* (Asteraceae). *Am. J. Bot.* **1998**, *85*, 235–244. [CrossRef] [PubMed]

15. Grombone-Guaratini, M.T.; Nascimento, A.A.D.; Santos-Gonçalves, A.P. Flowering and fruiting of *Aulonemia aristulata*: A gynomonoecious woody bamboo species from Atlantic Forest in Brazil. *Rev. Bras. Bot.* **2011**, *34*, 135–140. [CrossRef]
16. Lu, Y.; Huang, S.Q. Adaptive advantages of gynomonoecious species. *Acta Phytotaxon. Sin.* **2006**, *44*, 231–239, (In Chinese with English Abstract). [CrossRef]
17. Charnov, E.L.; Bull, J. When is sex environmentally determined? *Nature* **1977**, *266*, 828–830. [CrossRef] [PubMed]
18. Lloyd, D.G. Parental strategies in angiosperms. *N. Zeal. J. Bot.* **1979**, *17*, 595–606. [CrossRef]
19. Mamut, J.; Xiong, Y.Z.; Tan, D.Y.; Huang, S.Q. Flexibility of resource allocation in a hermaphroditic-gynomonoecious herb through deployment of female and male resources in perfect flowers. *Am. J. Bot.* **2017**, *104*, 461–467. [CrossRef]
20. Marshall, D.F.; Abbott, R.J. Polymorphism for outcrossing frequency at the ray floret locus in *Senecio vulgaris* L. III. Causes. *Heredity* **1984**, *53*, 145–149. [CrossRef]
21. Bertin, R.I.; Connors, D.B.; Kleinman, H.M. Differential herbivory on disk and ray flowers of gynomonoecious asters and goldenrods (Asteraceae). *Biol. J. Linn. Soc.* **2010**, *101*, 544–552. [CrossRef]
22. Zhang, G.F.; Xie, T.P.; Du, G.Z. Variation in floral sex allocation, female success, and seed predation within racemiform synflorescence in the gynomonoecious *Ligularia virgaurea* (Asteraceae). *J. Plant Res.* **2012**, *125*, 527–538. [CrossRef]
23. Collin, C.L.; Shykoff, J.A. Flowering phenology and female fitness: Impact of a pre-dispersal seed predator on a sexually polymorphic species. *Plant Ecol.* **2010**, *206*, 1–13. [CrossRef]
24. Lloyd, D.G. Breeding systems in *Cotula* L. (Compositae, Anthemideae). I. The array of monoclinous and diclinous systems. *New Phytol.* **1972**, *71*, 1181–1194. [CrossRef]
25. Álvarez, I.; Agudo, A.B.; Herrero, A.; Torices, R. The Mendelian inheritance of gynomonoecy: Insights from *Anacyclus* hybridizing species. *Am. J. Bot.* **2020**, *107*, 116–125. [CrossRef] [PubMed]
26. Bertin, R.I.; Gwisc, G.M. Floral sex ratios and gynomonoecy in *Solidago* (Asteraceae). *Biol. J. Linn. Soc.* **2002**, *77*, 413–422. [CrossRef]
27. Maurice, S. Gynomonoecy in *Silene italica* (Caryophyllaceae): Sexual phenotypes in natural populations. *Plant Biol.* **1999**, *1*, 346–350. [CrossRef]
28. Swenson, U.; Bremer, K. Patterns of floral evolution of four *Asteraceae* genera (Senecioneae, Blennospermatinae) and the origin of white flowers in New Zealand. *Syst. Biol.* **1997**, *46*, 407–425. [CrossRef]
29. Abbott, R.J.; Schmitt, J. Effect of environment on percentage female ray florets per capitulum and outcrossing potential in a self-compatible composite (*Senecio vulgaris* L. var. *hibernicus* Syme). *New Phytol.* **1985**, *101*, 219–229. [CrossRef]
30. Collin, C.L.; Pennings, P.S.; Rueffler, C.; Widmer, A.; Shykoff, J.A. Natural enemies and sex: How seed predators and pathogens contribute to sex-differential reproductive success in a gynodioecious plant. *Oecologia* **2002**, *131*, 94–102. [CrossRef]
31. Lande, R.S.; Schemske, D.W. The evolution of selffertilization and inbreeding depression in plants. *I. Genetic models. Evolution* **1985**, *39*, 24–40.
32. Charlesworth, D.; Morgan, M.T.; Charlesworth, B. Inbreeding depression, genetic load, and the evolution of outcrossing rates in a multilocus system with no linkage. *Evolution* **1990**, *44*, 1469–1489. [CrossRef]
33. Uyenoyama, M.K.; Holsinger, K.E.; Waller, D.M. Ecological and genetic factors directing the evolution of selffertilization. *Oxford Surv. Evol. Biol.* **1993**, *9*, 327–381.
34. Charlesworth, D.; Charlesworth, B. Inbreeding depression and its evolutionary consequences. *Annu. Rev. Ecol. Syst.* **1987**, *18*, 237–268. [CrossRef]
35. Hokanson, K.; Hancock, J. Early-acting inbreeding depression in three species of *Vaccinium* (Ericaceae). *Sex. Plant Reprod.* **2000**, *13*, 145–150. [CrossRef]
36. Keller, L.F.; Waller, D.M. Inbreeding effects in wild populations. *Trends Ecol. Evol.* **2002**, *17*, 231–241. [CrossRef]
37. Goodwillie, C.; Kalisz, S.; Eckert, C.G. The evolutionary enigma of mixed mating systems in plants: Occurrence, theoretical explanations, and empirical evidence. *Annu. Rev. Ecol. Evol.* **2005**, *36*, 47–79. [CrossRef]
38. Ashman, T.L. The relative importance of inbreeding and maternal sex in determining progeny fitness in *Sidalcea oregana* ssp. spicata, a gynodioecious plant. *Evolution* **1992**, *46*, 1862–1874. [CrossRef] [PubMed]
39. Sedlacek, J.; Schmid, B.; Matthies, D.; Albrecht, M. Inbreeding depression under drought stress in the rare endemic *Echium wildpretii* (Boraginaceae) on Tenerife, Canary Islands. *PLoS ONE* **2012**, *7*, e47415. [CrossRef]
40. Delph, L.F.; Mutikainen, P. Testing why the sex of the maternal parent affects seedling survival in a gynodioecious species. *Evolution* **2003**, *57*, 231–239. [CrossRef]
41. Dufay, M.; Lahiani, E.; Brachi, B. Gender variation and inbreeding depression in gynodioecious-gynomonoecious *Silene nutans* (Caryophyllaceae). *Int. J. Plant. Sci.* **2010**, *171*, 53–62. [CrossRef]
42. Chen, X.Q.; Liang, S.Y.; Xu, J.M.; Tamura, M.N. Liliaceae. In *Flora of China*; Wu, Z.Y., Raven, P.H., Eds.; Science Press: Beijing, China; Missouri Botanical Garden: St. Louis, MO, USA, 2000; Volume 24, pp. 87–88.
43. Shi, X.; Wang, J.C.; Zhang, D.Y.; Gaskin, J.F.; Pan, B.R. Pollination ecology of the rare desert species *Eremosparton songoricum* (Fabaceae). *Aust. J Bot.* **2010**, *58*, 35–41. [CrossRef]
44. Yuan, S.; Tang, H. Patterns of ephemeral plant communities and their adaptations to temperature and precipitation regimes in Junggar Desert, Xinjiang. *Biol. Sci.* **2010**, *18*, 346–354, (In Chinese with English Abstract).
45. Mamut, J.; Tan, D.Y.; Baskin, C.C.; Baskin, J.M. Intermediate complex morphophysiological dormancy in seeds of the cold desert sand dune geophyte *Eremurus anisopterus* (Xanthorrhoeaceae; Liliaceae *s.l.*). *Ann. Bot.* **2014**, *114*, 991–999. [CrossRef] [PubMed]

46. Kang, J.G.; van Iersel, M.W. Nutrient solution concentration affects shoot: Root ratio, leaf area ratio, and growth of subirrigated salvia (*Salvia splendens*). *Hortscience* **2004**, *39*, 49–54. [CrossRef]
47. Husband, B.C.; Schemske, D.W. Evolution of the magnitude and timing of inbreeding depression in plants. *Evolution* **1996**, *50*, 54–70. [CrossRef] [PubMed]
48. Baskin, J.M.; Baskin, C.C. Inbreeding depression and the cost of inbreeding on seed germination. *Seed Sci. Res.* **2015**, *25*, 355–385. [CrossRef]
49. Pinheiro, J.; Bates, D.; Debroy, S.; Sarkar, D.; R Core Team. nlme: Linear and Nonlinear Mixed Effects Models. R Package Version 3.1–140. 2019. Available online: https://CRAN.R-project.org/package=nlme (accessed on 12 May 2019).
50. Nihranz, C.T.; Walker, W.S.; Brown, S.J.; Mescher, M.C.; De Moraes, C.M.; Stephenson, A.G. Transgenerational impacts of herbivory and inbreeding on reproductive output in *Solanum carolinense*. *Am. J. Bot.* **2020**, *107*, 286–297. [CrossRef]
51. de Vries, F.T.; Manning, P.; Tallowin, J.R.B.; Mortimer, S.R.; Pilgrim, E.S.; Harrison, K.A.; Hobbs, P.J.; Quirk, H.; Shipley, B.; Cornelissen, J.H.C.; et al. Abiotic drivers and plant traits explain landscape-scale patterns in soil microbial communities. *Ecol. Lett.* **2012**, *15*, 1230–1239. [CrossRef] [PubMed]
52. Barton, K. MuMIn: Multi-Model Inference. R Package Version 1.7.2. 2012. Available online: http://CRAN.R-project.org/package=MuMIn (accessed on 7 March 2012).
53. Kalisz, S. Fitness consequences of mating system, seed weight, and emergence date in a winter annual, *Collinsia verna*. *Evolution* **1989**, *43*, 1263–1272. [CrossRef]
54. Ruane, L.G.; Dickens, M.E.; Wall, M.E. Fitness consequences of short- and long-distance pollinations in *Phlox hirsuta*, an endangered species. *Am. J. Bot.* **2016**, *102*, 1659–1665. [CrossRef]
55. Harper, J.L.; Lovell, P.H.; Moore, K.G. The shapes and sizes of seeds. *Annu. Rev. Ecol. Syst.* **1970**, *1*, 327–356. [CrossRef]
56. Seiwa, K. Effects of seed size and emergence time on tree seedling establishment: Importance of developmental constraints. *Oecologia* **2000**, *123*, 208–215. [CrossRef] [PubMed]
57. Koelewijn, H.P.; van Damme, J.M.M. Effects of seed size, inbreeding and maternal sex on offspring fitness in gynodioecious *Plantago coronopus*. *J. Ecol.* **2005**, *93*, 373–383. [CrossRef]
58. Bu, H.Y.; Chen, X.L.; Xu, X.L.; Liu, K.; Jia, P.; Du, G.Z. Seed mass and germination in an alpine meadow on the eastern Tsinghai–Tibet plateau. *Plant Ecol.* **2007**, *191*, 127–149. [CrossRef]
59. Dufay, M.; Billard, E. How much better are females? The occurrence of female advantage, its proximal causes and its variation within and among gynodioecious species. *Ann. Bot.* **2012**, *109*, 505–519. [PubMed]
60. Collin, C.L.; Shykoff, J.A. Outcrossing rates in the gynomonoecious-gynodioecious species *Dianthus sylvestris* (Caryophyllaceae). *Am. J. Bot.* **2003**, *90*, 579–585. [CrossRef] [PubMed]
61. Parker, W.C.; Noland, T.L.; Morneault, A.E. The effects of seed mass on germination, seedling emergence, and early seedling growth of eastern white pine (*Pinus strobus* L.). *New Forest.* **2006**, *32*, 33–49. [CrossRef]
62. Arnaud, M.; Jean-Philippe, B.; José, E.; Grégory, M.; Carlos, G.D.L. Rapid plant invasion in distinct climates involves different sources of phenotypic variation. *PLoS ONE* **2013**, *8*, e55627.
63. Zhang, D.Y.; Jiang, X.H. Maiting system evolution, resource allocation, and genetic diversity in plants. *Acta. Phytoecol. Sin.* **2001**, *25*, 130–143, (In Chinese with English Abstract).
64. Webb, C.J. Test of a model predicting equilibrium frequencies of females in populations of gynodioecious angiosperms. *Heredity* **1981**, *46*, 397–405. [CrossRef]
65. Li, J.M.; Koski, M.H.; Ashman, T.L. Functional characterization of gynodioecy in *Fragaria vesca* ssp. *bracteata* (Rosaceae). *Ann. Bot.* **2012**, *109*, 545–552. [CrossRef]
66. Lloyd, D.G. Self- and cross-fertilization in plants. II. The selection of self-fertilization. *Int. J. Plant Sci.* **1992**, *153*, 370–380. [CrossRef]
67. Ramsey, M.; Vaughton, G. Effect of environment on the magnitude of inbreeding depression in seed germination in a partially self-fertile perennial herb (*Blandfordia grandiflora*, Liliaceae). *Int. J. Plant Sci.* **1998**, *159*, 98–104. [CrossRef]
68. Barringer, B.C.; Geber, M.A. Mating system and ploidy influence levels of inbreeding depression in *Clarkia* (Onagraceae). *Evolution* **2008**, *62*, 1040–1051. [CrossRef] [PubMed]
69. Collin, C.L.; Penet, L.; Shykoff, J.A. Early inbreeding depression in the sexually polymorphic plant *Dianthus sylvestris* (Caryophyllaceae): Effects of selfing and biparental inbreeding among sex morphs. *Am. J. Bot.* **2009**, *96*, 2279–2287. [CrossRef] [PubMed]
70. Leppik, E.E. The evolution of capitulum types of the Compositae in the light of insect flower interaction. In *The Biology and Chemistry of the Compositae*; Heywood, V.H., Harbone, J.B., Turner, B.L., Eds.; Academic Press: London, UK, 1977; pp. 61–89.

diversity

MDPI

Article

Real-Time Classification of Invasive Plant Seeds Based on Improved YOLOv5 with Attention Mechanism

Lianghai Yang [1,2], Jing Yan [1], Huiru Li [1], Xinyue Cao [1], Binjie Ge [1], Zhechen Qi [2,3,*] and Xiaoling Yan [1,*]

1 Eastern China Conservation Centre for Wild Endangered Plant Resources, Shanghai Chenshan Botanical Garden, Shanghai 201602, China; 15150595752@163.com (L.Y.); yan.jing01@163.com (J.Y.); wwwlihuiru@163.com (H.L.); caoxinyue0417@163.com (X.C.); gebinjie@csnbgsh.cn (B.G.)
2 Zhejiang Province Key Laboratory of Plant Secondary Metabolism and Regulation, College of Life Sciences and Medicine, Zhejiang Sci-Tech University, Hangzhou 310018, China
3 Shaoxing Academy of Biomedicine, Zhejiang Sci-Tech University, Shaoxing 312366, China
* Correspondence: zqi@zstu.edu.cn (Z.Q.); xlyan@cemps.ac.cn (X.Y.)

Abstract: Seeds of exotic plants transported with imported goods pose a risk of alien species invasion in cross-border transportation and logistics. It is critical to develop stringent inspection and quarantine protocols with active management to control the import and export accompanied by exotic seeds. As a result, a method for promptly identifying exotic plant seeds is urgently needed. In this study, we built a database containing 3000 images of seeds of 12 invasive plants and proposed an improved YOLOv5 target detection algorithm that incorporates a channel attention mechanism. Given that certain seeds in the same family and genus are very similar in appearance and are thus difficult to differentiate, we improved the size model of the initial anchor box to converge better; moreover, we introduce three attention modules, SENet, CBAM, and ECA-Net, to enhance the extraction of globally important features while suppressing the weakening of irrelevant features, thereby effectively solving the problem of automated inspection of similar species. Experiments on an invasive alien plant seed data set reveal that the improved network model fused with ECA-Net requires only a small increase in parameters when compared to the original YOLOv5 network model and achieved greater classification and detection accuracy without affecting detection speed.

Keywords: weed seeds; seed identification; target detection; convolutional neural network; YOLOv5

Citation: Yang, L.; Yan, J.; Li, H.; Cao, X.; Ge, B.; Qi, Z.; Yan, X. Real-Time Classification of Invasive Plant Seeds Based on Improved YOLOv5 with Attention Mechanism. *Diversity* **2022**, *14*, 254. https://doi.org/10.3390/d14040254

Academic Editors: Jordi López-Pujol and Orsolya Valkó

Received: 29 January 2022
Accepted: 24 March 2022
Published: 29 March 2022

1. Introduction

Invasion by alien species refers to the introduction of certain organisms into a new ecological environment through natural or man-made activities from a different place of origin, which then reproduce and spread independently in the new environment and eventually exhibit a significant ecological impact, endangering the local biodiversity [1]. China is among the countries with the richest biodiversity in the world. The impact of alien species in China has become increasingly severe with the increased frequency of worldwide commerce and transit of products and persons, as well as the rapid expansion of international tourism and logistics industries. As of 2020, more than 660 invasive alien species have been discovered in China, with 370 of them being invasive plants [2]. Every year, invasive species cause massive environmental and economic damages totaling more than USD $30 billion.

In terms of biosecurity protocols, customs inspection and quarantine are considered first passes of defense against alien species invasion and the most critical part of invasive alien species prevention and control management. In 2020, the Chinese customs intercepted 69,500 batches of 384 species of quarantine pests and 4270 batches of 1258 alien species among inbound and outbound articles. As the world's largest grain importer, strict inspection and quarantine of imported grain are critical to the biosecurity of China. Taking Huangpu and Nansha ports in Guangdong province as an example, more than 4 million tons of sorghum imported from the United States alone were detected with a maximum quantity of 100 seeds per kilogram of 106 species of invasive plant seeds in 19 families from 2014 to 2016 [3]. The interception frequency and content of invasive plant seeds are high, and the huge quantity has brought great difficulties to the detection of customs staff. The rapid and accurate detection of the seeds of some notorious invasive plants is related to the biological security and food security of our country. Because some seeds in the same genus exhibit only slight differences in morphological characteristics, customs staff cannot successfully identify them without the assistance of specialists. Moreover, the traditional artificial detection and identification procedure is difficult, time-consuming, and inefficient. Therefore, it is critical to establish an invasive plant seed image data set and an automatic classification system.

With the rapid development of computer vision technology, the detection method based on image recognition technology of the species has been widely adopted, the secondary detection, identification, and classification of various species have achieved significant effect. The image-based target detection task involves identifying the target objects in the image, detecting their location, and determining their category. The target detection algorithm is divided into the traditional artificial feature extraction algorithm and the convolutional neural network (CNN) feature extraction algorithm. In the early stage, traditional machine learning methods were mainly used. Chtioui et al. [4] extracted the size, shape, and texture features of the four color seed images, respectively, and then used the stepwise discriminant analysis method and artificial neural network as the classifier to classify the four seeds and achieved the recognition rates of 92% and 99%. However, due to the small experimental data set and the prior knowledge required to extract features, this result has no practical generalization ability. The traditional feature extraction methods rely heavily on features designed based on prior knowledge. If the extracted features are insufficient, it will seriously affect the accuracy of recognition.

Thanks to the quick advances of GPU hardware and parallel computing performance, CNN (convolutional neural network) represented by AlexNet [5] algorithm provides new approaches for image recognition. In addition, the subsequent R-CNN (region with CNN features) [6] algorithm was widely adopted in image recognition due to its significantly improved performance of deep learning methods in target detection tasks. CNN-based deep learning algorithms are mainly divided into two categories: One is a two-stage target detection algorithm based on R-CNN and its improved algorithms, such as Fast R-CNN [7], Faster R-CNN [8], Mask R-CNN [9], etc. The other is a one-stage target detection algorithm, such as the SSD (Single Shot MultiBox Detector) [10] algorithm and the YOLO (You Only Look Once) [11–14] series algorithm. Compared with the two-stage target detection algorithm, one-stage does not need to map the anchor box to the feature map but directly classifies and regresses the anchor box. Overall, the classification detection accuracy of the one-stage target detection algorithm will be slightly lower than that of the two-stage target detection algorithm, but the detection speed will be enhanced. CNN-based deep learning algorithms have also been widely used in seed image recognition. Javanmardi et al. [15] employ a deep CNN as a general feature extractor for corn seed varieties. Luo et al. [16] compared the classification and detection capabilities of six popular CNN-based deep learning algorithms (AlexNet, VGG-16, Xception, GoogleNet, SqueezeNet, NasNet-Mobile) for weed seeds. Loddo et al. [17] proposed a new deep convolutional neural network, SeedNet, and compared it with other convolutional neural networks in two plant seed data sets. SeedNet can achieve the best results in terms of comprehensive performance. However, as a lightweight target detection algorithm, YOLOv5 has a faster classification and detection speed. Kundu et al. [18] developed a "Mixed Cropping Seed Classifier and Quality Tester (MCSCQT)" system using the YOLOv5 algorithm, which can classify the healthy and diseased seeds of pearl millet and maize. However, the application of the YOLOv5 algorithm in weed seed is scarce so far.

Therefore, we adopted the YOLOv5 algorithm to classify and detect invasive plant seeds in real time in this paper. The YOLO algorithm was proposed in 2016 and has undergone development from v1 to v5. YOLOv5 is a deep learning model based on PyTorch [19]. Benefiting from PyTorch's mature ecosystem, YOLOv5's environment support and model deployment are simpler and easier. YOLOv5 algorithm belongs to the one-stage target detection algorithm. Compared with the two-stage target detection algorithm, YOLOv5 has a smaller weight file and faster reasoning speed of the model [19]; meanwhile, it can still surpass many two-stage target detection algorithms in public data sets to obtain higher accuracy. In the present work, we apply YOLOv5 to classify invasive plant seeds in real time by refining the size model of the initial anchor box using a produced data set comprising captured seed images of 12 invasive plant species. In addition, we propose an improved YOLOv5 target detection algorithm, which combines the ECA attention mechanism. In addition, we also compared the improved model with the original YOLOv5 algorithm and several other attention mechanisms. The improved YOLOv5 method enhances the extraction ability of key features of plant seeds and achieves greater classification and detection accuracy without compromising detection speed.

2. Materials and Methods

2.1. *YOLOv5*

The YOLOv5 algorithm includes four network models: YOLOv5s, YOLOv5m, YOLOv5l, and YOLOv5x, listed in order of increasing network depth and weight file size. To realize high performance on real-time classification on handheld devices, such as cell phones and tablets, we chose the YOLOv5s model for experimental training from the perspective of minimizing computational cost and network weighting in this study. The improved YOLOv5s network model architecture incorporating the ECA-Net attention mechanism is shown in Figure 1. The entire network structure is divided into four parts: input, backbone, neck, and output. A part of the unit structure of the network model is shown in Figure 2.

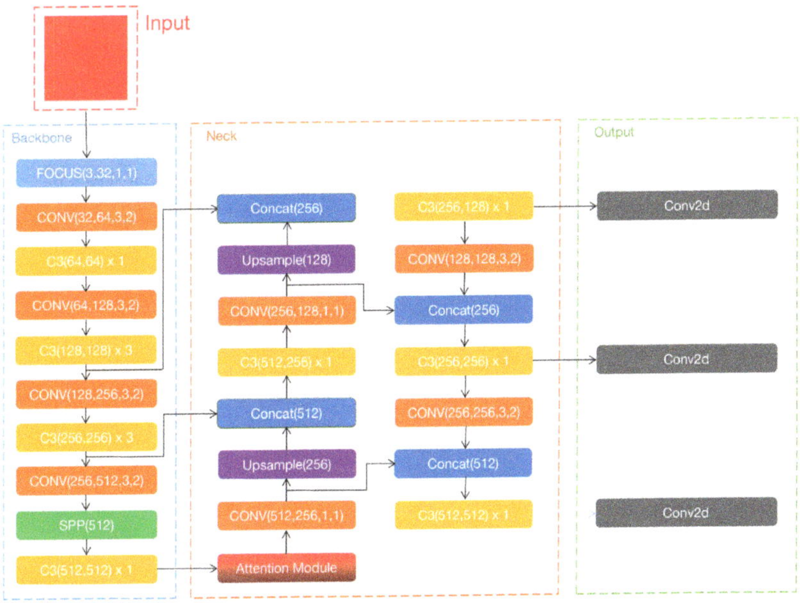

Figure 1. Improved structure of YOLOv5s network model. An attention module is added to the backbone part to capture the important feature information in the backbone and introduce the feature fusion layer.

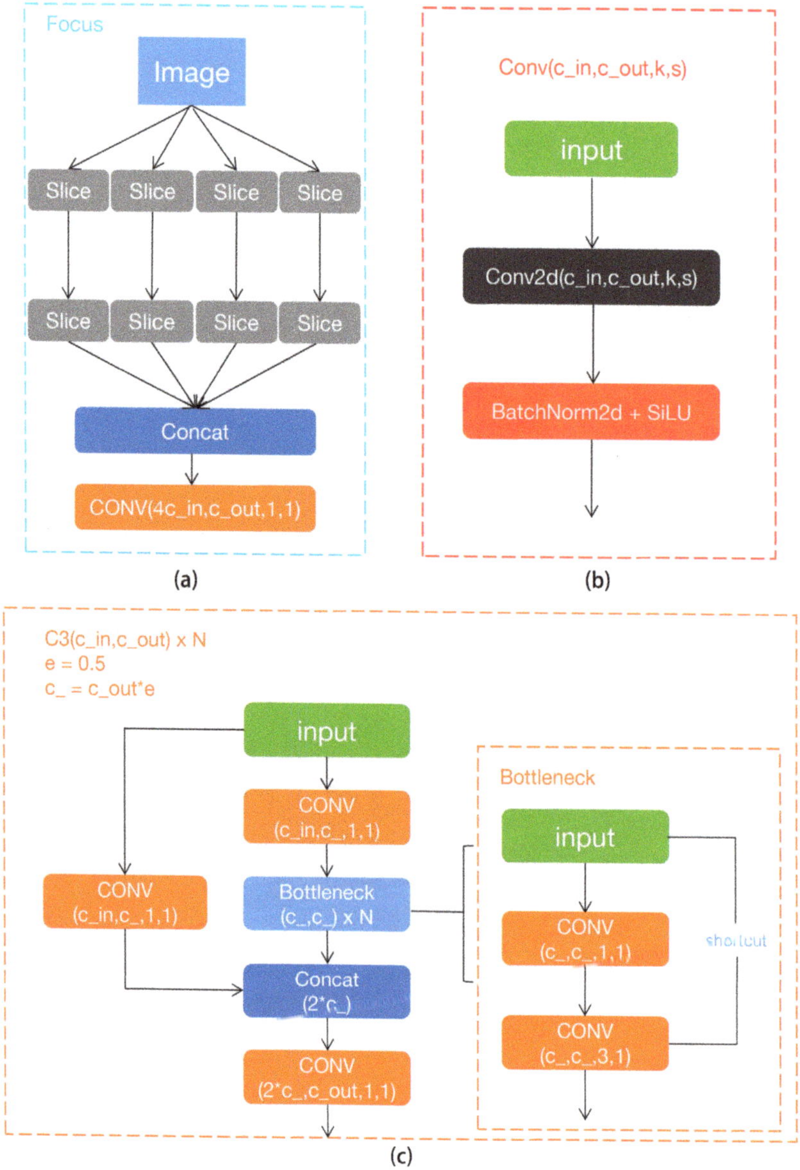

Figure 2. Partial cell structure in network mode: (**a**) structure of the Focus module; (**b**) structure of the Conv module; (**c**) structure of C3 module.

2.1.1. Input

There were fewer small targets in our self-constructed invasive plant seed data set owing to the capture of images with a macro lens, so we used the Mosaic [19] data enhancement method for the input part. Four images were scaled randomly and spliced into a picture to enhance the model's ability to detect small targets. The anchor box of YOLOv5 clustered out nine different sizes of a priori boxes under the three sizes of receptive fields based on the COCO [20] data set. Anchor boxes are statistics from all ground truth boxes in the training set, which are the most frequently occurring box shapes and sizes in the training set. The anchor box can effectively restrict the range of predicted objects in the training process and accelerate the convergence of the model. According to our data set, we use a genetic algorithm plus *k*-means [21] to re-cluster anchors, and the new anchor box is given as follows. [87, 88, 107, 107, 116, 137], [132, 136, 152, 155, 175, 197], [230, 244, 262, 266, 352, 372]. The training phase learns the offset parameters continuously along with the zoom ratio parameters through the anchor box and ground truth. The prediction phase uses these parameters to fine-tune the anchor box, and finally, a better prediction box is obtained from the test data.

2.1.2. Backbone

The backbone of the network architecture includes Focus, C3, Conv, SPP, and ECA attention modules.

Figure 2a shows the structure of the Focus module. Its main function is the slicing operation. As shown in Figure 3, the $4 \times 4 \times 3$ image is sliced into a $2 \times 2 \times 12$ feature map. After the slicing operation of the Focus module, the double down sampling feature map without information loss is obtained. The main purpose of the Focus layer is to reduce layers, reduce parameters, reduce FLOPS (floating-point operations per second), reduce CUDA memory, increase forward and backward speed while minimally impact on model accuracy [19].

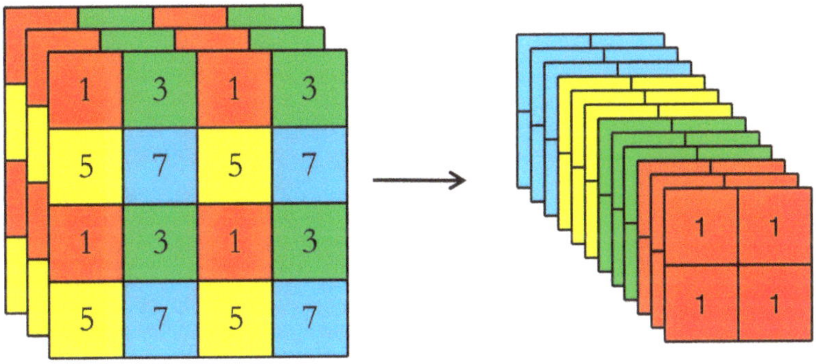

Figure 3. Slicing operation. $4 \times 4 \times 3$ image sliced into $2 \times 2 \times 12$ feature map.

The Conv module is a standard convolution module. A structure diagram is shown in Figure 2b. It consists of the Conv2d+BatchNorm2d+SiLU activation function. The SiLU (sigmoid linear unit) [22] activation function is used in the Conv module to further improve the detection accuracy of the algorithm and optimize the convergence effect of the model. SiLU is an improved, smoother version of the ReLU (rectified linear unit) [23] function. The formula of its activation function is as follows, and x represents the input of convolution.

$$\mathrm{ReLU} = \max(0, \mathrm{x}) \quad (1)$$

$$\mathrm{Sigmoid}(\mathrm{x}) = \frac{1}{1 + \mathrm{e}^{-\mathrm{x}}} \quad (2)$$

$$\mathrm{SiLU} = \mathrm{x} * \mathrm{Sigmoid}(\mathrm{x}) \quad (3)$$

The activation functions of SiLU and ReLU are similar in appearance, as shown in Figure 4. When the input value is less than 0, the output of the ReLU is always zero, as is the first-order derivative. As a result, some neurons may not be activated, and thus the corresponding parameters cannot be updated. The first-order derivative formulas of SiLU and ReLU are as follows.

$$\mathrm{ReLU}' = \max(0, 1) \quad (4)$$

$$\mathrm{SiLU}' = \mathrm{Sigmoid}(\mathrm{x}) + \mathrm{SiLU} * (1 - \mathrm{Sigmoid}(\mathrm{x})) \quad (5)$$

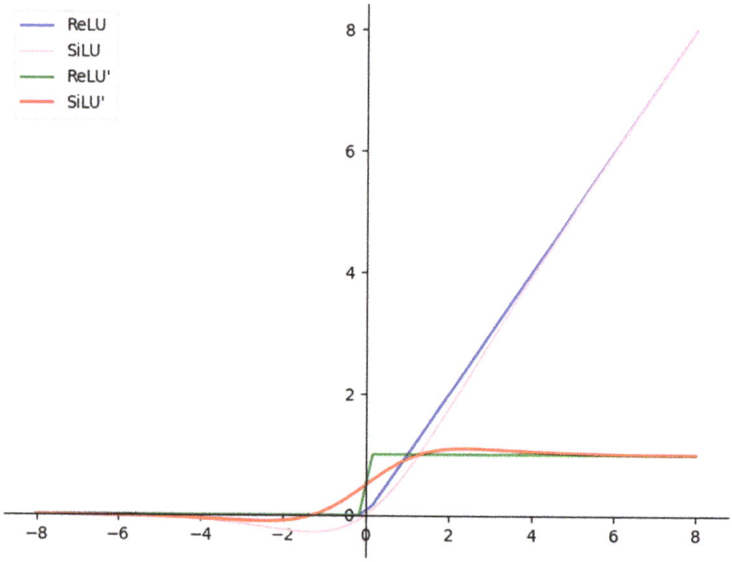

Figure 4. Activation functions and first derivative curves of ReLU and SiLU.

As shown by the curve marked ReLU′ in Figure 4, the ReLU activation function is not monotonically increasing, and the smooth ReLU has better stability, effectively avoiding neuron death and gradient explosion problems in the deep convolutional network.

The C3 module is a residual block based on cross-stage partial (CSP) networks and contains three convolution modules. As shown in Figure 2c, it is composed of a standard bottleneck residual module with three convolution modules. It draws on the design idea of the cross-stage local area network CSPNet [24], which can enhance the feature extraction ability and reduce the model parameters to reduce the memory cost.

As shown in Figure 5, in the spatial pyramid pooling (SPP) module [25], the feature output of different receptive fields is achieved by using three max-pooling layers with different kernel sizes. This can effectively solve the problem that the input image size of the convolutional neural network must be fixed and can avoid repeated extraction of image features to reduce computational costs.

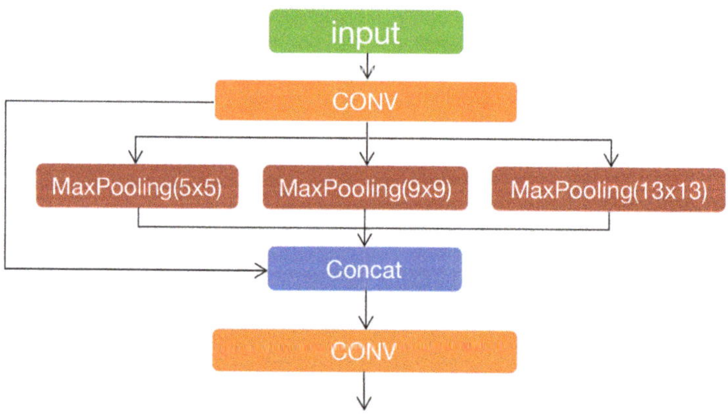

Figure 5. Spatial pyramid pooling structure.

2.1.3. Neck

The neck part adopts a structure combining an FPN (feature pyramid network) and a PAN (path augmentation network) framework, as shown in Figure 6. FPN transfers strong semantic features from top to bottom and then transfers strong positioning features upward through the bottom-up feature pyramid structure of the two PAN structures, and multiple feature fusions achieve full extraction of the three feature layers of the network.

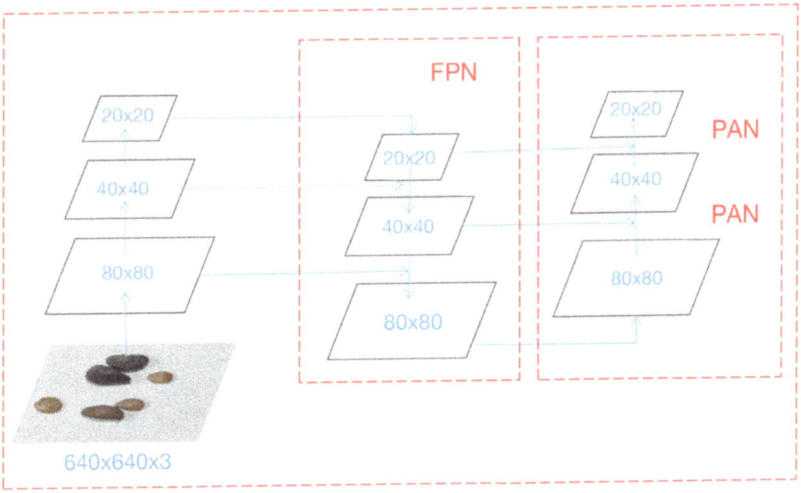

Figure 6. FPN+PAN structure.

2.1.4. Output

The output part uses GIoU_Loss (generalized intersection over union loss) function [26] for the bounding box. Compared with IoU_Loss (intersection over union loss), it increases the measurement of the intersection scale, which effectively alleviates the poor recognition caused by seed stacking in photos of invasive plants. GIoU finds a minimum closed box C to include them based on any two target boxes, A and B. It then calculates the ratio of the area of C that is not covered by A and B to the total area of C and then subtracts this ratio from the IoU of A and B, which is calculated as follows.

$$\mathrm{IoU} = \left|\frac{A \cap B}{A \cup B}\right| \tag{6}$$

$$\mathrm{GIoU} = \mathrm{IoU} - \frac{|C(A \cup B)|}{C} \tag{7}$$

2.2. Attention Mechanism

The concept of attention mechanisms was derived from the study of human vision. Hence, attention mechanisms in deep learning are essentially similar to the selective attention mechanism of humans. Their core purpose is to select more critical information on the current task goal from many information centers while ignoring other relatively unimportant information. Attention mechanisms are mainly divided into three types, including spatial, channel, and mixed mechanisms.

2.2.1. Spatial Attention Mechanism

The spatial attention mechanism only focuses on an area related to the task target and only looks for the most important part of the network for processing. A representative model is the spatial transformer network (STN) [27] proposed by Google DeepMind, and the network structure is shown in Figure 7. It contains three parts: a localization network, a grid generator, and a sampler.

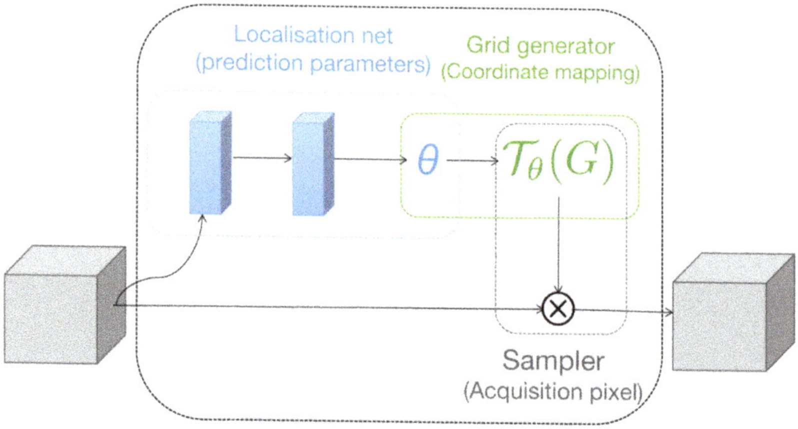

Figure 7. Spatial transformer network structure.

2.2.2. Channel Attention Mechanism

The channel attention mechanism learns the importance of each feature channel and then enhances the useful feature channels for different tasks and suppresses those that are less useful to realize the adaptive calibration of the feature channels.

As a representative model of the channel attention mechanism, Squeeze and Excitation Networks (SENet) [28] was the champion model of the 2017 ImageNet classification competition. The SE module proposed by SENet is shown in Figure 8, which mainly includes two operations: squeeze and excitation. The SE module first performs a squeeze operation on the feature map obtained by convolution to obtain the channel-level global features and then performs an excitation operation on the global features to obtain the weights of different channels and the relationship between the channels.

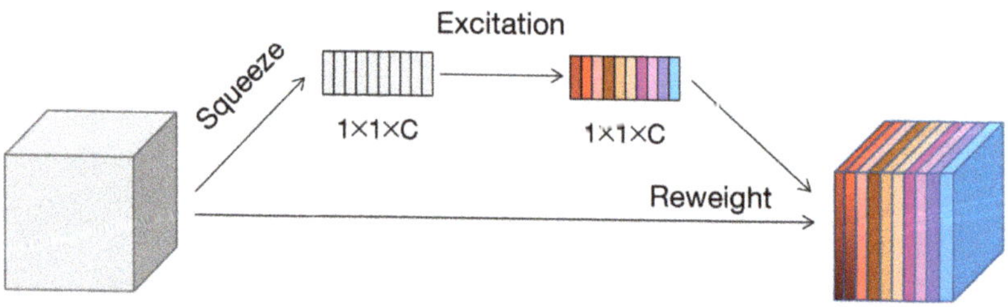

Figure 8. Squeeze and excitation structure.

Efficient channel attention networks (ECA-Net) [29] are introduced as an improved network based on SENet, which uses one-dimensional convolution to replace the bottleneck structure composed of two fully connected layers in SENet. The structure of the ECA module is illustrated in Figure 9. Given the aggregated features obtained by global average pooling (GAP), the local cross-channel interaction strategy without dimensionality reduction and the method of adaptively selecting the size of the convolution kernel is proposed to significantly reduce the complexity of the model while maintaining performance.

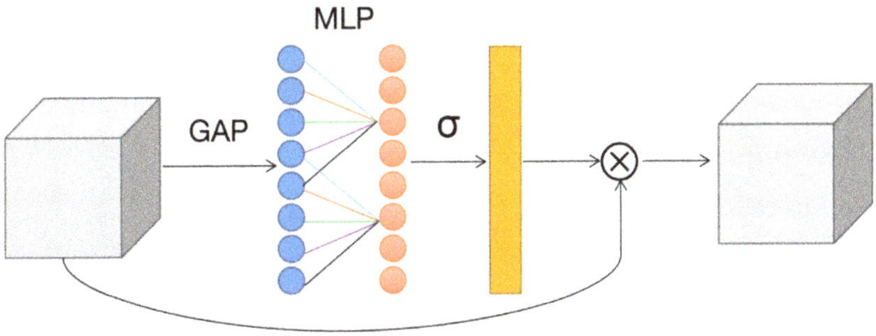

Figure 9. Efficient channel attention networks structure.

2.2.3. Mixed Attention Mechanism

The convolutional block attention module (CBAM) [30] was proposed to combine spatial and channel attention mechanisms. This model proposed a channel attention module and spatial attention module, as shown in Figure 10. The channel attention module processes the feature maps of different channels, and the spatial attention module processes the feature regions of the feature maps.

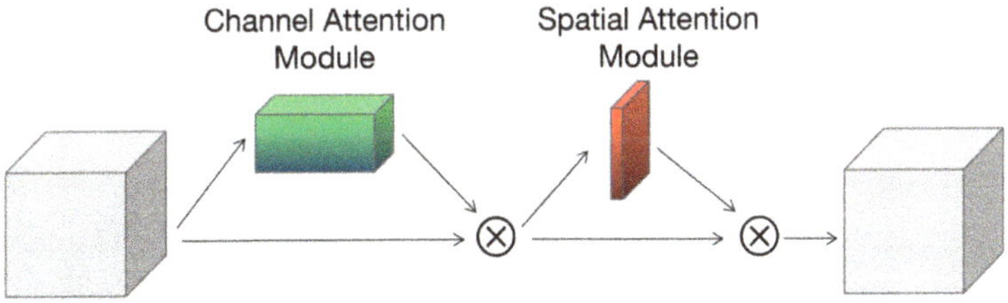

Figure 10. Convolutional block attention module structure.

3. Experiment

The hardware and software configuration of the experimental environment is shown in Table 1.

Table 1. Experimental environment configuration.

Item	Configuration
OS	CentOS 7.9
GPU	Tesla K80 (24GB)
CPU	Intel(R) Xeon(R) CPU E5-2690 v2@3.00GHz
Framework	Pytorch 1.7.1
Data annotation	LabelImg
Visualization	Tensorboard

3.1. Invasive Plant Seed Data Set

We selected 12 species of invasive plants in 10 genera and 7 families, and each species collected 50–80 seeds for shooting. To improve the reliability of the experiment content and results and to better fit the actual situation of customs inspection, we considered both the size of the seeds (the smallest seed of *Nicandra physalodes* (L.) Gaertner is less than 2 mm in average length and the largest seed of *Leucaena leucocephala* (Lam.) de Wit is more than 8 mm in average length) and their similarity (some sets of seeding having high similarity, such as *Solanum viarum* Dunal and *Solanum elaeagnifolium* Cav., as well as *Ipomoea lacunosa* L. and *Ipomoea triloba* L.) when selecting species to construct the invasive plant seeds data set. We did not choose seeds with a length of less than 1 mm, because too small seeds cannot better display surface features when shooting with mobile phones. Most of the selected seeds are between 3 and 5 mm in size, and only the length of Leucaena leucocephala is 8.5 mm. The number of invasive plant seeds larger than 5 mm intercepted by the customs is small, and it is easier to identify with the naked eye due to the large individual. The details of 12 species are shown in Table 2. Figure 11 shows the seed screenshot samples from the data set, which will be closer to the shooting in the actual environment.

(**a**) *Ipomoea lacunosa* (**b**) *Ipomoea triloba* (**c**) *Solanum viarum* (**d**) *Solanum pruinosum* (**e**) *Nicandra physalodes* (**f**) *Datura stramonium* (**g**) *Sida spinosa* (**h**) *Leucaena leucocephala* (**i**) *Sesbania cannabina* (**j**) *Veronica hederifolia* (**k**) *Hexasepalum teres* (**l**) *Paspalum urvillei*

Figure 11. Screenshots of 12 species in the data set.

When shooting the data set for the selected 12 species, we use the mobile phone macro lens (xiaomi10; 2 million pixels; F/2.4 aperture) to shoot the seeds under the light of 5500 K color temperature, which is more suitable for the application scenario of real-time quarantine of customs staff. To obtain more detailed features of seed appearance and improve the generalization ability of the data set, we used 5× optical zoom and 10× optical zoom to shoot seeds, respectively, taking 480 photos of single species and 520 photos of mixed species. Each picture contains a different number of targets, and some data sets are shown in Figure 12. Through random vertical and horizontal mirror flipping and random brightness adjustment, we expand the number of samples in the data set to three times the original image, a total of 3000 images and 14,682 targets. The target distribution of the 12 species is shown in Figure 13, and the target number of each species was evenly distributed. We used the LabelImg [31] data annotation tool to label the seed data, divide the data set into a training set, and test set according to a ratio of 4:1, and the picture resolution was 4344 × 3940.

Table 2. Details of 12 species in the data set: including family and genus information of the species and average data of seed length, width, and height of the species.

Scientific Name	Family	Genus	Length (mm)	Width (mm)	Height (mm)
Ipomoea lacunosa L.	Convolvulaceae	*Ipomoea*	4 ± 0.3	3.5 ± 0.3	2.9 ± 0.4
Ipomoea triloba L.	Convolvulaceae	*Ipomoea*	4.1 ± 0.5	3.1 ± 0.3	2.5 ± 0.3
Solanum viarum Dunal	Solanaceae	*Solanum*	2.3 ± 0.1	2 ± 0.1	0.7 ± 0.1
Solanum pruinosum Dunal	Solanaceae	*Solanum*	3 ± 0.2	2.3 ± 0.2	0.9 ± 0.1
Nicandra physalodes (L.) Gaertner	Solanaceae	*Nicandra*	1.7 ± 0.2	1.6 ± 0.1	0.6 ± 0.1
Datura stramonium L.	Solanaceae	*Datura*	3.5 ± 0.2	2.8 ± 0.1	1.4 ± 0.1
Sida spinosa L.	Malvaceae	*Sida*	2.2 ± 0.2	2 ± 0.2	1.5 ± 0.1
Leucaena leucocephala (Lam.) de Wit	Fabaceae	*Leucaena*	8.5 ± 0.5	5.4 ± 0.4	1.5 ± 0.2
Sesbania cannabina (Retz.) Poir.	Fabaceae	*Sesbania*	4 ± 0.5	2.4 ± 0.5	1.6 ± 0.2
Veronica hederifolia L.	Plantaginaceae	*Veronica*	2.2 ± 0.3	2 ± 0.3	1.6 ± 0.3
Hexasepalum teres (Walter) J. H. Kirkbr.	Rubiaceae	*Diodia*	3.4 ± 0.2	2.2 ± 0.2	1.6 ± 0.1
Paspalum urvillei Steud.	Poaceae	*Paspalum*	1.8 ± 0.3	1.3 ± 0.2	0.5 ± 0.1

Figure 12. Partial data set display: (**a**,**b**) single species images in data set; (**c**,**d**) mixed species images in data set.

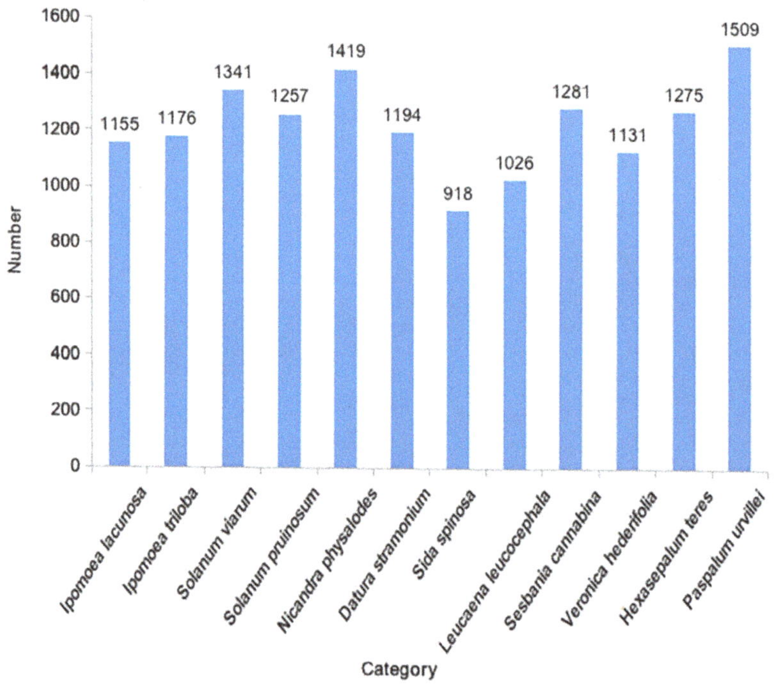

Figure 13. Target number statistics of each category in data set, the number of occurrences of the target seed in the data set.

3.2. Network Performance Evaluation

Common evaluation indicators for two-class problems are precision, recall, F1-score, mAP (mean average precision), FPS (frames per second), etc. A correctly identified sample is called a true positive (TP), whereas a negative sample incorrectly identified as a positive sample is called a false positive (FP), and a positive sample incorrectly identified as a negative sample is called a false negative (FN). Precision, recall, and F1-score are defined below.

$$\text{Precision} = \frac{\text{TP}}{\text{TP} + \text{FP}} \tag{8}$$

$$\text{Recall} = \frac{\text{FP}}{\text{FP} + \text{FN}} \tag{9}$$

$$\text{F1} - \text{score} = \frac{2 * \text{Precision} * \text{Recall}}{\text{Precision} + \text{Recall}} \tag{10}$$

According to the calculation formula of the evaluation index of the two-class problem, we derive the precision and recall rate of the multi-class problem. As shown in Table 3, we considered five classes (class 1–5) as an example (listing the details of class 1). In addition to the statistics of the five categories of predictions, we also show statistics that had content but did not predict the results and those that had no content but had the predicted results. The calculation formulas of Precision1 (P1), Recall1 (R1), and F1-score1 of class 1 are as follows.

$$\text{Precision1} = \frac{\text{TP1}}{\text{TP1} + \text{FP2} + \text{FP3} + \text{FP4} + \text{FP5} + \text{FP6}} \tag{11}$$

$$\text{Recall1} = \frac{\text{TP1}}{\text{TP1} + \text{FN2} + \text{FN3} + \text{FN4} + \text{FN5} + \text{FN6}} \tag{12}$$

$$\text{F1} - \text{score1} = \frac{2 * \text{Precision1} * \text{Recall1}}{\text{Precision1} + \text{Recall1}} \tag{13}$$

Table 3. Examples of five-classification precision and recall. TP (true positive), FP (false positive), and FN (false negative). In particular, when the real target is class1, but the prediction result is empty, we are defined as FN6. When there is no real target, but the prediction result is class1, we define it as FP6.

Multi-Class		Prediction					
		Class1	Class2	Class3	Class4	Class5	Null
Real	class1	TP1	FN2	FN3	FN4	FN5	FN6
	class2	FP2	TP2				
	class3	FP3		TP3			
	class4	FP4			TP4		
	class5	FP5				TP5	
	Null	FP6					

AP is the area under the P-R curve, indicating the accuracy in a category. The map represents the average accuracy of all categories and is used to measure the performance of the deep learning model in all categories. Among mAP@.5 indicates that the IoU (intersection over union) threshold of NMS (non-maximum suppression) is greater than the map value of 0.5, mAP@.5:.95 means that the IOU threshold is calculated every 0.05 from 0.5 to 0.95, and finally, the average value is calculated. AP and mAP are calculated as follows:

$$\text{AP} = \int_0^1 \text{P(R) dR} \tag{14}$$

$$\text{mAP} = \frac{\sum_{i=1}^{N} \text{AP}}{\text{N}} \tag{15}$$

where P is precision, R is recall, and N represents the number of classes in the data set.

Finally, the evaluation result was the average of all the classes. FPS represents the number of image frames that can be processed by the target detection method per second, which verifies the real-time performance of the detection method.

3.3. Experimental Implementations and Settings

This research integrates the SENet, CBAM, and ECA-Net modules in the backbone of YOLOv5s. We also conducted comparative experiments with the original YOLOv5s. The hyperparameter settings in the model are listed in Table 4. We refer to the application of the YOLOv5 target detection algorithm in other fields [32–34], and the network input size was 640 $\times$ 580, the batch size was set to 64 according to the performance of the GPU, and the initial learning rate was set to 0.01. The amount of data was not large; therefore, we used the Adam optimization algorithm to optimize the network parameters.

Table 4. Hyperparameter settings applied in the YOLOv5s.

Hyperparameter	Image Size	Batch Size	Epoch	Optimizer	Learning Rate	Beta1	Beta2
Value/Type	640	64	400	Adam	0.01	0.937	0.999

3.4. Experimental Results and Analysis

We tested YOLOv5s and three improved models that integrate attention mechanisms on the invasive alien plant seed data set. We recorded the differences between the models on the five aspects of Params, Precision, Recall, F1-score, and FPS. In terms of model parameters and detection speed, the original YOLOv5s has the best effect, but after integrating the ECA attention module, the three performance evaluation indicators of the model, Precision, Recall, F1-score, mAP@.5, and mAP@.5:.95 have been improved. The results are shown in Table 5, and in Figure 14, we show the prediction results of some test sets in the YOLOv5s+ECA model.

Table 5. Experimental comparison of YOLOv5s integrated attention module. Bold data represent the best result.

Models	Params	Precision/%	Recall/%	F1-Score/%	mAP@.5	mAP @.5:.95	FPS
YOLOv5s	**7,093,209**	93.02	89.28	91.06	90.65	80.40	**32**
YOLOv5s+SE	7,414,233	93.09	88.87	90.99	90.08	80.52	28
YOLOv5s+CBAM	7,137,121	92.83	89.52	91.22	90.83	81.16	29
YOLOv5s+ECA	7,283,164	**93.96**	**90.11**	**91.94**	**91.67**	**82.77**	29

Figure 14. Partial prediction results of YOLOv5s+ECA model.

Data analysis shows that in terms of detection speed when the inference size was 4344, the four network models were able to achieve millisecond-level real-time classification detection using an Nvidia Tesla K80 GPU. The number of parameters of all models that incorporate the attention module has different increases in parameters. Among them, the YOLOv5+SE model has a maximum increase of 4.53% in parameters, and FPS decreased by four frames, the final F1-score has a slight decrease. The minimum increment of the YOLOv5+CBAM model parameters is only 0.62%, and FPS is reduced by three frames, but it obtains a better detection effect than YOLOv5s+SE. The final F1-score is increased by

0.16% compared to YOLOv5s. YOLOv5s+ECA model has the best optimization results. Compared with YOLOv5s, the parameters of YOLOv5s+ECA increased by 2.68%, and FPS was only reduced by three frames, but the precision, recall, F1-score, mAP@.5, and mAP@.5:.95 increased by 0.94%, 0.83%, 0.88%, 1.02%, and 2.37%, respectively.

For a single category, the test results of the four algorithms on the alien invasive plant seed data set are shown in Table 6. Compared with other individuals, Sida spinosa showed obvious differences. The F1-scores of Sida spinosa in the four models were not high. We speculate that this is because the number of samples in the data set Sida spinosa is relatively small compared to other species. Ipomoea lacunosa and Ipomoea triloba belong to the same genus and are very similar in appearance, which may lead to poor recognition of the YOLOv5s network model. However, F1 scores were improved to different degrees after fusing the SENet, CBAM, and ECA-Net modules, and the ECA-Net module had the best effect. *Sesbania cannabina* also performs poorly in the models of YOLOv5s and integrated SENet. Because the seed surface of *Sesbania cannabina* is extremely smooth to reflect light and appear light spots in the case of flash photography. We cannot effectively solve this problem by adjusting the photographing method, which may lead to some models being unable to obtain more surface features and poor performance. The YOLOv5s+ECA model exhibited suitable detection results in 12 categories and achieved the highest F1-score among the four models in six categories.

Table 6. Comparison of F1-score detection results for 12 invasive plant seeds based on four algorithms with the produced data sets. Bold data represent the best result.

Category	YOLOv5s	YOLOv5s+SE	YOLOv5s+CBAM	YOLOv5s+ECA
Ipomoea lacunosa	0.8904	0.8906	0.8976	**0.9127**
Ipomoea triloba	0.9056	0.9106	**0.9214**	0.9186
Solanum viarum	**0.9440**	0.9383	0.9398	0.9405
Solanum pruinosum	0.9170	0.9134	0.9087	**0.9322**
Nicandra physalodes	0.9071	0.9164	**0.9180**	0.9164
Datura stramonium	0.9153	**0.9188**	0.9090	0.9168
Sida spinosa	0.8921	**0.8985**	0.8966	0.8980
Leucaena leucocephala	0.9136	0.9028	0.9108	**0.9182**
Sesbania cannabina	0.8978	0.8890	0.9030	**0.9076**
Veronica hederifolia	0.9284	0.9258	0.9207	**0.9406**
Hexasepalum teres	0.9000	0.8934	0.9000	**0.9120**
Paspalum urvillei	0.9156	**0.9214**	0.9217	0.9187

Based on the above analysis, we find that the integration of YOLOv5s with the ECA-Net attention module may be expected to increase the number of model parameters only slightly and not to reduce detection speed while improving detection accuracy. Moreover, it can achieve better performance in the real-time detection of invasive plant seeds.

4. Discussion

4.1. Potential Applications

In this paper, we establish a weed data set and offer a method for automatic detection of invasive plant seeds, which has a promising potential to assist customs staff in analyzing inbound alien plant seeds. We evaluate different improved methods on the invasive plant seed data set. The results show that the detection speed of the YOLOv5 method fused with ECA-Net is close to 30 FPS, which can meet the basic requirements of real-time detection of inbound plant seeds by customs staff. However, it should be noted that many species have small differences due to the particularity of plant seeds. Therefore, in the process of real-time classification and detection of plant seeds, it is necessary to achieve sufficient illumination intensity and high enough shooting resolution so that the deep learning model can effectively read the subtle characteristics of seeds.

4.2. Hyperparameter Exploration

In the previous experimental implementations and settings part, we used Adam optimizer to train the model. The image input size is resized to 640, each batch of model input was 64, the initial learning rate is 0.01, and the loss stabilized after 400 training iterations. To verify whether the settings of these hyperparameters are optimal, we observe the performance changes of the YOLOv5s model fused with ECA-Net on the invasive plant seed data set by adjusting these hyperparameters. The experimental results are shown in Table 7. The six experiments of "exp1–exp6" modified different hyperparameters, and the results showed that the F1-score of exp1 was the best. Because our invasive plant seed data set is a small data set, the Adam optimizer is obviously more suitable for our data set. This also verifies that our hyperparameter settings are optimal. The F1-score of exp4 is the worst. We speculate that the learning rate is too high, which makes the depth model unable to converge effectively. An appropriate learning rate is very important for the deep learning model.

Table 7. Hyperparameter adjustment and results. Exp1–6 represents six experiments performed.

Number	Image Size	Batch Size	Learning Rate	Optimizer	Epoch	F1-Score
Exp1	640	64	0.01	Adam	400	91.94%
Exp2	320	64	0.01	Adam	400	90.54%
Exp3	640	32	0.01	Adam	400	91.26%
Exp4	640	64	0.1	Adam	400	71.23%
Exp5	640	64	0.001	Adam	400	91.10%
Exp6	640	64	0.01	SGD	400	90.92%

5. Conclusions

Because of the difficulty of conventional customs biosecurity protocols based on human labor in detecting invasive plant seeds and the long detection process required, in this study, we proposed a real-time classification and detection network based on YOLOv5s fusion attention modules and constructed an image data set consisting of 12 types of invasive plant seeds. In order to better extract the important characteristics of invasive plant seeds and strengthen the comparison of similar species in the model, we combined the YOLOv5s network with the three attention modules of SENet, CBAM, and ECA-Net, and then conducted a comparative experiment with the original YOLOv5s network on the invasive plant seed data set. Using the invasive plant seed data set, the network model composed of the YOLOv5s fusion ECA-Net module achieved higher classification detection accuracy while only adding a small number of parameters and without loss of detection speed. By fusing the ECA-Net attention module, the ability to extract important features of similar species is enhanced, and the classification and detection accuracy of similar species are improved. In the experimental environment, the model FPS integrated with the ECA network can reach 32 frames, which can meet the real-time detection conditions of customs staff. The overall experimental results show that the modified model can achieve better results in the actual application of invasive plant seed detection.

In future studies, expanding the species and target number of the data set is crucial to improve the classification impact and practicability of the model on invasive plant seeds. Furthermore, we will continue to investigate the strategy of fusing the characteristic information of other modalities to improve the seed classification and detection accuracy, especially for the seeds of closely related species with high morphological similarity.

Author Contributions: Conceptualization, X.Y. and Z.Q.; methodology, L.Y. and X.C.; investigation, J.Y. and H.L.; resources, D.G., writing—original draft preparation, L.Y.; writing—review and editing, X.Y. and Z.Q.; project administration, X.Y. and Z.Q. All authors have read and agreed to the published version of the manuscript.

Funding: This research was funded by the Special Fund for Scientific Research of Shanghai Landscaping & City Appearance Administrative Bureau, grant numbers G212405, G222403; National Wild Plant Germplasm Resource Center, grant number ZWGX1902; Special Fundamental Work of the Ministry of Science and Technology, grant number 2014FY120400; the Natural Science Foundation of Zhejiang Province, grant number LY21C030008; and the Open Fund of Shaoxing Academy of Biomedicine of Zhejiang Sci-Tech University, grant number SXAB202020.

Institutional Review Board Statement: Not applicable.

Informed Consent Statement: Not applicable.

Data Availability Statement: Not applicable.

Acknowledgments: We are grateful to the reviewers for their thorough reviews and suggestions that helped to improve this paper.

Conflicts of Interest: The authors declare no conflict of interest.

References

1. Xu, H.; Qiang, S.; Han, Z.; Guo, J.; Huang, Z.; Sun, H.; He, S.; Ding, H.; Wu, H.; Wan, F. The status and causes of alien species invasion in China. *Biodivers. Conserv.* **2006**, *15*, 2893–2904. [CrossRef]
2. Bulletin on China's Ecological Environment in 2019. Available online: http://www.mee.gov.cn/hjzl/sthjzk/zghjzkgb/202006/P020200602509464172096.pdf (accessed on 1 April 2021).
3. Zuo, Y.; Jiang, X.; Li, P.; Yu, W.D. Analysis of the present situation of intercepted weed seeds from American sorghum. *J. Biosaf.* **2017**, *26*, 307–310.
4. Chtioui, Y.; Bertrand, D.; Dattée, Y.; Devaux, M.F. Identification of seeds by colour imaging: Comparison of discriminant analysis and artificial neural network. *J. Sci. Food Agric.* **1996**, *71*, 433–441. [CrossRef]
5. Krizhevsky, B.A.; Sutskever, I.; Hinton, G.E. ImageNet Classification with Deep Convolutional Neural Networks. *Commun. ACM* **2012**, *60*, 84–90. [CrossRef]
6. Girshick, R.; Donahue, J.; Darrell, T.; Malik, J. Rich feature hierarchies for accurate object detection and semantic segmentation. In Proceedings of the IEEE Conference on Computer Vision and Pattern Recognition, Columbus, OH, USA, 23–28 June 2014; pp. 580–587.
7. Girshick, R. Fast R-CNN. In Proceedings of the IEEE International Conference on Computer Vision, Santiago, Chile, 7–13 December 2015; pp. 1440–1448.
8. Ren, S.; He, K.; Girshick, R.; Sun, J. Faster R-CNN: Towards Real-Time Object Detection with Region Proposal Networks. *Adv. Neural Inf. Process. Syst.* **2015**, *28*, 91–99. [CrossRef] [PubMed]
9. He, K.; Gkioxari, G.; Dollar, P.; Girshick, R. Mask R-CNN. In Proceedings of the IEEE International Conference on Computer Vision, Venice, Italy, 22–29 October 2017; pp. 2961–2969.
10. Liu, W.; Anguelov, D.; Erhan, D.; Szegedy, C.; Reed, S.; Fu, C.Y.; Berg, A.C. SSD: Single shot multibox detector. In Proceedings of the European Conference on Computer Vision, Amsterdam, The Netherlands, 11–14 October 2016; pp. 21–37.
11. Redmon, J.; Divvala, S.; Girshick, R. You only look once: Unified, real-time object detection. In Proceedings of the IEEE Conference on Computer Vision and Pattern Recognition, Las Vegas, NV, USA, 12 December 2016; pp. 779–788.
12. Redmon, J.; Farhadi, A. YOLO9000: Better, Faster, Stronger. In Proceedings of the IEEE Conference on Computer Vision and Pattern Recognition, Hawaii, HI, USA, 21–26 July 2017; pp. 6517–6525.
13. Redmon, J.; Farhadi, A. Yolov3: An incremental improvement. *arXiv* **2018**, arXiv:1804.02767.
14. Bochkovskiy, A.; Wang, C.Y.; Liao, H.Y.M. Yolov4: Optimal speed and accuracy of object detection. *arXiv* **2020**, arXiv:2004.10934.
15. Javanmardi, S.; Miraei Ashtiani, S.H.; Verbeek, F.J.; Martynenko, A. Computer-vision classification of corn seed varieties using deep convolutional neural network. *J. Stored Prod. Res.* **2021**, *92*, 101800. [CrossRef]
16. Luo, T.; Zhao, J.; Gu, Y.; Zhang, S.; Qiao, X.; Tian, W.; Han, Y. Classification of weed seeds based on visual images and deep learning. *Inf. Process. Agric.* **2021**, 2214–3173. [CrossRef]
17. Loddo, A.; Loddo, M.; Di Ruberto, C. A novel deep learning based approach for seed image classification and retrieval. *Comput. Electron. Agric.* **2021**, *187*, 106269. [CrossRef]
18. Kundu, N.; Rani, G.; Dhaka, V.S. Seeds Classification and Quality Testing Using Deep Learning and YOLO v5. In Proceedings of the International Conference on Data Science, Machine Learning and Artificial Intelligence, New York, NY, USA, 9–12 August 2021; pp. 153–160.
19. GitHub. YOLOV5-Master. 2021. Available online: https://github.com/ultralytics/yolov5.git/ (accessed on 1 January 2021).
20. Common Objects in Context Dataset. Available online: https://cocodataset.org (accessed on 1 February 2021).
21. Hamerly, G.; Elkan, C. Learning the k in k-means. *Adv. Neural Inf. Process. Syst.* **2004**, *16*, 281–288.
22. Elfwing, S.; Uchibe, E.; Doya, K. Sigmoid-weighted linear units for neural network function approximation in reinforcement learning. *Neural Netw.* **2018**, *107*, 3–11. [CrossRef]

23. Xavier, G.; Antoine, B.; Yoshua, B. Deep Sparse Rectifier Neural Networks Xavier. In Proceedings of the Fourteenth International Conference on Artificial Intelligence and Statistics, Ft. Lauderdale, FL, USA, 11–13 April 2011; pp. 315–323.
24. Wang, C.Y.; Mark Liao, H.Y.; Wu, Y.H.; Chen, P.Y.; Hsieh, J.W.; Yeh, I.H. CSPNet: A new backbone that can enhance learning capability of CNN. In Proceedings of the 2020 IEEE CVF Conference on Computer Vision and Pattern Recognition Workshops (CVPRW), Seattle, WA, USA, 14–19 June 2020; pp. 390–391.
25. He, K.; Zhang, X.; Ren, S.; Sun, J. Spatial Pyramid Pooling in Deep Convolutional Networks for Visual Recognition. In Proceedings of the European Conference on Computer Vision (ECCV), Zurich, Hong Kong, 5–12 September 2014; pp. 346–361.
26. Rezatofighi, S.H.; Tsoi, N.; Gwak, J.; Sadeghian, A.; Reid, I.D.; Savarese, S. Generalized Intersection over Union: A Metric and A Loss for Bounding Box Regression. *arXiv* **2019**, arXiv:1902.09630.
27. Sønderby, S.K.; Sønderby, C.K.; Maaløe, L.; Winther, O. Recurrent Spatial Transformer Networks. *arXiv* **2015**, arXiv:1509.05329.
28. Hu, J.; Shen, L.; Albanie, S.; Sun, G.; Wu, E. Squeeze-and-Excitation Networks. In Proceedings of the IEEE Conference on Computer Vision and Pattern Recognition, Providence, RI, USA, 18–21 June 2018; pp. 7132–7148.
29. Wang, Q.; Wu, B.; Zhu, P.; Li, P.; Zuo, W.; Hu, Q. ECA-Net: Efficient channel attention for deep convolutional neural networks. In Proceedings of the Conference on Computer Vision and Pattern Recognition, Seattle, WA, USA, 14 June 2020.
30. Woo, S.; Park, J.; Lee, J.Y.; Kweon, I.S. CBAM: Convolutional block attention module. In Proceedings of the European Conference on Computer Vision, Munich, Germany, 6 October 2018; pp. 3–19.
31. GitHub. labelImg. 2021. Available online: https://github.com/tzutalin/labelImg.git/ (accessed on 1 February 2021).
32. Zhu, L.; Geng, X.; Li, Z.; Liu, C. Improving yolov5 with attention mechanism for detecting boulders from planetary images. *Remote Sens.* **2021**, *13*, 3776. [CrossRef]
33. Zhao, J.; Zhang, X.; Yan, J.; Qiu, X.; Yao, X.; Tian, Y.; Zhu, Y.; Cao, W. A Wheat Spike Detection Method in UAV Images Based on Improved YOLOv5. *Remote Sens.* **2021**, *13*, 3095. [CrossRef]
34. Yan, B.; Fan, P.; Lei, X.; Liu, Z.; Yang, F. A Real-Time Apple Targets Detection Method for Picking Robot Based on Improved YOLOv5. *Remote Sens.* **2021**, *13*, 1619. [CrossRef]

diversity

MDPI

Article

Reproductive Ecology of Distylous Shoreside *Polygonum criopolitanum* Hance

Ming-Lin Chen *, Meng-Ying Qi, Bei-Bei Bai and Xue Han

Provincial Key Laboratory of Conservation and Utilization of Biological Resources in Anhui, Anhui Normal University, Wuhu 241000, China; qmengying2022@126.com (M.-Y.Q.); bei12457@163.com (B.-B.B.); hanxue200403@163.com (X.H.)
* Correspondence: chenminglin@ahnu.edu.cn

Abstract: In this study, distyly was clearly confirmed in *Polygonum criopolitanum* Hance, which exhibited strict self-incompatibility. Unlike other distylous species, style-morph ratios of *P. criopolitanum* often deviated obviously from 1:1, and many populations were solely composed of long or short stylous flowers; the 1:1 style-morph ratio was occasionally found in very large populations. *P. criopolitanum* was dimorphic for intrinsic features such as style height and anther height and ancillary features such as pollen size and number. The L-morph flowers produced a significantly smaller and higher number of pollen grains than the S-morph flowers, and the stigma papillae of both morphs were not significantly different. We nearly found no seed sets in most wild populations and very low seed sets occasionally occurred in large populations, which was different from other species of Polygonaceae. Mating experiments showed that *P. criopolitanum* has a strict self-incompatibility system and clonal propagation was more common than sexual propagation, which was adaptive with the unisexual wild populations. Hygrocolous habitat, 20–60% soil water content, and height gap less than 4 m to the adjacent water were the main limiting factors for the distribution of *P. criopolitanum*.

Keywords: *Polygonum criopolitanum*; distyly; style-morph ratios; seed sets; strict self-incompatibility

Citation: Chen, M.-L.; Qi, M.-Y.; Bai, B.-B.; Han, X. Reproductive Ecology of Distylous Shoreside *Polygonum criopolitanum* Hance. *Diversity* **2022**, *14*, 222. https://doi.org/10.3390/d14030222

Academic Editors: Hong-Hu Meng and Yi-Gang Song

Received: 27 January 2022
Accepted: 8 March 2022
Published: 18 March 2022

1. Introduction

Heterostyly is a genetic polymorphism in which plant populations are composed of 2 (distyly) or 3 (tristyly) floral morphs. The morphs exhibit reciprocal positioning of anthers and stigmas. Flowers with the L-styled morphology have a stigma(s) positioned above the anthers, whereas flowers with the S-styled morphology have anthers placed above the stigma(s) [1,2]. Besides the differences in floral morphology, distyly is often linked to a sporophytically controlled, diallelic incompatibility system that results in intramorph incompatibility [3].

Heterostyly has been documented in at least 193 genera of 30 angiosperm families [2,4]. Some families-genera possess hundreds of heterostylous species, e.g., Oxalidaceae-*Oxalis*, Primulaceae-*Primula*, and Rubiaceae [3], and the occurrence of heterostyly has been recently reported in *Perovskia* [5]. Distyly in Polygonaceae was first described by Hildebrand in *Fagopyrum esculentum* over 100 years ago [6]; subsequently, many distylous species have been reported for the genera *Oxygonum* and *Aconogonum* [7–10]. The type genus for the family, *Polygonum* (which has more than 300 species) [11], has a worldwide distribution, but it has been seldom reported to be heterostylous. The first distylous *Polygonum chinense* was documented in detail by Reddy et al. (1977) [8]. To date, some other distylous species in *Polygonum* have been reported [12–17].

Polygonum criopolitanum Hance is an annual herb with a tufted, prostrate stem at the base that grows to 10–15 cm. Its inflorescence is terminal and capitate, and the perianth is composed of 5-parted purplish-red tepals. The species is always distributed in the sand by riversides and wet ditches. Most of the previous studies on this species focused mainly on its ecology [18], mating system [19], and plant resources [20], and, to the best of our knowledge, the reproductive characters of *P. criopolitanum* have never been reported in detail.

In this study, we examined variations in floral and reproductive traits and incompatibility systems in natural populations of *P. criopolitanum*. The aims of this study were as follows: (1) to determine whether *P. criopolitanum* is typically distylous as other documented distylous species; (2) to record the morph ratios of wild populations of *P. criopolitanum*; (3) to test the compatibility system after pollen germination, stigma receptivity, and seed set tests; and (4) to discuss the relationship between ecological factors and environmental factors by using canonical correlation analysis (CCA).

2. Materials and Methods

2.1. Study Materials

This study was conducted along the Yangzi River and in Anhui, Jiangxi Province (Figure 1). Herbarium specimens or living materials of *P. criopolitanum* from Anhui and Jiangxi Province, China, were used for this study. All vouchers were deposited at the herbarium of Anhui Normal University (ANUB), China.

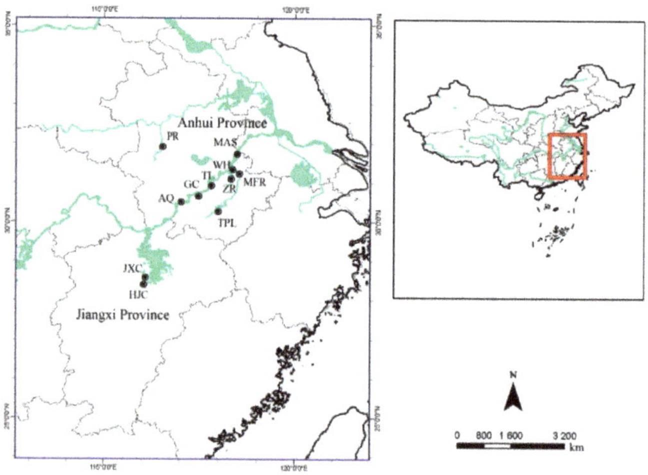

Figure 1. Study sites of *Polygonum criopolitanum* (•: Population. JXC = Jinxian County; HJC = Huangjiacun; TPL = Taiping Lake; MFR = Mafeng River; ZR = Zhang River; AQ = Anqing; GC = GuiChi; TL = Tongling; WH = Wuhu; MAS = Maanshan; PR =Pi River).

2.2. Study Methods

2.2.1. Floral Characters

To document distyly in *P. criopolitanum*, the height of stigma, height of anther, stigma-anther separation, length of tepal, and flower diameter (sketched in Figure 2) were recorded for 50 or 100 flowers of each morph at random from the Wuhu Mafeng River population (Figure 1), using a vernier caliper with a resolution of 0.01 mm.

To count the number of pollen produced, anthers from 100 flowers per morph of the Wuhu populations were placed on individual microscope slides. The numbers of pollen grains per flower were counted using a light microscope, and the diameter of 10 moderate pollen grains from each flower was measured [12,21]. Pollen counts for the style morphs were compared using Student's *t*-test.

2.2.2. Scanning Electron Microscopy

According to the standard acetolysis method [22], pollen grains were mounted in glycerin jelly and sealed with paraffin. The size of well-formed pollen grains from each sample was measured. Scanning electron microscopy was performed using acetolyzed pollen grains coated with Au/Pd under a Hitachi S4800 scanning electron microscope (SEM). Fresh stigmas were mounted in a slow-drying glue and observed under the SEM.

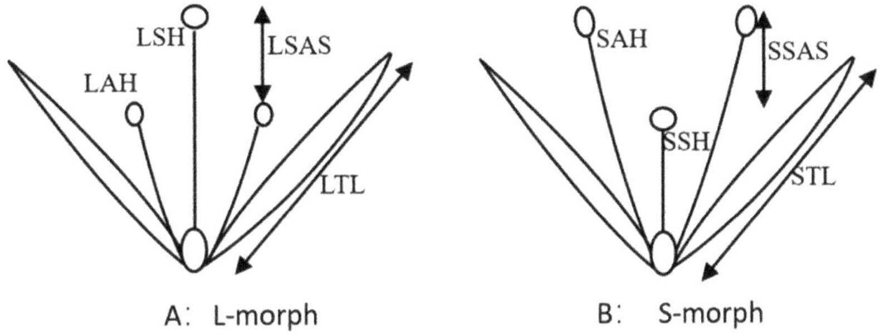

Figure 2. Pattern diagram of distylous flower of *P. criopolitanum* ((**A**): L-morph; (**B**): S-morph). LTL: tepal length of L-morph; LAH: anther height of L-morph; LSH: stigma height of L-morph; LSAS: stigma-anther separation of L-morph; STL: tepal length of S-morph; SAH: anther height of S-morph; SSH: stigma height of S-morph; SSAS: stigma-anther separation of S-morph.

2.2.3. Population Structure Survey

To determine the relative frequencies of the 2 style morphs in populations of *P. criopolitanum*, we surveyed 44 samples to test whether L- and S-morphs occurred at equal frequencies. The chi-square test was performed in R software [23]. We sampled all plants in sporadic populations and 1 × 1 m in patches of populations; for larger samples, and we performed 10 × 10 m sampling.

2.2.4. Environmental and Ecological Factors

Nineteen samples were obtained in Huangshan Taiping Lake, AnqingYingjiang Tower, Guichi Shibasuo, Tongling Shizishan, Yangzi River in Wuhu, Longwo Lake, and Zhang River. Four samples were obtained along the Yangzi River, Longwo Lake, and Zhang River. The first sample was located at the water side, and the other three samples were located along a line drawn perpendicular to the shore at 1-m intervals in altitude up the slope of the riverbank. The number of plants, height, relative coverage, species richness, and presence or absence of *P. criopolitanum* was recorded. The altitude, height gap to the adjacent water, habitat, annual precipitation, annual average temperature, slope, and canopy density of each sample were also documented, and1 kg of soil from each sample was transferred to the laboratory to test ecological factors such as soil water content, pH, organic matter, total nitrogen (TN), and total phosphorus (TP). The soil water content was measured using the GB9834-88 method; soil TN, Kjeldahl method; and soil TP, Mo-Sb colorimetric method [24].

2.2.5. CCA

CCA was performed using Canoco for Windows 4.5. The graph settings were confirmed using the data for the environmental and ecological factors.

2.2.6. Fluorescent Microscopy

To identify the extent of incompatibility of the breeding system of *P. criopolitanum*, pollen germination, and pollen tube growth were determined. The style morphs from each population were pollinated; we waited for 12 h to observe growth in the pollen tube, then they were stored in FAA until staining. To estimate the pollen germination, theharvested styles were softened in 8 mol/L NaOH for 24 h rinsed with distilled water. Then, they were stained for 4 h in 0.4 mg/mL aniline blue solution in phosphate buffer (pH 8.0) and gently squashed in a small drop of glycerol mounting medium under a cover slip. The pistil and pollen were scanned with an Olympus (BX61) epifluorescence microscope (420–470 nm excitation, 490–535 nm emission). For each style, pollen germination was detected by the presence of a pollen tube projecting from the grain, and 10 replicates were performed.

2.2.7. Seed Set Experiments

The seed set experiments were performed using natural populations and controlled pollination populations. For the natural populations, we set 14 1 × 1 m samples in Taiping Lake (2 samples), Luan Pi River (1), Guichi (1), Anqing (1), Tongling (1), Wuhu (5), and Jinxian (3). We calculated the style-morph ratio and then obtained all infructescences of one plant to count the number of seeds (n = 45).

The hybrid experiments were conducted in the laboratory. Legitimate and illegitimate pollination (including selfing) were performed; the seed sets were recorded in every treatment mode: emasculation and bagging, selfing, illegitimate (intramorph) pollination, and legitimate (intermorph) pollination. Each mode was conducted using 100 flowers.

2.2.8. Seed Germination Experiments

Normally developed seeds were soaked in water for 48 h and then disinfected in 75% ethyl alcohol for 30 s. Thirty treated seeds were placed in a culture dish in an illumination incubator to observe the seed germination, with 3 replicates.

3. Results

3.1. Floral Biology

P. criopolitanum is an annual herb with terminal and capitate inflorescence. The peduncle is covered by dense glandular hair, and each bract is 1-flowered. The pattern of floral variation in the wild populations demonstrates that *P. criopolitanum* has conventional distylous floral syndrome (Figure 3A–D). The tepal lengths of the long stylous flower (hereafter L-morph) and short stylous flower (hereafter S-morph) were 2.53 ± 0.44 and 2.47 ± 0.29 mm, respectively, and the tepal diameters of the L-morph and S-morph were 6.58 ± 0.47 and 6.42 ± 0.25 mm, respectively. No significant differences were observed between the tepal lengths as well tepal diameters of the L-morph and S-morph (Table 1, $p > 0.05$). The anthers and stigma are reciprocally positioned in the flowers of the two morphs (Figures 4 and 5). Five stamens are situated between the base of adjacent tepals, and the anthers are purple. Styles 2, seldom 3, connate at the middle-upper part (Figure 5A–D). In addition, the two morphs have five nectaries arranged at the base of each ovary (Figure 3C,D). The stigma and anther heights of the L-morph were 4.02 ± 0.66 and 2.19 ± 0.42 mm, respectively, and the stigma and anther heights of the S-morph were3.90 ± 0.42 and 1.83 ± 0.55 mm, respectively; significant differences were observed between the heights of the stigma and anther of both morphs ($p < 0.05$, Table 1). We found that the stigma-anther separation of the L-morph (2.06 ± 0.39) was longer than that of the S-morph (1.83 ± 0.55; $p < 0.05$, Table 1).

The stigma, which is capitate or spherical with globular papillae on the stigma surface, is similar in the L-morph and S-morph. No significant differences were observed in the size and shape of the stigma papillae (Figure 5C–F).

The pollen grains of *P. criopolitanum* are spheroidal in both morphs, and the pollen grain surface is covered with reticulate exine structures that are pentagonal or hexagonal. Each reticular mesh in the pollen contains many smooth papillae (Figure 5G–J).

The pollen size and number of the two morphs of *P. criopolitanum* showed significant differences. The mean pollen diameters of the L-morph and S-morph were 51 ± 1.92 μm and 62 ± 2.51 μm, respectively, and the mean pollen number per flower of the L-morph and S-morph was 647 ± 40 and 526 ± 38, respectively (Table 1). Although an overlap was detected in the pollen sizes of both morphs, flowers of the L-morph produced significantly more and smaller pollen grains than the ones of the S-morph (Table 1, $p < 0.01$).

Figure 3. Flowers of the L- and S-morphs of *P. criopolitanum* ((**A**,**C**): L-morph; (**B**,**D**): S-morph).

Table 1. Morphological features of the long- and short-morph flowers of *P.criopolitanum*. Differences between the means were analyzed using one-way analysis of variance (ANOVA; mean ± standard error).

Flower Characteristic	L-Morph	S-Morph	*p*-Value	Sample Number
Length of tepal(mm)	2.52 ± 0.44	2.47 ± 0.29	>0.05	50
Flower diameter(mm)	6.58 ± 0.47	6.42 ± 0.25	>0.05	50
Height of stigma(mm)	4.02 ± 0.66	1.84 ± 0.26	<0.001	50
Height of anther(mm)	2.19 ± 0.42	3.90 ± 0.42	<0.001	50
Stigma-anther separation(mm)	2.06 ± 0.39	1.83 ± 0.55	<0.05	100
Pollen number	647 ± 40	526 ± 38	<0.01	100
Pollen diameter(μm)	51 ± 1.92	62 ± 2.51	<0.001	100

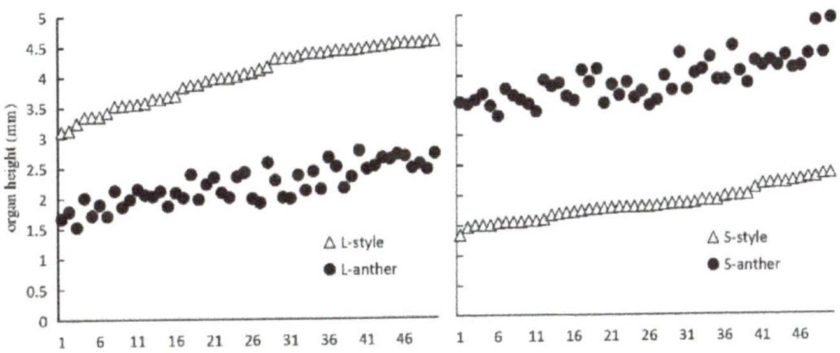

Figure 4. Style and anther length of *P.criopolitanum* (ranked by style length to illustrate the reciprocal correspondence of stigma and anther positions in the long- and short-styled morphs. Positions of the stigmas and anthers are indicated by open triangle (△) and solid circle (•), respectively).

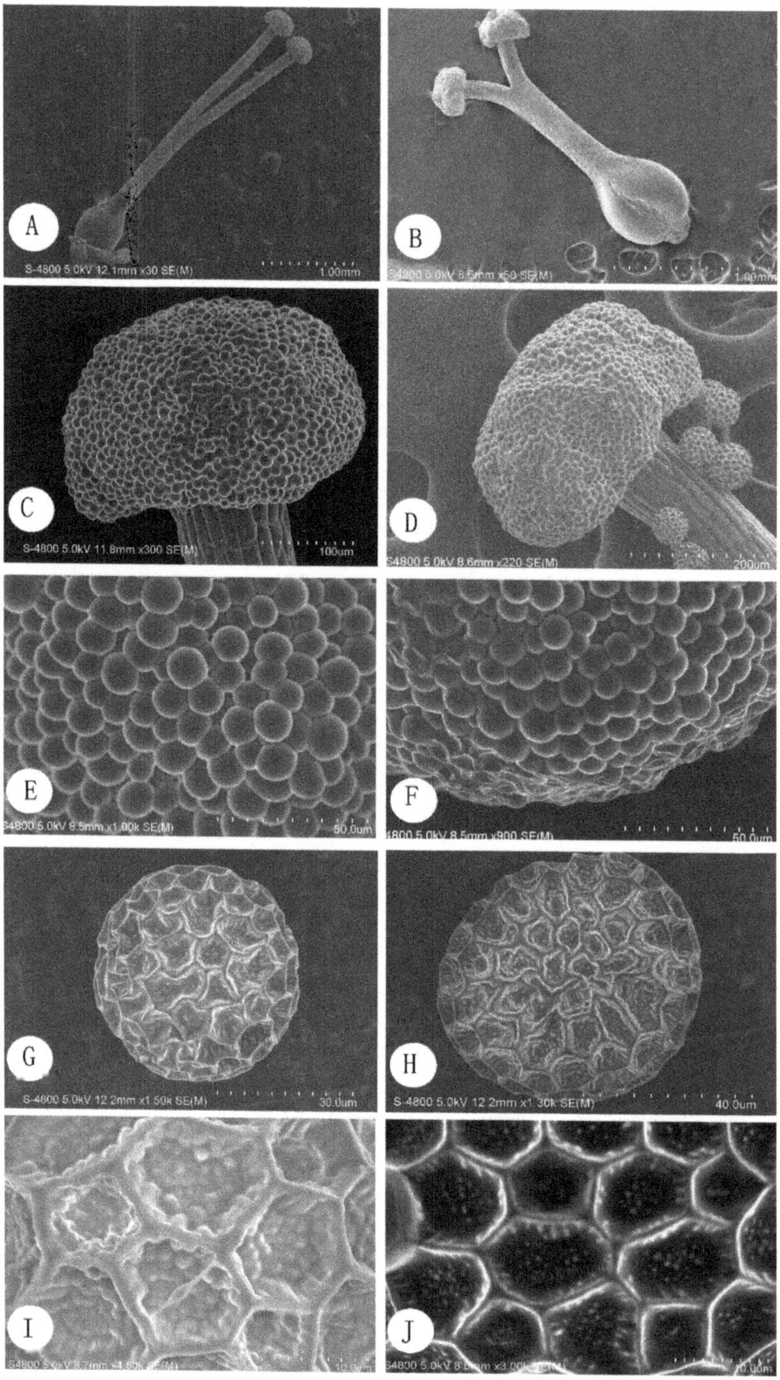

Figure 5. Micromorphological characteristics of *P. criopolitanum* (SEM). (**A**) pistil of L-morph; (**B**) pistil of S-morph; (**C**) stigma of L-morph; (**D**) stigma of S-morph; (**E**) stigma papillae of L-morph; (**F**) stigma papillae of S-morph; (**G**) pollen of L-morph; (**H**) pollen of S-morph; (**I**) pollen epidermal ornamentation of L-morph; (**J**) pollen epidermal ornamentation of S-morph.

3.2. *Style-Morph Ratios*

The chi-square test is a measure of whether the style-morph ratios deviate from 1:1. The survey results showed that the style-morph ratios of *P. criopolitanum* obviously deviated from 1:1, and parthenogenetic populations were always detected in the wild populations. We found that many populations were solely composed of long or short stylous flowers (e.g., ZhengfengTower or Jinxian County populations), whereas the 1:1 morph ratio was seldom found in larger populations (Jinxian County, Taiping Lake, and Mafeng River populations; Table 2).

Table 2. Style-morph ratios in 44 natural populations of *P. criopolitanum*.

Geographic Position	Sample Sites	Sample Size (m^2)	Flower No. of L-Morph	Flower No. of S-Morph	L-Morph:S-Morph	χ^2	*p*-Value	Deviate from 1:1
Anqing City	ZhengfengTower	1	211	0	-	211	2.2×10^{-16}	Y
	Zongyang County	1	185	0	-	185	2.2×10^{-16}	Y
Guichi City	Huamiao	1	12	0	-	12	5.32×10^{-4}	Y
	Shibasuo	1	0	97	-	97	2.2×10^{-16}	Y
Tongling City	Shizishan 1	1	0	37	-	37	1.18×10^{-9}	Y
	Shizishan 2	1	0	285	-	285	2.2×10^{-16}	Y
Wuhu City	MafengRiver	100	785	721	1.08	2.72	9.91×10^{-2}	N
	MafengRiver 1	1	0	63	-	63	2.07×10^{-15}	Y
	MafengRiver 2	1	82	0	-	82	2.2×10^{-16}	Y
	MafengRiver 3	1	46	0	-	46	1.18×10^{-11}	Y
	QingyiRiver	1	0	81	-	81	2.2×10^{-16}	Y
	Zhang River 1	100	102	0	-	102	2.2×10^{-16}	Y
	Zhang River 2	1	150	162	0.93	0.46	0.50	Y
	ZhaojiaRiver	1	72	0	-	72	2.2×10^{-16}	Y
	LongwoLake	1	0	95	-	95	2.2×10^{-16}	Y
	Wanzhi	1	0	89	-	89	2.2×10^{-16}	Y
Manshan City	Yushanqu 1	1	19	0	1	19	1.31×10^{-5}	Y
	Yushanqu 2	1	36	0	1	36	1.97×10^{-9}	Y
	Yushanqu 3	1	22	16	1.38	0.95	0.33	N
	Yushanqu 4	1	21	0	1	21	4.59×10^{-6}	Y
	Yushanqu 5	1	18	92	0.20	49.78	1.72×10^{-12}	Y
Tunxi City	Taiping Lake	100	340	316	1.08	0.88	0.35	N
	Taiping Lake 1	100	54	0	-	54	2×10^{-13}	Y
	Taiping Lake 2	1	31	0	-	31	2.58×10^{-8}	Y
	Taiping Lake 3	1	0	49	-	49	2.56×10^{-12}	Y
	Taiping Lake 4	1	11	0	-	11	9.11×10^{-4}	Y
Luan City	Luan Pi River	100	345	67	5.15	187.58	2.2×10^{-16}	Y
	Luan Pi River1	1	121	0	-	121	2.2×10^{-16}	Y
	Luan Pi River2	1	56	0	-	56	2.2×10^{-16}	Y
	Luan Pi River3	1	0	160	-	160	2.2×10^{-16}	Y
	Luan 1	1	42	24	1.75	4.91	0.03	Y
	Luan 2	1	35	0	1	35	3.30×10^{-9}	Y
	Luan 3	1	33	18	1.83	4.41	0.04	Y
Nanchang City	Jinxian County	100	895	842	1.06	1.62	0.20	N
	Jinxian County 1	100	112	0	-	112	2.2×10^{-16}	Y
	Jinxian County 2	1	56	12	4.67	28.47	9.51×10^{-9}	Y
	Jinxian County 3	1	0	90	-	90	2.2×10^{-16}	Y
	Huangjiacun 1	1	34	11	3.09	11.76	6.1×10^{-4}	Y
	Huangjiacun 2	1	0	34	1	34	2.2×10^{-16}	Y
	Huangjiacun 3	1	26	0	1	26	$3.41e^{-7}$	Y
	Huangjiacun 4	1	43	0	1	43	5.47×10^{-11}	Y
	Huangjiacun 5	1	27	30	0.90	0.16	0.69	N
	Huangjiacun 6	1	0	30	1	30	4.32×10^{-8}	Y
	Huangjiacun 7	1	29	27	1.07	0.07	0.79	N

3.3. *CCA Results*

3.3.1. Characteristics of the Environmental Factors

The environmental factor indices of 19 samples are listed in Table 3.

Table 3. Characteristics of the environmental factors of *P. criopolitanum*.

Sample	Population No.	Altitude (m)	Height Gap to the Adjacent Water (m)	Soil Water Content	Habitat (1: Aquatic, 2: Hygrocolous, 3: Xeromorphic)	Annual Precipitation (mm)	Annual Average Temperature (°C)	Slope (°C)	Canopy Density	pH	Organic Matter (g/kg)	Total Nitrogen (g/kg)	Total Phosphorus (g/kg)
		alt	heig	wat	hab	pre	tem	slo	cad	pH	orm	ton	top
WushiTown	1	18	0.5	0.39	2	1564.5	15.4	6	0.7	6.56	38.5	0.409	0.208
Taiping Lake	2	17	0.8	0.38	2	1564.5	15.4	12	0.56	6	50.4	0.254	0.45
AnqingYingjiangTower	3	21	1	0.32	2	1400	15.5	49	0.1	5.6	48.6	0.102	0.345
Guichishibasuo	4	17	1.2	0.25	2	1578.3	15.9	8	0.32	5.72	19.5	0.393	0.41
Tonglingshizishan	5	156	2.1	0.34	2	1589.1	15.8	32	0.38	7	22.7	0.327	0.208
Yangzi River 1	6	6.5	0	1	1	1054.8	16.2	12	0	7.32	33.5	0.256	0.467
Yangzi River 2	7	7.5	1	0.38	2	1054.8	16.2	25	0.75	6.33	21.9	0.278	0.285
Yangzi River 3	8	9.5	3	0.32	2	1054.8	16.2	25	0.76	6.38	34.6	0.346	0.156
Yangzi River 4	9	11.5	5	0.09	3	1054.8	16.2	25	0.56	6.36	34.9	0.165	0.234
Zhang River 1	10	9	0	1	1	1054.8	16.2	6	0	7.39	44.6	0.287	0.35
Zhang River 2	11	11	2	0.519	2	1054.8	16.2	23	0.63	7.17	29.3	0.158	0.4
Zhang River 3	12	12	3	0.485	2	1054.8	16.2	25	0.42	7.14	28.6	0.402	0.321
Zhang River 4	13	13	4	0.123	3	1054.8	16.2	26	0.35	7.52	44.7	0.483	0.235
LongwoLake 1	14	10	0	1	1	1054.8	16.2	42	0	7.23	45.6	0.306	0.408
LongwoLake 2	15	11	1	0.34	2	1054.8	16.2	12	0.23	7.26	48.9	0.415	0.296
LongwoLake 3	16	13	3	0.26	2	1054.8	16.2	28	0.25	7.42	21.3	0.439	0.318
LongwoLake 4	17	15	5	0.45	3	1054.8	16.2	12	0.27	7.05	20.5	0.308	0.375
Pi River 1	18	56	1.5	0.29	2	1100	16.3	21	0.32	5.67	42.6	0.446	0.284
Pi River 2	19	62	5	0.03	3	1100	16.3	19	0.29	5.62	38.6	0.479	0.368

3.3.2. Characteristics of Ecological Factors

The ecological factor indices of 19 samples are listed in Table 4.

Table 4. Characteristics of ecological factors of *P. criopolitanum*.

Samples	Sample Number	Plants Number	Relative Coverage	Height(cm)	Relative Frequency	Species Richness	Exist or Not
Wushi Town	1	118	0.85	16	0.76	3	1
Taiping Lake	2	12	0.34	15	0.54	4	1
AnqingYingjiangTower	3	4	0.16	14	0.12	1	1
GuichiShibasuo	4	3	0.12	11	0.21	2	1
TonglingShizishan	5	4	0.15	13	0.18	5	1
Yangzi River 1	6	0	0	0	0	0	0
Yangzi River 2	7	28	0.8	15	0.76	2	1
Yangzi River 3	8	39	0.82	14	0.82	5	1
Yangzi River 4	9	0	0	0	0	8	0
Zhang River 1	10	0	0	0	0	1	0
Zhang River 2	11	105	0.82	17	0.76	4	1
Zhang River 3	12	90	78	16	0.74	8	1
Zhang River 4	13	0	0	0	0	1	0
Longwo Lake 1	14	0	0	0	0	0	0
Longwo Lake 2	15	7	0.13	15	0.26	5	1
Longwo Lake 3	16	4	0.08	15	0.21	4	1
Longwo Lake 4	17	0	0	0	0	4	0
Pi River 1	18	29	0.76	16	0.35	6	1
Pi River 2	19	0	0	0	0	12	0

Note: pln: plants number; rec: relative coverage; heig: height; ref: relative frequency; spr: species richness; exn: exist or not; 1 = exist, 0 = absence.

3.3.3. CCA of *P. criopolitanum* Samples

The relationship between ecological factors and environmental factors is shown in Figure 6.

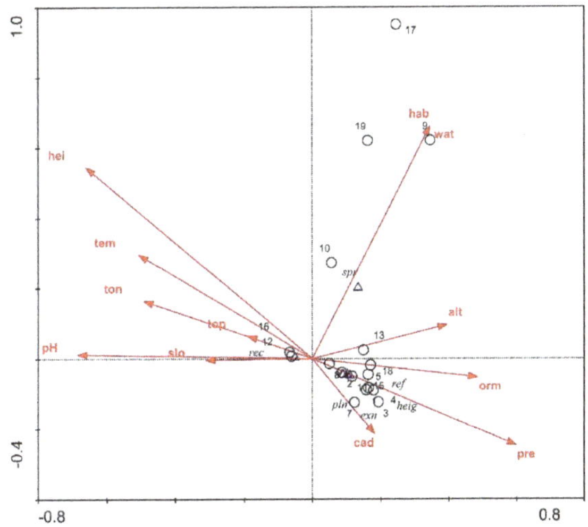

Figure 6. CCA ordination diagram of *P. criopolitanum*. Note: alt:altitude; hei: height gap; wat: water; hab: habitat; pre: precipitation; tem: temperature; slo: slope; cad: canopy density; pH: pH; orm: organic matter; ton: total nitrogen; top: total phosphorus. Positions of the samples and environmental factors are indicated by open circle (○) and open triangle (△), respectively.

In the CCA ordination diagram, the red lines and arrowheads refer to the environmental factors, and the length of the segment refers to the relationship degree between the sample distribution and environmental factors. The angle between the ordination axes and arrowhead connecting the line indicates the correlation degree between the environmental factor and its ordination axes, and the quadrant where the arrowhead is distributed indicates the positive or negative relationship between the environmental factor and its ordination axes. Figure 6 shows the major ecological factors that affect the population distribution of *P. criopolitanum*: habitat, soil water content, and height gap to the adjacent water. Hygrocolous habitat, 20–60% soil water content, and height gap less than 4 m to the adjacent water were the main limiting factors for the distribution of *P. criopolitanum*. Annual precipitation maydirectly influence the soil water content and hygrocoloushabitat. The population distribution of *P. criopolitanum* had no obvious relationship with the slope, pH, altitude, organic matter, TN, and TP. The axes showed that the soil water content increased progressively from top to bottom.

3.4. Mating System Relationships

The fluorescent experiments showed that the rate of pollen germination in the style of the intermorph was obviously higher than that of the self- and intramorph (Figure 7A–F). The statistics showed that the pollen germination rate of pin style × thrum pollen was 72.0 ± 4.0% (n = 10) and that of thrum style × pin pollen was 73.0 ± 3.5% (n = 10). During intramorph pollination, the pollen germination rate of pin style × pin pollen was 43 ± 2.6% (n = 10) and that of thrum style × thrum pollen was 42.0 ± 3.1% (n = 10).

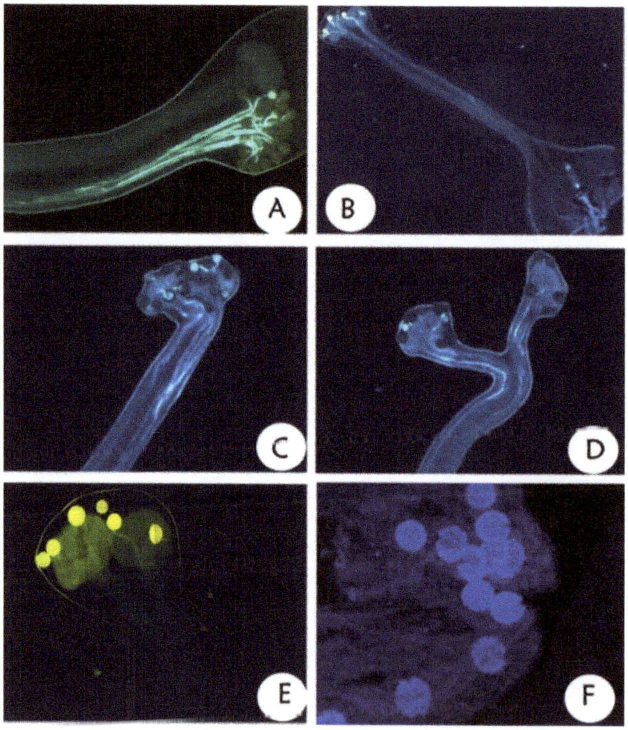

Figure 7. Fluorescent experiments of pollen germination of *P. criopolitanum*. (**A**) L-morph intermorph pollinations; (**B**) S-morph legitimate pollinations; (**C**) L-morph illegitimate pollination; (**D**) S-morph illegitimate pollination. (**E**) Selfing of L-morph; (**F**) Selfing of S-morph.

During self-pollination, the rate of pollen germination of the pin self was 9.0 ± 0.7%, whereas no pollen germination was identified in the case of the thrum self.

3.5. Seed Sets

The statistics showed that the seed sets of *P. criopolitanum* were very low in the wild populations; moreover, the seed sets in the population were often 0, and, even in larger populations with both L- and S-morph, the seed sets were always less than 10% (Table 5).

Table 5. Seed sets of *P.criopolitanum* in different populations under natural conditions.

Population	Population		Fruit Sets
	L-Morph	S-Morph	
Taipinghu 1	156	32	4.17
Taipinghu2	58	0	0
Luan Pi River	145	0	0
Guichi Shibasuo	0	45	0
Anqing Zhengfeng Tower	32	0	0
Tongling	49	0	0
Wuhu MafengRiver	176	152	3.16
Wuhu Qingyijiang	65	0	0
Wuhu Longwo Lake	0	47	0
Wuhu Zhang River 1	89	0	0
Wuhu Zhang River 2	80	11	0
Jinxian County 1	215	208	6.67
Jinxian County 2	0	73	0
Jinxian County 3	87	0	0

The statistics showed that the seed set rate was 0 under emasculation, selfing, and illegitimate pollination conditions. Moreover, the seed set rate was low under legitimate pollination (Table 6).

Table 6. Seed set rate under different treatment conditions.

Process Mode	Pin			Thrum		
	Flowers	Seeds	Seed Sets (%)	Flowers	Seeds	Seed Sets (%)
Emasculation, bagged	100	0	0	100	0	0
Selfing	100	0	0	100	0	0
Illegitimate pollination (intramorph)	100	0	0	100	0	0
Legitimate pollination	100	3	3%	100	5	5%

3.6. Seed Germination

The statistics showed that the seed germination of L-morph and S-morph was 13.33 ± 3.87% and 16.67 ± 3.87%, respectively, with no significant differences between the seed germination of L-morph and S-morph ($p < 0.01$). We found widespread asexual clonal reproduction phenomena in the fields.

4. Discussion

This study revealed that *P. criopolitanum* has all the polymorphic intrinsic features with respect to style and stamen heights and accessorial characteristics with respect to pollen size and number. The polymorphic density of stigma papillae has recently been reported for the tristylous *Lythrumsalicaria* [25] and distylous *Polygonum jucundum* [12]. However, in this study, we found no significant differences between the L-morph and S-morph of

P. criopolitanum, including tepal size; thus, *P. criopolitanum* is typically distylous in intrinsic features and not in all ancillary features.

Flower morph frequencies have always received much attention [26], especially with respect to the consequences for inbreeding [27] or long-term population persistence [28]. In addition, discrepancies in the morph ratio have been found to be much higher in small rather than large populations of *Primula veris* [29] and *Primula elatior* [30], but populations solely composed of the L-morph or S-morph were seldom reported. Interestingly, we observed that the monotypic populations of *P. criopolitanum* were always found in wild populations, which perfectly accounted for the particularly low seed sets of the species.

Fluorescent microscopy showed that pin × pin crosses do occur at a low rate, and no pollen germination was identified in the case of the thrum self, which can explain slight deviations from the 1:1 ratio in a large population. Very common clonal growth in *P. criopolitanum* can explain why the populations fixed for the L-morph or S-morph were always distributed in the fields.

In distylous species, different taxa havebreeding systems with different compatibility levels. Most heterostylous species have a mating system with strict self-incompatibility; for example, a diallelic self-incompatibility system was reported in *Tylosema esculentum* through in vivo and in vitro diallel crossing experiments. The major site of pollen tube inhibition in the intramorph crosses was found to be in style [31]. *Arnebia szechenyi* has also been recorded to show heteromorphic self-incompatibility, which was further supported by the fact that no fruit was produced by flowers subjected to self-pollination or intramorph pollination [32]. In contrast, some species do not exhibit a strict self-compatibility system. In many species of *Primula*, selfing or crossing with a plant of the same morph will also produce a small numberof seeds [33]; for example, the fertility of a legitimate crossof *Primula merrilliana* was high, whereas the fertility of an illegitimate cross waslow [34], and *Pulmonaria officinalis* and *Ceratostigma willmottianum* were found to bepartially self-compatible [35,36]. Unlike most heterostylous species, *Primula oreodoxa* was found to be fully self-compatible under controlled self- and cross-pollinations [37]; flowers of *Psychotria carthagenensis* were also self-compatible [38], and atypical distylous *Psychotria goyazensis* was proven to be an intramorph self-compatible species [39]. In Europe, *Armeria maritima* is completely self-incompatible, but it lost its self-incompatibility during its migration to the New World through the Arctic regions [40]. Seed production is thought to be more sensitive to habitat fragmentation in heterostylous plants than in plants with other breeding systems because potential mating partners are more limited [41].

In this study, we nearly found no seed sets in most wild populations and very low seed sets occasionally occurred in large populations, which was different from other species of Polygonaceae; for example, the seed set of *Polygonum perfoliatum* has a high seed set rate (even up to 84%) [42], and bagging experiments have shown that 47% flowers of *Polygonum thunbergii* are self-pollinated because of no pollinator visits. Despite a high probability of cross-pollination, the probability of fruit set within the ramet was 0.30 because of resource limitations [43]. What interested us the most was that we found a similar scenario in *Polygonum viviparum*. The fruit set of *P. viviparum* has never been observed in North American populations, and sexual reproduction is clearly a rare event in this species [44]. The lack of viable seed production in *P. viviparum* has no single developmental explanation. A similar adaptive reproduction mechanism may exist in these species. On the basis of a common monotypic distribution, pollen germination during the stigma experiments, and the absence of seed sets in the wild populations, we inferred that *P. criopolitanum* has a strict self-incompatibility system.

Previous surveys of the incompatibility status of island flora such as the flora of New Zealand, Hawaii, and the Galápagos have shown a deficit in taxa with heteromorphic incompatibility when compared with continental areas [45]. We found that the dependence on water and environmental characteristics of the hydro-fluctuation belt may be important factors for the establishment of thereproduction system of *P. criopolitanum*. On the basis of the seed sets (natural and artificial conditions) and shoreside distribution along the

Yangzi River or other lakes, the strict self-incompatibility, as well as the sex reproduction efficiency, of *P. criopolitanum* was less dominant than asexual clonal reproduction and uniparental reproduction occurred by clone and not by self-compatible sex reproduction. Baker (1955) referred mainly to self-fertilization as a trait that would confer reproductive assurance during colonization, but we found a different scenario in *P.criopolitanum* [40]. Somes pecies have been reported to abandon sexual reproduction for some form of clonal reproduction, at least in some habitats or parts of their geographic range [46,47].

In conclusion, *P. criopolitanum* was typical distylous species with a strict self-incompatibility reproductive system. The wild population of *P. criopolitanum* deviated obviously from 1:1, and 1:1 style-morph ratios were occasionally found in very large populations. The strict self-incompatibility reproductive system, main environmental factors stress, and common monomorphic populations resulted in the low seed sets, which can explain the general asexual clonal reproduction instead of sexual reproduction in the species. To understand the molecular adaptation mechanism of *P. criopolitanum*, further studies on pollen flow and gene flow at the molecular level need to be performed.

Author Contributions: Project administration and writing, M.-L.C.; investigation, M.-Y.Q.; data curation, B.-B.B. and X.H. All authors have read and agreed to the published version of the manuscript.

Funding: The Natural Science Foundation of Anhui Province (1808085MC76) and the opening fund for provincial key laboratory and key discipline of Colleges of Life Sciences in AHNU, China.

Institutional Review Board Statement: Not applicable.

Informed Consent Statement: Not applicable.

Data Availability Statement: Population data and environmental and ecological factors data were gotten in 2021. Morphological data were gotten from February 2019 to September 2021. The data from the study is available upon request from authors.

Conflicts of Interest: The authors declare no conflict of interest.

References

1. Ganders, E. The biology of heterostyly. *N. Z. J. Bot.* **1979**, *17*, 607–635. [CrossRef]
2. Lloyd, D.G.; Webb, C.J. The evolution of heterostyly. In *Evolution and Function of Heterostyly*; Barrett, S.C.H., Ed.; Springer: Berlin, Germany, 1992; pp. 151–178.
3. Barrett, S.C.H.; Jesson, L.K.; Baker, A.M. The evolution and function of stylar polymorphisms in flowering plants. *Ann. Bot.* **2000**, *85* (Suppl. A), 253–265. [CrossRef]
4. Chen, M.L.; You, Y.L.; Zhang, X.P. Advances in the research of heterostyly. *Acta Pratacult. Sin.* **2013**, *19*, 226–239.
5. Moon, H.K. Confirmation of distyly in *Perovskia* and floral dimorphism of *P. abrotanoides* (Salviinae: Lamiaceae). *Flora* **2021**, *283*, 151905. [CrossRef]
6. Darwin, C. *The Different Forms of Flowers on Plants of the Same Species*; Appleton & Co.: New York, NY, USA, 1877.
7. Graham, R.A. Polygonaceae. In *Flora of Tropical East*; Turril, W.B., Redhead, E.M., Eds.; Africa Crown Agents: London, UK, 1958; pp. 1–40.
8. Reddy, N.P.; Bahadur, B.; Kumar, P.V. Heterostyly in *Polygonum chinense*. *J. Genet.* **1977**, *63*, 79–82. [CrossRef]
9. Hong, S.P. The dimorphic heterostyly in *Aconogonon campanulatum* (Polygonaceae). *Plant Syst. Evol.* **1991**, *176*, 125–131. [CrossRef]
10. Hong, S.P. Pollen dimorphism in heterostylous species of *Oxygonum*. *Syst. Geogr. Plants* **1999**, *68*, 245–252. [CrossRef]
11. Li, A.J.; Bao, B.J. Polygonaceae. In *Flora of China*; Science Press: Beijing, China; Missouri Botanical Garden Press: St. Louis, MO, USA, 2003; Volume 5, pp. 277–350.
12. Chen, M.L.; Zhang, X.P. Distyly in *Polygonum jucundum* Meisn. *Plant Syst. Evol.* **2010**, *288*, 139–148. [CrossRef]
13. Chen, M.L. Floral morphology and breeding system in *Polygonum hastato-sagittatum* Mak. (Polygonaceae). *Flora* **2012**, *207*, 365–371. [CrossRef]
14. Lv, F.J.; Cui, M.C.; Chen, M.L. Reproductive biology of *Polygonum japonicum*. *Acta Pratacult. Sin.* **2013**, *22*, 196–203.
15. Wen, H.H.; Chen, M.L.; Zhang, Z.S. Reproductive biology in the distylous species of *Polygonum orientale*. *Acta Pratacult. Sin.* **2015**, *24*, 155–162.
16. Wang, C.H.; Du, W.; Wang, X.F. Reproductive investment in a cleistogamous morph of *Polygonum jucundum* (Polygonaceae). *PlantSyst. Evol.* **2017**, *303*, 559–563. [CrossRef]
17. Guo, Y.N.; Chen, S.F.; Chen, M.L.; Li, B. *Persicariajucunda* var. *rotunda* (Polygonaceae, Persicarieae), a distinct distylous taxa raised to specific rank. *PhytoKeys* **2019**, *126*, 127–138. [CrossRef] [PubMed]
18. Jin, J.W. *Reproductive Ecology of Polygonum criopolitanum*; Jiangxi Agricultural University: Nanchang, China, 2019.

19. Fang, Y.B.; Shao, J.W.; Wei, Y.; Zhang, X.P. Pilot study on the breeding system of seven species of *Polygonum*. *J. Biol.* **2009**, *26*, 38–40.
20. Xu, S.Q. Studies on *Polygonum* resources and utilization in Cixi. *Chin. Wild Plant Resour.* **2013**, *32*, 32–36.
21. Karise, R.; Mänd, M.; Ivask, M.; Koskor, E.; Bender, A. The effect of pollen amount and its caloric value in hybrid lucerne (*Medicago xvaria*) on its attractiveness to bumble bees (*Bombus terrestris*). *Agron. Res.* **2006**, *4*, 211–216.
22. Erdtman, G. The acetolysis technique: A revised description. *Sven. Bot. Tisdkr.* **1960**, *54*, 561–564.
23. R Core Team. R: A Language and Environment for Statistical Computing, Vienna. 2019. Available online: https://www.R-Project.org (accessed on 15 October 2021).
24. Bao, S.D. *Soil Agrochemical Analysis*, 3rd ed.; China Agriculture Press: Beijing, China, 2000.
25. Hermann, B.P.; Mal, T.K.; Williams, R.J.; Dollahon, N.R. Quantitative evaluation of stigma polymorphism in a tristylous weed, *Lythrumsalicaria* (Lythraceae). *Am. J. Bot.* **1999**, *86*, 1121–1129. [CrossRef]
26. Endels, P.; Jacquemyn, H.; Brys, R.; Hermy, M.; De Blust, G. Temporal changes (1986–1999) in populations of primrose (*Primula vulgaris* Huds.) in an agricultural landscape and implications for conservation. *Biol. Conserv.* **2002**, *105*, 11–25. [CrossRef]
27. Goodwillie, C. Inbreeding depression and mating systems in two species of *Linanthus* (Polemoniaceae). *Heredity* **2000**, *84*, 283–293. [CrossRef]
28. Byers, D.L.; Meagher, T.R. Mate availability in small populations of plant species with homomorphic sporophitic self-incompatibility. *Heredity* **1992**, *68*, 353–359. [CrossRef]
29. Kéry, M.; Matthies, D.; Spillmann, H.H. Reduced fecundity and offspring performance in small populations of the declining grassland plants *Primula veris* and *Gentiana lutea*. *J. Ecol.* **2000**, *88*, 17–30. [CrossRef]
30. Jacquemyn, H.; Brys, R.; Hermy, M. Patch occupancy, population size and reproductive success of a forest herb (*Primula elatior*) in a fragmented landscape. *Oecologia* **2002**, *130*, 617–625. [CrossRef] [PubMed]
31. Hartley, M.L.; Tshamekeng, E.; Thomas, S.M. Functional heterostyly in *Tylosema esculentum* (Caesalpinioideae). *Ann. Bot.* **2002**, *89*, 67–76. [CrossRef] [PubMed]
32. Zhang, C.; Wang, L.L.; Lan, D.; Yang, Y.P.; Duan, Y.W. Pollination ecology of *Arnebiaszechenyi*(Boraginaceae), a Chinese endemic perennial characterized by distyly and heteromorphic self-incompatibility. *Ann. Bot. Fenn.* **2014**, *51*, 297–304. [CrossRef]
33. Arnold, E.S.; Richards, A.J. On the occurrence of unilateral imcompatibility in *Primula* section *Aleuritia* Duby and the origin of *Primula scotica* Hook. *Bot. J. Linn. Soc.* **1998**, *128*, 359–368.
34. Chen, M.L. Comparative reproductive biology of *Primula merrilliana* Schltr. and *P. cicutariifolia* Pax. *Plant Syst. Evol.* **2009**, *278*, 23–32. [CrossRef]
35. Brys, R.; Jacquemyn, H.; Beeckman, T. Morph-ratio variation, population size and female reproductive success in distylous *Pulmonaria officinalis* (Boraginaceae). *J. Evol. Biol.* **2008**, *21*, 1281–1289. [CrossRef]
36. Gao, S.P.; Li, W.J.; Hong, M.T.; Lei, T.; Shen, P.; Li, J.N.; Jiang, M.Y.; Duan, Y.F.; Shi, L.S. The nonreciprocal heterostyly and heterotypic self-incompatibility of *Ceratostigma willmottianum*. *J. Plant Res.* **2021**, *134*, 543–557. [CrossRef]
37. Yuan, S.; Barrett, S.C.H.; Zhang, D.X. Genetics of distyly and homostyly in a self-compatible *Primula*. *Heredity* **2019**, *122*, 110–119. [CrossRef]
38. Faria, R.R.; Ferrero, V.; Navarro, L.; Araujo, A.C. Flexible mating system in distylous populations of *Psychotria carthagenensis* Jacq. (Rubiaceae) in Brazilian Cerrado. *Plant Syst. Evol.* **2012**, *298*, 619–627. [CrossRef]
39. Rodrigues, E.B.; Consolaro, H. Atypical distyly in *Psychotria goyazensis* Mull. Arg. (Rubiaceae), an intramorph self-compatible species. *Acta Bot. Bras.* **2013**, *27*, 155–161. [CrossRef]
40. Baker, H.G. Self-compatibility and establishment after "long-distance" dispersal. *Evolution* **1955**, *9*, 347–348.
41. Ågren, J. Population size, pollinator limitation, and seed set in the self-incompatible herb *Lythrumsalicaria*. *Ecology* **1996**, *77*, 1779–1790. [CrossRef]
42. Smith, J.R.; Hough-Goldstein, J.; Lake, E.C. Variable Seed Viability of Mile-a-Minute Weed (Devil's Tearthumb, *Persicaria perfoliata*). *Invasive Plant Sci. Manag.* **2014**, *7*, 107–112. [CrossRef]
43. Momose, K.; Inoue, T. Pollination and factors limiting fruit set of chasmogamous flowers of an amphicarpic annual, *Polygonum thunbergii* (Polygonaceae). *Res. Popul. Ecol.* **1993**, *35*, 79–93. [CrossRef]
44. Diggle, P.K.; Meixner, M.A.; Carroll, A.B.; Aschwanden, C.F. Barriers to sexual reproduction in *Polygonum viviparum*: A comparative developmental analysis of *P. viviparum* and *P. bistortoides*. *Ann. Bot.* **2002**, *89*, 145–156. [CrossRef]
45. Barrett, S.C.H. The reproductive biology and genetics of island plants. *Philos. Trans. R. Soc. Lond. B* **1996**, *351*, 725–733.
46. Sculthorpe, C.D. *The Biology of Aquatic Vascular Plants*; Edward Arnold: London, UK, 1967.
47. Philbrick, C.T.; Les, D.H. Evolution of aquatic angiosperm reproductive system. *BioScience* **1996**, *46*, 815–826. [CrossRef]

diversity

MDPI

Article

Effects of Landscape and Local Factors on the Diversity of Flower-Visitor Groups under an Urbanization Gradient, a Case Study in Wuhan, China

Mengyu He [1,2], Nan Ran [1], Huiqian Jiang [1], Zemin Han [1], Yuanyong Dian [1], Xiaoxia Li [3], Dong Xie [4,5], Peter A. Bowler [6] and Hui Wang [1,*]

1 College of Horticulture & Forestry Sciences, Hubei Engineering Technology Research Center for Forestry Information, Huazhong Agricultural University, Wuhan 430070, China; hmyforestry@webmail.hzau.edu.cn (M.H.); rannan@webmail.hzau.edu.cn (N.R.); jianghuiqian1023@163.com (H.J.); hzm@webmail.hzau.edu.cn (Z.H.); dianyuanyong@126.com (Y.D.)
2 Wuhan Dijiao Park Management Office, Garden and Forestry Bureau of Jiang'an District, Wuhan 430014, China
3 Institute of Wetland Research, Chinese Academy of Forestry, Beijing 100091, China; kindxiaoxia@163.com
4 Co-Innovation Center for Sustainable Forestry in Southern China, College of Biology and the Environment, Nanjing Forestry University, Nanjing 210037, China; xiedong0123@gmail.com
5 The National Wetland Ecosystem Field Station of Taihu Lake, National Forestry Administration, Suzhou 215107, China
6 Department of Ecology and Evolutionary Biology, University of California, Irvine, CA 92697-2525, USA; padowler@uci.edu
* Correspondence: wanghui@mail.hzau.edu.cn; Tel.: +86-1877-113-2510

Citation: He, M.; Ran, N.; Jiang, H.; Han, Z.; Dian, Y.; Li, X.; Xie, D.; Bowler, P.A.; Wang, H. Effects of Landscape and Local Factors on the Diversity of Flower-Visitor Groups under an Urbanization Gradient, a Case Study in Wuhan, China. *Diversity* **2022**, *14*, 208. https://doi.org/10.3390/d14030208

Academic Editors: Yi-Gang Song and Hong-Hu Meng

Received: 25 January 2022
Accepted: 10 March 2022
Published: 11 March 2022

Abstract: Urbanization is one of the primary forces driving worldwide pollinator decline. Moderate urban expansion with appropriate green space planning can help in maintaining pollinator diversity and pollination service. We investigated the relative effects of landscape and local factors on the diversity of flower-visitor functional groups in a moderately urbanized city, Wuhan, located in central China. We found that the proportion of impervious surface had no significant effect on the number of visitations, but it was negatively associated with the diversity of flower-visitor groups. The number of visitations by Halictidae and Lepidoptera correlated positively with local flower density and flowering plant species richness, respectively. Flowering plant species richness was also positively correlated with the diversity of flower-visitor groups. The proportion of green space was negatively associated with the visitation number of Muscidae and the overall diversity of flower-visitor groups, revealing the potential influence of green space quality on pollinator assemblage. The pollination networks under three urbanization levels (with a total of 11 flower visitor groups and 43 plant species) were asymmetric, highly nested, and generalized. The suburb sites contained the highest diversity of interactions. Core flowering plants (*Oenothera speciosa*, *Coreopsis grandiflora* and *Cyanus segetum*) are exotic species with attractive flowers. Improving green space quality (high flower density and flowering plant species richness) and using attractive native flowering plants (*Nandina domestica*, *Rosa chinensis*, *Astragalus sinicus*, *Cirsium arvense* var. *integrifolium*, and *Zabelia biflora*) would enhance the function of urban green space to maintain pollinator diversity and ecosystem stability.

Keywords: flower-visitor; insect pollination; plant species; urbanization; diversity; functional group; pollination network; green space

1. Introduction

The sustainable development of human societies relies on ecosystem services provided by nature, among which pollination service is vital and vulnerable [1]. Pollinators, such as bees, butterflies, and hoverflies have declined globally [2–6]. Flying insects in the nature reserves of Germany have declined by more than 75% during 1989–2016 [7]. The relative abundance of four bumblebee species in North America has decreased by 96% from 1980s to 2009 [8]. The primary factors causing large-scale pollinator decline include habitat loss and fragmentation, agricultural intensification, use of agricultural chemicals, pathogen infestation, invasive alien species, climate change, light pollution, and their collective interactions [8–11].

Urbanization is a dynamic process involving dramatic and continuous changes in land use, including a decrease in bare land, an increase in the area covered by impervious surfaces, and the loss of natural vegetation [10,12,13]. In general, urbanization causes habitat loss and reduces the habitat suitability of pollinators, and consequently reduces the abundance and diversity of pollinators [14–17]. However, some studies have found that moderate urban expansion or artificial land-use may increase the number of certain pollinators and their pollination services. Landscape heterogeneity is improved at the landscape level, resulting in diverse habitat types for different pollinator groups [18]. On the other hand, urban green spaces can provide suitable alternative habitats and food sources for pollinators [19,20]. Some studies have shown that floricultural suburban gardens contribute more to bumblebee population growth and nesting than agro-ecosystems or other rural landscapes [21–23]. Gardens with high flower diversity and density are considered to be pollinator-friendly [24]. To increase the local nectar resources, the cutting frequency of the green space should be reduced, in order to retain herb and shrub layers [25]. Therefore, suitable urbanized areas may play an important role in pollinator conservation.

At the landscape scale, residual semi-natural habitats in urban environments can serve as habitats and shelters for pollinators [26]. The impact of urban environments on insects is largely determined by the number and distribution of semi-natural habitats such as green spaces, lawns, hedges, and so forth [25]. Urban night light may directly affect the nocturnal pollinators and pollination networks, and indirectly affect the diurnal pollinators. Artificial night light disturbs nocturnal pollination and reduces plant reproduction, which also provide food resources for diurnal pollinators [27]. On the other hand, the night light may decrease nectar depletion by nocturnal pollinators and favor foraging by diurnal pollinators [28]. At the local scale, vegetation type and abundance of flower resources affects the habitat suitability and food availability for pollinators. For example, a high abundance and species richness of flowering plants can provide nectar and other food resources in different seasons, which can help maintain high pollinator diversity [20,29]. The plant species differ markedly in their attractiveness to pollinators [30,31], and species of *Origanum*, *Agastache*, *Lavandula*, and *Nepeta* were reported to be highly attractive [30]. Pollinators differ in their preference of floral traits [32], so the species of flowering plants also affects the pollinator assemblage. Therefore, studies of both landscape factors and local factors in moderately urbanized areas can help reveal the impact of urbanization on pollinator diversity and provide guidance in the planning of urban green spaces.

In this study, we investigated the effects of landscape factors and local factors driven by urbanization on the diversity of flower visitors in urban and suburban green spaces in Wuhan, central China. The objectives of this study were: (1) to investigate the effects of landscape factors (percentage of impervious surface, percentage of green space, and night light intensity) and local factors (species richness of flowering plants, and flower density) on the abundance and diversity of flower-visitor groups; (2) to investigate the effects of urbanization level on interaction between plants and flower-visitor groups, and clarify the key plant species that may play an important role in maintaining urban pollinator diversity.

2. Materials and Methods

2.1. Study Area

The study system we surveyed was located in Wuhan City (113°41′–115°05′ E, 29°58′–31°22′ N, ranging between 19.2 and 873.7 m in elevation and with a total area of 856,915 hectares) in Hubei Province, central China. This is one of the most rapidly growing cities in central China, where urban and arable areas have largely expanded over natural vegetation.

According to Wuhan's General Plan for Land Use (2006–2020), construction land was expected to reach 185,000 ha, accounting for 21.58% of total land area in 2020 [33].

Field investigations were conducted from May to July in 2019 at 19 independent sites in urban parks, covering an urbanization gradient. The study sites were selected on the basis of the proportion of impervious surfaces around the park within a 2000 m radius (Table A1, Figure 1), using Google satellite imagery 2016 and Hubei land cover maps (resolution of 30 m) [34]. We used 2000 m as the research scale of landscape variables for the following reasons. Firstly, some flower-visitors are capable of flying long distances, for example, the foraging flight of bumblebee workers is greater than 1 km and their maximum measurement reaches approximately 2 km [35]. Consequently, the habitat and resource that pollinators can reach cannot be evaluated at a small scale, and the effects of impervious space on the pollinator abundance are more apparent on a larger scale [36]. Secondly, the area of urban parks usually exceeds 100 hectares (for example, the area of Shizishan Parkland is 495 hectares, and the area of Wuhan Garden Expo Park is 213.8 hectares). We excluded the park area in calculation of impervious surface, because the parks with large squares and pavements usually lead to overestimation of urbanization level in their settings, especially for exurb parks. The water cover was excluded in the calculation of land cover composition following the approach used by Xie et al. (2017) [37]. Study sites were situated at least 3 km from each other. In keeping with McKinney's categorization of urban landscapes [38], the "City core" represented the area in which the impervious surfaces were greater than 80%, in the "Suburb" the impervious area was between 50 and 80%, and the impervious surface cover of the "Exurb" comprised sites with less than 50% coverage.

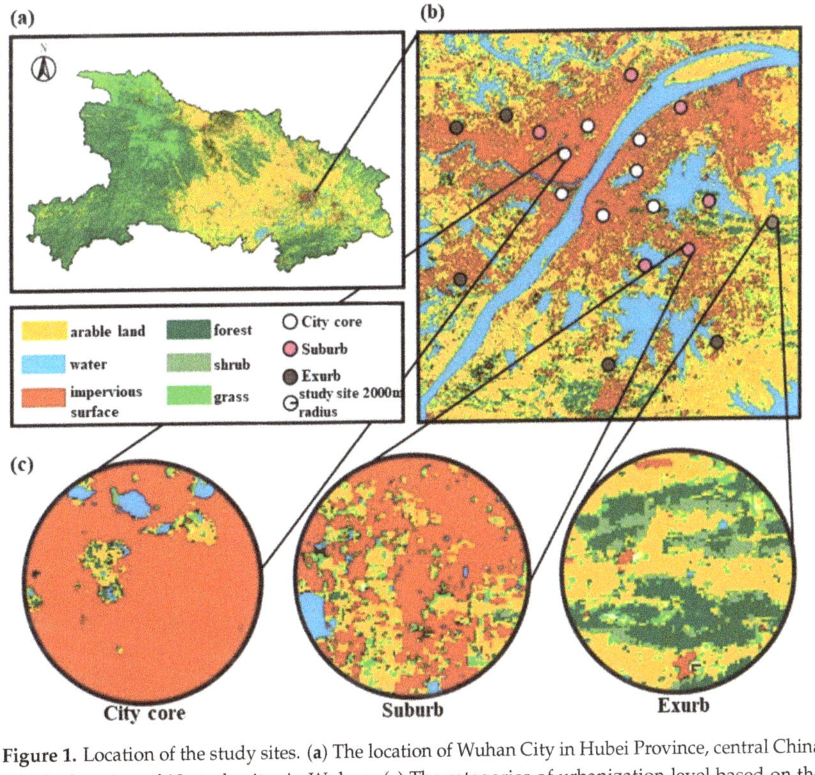

Figure 1. Location of the study sites. (**a**) The location of Wuhan City in Hubei Province, central China. (**b**) The location of 19 study sites in Wuhan. (**c**) The categories of urbanization level based on the proportion of impervious surface.

2.2. Landscape Variables

We calculated the proportion of green space within a 2000 m radius (excluding areas covered by water) around the study sites (Tables 1 and A1). Green space included parklands, protective green space, affiliated green space, regional green space, and other categories of non-developed settings.

Table 1. Description of landscape and local factors.

Factor	Scale	Description	Level
Proportion of impervious surface	r = 2000 m	Proportion of impervious surface around the park, as an indicator of urbanization level	Continuous variable, 0–100%
Proportion of green space	r = 2000 m	Proportion of green space around the study site, including parkland, protective green space, affiliated green space, regional green space, and other types	Continuous variable, 0–100%
Night light intensity	r = 2000 m	Average brightness of night light around the sites	Continuous variable
Density of flowers	20 m × 20 m plot	Number of flowers per m^2	Continuous variable
Species richness of flowering plant	20 m × 20 m plot	Number of flowering plant species	Discrete variable
Sampling time		The month of field survey	Nominal variable, May, June, July

The night light data were obtained from remote sensing images by the Luojia No. 1 night remote sensing satellite and have a resolution of 130 m. The average night light intensity within a 2000 m radius around the sites was calculated using Arcmap software (Version:10.6.1) (Tables 1 and A1).

2.3. Local Floral Resources and Flower-Visitors

At each site we selected a representative 20 × 20 m plot in which the floral resources were relatively abundant, and the level of human disturbance was low or absent. We conducted surveys between 1000 and 1700 h on sunny and calm days. Three survey rounds were performed in each site once a month.

Before our observations of flower-visitors, we recorded flowering plant species richness and the density of flowers (the number of flowers per meter squared) in each plot. The density of flowers was sampled in 10 randomly selected 1 × 1 m plots. The number of flowers were counted according to the procedure described by Gong and Huang (2009) [32]. Single flower plants (e.g., *Oenothera speciosa*) and plants with clustered florets in their inflorescence (e.g., *Trifolium repens* and composites) were counted as single floral units. For inflorescences with wide and sparse florets that can be visited independently by pollinators (e.g., *Althaea rosea* and *Oxalis articulate*), each floret was counted as a floral unit. Flowering trees higher than 2 m were not included because of the difficulty in surveying their flowers. The anemophilous flowers were not included.

In the 20 × 20 m plots, we recorded flower-visitors and visited plant species using line-transect methodology. We counted and recorded visits to flowers within 1 m of each side of the transect and 1 m in front of the observer at a steady walking speed. For each survey round, the transect was positioned across the 20 × 20 m plot and proceeded for 2 km (Figure 2). Due to the difficulty of species identification in the field, the flower-visitors were identified to genus (*Apis* spp., *Anthophora* spp.), family (Halictidae, Vespidae, Syrphidae), or order level (Lepidoptera, Coleoptera). To avoid interference with insect visitations, we caught individuals for further identification using a hand net as the survey round was completed. The flower visitors were transferred to separate vials and stored at −80 °C. Further identification was conducted utilizing DNA sequence comparisons

(COI-KC and ITS2) based upon insect leg tissue samples (Supplementary Material Table S1). The insect specimens were deposited in the lab of College of Horticulture & Forestry Sciences, Huazhong Agricultural University. The flower-visitors were classified into 11 functional groups: honeybee (*Apis cerana*, *A. dorsata*, *A. mellifera*, *A. nuluensis*), *Anthophora* spp., common wasp (Vespidae and Ichneumonidae), Halictidae (*Lasioglossum seillean*), Muscidae and Tachinidae (*Musca domestica*, *Drino inconspicua*), Lepidoptera (*Euxoa castanea*, *Gonepteryx cleopatra*, *Graphium sarpedon*, *Parnara batta*, *Pseudozizeeria maha*, etc.), Orthoptera (Oedipodidae), Coleoptera (*Dytiscus semisulcatus*), Odonata, ant (Formicidae), and Syrphidae (*Episyrphus balteatus*, *Eristalis tenax*, *Sphaerophoria philanthus*) (Supplementary Material Table S1). The functional groups were classified according to the insect body size and visiting behavior, for example, the short-tongued honeybee, long-tongued and large-sized bee (*Anthophora* spp.), solitary bee (Halictidae), long-tongued butterfly, and moth. We used Shannon's diversity index (H) to estimate the diversity of flower-visitor groups for each sample plot during each survey.

$$H = -\sum(Pi) \times (\ln Pi)$$

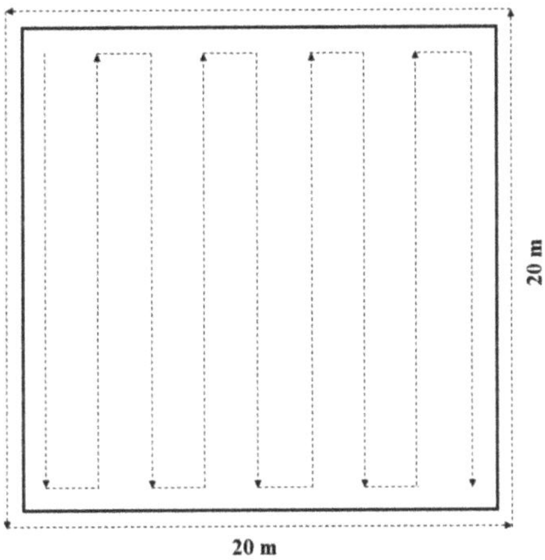

Figure 2. Transect route of flower-visitor survey.

Pi: the percentage of visits by each flower-visitor group to total visits.

Due to revegetation and plant replacement in green space areas, there were four times when there were no flowering plants present in the sample sites (Shouyi Cultural Park and Simeitang Park in the second round, Shizishan Parkland and Dijiao Park in the third round). The sites without flowering plants in the survey round were excluded from analysis.

2.4. Data Analysis

We examined the effect of the proportion of impervious surface, proportion of green space, night light intensity, local flower density, flowering plant species richness, and the sampling time (May, June, and July) on the number of visits by flower-visitors and the diversity of flower-visitor groups during each survey round, using the generalized linear mixed model (GLMM). The study site was included as a random effect. Variance inflation factors (VIFs) were checked to ensure that the factors met the assumption of independence. None of the factors had VIFs greater than 4, so the factors could be included independently in the model [31].

The number of visits by flower-visitor groups (discrete data with a large variance) were analyzed with a negative binomial distribution (with log link) [39]. We examined the number of visits by all groups and also by the main visitor groups (honeybee, Halictidae,

Muscidae, Lepidoptera). Honeybees are extremely abundant and widely distributed, and may drive the results. To address this possibility, we also analyzed the number of visits with honeybees excluded. The diversity of flower-visitor functional groups was analyzed with a normal distribution.

To illustrate the effect of urbanization level on the interaction between flowering plants and flower-visitor groups, we applied the Shannon's diversity index of interactions (*Pi* was calculated as the proportion of interactions) for each urbanization level. We also investigated the pollination networks for the three urbanization levels using R software (Version:4.0.3, 'bipartite' package).

3. Results

3.1. Survey Results

In total, we recorded 4115 floral visits on 53 plant species. The honeybees made 2534 visits and accounted for 61.58% of the total visitation, followed by Halictidae (625 visits, 15.19%), Lepidoptera (449 visits, 10.91%), and Muscidae and Tachinidae (335 visits, 8.14%). Other groups, including Formicidae (92 visits), *Anthophora* spp. (35 visits), Coleoptera (17 visits), Syrphidae (15 visits), Vespidae and Ichneumonidae (6 visits), Odonata (5 visits), and Orthoptera (2 visits), accounted for less than 5% of the total visits.

The study sites supported a diverse array of flowering plant species. We observed 53 plant species across 28 plant families. The most common plant families were the Asteraceae (13 species) and Leguminosae (4 species). Thirty-one plant species were exotic, of which nine species were invasive, including *Trifolium repens*, *Erigeron annuus*, *Trifolium pratense*, *Oenothera biennis*, *Veronica persica*, *Daucus carota*, *Geranium carolinianum*, *Silybum marianum*, and *Zephyranthes carinata* (Table A2) [40,41]. Each site contained a mean of 2.84 (±2.59) species of flowering plant during each round. The density of flowers ranged from 0.4 to 535.6 flower/m^2. The most common flowering plant species was *Rosa chinensis*, which was found in six sites, followed by *Oenothera speciosa* (five sites), *Erigeron annuus* (five sites), *Trifolium repens* L. (four sites), *Coreopsis grandiflora* (four sites), *Cosmos bipinnatus* (four sites), *Oxalis articulata* (four sites), and *Veronica persica* (four sites) (Table A2).

In this survey, a few plant species received the most visits (such as *Oenothera speciosa*, *Coreopsis grandiflora*, *Centaurea cyanus*, and *Oxalis articulata*). However, no visitor groups were observed for 10 plant species, *Weigela florida*, *Plantago depressa*, *Geranium carolinianum*, *Salvia miltiorrhiza*, *Silybum marianum*, *Oxalis corniculate*, *Ophiopogon japonicus*, *Platycodon grandiflorus*, *Canna indica*, and *Zephyranthes carinata*. These species were rare, found in only one survey site and in one survey round.

3.2. Factors Influencing the Number of Visits by Flower-Visitor Groups

The sampling time significantly affected the number of visits by all groups during each survey round (GLMM, $F_{2,45} = 3.365$, $p = 0.043$; Table 2). Visits by all groups were significantly higher in May (133.17 ± 125.00) than in June (54.80 ± 63.58) and July (41.53 ± 52.09) (Figure 3a).

None of the factors affected the number of visits when honeybees were excluded from the analysis (Table 2).

The number of visits by honeybees was significantly affected by the sampling time (GLMM, $F_{2,45} = 7.554$, $p = 0.001$; Table 2). During each survey round, honeybee visitations were significantly higher in May (104.17 ± 117.52) than that in June (26.00 ± 38.75) and July (9.27 ± 14.069). The visitation number did not differ between June and July (Figure 3c).

Visits by Halictidae were positively correlated with the flower density (GLMM, fixed coefficient = 0.040, $F_{1,45} = 47.175$, $p < 0.001$; Table 2).

Visits by Lepidoptera were positively associated with flowering plant species richness (GLMM, fixed coefficient = 0.298, $F_{1,45} = 6.317$, $p = 0.016$; Table 2).

Visits by Muscidae were affected by the sampling time (GLMM, $F_{2,45} = 10.366$, $p < 0.001$; Table 2, Figure 3f). During each round, the Muscidae made significantly more visits in May (13.56 ± 19.986), followed by June (4.35 ± 7.46), making least visits in July (1.27 ± 2.915) (Figure 2). The number of visits by Muscidae was negatively associated with the proportion of green space (GLMM, fixed coefficient = −5.384, $F_{1,45} = 4.393$, $p = 0.042$; Table 2).

Table 2. Effects of fixed factors on the number of visits by flower-visitor groups (GLMMs).

Term	Factor	Fixed Coefficient	F	*p*
All flower-visitors	Proportion of impervious surface	−1.579	0.666	0.419
	Sampling time	—	3.365	0.043 *
	Proportion of green space	−2.556	0.772	0.384
	Night light intensity	<−0.001	0.176	0.677
	Local flower density	0.001	0.155	0.696
	Flowering plant species richness	0.154	2.001	0.164
Non-honeybee visitors	Proportion of impervious surface	−2.309	1.410	0.241
	Sampling time	—	0.338	0.715
	Proportion of green space	−4.499	2.372	0.131
	Night light intensity	<−0.001	0.530	0.471
	Local flower density	0.003	1.021	0.318
	Flowering plant species richness	0.186	2.913	0.095
Honeybee	Proportion of impervious surface	−2.067	1.722	0.196
	Sampling time	—	7.554	0.001 ***
	Proportion of green space	0.250	0.011	0.917
	Night light intensity	<0.001	0.121	0.730
	Local flower density	−0.001	0.037	0.848
	Flowering plant species richness	0.119	1.601	0.212
Halictidae	Proportion of impervious surface	1.121	0.152	0.699
	Sampling time	—	1.288	0.286
	Proportion of green space	−1.607	0.134	0.716
	Night light intensity	<−0.001	0.043	0.837
	Local flower density	0.004	47.175	<0.001 ***
	Flowering plant species richness	−0.116	0.448	0.507
Lepidoptera	Proportion of impervious surface	−1.108	0.265	0.609
	Sampling time	—	0.525	0.595
	Proportion of green space	−3.400	1.101	0.300
	Night light intensity	<−0.001	0.641	0.428
	Local flower density	0.004	1.673	0.202
	Flowering plant species richness	0.298	6.317	0.016 *
Muscidae	Proportion of impervious surface	−1.023	0.348	0.558
	Sampling time	—	10.366	<0.001 ***
	Proportion of green space	−5.384	4.393	0.042 *
	Night light intensity	<−0.001	2.087	0.155
	Local flower density	0.002	0.364	0.549
	Flowering plant species richness	−0.005	0.003	0.956

Note: * $p < 0.05$, *** $p < 0.01$.

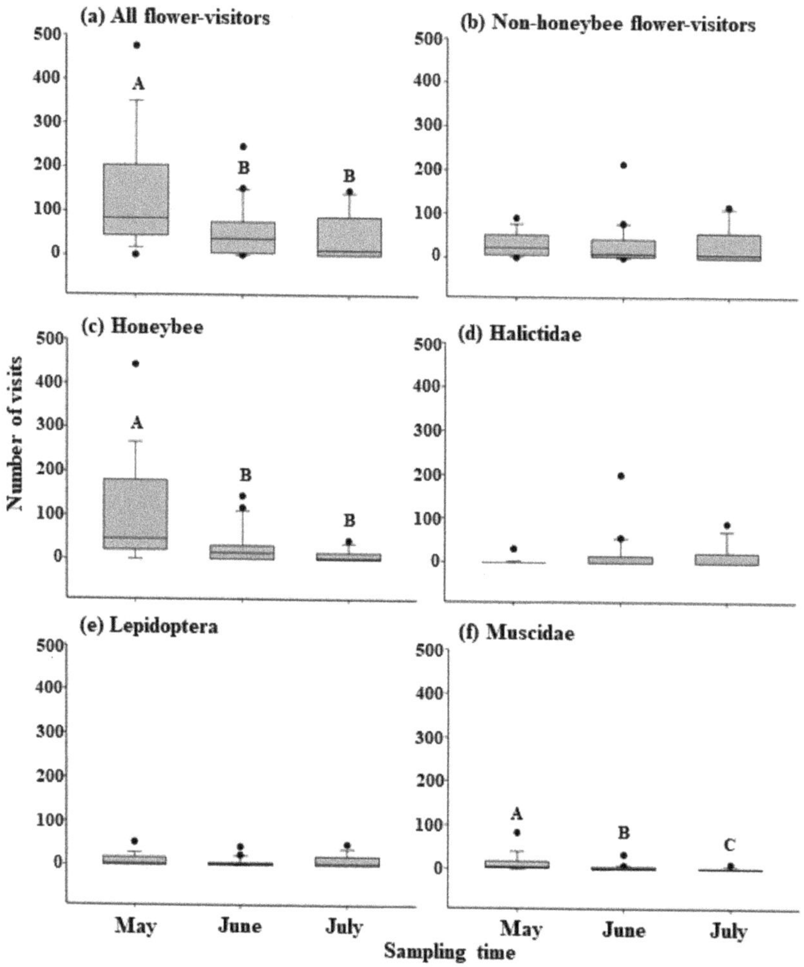

Figure 3. The effects of sampling time on the number of visits by (**a**) All flower visitors, (**b**) Non-honeybee visitors, (**c**) Honeybee, (**d**) Halictidae, (**e**) Lepidoptera, (**f**) Muscidae. Boxes with different capital letters indicate significant differences among three sampling months according to GLMMs ($p < 0.05$).

3.3. Factors Influencing the Composition and Diversity of Flower-Visitor Groups

There were six, ten, and eight functional groups of flower-visitors recorded in the City core, Suburb, and Exurb sites, respectively. Honeybees, Halictidae, Lepidoptera, Muscidae, and Tachinidae occurred under the three urbanization levels. Orthoptera only occurred in Suburb sites. Vespidae and Ichneumonidae were only found in Exurb sites. The visitor groups that appeared in City core sites were also recorded in the Suburb and Exurb sites.

The diversity of flower-visitor functional groups during each survey was negatively associated with the proportion of impervious surface and proportion of green space (GLMM, proportion of impervious surface, fixed coefficient = −1.179, $F_{1,45} = 4.847$, $p = 0.033$; proportion of green space, fixed coefficient = −1.864, $F_{1,45} = 5.223$, $p = 0.027$; Table 3, Figure 4a,b). The diversity of flower-visitor groups was positively correlated with the species richness of flowering plant (GLMM, fixed coefficient = 0.072, $F_{1,45} = 5.005$, $p = 0.030$; Table 3, Figure 4c).

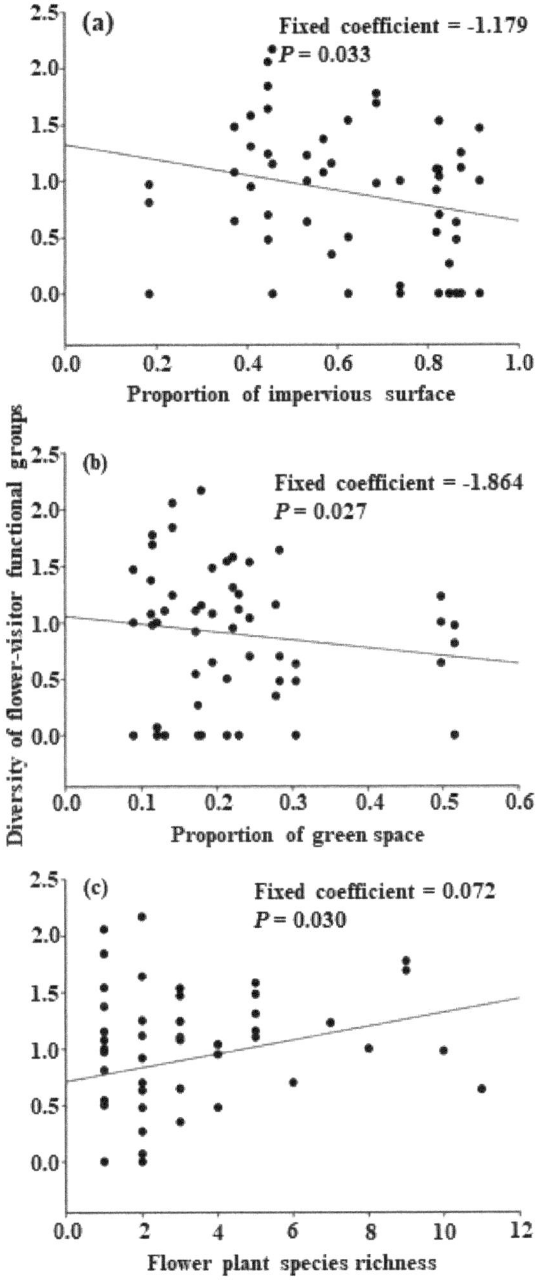

Figure 4. The correlation between diversity of flower-visitor functional groups and (**a**) Proportion of impervious surface, (**b**) Proportion of green space, (**c**) Flower plant species richness according to GLMMs.

Table 3. Effects of fixed factors on the diversity of flower-visitor functional group (GLMM).

Factor	Fixed Coefficient	F	p
Proportion of impervious surface	−1.179	4.847	0.033 *
Sampling time	—	0.172	0.842
Proportion of green space	−1.864	5.223	0.027 *
Night light intensity	<0.001	0.008	0.929
Local flower density	<0.001	0.199	0.658
Flowering plant species richness	0.072	5.005	0.030 *

Note: * $p < 0.05$, *** $p < 0.01$.

3.4. The Interaction between Flowering Plants and Flower-Visitor Groups under Three Urbanization Levels

The Shannon's diversity index results of interactions in the City core, Suburb, and Exurb sites were 2.8235, 2.9426, and 2.5174, respectively. The Suburb sites contained the most flower-visitor groups (10 groups) and flowering plant species (34 species), and it had the highest diversity of interactions.

The pollination networks were constructed for three urbanization levels, with a total of 11 flower-visitor groups (P1–P11) and 43 plant species which received visits (F1~F43) (Figure 5, Tables 4, A2 and A3). For the City core, Suburb, and Exurb sites, the Connectance (calculated as the abundance of links/abundance of potential links) was 0.31, 0.21, and 0.38, respectively, indicating that some potential links may have not been observed, especially at the Suburb sites (Table 4). The nestedness of the pollination network (calculated as (100-T)/100, T referring to the matrix temperature) was 0.89, 0.95, and 0.88 for three urbanization levels, indicating that the pollination networks were highly nested (Table 4). The pollination networks were highly generalized. The most connected visitor groups were honeybees, Lepidoptera and Muscidae associated with 30, 29, and 28 plant species, respectively (Table A3). The most connected plant species were *Coreopsis grandiflora* and *Centaurea cyanus*, which were visited by eight groups. The most connected native plant species was *Nandina domestica*, which received visits by four groups, followed by *Rosa chinensis*, *Astragalus sinicus*, *Cirsium arvense* var. *integrifolium*, *Zabelia biflora*, and *Medicago falcata*, which were visited by three visitor groups (Table 5).

Table 4. Metrics of pollination networks of three urbanization levels.

Metrics	City	Suburb	Exurb	Total
Functional groups of flower-visitors	6	10	8	11
Plant Species	18	34	13	43
Links	754	1433	1928	4115
Max links of flower visitor groups	462	994	1078	2534
Max links of plants	177	420	1009	1018
Abundance of links	39	70	40	117
Abundance of potential links	126	340	104	473
Connectance	0.31	0.21	0.38	0.24
Matrix temperature	10.80	5.16	12.12	6.25
Nestedness	0.89	0.95	0.88	0.94

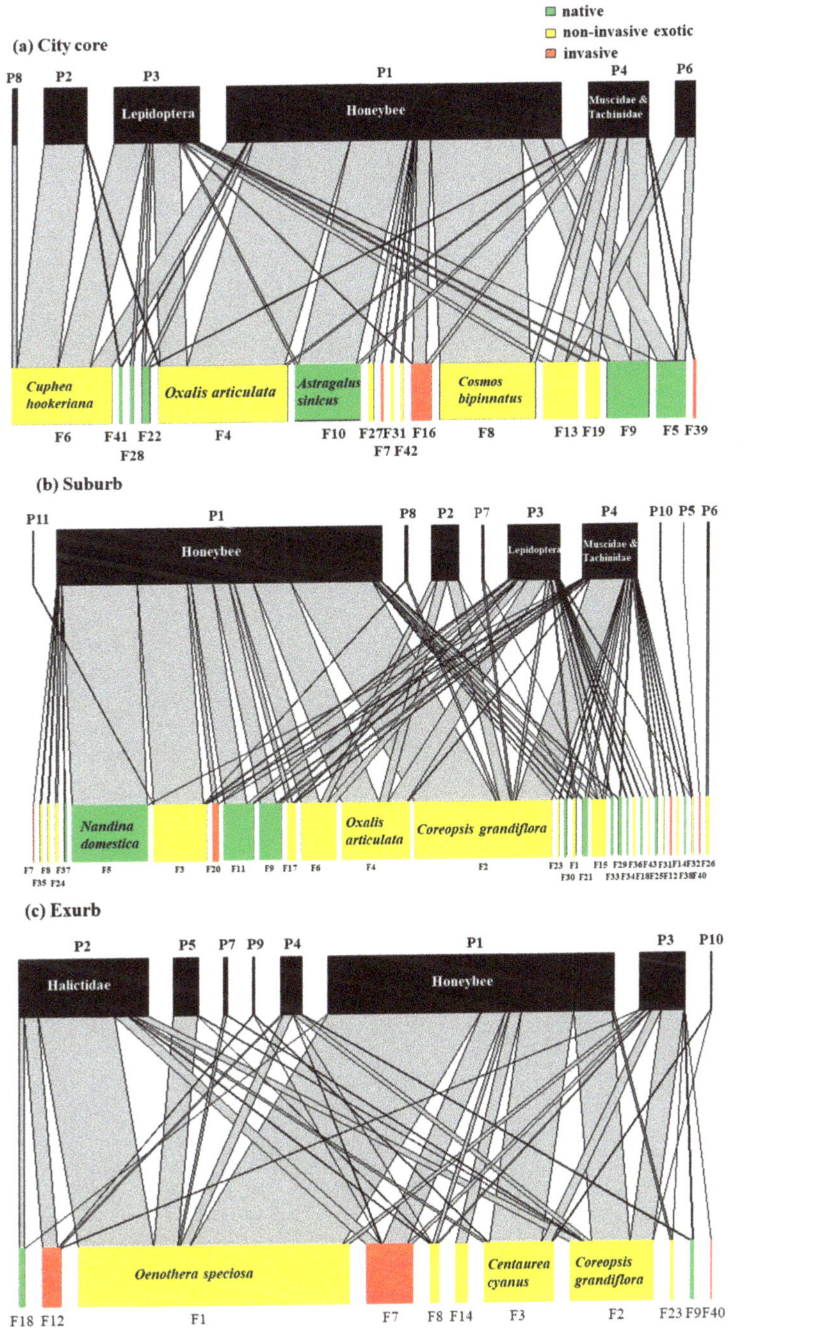

Figure 5. Visualization of the pollination networks of (**a**) City core, (**b**) Suburb, and (**c**) Exurb. The width of the lines connecting species is scaled to the number of links. Refer to Tables A2 and A3 for the code of plant species and flower-visitor groups in network.

Table 5. Code and links of native plant species in pollination network.

Code in Network	Plant Species	Number of Links	Number of Linked Pollinator Groups	Normalized Degree
F5	*Nandina domestica*	269	4	0.3636
F9	*Rosa chinensis*	135	3	0.2727
F10	*Astragalus sinicus*	89	3	0.2727
F11	*Cirsium arvense var. integrifolium*	87	3	0.2727
F18	*Zabelia biflora*	19	3	0.2727
F21	*Ligustrum sinense*	14	2	0.1818
F22	*Medicago falcata*	12	3	0.2727
F25	*Sedum lineare*	7	2	0.1818
F28	*Taraxacum mongolicum*	4	1	0.0909
F29	*Dianthus chinensis*	4	2	0.1818
F30	*Alcea rosea Linnaeus*	4	2	0.1818
F33	*Hemerocallis fulva*	3	2	0.1818
F34	*Acorus calamus*	3	1	0.0909
F37	*Orychophragmus violaceus*	2	1	0.0909
F38	*Vaccaria hispanica*	2	1	0.0909
F41	*Kerria japonica*	2	2	0.1818

4. Discussion

4.1. Effects of the Proportion of Impervious Surface to Flower-Visitor Groups

The results show that the proportion of impervious surface has no significant effect on the number of visits by flower-visitor groups (Table 2). Studies have reported that pollinators tend to prefer (semi-) natural habitats [42], but the visitor groups respond differently to urbanization. Social bees such as honeybees have different individuals specialized for pollen and nectar collection [43,44], and they are more likely to respond to the different spatial and temporal distribution of flower resources. The solitary bees experience landscapes at small scales [45,46], and may be particularly affected by landscape intensification and changing habitat [47]. The number of visits by other groups (Halictidae, Lepidoptera, Muscidae, Formicidae, etc.) was relatively less. The honeybee was the most abundant, and potentially drove the results.

The composition of flower-visitor groups differed in the three urbanization levels. The number of flower-visitor groups was the least in City core, where Vespidae and Ichneumonidae, Formicidae, Coleoptera, Odonata, and Orthoptera were not recorded. Since Vespidae, Formicidae, and Orthoptera are usually omnipresent and are as abundant in cities as in exurbs, this likely indicates a lack of detection due to low sampling effort in the study sites. The diversity of visitor groups negatively correlated with the proportion of impervious surface (Table 3, Figure 4a). This trend is consistent with previous studies. Abbate et al. (2019) found that species richness and diversity of bees were lower in highly urbanized areas [48]. Burdine and McCluney (2019) determined that bee diversity decreased as the proportion of impervious surface increased [49]. Collado et al. (2019) compared species richness and diversity of bees in natural habitats with those in urban core areas and found that natural habitat was better able to maintain bee diversity [42]. Persson et al. (2020) determined that the species richness of wild bees was decreased in densely built-up areas [50]. Our approach of categorizing flower-visitors to functional groups has limitations in clarifying the effect of urbanization on pollinator species, and thus may hardly be compared with findings of previous studies at species level. More studies at a species level would better clarify how different genera or species react to urbanization.

4.2. Effects of Proportion of Green Space on Flower-Visitor Groups

The proportion of green space caused no significant effect on visitation by honeybees, Halictidae, and Lepidoptera, but, contrary to previous studies, it was negatively associated with the number of visits by Muscidae and reduced the diversity of flower-visitor groups (Tables 2 and 3, Figure 4b). Green space is usually considered to be one of the important habitats for pollinators, and a high proportion of green space can significantly increase the number of pollinators [51,52]. However, studies also found that visitation by honeybees did not decrease as the distance between the study site and semi-natural habitat increased [53,54]. This was likely due to the wider foraging range available and a better adaptation to environmental change [53]. Honeybees were abundant and may have driven the results. Additionally, the low detection of visits by Halictidae, Lepidoptera, and other rare visitor groups increases the difficulty of analyzing the effect of the proportion of green space at our sample sites. The visits by Muscidae and the diversity of flower-visitor groups were negatively associated with the proportion of green space. The probable reason is that, in addition to the size or proportion of green space, the quality of green space also has a significant impact on pollinator visitation. Artificial green space in the city mainly exists in the form of park and residential green space, and only functions as a habitat connector or as temporary habitat. By contrast, natural habitat can better accommodate pollinators as their long-term habitat [42]. Various studies have investigated on how to enhance the habitat value of urban green space. First, 'Pollinator friendly gardens', with high flower diversity, abundance, and density, are considered to have a positive impact of maintaining pollinator diversity [24,55]. Second, as the pollinators diverge in active season, the selection of pollinator-attractive flowering plants should consider both the flowering period of plants and the seasonal activity of pollinators [31]. Third, management measures also play important role in pollinator conservation, including reducing the cutting frequency of meadows, retaining wide herbaceous margins and shrub layers with high nectar resources, and avoiding excessive growth in hedgerow height to reduce the barrier effect [25].

4.3. Effects of Night Light Intensity on Flower-Visitors

Night light intensity is caused by the level of urbanization and the magnitude of human activities, and it also has a significantly negative impact on nocturnal pollinators such as moths [56]. The artificial night lights reduce nocturnal pollination and the reproduction of plants. Diurnal pollinators also forage on these plants for food resource, so the effect of night light may extend through the plant–pollination network, and the diurnal pollinators are affected in the long run [27]. In this study, the night light intensity had no significant effect on the number of visits or the diversity of flower-visitor groups (Tables 2 and 3). This may be due to that no nocturnal pollinators were included in our study. In addition, the regeneration of flowering plants is frequent in urban parks, so the reproduction of certain plant species can't cause long-term effects on plant–pollinator interactions.

4.4. Effects of Local Flower Density and Flowering Plant Richness to Flower-Visitor Groups

Flower density and the species richness of flowering plants generally have positive effects on pollination, such as increasing pollinator visits [57], and diversity [19,58,59]. In this study, flower density positively correlated with the number of visits by Halictidae (Table 2). Flowering plant species richness was positively associated with the number of visits by Lepidoptera and the diversity of flower-visitor functional groups (Tables 2 and 3, Figure 4c). Diverse plant resources provide food and nesting resources for more pollinator groups. Studies have shown that high plant species richness has a positive effect on maintaining butterfly species richness due to their rapid response to changes in flowering plant abundance and species richness [60].

4.5. Effects of Urbanization Level on Flowering Plant–Visitor Group Interactions and Core Species

Suburb sites exhibited the highest diversity of interactions. However, the Suburb sites also contained high number of flower-visitor groups and plant species, and because of this, it is possible that some potential links may have been missed (Table 4). Among the three urbanization levels, the pollination networks were asymmetric, highly nested, and generalized. When compared with natural habitats and arable land, the urban areas are occupied by more generalists, which are supported by a higher number of plant species [61].

Core functional groups of flower-visitors (honeybee, Halictidae, Lepidoptera, and Muscidae) contribute the most links (Table A3). Core flowering plants (*Oenothera speciosa, Coreopsis grandiflora, Centaurea cyanus,* and *Oxalis articulata*) are exotic plant species and occurred at five, four, three, and four sites, respectively (Table A2). In the urban green space, flower-visitors may adapt and forage on exotic flowering plants, which somewhat enhances pollinator diversity by providing nectar or pollen at certain seasons, when native plants are not actively in flowering period [31,62,63]. However, the presence of attractive exotic and horticulturally modified flowering plants may displace endemic plants that pollinators are adapted to, and compete with native plants for pollinators, thus altering the plant–pollinator network [64–66]. The native plant species, *Nandina domestica, Rosa chinensis, Astragalus sinicus, Cirsium arvense* var. *integrifolium*, and *Zabelia biflora*, were relatively attractive to a diversity of visitor groups (Table 5). The use of these native flowering plants can help maintain the stability of flower–pollinator network and mitigate the negative effects of urbanization.

4.6. *Effects of Sampling Month to Flower-Visitor Groups*

The survey covered three months, during which the plant species, richness, and flower density changed with the seasonal climate. There are significant seasonal differences in the activity of various pollinator groups. Maintaining a diverse set of flower-visitor groups can help meet the pollination needs of different flowering plants.

5. Conclusions

This study has some limitations: Firstly, the approach of identifying flower-visitors to broad taxonomic levels (mostly genus or family) rather than species was inadequate for clarifying the responses of different flower-visitor species to urbanization. The conservation status of different flower-visitor species was ignored. Secondly, most of the interactions were between the abundant plant species and main flower-visitor groups, which are widely distributed and generalized. The sampling effort was not adequate to assess the flower-visitor assemblage of less abundant plants, as well as the effect of urbanization on rare and specialized flower-visitors. Thirdly, in the field survey of flower-visitors, transects in the plot were close, so the same individual may have been counted twice or more. To ensure data independence, transects should ideally be linear.

This study provides suggestions for the planning of urban green spaces and provides a theoretical reference for local vegetation configurations. Improvement of the quality of green spaces (high plant species richness, and flower density) and the use of attractive native flowering plants (such as *Nandina domestica, Rosa chinensis, Astragalus sinicus, Cirsium arvense* var. *integrifolium*, and *Zabelia biflora*) can help to reduce the negative effects of city expansion on pollinator diversity, and enhancing the function of urban green space in sustaining biodiversity and ecosystem stability.

Supplementary Materials: The following supporting information can be downloaded at: https://www.mdpi.com/article/10.3390/d14030208/s1, Table S1: Species list of flower-visitors according to DNA sequence comparisons; Table S2: The visits and diversity of flower-visitors; Table S3: The interaction between plants and flower-visitor groups under three urbanization levels.

Author Contributions: Development of the idea, H.W.; data collection, M.H., N.R. and H.J.; data analysis, M.H., Z.H. and Y.D.; writing-original draft preparation, M.H.; writing-review and editing, H.W., X.L. and P.A.B.; supervision of the project, D.X. All authors have read and agreed to the published version of the manuscript.

Funding: This research was funded by the National Natural Science Foundation of China, grant number 31600189, 32071683.

Institutional Review Board Statement: Not applicable.

Informed Consent Statement: Not applicable.

Data Availability Statement: Data is contained within the supplementary material.

Conflicts of Interest: The authors declare no conflict of interest.

Appendix A

Table A1. Information of Study Sites.

Study Site	Latitude	Longitude	Proportion of Impervious Surface	Proportion of Greenspace	Average Night Light Intensity (DN)
City core					
Jiefang Park	30.6091	114.2911	87.35%	22.87%	45141
Shahu Park	30.5758	114.3416	81.95%	17.13%	50630
Luojiashan Parkland	30.5379	114.3561	86.32%	30.47%	30709
Lianhuahu Park	30.5517	114.2741	82.58%	24.29%	38430
Shouyi Cultural Park	30.5433	114.3006	84.75%	17.45%	45896
Zhongshan Park	30.5858	114.2674	91.42%	8.93%	45881
Simeitang Park	30.5965	114.3313	82.46%	13.07%	47913
Suburb					
Shizishan Parkland	30.4750	114.3380	58.76%	27.81%	25860
Houxianghe Park	30.6115	114.2482	68.63%	11.45%	47191
Heping Park	30.6369	114.3814	62.38%	21.31%	28522
Wuhan Botanical Garden	30.5483	114.4160	53.32%	49.76%	10606
Dijiao Park	30.6655	114.3331	56.96%	11.27%	21320
Guanshan Holland Park	30.4964	114.3927	73.86%	12.05%	27054
Exurb					
Wuhan Garden Expo Park	30.6209	114.2156	44.67%	28.30%	38582
Shimenfeng Park	30.5166	114.4735	18.43%	51.55%	16231
Canglongdao Wetland Park	30.4069	114.4188	37.20%	19.38%	21667
Jiangxia Central Park	30.3796	114.3159	44.67%	14.10%	24925
Zhuyehai Park	30.6237	114.1607	45.68%	17.89%	29033
Tanghu Park	30.4727	114.1556	40.86%	22.10%	22490

Note: Proportion of impervious surface was calculated within a 2000 m radius (excluding water cover) around the parks. Proportion of green space was calculated within a 2000 m radius (excluding water cover) around the study sites. DN refers to digital number of pixel brightness.

Table A2. List of Surveyed Plant Species.

Family	Species	Region of Origin	Status	Distribution Sites	Code in Pollination Network
Amaryllidaceae	*Tulbaghia violacea* Harv.	South Africa	non-invasive exotic	Suburb (Houxianghe Park, Wuhan Botanical Garden)	F36
Amaryllidaceae	*Zephyranthes carinata* Herbert	South America	invasive	Suburb (Wuhan Botanical Garden)	F53
Araceae	*Acorus calamus* L.	China	native	Suburb (Wuhan Botanical Garden)	F34
Asteraceae	*Carthamus tinctorius* L.	Central Asia	non-invasive exotic	Suburb (Wuhan Botanical Garden)	F35
Asteraceae	*Centaurea cyanus* L.	Europe, Russia, north America	non-invasive exotic	Suburb, Exurb (Shizishan Parkland, Wuhan Botanical Garden, Canglongdao Wetland Park)	F3
Asteraceae	*Cirsium arvense* var. *integrifolium* Wimm. & Grab.	East and north Asia	native	Suburb (Wuhan Botanical Garden)	F11
Asteraceae	*Coreopsis grandiflora* Hogg ex Sw.	America	non-invasive exotic	Suburb, Exurb (Houxianghe Park, Dijiao Park, Jiangxia Central Park, Tanghu Park)	F2
Asteraceae	*Cosmos bipinnatus* Cav.	Mexico	non-invasive exotic	City, Suburb, Exurb (Shouyi Cultural Park, Shizishan Parkland, Houxianghe Park, Wuhan Garden Expo Park)	F8
Asteraceae	*Erigeron annuus* (L.) Pers.	North America	invasive	City, Suburb, Exurb (Jiefang Park, Houxianghe Park, Canglongdao Wetland Park, Jiangxia Central Park, Tanghu Park)	F12
Asteraceae	*Euryops pectinatus* (L.) Cass.	South Africa	non-invasive exotic	Suburb (Houxianghe Park)	F17
Asteraceae	*Helianthus annuus* L.	North America	non-invasive exotic	Suburb (Houxianghe Park)	F26
Asteraceae	*Leucanthemum vulgare* Lam.	West Europe	non-invasive exotic	Suburb, Exurb (Houxianghe Park, Jiangxia Central Park)	F15
Asteraceae	*Sanvitalia procumbens* Lam.	Mexico	non-invasive exotic	Suburb, Exurb (Shizishan Parkland, Canglongdao Wetland Park)	F23
Asteraceae	*Silybum marianum* (L.) Gaertn.	Europe, Mediterranean, north Africa, central Asia	invasive	Suburb (Wuhan Botanical Garden)	F48
Asteraceae	*Tagetes erecta* L.	Mexico	non-invasive exotic	City, Suburb (Shouyi Cultural Park, Houxianghe Park)	F31
Asteraceae	*Taraxacum mongolicum* Hand.-Mazz.	China, north Korea, Mongolia, Russia	native	City (Shahu Park, Lianhuahu Park)	F28
Berberidaceae	*Nandina domestica* Thunb.	China	native	City, Suburb (Jiefang Park, Lianhuahu Park, Guanshan Holland Park)	F5

Table A2. *Cont.*

Family	Species	Region of Origin	Status	Distribution Sites	Code in Pollination Network
Campanulaceae	*Platycodon grandiflorus* (Jacq.) A. DC.	East and north Asia	native	Suburb (Wuhan Botanical Garden)	F51
Cannaceae	*Canna indica* L.	Japan	non-invasive exotic	Suburb (Wuhan Botanical Garden)	F52
Caprifoliaceae	*Weigela florida* (Bunge) A. DC.	China, japan, india, america	native	City (Shahu Park)	F44
Caprifoliaceae	*Zabelia biflora* (Turcz.) Makino	China, korea	native	Suburb, Exurb (Wuhan Botanical Garden, Tanghu Park)	F18
Caryophyllaceae	*Dianthus chinensis* L.	China	native	Suburb (Houxianghe Park)	F29
Caryophyllaceae	*Vaccaria hispanica* (Miller) Rauschert	Europe, asia	native	Suburb (Shizishan Parkland)	F38
Crassulaceae	*Sedum lineare* Thunb.	China, japan, vietnam	native	Suburb (Houxianghe Park)	F25
Cruciferae	*Orychophragmus violaceus* (Linnaeus) O. E. Schulz	China, north Korea	native	Suburb (Shizishan Parkland)	F37
Geraniaceae	*Geranium carolinianum* L.	America	invasive	City (Lianhuahu Park)	F46
Lamiaceae	*Salvia miltiorrhiza* Bunge	China, japan	native	Suburb (Wuhan Botanical Garden)	F47
Lamiaceae	*Scutellaria barbata* D. Don	India, nepal, myanmar, laos, thailand	non-invasive exotic	City (Shouyi Cultural Park)	F42
Leguminosae	*Astragalus sinicus* L.	China	native	City (Luojiashan Parkland)	F10
Leguminosae	*Medicago falcata* L.	China	native	City (Shahu Park, Luojiashan Parkland, Lianhuahu Park)	F22
Leguminosae	*Trifolium pratense* L.	Central Europe	invasive	City (Shahu Park)	F16
Leguminosae	*Trifolium repens* L.	Europe, north Africa	invasive	City, Suburb, Exurb (Simeitang Park, Houxianghe Park, Zhuyehai Park, Tanghu Park)	F7
Liliaceae	*Hemerocallis fulva* (L.) L.	China, south Europe	native	Suburb (Wuhan Botanical Garden)	F33
Liliaceae	*Ophiopogon japonicus* (L. f.) Ker-Gawl.	China, japan, vietnam, india	native	City (Shahu Park)	F50
Lythraceae	*Cuphea hookeriana* Walp.	Mexico	non-invasive exotic	City, Suburb (Lianhuahu Park, Houxianghe Park)	F6
Malvaceae	*Alcea rosea* Linnaeus	China	native	Suburb (Wuhan Botanical Garden)	F30
Oleaceae	*Ligustrum sinense* Lour.	China	native	Suburb (Houxianghe Park)	F21
Onagraceae	*Oenothera biennis* L.	North America	invasive	Suburb (Wuhan Botanical Garden)	F20

Table A2. *Cont.*

Family	Species	Region of Origin	Status	Distribution Sites	Code in Pollination Network
Onagraceae	*Oenothera speciosa* Nutt.	America	non-invasive exotic	Suburb, Exurb (Shizishan Parkland, Wuhan Botanical Garden, Wuhan Garden Expo Park, Zhuyehai Park, Tanghu Park)	F1
Oxalidaceae	*Oxalis articulata* Savigny	South America	non-invasive exotic	City, Suburb (Shahu Park, Lianhuahu Park, Simeitang Park, Wuhan Botanical Garden	F4
Oxalidaceae	*Oxalis corniculata* L.	Temperate and subtropical Asia, Europe, Mediterranean, north America	native	City (Lianhuahu Park)	F49
Papaveraceae	*Papaver rhoeas* L.	Europe	non-invasive exotic	Suburb, Exurb (Shizishan Parkland, Wuhan Garden Expo Park, Canglongdao Wetland Park)	F14
Plantaginaceae	*Plantago depressa* Willd.	China, north Korea, Russia, Kazakhstan	native	Exurb (Canglongdao Wetland Park)	F45
Plantaginaceae	*Veronica persica* Poir.	Western Asia, Europe	invasive	City, Suburb, Exurb (Lianhuahu Park, Simeitang Park, Wuhan Botanical Garden, Wuhan Garden Expo Park)	F39
Portulacaceae	*Portulaca grandiflora* Hook.	Brazil	non-invasive exotic	City (Shouyi Cultural Park)	F27
Rosaceae	*Kerria japonica* (L.) DC.	China, japan	native	City (Lianhuahu Park)	F41
Rosaceae	*Rosa chinensis* Jacq.	China	native	City, Suburb, Exurb (Jiefang Park, Zhongshan Park, Houxianghe Park, Heping Park, Guanshan Holland Park, Shimenfeng Park)	F9
Saxifragaceae	*Heuchera micrantha* Douglas	Cenral america	non-invasive exotic	Suburb (Houxianghe Park)	F43
Solanaceae	*Petunia* × *atkinsiana* D. Don ex Loudon	South America	non-invasive exotic	City (Zhongshan Park	F19
Umbelliferae	*Coriandrum sativum* L.	Mediterranean	non-invasive exotic	Suburb (Wuhan Botanical Garden)	F24
Umbelliferae	*Daucus carota* L.	Europe	invasive	Suburb, Exurb (Wuhan Botanical Garden, Canglongdao Wetland Park)	F40
Verbenaceae	*Glandularia* × *hybrida* (Groenland & Rümpler) G. L.Nesom & Pruski	Panama, honduras, venezuela	non-invasive exotic	Suburb (Houxianghe Park)	F32
Violaceae	*Viola cornuta* Desf.	Europe	non-invasive exotic	City (Zhongshan Park)	F13

Table A3. Code and Links of Flower-Visitor Functional Groups in Pollination Network.

Code in Network	Functional Groups	Number of Links	Number of Linked Plant Species	Normalised Degree
P1	Honeybee	2534	30	0.6977
P2	Halictidae	625	10	0.2326
P3	Lepidoptera	449	29	0.6744
P4	Muscidae and Tachinidae	335	28	0.6512
P5	Formicidae	92	4	0.0930
P6	*Anthophora* spp.	35	3	0.0698
P7	Coleoptera	17	3	0.0698
P8	Syrphidae	15	3	0.0698
P9	Vespidae and Ichneumonidae	6	3	0.0698
P10	Odonata	5	3	0.0698
P11	Orthoptera	2	1	0.0233

References

1. Garibaldi, L.A.; Steffan-Dewenter, I.; Winfree, R.; Aizen, M.A.; Bommarco, R.; Cunningham, S.A.; Kremen, C.; Carvalheiro, L.G.; Harder, L.D.; Afik, O.; et al. Wild Pollinators Enhance Fruit Set of Crops Regardless of Honey Bee Abundance. *Science* **2013**, *339*, 1608–1611. [CrossRef] [PubMed]
2. Goulson, D.; Lye, G.C.; Darvill, B. Decline and Conservation of Bumble Bees. *Annu. Rev. Èntomol.* **2008**, *53*, 191–208. [CrossRef] [PubMed]
3. Bartomeus, I.; Ascher, J.S.; Gibbs, J.; Danforth, B.N.; Wagner, D.L.; Hedtke, S.M.; Winfree, R. Historical changes in northeastern US bee pollinators related to shared ecological traits. *Proc. Natl. Acad. Sci. USA* **2013**, *110*, 4656–4660. [CrossRef] [PubMed]
4. Carvalheiro, L.G.; Kunin, W.E.; Keil, P.; Aguirre-Gutiérrez, J.; Ellis, W.N.; Fox, R.; Groom, Q.; Hennekens, S.; Van Landuyt, W.; Maes, D.; et al. Species richness declines and biotic homogenisation have slowed down for NW-European pollinators and plants. *Ecol. Lett.* **2013**, *16*, 870–878. [CrossRef] [PubMed]
5. Goulson, D.; Nicholls, E.; Botías, C.; Rotheray, E.L. Bee declines driven by combined stress from parasites, pesticides, and lack of flowers. *Science* **2015**, *347*, 1255957. [CrossRef] [PubMed]
6. Johansson, V.; Gustafsson, L.; Andersson, P.; Hylander, K. Fewer butterflies and a different composition of bees, wasps and hoverflies on recently burned compared to unburned clear-cuts, regardless of burn severity. *For. Ecol. Manag.* **2020**, *463*, 118033. [CrossRef]
7. Hallmann, C.A.; Sorg, M.; Jongejans, E.; Siepel, H.; Hofland, N.; Schwan, H.; Stenmans, W.; Müller, A.; Sumser, H.; Hörren, T.; et al. More than 75 percent decline over 27 years in total flying insect biomass in protected areas. *PLoS ONE* **2017**, *12*, e0185809. [CrossRef]
8. Cameron, S.A.; Lozier, J.D.; Strange, J.P.; Koch, J.B.; Cordes, N.; Solter, L.F.; Griswold, T.L. Patterns of widespread decline in North American bumble bees. *Proc. Natl. Acad. Sci. USA* **2011**, *108*, 662–667. [CrossRef]
9. Potts, S.G.; Biesmeijer, J.C.; Kremen, C.; Neumann, P.; Schweiger, O.; Kunin, W.E. Global pollinator declines: Trends, impacts and drivers. *Trends Ecol. Evol.* **2010**, *25*, 345–353. [CrossRef]
10. González-Varo, J.P.; Biesmeijer, J.C.; Bommarco, R.; Potts, S.G.; Schweiger, O.; Smith, H.G.; Steffan-Dewenter, I.; Szentgyörgyi, H.; Woyciechowski, M.; Vilà, M. Combined effects of global change pressures on animal-mediated pollination. *Trends Ecol. Evol.* **2013**, *28*, 524–530. [CrossRef]
11. Wagner, D.L.; Van Driesche, R.G. Threats Posed to Rare or Endangered Insects by Invasions of Nonnative Species. *Annu. Rev. Èntomol.* **2010**, *55*, 547–568. [CrossRef] [PubMed]
12. Ahrné, K.; Bengtsson, J.; Elmqvist, T. Bumble Bees (*Bombus* spp.) along a gradient of increasing urbanization. *PLoS ONE* **2009**, *4*, 1–9. [CrossRef] [PubMed]
13. Wang, M.; Li, J.; Kuang, S.; He, Y.; Chen, G.; Huang, Y.; Song, C.; Anderson, P.; Łowicki, D. Plant Diversity along the Urban-Rural Gradient and Its Relationship with Urbanization Degree in Shanghai, China. *Forests* **2020**, *11*, 171. [CrossRef]
14. McKinney, M.L. Urbanization, Biodiversity, and Conservation. *Bioscience* **2002**, *52*, 883–888. [CrossRef]
15. Tscharntke, T.; Klein, A.M.; Kruess, A.; Steffan-Dewenter, I.; Thies, C. Landscape perspectives on agricultural intensification and biodiversity—Ecosystem service management. *Ecol. Lett.* **2005**, *8*, 857–874. [CrossRef]
16. Geslin, B.; Gauzens, B.; Thébault, E.; Dajoz, I. Plant Pollinator Networks along a Gradient of Urbanisation. *PLoS ONE* **2013**, *8*, e63421. [CrossRef]
17. Fortel, L.; Henry, M.; Guilbaud, L.; Guirao, A.L.; Kuhlmann, M.; Mouret, H.; Rollin, O.; Vaissière, B.E. Decreasing Abundance, Increasing Diversity and Changing Structure of the Wild Bee Community (Hymenoptera: Anthophila) along an Urbanization Gradient. *PLoS ONE* **2014**, *9*, e104679. [CrossRef]

18. Winfree, R.; Aguilar, R.; Vázquez, D.P.; Lebuhn, G.; Aizen, M.A. A meta-analysis of bees' responses to anthropogenic disturbance. *Ecology* **2009**, *90*, 2068–2076. [CrossRef] [PubMed]
19. Hausmann, S.L.; Petermann, J.; Rolff, J. Wild bees as pollinators of city trees. *Insect Conserv. Divers.* **2016**, *9*, 97–107. [CrossRef]
20. Theodorou, P.; Radzeviciute, R.; Settele, J.; Schweiger, O.; Murray, T.E.; Paxton, R.J. Pollination services enhanced with urbanization despite increasing pollinator parasitism. *Proc. R. Soc. B Boil. Sci.* **2016**, *283*, 20160561. [CrossRef]
21. Goulson, D.; Hughes, W.; Derwent, L.C.; Stout, J. Colony growth of the bumblebee, Bombus terrestris, in improved and conventional agricultural and suburban habitats. *Oecologia* **2002**, *130*, 267–273. [CrossRef] [PubMed]
22. Osborne, J.L.; Martin, A.P.; Shortall, C.R.; Todd, A.D.; Goulson, D.; Knight, M.E.; Hale, R.J.; Sanderson, R.A. Quantifying and comparing bumblebee nest densities in gardens and countryside habitats. *J. Appl. Ecol.* **2008**, *45*, 784–792. [CrossRef]
23. Goulson, D.; Lepais, O.; O'Connor, S.; Osborne, J.L.; Sanderson, R.A.; Cussans, J.; Goffe, L.; Darvill, B. Effects of land use at a landscape scale on bumblebee nest density and survival. *J. Appl. Ecol.* **2010**, *47*, 1207–1215. [CrossRef]
24. Majewska, A.A.; Altizer, S. Planting gardens to support insect pollinators. *Conserv. Biol.* **2018**, *34*, 15–25. [CrossRef]
25. Luppi, M.; Dondina, O.; Orioli, V.; Bani, L. Local and landscape drivers of butterfly richness and abundance in a human-dominated area. *Agric. Ecosyst. Environ.* **2018**, *254*, 138–148. [CrossRef]
26. Bubová, T.; Vrabec, V.; Kulma, M.; Nowicki, P. Land management impacts on European butterflies of conservation concern: A review. *J. Insect Conserv.* **2015**, *19*, 805–821. [CrossRef]
27. Knop, E.; Zoller, L.; Ryser, R.; Gerpe, C.; Hörler, M.; Fontaine, C. Artificial light at night as a new threat to pollination. *Nat.* **2017**, *548*, 206–209. [CrossRef]
28. Giavi, S.; Fontaine, C.; Knop, E. Impact of artificial light at night on diurnal plant-pollinator interactions. *Nat. Commun.* **2021**, *12*, 1690. [CrossRef]
29. Venjakob, C.; Klein, A.-M.; Ebeling, A.; Tscharntke, T.; Scherber, C. Plant diversity increases spatio-temporal niche complementarity in plant-pollinator interactions. *Ecol. Evol.* **2016**, *6*, 2249–2261. [CrossRef]
30. Garbuzov, M.; Ratnieks, F.L.W. Quantifying variation among garden plants in attractiveness to bees and other flower-visiting insects. *Funct. Ecol.* **2014**, *28*, 364–374. [CrossRef]
31. Tasker, P.; Reid, C.; Young, A.D.; Threlfall, C.G.; Latty, T. If you plant it, they will come: Quantifying attractiveness of exotic plants for winter-active flower visitors in community gardens. *Urban Ecosyst.* **2020**, *23*, 345–354. [CrossRef]
32. Gong, Y.-B.; Huang, S.-Q. Temporal stability of pollinator preference in an alpine plant community and its implications for the evolution of floral traits. *Oecologia* **2011**, *166*, 671–680. [CrossRef] [PubMed]
33. Wuhan Natural Resources and Planning Bureau. Wuhan's General Plan for Land Use (2006–2020). 2011. Available online: http://zrzyhgh.wuhan.gov.cn/ (accessed on 20 January 2022).
34. Gong, P.; Liu, H.; Zhang, M.; Li, C.; Wang, J.; Huang, H.; Clinton, N.; Ji, L.; Li, W.; Bai, Y.; et al. Stable classification with limited sample: Transferring a 30-m resolution sample set collected in 2015 to mapping 10-m resolution global land cover in 2017. *Sci. Bull.* **2019**, *64*, 370–373. [CrossRef]
35. Hagen, M.; Wikelski, M.; Kissling, W.D. Space Use of Bumblebees (*Bombus* spp.) Revealed by Radio-Tracking. *PLoS ONE* **2011**, *6*, e19997. [CrossRef] [PubMed]
36. Glaum, P.; Simao, M.-C.; Vaidya, C.; Fitch, G.; Iuliano, B. Big city Bombus: Using natural history and land-use history to find significant environmental drivers in bumble-bee declines in urban development. *R. Soc. Open Sci.* **2017**, *4*, 170156. [CrossRef]
37. Xie, Q.J.; Chen, K.L.; Jin, G. Quantifying urbanization and its impact on urban heat island effect in Wuhan. *Sci. Surv. Mapp.* **2017**, *42*, 71–76+87. (In Chinese)
38. McKinney, M.L. Effects of urbanization on species richness: A review of plants and animals. *Urban Ecosyst.* **2008**, *11*, 161–176. [CrossRef]
39. Baldock, K.C.R.; Goddard, M.; Hicks, D.M.; Kunin, W.E.; Mitschunas, N.; Osgathorpe, L.M.; Potts, S.G.; Robertson, K.M.; Scott, A.V.; Stone, G.; et al. Where is the UK's pollinator biodiversity? The importance of urban areas for flower-visiting insects. *Proc. R. Soc. B Boil. Sci.* **2015**, *282*, 20142849. [CrossRef]
40. Catalogue of Life. Available online: https://www.catalogueoflife.org/ (accessed on 20 January 2022).
41. Invasive Alien Species of China (ISAC). Available online: https://www.plantplus.cn/ias/protlist (accessed on 20 January 2022).
42. Collado, M.Á.; Sol, D.; Bartomeus, I. Bees use anthropogenic habitats despite strong natural habitat preferences. *Divers. Distrib.* **2019**, *25*, 924–935. [CrossRef]
43. Ollerton, J.; Waser, N.M. *Plant-Pollinator Interactions: From Specialization to Generalization*; University of Chicago Press: Chicago, IL, USA, 2006.
44. Klein, S.; Pasquaretta, C.; He, X.J.; Perry, C.; Søvik, E.; Devaud, J.-M.; Barron, A.B.; Lihoreau, M. Honey bees increase their foraging performance and frequency of pollen trips through experience. *Sci. Rep.* **2019**, *9*, 6778. [CrossRef]
45. Gathmann, A.; Tscharntke, T. Foraging ranges of solitary bees. *J. Anim. Ecol.* **2002**, *71*, 757–764. [CrossRef]
46. Zurbuchen, A.; Cheesman, S.; Klaiber, J.; Müller, A.; Hein, S.; Dorn, S. Long foraging distances impose high costs on offspring production in solitary bees. *J. Anim. Ecol.* **2010**, *79*, 674–681. [CrossRef]
47. Le Féon, V.; Schermann-Legionnet, A.; Delettre, Y.; Aviron, S.; Billeter, R.; Bugter, R.; Hendrickx, F.; Burel, F. Intensification of agriculture, landscape composition and wild bee communities: A large scale study in four European countries. *Agric. Ecosyst. Environ.* **2010**, *137*, 143–150. [CrossRef]

48. Abbate, A.; Campbell, J.W.; Kimmel, C.B.; Kern, W.H. Urban development decreases bee abundance and diversity within coastal dune systems. *Glob. Ecol. Conserv.* **2019**, *20*, e00711. [CrossRef]
49. Burdine, J.D.; McCluney, K.E. Interactive effects of urbanization and local habitat characteristics influence bee communities and flower visitation rates. *Oecologia* **2019**, *190*, 715–723. [CrossRef]
50. Persson, A.S.; Ekroos, J.; Olsson, P.; Smith, H.G. Wild bees and hoverflies respond differently to urbanisation, human population density and urban form. *Landsc. Urban Plan.* **2020**, *204*, 103901. [CrossRef]
51. Kratschmer, S.; Pachinger, B.; Schwantzer, M.; Paredes, D.; Guernion, M.; Burel, F.; Nicolai, A.; Strauss, P.; Bauer, T.; Kriechbaum, M.; et al. Tillage intensity or landscape features: What matters most for wild bee diversity in vineyards? *Agric. Ecosyst. Environ.* **2018**, *266*, 142–152. [CrossRef]
52. Beyer, N.; Gabriel, D.; Kirsch, F.; Schulz-Kesting, K.; Dauber, J.; Westphal, C. Functional groups of wild bees respond differently to faba bean Vicia faba L. cultivation at landscape scale. *J. Appl. Ecol.* **2020**, *57*, 2499–2508. [CrossRef]
53. Garibaldi, L.A.; Steffan-Dewenter, I.; Kremen, C.; Morales, J.M.; Bommarco, R.; Cunningham, S.; Carvalheiro, L.G.; Chacoff, N.P.; Dudenhöffer, J.-H.; Greenleaf, S.S.; et al. Stability of pollination services decreases with isolation from natural areas despite honey bee visits. *Ecol. Lett.* **2011**, *14*, 1062–1072. [CrossRef]
54. Rader, R.; Bartomeus, I.; Garibaldi, L.A.; Garratt, M.P.D.; Howlett, B.G.; Winfree, R.; Cunningham, S.A.; Mayfield, M.M.; Arthur, A.D.; Andersson, G.K.S.; et al. Non-bee insects are important contributors to global crop pollination. *Proc. Natl. Acad. Sci. USA* **2016**, *113*, 146–151. [CrossRef]
55. Threlfall, C.G.; Walker, K.; Williams, N.S.; Hahs, A.K.; Mata, L.; Stork, N.; Livesley, S.J. The conservation value of urban green space habitats for Australian native bee communities. *Biol. Conserv.* **2015**, *187*, 240–248. [CrossRef]
56. MacGregor, C.J.; Pocock, M.J.O.; Fox, R.; Evans, D.M. Pollination by nocturnal L epidoptera, and the effects of light pollution: A review. *Ecol. Èntomol.* **2015**, *40*, 187–198. [CrossRef] [PubMed]
57. Ghazoul, J. Floral diversity and the facilitation of pollination. *J. Ecol.* **2006**, *94*, 295–304. [CrossRef]
58. Blaauw, B.R.; Isaacs, R. Larger patches of diverse floral resources increase insect pollinator density, diversity, and their pollination of native wildflowers. *Basic Appl. Ecol.* **2014**, *15*, 701–711. [CrossRef]
59. Vrdoljak, S.M.; Samways, M.J.; Simaika, J.P. Pollinator conservation at the local scale: Flower density, diversity and community structure increase flower visiting insect activity to mixed floral stands. *J. Insect Conserv.* **2016**, *20*, 711–721. [CrossRef]
60. Myers, M.C.; Hoksch, B.J.; Mason, J.T. Butterfly response to floral resources during early establishment at a heterogeneous prairie biomass production site in Iowa, USA. *J. Insect Conserv.* **2012**, *16*, 457–472. [CrossRef]
61. Theodorou, P.; Albig, K.; Radzeviciute, R.; Settele, J.; Schweiger, O.; Murray, T.E.; Paxton, R. The structure of flower visitor networks in relation to pollination across an agricultural to urban gradient. *Funct. Ecol.* **2017**, *31*, 838–847. [CrossRef]
62. Senapathi, D.; Goddard, M.; Kunin, W.E.; Baldock, K.C.R. Landscape impacts on pollinator communities in temperate systems: Evidence and knowledge gaps. *Funct. Ecol.* **2017**, *31*, 26–37. [CrossRef]
63. Stelzer, R.J.; Chittka, L.; Carlton, M.; Ings, T.C. Winter Active Bumblebees (*Bombus terrestris*) Achieve High Foraging Rates in Urban Britain. *PLoS ONE* **2010**, *5*, e9559. [CrossRef]
64. Grass, I.; Berens, D.G.; Peter, F.; Farwig, N. Additive effects of exotic plant abundance and land-use intensity on plant–pollinator interactions. *Oecologia* **2013**, *173*, 913–923. [CrossRef]
65. Stout, J.C.; Tiedeken, E.J. Direct interactions between invasive plants and native pollinators: Evidence, impacts and approaches. *Funct. Ecol.* **2017**, *31*, 38–46. [CrossRef]
66. Hansen, S.; Roets, F.; Seymour, C.L.; Thébault, E.; Van Veen, F.F.; Pryke, J.S. Alien plants have greater impact than habitat fragmentation on native insect flower visitation networks. *Divers. Distrib.* **2018**, *24*, 58–68. [CrossRef]

diversity

MDPI

Article

Response of Phytoplankton Community Structure to Vegetation Restoration after Removal of Purse Seine in Shengjin Lake

Xudong Zheng, Jingwen Chen, Wenli Guo, Zhongze Zhou and Xiaoxin Ye *

School of Resources and Environmental Engineering, Anhui University, Hefei 230601, China; x19301057@stu.ahu.edu.cn (X.Z.); x19201018@stu.ahu.edu.cn (J.C.); x19301100@stu.ahu.edu.cn (W.G.); zhzz@ahu.edu.cn (Z.Z.)

* Correspondence: yexx@ahu.edu.cn

Abstract: Aquatic vegetation has been restored since the removal of seine nets from the lake surface of Shengjin Lake in 2018. Through four seasons of phytoplankton sampling surveys from 2019–2020, we analyzed spatial and temporal changes in phytoplankton communities, water quality, and aquatic plant recovery in conjunction with previous research literature to reveal the response mechanisms of phytoplankton community structure to rapidly recovering aquatic vegetation. The results showed that the Secchi depth increased (0.4 m to 0.7 m), the concentration of total phosphorus decreased (0.053 mg/L to 0.41 mg/L), the species of aquatic plants (5 species to 16 species), phytoplankton species (210 species to 254 species) and cell density increased after the removal of the seine. Since the removal of the seine of Shengjin Lake, the aquatic vegetation cover has exceeded 80%, the phytoplankton biodiversity has increased, and the water quality has recovered to II-III water status. Our results show that aquatic plants improve water quality through direct and indirect effects and influence phytoplankton community structure together with the water environment, which can provide guidance for the restoration situation of the middle and lower reaches of the Yangtze River through-river lake ecosystems.

Keywords: vegetation restoration; phytoplankton; Shengjin Lake; environmental factors

Citation: Zheng, X.; Chen, J.; Guo, W.; Zhou, Z.; Ye, X. Response of Phytoplankton Community Structure to Vegetation Restoration after Removal of Purse Seine in Shengjin Lake. *Diversity* **2022**, *14*, 178. https://doi.org/10.3390/d14030178

Academic Editor: Mario A. Pagnotta

Received: 15 January 2022
Accepted: 20 February 2022
Published: 28 February 2022

1. Introduction

Aquatic ecosystems are responsible for the energy flow and circulation in the food chain of lakes, and phytoplankton is the primary producer in the aquatic ecosystem (Plus et al., 2015). Because of the high sensitivity of phytoplankton to the environment, it is used as an indicator in ecological evaluation [1,2]. By studying the changes and characteristics of phytoplankton community structure, the water quality of lakes was analyzed, and the nutritional status of lakes was evaluated. Generally speaking, the change of phytoplankton community structure is mainly affected by water temperature (WT), pH, Secchi depth (SD), dissolved oxygen (DO), and other hydrological conditions. The grazing of fish and zooplankton in water is also one of the factors limiting the growth of phytoplankton [3–5]. Studies have shown that excessive inflow of nutrients such as nitrogen and phosphorus will lead to a sharp increase in the number of phytoplankton, especially Cyanophyta, which is also the cause of eutrophication in most lakes in China [6,7]. However, a single environmental factor can not effectively describe the complex dynamic changes of phytoplankton, and each environmental factor has a coordinated effect on the phytoplankton community. In order to analyze which environmental factors play a leading role in affecting the phytoplankton community structure, we conducted an in-depth study of the lake.

There is a large area of lakes in the middle and lower reaches of the Yangtze River in Anhui Province. These lakes are connected with the Yangtze River, forming the important and unique aquatic ecosystem of a river-connected lake. At the same time, these lakes constitute both an ecological barrier and an important ecological function of the economically developed area of the Yangtze River Delta. The unique hydrological conditions formed by the connection between river-connected lake and the Yangtze River cause the lake water level (WL) to change seasonally with the Yangtze River. The changing WL, velocity, and WT are the basic elements to maintain the vitality of the wetland ecosystem, which is also conducive to the feeding and reproduction of waterfowl and fish, and improve their biological species and density [8]. During the summer, rainfall is heavy. As a result, the Yangtze River water flow increases to the lakes, leading to a rise in WL. The WL reaches its highest point in August and remains fairly constant until after September. This creates a period of high water. After September, the amount of water in the Yangtze River decreases, the lake sluice opens, and the water flows back into the Yangtze River, replenishing it. The WL drops to its lowest point in the winter (after December), and the shoals around the lake are exposed. This is the low water period. At the beginning of April, the WL gradually returns to normal [9]. Shengjin Lake is a typical river-connected lake in the middle and lower reaches of the Yangtze River. It is a National Nature Reserve in Anhui Province that protects wintering migratory birds, and more than a million waterfowls gather to overwinter every year [10]. This is the most important wintering place for whiteheaded cranes in China, an important resting place for waterbirds on the migratory route from East Asia to Australasia, and a place for fish migration and breeding in the Yangtze River [11–13]. However, starting in 1997, the local residents developed a purse seine fishery based on river crab culture in order to meet the needs of the economy. This overexploited the natural resources. Consequently, there are a series of environmental problems in Shengjin Lake: the aquatic vegetation is seriously degraded, the submerged vegetation has almost disappeared, the ecological function is degraded, the water quality deteriorates, and the eutrophication has intensified [14]; the number of birds feeding on the roots of submerged plants has decreased due to the decrease of food; purse seine blocks the hydrological connectivity between Shengjin Lake and the Yangtze River, resulting in the loss of spawning and migration channels for migratory fish, the destruction of fish physiological habits and the reduction of fish species.

The Chinese government attaches great importance to the ecological restoration of Shengjin Lake, Huayang Lake, Caizi Lake, and other lakes in the Yangtze River Basin. Since 2016, the government has implemented the Yangtze River protection policy. The government has invested hundreds of millions of funds to dismantle the seine in the lake, ban fishing and compensate the fishermen, enforcing the "no seine, no fishing, no boat" of the lake, to restore the ecological diversity of the lake and maintain the stability of wetland ecosystem function. After the removal of purse seine in 2018–2019, the species of overwintering waterbirds increased to 29, and the diversity of fish and birds increased [15]. In the investigation of Huayang Lake, the diversity of fish showed similar changes [16]. Since the complete removal of the seine net on Shengjin Lake in 2018, aquatic vegetation has recovered rapidly. Through regular monitoring of the aquatic vegetation of Shengjin Lake, we have found that the emerged plant *Zizania caduuciflora* and the floating plant *Trapa incisa* has a large distribution in the lake area of Shengjin Lake as pioneer species, especially in some waters of the upper lake (UL), where the cover of *Z. caduuciflora* and *T. incisa* is as high as more than 90%, and the submerged plant *Vallisineria spirali*s and *Myriophyllum verticillatum* are the main companion species.

Studies have shown that aquatic vegetation will either directly or indirectly affect the phytoplankton community structure, depending on the type of vegetation. Submerged macrophytes absorb a lot of nitrogen, phosphorus, and inorganic carbon through their rhizomes, their leaves block light, and their bodies secrete allelochemicals; large areas of aquatic vegetation weaken the effect of water flow and wind and impacts the hydrological conditions such as WT and DO [17,18]. On the other hand, aquatic vegetation provides a living environment for zooplankton to limit predation by large carnivorous fish, which increases the biomass of zooplankton and limits the reproduction of phytoplankton [19,20]. Therefore, in order to study the effects of aquatic vegetation restoration and lake environmental factors on the phytoplankton community structure, long-term monitoring of Shengjin Lake is needed.

In this study, our research team regularly monitored the water environmental factors, aquatic vegetation distribution, and phytoplankton community structure of Shengjin Lake from April 2019 to January 2020 with the following purpose: (1) to find out the distribution of aquatic vegetation through field investigation and remote sensing monitoring data; (2) by analyzing the temporal and spatial changes of phytoplankton community structure and water quality data, combined with the previous research literature, the response mechanism of phytoplankton community structure to rapidly recovering aquatic vegetation was revealed. The results can provide a more comprehensive scientific basis for the management and protection of freshwater lakes in the middle and lower reaches of the Yangtze River.

2. Materials and Methods

2.1. Description of Study Area

The study area is Shengjin Lake National Nature Reserve (E116°55′~117°15′, N30°15′~30°30′), which is located in the South Bank of the middle and lower reaches of the Yangtze River in Anhui Province, China. It is connected with the Yangtze River through a waterway, Huangpeng Sluice. The main flood season of the Yangtze River is from July to September, which is basically the same as the rainy season due to the influence of rainfall in the basin. Therefore, the WL of Shengjin Lake rises and falls with the WL of the Yangtze River. The water source comes from Zhangxi River in the southeast and Tangtian River in the northeast. The lake area is 132.8 square kilometers. The WL ranges from 0.20 to 8.74 m, and the average WL is 5.21 m. The area has a subtropical monsoon climate with four distinct seasons, abundant rainfall, and sunshine. The annual average temperature was 16.1 °C, the average summer temperature was 28.8 °C, and the average winter temperature was 3.9 °C. Under the influence of the monsoon, the rainfall has obvious seasonal changes, with the largest rainfall in summer and an average annual rainfall of 1600 mm with abundant surface runoff [21].

2.2. Data Collection

We collected the samples upper (UL), middle, and lower (ML) of Shengjin Lake, with 8 sections and 24 sampling points (Figure 1). The sampling point SJ1~12 was located in the UL and SJ13~24 in the ML, including SJ1~2, SJ5~9 and SJ11~14, which was covered by the floating plant *T. incisa*, and SJ3, which was covered by the emerged plant *Z. caduuciflora*. From April 2019 to January 2020, we collected 4 samples that represented the water quality of the lake in spring (April), summer (August), autumn (October), and winter (January of the following year).

We used the SX751 portable water quality analyzer (Sanxin Instrument Co., Ltd. in Shanghai, China) to measure the water quality of Shengjin Lake on site, including WT, DO, Cond, and pH. We used a Hach 2100Q portable turbidimeter to calculate the turbidity (Turb). SD was measured with the Secchi disk, and WL was measured with a scale lead hammer. Chl a and nutrient concentrations were measured in the laboratory [22], and total phosphorus (TP) was determined by ammonium molybdate spectrophotometry, total nitrogen (TN) by alkaline potassium persulfate ablation, and Chl a by acetone extraction spectrophotometry under an ultraviolet spectrophotometer (UV 2450), respectively [23].

For quantitative analysis of phytoplankton, we collected 1 liter (L) of water sample and added 10 milliliter (mL) of 1% Lugol iodine solution for fixation. After letting it stand for 48 h, we slowly siphoned the supernatant to 30 mL in order to identify the phytoplankton [24,25]. Under the optical microscope, we counted 0.1 mL samples with 40 × 10 magnification and randomly selected 100 fields for the identification and counting of each sample [26]. We identified phytoplankton species based on morphology [27]. The biomass of phytoplankton is mainly measured by individual traits. Phytoplankton abundance and biomass were expressed as cell/L and mg/L, respectively [28].

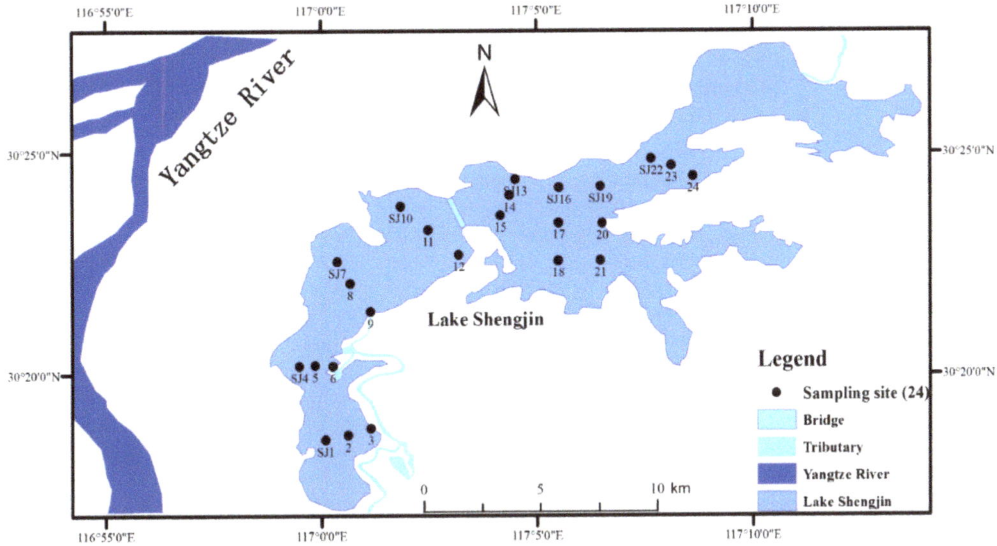

Figure 1. Distribution of sampling points in the study area of Shengjin Lake.

2.3. Data Analysis

The calculation formula of Mcnaughton dominance index (Y) is as follows:

$$Y = n_i/N \times f_i \tag{1}$$

In order to analyze the community structure of the phytoplankton, we applied the Shannon-Wiener index (H'), the Pielou index (J'), and the Margalef index (D). The calculation formulas are as follows [29]:

$$H' = -\sum_{i=1}^{S} \frac{n_i}{N} \ln \frac{n_i}{N} \tag{2}$$

$$J' = H' / \ln S \tag{3}$$

$$D = (S-1) / \ln N \tag{4}$$

where n_i is the number of individuals of species i, N is the total number of individuals of all species, S is the total number of phytoplankton species, and f_i is the frequency of individuals in the i species; following [30], we consider (Y) $\geq$ 0.02 as an indication of phytoplankton dominance.

Using IBM SPSS Statistics (Version 26.0), we conducted a one-way ANOVA method to test for significant differences between environmental parameters and phytoplankton communities' cell density between seasons. We evaluated the effect of water quality parameters on the phytoplankton density with the Pearson correlation analysis, using IBM SPSS26.0 Statistics, according to the correlation coefficient, the significant correlation between water quality parameters, and phytoplankton density was tested. We performed RDA utilizing Canoco 5 to explain the relationship between dominant phytoplankton communities and environmental factors. To reduce the effect of rare phytoplankton species, we transformed the phytoplankton data into log10(x + 1) format before data analysis. The results of RDA were visualized in the form of ordination plots based on AX 1 and AX 2 axes generated by Canoco for Windows 5 software. Environmental variables and species are designated by arrows. The position of each species in the ordination plot indicates the point corresponding to the optimum value of that species in the gradient. We produced the bar graphs with Microsoft Excel and the spatial interpolation of cell abundance and the distribution of aquatic vegetation with the ArcGIS 10.2 software applying the kriging method.

3. Results

3.1. Environmental Characteristics Parameters

The following table shows the main environmental factors of Shengjin Lake from 2019 to 2020 (Table 1). The results revealed that there were significant differences in the seasonal and spatial distribution of WT in Shengjin Lake ($p < 0.01$). During the study period, when the WT was 8~35.7 °C, the highest WT appeared in the summer and the lowest in the winter. The pH was between 7.47~9.26, the mean was 8.68, and the whole lake was weakly alkaline. DO is significantly higher in the winter than in the other seasons, with an average of 11.75 mg/L, and the lowest in summer, with an average of 7.69 mg/L. One-way ANOVA showed that WL and SD in summer were significantly higher than those in other seasons, ranging from 0.31 m to 6.30 m and 0.07 m to 2.45 m, respectively. Pearson correlation results there was a significant correlation between WL and SD (* $p < 0.05$). The concentration of nitrogen and phosphorus is one of the important indexes to evaluate the health and self-purification of lake water. The concentration of TN ranges from 0.08 mg/L to 1.62 mg/L. The concentration of TP ranged from 0.003 mg/L to 0.158 mg/L, and the concentration of nitrogen and phosphorus in the UL was lower than that in the ML. The concentration of nitrate nitrogen (NO_3-N) ranged from 0.001 mg/L to 3.41 mg/L. The average concentration of NO_3-N in the whole lake was the lowest in the summer when the plants were growing vigorously, and it gradually increased with the seasonal changes. The average concentration of NO_3-N in the spring was the highest. In general, the water quality of Shengjin Lake was in the II-III standard during the study period.

3.2. Distribution of Aquatic Vegetation

Shengjin Lake has abundant precipitation in July and August, and the lake is in a period of abundant water. A total of 16 aquatic plants were monitored during the study. The aquatic plants detected in the spring included *Ceratophyllum demersum, Hydrilla verticillata, Utricularia aurea, T. incisa, Euryale ferox, Nelumbo nucifera, Z. caduuciflora*, etc. Figure 2 illustrates the distribution of dominant associations of main aquatic plants' spread in the lake during the summer: (1) the aquatic plant communities in the lake at SJ1~2 of the UL District were Assoziation. *Trapa incisa-N. nucifera*, associated species are *Z. caduuciflora* and *V. spiralis*, and the coverage was 95%; (2) the aquatic plant community distributed on the lake surface at SJ3 is Ass. *Zizania caduuciflora*, the coverage was 90%, associated plants are *H. verticillata, C. demersum*; (3) SJ5~9 is the central water area of the UL, and the distribution of floating plant is Ass.*Trapa* sp., and the coverage was 55%, the associated species are the emerged plant *Z. caduuciflora* and the submerged plant *M. verticillatum, H. verticillata;* (4) Ass. *Polygonum lapathifolium* is an emerged plant distributed in the water area of SJ7, and the coverage was 15%; (5) there is a large area of Ass.*Trapa incisa* near the bridge, covering more than 90%. There is no aquatic vegetation with large coverage in the open water area of the ML and a small number of Ass.*Trapa incisa* is distributed near the shore. Compared with the UL, the coverage of aquatic vegetation in the ML is greatly reduced. Compared with the spring, vegetation coverage in the summer increased greatly. As the dominant aquatic plants in Shengjin Lake are *Z. caduuciflora* and *T. incisa*, both of which are thermophilic plants. The suitable temperature for growth is 10–25 °C, and they will die in autumn and winter. On the contrary, *Potamogeton crispus* is a typical submerged plant that germinates in autumn and grows over the winter. Its buds germinate in early October. In the autumn and winter, we monitor the newly grown submerged plants *Potamogeton crispus*. They were scattered in the shallow coastal waters while the other aquatic plants gradually disappeared.

Table 1. The average value of environmental parameters of Shengjin Lake.

Environmental Factors	Spring (April)		Summer (July)		Autumn (October)		Winter (November)		Seasonal *p*-Value	Spatial *p*-Value
	UL	ML	UL	ML	UL	ML	UL	ML		
WT (°C)	27.66 ± 0.73	27.25 ± 0.31	34.92 ± 0.48	34.43 ± 0.15	17.35 ± 2.65	16.27 ± 0.26	8.57 ± 0.25	8.13 ± 0.11	**	*
pH	8.47 ± 0.45	9.03 ± 0.15	8.6 ± 0.18	8.77 ± 0.1	8.64 ± 0.36	8.43 ± 0.13	8.77 ± 0.04	8.52 ± 0.18	-	-
DO (mg/L)	7.98 ± 1.02	8.39 ± 0.29	7.67 ± 0.39	7.72 ± 0.16	9.73 ± 0.17	9.45 ± 0.08	11.75 ± 0.11	11.77 ± 0.15	*	*
Cond (Us/cm)	116.05 ± 6.82	111.21 ± 2.16	190.57 ± 6.5	177.04 ± 3.36	247.5 ± 8.5	234.5 ± 7.76	157.73 ± 2.44	154.38 ± 3.52	*	-
SD (m)	1.08 ± 0.1	0.86 ± 0.12	1.37 ± 0.56	1.14 ± 0.19	0.20 ± 0.06	0.19 ± 0.06	0.16 ± 0.01	0.13 ± 0.02	**	*
WL (m)	2.54 ± 0.45	3.66 ± 0.38	4.19 ± 0.30	5.57 ± 0.62	0.45 ± 0.25	0.97 ± 0.27	1.08 ± 0.23	1.43 ± 0.13	**	*
Turb	4.06 ± 1.63	8.06 ± 1.08	5.54 ± 1.95	5.48 ± 1.68	41 ± 24.2	56.33 ± 44	203.72 ± 77	232 ± 89.37	**	*
TP (mg/L)	0.009 ± 0.1	0.013 ± 0.01	0.027 ± 0.01	0.036 ± 0.02	0.072 ± 0.06	0.077 ± 0.04	0.051 ± 0.01	0.073 ± 0.09	**	*
TN (mg/L)	0.97 ± 0.21	1.15 ± 0.36	0.57 ± 0.52	0.92 ± 0.35	1.07 ± 0.14	1.33 ± 0.82	1.12 ± 0.51	1.18 ± 1.08	*	*
NO_3-N (mg/L)	2.12 ± 0.99	0.82 ± 0.33	0.15 ± 0.05	0.06 ± 0.04	0.013 ± 0.01	0.14 ± 0.17	0.18 ± 0.02	0.21 ± 0.06	-	-
NH_3-N (mg/L)	0.09 ± 0.05	0.13 ± 0.1	0.09 ± 0.05	0.13 ± 0.09	0.89 ± 0.36	0.81 ± 0.23	0.24 ± 0.07	0.28 ± 0.14	*	-
Chl a (µg/L)	2.47 ± 4.72	3.42 ± 3.15	1.32 ± 0.71	2.58 ± 1.05	1.52 ± 1.1	0.87 ± 0.41	3.08 ± 0.78	2.92 ± 0.71	*	*

* $p < 0.05$ indicates significant difference at 0.05 level, ** $p < 0.01$ indicates a highly significant difference at the 0.01 level.

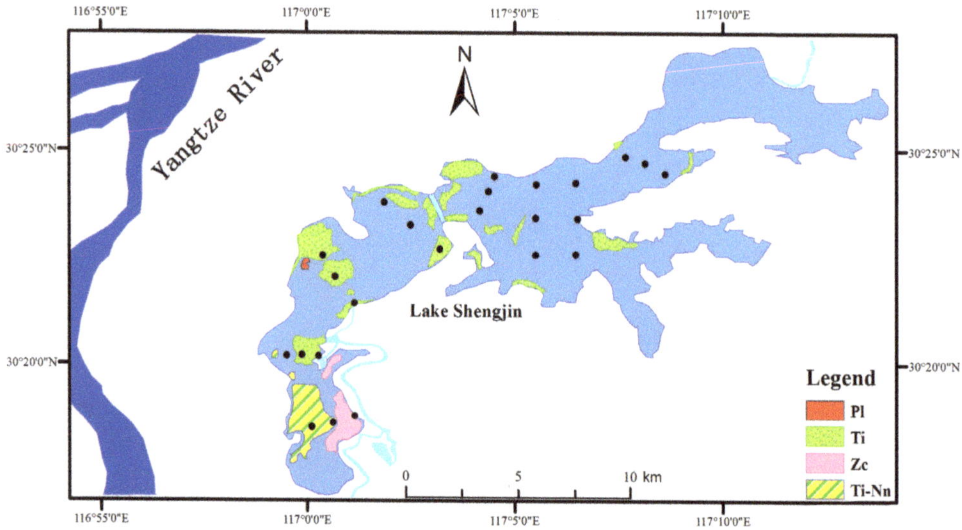

Figure 2. Distribution of dominant associations of main aquatic plants in summer. The following abbreviations used in the diagram are: Pl: *Polygonum lapathifolium*, Ti: *Trapa incisa*, Zc: *Zizania caduuciflora*, and Ti-Nn: *Trapa incisa-Nelumbo nucifera*.

3.3. Phytoplankton Community Diversity and Dominant Species

During the survey of this study (Figure 3), we identified a total of 254 species (including varieties) in 8 phyla and 137 genera of phytoplankton. The most numerous of these are Chlorophyta, with 109 species, accounting for 42.27% of the total species; this was followed by Bacillariophyta and Cyanophyta, with 56 and 34 species, accounting for 22.76% and 13.82%, respectively. In the summer survey, there were 174 phytoplankton species in the *T. incisa* range, 169 species in the *Z. caduuciflora* range, and 147 species in open water. Among them, we did not find *Dinobryon bavaricum*, *Pediastrum*, *Fragilaria capucina*, and *Melosira granulata* in most open waters, and we only detected *Actinastrum fluviatile* and *Strombomonas ensifera* in *Z. caduuciflora* distribution area.

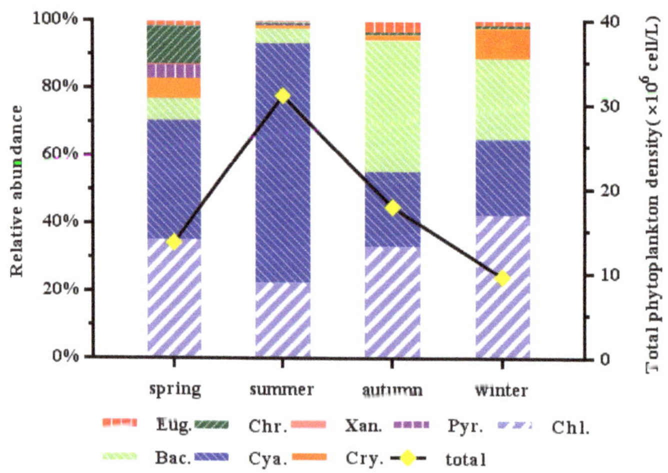

Figure 3. Total phytoplankton cell density and relative abundance ($\times 10^6$ cell/L) in Shengjin Lake.

The Shannon–Wiener index for phytoplankton in Shengjin Lake ranged from 1.45 to 3.52, with a mean value of 2.74. We recorded the high mean value, 2.96, in the winter. Spatially, the *T. incisa* range (3.06) > the *Z. caduuciflora* range (2.83) > the open water (2.41). The Pielou homogeneity index fluctuated from 0.55 to 0.87, with a mean value of 0.71. Temporally, the highest mean value was recorded in winter, at 0.73. Spatially, there was no significant trend. The Margalef species richness index ranged from 2.15 to 3.52, with a mean value of 3.15. Once again, we noted that, in the winter, the mean value was at its highest with 3.24. Spatially, the *T. incisa* distribution area was the highest, at 3.22. In summary, the results of the diversity indices illustrated that Shengjin Lake had the highest average in winter and the diversity index of the area with aquatic vegetation was higher than that of the open water area.

The analysis of the dominant phytoplankton species during the study period showed that they varied greatly among seasons (Table 2). In spring, the dominant species were *Merismopedia marssonii, P. pusillum,* and *D. bavaricum*. In summer, *Planktolyngbya subtilis, Phormidium tenuis,* and *Microcystis aeruginosa* dominated the Cyanophyta, *Selenastrum minutum*, and *Staurastrum indentatum* dominated the Chlorophyta. *P. subtilis* is the most dominant species, with the $Y = 0.22$, and the relative abundance of *P. subtilis* is more than 30% in the whole lake area with a detection frequency of 95.8% at each sampling. Aquatic vegetation grows vigorously in the summer, and there were significant differences in the distribution of aquatic vegetation in the lake area. Upon further analysis of the spatial distribution of the dominant species based on the results, we observed that there were spatial differences in the dominance index of *P. subtilis*, among which the *T. incisa* distribution area (0.24) > *Z. caduuciflora* distribution area (0.19) > open water area (0.11). In autumn, the dominant species shift from Cyanophyta to Bacillariophyta and Chlorophyta. This decreased the dominance of the *P. subtilis* ($Y = 0.03$), while *M. granulata* became the main dominant species with a relative abundance of 22%. *M. tenuissima, Synedra acus, M. granulata Her, Ankistrodesmus convolutus, Scenedesmus quadricauda, A. fluviatile* and *Cryptomonas ovata*, etc., were the dominant species in the winter. The dominant species differed among seasons, and we found the frequencies of the main dominant species to be similar in the UL and ML, but the dominant values of the dominant species varied among sampling sites.

Table 2. Dominant species and dominance of phytoplankton communities in each season.

Species		Dominance			
		Spring (April)	**Summer (July)**	**Autumn (October)**	**Winter (January)**
Cyanophyta	*P. subtilis*	0.012	0.228	0.032	0.019
	P. tenuis	0.003	0.035	0.014	0.01
	M. tenuissima	0.002	0.005	0.011	0.033
	M. marssonii	0.064	0.031	0.002	0.004
	M. aeruginosa	-	0.031	-	0.004
Chlorophyta	*S. minutum*	0.0007	0.035	0.006	0.017
	S. indentatum	0.005	0.025	0.009	-
	Chlorella vulgaris	0.015	0.011	0.059	0.019
	S. arcuatus	0.008	0.0006	0.035	0.002
	A. convolutus	0.006	0.017	0.012	0.043
	S. quadricauda	0.016	-	0.0031	0.081
	A. fluviatile	0.1	0.001	-	0.065
Bacillariophyta	*M. granulata Her*	0.002	-	0.134	0.034
	M. granulata	0.005	0.003	0.051	0.018
	S. acus	0.013	0.002	0.019	0.073
Cryptophyta	*C. ovata*	0.015	0.009	0.01	0.064
Pyrrophyta	*P. pusillum*	0.032	0.008	-	-
Chrysophyta	*D. bavaricum*	0.079	0.01	0.0002	-

3.4. Spatial and Temporal Variation of Phytoplankton Communities

The results showed (Figure 3) that there were seasonal differences ($p < 0.05$) in the mean cell density of phytoplankton at each sampling site, with the highest mean cell density of 3.11 $\times$ 10^7 cell/L in the summer, decreased gradually to 1.79 $\times$ 10^7 cell/L in the autumn, and the lowest of 0.96 $\times$ 10^7cell/L in the winter. Cyanophyta was a major component of the phytoplankton community, accounting for more than 20% of the relative abundance in all seasons. The relative abundance of Cyanophyta in the summer was more than 70%, 47.72% of Cyanophyta are *P. subtilis*. The relative abundance of Cyanophyta gradually decreased in the autumn and reached the lowest value of 22.54% in the winter, and recovered in the spring. The relative abundance of Chlorophyta increased from 22.15% in the summer to the highest value of 42.57% in the winter. The relative abundance of Bacillariophyta climbed to the highest value of 38.29% in the autumn. *M. granulata* accounted for 48.84% of the abundance of Bacillariophyta. The number of Cryptophyta increased in the winter, and the relative abundance rose to 8.8%. Although Cyanophyta had absolute dominance in the summer, it did not in all the other seasons, allowing the relative abundance of Chrysophyta, Euglenophyta, Xanthophyta, and Pyrrophyta to increase.

The phytoplankton biomass of Shengjin Lake spanned from 0.78–15.52 mg/L with seasonal variation, reaching a maximum in the summer and then gradually decreasing, with similar trends in phytoplankton cell abundance.

In the following figure (Figure 4), the spatial distribution of the annual mean phytoplankton cell density in Shengjin Lake was significantly different ($p < 0.05$), with the annual mean phytoplankton cell density in the UL area being 1.76 $\times$ 10^7 cell/L. The trend shows that the mean cell density is higher in the central open water area, with a mean cell density of 2.02 $\times$ 10^7 cell/L, and lower in the surrounding aquatic vegetation cover area, with the lowest value found in the SJ1, where the water-holding plant experienced a good recovery, with a cell density of 1.51 $\times$ 10^7 cell/L. The phytoplankton cell density in the ML was generally higher than that in the UL, with an annual average cell density of 1.95 $\times$ 10^7 cell/L and the highest value of 2.24 $\times$ 10^7 cell/L, which was found in the SJ23, showing an obvious gradient distribution characteristic.

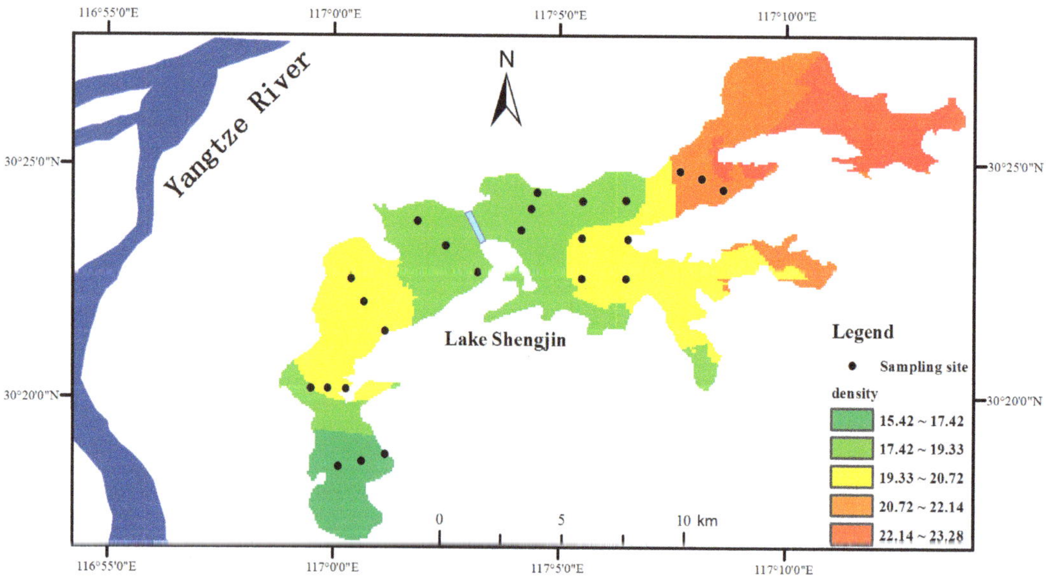

Figure 4. Distribution of annual mean phytoplankton cell density in Shengjin Lake ($\times 10^5$).

3.5. Relationship between Phytoplankton Communities and Environmental Parameters

In this study, we selected a total of 18 phytoplankton species (dominant species) and 14 environmental factors to analyze the relationship between the dominant phytoplankton species and environmental parameters throughout the year in Shengjin Lake [31,32]. We analyzed the relationship between the phytoplankton species, and the environmental parameters was analyzed using RDA. The adjusted variance explained 80.9%. Sort axes 1 and 2 explained the 22.9% and 31.5% of the variance in phytoplankton (Table 3).

Table 3. Results of RDA analysis of dominant phytoplankton species.

Sort Axis	1	2	3	4
Eigenvalue	0.329	0.211	0.126	0.101
Species-environment correlation	0.852	0.757	0.682	0.714
Cumulative percentage of species	22.9	31.5	36.3	38.6
Cumulative percentage of species-environment relationships	43.4	71.9	83.8	91.2

We conducted Monte Carlo tests conducted on phytoplankton communities and the environmental parameters, where WL, WT, TN, and Cond were the main parameters ($p < 0.05$). The results of the analysis showed that (Figure 5) the abundance of cells of the dominant species, *P. pusillum* and *D. bavaricum*, was significantly and positively correlated with NO_3-N; *P. subtilis* was positively correlated with *T. incisa*, SD and WT; *S. indentatum* and *P. tenuis* were positively correlated with *Z. caduuciflora*, WT and WL; *M. granulata* and *S. acus* were positively correlated with TN, DO, and the significant negative correlations were found between SD and *T. incisa*.

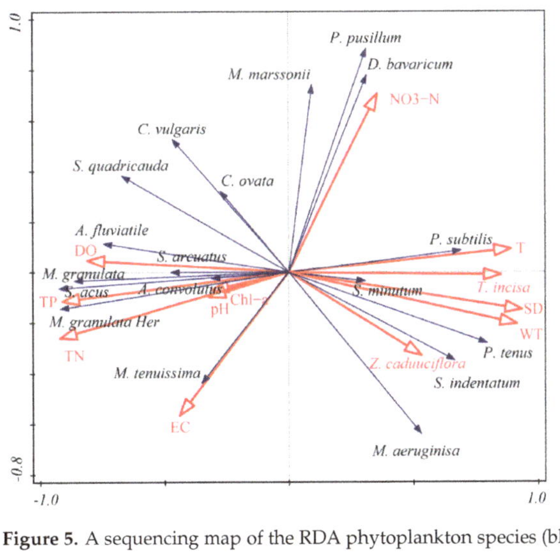

Figure 5. A sequencing map of the RDA phytoplankton species (blue lines with arrowhead) and the environmental parameters (red lines with arrowhead).

4. Discussion

4.1. Aquatic Vegetation Restored after Removal of Purse Seine

Since the removal of the seine net in 2018, the aquatic vegetation of the whole lake surface increased significantly. Compared to 2015, the area of aquatic vegetation increased to 26 km^2, and the coverage of aquatic vegetation was more than 50%, especially the coverage of the emerged plant *Z. caduuciflora* in the UL waters of sampling points SJ1~3 in the UL was more than 90%. The purse seine fishery, which began in 1997, led to severe degradation of aquatic vegetation. In the 2007 monitoring survey, the area of purse seine farming had reached 100%, and the aquatic vegetation cover was only 30%. There were large areas of water that were vegetation-free in the UL, and only the channel in the ML contained small amounts of the floating plant growth [33,34]. With the advent of the fishing ban policy in 2015 in the UL, the submerged plants *C. demersum* and *M. verticillatum* have begun to appear in patches, the distribution area has expanded from the central to the northern part, and we found only a small amount of the emerged plant *Z. caduuciflora* between the purse seine in the ML [35]. According to the data, Hangzhou West Lake and Wuhan East Lake also experienced serious damage to aquatic vegetation caused by the over-farming of fish and later restored aquatic vegetation by reducing fish density.

According to the available data, the early stage of the commercial use of Shengjin Lake was Chinese velvet chelonia farming, and the middle and late stages developed mainly into fishery seine farming. With the bridge as the boundary, seine nets were deployed only in the junction area of the UL and the ML. Extensive contract fishery culture was implemented in the UL, while numerous artificial dikes and culture nets were constructed in the ML. The lake was divided into multiple culture areas with intensive seine nets for high-density aquaculture throughout the year. Filter-feeding fish such as *Hypophthalmichthys molitrix* and *Aristichthys nobilis* were the main dominant species [36]. The small number of seine nets in the UL and smaller catches than in the ML may be the reason for the presence of aquatic plants. After the removal of seine nets in Shengjin Lake in 2018, the density of fish decreased, the feeding of herbivorous fish on aquatic plants decreased, and aquatic plants were able to grow and renew. At the same time, after the ban on aquaculture, pollution from bait placement was alleviated to a greater extent, the lake water quality improved, aquatic plants adapted to water quality conditions, and aquatic vegetation recovered rapidly, especially submerged vegetation.

4.2. Phytoplankton Community Composition and Spatial and Temporal Variation

The findings of the current study revealed large seasonal differences in phytoplankton communities, and similar variations have been identified in previous surveys [37,38]. Compared to the 2014 survey data, the number of phytoplankton species increased from 210 to 254, with an increase in the number of species of the Chlorophyta and Euglenophyta. The decrease in phytoplankton cell abundance and the increase in biodiversity was due to the improvement in water quality after the removal of the purse seine ban.

According to the data of previous surveys [39], the TP concentration in Shengjin Lake decreased by 20% (0.053 mg/L to 0.41 mg/L). The average transparency increased from 0.4 m to 0.7 m, and some waters covered by aquatic plants can reach 1.2 m. Phytoplankton is highly sensitive to environmental changes. The changes in environmental factors after the removal of purse seine affect the structural distribution of the phytoplankton community.

Compared with before and after the removal of the purse seine, the species of Chlorophyta are the most abundant, with the highest Cyanophyta cell density after purse seine removal in summer. Shengjin Lake has a high water level in the wet season in summer, and the water temperature exceeds 35 °C, which is most suitable for the growth of Cyanophyta. At the same time, after the purse seine is removed, the coverage area of aquatic vegetation is greatly increased, which has enhanced the purification of lake water quality, high water transparency, and strong photosynthesis of Cyanophyta. The main dominant species in the summer after the removal of the purse seine is *P. subtilis*, which is significantly less than that of other cyanobacteria before the removal of the purse seine. *P. subtilis* belongs to *Lyngbya*, and there is a gelatinous sheath inside the cells that regulate the intensity of light received. This may be the reason why *P. subtilis* is the dominant species in the summer [40].

As the number of aquatic plants increased significantly after the removal of the seine, almost all aquatic plants died in autumn and winter, and the nitrogen and phosphorus from the plant body entered the water body, resulting in higher nitrogen and phosphorus concentrations compared to the seine period. After the removal of the purse seine, the *M. granulata* replaced Chlorophyta as the dominant species in autumn due to the appropriate temperature and high nutrient level. The RDA results showed that Bacillariophyta cell abundance correlated negatively with WT and SD. When the water of Shengjin Lake flowed into Yangtze River, the WL in most of the lake area was at 1 m, the WT was below 15 °C, the aquatic vegetation disappeared, and the water environment favored favorable for the growth of Bacillariophyta [41]. Compared with before the removal of the purse seine, the number of phytoplankton species increased, and *Scenedesmus* became the dominant species in winter, probably because *Scenedesmus* thrives in a low-temperature environment, the chemosensory effect of *Scenedesmus* decreased with the disappearance of aquatic plants [42,43].

4.3. Response Mechanisms of Phytoplankton Communities to Aquatic Vegetation Restoration

Studies have shown that aquatic vegetation indirectly affects the community structure of phytoplankton by altering the lake water environment. The restoration of floating and submerged aquatic vegetation in Shengjin Lake provides diverse habitats for the growth and development of phytoplankton, resulting in increased phytoplankton diversity. The results of RDA analysis that the content of nitrogen and phosphorus correlated positively with the growth of *M. granulata* and most Chlorophyta, and the nutrient concentration directly affected the phytoplankton community structure. The death of aquatic plants in the autumn and winter to produce organic matter explained the high cell abundance of Bacillariophyta and Chlorophyta during those seasons, while the fixation of suspended matter on the lake surface by aquatic vegetation disappeared, and wind and rainfall caused the release of nitrogen and phosphorus from the bottom mud [44]. The cell abundance of the dominant species of *P. tenuis* and *P. subtilis*, was negatively correlated with DO and positively correlated with the floating plant *T. incisa*. Several sites covered with *T. incisa* in the summer identified that the relative abundance of the algae *P. subtilis* was higher than at the open water sites. Aquatic plants consume a large amount of DO and reduce pH when decomposing organic matter and nutrients [45]. As a result, the available light of phytoplankton in the area with high *T. incisa* coverage in the UL is less, the respiration rate of phytoplankton is greater than that of light cooperation and the photosynthesis and respiration of aquatic plants and phytoplankton jointly affect the trend of DO.

Aquatic plants also directly influence the community structure of phytoplankton through themselves. In the summer, we identified a high abundance of *D. bavaricum* and Pyrrophyta in the southern sampling sites covered by *T. incisa* and *Z. caduuciflora* in the UL. The fixation type algae such as *P. subtilis* and *Dinobryon* have flagella or fixation filaments. Aquatic plants provide a habitat for the large fixation type phytoplankton, and algae attach to aquatic plants to avoid predation by fish [46]. The phytoplankton identified in the distribution waters of aquatic vegetation, such as *Pediastrum* and Euglenophyta, are mostly above 10 μm in diameter, have a fresh weight of more than 0.01×10^{-6} mg, and are presumably shade and cold-tolerant species.

Aquatic plants improve water quality and also compete with phytoplankton for light and nutrients. The three UL sampling sites (SJ1~3) with high aquatic vegetation cover in the summer have low phytoplankton abundance, with an average cell density of 1.47×10^7 cell/L. Aquatic plants provide a sheltered environment for zooplankton and filter-feeding fish, causing smaller phytoplankton to be more easily predated and the abundance of larger Chrysophyta and Pyrrophyta to increase relatively. Some studies have shown that filter-feeding fish tend to make phytoplankton larger under low nutrient conditions. On the other hand, aquatic plants and phytoplankton have similar ecological niches, and both may secrete chemosensitive substances to inhibit each other's growth [47]. There are other studies detailing how the main components of chemosensitive substances in aquatic plants, such as *N. nucifera* and *C. demersum*, include secondary metabolites such as flavonoids and organic acids [48,49], which can damage the cell structure of Cyanophyta and other phytoplankton, affect intracellular enzyme activities, and prevent their photosynthesis [50,51].

5. Conclusions

The results showed that the spatial and temporal changes in the phytoplankton community structure of river-connected lake were in response to the restoration of aquatic vegetation and changes in environmental factors in the lake. After the seine was removed, the aquatic plants re-grew. The direct and indirect effects of aquatic plants are one of the important causes of changes in the structure of phytoplankton communities. Monitoring phytoplankton to evaluate the restoration of the middle and lower reaches of Yangtze River through-river lake ecosystem is an important tool in the current lake ecological restoration project, and conducting research related to aquatic plants and phytoplankton has important ecological significance and value.

Author Contributions: Conceptualization: Z.Z. and X.Y.; data curation: X.Z., J.C. and W.G.; investigation: X.Z., J.C. and W.G.; methodology: Z.Z.; writing—original draft: X.Z.; writing—review and editing: X.Z., Z.Z. and X.Y.; funding acquisition: Z.Z. All authors have read and agreed to the published version of the manuscript.

Funding: This research was supported by Joint Research Project for the Yangtze River Conservation (Phase I), China (No. 2019-LHYJ-01-0212, 2019-LHYJ-01-0212-17).

Institutional Review Board Statement: Not applicable.

Data Availability Statement: The data presented in this study are available on request from the corresponding author Xiaoxin Ye. E-mail: yexx@ahu.edu.cn.

Acknowledgments: We would like to thank Jingwen Chen, Wenli Guo for their assistance with data collection. We would also like to thank Zhongze Zhou and Xiaoxin Ye for their advice on the paper. We would like to thank Marci Baun from the University of California, Los Angeles, for editing the paper.

Conflicts of Interest: The authors declare no conflict of interest.

References

1. Kallfa, A.K.; Babani, F.; Kraja, A.Y. Evaluation of the Trophic Level of Kune and Vain Lagoons in Albania, Using Phytoplankton as a Bioindicator. *J. Sustain. Dev. Energy Water Environ. Syst.* **2014**, *2*, 61–67. [CrossRef]
2. Gao, S.R.; Pan, L.J.; Sun, F.Y. Assessment of environmental water pollution and eutrophication by aquatic organisms. *Environ. Sci. Manag.* **2006**, *31*, 174–176.
3. Wei, Y.Q.; Sun, J.; Zhang, G.C.; Wang, Z.X.; Wang, F. Environmental factors controlling the dynamics of phytoplankton communities during spring and fall seasons in the southern Sunda Shelf. *Environ. Sci. Pollut. Res.* **2020**, *27*, 23222–23233. [CrossRef] [PubMed]
4. Miao, X.M.; Wang, S.K.; Liu, M.Q.; Ma, J.; Hu, J.; Li, T.L.; Chen, L.J. Changes in the phytoplankton community structure of the Backshore Wetland of Expo Garden, Shanghai from 2009 to 2010. *Aquacult. Fish.* **2019**, *4*, 198–204. [CrossRef]
5. Liu, C.Q.; Liu, L.S.; Shen, H.T. Seasonal variations of phytoplankton community structure in relation to physico-chemical factors in Lake Baiyangdian, China. *Procedia Environ. Sci.* **2010**, *2*, 1622–1631. [CrossRef]
6. Loza, V.; Perona, E.; Mateo, P. Specific responses to nitrogen and phosphorus enrichment in cyanobacteria: Factors influencing changes in species dominance along eutrophic gradients. *Water Res.* **2014**, *48*, 622–631. [CrossRef]
7. Ma, J.R.; Qin, B.Q.; Paerl, H.W.; Brookes, J.D.; Wu, P.; Zhou, J.; Deng, J.M. Green algal over cyanobacterial dominance promoted with nitrogen and phosphorus additions in a mesocosm study at Lake Taihu, China. *Environ. Sci. Pollut. Res.* **2015**, *22*, 5041–5049. [CrossRef]
8. Wu, B.; Zhao, Q.; Ma, F.K. Study on Restoration of the Relationship between Blocking Lakes and Yangtze River. *Environ. Sci. Surv.* **2019**, *38*, 10–14.
9. Cui, Y.H.; Wang, J. Influence of the sluice on water level and area of Yangtze river-connected lakes: A case study in Shengjin Lake. *J. Water Resour. Water Eng.* **2018**, *29*, 47–52.
10. Song, X.W.; Zhou, L.Z. Effects of habitat changes on spatiotemporal pattern of the wintering waterbird community at Shengjin Lake. *J. Anhui Agric. Univ.* **2019**, *46*, 610–617.
11. Chen, J.Y.; Zhou, L.Z.; Zhou, B.; Xu, R.X.; Xu, W.B. Seasonal dynamics of wintering waterbirds in two shallow lakes along Yangtze River in Anhui Province. *Zool. Res.* **2011**, *32*, 540–548. [CrossRef]
12. Cao, L.; Zhang, Y.; Barter, M.; Lei, G. Anatidae in eastern China during the non-breeding season: Geographical distributions and protection status. *Biol. Conserv.* **2009**, *143*, 650–659. [CrossRef]
13. Li, C.L.; Yang, Y.; Wang, Z.; Zhou, L.Z. The relationship between seasonal water level fluctuation and habitat availability for wintering waterbirds at Shengjin Lake, China. *Bird Conserv. Int.* **2018**, *29*, 100–114. [CrossRef]

14. Xu, H.Q. Study on phytoplankton community structure characteristics in the Shengjin Lake of Anhui Province. Master's Thesis, Anhui University, Hefei, China, 2011.
15. Fox, A.D.; Cao, L.; Zhang, Y.; Barter, M.; Meng, F.J.; Wang, S.L. Declines in the tuber-feeding waterbird guild at Shengjin Lake National Nature Reserve, China—A barometer of submerged macrophyte collapse. *Aquat. Conserv. Mar. Freshw. Ecosyst.* **2011**, *21*, 82–91. [CrossRef]
16. Zhang, K.; Jiang, F.Y.; Chen, H.; Wu, Q.L.; Zhou, Z.Z. Temporal and spatial variations in zooplankton communities in relation to environmental factors in four floodplain lakes located in the middle reach of the Yangtze River, China. *Environ. Pollut.* **2019**, *251*, 277–284. [CrossRef]
17. Bian, G.G.; Zhao, W.D.; Da, L. The research and application of the allelopathic effects of submerged plants inhibit the growth of algae. *Environ. Dev.* **2012**, *24*, 59–64.
18. Scheffer, M.; Hosper, S.H.; Meijer, M.L. Alternative equilibria in shallow lakes. *Trends Ecol. Evol.* **1993**, *8*, 275–279. [CrossRef]
19. Carlos, I.; Guillermo, G.; Nestor, M.; Mariana, M.; Elena, R.; Erik, J. Horizontal dynamics of zooplankton in subtropical Lake Blanca (Uruguay) hosting multiple zooplankton predators and aquatic plant refuges. *Hydrobiologia* **2007**, *584*, 179–189. [CrossRef]
20. Ellen, D.; Wouter, J.B. Impact of submerged macrophytes including charophytes on phyto- and zooplankton communities: Allelopathy versus other mechanisms. *Aquat. Bot.* **2002**, *72*, 261–274. [CrossRef]
21. Hu, W.H.; Yu, G.Q.; Liu, Z.Y. The ecology environment and its protection of Shengjin Lake wetland. *Territ. Nat. Resour. Study.* **2004**, *04*, 70–71.
22. State Environmental Protection Bureau (SEPB). *Methods of Monitoring and Analysis for Water and Wastewater*, 4th ed.; SEPB: Beijing, China, 2002.
23. *GB 3838-2002*; Environmental Quality Standards for Surface Water. Ministry of Ecology and Environment of the People's Republic of China: Beijing, China, 2002.
24. Xu, Y.G.; Li, J.A.; Qin, J.H.; Li, Q. Seasonal patterns of water quality and phytoplankton dynamics in surface waters in Guangzhou and Foshan, China. *Sci. Total Environ.* **2017**, *590–591*, 361–369. [CrossRef] [PubMed]
25. Guesdon, S.; Stachowski-Haberkorn, S.; Lambert, C.; Beker, B.; Brach-Papa, C.; Auger, D.; Béchemin, C. Effect of local hydroclimate on phytoplankton groups in the Charente estuary. *Estuar. Coast. Shelf Sci.* **2016**, *181*, 325–337. [CrossRef]
26. Guo, J.S.; Sheng, J.P.; Li, Z.; Hong, Z. Seasonal variation of phytoplankton community in Xiaojiang backwater area during the preliminary operation stage of the Three Gorges Reservoir. *Environ. Sci.* **2010**, *31*, 1492–1497.
27. Hu, H.J.; Wei, Y.X. *The Freshwater Algae of China—Systematics, Taxonomy and Ecology*; Science Press: Beijing, China, 2006.
28. Zhang, Z.S.; Huang, X.F. *Research Methods of Freshwater Plankton*; Science Press: Beijing, China, 1991; Volume 10, pp. 336–339.
29. Wu, C.; Zhang, Q.G. Research Progress on diversity and quantitative analysis methods of freshwater plankton. *Anhui Agric. Sci. Bull.* **2009**, *15*, 41–42.
30. Lampitt, R.S.; Wishner, K.F.; Turley, C.M.; Angel, M.V. Marine snow studies in the Northeast Atlantic Ocean: Distribution, composition and role as a food source for migrating plankton. *Mar. Biol.* **1993**, *116*, 689–702. [CrossRef]
31. Jiang, Z.B.; Liu, J.J.; Chen, J.F.; Chen, Q.Z.; Yan, X.J.; Xuan, J.L.; Zeng, J.N. Responses of summer phytoplankton community to drastic environmental changes in the Changjiang (Yangtze River) estuary during the past 50 years. *Water Res.* **2014**, *54*, 1–11. [CrossRef]
32. Wendroth, O. Multivariate Analysis of Ecological Data using CANOCO. *Vadose Zone J.* **2004**, *3*, 1057. [CrossRef]
33. Chen, B.P.; Wu, L.Q. Ecological security and sustainable utilization of wetland in Shengjin Lake National Nature Reserve. *J. Chizhou Univ.* **2006**, *20*, 78–80, 95.
34. Xu, L.L.; Xu, W.B.; Sun, Q.Y. Flora and Vegetation in Shengjin Lake. *Plant Sci. J.* **2008**, *27*, 264–270.
35. Liu, J.J.; Zhou, Z.Z.; Tian, H.X. Vascular plant community types and flora in Shengjin Lake National Nature Reserve. *J. Biol.* **2016**, *33*, 40–46.
36. Yang, L. Effects of Lake Wetland Degradation on foraging activities of overwintering white headed crane (Grus monacha). Ph.D. Thesis, Anhui University, Hefei, China, 2015.
37. Wu, F.Y.; Wang, L.; Xu, M. Community dynamics of phytoplankton and related affecting factors in Shengjin Lake. *J. Biol.* **2016**, *33*, 34–39.
38. Wang, L.; Wang, C.; Deng, D.G.; Zhao, X.X.; Zhou, Z.Z. Temporal and spatial variations in phytoplankton: Correlations with environmental factors in Shengjin Lake, China. *Environ. Sci. Pollut. Res.* **2015**, *22*, 14144–14156. [CrossRef] [PubMed]
39. Wang, L.; Wang, C.; Zhao, X.X.; Zhou, Z.Z. Seasonal and spatial variation of chlorophyll-a concentration in Shengjin Lake. *J. Biol.* **2016**, *33*, 49–52.
40. Ma, J.R.; Deng, J.M.; Qin, B.Q. Progress and prospects on Cyanobacteria bloom-forming mechanism in lakes. *Acta Ecol. Sin.* **2013**, *33*, 3020–3030.
41. Silva, C.; Torgan, L.C.; Schneck, F. Temperature and surface runoff affect the community of periphytic diatoms and have distinct effects on functional groups: Evidence of a mesocosms experiment. *Hydrobiologia* **2019**, *839*, 37–50. [CrossRef]
42. Ma, H.T.; Zhang, F.; Zhang, H.J. The inhibition effect of two common limnetic submerged macrophytes on Scenedesmus obliquus under co-culture condition. *J. Biol.* **2020**, *37*, 72–75.
43. Xie, X.L.; Zhou, R.; Deng, Z.F. Overcompensation and competitive effects of Microcystis aeruginosa and Scenedesmus obliquus after low temperature and light stresses. *Acta Ecol. Sin.* **2014**, *34*, 1224–1234.

44. Figueredo, C.C.; Giani, A. Seasonal variation in the diversity and species richness of phytoplankton in a tropical eutrophid reservoir. *Hydrobiologia* **2001**, *445*, 165–174. [CrossRef]
45. Zhu, Q.C.; Yang, W.D.; Cheng, H. Test of Potamogeton crispus Decomposing in Water. *Mod. Agric. Sci. Technol.* **2013**, *23*, 224–226, 229.
46. Liu, C.G.; Qiu, J.Q.; Wang, W. Advances on theory of biomanipulation in control of eutrophicated lakes. *J. Agro-Environ. Scie.* **2004**, *23*, 198–201.
47. Wang, L.X.; Wu, G.R. Inhibitory effect of Hydrilla verticillata on Microcystis aeruginosa. *J. Lake Sci.* **2004**, *16*, 337–342.
48. Hu, H.Y.; Men, Y.J.; Li, F.M.; Men, Y. Research progress on phyto-allelopathic algae control. *Ecol. Environ. Sci.* **2006**, *15*, 153–157.
49. Park, M.H.; Chung, I.M.; Ahmad, A.; Hwang, S.J. Growth inhibition of unicellular and colonial Microcystis strains (Cyanophyceae) by compounds isolated from rice (*Oryza Sativa*) hulls. *Aquat. Bot.* **2008**, *90*, 309–314. [CrossRef]
50. Elisabeth, M. Allelopathy of Aquatic Autotrophs. *Crit. Rev. Plant Sci.* **2003**, *22*, 313–339. [CrossRef]
51. Friebe, A.; Roth, U.; Kück, P.; Schnabl, H.; Schulz, M. Effects of 2, 4-dihydroxy-1, 4-benzoxazin-3-ones on the activity of plasma membrane H +− ATPase. *Phytochemistry* **1997**, *44*, 979–983. [CrossRef]

diversity
MDPI

Article

Bauhinia (Leguminosae) Fossils from the Paleogene of Southwestern China and Its Species Accumulation in Asia

Lin-Bo Jia [1,*,†], Jin-Jin Hu [1,†], Shi-Tao Zhang [2], Tao Su [3], Robert A. Spicer [3,4], Jia Liu [3], Jiu-Cheng Yang [2], Pu Zou [5,*], Yong-Jiang Huang [1] and Zhe-Kun Zhou [3]

1 CAS Key Laboratory for Plant Diversity and Biogeography of East Asia, Kunming Institute of Botany, Chinese Academy of Sciences, Kunming 650201, China; hujinjin@mail.kib.ac.cn (J.-J.H.); huangyongjiang@mail.kib.ac.cn (Y.-J.H.)
2 Faculty of Land Resource Engineering, Kunming University of Science and Technology, Kunming 650093, China; taogezhang@hotmail.com (S.-T.Z.); 18395636983@163.com (J.-C.Y.)
3 CAS Key Laboratory of Tropical Forest Ecology, Xishuangbanna Tropical Botanical Garden, Chinese Academy of Sciences, Mengla 666303, China; sutao@xtbg.org.cn (T.S.); ras6@open.ac.uk (R.A.S.); liujia@xtbg.ac.cn (J.L.); zhouzk@xtbg.ac.cn (Z.-K.Z.)
4 School of Environment, Earth and Ecosystem Sciences, The Open University, Milton Keynes MK7 6AA, UK
5 South China Botanical Garden, Chinese Academy of Sciences, Guangzhou 510650, China
* Correspondence: jialinbo@mail.kib.ac.cn (L.-B.J.); zoupu@scbg.ac.cn (P.Z.)
† These authors contributed equally to this work.

Abstract: Extant *Bauhinia* (Leguminosae) is a genus of 300 species of trees, shrubs, and lianas, widely distributed in pantropical areas, but its diversification history in southeastern Asia, one of its centers of highest diversity, remains unclear. We report new fossils of three *Bauhinia* species with cuticular preservation from the Paleogene of Puyang Basin, southwestern China. Our finding likely extends the emergence of *Bauhinia* in Asia to the late Eocene. Together with previously reported fossil records, we show that the diversification of *Bauhina* in Asia and the phenomenon of a small region harboring multiple *Bauhinia* species in southwestern China could be traced back to the Paleogene.

Keywords: Asia; Biogeography; *Bauhinia*; Fabaceae; late Eocene

Citation: Jia, L.-B.; Hu, J.-J.; Zhang, S.-T.; Su, T.; Spicer, R.A.; Liu, J.; Yang, J.-C.; Zou, P.; Huang, Y.-J.; Zhou, Z.-K. *Bauhinia* (Leguminosae) Fossils from the Paleogene of Southwestern China and Its Species Accumulation in Asia. *Diversity* **2022**, *14*, 173. https://doi.org/10.3390/d14030173

Academic Editor: Ben-Erik Van Wyk

Received: 30 January 2022
Accepted: 24 February 2022
Published: 27 February 2022

1. Introduction

Bauhinia L. (Leguminosae) today comprises about 300 species of trees, shrubs, and lianas and is widely distributed in pantropical areas, with the largest diversity center being in the neotropics, and the second largest in southeastern Asia [1–3] (Figure 1). A typical leaf of this genus is simple and bilobed, rarely entire or two-foliolate, with pulvinus on both ends of the petiole [1]. Its fruit is flat, elliptic, oblong, or linear, woody or thinly valved. Species of *Bauhinia* are widely cultivated as ornamentals [1]. For example, the orchid tree (*Bauhinia* × *blakeana* Dunn) was chosen as the city flower of Hongkong. Several species of *Bauhinia* (e.g., *B. purpurea* L.) are used in local medicine and seeds of *B. petersiana* Bolle can be used as a coffee substitute [3].

Recent phylogenetic studies show that *Bauhinia* is an early-diverged member of Leguminosae [2,4–7]. However, due to the nesting of *Griffonia* and *Brenierea* within the genus, *Bauhinia* is not monophyletic [2,5]. The phylogenetic relationships of the *Bauhina* + *Griffonia* + *Brenierea* clade has not yet been well resolved [2,5]. In some recent treatments, *Bauhinia* is divided into eight genera including *Bauhinia* L. s.s., *Barklya* F. Muell., *Gigasiphon* Drake, *Lysiphyllum* (Benth.) de Wit, *Phanera* Lour., *Piliostigma* Hochst., *Schnella* Raddi, and *Tylosema* (Schweinf.) Torre et Hillc. [5,8]. Here we adopt the traditional broad treatment of *Bauhinia* because this study mainly concerns plant morphology and the character suite available in *Bauhinia* leaf fossils limits taxonomic resolution.

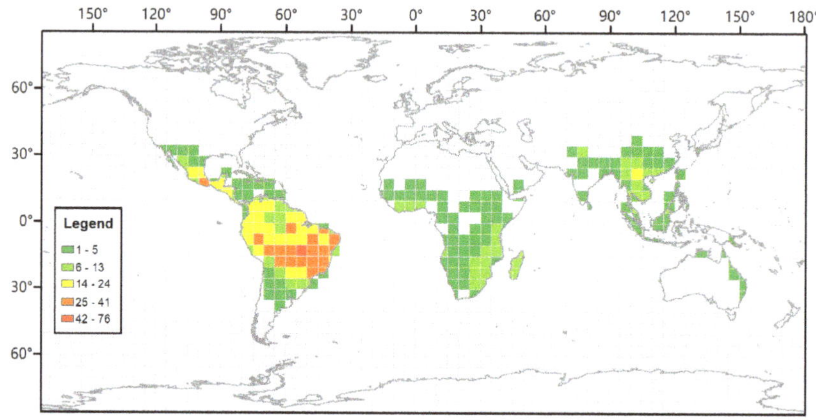

Figure 1. Extant distribution of *Bauhinia*. Different colors in the map indicate the number of species in each grid square. Extant occurrence data of *Bauhinia* are from the Global Biodiversity Information Facility (GBIF).

Fossils of *Bauhinia* have been documented in various forms of wood, leaf, and twig with attached fruit [9–12]. The earliest reliable fossils are leaves from the early Oligocene of China [11,13]. Later fossils of the genus are documented from the late Oligocene and Middle Miocene of China, the Oligocene of Mexico, and the early Miocene and middle Miocene–middle Pleistocene of India, the Pliocene of Uganda, and the Miocene of Ecuador [9,10,12,14–18].

This study reports new *Bauhinia* fossils from the late Eocene of southeastern China. First, we morphologically compared the macroscopic morphology and cuticle features of the fossils with those of the extant and fossil species in the genus. Then we discussed the implications of the fossils in the context of our current understanding of the evolutionary history of *Bauhinia* in Asia.

2. Materials and Methods

2.1. Geological Setting

The Puyang Basin (105.26° E, 23.48° N; 825 m asl) is a wedge-shaped strike-slip basin located in the southeastern Yunnan province, China [19–21] (Figure 2). The base of the basin is Cambrian limestone, with Cenozoic sediments unconformably lain above [20] (Figure 3). The lower part of the Cenozoic basin fill is dominated by lignite beds representing swamp facies, and the upper part is mainly lacustrine grey to yellow mudstone [20]. Recently, a mammal fossil, belonging to Anthracotheriidae Leidy, similar to the late Eocene *Bothriogenys hui* from Yunnan and *B. orientalis* from Thailand [22,23], was recovered from the lignite (Figure 3). Pollen analysis also suggests a late Eocene age for the lignite bed (Yang et al. under review) and suggests that basin formation was roughly coeval with other regional basins such as Wenshan [13] and Lühe [24] which have been dated radiometrically. Our fossils are collected from the lacustrine mudstone above the lignite bed (Figure 3) and are most likely also late Eocene in age.

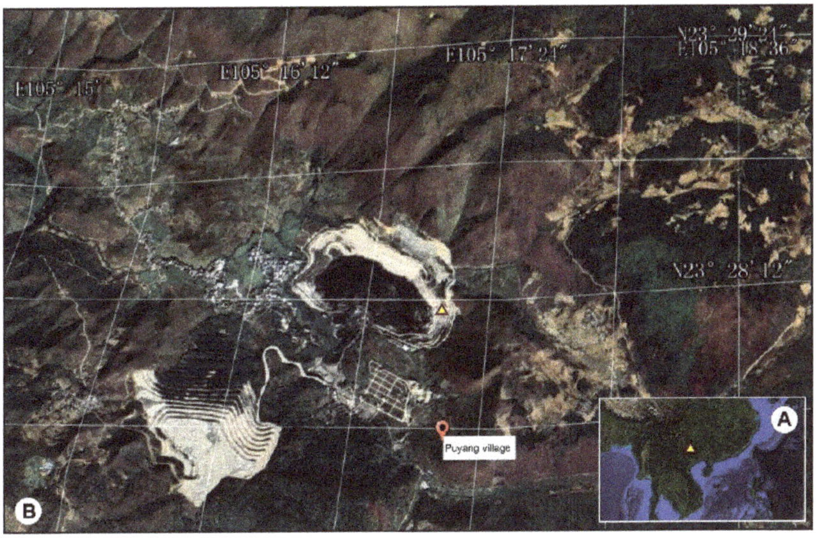

Figure 2. The position of the fossil locality in a broad view of southeastern Asia (**A**) and in a magnified view of Puyang Basin, Yunnan province, China (**B**).

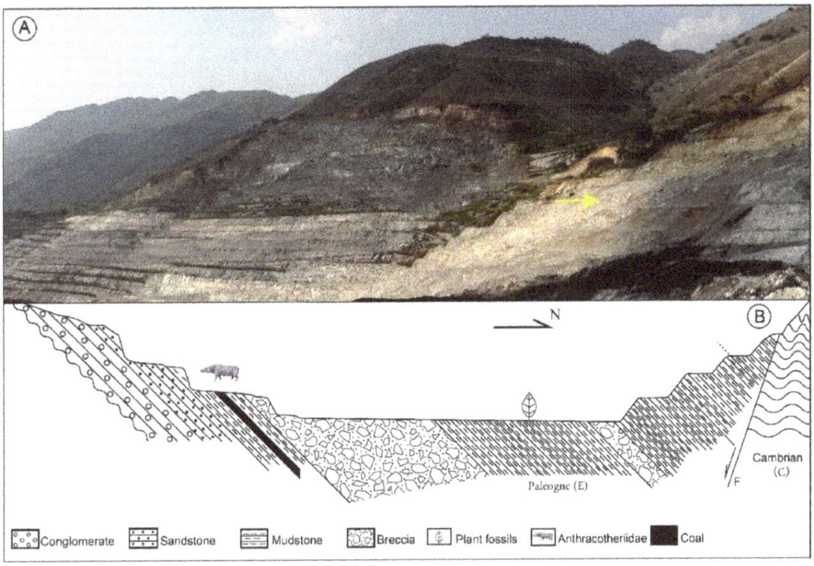

Figure 3. Cross section of the fossil locality in Puyang Basin, southeastern Yunnan Province, China. (**A**) The fossiliferous outcrop. The light green arrow indicates the layers where our fossils were collected. (**B**) Cross section of the fossil locality.

2.2. *Macroscopic Feature Observations and the Modern Distribution of Bauhinia*

Fifteen leaf fossils and one fruit compression were recovered and photographed using a digital camera (Nikon D750, Kanagawa, Japan). Fine-scale details of the fossils were further examined under a stereo microscope (Leica S8APO, Wetzlar, Germany), and images were taken. The raw data for the extant occurrence of *Bauhinia* was download from Gbif [25], and first cleaned using an R program and then checked manually [26]. Finally, the cleaned data were imported into Arcgis 10.0 to prepare the distributional heat map.

2.3. Cuticle Preparation for Fossil and Extant Materials

Fossil leaf fragments were treated with HCl and HF to remove calcareous and siliceous materials, and then macerated using 3% NaClO solution for 30 min to one hour until they became translucent [27–29]. For extant materials, fragments from mature leaves were macerated using a 1:1 solution of CH_3COOH and 30% H_2O_2 at 80 °C for about one hour [30,31]. After the mesophyll tissue was removed, the adaxial and abaxial cuticles for both fossil and extant materials were stained for about 30 min using Safranin O, mounted in glycerine on glass slides, and then photographed using a light microscope (Leica DM 750 with a Leica DFC 295 camera). All cuticular slides are stored at Kunming Institute of Botany, Chinese Academy of Sciences.

3. Results

Family: Leguminosae Juss. (or Fabaceae Lindl).
Subfamily: Caesalpinioideae DC.
Genus: *Bauhinia* L.
Locality: The Puyang Basin, Funing county, Yunnan province, China.
Age: The late Eocene.
Leaf.
Species: *Bauhinia wenshanensis* H.H. Meng et Z.K. Zhou (morphotype 1).
2014 *Bauhinia wenshanensis* H.H. Meng et Z.K. Zhou, Figure 4A–D.
Specimens: FN0403 (Figure 4A); FN0399 (Figure 4B); FN0106 (Figure 4C); FN06005 (Figure 4D).

Description: Leaf is entire and bilobed, 28–34 mm long and 18–26 mm wide, ovate to elliptical in outline (Figure 4A–D). The basal portion is cordate, slightly asymmetrical (Figure 4A,B). The widest part is in the lower third of the leaf, and the lamina gradually tapers toward the apex (Figure 4A,B). The apex is bifid to form two acute lobes at an angle of 31°–49° (Figure 4A,B). The primary vein framework is palmate with nine basal veins (Figure 4A,B). Primary veins near the midvein extend into the apices of lobes (Figure 4A–C). Additional primary veins extend toward the adjacent primary vein at the inner side (Figure 4A,B). Major secondaries originate from the primary veins and extend toward the apex of the leaf (Figure 4A,B).

The adaxial cuticle consists of irregular epidermal cells with sinuolate epidermal walls (Figure 4E,F) and a few single-celled trichome bases (Figure 4E,G,H). No stomata were observed in the adaxial cuticle. The abaxial epidermal cells are polygonal or irregularly shaped. Stomatal complexes are paracytic and tetracytic with sunken guard cells, and the stomatal rim is single-layered (Figure 4I–K) Subsidiary cells are crescent, polygonal, or irregularly shaped. Many single-celled trichome bases exist in the abaxial cuticle (Figure 4I,L).

Species: *Bauhinia* sp. (morphotype 2).
Specimens: FN0411a (Figure 5A); FN0411b (Figure 5B); FN0292 (Figure 5C).

Description: Leaf is simple, petiolate and bilobed, 25–27 mm long and 17–31 mm wide, elliptical to obovate in outline (Figure 5A–C). Base is almost straight in open leaf (Figure 5A–C). Apex is round (Figure 5A,B). The primary vein framework is palmate with seven basal veins (Figure 5A–C).

The adaxial cuticle consists of polygonal epidermal cells with straight arched epidermal walls (Figure 5D–F) and few single-celled trichome bases (Figure 5G). No stomata were observed in the adaxial cuticle. The abaxial epidermal cells are similar in shape and size to those in the adaxial cuticle. Stomatal complexes are paracytic and tetracytic with sunken guard cells, and the stomatal rim is double-layered (Figure 5H–J). Subsidiary cells are polygonal or irregularly shaped. Many single-celled trichome bases exist in the abaxial cuticle (Figure 5H–K).

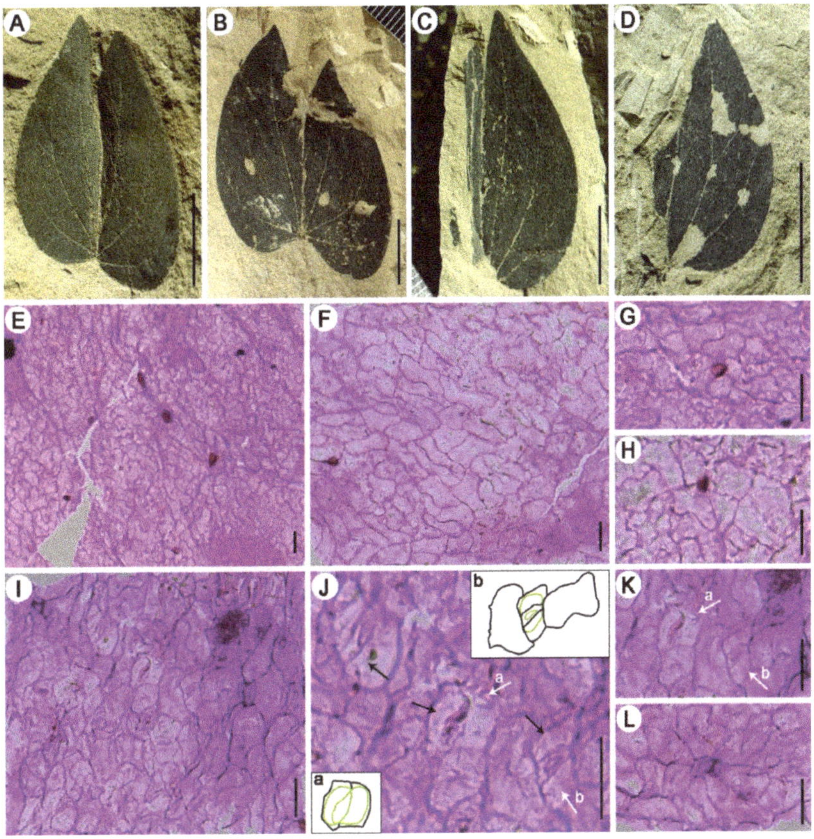

Figure 4. Leaf morphologies and cuticular structures of *Bauhinia wenshanensis* H.H. Meng et Z.K. Zhou (morphotype 1) from the Puyang Basin, Funing, Yunnan province. The scale bars represent 1 cm for leaf fossils and 20 μm for cuticle images. (**A–D**) Leaves of *B. wenshanensis*. All the cuticle images are from the fossil specimen in (**D**). (**E,F**) Adaxial cuticle showing sinuolate epidermal walls. (**G,H**) Single-celled trichome bases of adaxial cuticle. (**I–K**) Abaxial cuticle showing the orientation of stomata. Black arrows indicate sunken guard cells; white arrows indicate (a) paracytic stomatal complex, (b) tetracytic stomatal complex, and that the stomatal rim is single-layered. Line drawings illustrate (a) the paracytic stomatal complex indicated with white arrow a, (b) the tetracytic stomatal complex indicated with white arrow b. The green lines indicate stomata, and the black lines indicate subsidiary cells. (**L**) Single-celled trichome base of abaxial cuticle. (**A**), FN0403; (**B**), FN0399; (**C**), FN0106; (**D**), FN06005.

Species: *Bauhinia* sp. (morphotype 3).

Specimens: FN0189 (Figure 6A); FN0616 (Figure 6B).

Description: Leaf is entire and bilobed, 18 mm long and 22 mm wide, elliptical to oblong in outline (Figure 6A,B). The basal portion is cordate and weakly asymmetrical (Figure 6B). The widest part is in the middle of the leaf (Figure 6A,B). Primary vein framework is palmate with seven basal veins (Figure 6B).

Fruit.

Species: cf. *Bauhinia* sp. (morphotype 4).

Specimens: FN0465 (Figure 6C–E).

Description: Fruit is flat, elliptic to oblong, 42 mm long and 17 mm wide (Figure 6C). The left flank of the proximal end is nearly straight, and the right flank is convex (Figure 6D).

The distal end is acuminate (Figure 6E). The stigmatic remain is short and persistent (Figure 6E). There is a constriction in the middle of the fruit (Figure 6C). The suture lines are prominent, about 0.5 mm wide (Figure 6D,E). The seed chambers are elliptic, 4.7–10.4 mm long and 3.6–4.6 mm wide (Figure 6C,E). The angle between the long axis of the seed chambers and those of fruit is 94–100° (Figure 6C).

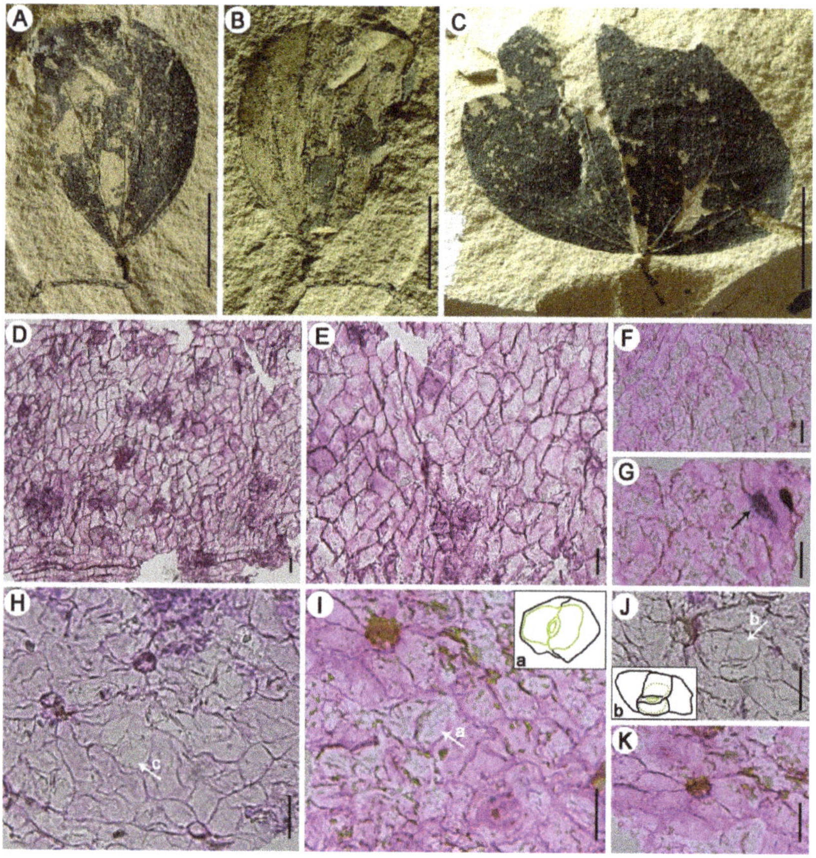

Figure 5. Leaf morphologies and cuticular structures of *Bauhinia* sp. (morphotype 2) from the Puyang Basin, Funing, Yunnan province. The scale bars represent 1 cm for leaf fossils and 20 μm for cuticle images. (**A–C**) Leaves of *Bauhinia* sp. The cuticle images in (**D,E,H,J**) are from the fossil specimen in (**C**); (**F,G,I,K**) are from the fossil specimen in (**A**). (**D–G**) Adaxial cuticle showing straight arched epidermal walls. The black arrow in (**G**) indicates a single-celled trichome base. (**H–J**) Abaxial cuticle showing the orientation of stomata. White arrows indicate (a) paracytic stomatal complex with double-layered stomatal rim, (b) tetracytic stomatal complex, and (c) stomatal complex with double-layered stomatal rim. Line drawings illustrate (a) the paracytic stomatal complex indicated with white arrow a, (b) the tetracytic stomatal complex indicated with white arrows b. The green lines indicate stomata, the dashed green lines indicate indistinct sunken guard cells, and the black lines indicate subsidiary cells. (**K**) Single-celled trichome base of abaxial cuticle. (**A**), FN0411a; (**B**), FN0411b; (**C**), FN0292.

Figure 6. Leaf morphologies of *Bauhinia* sp. (morphotype 3) and fruit morphology of cf. *Bauhinia* sp. (morphotype 4). The scale bars represent 1 cm. (**A**) FN0189; (**B**) FN0616; (**C–E**) FN0465.

4. Discussion

4.1. Morphological Comparison

The fossil leaves are characterized by simple and bilobed leaves. As far as we know, such kinds of leaves are seen in several families including Ginkgoaceae Engl., Lauraceae Juss., Passifloraceae Juss. ex Roussel, Proteaceae Juss., and Leguminosae Juss. (Figure 7). However, venation of the Ginkgoaceae leaves is dichotomous, and so different from our fossils where the venation is reticulate. The leaves of Passifloraceae have a small middle lobe or a broad angle (larger than 90°) between the lobes, distinguishing them from our fossils, which are strictly bilobed and diverge at an angle less than 90°. The palmate venation of *Dilobia* Thours. in the Proteaceae (see cleared leaf in Pole and Bowman (1996) [32]) and some species in Lauraceae such as *Sassafras albidum* (Nutt.) Nees is suprabasal but that of our fossils is basal. In Leguminosae, *Pueraria* DC., *Desmodium* Desv., *Christia* Moench, and *Bauhinia* L. have bilobed leaves, but leaves of *Pueraria* are prominently deflective and asymmetric, differing from our fossils, which are nearly symmetric. Secondary veins of *Desmodium* leaves are parallel whereas those of our fossils extend towards the leaf apices. Leaves of *Christia* have pinnate venation, distinguishing them from our fossils which possess palmate venation. Overall, our fossil leaves are a close match with *Bauhinia*.

Based on macroscopic and cuticular morphology, the fossil leaves can be divided into three morphotypes. Morphotype 1 has a cordate base, acute apex, single-layered stomatal rim, and sinuolate adaxial epidermal walls. Morphotype 2 has a straight base, round apex, double-layered stomatal rim, and straight arched epidermal walls, whereas morphotype 3 has a cordate base and a round apex. Although leaf and epidermal cell shape display intraspecific variability, the characteristics of the stomatal rim are considered stable at intraspecific level [33,34], so morphotypes 1 and 2 should represent different species. Whether morphotype 3 is another species will be discussed below. The three morphotypes are further compared with 46 extant species based on macroscopic morphology and cuticle features (see images and tables in Zou [35]), and then they are compared with fossil species.

Morphotype 1 is similar to *B. acuminata* L., *B. comosa* Craib, and *B. esquirolii*; Gagnep. in gross macroscopic morphology (Figures 4 and 8). However, the abaxial cuticle of morphotype 1 has paracytic and tetracytic stomatal complexes with sunken guard cells (Figure 4J,K), and so is different from those of *P. comosa* and *P. esquirolii* that have paracytic stomatal complexes and the guard cells are not sunken (Figure 8F,I). The abaxial cuticle of morphotype 1 is similar to that of *B. acuminata* in features including paracytic and tetracytic (some atypical) stomatal complexes with sunken guard cells and single-layered stomatal rim (Figure 8C). A combination of macroscopic and cuticular features suggests morphotype 1 is possibly a close relative of *B. acuminata*. However, it is worth noting that stomata also appear on the adaxial epidermis of *B. acuminata* in small number (Figure 8B), but we did not observe any from the adaxial epidermis of morphotype 1 (Figure 4E,F). This may be because the region from which we successfully extracted epidermis lacks stomata while the rest of leaf has them, or that morphotype 1 is a hypostomatic leaf. When compared with the reported fossil species (Table 1), morphotype 1 is similar to *B. wenshanensis* H. H. Meng et Z. K. Zhou found from the early Oligocene of Yunan, China [11]. In consideration of its near identical age, we assign morphotype 1 to *B. wenshanensis*. However, cuticular features have not yet been reported for *B. wenshanensis*, so our description of the cuticle for morphotype 1 may be taken as a tentative description for *B. wenshanensis*, but this should be used with caution. Future studies may obtain cuticle for *B. wenshanensis* from its type locality.

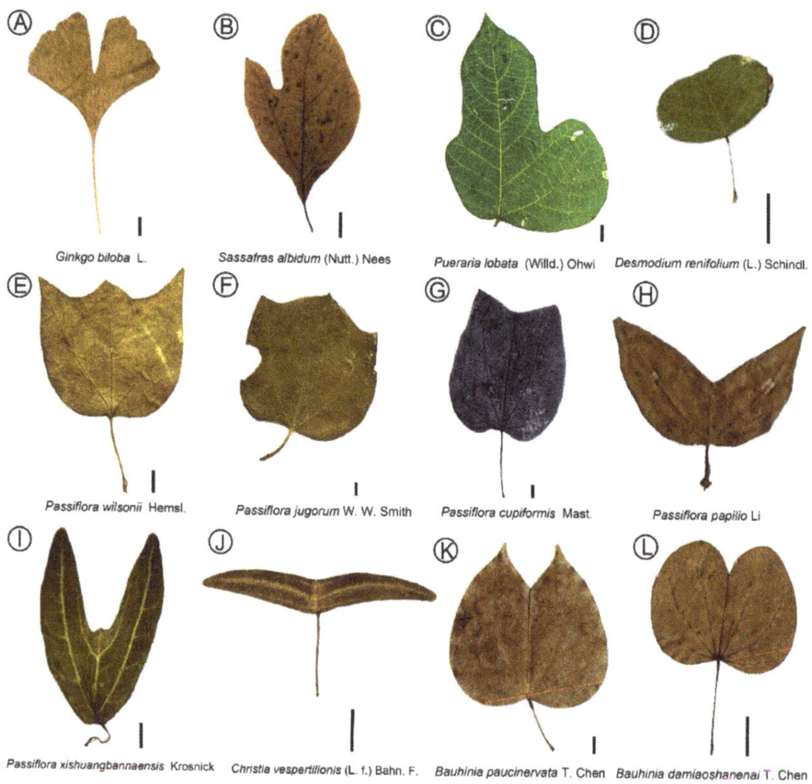

Figure 7. Leaf morphologies of bilobed or bilobe-like leaves in Ginkgoaceae Engl. (**A**) Lauraceae Juss (**B**), Passifloraceae Juss. ex Roussel (**E–I**), and Leguminosae Juss. (**C**,**D**,**J–L**). The scale bars represent 1 cm.

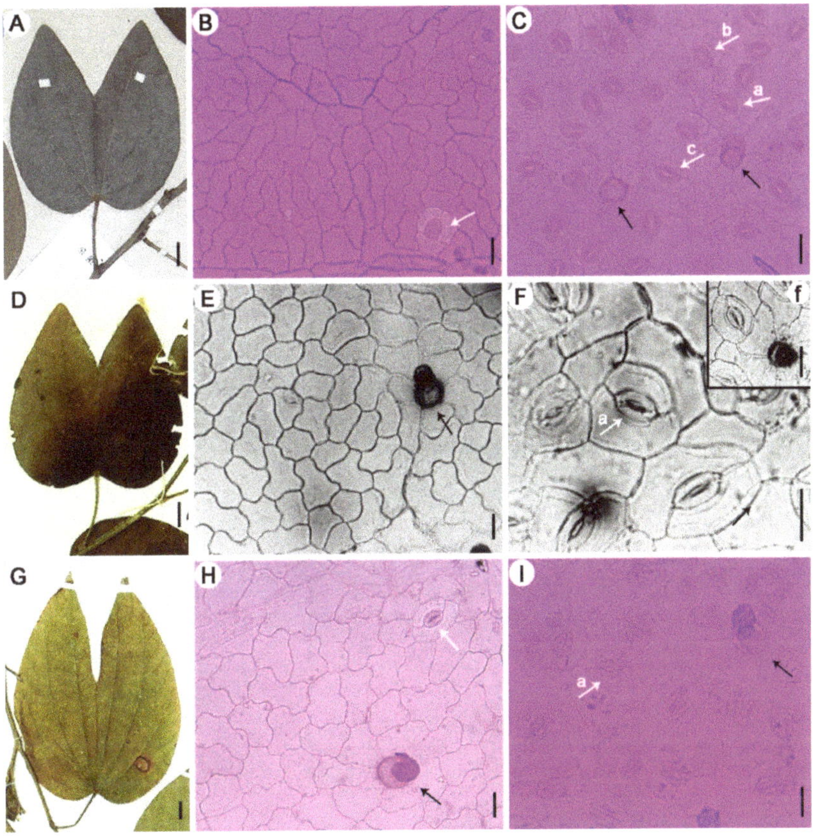

Figure 8. Leaf morphologies, adaxial (**B**,**E**,**H**) and abaxial (**C**,**F**,**I**) epidermises of *Bauhinia acuminata* L. (**A**–**C**) (IBSC, 0166379), *B. comosa* Craib (**D**–**F**) (HITBC, 0009647), and *B. esquirolii* Gagnep. (**G**–**I**) (HITBC, 0021422). The scale bars represent 1 cm for leaf specimens and 20 μm for epidermal images. The image f in (**F**) indicates a paracytic stomatal complex and a single-celled trichome base in abaxial epidermis. White arrows in (**B**,**H**) indicate few stomata in adaxial epidermis; white arrows in (**C**,**F**,**I**) indicate (a) paracytic stomatal complex, (b) tetracytic stomatal complex, and (c) sunken guard cells; black arrows indicate single-celled trichome bases.

Table 1. Leaf and fruit fossil records of *Bauhinia*.

No.	Species	Type	Age	Locality	Reference
1	*Bauhinia wenshanensis* H.H. Meng et Z.K. Zhou	Leaf	Early Oligocene	Dashidong Town, Wenshan County, Yunnan Province, China	[11]
2	*Bauhinia larsenii* D.X. Zhang et Y.F. Chen	Leaf and fruit	Late Oligocene	Ningming County, Guangxi, China	[10,12]
3	*Bauhinia ningmingensis* Q. Wang	Leaf	Late Oligocene	Ningming, Guangxi, China	[10]
4	*Bauhinia cheniae* Q. Wang	Leaf	Late Oligocene	Ningming, Guangxi, China	[10]
5	*Bauhinia ningmingensis* Q. Wang	Leaf	Late Oligocene	Ningming, Guangxi, China	[10]
6	*Bauhcis moranii* Calvillo-Canadell et Cevallos-Ferriz	Leaf	Oligocene	Los Ahuehuetes, Tepexi de Rodríguez, Puebla, Mexico	[14]
7	*Bauhinia krishnanunnii* A. K. Mathur	Leaf	Early Miocene	Unmetalled way to Babu Mohalla, Dagshai Cantonment, Solan District, Himachal Pradesh, India	[16]
8	*Bauhinia fotana* F. M. B. Jacques	Leaf	Middle Miocene	Zhangpu, County, Zhangzhou City, Fujian Province, Southeast China	[15]
9	*Bauhinia ungulatoides* Y. X. Lin, W. O. Wong, G. L. Shi, S. Shen et Z. Y. Li	Leaf	Middle Miocene	Zhangpu, Fujian, China	[9]
10	*Bauhinia ecuadorensis* E.W. Berry	Leaf	Miocene	Loja Basin, Ecuador	[17]
11	*Bauhinia siwalika* R.N. Lakh. et N. Awasthi	Leaf	Middle Miocene-Pleistocene	Bhikhnathoree, West Champaran District, Bihar, India	[36]
12	*Bauhinia nepalensis* N. Awasthi et N. Prasad	Leaf	Middle Miocene-Pleistocene	Bhikhnathoree, West Champaran District, Bihar, India	[36]
13	*Bauhinia nepalensis* N. Awasthi et N. Prasad	Leaf	Middle Miocene-Pleistocene	Surai Khola beds, near SuraiKhola bridge, Surai Khola area, India	[37]
14	*Bauhinia siwalika* R.N. Lakh. et N. Awasthi	Leaf	Middle Miocene-Middle Pleistocene	Bhikhnathoree, West Champaran District, Bihar, India	[38]
15	*Bauhinia waylandii* R.W. Chaney	Leaf	Pliocene	Busano, Bugishu, District, Eastern, Province, Uganda	[18]
16	*Bauhinia potosiana* E.W. Berry	Leaf	Pliocene-Early Pleistocene	Potosi, Bolivia	[39]
17	*Bauhinia* sp. cf. *B. purpurea* L.	Leaf	Late Cenozoic	Mahuadanr, Palamau District, Bihar, India	[40]

Note. Carpenter et al. [41] reported bilobed leaves from the Cenozoic of Australia and assigned these fossils to cf. Cercideae/Detarieae. It is clear that the veins diverge from the midvein and extend into the apex of the lobes in these fossils, which distinguishes them from *Bauhinia* in which two of the basal veins extend into the lobe apices. Moreover, Biagolini et al. [42] documented a fragment of a leaf, lacking apex and cuticle, from the Paleogene of Brazil. Their fossil seems to have palmate venation, a kind of venation pattern that exists in many families such as Malvaceae, Euphorbiaceae and Lauraceae. This makes the assignment of the leaf to *Bauhinia* superficial.

Morphotype 2 is similar to *B. purpurea* L., *B. viridescens* Desv., *B. tomentosa* L., and *B. racemosa* Lam. in terms of macroscopic morphology (Figure 9). However, the adaxial and abaxial epidermis of *B. purpurea* do not have trichome bases (Figure 9C,D), and so are different from those of morphotype 2 that possesses single-celled trichome bases (Figure 5G,K). The adaxial epidermal walls of *B. viridescens* are sinuolate (Figure 9G) whereas those of morphotype 2 are straight (Figure 5D–F). The abaxial epidermis of *B. tomentosa* has single-celled glandular trichome bases (Figure 9L), distinguishing it from morphotype 2 that has regular single-celled trichome bases (Figure 5G,K). The trichome bases for adaxial and abaxial epidermises of *B. racemose* are multicellular (Figure 9O,P) whereas those of morphotype 2 are unicellular (Figure 5G,K). In addition, *B. purpurea*, *B. viridescens*, and *B. tomentosa* have paracytic stomatal complexes (Figure 9D,H,L), and so are different from morphotype 2 with paracytic and tetracytic stomatal complexes (Figure 5H–J). Moreover, these four extant *Bauhinia* species exhibit a single-layered stomatal rim (Figure 9C,D,G,H,K,L,O,P), whereas morphotype 2 has a double-layered stomatal rim (Figure 5H–I). Overall, the four extant species above are different from morphotype 2 in their cuticular features. When compared to the fossil species, morphotype 2 is similar to *Bauhcis moranii* Calvillo-Canadell et Cevallos-Ferriz from the Oligocene of Mexico [14]. Due to no cuticular information in *Bauhcis moranii* and limited preservation of morphotype 2, we leave the nomenclature open for discussion.

Morphotypes 3 and 4, are similar to leaves and fruits of two species, i.e., *Bauhinia touranensis* Gagnep. and *B. damiaoshanensis* T. Chen (Figure 10). The two morphotypes possibly represent the same species, but this species is distinguished from those represented by morphotypes 1 and 2 because the fruit (morphotype 4) is different from those of the close recent relatives of morphotypes 1 and 2. When compared to fossils, morphotype 3 is different from any previously reported species. *Bauhinia larsenii* D.X. Zhang et Y.F. Chen from the late Oligocene of Ningming Basin, southern China [12], is the only fruit fossil assigned to the genus so far. However, morphotype 4 is banded with an acuminate stigmatic remnant, distinguishing it from *B. larsenii* which is elliptical with an acute stigmatic remnant. We here treat the morphotypes 3 and 4 as undetermined species.

To conclude, our fossils represent at least three species. Morphotype 1 is assigned to *B. wenshanensis*. Morphotype 2 constitutes the second species (*B.* sp.), and morphotypes 3 and 4 are possibly the third one (*B.* sp.).

4.2. The Diversification of Bauhinia in Southeastern Asia

Southeastern Asia is one of the diversity centers of *Bauhinia* (Figure 1). A recent molecular phylogenetic study suggests that the diversification of Asian *Bauhinia* can be traced back to the Paleocene (~60 Ma) [11]. However, Asian *Bauhinia* fossils are only known from the early Oligocene so far (Table 1). This forms a gap between the fossil evidence and molecular dating. Our finding is most likely the earliest reliable fossil records of *Bauhinia* in Asia, extending the existence of the genus to the late Eocene. *Bauhinia wenshanensis* has been reported from the early Oligocene of Wenshan, and four species, i.e., *B. larsenii* D.X. Zhang et Y.F. Chen, *B. ningmingensis* Q. Wang, *B. cheniae* Q. Wang, and *Bauhcis moranii* have also been found from the late Oligocene of Ningming Basin, Guangxi, China [9,11,12]. This provides evidence that *Bauhinia* apparently diversified in Asia in the Oligocene. In the Miocene and later periods, *B. krishnanunnii* A. K. Mathur comes from the early Miocene of India, *B. fotana* F. M. B. Jacques and *B. ungulatoides* Y. X. Lin, W. O. Wong, G. L. Shi, S. Shen et Z. Y. Li have been documented from the Middle Miocene Fujian province of China, and *B. siwalika* R.N. Lakh. et N. Awasthi and *B. nepalensis* N. Awasthi et N. Prasad from the middle Miocene to Pleistocene of India. This suggests further accumulation of species diversity occurred within the genus.

An interesting phenomenon for the extant distributional pattern of *Bauhinia* in southwestern China is that a small area can harbor many species. For example, 10 species have been found living in the Laojun Mountain area while 14 species have been recorded from Dawei Mountain, southeastern Yunnan [43,44], close to the locality that yielded fossils in this study. Of primary interest is when this kind of pattern formed. The discovery of three *Bauhinia* species from the late Eocene Puyang Basin and four species in the late Oligocene Ningming Basin (Table 1) shows that the two basins once harbored multiple *Bauhinia* species. Therefore, the phenomenon of many *Bauhinia* species coexisting in a small area can now be traced back to at least the Paleogene.

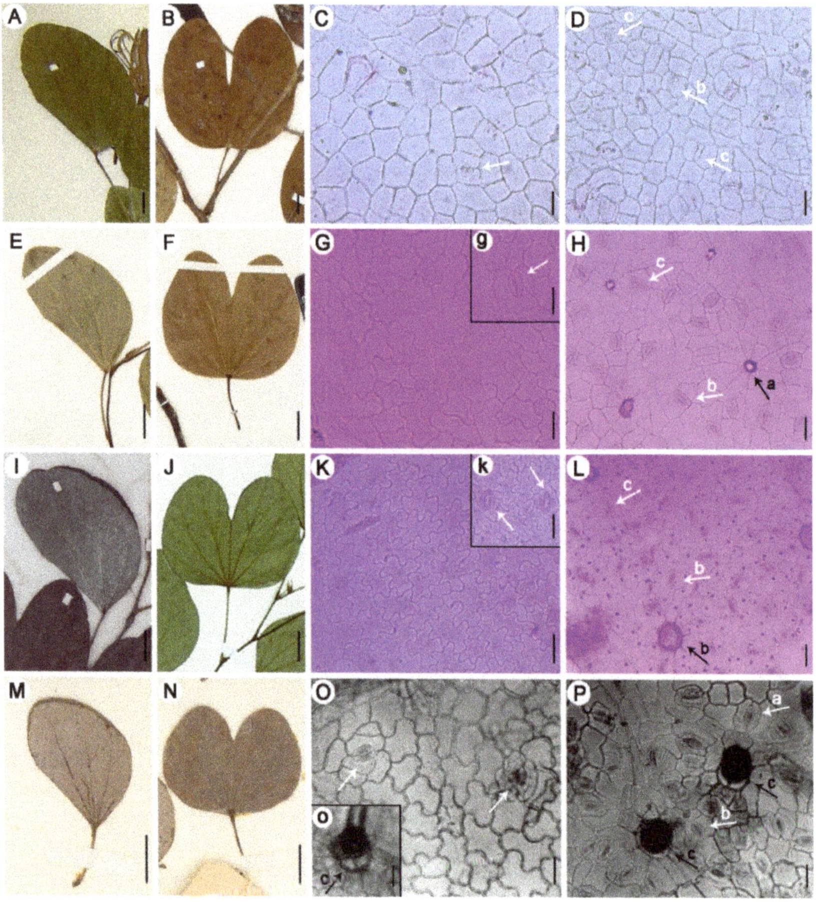

Figure 9. Leaf morphologies, together with adaxial (**C**,**G**,**K**,**O**) and abaxial (**D**,**H**,**L**,**P**) epidermises of *Bauhinia purpurea* L. (**A**–**D**) ((A) PE, 00327078; (B) KUN, 0125154), *B. viridescens* Desv. (**E**–**H**) (KUN 0169829), *B. tomentosa* L. (**I**–**L**) ((**I**) SYS, 00044882; (**J**) HITBC, 0021440), and *B. racemosa* Lam. (**M**–**P**) (KUN, 0125165). The scale bars represent 1 cm for leaf specimens and 20 μm for epidermal images. The image g in (**G**) indicates a stomatal complex of an adaxial epidermis. The image k in (**K**) indicates stomata of an adaxial epidermis. The image o in (**O**) indicates a trichome base of an adaxial epidermis. White arrows in (**C**), g, k, and (**O**) indicate stomata of adaxial epidermises; white arrows in (**D**,**H**,**L**,**P**) indicate (a) paracytic stomatal complex, (b) tetracytic stomatal complexes, and (c) sunken guard cells; black arrows indicate (a) a single-celled trichome base, (b) a single-celled glandular trichome base, and (c) multicellular trichome bases.

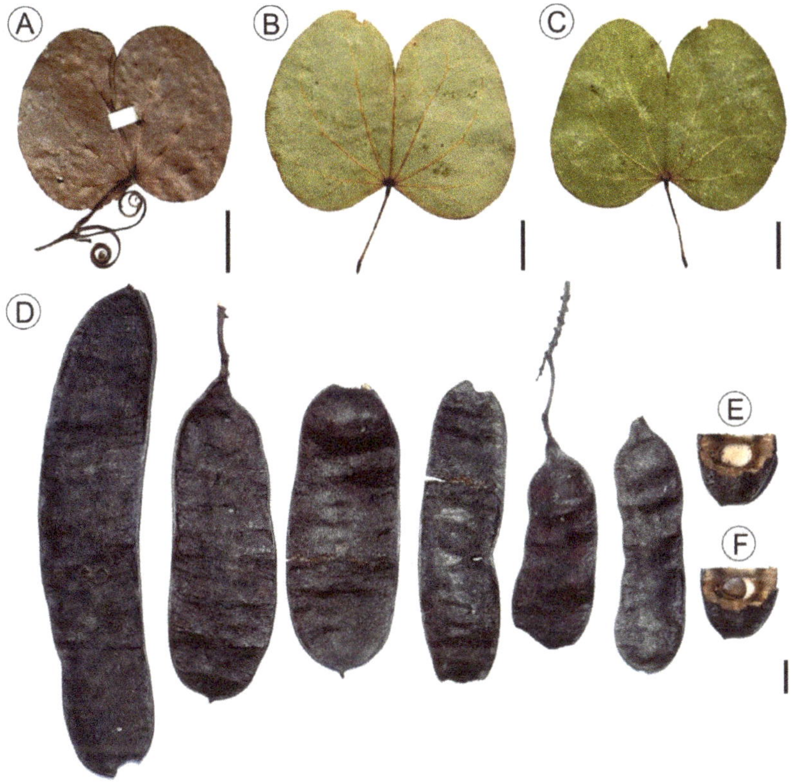

Figure 10. Morphology of *Bauhinia damiaoshanensis* T. Chen leaf (**A**) and *Bauhinia touranensis* Gagnep. leaves and fruits (**B–F**). The bars are 1 cm.

It is worth noting that although the Neotropics today host the largest diversity of *Bauhinia* species, few early fossil records have been documented there (Table 1 and note therein). Moreover, a recent molecular work points to a Neogene diversification of Neotropical *Bauhinia* species [11]. This scenario suggests that Asia is probably an ancient diversification center of *Bauhinia*, while the Neotropics is a more recent one, although this could result from under investigation of fossil records.

Author Contributions: Conceptualization, L.-B.J. and Z.-K.Z.; methodology, J.-C.Y. and J.-J.H.; software, J.L. and J.-C.Y.; validation, R.A.S., S.-T.Z. and Z.-K.Z.; formal analysis, J.-J.H. and L.-B.J.; investigation, J.-J.H. and Y.-J.H.; resources, P.Z.; data curation, P.Z.; writing—original draft preparation, L.-B.J. and J.-J.H.; writing—review and editing, R.A.S.; visualization, J.-J.H.; supervision, R.A.S., T.S. and Z.-K.Z.; project administration, L.-B.J.; funding acquisition, L.-B.J. and Z.-K.Z. All authors have read and agreed to the published version of the manuscript.

Funding: This work was supported by the National Natural Science Foundation of China (No. 31900194), the Science and technology Project of Yunnan Science and Technology Department (No. 202101AT070163, 202001AU070058), and the Chinese Academy of Sciences "Light of West China" Program and Yunnan Basic Research Projects (No.2019FB026).

Institutional Review Board Statement: Not applicable.

Data Availability Statement: All data are available in the text.

Acknowledgments: The authors are indebted to Tie-Yao Tu for identification of some extant *Bauhinia* species, Li Chen for collecting extant specimens of *Bauhinia* from southwestern Yunnan, China, and

Tao Deng, Xue-Dan Xie, Jing-Hua Wang, Ping Li, Xia Yang, Shi-Wei Guo, and Yu-Lan Yang for providing pictures of extant *Bauhinia* and help in morphological comparisons. We are also grateful to the Herbarium of Kunming Institute of Botany, Chinese Academy of Sciences (KUN) for providing access to the extant specimens.

Conflicts of Interest: The authors declare no conflict of interest.

References

1. Chen, D.; Zhang, D.; Larsen, K.; Larsen, S.S. *Bauhinia*; Wu, Z.Y., Raven, P.H., Hong, D.Y., Eds.; Science Press: Beijing, China; Missouri Botanical Garden Press: St. Louis, MO, USA, 2010; pp. 6–21.
2. Sinou, C.; Forest, F.; Lewis, G.P.; Bruneau, A. The genus *Bauhinia* s.l. (Leguminosae): A phylogeny based on the plastid *trnL–trnF* region. *Botany* **2009**, *87*, 947–960. [CrossRef]
3. Lewis, G.; Schrire, B.; Mackinder, B.; Lock, M. *Legumes of the World*; Royal Botanic Gardens, Kew: London, UK, 2005.
4. Zhang, R.; Wang, Y.H.; Jin, J.J.; Stull, G.W.; Bruneau, A.; Cardoso, D.; de Queiroz, L.P.; Moore, M.J.; Zhang, S.D.; Chen, S.Y.; et al. Exploration of plastid phylogenomic conflict yields new insights into the deep relationships of Leguminosae. *Syst. Biol.* **2020**, *69*, 613–622. [CrossRef]
5. The Legume Phylogeny Working Group. A new subfamily classification of the Leguminosae based on a taxonomically comprehensive phylogeny. *Taxon* **2017**, *66*, 44–77. [CrossRef]
6. Koenen, E.J.M.; Ojeda, D.I.; Steeves, R.; Migliore, J.; Bakker, F.T.; Wieringa, J.J.; Kidner, C.; Hardy, O.J.; Pennington, R.T.; Bruneau, A.; et al. Large-scale genomic sequence data resolve the deepest divergences in the legume phylogeny and support a near-simultaneous evolutionary origin of all six subfamilies. *New Phytol.* **2020**, *225*, 1355–1369. [CrossRef]
7. Zhao, Y.; Zhang, R.; Jiang, K.W.; Qi, J.; Hu, Y.; Guo, J.; Zhu, R.; Zhang, T.; Egan, A.N.; Yi, T.S.; et al. Nuclear phylotranscriptomics and phylogenomics support numerous polyploidization events and hypotheses for the evolution of rhizobial nitrogen-fixing symbiosis in Fabaceae. *Mol. Plant* **2021**, *14*, 748–773. [CrossRef] [PubMed]
8. Mackinder, B.A.; Clark, R. A synopsis of the Asian and Australasian genus *Phanera* Lour. (Cercideae: Caesalpinioideae: Leguminosae) including 19 new combinations. *Phytotaxa* **2014**, *166*, 49–68. [CrossRef]
9. Lin, Y.; Wong, W.O.; Shi, G.; Shen, S.; Li, Z. Bilobate leaves of *Bauhinia* (Leguminosae, Caesalpinioideae, Cercideae) from the middle Miocene of Fujian Province, southeastern China and their biogeographic implications. *BMC Evol. Biol.* **2015**, *15*, 252. [CrossRef]
10. Wang, Q.; Song, Z.; Chen, Y.; Shen, S.; Li, Z. Leaves and fruits of *Bauhinia* (Leguminosae, Caesalpinioideae, Cercideae) from the Oligocene Ningming Formation of Guangxi, South China and their biogeographic implications. *BMC Evol. Biol.* **2014**, *14*, 88. [CrossRef] [PubMed]
11. Meng, H.H.; Jacques, F.M.B.; Su, T.; Huang, Y.J.; Zhang, S.T.; Ma, H.J.; Zhou, Z.K. New Biogeographic insight into *Bauhinia* s.l. (Leguminosae): Integration from fossil records and molecular analyses. *BMC Evol. Biol.* **2014**, *14*, 181. [CrossRef]
12. Chen, Y.-F.; Zhang, D.-X. *Bauhinia larsenii*, a fossil legume from Guangxi, China. *Bot. J. Linn. Soc.* **2005**, *147*, 437–440. [CrossRef]
13. Tian, Y.; Spicer, R.A.; Huang, J.; Zhou, Z.; Su, T.; Widdowson, M.; Jia, L.; Li, S.; Wu, W.; Xue, L.; et al. New early oligocene zircon U-Pb dates for the 'Miocene' Wenshan Basin, Yunnan, China: Biodiversity and paleoenvironment. *Earth Planet. Sci. Lett.* **2021**, *565*, 116929. [CrossRef]
14. Calvillo-Canadell, L.; Cevallos-Ferriz, S.R.S. *Bauhcis moranii* gen. et sp. nov. (Cercideae, Caesalpinieae), an Oligocene plant from Tepexi de Rodríguez, Puebla, Mex., with leaf architecture similar to *Bauhinia* and *Cercis*. *Rev. Palaeobot. Palynol.* **2002**, *122*, 171–184. [CrossRef]
15. Jacques, F.M.B.; Shi, G.; Su, T.; Zhou, Z. A tropical forest of the middle Miocene of Fujian (SE China) reveals Sino-Indian biogeographic affinities. *Rev. Palaeobot. Palynol.* **2015**, *216*, 76–91. [CrossRef]
16. Guleria, J., Rashmi, S., Prasad, M. Some fossil leaves from the Kasauli Formation of Himachal Pradesh, North-West India. *Himal. Geol.* **2000**, *21*, 43–52.
17. Berry, E.W. Fossil floras from southern Ecuador. *Johns Hopkins Univ. Stud. Geol.* **1945**, *14*, 93–150.
18. Chaney, R.W. A tertiary flora from Uganda. *J. Geol.* **1933**, *41*, 702–709. [CrossRef]
19. Bureau of Geology and Mineral Resources. *Regional Geology of Yunnan Province*; Geology Press: Beijing, China, 1990.
20. Guo, Q.; Littke, R.; Sun, Y.; Zieger, L. Depositional history of low-mature coals from the Puyang Basin, Yunnan Province, China. *Int. J. Coal Geol.* **2020**, *221*, 103428. [CrossRef]
21. Luo, X.Y.; Zhang, Y.H. Neogene coal-accumulation basin characteristics and genetic types in Yunnan Province. *Coal Geol. China* **2013**, *25*, 10–17.
22. Ducrocq, S. The anthracotheriid genus *Bothriogenys* (Mammalia, Artiodactyla) in Africa and Asia during the Paleogene: Phylogenetical and paleobiogeographical relationships. *Stuttg. Beiträge Nat. B Geol. Paläontol.* **1997**, *250*, 1–44.
23. Chow, M. Some Oligocene mammals from Lunan, Yunnan. *Vertebr. PalAsiatica* **1958**, *2*, 263–267.
24. Linnemann, U.; Su, T.; Kunzmann, L.; Spicer, R.A.; Ding, W.N.; Spicer, T.E.V.; Zieger, J.; Hofmann, M.; Moraweck, K.; Gärtner, A.; et al. New U-Pb dates show a Paleogene origin for the modern Asian biodiversity hot spots. *Geology* **2017**, *41*, 3–6. [CrossRef]
25. GBIF.org. Available online: https://www.gbif.org/ (accessed on 29 December 2021).

26. Jin, J.; Yang, J. BDcleaner: A workflow for cleaning taxonomic and geographic errors in occurrence data archived in biodiversity databases. *Glob. Ecol. Conserv.* **2020**, *21*, e00852. [CrossRef]
27. Kerp, H. The study of fossil gymnosperms by means of cuticular analysis. *Palaios* **1990**, *5*, 548–569. [CrossRef]
28. Leng, Q. An effective method of observing fine venation from compressed angiosperm fossil leaves. *Acta Palaeontol. Sin.* **2000**, *39*, 157–158, (In Chinese with English Abstract).
29. Hu, J.J.; Xing, Y.W.; Turkington, R.; Jacques, F.M.B.; Su, T.; Huang, Y.J.; Zhou, Z.K. A new positive relationship between pCO_2 and stomatal frequency in *Quercus guyavifolia* (Fagaceae): A potential proxy for palaeo-CO_2 levels. *Ann. Bot.* **2015**, *115*, 777–788. [CrossRef]
30. Poole, I.; Kürschner, W.M. Stomatal Density and Index: The Practice. Fossil Plants and Spores: Modern Techniques. In *Fossil Plants and Spores: Modern Techniques*; Jones, T.P., Rowe, N.P., Eds.; Geological Society: London, UK, 1999; pp. 257–260.
31. Stace, C.A. Cuticular studies as an aid to plant taxonomy. Bulletin of the British Museum (Natural History). *Bot. Ser.* **1965**, *4*, 3–78.
32. Pole, M.S.; Bowman, D.M.J.S. Tertiary Plant Fossils from Australia's 'Top End'. *Aust. Syst. Bot.* **1996**, *9*, 113–126. [CrossRef]
33. Zeng, N.; Zhang, J.R.; Chang, Z.Y. Micromorphological characteristics of leaf epidemis and systematic significance of *Rosa* L. from China. *Guihaia* **2017**, *37*, 169–185, (In Chinese with English Abstract).
34. Yang, Z.R.; Lin, Q. Comparative morphology of the leaf epidermis in *Schisandra* (Schisandraceae). *Bot. J. Linn. Soc.* **2005**, *148*, 39–56. [CrossRef]
35. Zou, P. *Comparative Morphology of Cercideae (Leguminosae: Caesalpinioideae)*; South China Botanical Garden, Chinese Academy of Sciences: Guangzhou, China, 2006; p. 109.
36. Awasthi, N. India Fossil Legumes. In *Advances in Legume Systematics: Part 4 The fossil records*; Herendeen, P.S., Dilcher, D.L., Eds.; Royal Botanic Gardens, Kew: London, UK, 1992; pp. 225–250.
37. Awasthi, N.; Prasad, M. Siwalik plant fossils from Surai Khola area, western Nepal. *Palaeobotanist* **1990**, *38*, 298–318.
38. Lakhanpal, R.; Awasthi, N. A late Tertiary Florule from near Bhikhnathoree in West Champaran District, Bihar. In *Proceedings of the Symposium on Evolutionary Botany and Biostratigraphy, Current Trends in Life Sciences 10 (A. K. Ghosh Commem-oration Volume)*; Sharma AK Mitra, G.C., Banerjee, M., Eds.; Today and Tomorrow's Printers & Publishers: New Delhi, India, 1984; pp. 587–596.
39. Berry, E.W. Fossil plants from Bolivia and their bearing upon the age of uplift of the eastern Andes. *Proc. U. S. Natl. Mus.* **1917**, *54*, 103–164. [CrossRef]
40. Bande, M.; Srivastava, G. Late Cenozoic plant impressions from Mahuadanr Valley, Palamu District, Bihar. *Palaeobotanist* **1990**, *37*, 331–336.
41. Carpenter, R.J.; Goodwin, M.P.; Hill, R.S.; Kanold, K. Silcrete plant fossils from Lightning Ridge, New South Wales: New evidence for climate change and monsoon elements in the Australian Cenozoic. *Aust. J. Bot.* **2011**, *59*, 399–425. [CrossRef]
42. Biagolini, C.H.; Bernardes-de-Oliveira, M.E.C.; Caramês, A.G. Itaquaquecetuba Formation, São Paulo basin, Brazil: New angiosperm components of Paleogene Taphoflora. *Braz. J. Geol.* **2013**, *43*, 639–652. [CrossRef]
43. Yang, Y.; Tian, K.; He, S. *Study on the Scientific Survey of Wenshan National Reserve in China*; Science Press: Beijing, China, 2008.
44. Shui, Y.-M. *The Phytogeography of Seed Plants from Dawei Mountain*; Kunming Institute of Botany: Kunming, China, 2000; p. 211.

diversity

MDPI

Article

Population Genetic Structure and Biodiversity Conservation of a Relict and Medicinal Subshrub *Capparis spinosa* in Arid Central Asia

Qian Wang and Hong-Xiang Zhang *

State Key Laboratory of Desert and Oasis Ecology, Xinjiang Institute of Ecology and Geography, Chinese Academy of Sciences, Urumqi 830011, China; wangqian@ms.xjb.ac.cn
* Correspondence: zhanghx561@ms.xjb.ac.cn

Abstract: As a Tertiary Tethyan relict, *Capparis spinosa* is a typical wind-preventing and sand-fixing deciduous subshrub in arid central Asia. Due to its medicinal and energy value, this species is at risk of potential threat from human overexploitation, habitat destruction and resource depletion. In this study, our purpose was to evaluate the conservation strategies of *C. spinosa* according to its genetic structure characteristics and genetic diversity pattern among 37 natural distributional populations. Based on genomic SNP data generated from dd-RAD sequencing, genetic diversity analysis, principal component analysis, maximum likelihood phylogenetic trees and ADMIXTURE clustering, the significant population structure and differentiation were explored. The results showed the following: (1) Six distinct lineages were identified corresponding to geographic locations, and various levels of genetic diversity existed among the lineages for the natural habitat heterogeneity or human interferences; (2) The lineage divergences were influenced by isolation by distances, vicariance and restricted gene flow under complex topographic and climatic conditions. Finally, for the preservation of the genetic integrity of *C. spinosa*, we suggest that conservation units should be established corresponding to different geographic groups, and that attention should be paid to isolated and peripheral populations that are experiencing biodiversity loss. Simultaneously, monitoring and reducing anthropogenic disturbances in addition to rationally and sustainably utilizing wild resources would be beneficial to guarantee population resilience and evolutionary potential of this xerophyte in response to future environmental changes.

Keywords: genetic divergence; isolation by distance; biodiversity conservation; xerophytic relict plant; source-sink metapopulations; ddRAD-seq

Citation: Wang, Q.; Zhang, H.-X. Population Genetic Structure and Biodiversity Conservation of a Relict and Medicinal Subshrub *Capparis spinosa* in Arid Central Asia. *Diversity* **2022**, *14*, 146. https://doi.org/10.3390/d14020146

Academic Editors: Mario A. Pagnotta, Hong-Hu Meng and Yi-Gang Song

Received: 24 November 2021
Accepted: 14 February 2022
Published: 17 February 2022

1. Introduction

Arid central Asia is deemed to be the largest arid region in the temperate zones of the northern hemisphere and even the world [1]. The floristic compositions here are mostly descended from Tethyan xerophytic vegetation flora and have been phenoecological plant taxa shaped by the long-term continental–arid desert climate [2,3]. Because of the specific geomorphologic landscapes comprising the vertical and horizontal convergence of several major Asian tectonic mountains (Himalayas, Karakoram, Kunlun and Tianshan Mountains) and plateaus (Pamirs and Tibet Plateau), as well as the mosaic distribution of many large basins (Tarim, Turpan-Hami and Junggar basins) and deserts (Taklimakan, Kumtag and Gurbantunggut deserts), the monsoon current and moisture from the Indian and Pacific Oceans have been blocked and intercepted, and the amount of precipitation is low and unevenly distributed, which has constituted the complexity of the extreme geographic and climatic conditions, consequently forming a very fragile ecological environment [1]. The types of vegetation are relatively sparse and simple in this region. To adapt to environment variability, desert species from oligotypic genera and a monotypic genus with drought resistance and barren tolerance have mainly occurred and been discontinuously distributed [2,4,5]. It is speculated that the formation of this pattern is the result of intense orogeny during the geological period, coupled with extensive Quaternary glaciation, as well as the long-term process of aridification in central Asia [5–7]. Most xerophytic plants have exerted their advantages in the maintenance and restoration of desert vegetation ecosystems [8]. Given this, scientists and departments of environmental protection have paid increasing attention to studying and persisting the complex and unique community structure characteristics, as well as the vegetation diversification pattern in arid central Asia in recent years [9–11].

Capparis spinosa L. is considered to be a relict taxon that originated from arid Tethyan flora during the middle to late Tertiary [3,4]. It is widely distributed in southern Europe, northern Africa, western and southern Asia in addition to arid central Asia. The family, Capparaceae, to which it belongs, currently has polytypic genera mostly in the tropics and subtropics [12], with oligotypic genera in the transitional regions from western to central Asia, but a residual monotypic genus (*C. spinosa*) remains in eastern central Asia (arid region of western China) [13,14]. In arid central Asia, its natural populations inhabit heterogeneous quality landscapes with higher desert plateaus, mid-altitudinal warm dry valleys or proluvial fans and lower desert basins (Figure 1). As a representative xerophyte in desert regions, its perennial plants have shown excellent ecological value in terms of wind prevention, sand fixation and water as well as soil preservation [8]. However, due to its commercial and economic value in view of its medicinal and oil-bearing fruits and seeds [8], this wild subshrub has been suffering from long-term over-exploitation, such as excessive harvesting and grazing, without being reasonably and effectively protected, combined with complex natural habitat isolation, agricultural reclamation and industrial pollution, putting it under multiple pressures, resulting in range shrinkage with concomitant local biodiversity loss. Hence, to avoid the hidden danger of resource exhaustion or even the verge of extinction, it is urgent to formulate appropriate conservation strategies to reduce resource plundering and habitat destruction caused by anthropogenic activity.

Figure 1. Different habitat requirements of *C. spinosa*: (**a**) high-altitudinal desert plateau, (**b**) medium-altitudinal valleys and (**c**) low-altitudinal desert basins.

A number of previous studies have hitherto been carried out on the genetic variation and diversity of *C. spinosa* [15–21]. Earlier on, scholars mainly devoted their efforts to morphological taxonomy and subspecies or variety identification [15–17]. In recent years, with the development of molecular marker technology, researchers have achieved certain advancements in intraspecific diversity and infraspecific differentiation by virtue of methodologies such as RAPD, AFLP, ISSR, IRAP and EST-SSR [18–21]. Nonetheless, most of these studies have focused on Mediterranean coastal regions with a subtropical dry summer climate. However, the relict wild populations remaining in arid central Asia, dominated by a temperate continental desert climate, have been shown a lack of concern. Valuable information on the genetic structure and diversity conservation of *C. spinosa* in this region is relatively scarce and is of significant biogeographical and conservation interest to phytologists in the fields of plant genetic diversity and biodiversity conservation.

The conservation of genetic diversity is of great importance to the maintenance of biological species and ecosystem diversification, which provides a crucial foundation for the survival and development of species and enhances their long-term adaptive evolution to environmental changes, reducing the risk of extinction [22–24]. The evaluation of levels of genetic diversity based on integrated population structures is a critical prerequisite for species protection [25]. For source-sink metapopulations of terrestrial plants, the genetic structures and divergences are not only dominated by their own species and population characteristics, such as growth rates, effective population sizes and dispersal capacities, but are also closely related to their demographic responses to external spatial and temporal variability of environment [26,27]. Drivers of isolation by distances (IBD), vicariance and landscape heterogeneity influence the species' evolutionary processes (genetic drift effects, gene exchanges, natural selection and local adaptation), thereby determining the genetic diversity patterns of metapopulations [28,29]. Human disturbance is also a contributory external factor influencing biodiversity. Since the beginning of the Anthropocene in the latter part of the 18th century, hazards of wild habitat destruction and loss caused by expansions in the range of anthropogenic activities have been becoming increasingly serious [30,31]. Under the background of global environmental and climatic changes, mankind's excessive pursuits of economic benefits are causing the growing reduction in biological resources, the extinction of species and populations, as well as the modification of natural landscapes [32]. Notably for small-scale isolated or peripheral populations, due to lack of connectivity with source populations, they are particularly prone to genetic depletion through genetic drift and inbreeding [33], whereas anthropogenic interference undoubtedly further weakens their fitness, decreasing survivorship and reproduction, and increasing extinction risks [34]. Therefore, research on and the protection of these sink populations are considered to be essential for maintaining the genetic integrity of large-scale metapopulations. Hitherto, the conservation of genetic resources has mainly focused on native or endemic rare and endangered species in global biodiversity hotspots, such as the Andes Mountains, Mediterranean Basin, eastern Himalayas and Hengduan Mountains in the tropics and subtropics [35–38]. However, some widespread natural resources, typically with medicinal or economic value, experiencing diversity reduction in temperate arid regions have not been sufficiently recognized and effectively protected [39]. If this continues, these species would not escape from the hidden dangers of resource exhaustion or even extinction.

Compared with traditional molecular genetic markers, genome-wide single-nucleotide polymorphisms (SNPs) have the characteristics of higher genotyping efficiency and data quality, analytical simplicity, broader genome coverage and sheer abundance, as well as advantages in further analyzing complex population structures and distinguishing the evolutionary relationship of genetic polymorphisms [40,41]. In recent years, they have been increasingly widely applied in research fields of phylogenetics, molecular ecology, conservation genetics and biogeography [41–43]. As a high-efficiency next-generation high-throughput sequencing technology (NGS) that has been developed and commonly used, restriction-site-associated DNA sequencing (RAD-Seq) can rapidly identify and score high-resolution SNP data from thousands of orthologous sites, while reducing the complexity of genomic sequences, even in the absence of reference genomes [44,45], becoming one of the main effective methods to address critical issues in the fields of evolution and genetics at the genomic level [46], especially for studies of species groups with ancient origins and complex evolutionary and genetic structures [47,48].

In the present study, for the first time, we used the SNP marker from dd-RAD sequencing, ADMIXTURE clustering, principal component analysis (PCoA) and maximum likelihood (ML) phylogenetic analysis methods to fully explore the integrated genetic diversity mechanisms and conservation significance of *C. spinosa* across large-scale environmental gradients in arid central Asia, based on spatial source-sink theory of metapopulation dynamics. Thus, we aimed to: (1) elucidate the population structure and intraspecific divergence of the lineages of this species; (2) to adequately reveal the spatial pattern of genetic diversity among identified geographical groups and, accordingly, formulate feasible conservation strategies for maintaining the genetic integrity and population resilience of this xerophytic subshrub; and (3) to provide a referential understanding for future research on the diversity conservation of relict plants in arid regions.

2. Materials and Methods

2.1. Sampling, DNA Extraction and dd-RAD Library Construction

We located and collected a total of 37 natural populations (5–12 individuals for each population) of *C. spinosa* in arid central Asia during 2019–2020 (Figure 2, Table S1), according to the distribution described in *Flora of China*, *Flora Xinjiangensis* and *Flora of Tibet* as references [12–14]. The sampling spacing was at least 100 m between the neighboring individuals within each population. We also collected *Capparis erythrocarpos* Isert (northeastern Kenya, Africa) and *Capparis bodinieri* Lévl. (Xishuangbanna, China) as outgroups for the subsequent phylogenetic analysis. Total genomic DNA was extracted from silica-dried leaf tissues using a DNeasy Plant Mini Kit (Qiagen, Valencia, CA, USA), following the manufacturer's protocol. The cleaned DNA had previously conducted simulated double digestion with 15 combinations among seven common or rare restriction enzymes to predict enzyme cutting sites (Table S2). Considering the yielded tag numbers and average sequencing depths, we finally selected the optimal combination of *Bfa*I (5′-C↓TAG-3′) and *Dpn*II (5′-↓GATC-3′) (New England Biolabs, Ipswich, MA, USA) to construct the dd-RAD library, following a modified approach of Peterson et al. [45]. The library with an inserted fragment between 200 and 600 bp was recovered for sequencing. All the qualified samples' sequencing was performed on the Illumina Hiseq X Ten platform (Illumina, San Diego, CA, USA) with paired-end reads of 2 × 150 bp in length. Library preparation and sequencing were performed by Shanghai Personalbio Biotechnology Co., Ltd.

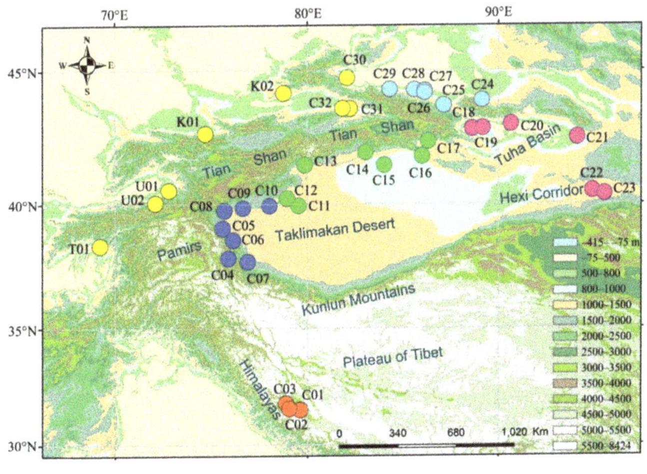

Figure 2. Natural geographical distribution of *C. spinosa* populations in arid central Asia. Different colors distinguish different geographic groups with red (western Himalayas), blue (eastern Pamirs), yellow (western Tianshan), azure (northern piedmont of Tianshan), green (southern piedmont of Tianshan) and purple (eastern Tianshan). The SRTM elevation data at a resolution of 30 s was downloaded from the WorldClim (https://www.worldclim.org/ accessed on 22 October 2021).

2.2. Data Processing and SNP Calling

Raw data produced by Illumina were processed to yield RAD tags and to obtain SNPs using the Stacks v1.4.8 program [49]. Initially, paired-end reads were cleaned and filtered for quality control and then trimmed to 140 bp × 2 in length using the *process_radtags* module. Individuals with sufficient RAD tags and coverage depth were used for SNP calling. Since there were no available reference genome data for *C. spinosa*, we next chose the *ustack* module to assemble reads per individual into RAD tags following the protocol of de novo workflow [49]. To remove high-repetitive stacks via a deleveraging algorithm, we set coverage depth to at least $m = 3$, the maximum distance allowed between stacks as $M = 2$ and the maximum number of mismatches allowed between loci as $n = 1$. Then, we combined the consensus loci among all 263 samples to build a catalog file in a *cstacks* module and aligned each loci sequence with the catalog to generate alleles in an *sstacks* module. Finally, filtered SNP matrices were identified and exported using a *populations* module, where the optimal operation parameters were set as follows: complete degree > 0.5 (individuals where the same RAD locus was detected accounted for at least 50% of each population), minor allele frequency (*MAF*) > 0.05 and minimum stack depth $m = 4$, $p = 100$ (one locus appeared in at least 100 populations). Based on the above calculations, the resulting high-quality SNP dataset was used for further analysis.

2.3. Genetic Structure Analysis

Population genetic structure, phylogenetic analysis and principal component analysis (PCoA) were comprehensively considered to evaluate the spatial distribution pattern of genetic variation. We performed a maximum likelihood estimation to analyze the genetic structure of *C. spinosa* using the ADMIXTURE program, version 1.3.0 [50], in order to identify the group clustering of each individual based on unlinked SNP datasets via a block relaxation algorithm. When the numbers of grouping (*K* value) were preset as $1 \leq K \leq 10$, ten replicated runs were implemented in ADMIXTURE to compute the mean cross-validation (*CV*) error for each *K*. The optimal *K* value was determined in conditions of the lowest *CV* error. For the analysis of phylogenetic relationships, we then reconstructed a maximum likelihood (ML) tree based on the unlinked genomic SNPs among these 263 individuals using the PhyML program, version 3.0 [51]. *Capparis erythrocarpos* Isert and *Capparis bodinieri* Lévl were chosen as the outgroups to root the tree. After the tree topology with maximum likelihood was established, the reliability of the branches was verified through a bootstrap (BP) with 1000 replications. Lastly, principal component analysis (PCoA) was performed to exhibit the first two axes of population variation in the species using GCTA software, version 1.93.2 [52]. After removing SNPs with *MAF* values of less than 0.05, genomic SNP data of different individuals were formed into a genetic matrix where the feature vectors that made a major contribution to variance were extracted, and then a scatter plot was drawn using the first two of the eigenvectors. According to the distribution characteristics of the principal components, the clustering relationship of individuals could be inferred.

2.4. Genetic Diversity and IBD Analysis

Based on the results of genetic grouping, population genetic statistics, including observed heterozygosity (H_o), expected heterozygosity (H_e), nucleotide diversity (P_i) and inbreeding coefficient (F_{IS}), were computed using the *populations* command in the Stacks v1.4.8 package [49] at the group and population levels of the species. Degrees of deviations from the Hardy–Weinberg equilibrium were measured for all 37 populations. Hierarchical AMOVA (analysis of molecular variance) in the ARLEQUIN version 3.5.2.2 program was used to classify the total genetic variation partitioning within and across groups by calculating the evolutionary distance between alleles or genotypes [53]. Pairwise genetic differentiation indices (F_{ST}) among populations were also calculated in the *populations* module in the Stacks. Additionally, the significance of isolation by distance (IBD) was further tested to compare Pearson's correlation between the genetic distance (F_{ST}) matrix and geographic distances matrix among these 37 populations using the *cor.test* function in the R v3.3.3 package [54].

3. Results

3.1. SNPs from dd-RAD Analysis of *C. spinosa*

A total of 263 *C. spinosa* individuals from 37 populations were sequenced using the dd-RAD library construction methodology. Both the base content and quality distribution across all the sequences were uniform and optimal, considering that the average proportion of bases whose recognition accuracy exceeded 99.9% (Q30) was greater than 90% and that the peak value was around Q36. The numbers of raw reads for each sample ranged from 6,455,458 to 26,065,222. After filtering out low-quality reads, the numbers of high-quality reads ranged from 5,995,376 to 24,640,428, and the mapping rate ranged from 90.24% to 97.98%. According to the clustering algorithm through the Stacks procedure, the obtained tag number per sample ranged from 247,350 to 823,581, and the average depth of sequencing was from 10× to 31×. Via mutation spectrum analysis, C:G > T:A and T:A > C:G account for the overwhelming majority of mutation types of SNPs (Figure S1). The final dataset, which contained 559,586 high-integrity SNPs, could be used for the subsequent population genetics analyses.

3.2. Population Genetic Structure of *C. spinosa*

The ADMIXTURE analysis showed that, when $K = 6$, the minimal value of the cross-validation (CV) error was detected (Figure S2), indicating that the best-fit number of genetic groupings of *C. spinoca* was six: (1) group WH (C01-C03, western Himalayas); (2) group EP (C04-C10, eastern Pamirs); (3) group WT (K01, K02, U01, U02, T01, C30-C32, western Tianshan Mountains); (4) group ST (C11-C17, the southern foothills of Tianshan); (5) group NT (C24-C29, the northern piedmont of Tianshan); and (6) group ET (C18-C23, eastern Tianshan) (Figures 2 and 3, Table S1). This pattern of population genetic structure was also supported by the topological relationship of the ML phylogenetic tree, with the main nodal bootstrap values exceeding 89.3% at the intraspecific level (Figure 4). Populations from group WH (red) always diverged first from all other populations, whether in the ADMIXTURE clustering or in the phylogenetic tree (Figures 3 and 4). Then, group EP (blue) branched off, followed by group WT (yellow) and group ST (green). Group NT (azure) was nested within group ST, and populations from group ET (purple) formed one clade (Figure 4).

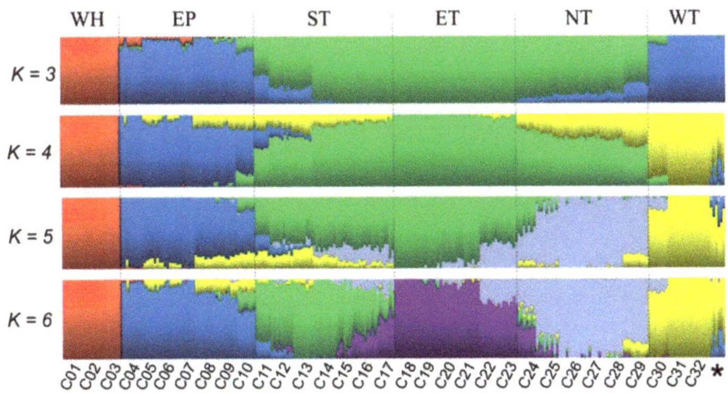

Figure 3. Population genetic structure of *C. spinosa* according to posterior probabilities assigned to the Bayesian clusters in the ADMIXTURE analysis. The patterns are shown with different K values (K = 3–6). The six groups are delimited by gray lines. Group WH, western Himalayas; group EP, eastern Pamirs; group ST, the southern piedmont of Tianshan; group NT, the northern piedmont of Tianshan; group ET, eastern Tianshan; group WT, western Tianshan. *, K02 (Almaty population), K01 (Bishkek population), U01 (Osh population), U02 (Batken population), T01 (Nurek population).

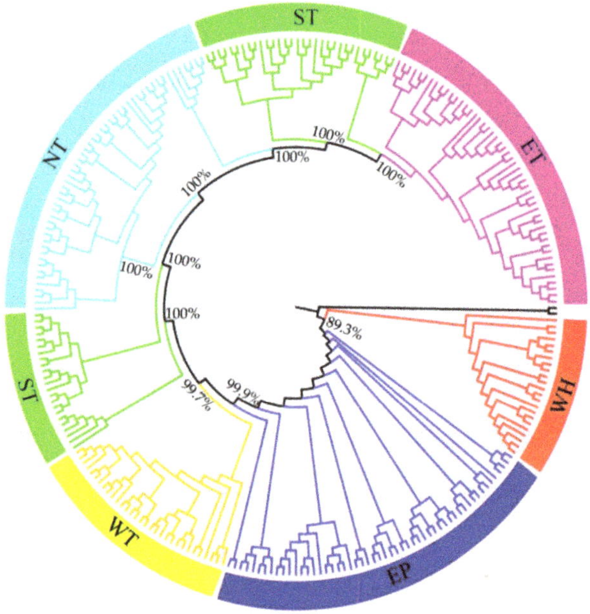

Figure 4. Maximum likelihood (ML) phylogenetic tree of 263 individuals from 37 sampled populations and two outgroup species (*C. erythrocarpos* and *C. bodinieri*) constructed with the genome-wide SNP dataset. The color-coded clades indicate different geographic groups of *C. spinosa* with red (western Himalayas), blue (eastern Pamirs), yellow (western Tianshan), azure (the northern piedmont of Tianshan), green (the southern piedmont of Tianshan) and purple (eastern Tianshan).

The first two coordinate axes of PCoA identified three major clusters, and corresponding principal components explained 16.55% (PC1) and 6.99% (PC2) of the total genetic variation, respectively (Figure 5a). Considering the preliminary results, group WH formed the first cluster, which was completely separated from the rest of the groups; group EP and group WT were generally merged into the second cluster; and group ST, group NT and group ET were broadly integrated into the third cluster. Then, we conducted further PCoA analyses on the latter two clusters, separately. The substructure within the second cluster showed that group EP was gradually split from group WT, with 16.15% (PC1) and 6.99% (PC2) of the total variation (Figure 5b). The substructure within the third cluster showed that populations in groups ST, NT and ET were mostly divided from each other (PC1, 16.15% vs. PC2, 6.99%), whereas admixed populations crossed among them (Figure 5c), which was also evident in the ADMIXTURE diagram (Figure 3).

3.3. Genetic Diversity Pattern and IBD

The results of the genetic diversity analysis showed that the overall genetic diversity among geographical groups was ranked as follows: EP > ST > NT > WT > ET > WH, which exhibited distinct differences in various degrees (Table 1). At the population level, the maximum observed heterozygosity (H_o), expected heterozygosity (H_e) and nucleotide diversity indices (P_i) were in Manas (C26) and Jiashi (C10), while the minimum H_o, H_e and H_e values were in Zabrang (C01) (Table 1). Inbreeding coefficients (F_{IS}) that measured degrees of deviation from the Hardy–Weinberg equilibrium were positive for all 37 populations. The results of the AMOVA showed that the percentage of variation among the six geographical groups was relatively low; the variation was mainly from individuals within the populations, and the variation rate across populations within groups was the lowest (Table 2). Significant genetic differentiation existed among geographic populations (Figure S3, Table S3), with F_{ST} values on a scale from 0.076 to 0.707 (Table S3). The maximum genetic differentiation was detected between C02 (Sarang population) and K01 (Bishkek population) (F_{ST} = 0.707), while the minimum genetic differentiation existed between C27 (Shihezi population) and C29 (Wusu population) (F_{ST} = 0.076) (Table S3). *Nei*'s genetic distance was roughly synchronous with pairwise F_{ST} between populations (Figure S3). Furthermore, the result of isolation by distance (IBD) analysis displayed a considerable correlated pattern between genetic divergence and geographical distance among these 37 populations, given that the two-tailed test of the Pearson correlation coefficient was of significance (r = 0.499, p < 0.001) (Figure 6).

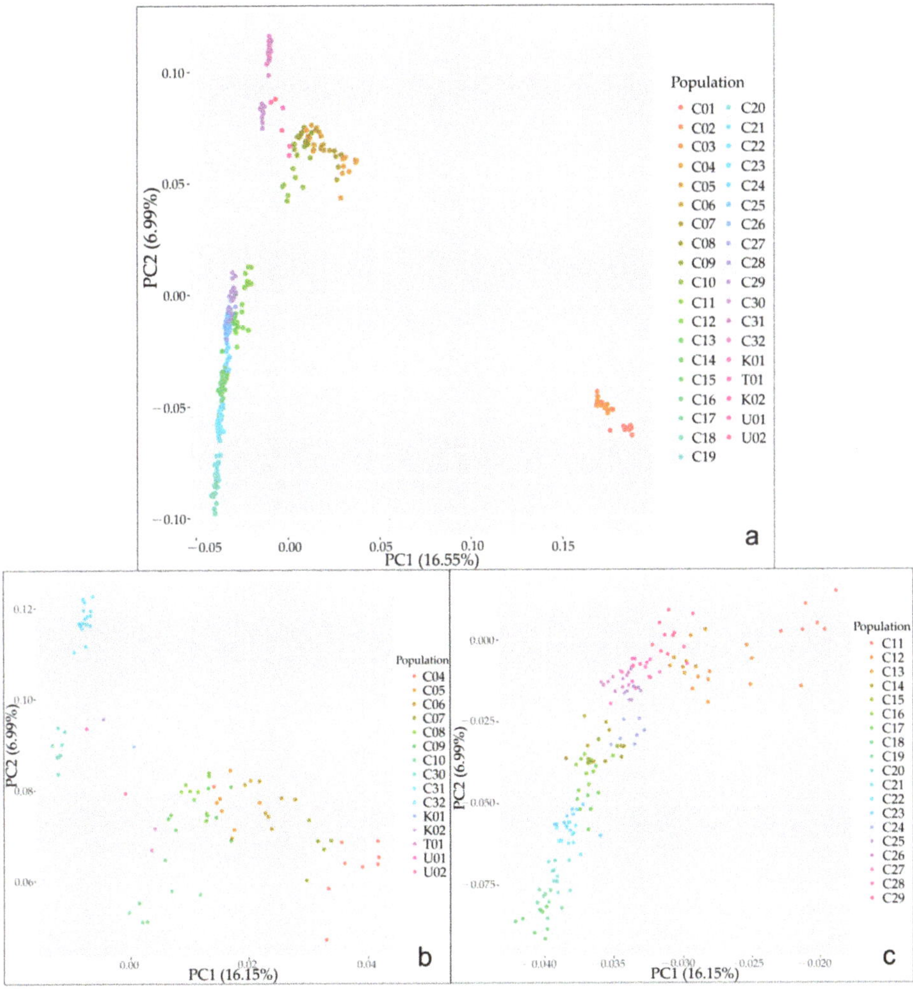

Figure 5. PCA analysis based on SNP dataset. (**a**) PC1 and PC2 (16.55% vs. 6.99%) are the distributions of the two main principal components of the genetic variation of *C. spinosa*, (**b**) substructures within group EP and group WT (PC1, 16.15% vs. PC2, 6.99%) in addition to (**c**) substructures within groups ST (the southern piedmont of Tianshan), NT (the northern piedmont of Tianshan) and ET (eastern Tianshan Mountains) (PC1, 16.15% vs. PC2, 6.99%). The individuals are color coded according to different populations.

Table 1. Genetic diversity indices for natural populations of *C. spinosa*.

Code	Pop. Locality	H_o	H_e	P_i
Group WH	(Western Himalayas)	0.0455	0.0648	0.0671
C01	Zabrang, Tibet, China	0.0394	0.0429	0.0465
C02	Sarang, Tibet, China	0.0512	0.0448	0.0594
C03	Diya, Tibet, China	0.0528	0.0637	0.0687
Group EP	(Eastern Pamirs)	0.1169	0.2283	0.2318
C04	Taxkorgan, Xinjiang, China	0.1109	0.1822	0.2003
C05	Akto, Xinjiang, China	0.1154	0.1665	0.1932
C06	Yengisar, Xinjiang, China	0.1168	0.1805	0.1989
C07	Yecheng, Xinjiang, China	0.1220	0.1420	0.1703
C08	Wuqia, Xinjiang, China	0.1026	0.1665	0.1858
C09	Artux, Xinjiang, China	0.1234	0.1946	0.2134
C10	Jiashi, Xinjiang, China	0.1273	0.1991	0.2234
Group WT	(Western Tianshan)	0.1025	0.1914	0.1965
K02	Almaty, Kazakhstan	0.0932	0.0466	0.0932
K01	Bishkek, Kyrgyzstan	0.0561	0.0281	0.0561
U01	Osh, Kyrgyzstan	0.1057	0.0529	0.1057
U02	Batken, Kyrgyzstan	0.1115	0.0558	0.1115
T01	Nurek, Khatlon, Tajikistan	0.0940	0.0769	0.1154
C30	Bole, Xinjiang, China	0.1150	0.1693	0.1866
C31	Yining, Xinjiang, China	0.1021	0.1367	0.1539
C32	Gongliu, Xinjiang, China	0.1020	0.1475	0.1602
Group ST	(southern piedmont of Tianshan)	0.1125	0.2223	0.2255
C11	Tumxuk, Xinjiang, China	0.0911	0.1372	0.1585
C12	Kalpin, Xinjiang, China	0.1178	0.1840	0.2042
C13	Wensu, Xinjiang, China	0.1145	0.1739	0.1945
C14	Baicheng, Xinjiang, China	0.1208	0.1745	0.1885
C15	Lunnan, Xinjiang, China	0.0956	0.1321	0.1555
C16	Korla, Xinjiang, China	0.1135	0.1714	0.1897
C17	Hejing, Xinjiang, China	0.1178	0.1691	0.1904
Group NT	(northern piedmont of Tianshan)	0.1206	0.2214	0.2245
C24	Jimsar, Xinjiang, China	0.1155	0.1623	0.1827
C25	Urumqi, Xinjiang, China	0.1120	0.1850	0.2020
C26	Manas, Xinjiang, China	0.1288	0.1783	0.1925
C27	Shihezi, Xinjiang, China	0.1161	0.1873	0.2046
C28	Shawan, Xinjiang, China	0.1298	0.1692	0.1940
C29	Wusu, Xinjiang, China	0.1198	0.1915	0.2069
Group ET	(Eastern Tianshan)	0.0999	0.1910	0.1944
C18	Toksun, Xinjiang, China	0.0973	0.1270	0.1455
C19	Turpan, Xinjiang, China	0.1060	0.1479	0.1631
C20	Shanshan, Xinjiang, China	0.1059	0.1549	0.1699
C21	Hami, Xinjiang, China	0.0920	0.1310	0.1473
C22	Dunhuang, Gansu, China	0.1029	0.1286	0.1516
C23	Guazhou, Gansu, China	0.0942	0.1595	0.1749

H_o, observed heterozygosity; H_e, expected heterozygosity; P_i, nucleotide diversity.

Table 2. AMOVA among different geographic groups, inter- and intra-populations of *C. spinosa*.

Source of Variation	*df*	Sum of Squares	Variance Components	Percentage of Variation (%)	*Φ*-Statistics
Among groups	5	99,228.136	212.66396	27.36	$\Phi_{CT} = 0.27$ **
Among populations within groups	29	39,934.919	59.33213	7.63	$\Phi_{SC} = 0.11$ **
Within populations	491	248,090.120	505.27519	65.01	$\Phi_{ST} = 0.35$ **
Total	525	387,253.175	777.27128		

df, degrees of freedom; **, $p < 0.001$.

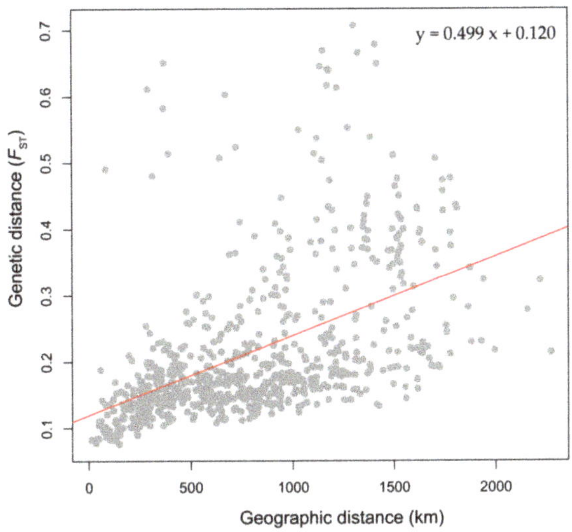

Figure 6. Isolation by distance (IBD) analysis: correlation between genetic distance and geographical distance among 37 populations ($r = 0.499$, $p < 0.001$).

4. Discussion

4.1. Relationships between Genetic Structure, Lineage Differentiation and Geographic Distribution

In this study, we obtained a clear genetic structure of *C. spinosa* through RAD-seq and identified six clustering results corresponding to geographic locations (Figures 2 and 3, Table S1). For this Tertiary relic plant, the effects of isolation by distance and vicariance has caused long-term shaping of local ecological environments, which had a profound influence on population structure and lineage differentiation among different geographic units. It is speculated that the formation of this disjunction is the result of intense and frequent orogeny during the geological period, coupled with extensive Quaternary glaciation, as well as the long-term process of aridification in central Asia [5–7]. The allopatric divergence and local adaptive evolution among distant geographical groups, especially for the isolated populations, are probably attributed to the incapability of long-distance genetic exchange and being hindered by mountains and immense deserts. Similar disjunctive or fragmented population structure patterns have also been reported in other Tethyan relicts that occur in this area, such as *Gymnocarpos przewalskii*, *Amygdalus mongolica* and *Ammopiptanthus*, owing to the influence of long-term aridification [9–11]. The constantly extended deserts and climatic changes have led to the fragmentation and separation of suitable habitats for those xerophytic taxa, thereby restricting the migration among distant geographic populations [55].

The relatively isolated populations distributed in the northern piedmont of the western Himalayas (Group WH, C01–C03) have the maximum genetic distance from other populations. Plant taxon located in this desert plateau region is related to the sub-frigid arid climate, precipitous terrain and topography, as well as abnormally barren soil conditions (Figure 1a). Due to the complex and discontinuous natural environment, the sink populations here, to a certain extent, are differentiated by hereditary and habitat variability. Compared to populations distributed at high altitudes, populations in medium-altitudinal valleys and low-altitudinal desert basins have different habitat requirements (Figure 1). Among the eastern Pamir region and the southern piedmonts of the Tianshan Mountains, we found that population structures were associated with spatial geographic distribution, as well as heterogeneous environments [29]. For instance, the Jiashi population (C10) is geographically nearby the Kalpin (C12) and Tumxuk (C11) populations, but due to the hereditary differences and landscape heterogeneity (gravel Gobi vs. dried clay desert) they belong, respectively, to group EP and group ST, although some of the mixed genotypes were found in these divergent populations (Figure 3). On the contrary, according to research results from Mims et al. (2016), among populations within the same geographic unit, the habitat preferences are of continuity and similarity, where high degree of gene flow could exist among them in a relatively flatter and more expansive range [56]. For example, source populations that inhabit piedmont proluvial fans on the southern side of the Tianshan Mountains (Group ST, C11–C17) are more closely distributed, and the connectivity of high-quality breeding habitats may play a role in facilitating closer reduced genetic distance between these populations.

4.2. Genetic Diversity Pattern of Species Metapopulations

The genetic diversity of *C. spinosa* (Pi = 0.2644) (Table 1) through RAD-seq is higher than that of other relic species (such as *Euptelea pleiosperma* and *Gymnocarpos przewalskii*) [11,48] because widespread species have a higher genetic diversity than species with endemic distribution. The total diversity index in arid central Asia is also different from that obtained by previous research using AFLP, RAPD and EST-SSR for *C. spinosa* in the Mediterranean coast (southern Europe, north Africa and west Asia) [18,20,21]. The primary reasons may be the different systematic molecular markers and the varying regional environments and weather conditions. In the arid region of central Asia, the distribution of *C. spinosa* covers most of the mountain piedmont, warm dry valley and Gobi desert terrains. Based on the results of SNPs, the range of this species is divided into six geographic groups with various levels of genetic diversity, indicating the levels of geographic isolation and environment gradients. According to the observed heterozygosity, the genetic diversity of source populations in groups ST and NT remained at relatively high and stable levels (Table 1), indicating that populations within these two groups are both likely to have experienced long-term migration of gene flow. In these geographic units, the high-quality piedmont alluvial or diluvial fan habitats are usually continuous, which is likely conducive to maintaining a high level of genetic diversity throughout these regions. In contrast, sink populations in group WH (western Himalayas) inhabit low-quality desert plateaus with high altitudes and low temperatures, especially the isolated Zabrang population (C01) with lowest genetic diversity (Table 1) adjacent to the middle reaches of the Sutlej River Basin, which grows on steep rocky slopes and seems a bit far away from the water source (Figure 1a); even worse, natural disasters, such as landslides and mudslides, occur frequently during the annual flowering and fruiting phases. In addition, because of insufficient visits by pollinators, inbreeding or self-breeding dominate the reproduction of *C. spinosa* under this situation, resulting in significant population depression (current effective size of <50 individuals). Many previous studies have demonstrated that, in nature, isolated populations with smaller effective sizes often have more genetic depletion and extinction risks due to genetic drift and inbreeding, which are, to a large extent, disadvantageous to the persistence of population diversity [57,58]. In the process of evolutionary history, the Zabrang population survives in a separate habitat but has not adapted to the extreme environment, which may be one of the reasons why its population scale is declining and even endangered. Nevertheless, it is noteworthy that the Taxkorgan (C04, 2855 m alt) and Akto (C05, 2845 m alt) populations, which are also located at high altitudes, could maintain certain high levels of genetic diversity because they are not completely isolated in the eastern Pamirs region (group EP). These populations have continuous extension and buffer along the descending elevation gradients, and they also have a certain degree of gene exchange with the Yegisar population (C06) in the piedmont belt.

Researchers have proved that, compared with diffusion within short distances, the long-distance dispersal of gene flow plays a more pivotal role in the persistence of the entire connectivity and genetic diversity of plant metapopulations [59,60]. Simultaneously, the contributions of wind, pollinators and river water help the pollen flow to spread farther. However, within some geographic groups (group ET, group WT), moderate degrees of genetic differentiation (F_{ST} > 0.1) (Table S3) have appeared in neighboring populations, indicating that anthropogenic activities may have encroached into these natural territories, leading to the discontinuous gene flow. Driven by economic interests and commercial purposes, human over-exploitation has weakened the reproductive capacity of *C. spinosa* populations and has arbitrarily cut off the long-distance dispersal chains of gene flow, making short-distance spreading predominant, increasing the rates of inbreeding depression and resulting in population decline, thereby making it difficult for them to maintain their genetic diversity and evolutionary potential. Some populations near human activity areas, such as Toksun (C18), Hami (C21), Dunhuang (C22), Yining (C31) and Gongliu (C32), are frequently threatened by natural resource plundering and other disturbances, such as riverway- or highway-widening, agricultural reclamation and industrial expansion, etc. The relatively low genetic heterozygosity and diversity levels (Table 1) found in these populations may be primarily caused by the results of the above depredations.

4.3. *Implications for Conservation*

Evaluating the spatial genetic structure and survival situation of metapopulations across regional environment variability could efficiently and effectively support and improve the management measures of biodiversity conservation, thereby conducing to maintain the evolutionary potential of widespread species in response to changing environments [33,56]. In this study, most of the central source populations with large effective sizes are located in the expansive piedmonts of north and south Tianshan Mountains, where multiple hotspots of biodiversity have been shaped in their long evolution history in the arid region. In contrast, isolated populations on the desert plateau of western Himalayas, as well as peripheral populations in the desert basin adjacent to eastern Tianshan, are experiencing a decrease in or even disappearance of biodiversity. As a result of the adverse effects of hereditary factors or human disturbance, these sink populations are extremely vulnerable to habitat fragmentation and resource exhaustion [34], so it is particularly important to monitor population trajectories, minimize human disturbances and restore natural landscapes [27]. Given the limited manpower and material resources, in addition to relevant policies for protecting xeromorphic vegetations in arid regions, we recommend that conservation management units should be established that are consistent with the six effective evolutionary groups and that the emphasis of preservation work should be laid on the genetic rescue of isolated populations and habitat restoration of peripheral populations in order to persist the genetic integrity of this plant species.

Conservation of sink populations has been described as an important part of biodiversity persistence of integrated metapopulations because they often contain certain rare alleles or genotypes in the process of historic immigration [61]. The Zabrang population (C01) is distributed in more sparse and localized low-quality habitats, resulting in its greater genetic differentiation with other populations within group WH (F_{ST} = 0.1357–0.1878) (Table S3). However, due to a lack of corresponding attention and conservation, the genetic diversity and uniqueness of such isolated population are being threatened and almost disappearing. Empirical studies have illustrated that inputting a certain level of stable gene flow from large source populations to small sinks may counter the negative effects of inbreeding depression and genetic drift [26,33]. Currently, we suggest that genetic rescue such as immigration, exchange of pollen flow and hybridization, might reduce the extinction risk of isolated populations of *C. spinosa*. If conditions permit, ex situ preservation and reintroduction instead of natural dispersal may be more effective protection strategies to increase local effective population sizes.

The disturbance of human activities, such as the expansion of farmland, makes this subshrub lose its competitive advantage, which could consequently lead to its habitat shrinking. Additionally, the industrial waste generated by newly built factories in the suburbs has caused serious pollution to the soil environment and increased the saline–alkali stresses on its habitat, whereupon the mortality rates of plant individuals have also risen. If these trends persist, the relic species would scarcely escape from the hidden danger of resource exhaustion or even the verge of extinction. Populations in the Tuha

Basin–Hexi Corridor (Group EP) are located on the edge of the Kumtag Desert, where water shortages and soil salinization are becoming increasingly serious. These findings are the consequences of continuous increasing human pressure on the barren habitats of xerophytes. According to our field investigation conducted from 2011 to 2020, the population sizes within group EP in the desert basin were originally large (>300 individuals per population in 2011). Nevertheless, over the past decade, due to humans' immoderate harvesting and unceasing expansion of agriculture and industry, the living space of *C. spinosa* in this region has decreased sharply. In particular, the genetic diversity of the peripheral Hami and Dunhuang populations (C21 and C22) has been gradually decreasing, and the plant morphology and population sizes (<100 individuals per population in 2020) here have also been shrinking under these dual pressures, which weaken the plant's ability to prevent wind and sand and preserve water and soil (Figure 1c). This region has been trapped in a vicious circle of natural desert vegetation destruction, water loss and soil erosion, as well as aggravated saline–alkali desertification. In order to realize the sustainable development of the ecological economy in arid regions, we should rationally conduct agricultural production on the basis of protecting natural desert landscapes, reduce the pollution of industrial waste to primitive habitats and limit the predatory exploitation of germplasm resources, therefore protecting the biodiversity of wild drought-resistant species to ultimately slow down the processes of desertification and salinization.

5. Conclusions

In this study, we analyzed the population genetic structure and biodiversity conservation management of *C. spinosa*, a Tethyan relic medicinal subshrub in arid central Asia. Six geographical units with differences in genetic diversity and lineage differentiation were distinguished. This excellent wind-preventing, sand-fixing and soil- and water-conserving xerophyte is at potential risk of natural resource depletion. Populations at high altitudes of the western Himalayas have been experiencing a gradual decline in genetic diversity and effective scales as a result of inbreeding depression by IBD and habitat isolation. On the other hand, peripheral populations at low altitudes in eastern desert basins with original rich genetic diversity are suffering from the plundering of germplasm resources and the destruction of survival environment. Therefore, to restore the genetic integrity of the metapopulations, we suggest that central source populations should be preserved in situ corresponding to different geographical units and that conservation priority should be focused on the genetic rescue of isolated populations and the habitat restoration of peripheral populations. Meanwhile, minimizing human activities in addition to the rational and sustainable utilization of natural resources will be of significance to the population resilience and evolutionary potential of *C. spinosa* in response to the long-term aridification and future changing environment, and it will eventually maintain the ecosystem balance, as well as slow the desertification process in arid regions.

Supplementary Materials: The following are available online at https://www.mdpi.com/article/10.3390/d14020146/s1, Table S1: Details of geographical locations, altitude and effective sizes of natural populations of *C. spinosa*. Table S2: Digestion sites of several common or rare restriction enzymes, as well as their combinations for screening schemes in this study. Table S3: Pairwise genetic differentiation indices (F_{ST}) among the 37 populations. Figure S1: Mutation spectrum of genome-wide single-nucleotide polymorphism (SNP) dataset of *C. spinosa*. Mutation types of SNPs are distinguished by different colors. T:A > C:G and C:G > T:A account for the overwhelming majority. Figure S2: Distribution of cross-validation (CV) error in the ADMIXTURE analysis. $K = 6$ is the optimal K value in condition of the lowest CV error. Figure S3: Pairwise *Nei*'s genetic distance among populations of *C. spinosa*.

Author Contributions: Conceptualization, Q.W.; data curation, H.-X.Z. and Q.W.; formal analysis, H.-X.Z. and Q.W.; funding acquisition, Q.W.; investigation, Q.W.; methodology, H.-X.Z.; project administration, Q.W.; visualization, H.-X.Z. and Q.W.; writing—original draft, Q.W.; and writing—review and editing, Q.W. and H.-X.Z. All authors have read and agreed to the published version of the manuscript.

Funding: This research was funded by National Natural Science Foundation of China (31700287, 31870323, 32170391 and 31770227).

Institutional Review Board Statement: Not applicable.

Data Availability Statement: All RAD-seq data of *C. spinosa* were submitted to the NCBI (SRA accession: SAMN24698411–SAMN24698673).

Acknowledgments: The authors are grateful to Qing-Song Pu, Ke-Jing Ouyang and Ze-Run Zhu for their kind help and guidance in the field investigation in Tibet Autonomous Region for this paper.

Conflicts of Interest: The authors declare no conflict of interest.

References

1. Chen, X.; Jiang, F.Q.; Wang, Y.J.; Li, Y.M.; Hu, R.J. Characteristics of the eco-geographical pattern in arid land of Central Asia. *Arid Zone Res.* **2013**, *30*, 385–390, (In Chinese with English Abstract).
2. Yong, S.P.; Zhu, Z.Y. A certain fundamental characteristics of gobi desert vegetation in the Centre Asia. *Acta Sci. Nat. Univ. Intramongol.* **1992**, *23*, 235–244, (In Chinese with English Abstract).
3. Wu, Z.Y.; Sun, H.; Zhou, Z.K.; Li, D.Z.; Peng, H. *Floristics of Seed Plants from China*; Science Press: Beijing, China, 2010. (In Chinese)
4. Liu, Y.X. A study on origin and formation of the Chinese desert floras. *Acta Phyt. Sin.* **1995**, *33*, 131–143, (In Chinese with English Abstract).
5. Sun, H.; Li, Z.M. Qinghai-Tibet Plateau uplift and its impact on Tethys Flora. *Adv. Earth Sci.* **2003**, *18*, 852–862.
6. Fang, X.M.; Lv, L.Q.; Yang, S.L.; Li, J.J.; An, Z.S.; Jiang, P.A.; Chen, X.L. Loess in Kunlun Mountain and its implications on desert development and Tibetan Plateau uplift in west China. *Sci. China Ser. D* **2002**, *45*, 289–299. [CrossRef]
7. Guan, Q.Y.; Pan, B.T.; Li, N.; Zhang, J.D.; Xue, L.J. Timing and significance of the initiation of present day deserts in the northeastern Hexi Corridor, China. *Palaeogeogr. Palaeoclimatol. Palaeoecol.* **2011**, *306*, 70–74.
8. Zhang, L.Y.; Hai, Y. Plant communities excluded in the book of "The Vegetation and its Utilization in Xinjiang": I. The desert plant communities. *Arid Land Geogr.* **2002**, *25*, 84–89.
9. Su, Z.H.; Pan, B.R.; Zhang, M.L.; Shi, W. Conservation genetics and geographic patterns of genetic variation of endangered shrub *Ammopiptanthus* (Fabaceae) in northwestern China. *Conserv. Genet.* **2016**, *17*, 485–496. [CrossRef]
10. Ma, S.M.; Nie, Y.B.; Jiang, X.L.; Xu, Z.; Ji, W.Q. Genetic structure of the endangered, relict shrub *Amygdalus mongolica* (Rosaceae) in arid northwest China. *Aust. J. Botany* **2019**, *67*, 128–139. [CrossRef]
11. Zhang, H.X.; Wang, Q.; Jia, S.W. Genomic phylogeography of *Gymnocarpos przewalskii* (Caryophyllaceae): Insights into habitat fragmentation in arid Northwestern China. *Diversity* **2020**, *12*, 335. [CrossRef]
12. Zhang, M.L.; Tucker, G.C. Capparaceae. In *Flora of China*; Wu, Z.Y., Raven, P.H., Eds.; Science Press: Beijing, China; Missouri Botanical Garden Press: St. Louis, MO, USA, 2008; Volume 7, pp. 433–450.
13. Wu, S.K. Capparidaceae. In *Flora of Tibet*; Wu, Z.Y., Ed.; Science Press: Beijing, China, 1985; Volume 2, pp. 322–323. (In Chinese)
14. An, Z.X. Capparidaceae. In *Flora Xinjiangensis*; Mao, Z.M., Ed.; Xinjiang Science, Technology and Hygiene Publishing House: Urumqi, China, 1995; Volume 2, pp. 35–36. (In Chinese)
15. Fici, S.; Gianguzzi, L. Diversity and conservation in wild and cultivated *Capparis* in Sicily. *Bocconea* **1997**, *7*, 437–443.
16. Saadaoui, E.; Khaldi, A.; Khouja, M.L.; Mohamed, E.G. Intraspecific variation of *Capparis spinosa* L. in Tunisia. *J. Herbs Spices Med. Plants* **2009**, *15*, 9–15. [CrossRef]
17. Fici, S. A taxonomic revision of the *Capparis spinosa* group (Capparaceae) from the Mediterranean to Central Asia. *Phytotaxa* **2014**, *174*, 001–024. [CrossRef]
18. Özbek, Ö.; Kara, A. Genetic variation in natural populations of *Capparis* from Turkey, as revealed by RAPD analysis. *Plant Syst. Evol.* **2013**, *299*, 1911–1933. [CrossRef]
19. Al-Safadi, B.; Faouri, H.; Elias, R. Genetic diversity of some *Capparis* L. species growing in Syria. *Braz. Arch. Biol. Technol.* **2014**, *57*, 916–926. [CrossRef]
20. Aichi-Yousfi, H.; Bahri, B.A.; Medini, M.; Rouz, S.; Rejeb, M.N.; Ghrabi-Gammar, Z. Genetic diversity and population structure of six species of *Capparis* in Tunisia using AFLP markers. *Comptes Rendus Biol.* **2016**, *339*, 442–453. [CrossRef]
21. Mercati, F.; Fontana, I.; Gristina, A.S.; Martorana, A.; El Nagar, M.; De Michele, R.; Fici, S.; Carimi, F. Transcriptome analysis and codominant markers development in caper, a drought tolerant orphan crop with medicinal value. *Sci. Rep.* **2019**, *9*, 10411. [CrossRef]
22. Futuyma, D.J. *Evolutionary Biology*, 2nd ed.; Sinauer Associates: Sunderland, MA, USA, 1986.
23. Vellend, M.; Geber, M.A. Connections between species diversity and genetic diversity. *Ecol. Lett.* **2005**, *8*, 767–781. [CrossRef]
24. Frankham, R. Genetics and extinction. *Biol. Conserv.* **2005**, *126*, 131–140. [CrossRef]
25. Sakai, A.K.; Allendorf, F.W.; Holt, J.S.; Lodge, D.M.; Molofsky, J.; With, K.A.; Baughman, S.; Cabin, R.J.; Cohen, J.E.; Ellstrand, N.C.; et al. The population biology of invasive species. *Annu. Rev. Ecol. Syst.* **2001**, *32*, 305–332. [CrossRef]
26. Pulliam, R.H. Sources, sinks, and population regulation. *Am. Nat.* **1988**, *132*, 652–661. [CrossRef]
27. Heinrichs, J.A.; Lawler, J.J.; Schumaker, N.H. Intrinsic and extrinsic drivers of source–sink dynamics. *Ecol. Evol.* **2016**, *6*, 892–904. [CrossRef] [PubMed]

28. Wright, S. Isolation by distance. *Genetics* **1943**, *28*, 114–138. [CrossRef] [PubMed]
29. Walters, S. Landscape pattern and productivity effects on source–sink dynamics of deer populations. *Ecol. Model.* **2001**, *143*, 17–32. [CrossRef]
30. Crutzen, P.J. Geology of mankind. *Nature* **2002**, *415*, 23. [CrossRef]
31. Lewis, S.L.; Maslin, M.A. Defining the Anthropocene. *Nature* **2015**, *519*, 171–180. [CrossRef]
32. Meng, H.H.; Zhou, S.S.; Li, L.; Tan, Y.H.; Li, J.W.; Li, J. Confict between biodiversity conservation and economic growth: Insight into rare plants in tropical China. *Biodivers. Conserv.* **2019**, *28*, 523–537. [CrossRef]
33. Hanski, I. Metapopulation dynamics. *Nature* **1998**, *396*, 41–49. [CrossRef]
34. Nunes, F.; Norris, R.D.; Knowlton, N. Implications of isolation and low genetic diversity in peripheral populations of an amphi-Atlantic coral. *Mol. Ecol.* **2009**, *18*, 4283–4297. [CrossRef]
35. Myers, N.; Mittermeier, R.A.; Mittermeier, C.G.; da Fonseca, G.A.B.; Kent, J. Biodiversity hotspots for conservation priorities. *Nature* **2000**, *403*, 853–858. [CrossRef]
36. Kessler, M. Patterns of diversity and range size of selected plant groups along an elevational transect in the Bolivian Andes. *Biodivers. Conserv.* **2001**, *10*, 1897–1921. [CrossRef]
37. Liu, J.; Moller, M.; Provan, J.; Gao, L.M.; Poudel, R.M.; Li, D.Z. Geological and ecological factors drive cryptic speciation of yews in a biodiversity hotspot. *New Phytol.* **2013**, *199*, 1093–1108. [CrossRef] [PubMed]
38. Fernández-Mazuecos, M.; Jiménez-Mejías, P.; Rotllan-Puig, X.; Vargas, P. Narrow endemics to Mediterranean islands: Moderate genetic diversity but narrow climatic niche of the ancient, critically endangered *Naufraga* (Apiaceae). *Perspect. Plant Ecol. Evol. Syst.* **2014**, *16*, 190–202. [CrossRef]
39. Millar, C.I.; Libby, W.J. Strategies for conserving clinal, ecotypic, and disjunct population diversity in widespread species. In *Genetics and Conservation of Rare Plants*; Falk, D.A., Holsinger, K.E., Eds.; Oxford University Press: New York, NY, USA, 1991; Volume 10, pp. 149–170.
40. Brumfield, R.T.; Beerli, P.; Nickerson, D.A.; Edwards, S.V. The utility of single nucleotide polymorphisms in inferences of population history. *Trends Ecol. Evol.* **2003**, *18*, 249–256. [CrossRef]
41. Morin, P.A.; Luikart, G.; Wayne, R.K.; The SNP Workshop Group. SNPs in ecology, evolution and conservation. *Trends Ecol. Evol.* **2004**, *19*, 208–216. [CrossRef]
42. Leaché, A.D.; Oaks, J.R. The utility of single nucleotide polymorphism (SNP) data in phylogenetics. *Annu. Rev. Ecol. Evol. Syst.* **2017**, *48*, 69–84. [CrossRef]
43. Zhang, H.X.; Li, X.S.; Wang, J.C.; Zhang, D.Y. Insights into the aridification history of Central Asian Mountains and international conservation strategy from the endangered wild apple tree. *J. Biogeogr.* **2021**, *48*, 332–344. [CrossRef]
44. Davey, J.W.; Blaxter, M.L. RADSeq: Next-generation population genetics. *Brief. Funct. Genom.* **2011**, *9*, 416–423. [CrossRef]
45. Peterson, B.K.; Weber, J.N.; Kay, E.H.; Fisher, H.S.; Hoekstra, H.E. Double digest RADseq: An inexpensive method for *de novo* SNP discovery and genotyping in model and non-model species. *PLoS ONE* **2012**, *7*, e37135. [CrossRef]
46. Andrews, K.R.; Good, J.M.; Miller, M.R.; Luikart, G.; Hohenlohe, P.A. Harnessing the power of RADseq for ecological and evolutionary genomics. *Nat. Rev. Genet.* **2016**, *17*, 81–92. [CrossRef]
47. Jiang, X.L.; Gardner, E.M.; Meng, H.H.; Deng, M.; Xu, G.B. Land bridges in the Pleistocene contributed to flora assembly on the continental islands of South China: Insights from the evolutionary history of *Quercus championii*. *Mol. Phylogenet. Evol.* **2019**, *132*, 36–45. [CrossRef] [PubMed]
48. Cao, Y.N.; Zhu, S.S.; Chen, J.; Comes, H.P.; Wang, I.J.; Chen, L.Y.; Sakaguchi, S.; Qiu, Y.X. Genomic insights into historical population dynamics, local adaptation, and climate change vulnerability of the East Asian Tertiary relict *Euptelea* (Eupteleaceae). *Evol. Appl.* **2020**, *13*, 2038–2055. [CrossRef] [PubMed]
49. Catchen, J.; Hohenlohe, P.A.; Bassham, S.; Amores, A.; Cresko, W.A. Stacks: An analysis tool set for population genomics. *Mol. Ecol.* **2013**, *22*, 3124–3140. [CrossRef] [PubMed]
50. Alexander, D.H.; Novembre, J.; Lange, K. Fast model-based estimation of ancestry in unrelated individuals. *Genome Res.* **2009**, *19*, 1655–1664. [CrossRef] [PubMed]
51. Guindon, S.; Dufayard, J.-F.; Lefort, V.; Anisimova, M.; Hordijk, W.; Gascuel, O. New algorithms and methods to estimate maximum-likelihood phylogenies: Assessing the performance of PhyML 3.0. *Syst. Biol.* **2010**, *59*, 307–321. [CrossRef] [PubMed]
52. Yang, J.; Lee, S.H.; Goddard, M.E.; Visscher, P.M. GCTA: A tool for genome-wide complex trait analysis. *Am. J. Hum. Genet.* **2011**, *88*, 76–82. [CrossRef]
53. Excoffier, L.; Lischer, H.E. Arlequin suite ver 3.5: A new series of programs to perform population genetics analyses under Linux and Windows. *Mol. Ecol. Res.* **2010**, *10*, 564–567. [CrossRef]
54. R Core Team. *R: A Language and Environment for Statistical Computing*; R Foundation for Statistical Computing: Vienna, Austria, 2013.
55. Meng, H.H.; Gao, X.Y.; Huang, J.F.; Zhang, M.L. Plant phylogeography in arid Northwest China: Retrospectives and perspectives. *J. Syst. Evol.* **2015**, *53*, 33–46. [CrossRef]
56. Mims, M.C.; Hauser, L.; Goldberg, C.S.; Olden, J.D. Distance, and metapopulation dynamics of the Arizona Treefrog (*Hyla wrightorum*) in an isolated portion of its range. *PLoS ONE* **2016**, *11*, e0160655. [CrossRef]
57. Frankham, R.; Ballou, J.D.; Briscoe, D.A. *Introduction to Conservation Genetics*; Cambridge University Press: Cambridge, UK, 2002.
58. Beebee, T.J.C. Conservation genetics of amphibians. *Heredity* **2005**, *95*, 423–427. [CrossRef]

59. Tero, N.; Aspi, J.; Siikamäki, P.; Jäkäläniemi, A.; Tuomi, J. Genetic structure and gene flow in a metapopulation of an endangered plant species, *Silene tatarica*. *Mol. Ecol.* **2003**, *12*, 2073–2085. [CrossRef] [PubMed]
60. O'Connell, L.M.; Mosseler, A.; Rajora, O.P. Extensive long-distance pollen dispersal in a fragmented landscape maintains genetic diversity in white spruce. *J. Hered.* **2007**, *98*, 640–645. [CrossRef] [PubMed]
61. Howe, R.W.; Davis, G.J.; Mosca, G.J. The demographic significance of sink populations. *Biol. Conserv.* **1991**, *57*, 239–255. [CrossRef]

diversity

MDPI

Article

Phylogenetics and Biogeography of *Lilium ledebourii* from the Hyrcanian Forest

Shekoofeh Shokrollahi [1], Hamed Yousefzadeh [2,3,*], Christian Parisod [4], Gholamali Heshmati [1], Hamid Bina [3], Shujait Ali [5], Narjes Amirchakhmaghi [3] and Yigang Song [6,*]

[1] Department of Rangeland, Gorgan University of Agricultural Sciences and Natural Resources, Gorgan 49138-15749, Iran; shokrollahi.sh93@yahoo.com (S.S.); heshmati@gau.ac.ir (G.H.)
[2] Department of Environment, Faculty of Natural Resources, Tarbiat Modares University, Tehran 14115-111, Iran
[3] Department of Forestry, Faculty of Natural Resources, Tarbiat Modares University, Tehran 14115-111, Iran; hamid.bina@ymail.com (H.B.); n_amirchakhmaghi@yahoo.com (N.A.)
[4] Institute of Plant Sciences, University of Bern, Altenbergrain 21, CH-3013 Bern, Switzerland; christian.parisod@ips.unibe.ch
[5] Center for Biotechnology and Microbiology (CBM), University of Swat, Odigram, Mingora 19130, Pakistan; shujaitswati@yahoo.com
[6] Eastern China Conservation Centre for Wild Endangered Plant Resources, Shanghai Chenshan Botanical Garden, Shanghai 201602, China
* Correspondence: h.yousefzadeh@modares.ac.ir (H.Y.); ygsong@cemps.ac.cn (Y.S.)

Abstract: *Lilium ledebourii* (Baker) Boiss is one of the most endangered lilies, restricted to only a few small and fragmented areas in the Hyrcanian forest. This study aimed at evaluating the taxonomy of this unique Iranian lily and reconstructing divergence time from other species of the genus *Lilium* to address the role of this region in its diversification. Phylogenetic trees based on nuclear ITS and chloroplastic matK strongly supported the monophyly of the genus *Lilium* and division into subclades hardly matching prior morphological classifications. Biogeographic analyses using S-DIVA revealed East Asia as the ancestral range from where *Lilium* presented a multidirectional expansion towards North America, West-Central Asia, North Asia, and Europe. Diverging from ancestral *Lilium* during the beginning of Eocene (50 Ma; 95% HDP: 68.8–36.8). Specific members of *Lilium* colonized Iran (Western Asia) separated from the Clade IV (West-Central Asia and Europe lineage), and then yielded the Iranian *L. ledebourri*. Accordingly, the north of Iran appears to have promoted both long-term persistence and migration of Lily species from Asia to the Europe.

Keywords: phylogeny; biogeography; divergence; critically endangered; rare species

Citation: Shokrollahi, S.; Yousefzadeh, H.; Parisod, C.; Heshmati, G.; Bina, H.; Ali, S.; Amirchakhmaghi, N.; Song, Y. Phylogenetics and Biogeography of *Lilium ledebourii* from the Hyrcanian Forest. *Diversity* **2022**, *14*, 137. https://doi.org/10.3390/d14020137

Academic Editors: Mario A. Pagnotta and Michael Wink

Received: 24 November 2021
Accepted: 9 February 2022
Published: 15 February 2022

1. Introduction

Lilium L. (Liliaceae) is one of the most fascinating genera among the plant kingdom, and it adds to its importance on horticulture, medicine and food [1–3]. *Lilium* is a perennial herbs with subterranean bulbs and mostly spring-flowering that grow in steppes and mountain meadows [4]. This genus occurs in Eurasia and North America and has three main ranges including the Qinghai–Tibetan Plateau in East Asia, North America, and the Caucasus [5,6]. The genus of *Lilium*, with approximately 100 species, has been classified into five to eleven sections [7–9]. Due to the frequent gene flow among sections, the major phylogenetic clades of *Lilium* are still controversial but have been clear [2,3,8,10]. Traditionally, seven sections were recognized based on morphological taxonomy and molecular phylogenetic methods [2,8,10].

The Hyrcanian area in the southern shores of the Caspian Sea presents remnants of natural deciduous forests [11] and forms a unique vegetation belt from the Talish region in Azerbaijan to Golestan National Park in Iran (between 48°–56° E and 38°55′–35°05′ N) [12]. Decreasing water levels during the Quaternary, coupled with the rising of the Earth's crust in the northern Caucasus, led to the separation of the Black Sea and the Caspian Sea, promoting the isolation of the Hyrcanian region [13]. The presence of many Arcto-Tertiary relict species in the area was generally viewed as coherent with the Hyrcanian forest having acted as major refugia sheltering the long-term persistence of plant species [14,15]. To confirm this hypothesis, additional data are necessary.

Lilium ledebourii (Baker) Boiss (Iranian lily) is a rare and endemic lily species distributed only in a few small areas of the Hyrcanian forest [16]. Its habitat was recorded as a natural monument in Iran's protected areas in 1976 [17]. This species is attractive and widely used in ornamental breeding programs, and some biochemical features were also studied [18,19]. Limited information is available regarding its phylogenetic position; however, there was one recent study based on the nuclear rDNA-internal transcribed spacers (ITS) and another based on the 5.8 S ribosomal DNA sequence [19,20]. However, due to the incomplete sampling on the distribution range of *L. ledebourii* and single sequence, its phylogenetic position is still unclear. Furthermore, there are several biogeographic studies of the genus *Lilium* missing this species. As the famous species is restricted to the biodiversity hotspots, *L. ledebourii* presents a less known biogeographic synthesis, and deserves further attention [21,22].

Despite the advent of genomics, DNA barcoding based on nucleotide sequences from variable chloroplastic and nuclear loci is still facilitated to monitor the phylogenetic position accurately and economically [23–25]. Here, we thus address the phylogenetic position of *L. ledebourii* concerning worldwide *Lilium* species and shed light on its role for the biogeography and diversification of the genus. The first purpose of this study is to present the phylogenetic relationships of the Iranian Lily based on nuclear and plastid DNA markers; the second goal is to estimate the divergence time of the Iranian Lily and discuss the biogeography of *Lilium* after adding the Iranian Lily.

2. Materials and Methods

2.1. Sampling and DNA Extraction

Leaf samples of *L. ledebourii* were collected from four natural populations (all reported sites for this species) of the Hyrcanian forest (Figure 1). Total genomic DNA was extracted from at least 10 individuals from each population following Murray and Thompson (1980), with modifications according to Janfaza et al. [26].

2.2. PCR Amplification and Sequencing

For amplification and sequencing, four candidate DNA barcoding regions were selected: the internal transcribed spacer regions (ITS) from nuclear genomes and three plastid regions (matK, trnL-F and psbA-trnH). PCR amplifications were performed in 20 μL using the AccuPower HotStart PCR PreMix kit (Bioneer, Daejeon, Korea), with either universal ITS-1 and ITS-4 primers from White et al. [27] or matK primers from Johnson et al. [28] or primers for the trnL-F intergenic spacer from Taberlet et al. [29] or primers for the psbA-trnH intergenic spacer from Sang et al. [30]. For ITS, psbA-trnH and trnL-F regions, PCR cycles consisted of an initial denaturation for 6 min at 95 °C, followed by 32 cycles of 60 s at 95 °C, 45 s at 56 °C, 90 s at 72 °C, and a final extension of 5 min 72 °C. The matK region was amplified following Hayashi and Kawano (2000) [31], including 35 cycles of 60 s at 94 °C, 120 s at 50 °C, 180 s at 72 °C and a final extension at for 7 min at 72 °C). After a check through electrophoresis, PCR products were sent to Bioneer service (Bioneer, Daejeon, Korea) for sequencing, and the resulting electropherograms were manually inspected using Chromas version 2.3 (www.technelysium.com.au (15 February 2020) to yield accurate sequence data.

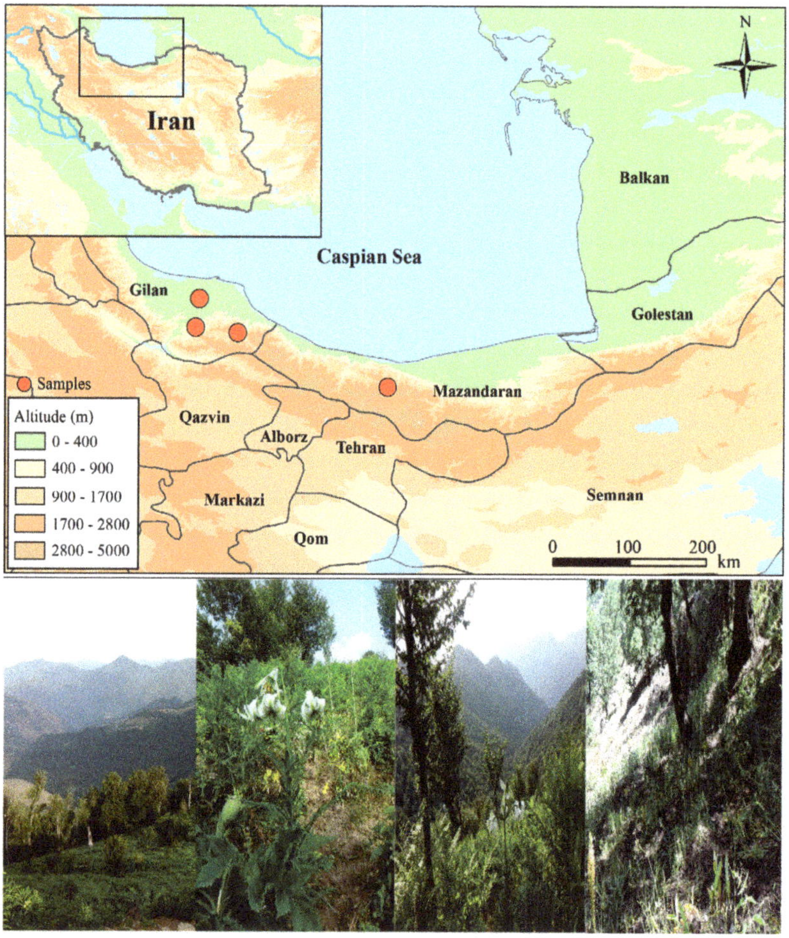

Figure 1. Geographical sampling sites of *Lilium ledebourii* in the Hyrcanian forest, part of Iran.

2.3. Phylogenetic Analysis and ITS2 Secondary Structure

There are eight individuals of *L. ledebourii* were used in the phylogenetic analysis in this study. The analysis based on ITS sequences used 35 *Lilium* species and took five *Tulipa* species as outgroups. The analysis based on matK sequences used 21 *Lilium* species and taken two *Tulipa* species as outgroups.

Alignments of sequences were performed for each locus using MUSCLE and nucleotide composition, the number of variables, parsimony and conserved sites among studied taxa were estimated in MEGA 6 [32]. A median-joining (MJ) network was inferred from all variable characters of the complete alignment of concatenated sequences using NETWORK 3.1.1.1 [33]. The phylogenetic tree was constructed based on maximum likelihood (ML) and Bayesian inference (BI) for all taxa.

Bayesian inference tree was constructed by MrBayes v. 3.2.6 with a cold chain and three incrementally heated chains (T = 0.2), running for 10,000,000 generations with a sampling frequency of 1000. The first 25% of the trees were discarded (burnin) and the remaining trees were used to build a consensus tree and estimate Bayesian posterior probabilities. The MEGA (version 6) [32] software was used to choose the best model and create phylogenetic trees using maximum likelihood (ML) methods, [34]. Bootstrap analysis with 1000 replications was used to analyze the reliability of each branch. ITS2 secondary structure for each taxon was inferred by minimizing free energy and homology search across the ITS2 database (http://its2.bioapps.biozentrum.uni-wuerzburg.de/ (27 February 2020).

2.4. Divergence Time Estimate and Biogeographic Analysis

To estimate the divergence time of *Lilium* and reconstruct the ancestral geographical range of this genus, we used 31 *Lilium* species covering its entire distribution area. Partition homogeneity or incongruence length difference (ILD) [35] test was conducted in PAUP v4.0b10 [36] following a heuristic search approach of 1000 replicates with 100 random stepwise additions and tree bisection reconstruction (TBR) branch swapping supported congruity of ITS and matK datasets ($p = 0.68$). Similarly, the likelihood ratio test (LRT) carried out in PAUP v4.0b10 [36] supported a relaxed molecular clock approach for our combined dataset ($p = 0.00001$).

Divergence times were estimated by calibrating the stem node of Ripogonaceae, Luzuriaga, according to Kim and Kim (2018) [37], and the crown node of *Smilax* (the outgroup) to a mean age of 55 million years ago (Ma), 24 Ma and 46 Ma respectively, using the log- normal distribution prior with standard deviation of 1.0. The phylogenetic analysis was conducted for 50,000,000 generations using BEAST ver. 1.6.1 [38] under uncorrelated lognormal relaxed clock parameter and the Yule model of speciation as tree prior. The GTR + G substitution model was selected as the best fit model using jModel test under Akaike's information criterion (AIC). Tree Annotatorver.1.7.5 [39] and FigTreever.1.4 [40] were used for the generation and visualization of the maximum clade credibility tree.

The distribution of *Lilium* was divided into six regions based on the available samples: A (East Asia), B (North Asia), C (West and Central Asia), D (Europe), E (North America), and F (South America). Reconstruct Ancestral State in Phylogenies (RASP) software [41] was used for biogeographic inferences using Statistical Dispersal-Vicariance Analysis (S-DIVA) approaches. To overcome uncertainties associated with S-DIVA analysis out of 10,000 trees from BEAST, the maximum clade credibility tree and distribution file were uploaded into the RASP software for biogeographic reconstructions. In the BBM analysis, only the maximum clade credibility tree was used along with the distribution file to obtain a reconstruction and the Markov chain Monte Carlo (MCMC) was run under the JC + G (Jukes-Cantor + Gamma) model for 5,000,000 generations.

3. Results

3.1. ITS and Plastid Regions (MatK, TrnL-F and TrnH-PsbA)

The nucleotide composition of the Iranian samples of *L. ledebourii* from four different regions is summarized and compared with available samples of lilies in Table 1. ITS sequences from Iranian samples were longer (644–654 bp) than those retrieved from NCBI (465–485 bp). The matK region showed low variability among *Lilium* species (length = 853–881 bp, conserved sites = 849). Iranian samples presented 19 variable sites, including nine singletons, whereas species available in NCBI presented 36 variable sites. The trnL-F and trnH-psbA regions showed the high conservation sites (length = 270–255, conserved sites = 246) and nucleotide percent of Temin (Iranian sample = 34.5, sample from NCBI = 35.5).

Table 1. Sequences characterization and nucleotide composition four regions trnH-psbA, trnL-F, matK and ITS without the outgroups.

	Region	Conserved Sites	Variable Sites	Parsim Informative Sites	Singleton Sites	Length (bp)	A%	T%	C%	G%	Model
Iranian Sample	ITS	549	80	27	52	644–654	18.8	21.1	27.2	32.9	T93 + G
	trnH-psbA	190	244	10	234	465–485	36.8	38.7	16.5	18	T92
	matk	849	19	-	9	853–862	31.1	38.5	15.8	14.6	T92
	trnL-F	246	13	-	13	270	30.5	34.5	18.5	16.6	T92
Other species	ITS	436	171	108	63	644–660	18.4	20.7	27.8	33	T93 + G
	trnH-psbA	251	170	12	153	460–480	31	36.1	15.9	17	T92
	matk	882	36	19	17	872–881	31	38.6	15.8	14.6	T92
	trnL-F	255	4	2	2	260–270	30.4	35.5	13.8	15.9	T92

3.2. Phylogenetic Position of L. ledebourii

Bayesian tree and ML trees based on ITS strongly supported a monophyletic genus Lilium and divided the species into six clades (Figure 2A), consistent with the MJ network (Supplementary Material Figure S1). These clades are somewhat consistent with the four main groups defined by Baker et al. [42] based on flower morphology: i.e., L. sect. *Eulirion* Rchb. (funnel-flowered lilies), L. sect. *Archelirion* Baker (open-flowered lilies), L. sect. *Isolirion* Baker (erect-flowered lilies) and L. sect. *Martagon* Rchb. (Turk's cap lilies). The results of the phylogenetic tree showed that the Caucasian and European species were placed in three separate clade (Figure 2A). All species of the European region were grouped in one clade, while seven species of the Caucasus were grouped in three separate clades. Surprisingly, *L. ledebourii* was placed in a clade with European species, and the *L. ciliatum* itself formed a separate clade, while other Caucasian species formed the third clade.

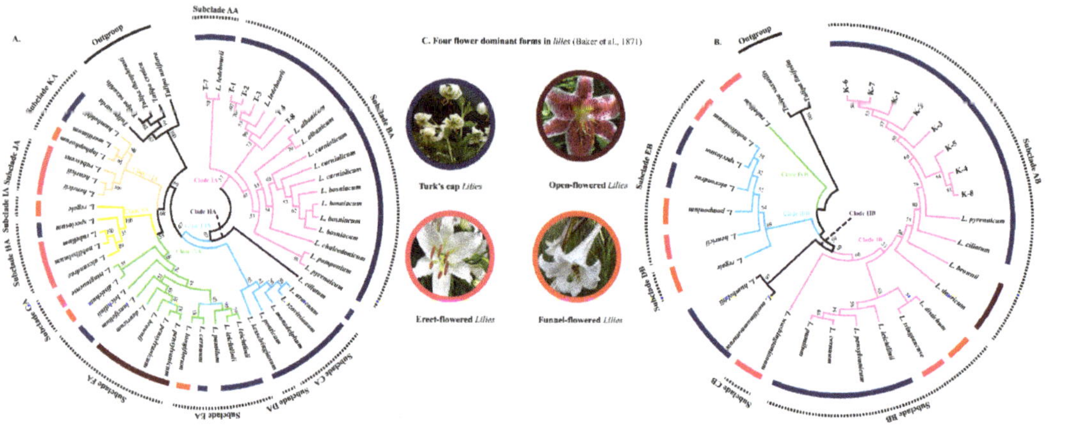

Figure 2. (**A**) Phylogenetic tree of *Lilium* based on ITS sequences (ITS1, 5.8 s, ITS2) with Maximum Likelihood method (model: Tumara3 parameter + Gamma distributed) and bootstrap 1000, (**B**) Phylogenetic tree of *Lilium* based on matK sequences with Maximum Likelihood method (model: Tumara3 parameter) and bootstrap 1000. (T and K legends are represented repeated sequences of ITS and matK regions from the Hyrcanian forest, respectively). (**C**) Four morphological flower types of the genus *Lilium* based on classification of Baker et al., 1871).

The Bayesian matK tree has a backbone polytomy and four major lineages, with significant differences in composition between these lineages and the six resolved in the ITS phylogeny. The phylogenetic tree based on this region showed a high inconsistency with the Baker et al., 1871 classification, but divided all Hyrcanian specimens along with *L. ciliatum* from the Caucasus and *L. pyrnaicum* from Europe into one sub clade.

3.3. Biogeography of the Genus Lilium

The results of S-DIVA analyses suggest that the current distribution of *Lilium* species is the outcome of multiple dispersal and vicariance events (Figure 3). S-DIVA suggests 9 dispersal and three vicariance events. They congruently indicate that *Lilium* ancestors (node I) originated in East Asia, or East-West-Central Asia (Figure 3) at the beginning of Eocene (c.a. 50 MYA, 95% HDP: 68.8–36.8, Table 2). Then, the ancestral *Lilium* from East Asia began to diverge into two clades (Clade II and Clade III) during the early and middle Eocene (Figure 3). During the Late Eocene and Oligocene, the *Lilium* began to spread to North America (Clade II) and West and Central Asia (Clade III). The Clade II have been included two obvious lineages (East Asia and North America lineage), and they carried out an independent evolution and diffusion from Miocene. In Clade III, the *Lilium* species in East Asia have four times the dispersal after Pliocene. Ancestral reconstructions of Clades IV and V further favor an ancestral range in western Asia and Europe (with 93.30% and 78% marginal probability, respectively) in Clade III, including the Iranian *L. ledebourri*. *Lilium ledebourii* in West Asia looks like the first lineage separate from Clade IV.

Table 2. Age estimation in million year ago (MYA) of nodes of divergence time clade based on S-DIVA and BBM.

	Age Estimation (MYA)			S-DIVA		BBM		
Nodes	**Mean**	**95% HPDlower**	**95% HPDupper**	**AR**	**MP(%)**	**AR**	**MP**	**Support (PP)**
I	50.71	36.8	68.8	A/AF	49/36	A/AD	60/16	0.90
II	40.84	23.0	63.2	AF	99.71	AF/A	34/26	0.80
III	40.45	28.2	56.5	AC	93.40	A/AD	42/23	0.95
IV	33.25	20.5	49.8	C	93.30	C/CD	46/38	1.00
V	27.08	15.7	44.0	CD	78.00	C/CD	49/42	0.90

The overall scenario favored by S-DIVA (Table 3) supports stepping-stone radiation of the genus *Lilium*, with a primary center of diversification in eastern Asia (i.e., region A). Our inclusion of the Iranian samples here looks critical, as it supports an expansion and secondary diversification in western Asia (i.e., region C) before further range expansion and diversification towards Europe (i.e., region D).

Table 3. Dispersal details of different distribution area based on S-DIVA and BBM.

S-DIVA				BBM			
Distribution Range	**Dispersal from**	**Dispersal to**	**Within**	**Distribution Range**	**Dispersal from**	**Dispersal to**	**Within**
A	44.00	0.00	28.0	A	64.0	0.00	28.0
B	0.00	5.00	1.00	B	0.00	9.00	1.00
C	2.00	5.00	8.00	C	2.00	7.00	9.00
D	1.00	19.0	15.0	D	2.00	28.00	26.0
E	0.00	3.00	0.00	E	0.00	4.00	0.00
F	1.00	16.0	9.00	F	4.00	24.00	18.0

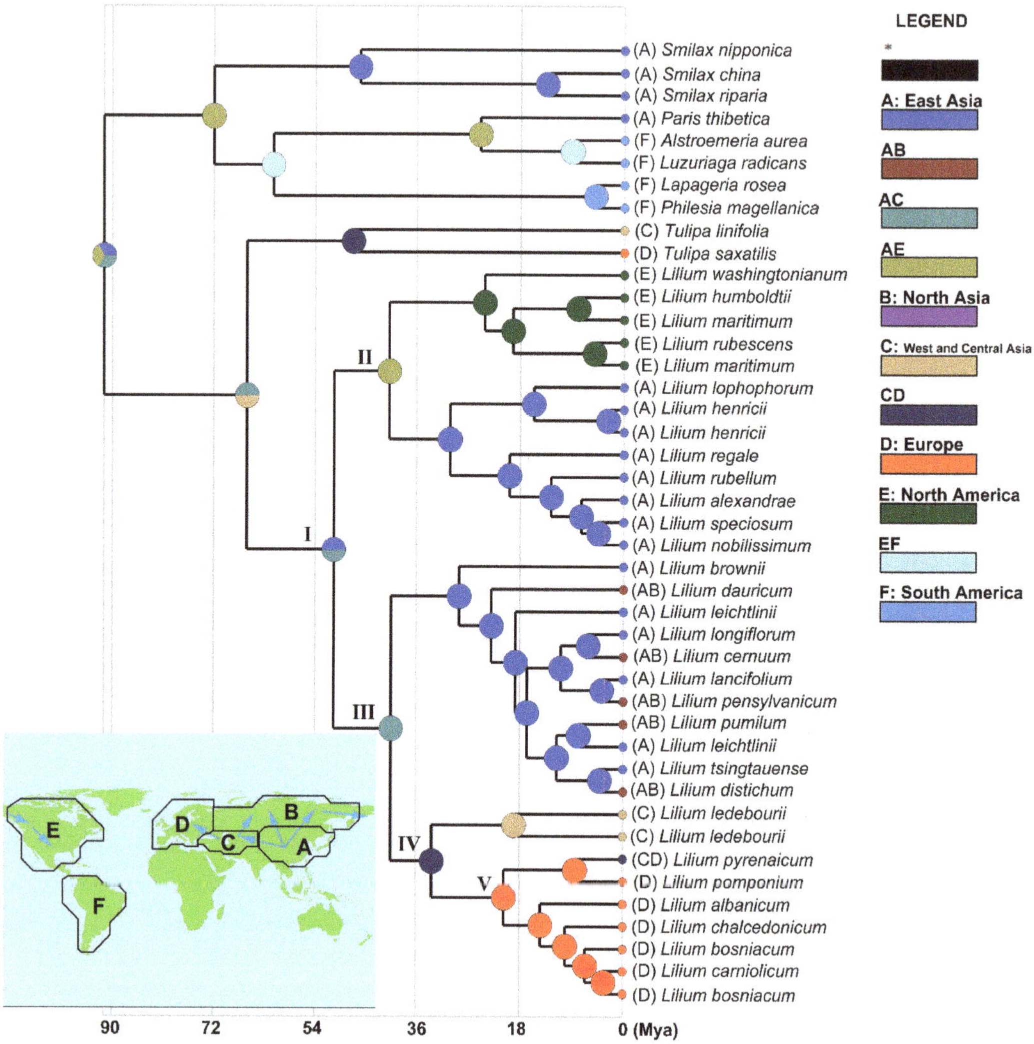

Figure 3. Divergence time estimations and the major divergence events among the *Lilium* species based on region combined data (ITS + matK regions) and Dispersal scenario of *Lilium* in the world. Pie charts depict ancestral area reconstruction probability, with the colors of pie slices defined in the legend. (The reader is directed to the web version of this article for an explanation of the color references in this figure legend).

4. Discussion

4.1. Phylogeny and Biogeography of Lilies

Our research sheds light on the natural divergence histories of *L. ledebourii* in the Caucasian Hyrcanian forest of Iran, which is a hotspot for biodiversity. The results showed that *L. ledebourii* from this region belonging to the Caucasians are older than the European samples, and thus Iran could be a bridge for transforming *L. ledebourii* into Europe. Furthermore, we discovered that the phylogenetic position of the *Lilium* genus did not follow the flower shape.

Phylogenetic inferences based on ITS and matK loci here largely confirmed earliest classifications based on morphology to distinguish either five sections [9] or four main groups [42]. Iranian lilies (*L. ledebourii*), analyzed here for the first time, were closely related to European species (*L. pyrenaicum*) as well as species from Caucasus [43] to be part of the *Martagon* section composed of species forming mostly Turk's cap flowers.

The inclusion of the Iranian samples offers a comprehensive sampling with representative species from most regions colonized by lilies to bring fresh insights into their evolution. For instance, the controversial *L. lophophorum* here appears consistent with the classification of Wilson [44] in a *Lophophorum* subsection, whereas close relationships with *L. humboldtii* further supports an early origin out of Chinese lineages. All species comprised here in clade V (sections *Archelirion* and *Martagon* of Baker's classification) share an origin in east Asia and support Comber [8] classification. Although the impact of hybridization remains controversial in the genus *Lilium* [45], present results indicate close relationships between species of *Sinomartagon* and *Archelirion* sections, supporting the view of Leslie [46] about hybridization between corresponding species. In particular, *L. henrici* here is shown to be very close to *L. alexandrae* and these species may have undergone gene flow. However, inference of reticulate evolution across the genus *Lilium* is beyond the scope of this work.

4.2. Biogeography of *Lilium* Emphasis on West Asia

Even though *Lilium* is one of the most famous genera all over the world, the biogeographic study of this genus is limited [5]. Gao et al. (2013) tried to analyze the evolutionary events in *Lilium*. Due to the sampling, this study only focused on the biogeography of the Q-T plateau and the Hengduan Mountains. In this study, we try to cover the distribution area of the genus to explore the biogeography of this genus, adding some species from West Asia.

Biogeographic reconstructions using S-DIVA indicated an ancestral origin of *Lilium* in Eastern Asia or the extensive area of East-Central-West Asia during the Eocene around 50 MYA (95% HDP: 68.8–36.8). This contrasts with prior dating based on the plastid dataset that concluded on a last common ancestor of *Lilium* some 13.19 MYA [5]. Thus, our results suggest that *Lilium* may have early origins in the Pan-Asian region. The genus began to spread outside of Asia during the Eocene. Two early radiation events from the ancestral area have particularly supported independent colonization toward Western Asia and towards North America. This event may have benefited from the warm climatic conditions in this Geologic period. This is in keeping with prior hypotheses of migration for diverse plants and animals between the Old and the New Worlds through either the North Atlantic Land Bridge (NALB) or Beringia [47], rather than the suggestion in some works of literature that the Himalayas were the center of origin of the genus *Lilium* [5,6]. This pattern also contrasts with the proposal of Ikinci et al. [48] that European species derive from multiple routes, with L. *martagon* first colonizing Europe, whereas a second route later gave rise to *L. bulbiferum* and further diversification of all other European species. *Lilium* species of Iran (Western Asia) indeed involved radiation of ancestors from East Asia to Europe, with early colonization of Western Asia from East Asia (50–40 MYA) followed by additional events some 20 MYA and 10 MYA. In particular, three radiations were apparent between Western Asia and Europe in our analysis, supporting dispersal from Western Asia to Europe during the Oligocene (28 MYA) and again during the Miocene (18 MYA).

Due to the limited impact of the ice ages during the Pleistocene, it indeed hosts an impressive number of endemic and relict species [49]. Our analysis showed that the rare *Lilium ledebourii* is such a relict species that currently scattered among small populations in cliffs above the tree line. These populations seemingly represent survivors of previously large populations having migrated to the area before the ice age. Accordingly, the northern forests of Iran likely acted as refugia for such species. The presence of tree species from typically high latitudes such as *Betula pendula* and *Sorbus aucaparia* in the Hyrcanian forest further corroborates the case of *L. ledebourii*.

A non-mutually exclusive hypothesis is that the north of Iran represents a path for migration of species from East Asia to Europe during periods of drastic environmental changes. Khalilzadeh et al. [50] indeed concluded that the north west of Iran (specifically Southwest Caspian Sea) is a major contact zone between Asian and European clades of Wild Boar and biogeographic inferences on several plant species [51,52] supported the north of Iran (Hyrcanian forest) as a main corridor for plant migration between the East and the West of the Eurasian continent. As expressed by Manafzadeh et al. [53], the Irano-Turanian floristic region can serve as a 'donor' of xerophytic taxa to 'recipient' neighboring regions, including the Mediterranean floristic region. Additional studies appear necessary to accurately quantify the refugial vs. corridor role of the Hyrcanian forest.

In conclusion, the phylogenetic position and biogeography of endemic lily species in the Hyrcanian forest were studied using the barcoding technique. The phylogenetic tree utilizing ITS rather than matK produced results that were more consistent with morphological classification and placed *L. ledebourii* in a phylogenetic group with other Caucasian species and section *Liriotypus*. The divergence time of *L. ledebourii* determined that *L. ledebourii* diverged from the radiation of ancestors in East Asia, and then formed a West-Central Asia and Europe lineage during the end of the Eocene. Accordingly, the north of Iran appears to act as long-term persistence and migration of Lily species from Asia to Europe.

Supplementary Materials: The following are available online at https://www.mdpi.com/article/10.3390/d14020137/s1, Figure S1: Network tree of the genus *Lilium* based on ITS region (A) and mat K region (B).

Author Contributions: H.Y., G.H. and S.S. designed the study, H.Y. and S.S. collected the plant material, H.Y., H.B., S.A., N.A. and Y.S. performed analysis of data and H.Y., C.P. and Y.S. wrote the paper. All authors have read and agreed to the published version of the manuscript.

Funding: This research received no external funding.

Institutional Review Board Statement: Not applicable.

Informed Consent Statement: Not applicable.

Data Availability Statement: The data reported here are archived as Supplemental Material in PSE. Compliance with ethical standards.

Conflicts of Interest: The authors declare that they have no conflict of interest.

References

1. Woodcock, H.; Stearn, W.T. *Lilies of the World, Their Cultivation and Classification*; Country Life Limited: London, UK, 1950.
2. Gong, X.; Huang, H.H.; Ting, Y.W.; Hsu, T.W.; Malikova, L.; Tran, H.T.; Huang, C.L.; Liu, S.H.; Chiang, T.Y. Frequent gene flow blurred taxonomic boundaries of sections in *Lilium* L. (*Liliaceae*). *PLoS ONE* **2017**, *12*, e0183209. [CrossRef] [PubMed]
3. Kim, H.T.; Lim, K.B.; Kim, J.S. New insights on *Lilium* phylogeny based on a comparative phylogenomic study using complete plastome sequences. *Plants* **2019**, *8*, 547. [CrossRef] [PubMed]
4. Lee, C.S.; Kim, S.C.; Yeau, S.H. Major lineages of the genus *Lilium* (*Liliaceae*) based on nrDNA ITS sequences, with special emphasis on the Korean species. *J. Plant Biol.* **2011**, *54*, 159–171. [CrossRef]
5. Gao, Y.-D.; Harris, A.; Zhou, S.-D.; He, X.-J. Evolutionary events in *Lilium* (including Nomocharis, Liliaceae) are temporally correlated with orogenies of the Q–T plateau and the Hengduan Mountains. *Mol. Phylogenet. Evol.* **2013**, *68*, 443–460. [CrossRef]
6. Patterson, T.B.; Givnish, T.J. Phylogeny, concerted convergence, and phylogenetic niche conservatism in the core *Liliales*: Insights from rbcL and ndhF sequence data. *Evolution* **2002**, *56*, 233–252. [CrossRef]
7. Baranova, M. The lily species in the flora of the former Soviet Union and their classification within the genus Lilium. In *International Symposium on the Genus Lilium 414*; ISHS: Leuven, Belgium, 1994; pp. 133–136.
8. Comber, H.F. A new classification of the genus *Lilium*. *Lily Year Book RHS* **1949**, 86–105.
9. Endlicher, S.L. *Genera Plantarum Secundum Ordines Naturales Disposita: Genera Plantarum ... 1836–1840*; Beck: Munich, Germany, 1840; Volume 2.
10. Ikinci, N. Molecular phylogeny and divergence times estimates of *Lilium* section *Liriotypus* (*Liliaceae*) based on plastid and nuclear ribosomal ITS DNA sequence data. *Turk. J. Bot.* **2011**, *35*, 319–330.
11. Naqinezhad, A.; Hamzeh'ee, B.; Attar, F. Vegetation–environment relationships in the alderwood communities of Caspian lowlands, N. Iran (toward an ecological classification). *Flora-Morphol. Distrib. Funct. Ecol. Plants* **2008**, *203*, 567–577. [CrossRef]

12. Miller, A.G. Hyrcanian forests, Iran and Azerbaijan. Centres of Plant Diversity, A Guide and Strategy for their Conservation Oxford. *IUCN* **1994**, *1*, 343–344.
13. Olteanu, R.; Jipa, D. Dacian Basin environmental evolution during Upper Neogene within the Paratethys domain. *Geo-Eco-Marina* **2006**, *12*, 91–105.
14. Tralau, H. *Asiatic Dicotyledonous Affinities in the Cainozoic Flora of Europe*; Almqvist & Wiksell: Tofnes, UK, 1963.
15. Zohary, M. *Geobotanical Foundation of the Middle East*; Fischer-Verlag: Stuttgart, Germany, 1983.
16. Mimaki, Y.; Satou, T.; Kuroda, M.; Sashida, Y.; Hatakeyama, Y. Steroidal saponins from the bulbs of *Lilium candidum*. *Phytochemistry* **1999**, *51*, 567–573. [CrossRef]
17. Fataei, E. Recognition of new habitats of Lilium ledebourii Baker in Northwest Ardabil Province of Iran using GIS technology. In *II International Symposium on the Genus Lilium 900*; ISHS: Leuven, Belgium, 2010; pp. 65–70.
18. Azadi, P.; Khosh-Khui, M. Micropropagation of *Lilium ledebourii* (Baker) Boiss as affected by plant growth regulator, sucrose concentration, harvesting season and cold treatments. *Electron. J. Biotechnol.* **2007**, *10*, 582–591. [CrossRef]
19. Salehi, M.; Hatamzadeh, A.; Jafarian, V.; Zarre, S. New molecular record and some biochemical features of the rare plant species of Iranian lily (*Lilium ledebourii* Boiss.). *Hortic. Environ. Biotechnol.* **2019**, *60*, 585–593. [CrossRef]
20. Ghanbari, S.; Fakheri, B.A.; Naghavi, M.R.; Mahdinezhad, N. Evaluating phylogenetic relationships in the *Lilium* family using the ITS marker. *J. Plant Biotechnol.* **2018**, *45*, 236–241. [CrossRef]
21. Rockwell, F.F. *Complete Book of Lilies*; Doubleday: New York, NY, USA, 1961.
22. Woodcock, H.; Stearn, W.T. *Lilies of the World*; Country Life Limited: London, UK, 1950.
23. Vassou, S.L.; Kusuma, G.; Parani, M. DNA barcoding for species identification from dried and powdered plant parts: A case study with authentication of the raw drug market samples of *Sida cordifolia*. *Gene* **2015**, *559*, 86–93. [CrossRef]
24. Tripathi, A.M.; Tyagi, A.; Kumar, A.; Singh, A.; Singh, S.; Chaudhary, L.B.; Roy, S. The internal transcribed spacer (ITS) region and trnhH-psbA are suitable candidate loci for DNA barcoding of tropical tree species of India. *PLoS ONE* **2013**, *8*, e57934. [CrossRef]
25. Chen, C.W.; Huang, Y.M.; Kuo, L.Y.; Nguyen, Q.D.; Luu, H.T.; Callado, J.R.; Farrar, D.R.; Chiou, W.L. trnL-F is a powerful marker for DNA identification of field vittarioid gametophytes (Pteridaceae). *Ann. Bot.* **2013**, *111*, 663–673. [CrossRef]
26. Janfaza, S.; Yousefzadeh, H.; Hosseini Nasr, S.M.; Botta, R.; Asadi Abkenar, A.; Torello Marinoni, D. Genetic diversity of *Castanea sativa* an endangered species in the Hyrcanian forest. *Silva Fenn.* **2017**, *51*, 1–15. [CrossRef]
27. White, T.J.; Bruns, T.; Lee, S.; Taylor, J. Amplification and direct sequencing of fungal ribosomal RNA genes for phylogenetics. *PCR Protoc. Guide Methods Appl.* **1990**, *18*, 315–322.
28. Johnson, L.A.; Soltis, D.E. matK DNA sequences and phylogenetic reconstruction in *Saxifragaceae* s. str. *Syst. Bot.* **1994**, 143–156. [CrossRef]
29. Taberlet, P.; Gielly, L.; Pautou, G.; Bouvet, J. Universal primers for amplification of three non-coding regions of chloroplast DNA. *Plant Mol. Biol.* **1991**, *17*, 1105–1109. [CrossRef] [PubMed]
30. Sang, T.; Crawford, D.J.; Stuessy, T.F. Chloroplast DNA phylogeny, reticulate evolution, and biogeography of *Paeonia* (Paeoniaceae). *Am. J. Bot.* **1997**, *84*, 1120–1136. [CrossRef] [PubMed]
31. Hayashi1, K.; Kawano2, S. Molecular systematics of *Lilium* and *allied* genera (*Liliaceae*): Phylogenetic relationships among *Lilium* and related genera based on the rbcL and matK gene sequence data. *Plant Species Biol.* **2000**, *15*, 73–93. [CrossRef]
32. Tamura, K.; Stecher, G.; Peterson, D.; Filipski, A.; Kumar, S. MEGA6: Molecular evolutionary genetics analysis version 6.0. *Mol. Biol. Evol.* **2013**, *30*, 2725–2729. [CrossRef] [PubMed]
33. Bandelt, H.-J.; Forster, P.; Röhl, A. Median-joining networks for inferring intraspecific phylogenies. *Mol. Biol. Evol.* **1999**, *16*, 37–48. [CrossRef] [PubMed]
34. Tamura, K.; Stecher, G.; Kumar, S. MEGA11: Molecular evolutionary genetics analysis version 11. *Mol. Biol. Evol.* **2021**, *38*, 3022–3027. [CrossRef] [PubMed]
35. Farris, J.S.; Kallersjo, M.; Kluge, A.G.; Bult, C. *Testing Significance of Incongruence*; Elsevier: Amsterdam, The Netherlands, 1994.
36. Swofford, D.L. PAUP: Phylogenetic analysis using parsimony version 3.11. In *Computer Program Distributed by the Illinois Natural History Survey*; University of Illinois at Urbana-Champaign: Champaign, IL, USA, 1993.
37. Kim, J.S.; Kim, J.-H. Updated molecular phylogenetic analysis, dating and biogeographical history of the *lily* family (*Liliaceae: Liliales*). *Bot. J. Linn. Soc.* **2018**, *187*, 579–593. [CrossRef]
38. Drummond, A.J.; Rambaut, A. BEAST: Bayesian evolutionary analysis by sampling trees. *BMC Evol. Biol.* **2007**, *7*, 214. [CrossRef]
39. Drummond, A.J.; Suchard, M.A.; Xie, D.; Rambaut, A. Bayesian phylogenetics with BEAUti and the BEAST 1.7. *Mol. Biol. Evol.* **2012**, *29*, 1969–1973. [CrossRef]
40. Rambaut, A. Available online: FigTree.tree.bio.ed.ac.uk/software/figtree (accessed on 12 December 2014).
41. Yu, Y.; Harris, A.J.; Blair, C.; He, X. RASP (Reconstruct Ancestral State in Phylogenies): A tool for historical biogeography. *Mol. Phylogenet. Evol.* **2015**, *87*, 46–49. [CrossRef]
42. Baker, J.G. A new synopsis of all the known lilies. *Gard. Chron.* **1871**, *104*, 1650.
43. Nishikawa, T.; Okazaki, K.; Uchino, T.; Arakawa, K.; Nagamine, T. A molecular phylogeny of *Lilium* in the internal transcribed spacer region of nuclear ribosomal DNA. *J. Mol. Evol.* **1999**, *49*, 238–249. [CrossRef] [PubMed]
44. Wilson, C.T. The Acceleration of β-Particles in Strong Electric Fields Such as Those of Thunderclouds. *Math. Proc. Camb. Philos. Soc.* **1925**, *22*, 534–538. [CrossRef]
45. McRae, K.; Boisvert, S. Automatic semantic similarity priming. *J. Exp. Psychol. Learn. Mem. Cogn.* **1998**, *24*, 558. [CrossRef]

46. Leslie, A.C. *The International Lily Register, Including 17 Additions (1984–1998)*; Royal Horticultural Society: London, UK, 1982.
47. Tiffney, B.H.; Manchester, S.R. The use of geological and paleontological evidence in evaluating plant phylogeographic hypotheses in the Northern Hemisphere Tertiary. *Int. J. Plant Sci.* **2001**, *162* (Suppl. 6), S3–S17. [CrossRef]
48. İkinci, N.; Oberprieler, C.; Güner, A. On the origin of European lilies: Phylogenetic analysis of *Lilium* section *Liriotypus* (*Liliaceae*) using sequences of the nuclear ribosomal transcribed spacers. *Willdenowia* **2006**, *36*, 647–656. [CrossRef]
49. Joger, U.; Fritz, U.; Guicking, D.; Kalyabina-Hauf, S.; Nagy, Z.T.; Wink, M. Relict populations and endemic clades in palearctic reptiles: Evolutionary history and implications for conservation. In *Relict Species*; Springer: Cham, Switzerland, 2010; pp. 119–143.
50. Khalilzadeh, P.; Rezaei, H.R.; Fadakar, D.; Serati, M.; Aliabadian, M.; Haile, J.; Goshtasb, H. Contact zone of Asian and European wild boar at North West of Iran. *PLoS ONE* **2016**, *11*, e0159499.
51. Colagar, A.H.; Yousefzadeh, H.; Shayanmehr, F.; Jalali, S.G.; Zare, H.; Tippery, N.P. Molecular taxonomy of Hyrcanian *Alnus* using nuclear ribosomal ITS and chloroplast trnH-psbA DNA barcode markers. *Syst. Biodivers.* **2016**, *14*, 88–101. [CrossRef]
52. Mostajeran, F.; Yousefzadeh, H.; Davitashvili, N.; Kozlowski, G.; Akbarinia, M. Phylogenetic relationships of *Pterocarya* (*Juglandaceae*) with an emphasis on the taxonomic status of Iranian populations using ITS and trnH-psb A sequence data. *Plant Biosyst.* **2017**, *151*, 1012–1021. [CrossRef]
53. Manafzadeh, S.; Salvo, G.; Conti, E. A tale of migrations from east to west: The Irano-Turanian floristic region as a source of Mediterranean xerophytes. *J. Biogeogr.* **2014**, *41*, 366–379. [CrossRef]

diversity

Article

Genetic Diversity and Mating System of Two Mangrove Species (*Rhizophora apiculata* and *Avicennia marina*) in a Heavily Disturbed Area of China

Wenxun Lu [1,2], Zhen Zou [2], Xueying Hu [1] and Shengchang Yang [2,3,*]

[1] College of Life Sciences, Peking University, Beijing 100871, China; wenxunlu@pku.edu.cn (W.L.); huxueying@pku.edu.cn (X.H.)
[2] College of the Environment and Ecology, Xiamen University, Xiamen 361102, China; zouzhen29@163.com
[3] Key Laboratory of the Ministry of Education for Coastal and Wetland Ecosystems, Xiamen University, Xiamen 361102, China
* Correspondence: scyang@xmu.edu.cn

Abstract: Mangrove forests are distributed in the intertidal zones of tropical and subtropical regions, and have been severely damaged by anthropogenic activities, climate change, and stochastic events. Although much progress has been made in the conservation and restoration of mangroves in China, studies of the genetic diversity of mangroves are lacking, especially for isolated populations, yet such studies are essential for guiding conservation and restoration efforts. Here, we evaluated the genetic diversity, spatial genetic structure, and mating system of two mangrove species, *Rhizophora apiculata* and *Avicennia marina*, in a heavily disturbed area in Tielu Harbor, Sanya City, Hainan Island, China, using 18 nuclear microsatellite markers. We found that the genetic diversity of *R. apiculata*, which is classified as 'Vulnerable' in the China Red List categories, was high and similar compared with the genetic diversity estimates of other populations reported in previous studies. In contrast, the genetic diversity of *A. marina*, which is classified as a species of 'Least Concern', was low compared with the genetic diversity estimates of other populations. We then evaluated the presence of genetic bottlenecks, spatial genetic structure, and the mating system to determine the effects that habitat destruction has had on these two species. Our findings indicate that distinct conservation and restoration approaches are needed for these two species. Generally, our results provide valuable information that will aid the development of conservation and restoration strategies for the mangroves of Tielu Harbor.

Keywords: genetic diversity; spatial genetic structure; mating system; *Rhizophora apiculata*; *Avicennia marina*; habitat destruction; conservation; restoration

Citation: Lu, W.; Zou, Z.; Hu, X.; Yang, S. Genetic Diversity and Mating System of Two Mangrove Species (*Rhizophora apiculata* and *Avicennia marina*) in a Heavily Disturbed Area of China. *Diversity* **2022**, *14*, 115. https://doi.org/10.3390/d14020115

Academic Editors: Hong-Hu Meng and Yi-Gang Song

Received: 28 December 2021
Accepted: 2 February 2022
Published: 6 February 2022

1. Introduction

Mangrove forests, which are dominated by plants in the Rhizophoraceae, Verbenaceae, and Combretaceae families, occur in the intertidal zones of tropical and subtropical regions [1,2]. Mangrove ecosystems are some of the world's most biodiverse and productive forest ecosystems [3], and they provide important ecosystem services, such as carbon sequestration, wave attenuation, and refuges for organisms [4,5]. Despite the high social, economic, and ecological value of mangroves, they have been severely degraded due to anthropogenic activities [6,7], climate change [8], and stochastic events [9]. For example, many mangroves have been converted to nursery areas for fishery species and play a critically important role in sustaining production in coastal fisheries [10]. Mangrove conservation and restoration strategies have been widely implemented given the alarming rate of decline in the area of mangrove forests worldwide [10–13].

Mangrove forests in China are mainly distributed along the southeastern coast, including Hainan, Guangxi, Guangdong, Fujian, and Taiwan Provinces [14]. China's mangrove forests have been severely fragmented in previous decades, and China has historically had one of the highest rates of mangrove loss worldwide [14–16]. The area of mangrove forests in China declined from 48,300 hm^2 in 1950 to 22,025 hm^2 in 2001 [16–18]. The implementation of strict protection and large-scale restoration measures after 2001 has resulted in an increase in the area of mangrove forests; by 2019, the area of mangroves in China reached 30,000 hm^2 [19,20]. Despite this conservation success, an increase in the area of mangroves does not, by itself, result in the restoration of a healthy mangrove ecosystem [16], as habitat degradation, decreases in biodiversity, and biological invasions are still major problems affecting mangroves in China [21–23].

Reductions in population size are the most direct effect of anthropogenic activities and climate change on populations [24]; however, anthropogenic activities and climate change can also affect other characteristics of populations, such as genetic diversity, the mating system, and spatial genetic structure. Genetic diversity refers to the genetic variation among different individuals or populations within a species [25], and it buffers populations against variability in environmental conditions [26]. Thus, characterizing the genetic diversity of populations is a major goal of research in biodiversity conservation [27,28]. The effects of anthropogenically driven habitat fragmentation and destruction on the genetic diversity of populations of species have been well studied [29,30]. Habitat fragmentation reduces the size of populations and increases their isolation, which results in reduced heterozygosity, the loss of alleles, and reduced gene flow between populations. This can lead to an increase in inbreeding and reductions in effective population size and can have deleterious effects on the long-term persistence of populations [31,32]. Small population size and the isolation of populations can have negative effects on the mating system of plants [33]. Some species also possess ecological and genetic traits, such as those that prevent self-pollination, that counteract the deleterious effects of small population size and isolation on mating systems [34]. Spatial genetic structure is another important component of the population genetics of a species, as a thorough understanding of the spatial genetic structure within and among populations can aid in the development of conservation and restoration strategies [35,36]. An increasing number of studies have examined the population genetics of mangroves in recent years, and investigations of the population genetic diversity and spatial genetic structure of different species in mangroves have been conducted to guide mangrove conservation and restoration efforts [9,37,38].

Tielu Harbor Mangrove Nature Reserve is located in Sanya City in southern Hainan Island. A total of 13 true mangrove species in nine genera and six semi-mangrove species in six genera have been documented in the reserve [39]. Mangrove resources are abundant in Tielu Harbor, and the mangroves in this area are ancient and endangered [14]. Previous studies indicate that many ancient trees of the mangrove species, such as *Bruguiera sexangular*, *Bruguiera gymnorhiza*, *Lumnitzera racemose*, *Lumnitzera littorea*, and *Xylocarpus granatum*, occur in Tielu Harbor [14,39,40]. Increases in human activities have reduced the area of mangroves in Tielu Harbor, which is currently approximately 3–4 hm^2 [14,40], and degraded existing mangrove habitat. The mangroves of Tielu Harbor are thus in a precarious state and require urgent conservation attention [14,39].

Mangroves in Tielu Harbor can be divided into four major types: *Rhizophora apiculata* communities, *Avicennia marina* communities, *Lumnitzera racemosa* communities, and *Xylocarpus granatum* communities [14,39]. Of these, *R. apiculata* is the dominant species, and *A. marina* is generally considered a pioneer species (Figure 1). *R. apiculata* and *A. marina* are listed as 'Vulnerable' and of 'Least Concern', respectively, in the China Red List categories [16]. Study of the genetic diversity, mating systems, and population genetic structure of these two important mangrove species is needed to ensure that effective conservation measures are taken for the mangrove forests in Tielu Harbor.

Figure 1. Images of *Rhizophora apiculata* and *Avicennia marina* in Tielu Harbor Mangrove Nature Reserve. Individuals of *R. apiculata* on land (**a**) and near the sea (**b**) and flowers of *R. apiculata* (**c**). Individuals of *A. marina* on land (**d**) and near the sea (**e**) and flowers of *A. marina* (**f**). The red arrows indicate pieces of garbage in the mangroves.

Here, we investigated the genetic diversity, spatial genetic structure, and mating systems of two mangrove species, *R. apiculata* and *A. marina*, in Tielu Harbor Mangrove Nature Reserve. We also conducted a comprehensive analysis of our results and recommended specific conservation measures that could be taken to guide ongoing mangrove conservation and restoration efforts in Tielu Harbor.

2. Materials and Methods

2.1. Study Area and Plant Sampling

Plant materials for this study were collected from Tielu Harbor Mangrove Nature Reserve, Sanya City, Hainan Province, China (18°15′–18°17′ N, 109°42′–109°42′ E, Figure 2). Tielu Harbor features a tropical monsoon climate, and the mean annual precipitation and temperature are 1255 mm and 25.5 °C, respectively [39]. Sampling was conducted from June to August 2016. Two quadrats were established for *R. apiculata* (80×130 m^2) and *A. marina* (110×180 m^2). Leaves of all adult individuals and some seedlings of the two species in quadrats were sampled. The coordinates of each sample were recorded with a Garmin GPSmap 60CSx (Garmin Ltd., Lenexa, KA, USA). A total of 167 samples (139 adults and 28 seedlings) and 210 samples (152 adults and 58 seedlings) were collected for *R. apiculata* and *A. marina*, respectively. In addition, three individuals of *R. apiculata* and five individuals of *A. marina* with propagules were randomly selected and used as mother trees for paternity analysis. To minimize the impact on the capacity for population renewal, approximately 10 propagules were randomly sampled from each mother tree. Leaves and propagules collected were desiccated in plastic zip-lock bags with silica gel and stored at room temperature until DNA extraction.

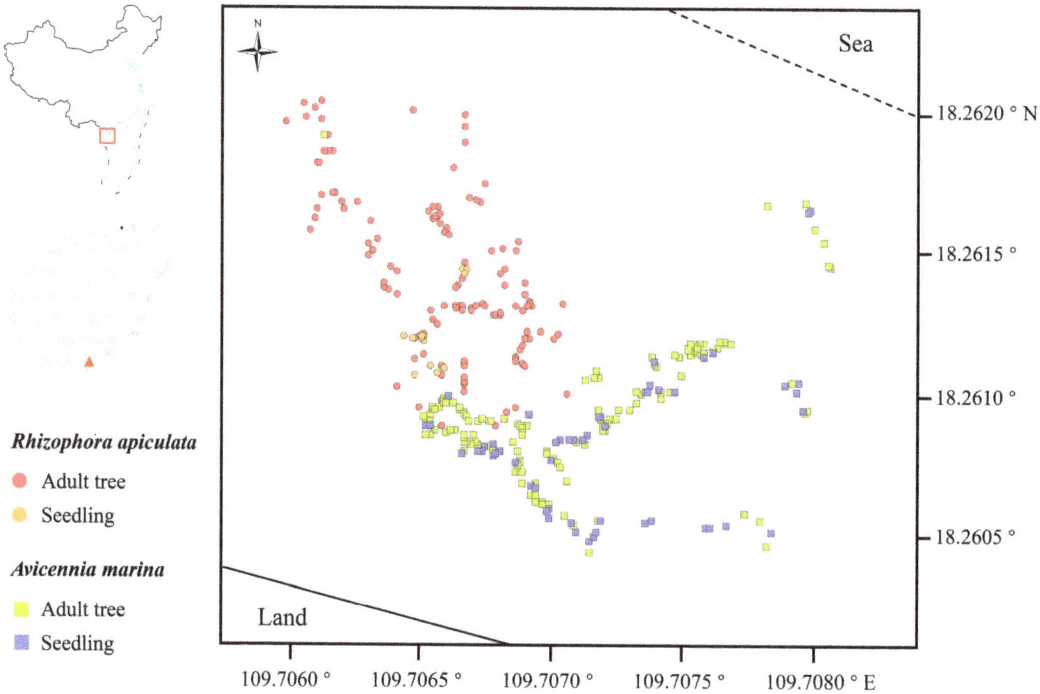

Figure 2. Map of *Rhizophora apiculata* (circles) and *Avicennia marina* (squares) individuals in quadrats.

2.2. DNA Extraction, PCR Amplification, and Microsatellite Screening

The total genomic DNA was extracted from the dried leaves and propagules of each sample using a Plant Genomic DNA Kit DP305 (Tiangen Biotech, Beijing, China). Specifically, for *A. marina*, we collected their propagules and cultured them until new leaves appeared in the laboratory, and then collected leaves of each seedling for DNA extraction. For *R. apiculate*, we directly collected a part of each propagule for subsequent DNA extraction because of its large propagule. After the concentration and purity were determined, each DNA sample was diluted with TE buffer to a concentration of 5 to 20 ng/μL and stored at −20 °C until use. Eighteen nuclear microsatellite markers (nine nuclear simple-sequence repeat (nSSR) markers for each species) developed by previous studies [41–44] were amplified for *R. apiculata* and *A. marina* (Supplementary Table S1). Fluorescently labeled primers (with HEX or ROX) were synthesized (Sangon Biotech [Shanghai] Co., Ltd.), and the PCR amplification conditions used were based on a previously published protocol [41–44]. All amplified PCR products were screened using capillary electrophoresis (Sangon Biotech Co., Ltd., Shanghai, China).

2.3. Statistical Analysis of Genetic Parameters

For each nuclear locus, the Hardy–Weinberg equilibrium (*HWE*), linkage disequilibrium (*LD*), and null alleles were tested using GenePop Volume 3.4 [45]. Genetic diversity parameters, such as the number of alleles (*Na*), number of effective alleles (*Ne*), Shannon diversity index (*I*), observed heterozygosity (*Ho*), expected heterozygosity (*He*), and unbiased expected heterozygosity (*uHE*), were calculated using GenAlex Version 6.5 [46]; ADZE Version 1.0 [47] was used to determine the allele abundance (*AR*) and private allele abundance (*PAR*). The inbreeding coefficient (*Fis*) and its 95% confidence interval was calculated using GDA v1.1 (http://en.bio-soft.net/dna/gda.html, accessed on 28 December 2021). The total paternity exclusion probability of the first [Pr(*Ex1*)] and second parent [Pr(*Ex2*)] was estimated using CERVUS Version 3.0 [48]. In addition, we performed the analysis of molecular variance (AMOVA) to estimate the distribution of genetic variance among adults and seedlings using Φ-statistics with GenAlEx Version 6.5 [46] for both two species.

2.4. Analysis of Bottlenecks and Spatial Genetic Structure

The likelihood of prior population bottlenecks in the two species was estimated using BOTTLENECK Version 1.2.02 [49] using three models, the two-phase model (TPM), the infinite allele model (IAA), and the stepwise mutation model (SMM), under default settings. The significance of bottlenecks was estimated using the sign test and one-tailed Wilcoxon sign-rank test.

Spatial autocorrelation analyses were carried out for the entire population and for seedlings and adults using SPAGeDi Version 1.2 [50]. Average multilocus kinship coefficients (*Fij*) were calculated for nine distance classes according to size of quadrats and a previous study [51]. Specifically, for *A. marina*, distance classes were 1, 2, 4, 8, 16, 32, 64, 128 and 300 m; and we set distance classes as 1, 2, 4, 8, 16, 32, 64, 128 and 200 m for *R. apiculate*. The 95% confidence interval of the different distance classes was tested using 10,000 random permutations. The *Sp* statistic was calculated for individuals from the first distance class following the method of Vekemans and Hardy [52].

2.5. Analysis of Mating System Parameters and Pollen Dispersal

MLTR Version 3.2 [53] was used to analyze the multilocus outcrossing rate (*tm*), the single-locus outcrossing rate (*ts*), and the biparental inbreeding rate (*tm*−*ts*) under the mixed mating model with 1000 bootstrap replications to assess the 95% confidence intervals for standard errors. The most likely pollen donor for each propagule was determined via maximum-likelihood assignment in CERVUS Version 3.0 [48]. For paternity analyses, all adult individuals were considered as candidate parents, and 10,000 simulations were conducted with 0.01 as the mistyped rate, 0.9 as the sampled candidate parent proportion, and 95% as the strict and 80% as the relaxed confidence level.

3. Results

3.1. Genetic Diversity of R. apiculata and A. marina

Two nSSR loci (RM102 and Am98) were not included in subsequent analyses because they were not polymorphic. A total of 37 alleles of eight nSSR loci and 23 alleles of eight nSSR loci were detected in both adults and seedlings of *R. apiculata* and *A. marina*, respectively. The genetic diversity parameters of each nSSR locus for all individuals of the two species are shown in Table 1. The average frequency of null alleles in *R. apiculata* and *A. marina* was 18.8% (3.5–47.1%) and 20.7% (8.4–29.3%), respectively. Only four out of 28 pairwise comparisons showed significant *LD* in the *R. apiculata* population (including seedlings, $p < 0.01$), whereas approximately half (13 out of 28) of the pairwise comparisons showed significant *LD* in the *A. marina* population (including seedlings, $p < 0.01$). Six nSSR loci (Rhst01, Rhst11, Rhst13, Rhst20, RM114, and RM116) deviated significantly from *HWE* in the *R. apiculata* population (heterozygote deficiency, $p < 0.01$), and all nSSR loci in the *A. marina* population showed heterozygote deficiency ($p < 0.01$).

Table 1. Genetic diversity of the selected nSSR primer pairs in all individuals of *R. apiculata* and *A. marina*.

Locus	*n*	*Na*	*Ne*	*Ho*	*He*	*uHe*	*I*	*Pr(Ex1)*	*Pr(Ex2)*
R. apiculata									
Rhst01	167	2	1.519	0.018	0.342	0.343	0.525	0.962	0.880
Rhst02	167	4	1.877	0.665	0.467	0.469	0.747	0.871	0.755
Rhst11	167	2	1.916	0.347	0.478	0.479	0.671	0.903	0.828
Rhst13	167	7	2.065	0.413	0.516	0.517	0.920	0.861	0.743
Rhst20	167	8	4.322	0.311	0.769	0.771	1.589	0.630	0.452
RM113	167	5	2.453	0.593	0.592	0.594	1.056	0.814	0.671
RM114	167	6	2.207	0.467	0.547	0.548	0.999	0.839	0.692
RM116	167	3	1.068	0.042	0.064	0.064	0.155	0.997	0.960
Mean	167	4.6	2.178	0.357	0.472	0.473	0.833	/	/
A. marina									
Avma01	210	3	1.840	0.262	0.457	0.458	0.662	0.881	0.813
Avma02	210	5	1.479	0.248	0.324	0.325	0.596	0.930	0.823
Avma16	210	4	1.975	0.238	0.494	0.495	0.722	0.874	0.804
Avma17	210	5	2.795	0.476	0.642	0.644	1.201	0.797	0.634
Am3	210	2	1.379	0.157	0.275	0.275	0.447	0.975	0.901
Am32	210	3	1.100	0.010	0.091	0.091	0.212	0.999	0.976
Am40	210	4	1.776	0.152	0.437	0.438	0.671	0.903	0.819
Am81	210	4	1.547	0.014	0.354	0.354	0.594	0.908	0.818
Mean	210	3.5	1.750	0.197	0.387	0.389	0636	/	/

Note: Genetic diversity parameters were estimated for combined adult and seedling plants; **Abbreviations**: ***n***, number of individuals; ***Na***, number of alleles; ***Ne***, effective number of alleles; ***Ho***, observed heterozygosity; ***He***, expected heterozygosity; ***uHe***, unbiased expected heterozygosity; ***I***, Shannon's index of diversity; **[Pr(*Ex1*)]** and **[Pr(*Ex2*)]**, exclusion probability of the first and second parent, respectively.

The results of the genetic diversity analysis of different age groups (adults and seedlings) of the two species are shown in Table 2. Because the number of seedlings sampled was small, several genetic diversity parameters (*Na*, *I*) were lower in seedlings than in adults in the *R. apiculata* population. A similar pattern was observed in the *A. marina* population. Because *Na* is sensitive to sample size, we used ADZE Version 1.0 to calculate allele richness. In the *R. apiculata* population, the *AR* of seedlings and adults was similar, but the *PAR* of seedlings was lower than that of adults. The *AR* and *PAR* of seedlings were higher than that of adults in the *A. marina* population (Table 2). AMOVA analysis for adults and seedlings of *R. apiculata* revealed that the genetic differences mainly existed among different individuals (99.00%), only 1.00% occurred between different age groups (Supplementary Table S2). Similarly, AMOVA analysis in *A. marina* population revealed that 100% genetic variation existed within different individuals (Supplementary Table S3).

Table 2. Genetic diversity parameters for adults and seedlings of *R. apiculata* and *A. marina*.

Species	*n*	*Na*	*Ne*	*Ho*	*He*	*uHe*	*I*	*Fis*	*AR*	*PAR*
R. apiculata										
Adults	139	4.625	2.165	0.366	0.464	0.466	0.828	0.215	3.046	0.433
		(0.800)	(0.365)	(0.084)	(0.073)	(0.074)	(0.152)	/	(0.441)	(0.130)
Seedlings	28	3.375	2.169	0.313	0.486	0.495	0.813	0.373 *	2.941	0.327
		(0.460)	(0.262)	(0.081)	(0.071)	(0.072)	(0.135)	/	(0.402)	(0.155)
Total	167	4.625	2.178	0.357	0.472	0.473	0.833	0.246	/	/
		(0.800)	(0.341)	(0.083)	(0.072)	(0.073)	(0.149)	/	/	/

Table 2. *Cont.*

Species	*n*	*Na*	*Ne*	*Ho*	*He*	*uHe*	*I*	*Fis*	*AR*	*PAR*
A. marina										
Adults	152	3.500	1.750	0.197	0.387	0.389	0.636	0.497 *	2.623	0.277
		(0.267)	(0.181)	(0.056)	(0.060)	(0.061)	(0.102)	/	(0.233)	(0.111)
Seedlings	58	2.875	1.699	0.190	0.372	0.376	0.625	0.497 *	2.704	0.358
		(0.389)	(0.181)	(0.049)	(0.055)	(0.055)	(0.097)	/	(0.329)	(0.143)
Total	210	3.750	1.736	0.195	0.384	0.385	0.638	0.497 *	/	/
		(0.366)	(0.180)	(0.053)	(0.058)	(0.058)	(0.099)	/	/	/

Note: The standard error of the corresponding parameter is shown in parentheses; * Significant at the 95% confidence level. **Abbreviations:** ***n***, number of individuals; ***Na***, number of alleles; ***Ne***, effective number of alleles; ***Ho***, observed heterozygosity; ***He***, expected heterozygosity; ***uHe***, unbiased expected heterozygosity; ***I***, Shannon's index of diversity; ***Fis***, coefficient of local inbreeding; ***AR***, allele abundance; ***PAR***, private allele abundance.

3.2. Bottlenecks and Spatial Genetic Structure

We next conducted bottleneck analysis to determine whether these two populations have undergone population bottlenecks. We detected a significant excess of heterozygotes in the seedlings of *R. apiculata* under the IAA model according to both the sign test and the one-tailed Wilcoxon sign-rank test, suggesting that the seedlings of this population may have experienced a reduction in population size (Table 3). Our results indicated that the total population (adults + seedlings) of *R. apiculata* has experienced a bottleneck under the IAA model (Table 3).

Table 3. The probability of population bottlenecks of *R. apiculata* and *A. marina* under three models.

	Model	*R. apiculata*			*A. marina*		
		Adults	**Seedlings**	**Total**	**Adults**	**Seedlings**	**Total**
Sign test	TPM	0.450	0.200	0.430	0.531	0.142	0.534
	IAA	0.154	**0.042**	0.145	0.130	0.096	0.138
	SMM	0.086	0.269	0.083	0.225	0.593	0.069
	mean	0.230	0.170	0.219	0.295	0.277	0.247
One-tailed Wilcoxon test	TPM	0.422	0.125	0.422	0.527	0.230	0.578
	IAA	0.098	**0.020**	**0.027**	0.098	0.125	0.125
	SMM	0.963	0.473	0.902	0.973	0.527	0.980
	mean	0.494	0.206	0.451	0.533	0.294	0.561

Note: The one-tailed Wilcoxon sign-rank test was only used to determine heterozygosity excess. Values indicating significant bottlenecks are in bold ($p < 0.05$). **Abbreviations**: **TPM**, two-phase model; **IAA**, infinite allele model; **SMM**, stepwise mutation model.

The spatial autocorrelation analysis indicated significant spatial genetic structure at short distance classes for both seedlings and adults in the *R. apiculata* population (2–4, 4–8, 8–16, and 16–32 m; Figure 3a). Significant positive *Fij* values were observed at short distance classes among *R. apiculata* adults (2–4, 4–8, 8–16, and 16–32 m; Figure 3b). *Fij* values decreased with distance in the *R. apiculata* population. Significant positive *Fij* values were also observed for both seedlings and adults of *A. marina* (until 32 m; Figure 3c), and a similar pattern was observed for *A. marina* adults. However, *Fij* values among adults were lower than those among the total *A. marina* population, and correlations between *Fij* values and distance class were weak (Figure 3d). *Sp* statistics for each analysis are shown in Figure 2 and reveal a strong spatial genetic structure for these two species at short distances.

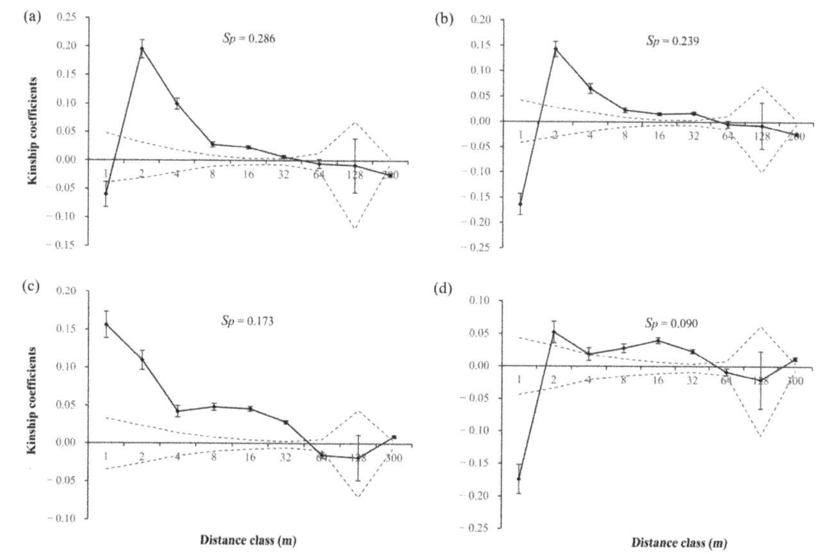

Figure 3. Kinship coefficients (*Fij*) in eight distance classes for all individuals (**a**) and adults (**b**) of the *R. apiculata* population and all individuals (**c**) and adults (**d**) of the *A. marina* population. The dashed lines indicate 95% confidence limits.

3.3. Mating System Parameters

The *tm*, *ts*, and *tm*−*ts* for *R. apiculata* at the population level were 1.135 (0.029), 1.111 (0.051), and 0.024 (0.042), respectively. Bootstrap analysis suggested that both *tm* (95% CI = 1.007–1.193) and *ts* (95% CI = 1.009–1.213) were greater than 1, and *tm*−*ts* (95% CI = −0.06–0.108) did not differ significantly from 0. This suggested the presence of high outcrossing and a relatively low proportion of biparental inbreeding in *R. apiculata*. *tm* (1.001–1.116) and *ts* (1.082–1.551) were also greater than 1 for various mother trees of *R. apiculata* (Table 4). For *A. marina*, *tm* (95% CI = 0.448–0.980) and *ts* (95% CI = 0.318–0.874) were lower than 1 at the population level, which indicated that selfing has occurred in this population. *tm* − *ts* (95% CI = 0.020–0.216) significantly differed from 0. This suggested the presence of significant biparental inbreeding. There was a high degree of variation in the mating system parameters of each mother tree in the *A. marina* population (Table 4).

Table 4. Mating system parameters for each mother tree of *R. apiculata* and *A. marina*.

Mother Tree	Propagules	*tm* (SE)	*ts* (SE)	*tm*−*ts*
R. apiculata				
Z12	6	1.001 (0.001)	1.138 (0.053)	−0.137
Z85	11	1.116 (0.041)	1.551 (0.165)	−0.435
Z110	6	1.004 (0.001)	1.082 (0.112)	−0.078
A. marina				
A1	17	0.822 (0.209)	1.027 (0.203)	−0.205
A2	10	0.186 (0.151)	0.094 (0.073)	0.092
A21	9	0.755 (0.170)	0.691 (0.199)	0.064
A48	9	1.024 (0.001)	0.589 (0.079)	0.435
A126	6	0.849 (0.162)	0.818 (0.228)	0.031

Abbreviations: *tm*, multilocus outcrossing rates; *ts*, single-locus outcrossing rates; *tm*−*ts*, biparental inbreeding rates.

We identified pollen donors within the *R. apiculata* quadrat for 35% (8 of 23) of the propagules at the 80% confidence level. The remaining 65% of the propagules were likely derived from pollen donors outside the quadrat. All eight propagules originated from outcrossing. These results were consistent with the mating system parameters estimated using MLTR software. However, only three (6%) propagules were identified as pollen donors within the quadrat at the 80% confidence level for *A. marina*, indicating that nearly all propagules originated from outcrossing. Mother trees of these three propagules were A48 (two propagules) and A126 (one propagule), and their *tm* was higher compared with that of other trees in the *A. marina* quadrat. The above findings suggest that most of the pollen donors were located outside of the *A. marina* quadrat. Because most pollen donors were from outside the quadrat, we did not calculate the pollen dispersal distance.

4. Discussion

Protections for mangroves have strengthened in China over the past two decades, and many mangrove reserves have been established [16–18]. Although much conservation effort has been devoted to increasing the area of mangrove forests, the genetic health of mangroves has been largely ignored by comparison. Small and isolated populations are often characterized by lower genetic diversity, engage in higher levels of inbreeding, and experience bottlenecks, all of which affect their resilience against a backdrop of anthropogenic threats and pressures and climate change [9,54]. Assessing the genetic diversity and mating system of such small and isolated populations of mangroves is critically important for setting reasonable restoration goals for maintaining and improving levels of genetic diversity of remaining mangrove populations [55–57].

We evaluated the genetic diversity of two mangrove species, *R. apiculata* and *A. marina*, in Tielu Harbor Mangrove Nature Reserve. There were no significant differences in the genetic diversity (*Ho* = 0.357, *He* = 0.472) of the *R. apiculata* population in Tielu Harbor with other *R. apiculata* populations. Yahya et al. [58] characterized the genetic variation of 15 *R. apiculata* populations in the Greater Sunda Islands of Indonesia using five microsatellite loci and found that *Ho* and *He* values were 0.338 (0.117–0.457) and 0.378 (0.123–0.482), respectively. The genetic diversity of *R. apiculata* in Malaysia was found to be particularly low, as the *Ho* and *He* values were 0.299 (0.194–0.483) and 0.325 (0.247–0.503), respectively [59]. Although the genetic diversity of *R. apiculata* in our study population was not low, we found that the *AR* and *PAR* values of seedlings were lower than those of adults (Table 2). Given that it takes nearly three years for propagules of *R. apiculata* to develop from flower bud primordia to maturity [60], the low genetic diversity of these seedlings may reflect the impact of recent anthropogenic activities on the population.

The genetic diversity of the *A. marina* population in our study (*Ho* = 0.195, *He* = 0.384) was lower compared with that of other populations examined in a previous study [61]. Maguire et al. [58] detected high levels of genetic diversity of *A. marina* in 14 populations across its global range using three microsatellite loci. Nearly all of the populations examined in Maguire et al. [61] had higher *Ho* values compared with the *A. marina* population in Tielu Harbor; the one exception was a Japanese population (*Ho* = 0.000). Hou et al. (unpublished data) investigated the genetic diversity of an *A. marina* population along the Sanya River, which is located close to Tielu Harbor, and the *Ho* and *He* values of this population were 0.532 and 0.575, respectively. Our data indicate that habitat degradation might have had a significant effect on the genetic diversity of the *A. marina* population in Tielu Harbor. Similarly, Salas-Leiva et al. [55] reported that the low genetic diversity (*Ho* = 0.277) of another *Avicennia* species (*A. germinans*) along the Colombian Pacific coast was associated with habitat degradation and fragmentation processes. However, we found that the *AR* and *PAR* of the seedlings of *A. marina* were higher than that of adults. The propagules of *A. marina* are small and light and can spread by sea currents [62]; thus, some of these seedlings might have originated from outside the quadrat, such as populations near Tielu Harbor (e.g., the Sanya River population and Qingmei Harbor population [39]).

We found that *A. marina* populations in Tielu Harbor had significant positive *Fis* values. High levels of inbreeding have been documented in several mangrove species, such as *Rhizophora mangle* [63], *Rhizophora stylosa* [43], *A. marina* [61], and *Kandelia candel* [64]. The high level of inbreeding observed in *A. marina* can be explained by the habitat degradation and fragmentation [55]. Considering that the population was not detected to have experienced bottleneck (Table 3) and results of the mating system and paternity analyses, we speculate that the significant positive *Fis* values and low genetic diversity in *A. marina* reflect mating between close relatives rather than a reduction in effective population size. The results of the genetic bottleneck analysis indicate that the *R. apiculata* population in Tielu Harbor may have experienced a reduction in population size induced by habitat fragmentation. Although previous studies have shown that habitat fragmentation can have significant deleterious effects on the genetic health and mating system of wind-pollinated trees [65], the outcrossing rates observed for *R. apiculata* in our study were higher than those reported for other mangrove trees [51,66]. The continuous distribution of *R. apiculata* trees in Tielu Harbor might facilitate the exchange of pollen among plants and explain the high outcrossing rate [67].

The effect of seedlings was greater on the spatial genetic structure of *A. marina* than that of *R. apiculata* (Figure 3). Pronounced spatial genetic structure has been detected over short distances within populations of mangrove species given that many propagules become established around the mother tree [51]. The pollen and propagule dispersal of *R. apiculata*, a viviparous species, is limited; similar observations have been made for another viviparous mangrove plant, *K. candel* [51,68]. However, because each *R. apiculata* individual occupied a large space, virtually no seedlings can grow further up within 2 m of the mother tree. As discussed above, this creates significant genetic structure over short distances, as was observed in adults of *R. apiculata* (2–32 m, Figure 3b). The spatial genetic structure of *A. marina* within 2 m changed significantly depending on whether seedlings were included in the analysis. This can be explained by the amount of space occupied by adult trees. Adult *A. marina* trees are small, and seedlings can grow around their mother trees. Thus, when seedlings were excluded from the analysis, genetic structure was not significant within 2 m, but when they were included, genetic structure was significant.

We studied the genetic diversity, spatial genetic structure, and mating system of two mangrove species, *R. apiculata* and *A. marina*, in a heavily disturbed area in Tielu Harbor, China, using nuclear microsatellite markers. Microsatellite markers are widely used in population genetic studies of mangrove plants, such as *R. apiculata* [58,59], *K. candel* [51,68], *A. germinans* [55–57] and *A. marina* [61], to estimate genetic diversity, genetic structure and mating system. This indicates the usefulness of microsatellites in the genetic analysis of mangrove plants. The findings from our studies and field investigations suggest that these two mangrove species require distinct restoration and rehabilitation approaches. Given that the diversity of the *R. apiculata* population in Tielu Harbor is similar to that of other populations [58,59], the propagules produced by local adult trees could be used for mangrove restoration. However, for this approach to succeed, the connectivity among individuals needs to be maintained, and this includes the connectivity between newly restored mangroves with existing mangroves, to minimize the effects of inbreeding on future generations. Connectivity among individuals is also considered important in the in-situ conservation of another mangrove plant, *A. germinans* [57]. The low genetic diversity and significant inbreeding of *A. marina* in Tielu Harbor suggest that the genetic health of the *A. marina* population is poorer than that of the *R. apiculata* population [61]. Consequently, the use of propagules from the *A. marina* population in Tielu Harbor for mangrove restoration is likely insufficient for ensuring the long-term persistence of restored *A. marina* populations. In a previous study, Salas-Leiva et al. [56] considered that reforestation using propagules from different populations would improve the maintenance of genetic diversity and the viability of the reforested population in the short and medium term. Propagules from numerous populations with high genetic diversity are needed to ensure that restored

A. marina populations are resilient enough to persist in the face of anthropogenically driven habitat degradation and climate change [28,69].

Mangrove conservation and restoration efforts in China have generally been successful over the last several decades. Approximately 67% of mangroves are now within nature reserves, and 43 mangrove protected areas have been established [16]. Studies of the genetic diversity, fine-scale genetic structure, and mating system of mangroves are needed to ensure the success of mangrove conservation efforts. In this study, we only selected limited seedlings and propagules to minimize the impact on the capacity for population renewal. We believed that if more seedling and propagule samples can be taken, the results should be more reliable. Furthermore, additional studies are needed to evaluate the genetic diversity of populations following mangrove restoration, as well as monitor the status of populations. Such studies of genetic diversity and structure would provide valuable insight into whether restoration efforts are having their intended effect of increasing the resilience of restored areas.

Supplementary Materials: The following are available online at https://www.mdpi.com/article/10.3390/d14020115/s1, Table S1: Information of 18 nSSR primer pairs, Table S2: Summary of the AMOVA results for adults and seedlings in *R. apiculata* population, Table S3: Summary of the AMOVA results for adults and seedlings in *A. marina* population.

Author Contributions: Conceptualization, W.L. and S.Y.; methodology, W.L. and S.Y.; validation, W.L. and Z.Z.; formal analysis, W.L.; investigation, W.L. and S.Y.; data curation, W.L. and Z.Z.; writing—original draft preparation, W.L.; writing—review and editing, W.L., X.H. and S.Y.; visualization, W.L.; supervision, S.Y.; project administration, S.Y.; funding acquisition, S.Y. All authors have read and agreed to the published version of the manuscript.

Funding: This research was funded by the National Key Research and Development Program of China (2017YFC0506103); Ocean Public Fund Research Project Grants of State Oceanic Administration (No. 201305021-2); and National Natural Science Foundation of China (No. 30972334).

Institutional Review Board Statement: Not applicable.

Data Availability Statement: Not applicable.

Acknowledgments: We thank editors and three anonymous reviewers for their suggestions.

Conflicts of Interest: The authors declare no conflict of interest.

References

1. Alongi, D.M. Present state and future of the world's mangrove forests. *Environ. Conserv.* **2002**, *29*, 331–349. [CrossRef]
2. Kathiresan, K.; Bingham, B.L. Biology of mangroves and mangrove ecosystems. *Adv. Mar. Biol.* **2001**, *40*, 81–251.
3. Tomlinson, P.B. *The Botany of Mangroves*, 2nd ed.; Cambridge University Press: Cambridge, UK, 2016; pp. 11–28.
4. Costanza, R.; d'Arge, R.; de Groot, R.; Farber, S.; Grasso, M.; Hannon, B.; Limburg, K.; Naeem, S.; O'Neill, R.V.; Paruelo, J.; et al. The value of ecosystem service and nature capital in the world. *Nature* **1997**, *387*, 235–260. [CrossRef]
5. Alongi, D.M. Carbon sequestration in mangrove forests. *Carbon. Manag.* **2012**, *3*, 313–322. [CrossRef]
6. Polidoro, B.A.; Carpenter, K.E.; Collins, L.; Duke, N.C.; Ellison, A.M.; Ellison, J.C.; Farnsworth, E.J.; Fernando, E.S.; Kathiresan, K.; Koedam, N.E.; et al. The loss of species: Mangrove extinction risk and geographic areas of global concern. *PLoS ONE* **2010**, *5*, e10095. [CrossRef] [PubMed]
7. Eddy, S.; Milantara, N.; Sasmito, S.D.; Kajita, T.; Basyuni, M. Anthropogenic drivers of mangrove loss and associated carbon emissions in South Sumatra, Indonesia. *Forests* **2021**, *12*, 187. [CrossRef]
8. Field, C.D. Impact of expected climate change on mangroves. *Hydrobiologia* **1995**, *295*, 75–81. [CrossRef]
9. Bologna, P.; Campanella, J.J.; Restaino, D.J.; Fetske, Z.A.; Lourenco, M.; Smalley, J.V. Lingering impacts of hurricane Hugo on *Rhizophora mangle* (red mangrove) population genetics on St. John, USVI. *Diversity* **2019**, *11*, 65. [CrossRef]
10. Lee, S.-Y.; Primavera, J.H.; Dahdouh-Guebas, F.; McKee, K.; Bosire, J.O.; Cannicci, S.; Diele, K.; Fromard, F.; Koedam, N.; Marchand, C.; et al. Ecological role and services of tropical mangrove ecosystems: A reassessment. Glob. *Ecol. Biogeogr.* **2014**, *23*, 726–743. [CrossRef]
11. FAO. *The World's Mangroves 1980–2005*; Food and Agriculture Organization of the United Nations: Rome, Italy, 2007.
12. Fan, H.; Wang, W. Some thematic issues for mangrove conservation in China (in Chinese). *J. Xiamen Univ. Nat. Sci.* **2017**, *56*, 323–330.

13. Duke, N.C.; Meynecke, J.-O.; Dittmann, S.; Ellison, A.M.; Anger, K.; Berger, U.; Cannicci, S.; Diele, K.; Ewel, K.C.; Field, C.D.; et al. A world without mangroves? *Science* **2007**, *317*, 41–42. [CrossRef]
14. Wang, W.Q.; Wang, M. *The Mangrove of China*, 1st ed.; Science Press: Beijing, China, 2007; pp. 1–186.
15. SFA. *National Mangrove Resource Inventory Report*; The State Forestry Administration of the People's Republic of China: Beijing, China, 2002.
16. Wang, W.; Fu, H.; Lee, S.-Y.; Fan, H.; Wang, M. Can strict protection stop the decline of mangrove ecosystems in China? From rapid destruction to rampant degradation. *Forests* **2020**, *11*, 55. [CrossRef]
17. Yang, S.-C.; Lu, W.-X.; Zou, Z.; Li, S. Mangrove Wetlands: Distribution, Species Composition and Protection in China (in Chinese). *Subtrop. Plant Sci.* **2017**, *46*, 301–310.
18. Jia, M.; Wang, Z.; Zhang, Y.; Mao, D.; Wang, C. Monitoring loss and recovery of mangrove forests during 42 years: The achievements of mangrove conservation in China. *Int. J. Appl. Earth Obs. Geoinf.* **2018**, *73*, 535–545. [CrossRef]
19. Hamilton, S.E.; Casey, D. Creation of a high spatio-temporal resolution global database of continuous mangrove forest cover for the 21st century (CGMFC-21). *Glob. Ecol. Biogeogr.* **2016**, *25*, 729–738. [CrossRef]
20. Wang, M.; Wang, W.; Lin, G.; Ma, W.; Fu, R. *The Mangroves of Sanya*, 1st ed.; Science Press: Beijing, China, 2019; pp. 1–212.
21. Ma, Z.; Melville, D.S.; Liu, J.; Chen, Y.; Yang, H.; Ren, W.; Zhang, Z.; Piersma, T.; Li, B. Rethinking China's new great wall. *Science* **2014**, *346*, 912–914. [CrossRef] [PubMed]
22. Wang, Q.; Song, L.; Agusti, S.; Duarte, C.; Christakos, G.; Wu, J. Changes of the Macrobenthos Community with Non-native Mangrove Rehabilitation (*Kandelia obovata*) and Salt Marsh Invasion (*Spartina alterniflora*) in Ximen Island, Zhejiang, China. *Ocean Sci. J.* **2021**, *56*, 395–405. [CrossRef]
23. Maiti, S.K.; Chowdhury, A. Effects of anthropogenic pollution on mangrove biodiversity: A review. *J. Environ. Prot.* **2013**, *4*, 1428–1434. [CrossRef]
24. Feeley, K.J.; Silman, M.R. Land-use and climate change effects on population size and extinction risk of Andean plants. *Glob. Chang. Biol.* **2010**, *16*, 3215–3222. [CrossRef]
25. Çalişkan, M. *Genetic Diversity in Plants*, 1st ed.; InTech: Rijeka, Croatia, 2012; pp. 1–493.
26. Verma, A.K. Genetic diversity as buffer in biodiversity. *Indian J. Biol.* **2017**, *4*, 61–63.
27. Rao, V.R.; Hodgkin, T. Genetic diversity and conservation and utilization of plant genetic resources. *Plant Cell Tissue Organ Cult.* **2002**, *68*, 1–19.
28. Kahilainen, A.; Puurtinen, M.; Kotiaho, J.S. Conservation implications of species–genetic diversity correlations. *Glob. Ecol. Conserv.* **2014**, *2*, 315–323. [CrossRef]
29. Dixo, M.; Metzger, J.P.; Morgante, J.S.; Zamudio, K.R. Habitat fragmentation reduces genetic diversity and connectivity among toad populations in the Brazilian Atlantic Coastal Forest. *Biol. Conserv.* **2009**, *142*, 1560–1569. [CrossRef]
30. Fahrig, L. Effects of habitat fragmentation on biodiversity. *Annu. Rev. Ecol. Evol. Syst.* **2003**, *34*, 487–515. [CrossRef]
31. Tambarussi, E.V.; Boshier, D.H.; Vencovsky, R.; Freitas, M.L.M.; Di-Dio, O.J.; Sebbenn, A.M. Several small: How inbreeding affects conservation of *Cariniana legalis* Mart. Kuntze (Lecythidaceae) the Brazilian Atlantic Forest's largest tree. *Int. For. Rev.* **2016**, *18*, 502–510. [CrossRef]
32. Coelho, N.H.P.; Tambarussi, E.V.; Aguiar, B.I.; Roque, R.H.; Portela, R.M.; Braga, R.C.; Sanson, D.; Silva, R.A.R.; Ferraz, E.M.; Moreno, M.A.; et al. Understanding genetic diversity, spatial genetic structure, and mating system through microsatellite markers for the conservation and sustainable use of *Acrocomia aculeata* (Jacq.) Lodd. Ex Mart. *Conserv. Genet.* **2018**, *19*, 879–891. [CrossRef]
33. Eckert, C.G.; Kalisz, S.; Geber, M.A.; Sargent, R.; Elle, E.; Cheptou, P.O.; Goodwillie, C.; Johnston, M.O.; Kelly, J.K.; Moeller, D.A.; et al. Plant mating systems in a changing world. *Trends Ecol. Evol.* **2010**, *25*, 35–43. [CrossRef]
34. Hopper, S.D. OCBIL theory: Towards an integrated understanding of the evolution, ecology and conservation of biodiversity on old, climatically buffered, infertile landscapes. *Plant Soil* **2009**, *322*, 49–86. [CrossRef]
35. Miller, M.P.; Haig, S.M.; Wagner, R.S. Phylogeography and spatial genetic structure of the southern torrent salamander: Implications for conservation and management. *J. Hered.* **2006**, *97*, 561–570. [CrossRef]
36. Caballero, A.; Rodríguez-Ramilo, S.T.; Ávila, V.; Fernández, J. Management of genetic diversity of subdivided populations in conservation programmes. *Conserv. Genet.* **2010**, *11*, 409–419. [CrossRef]
37. Al-Qthanin, R.N.; Alharbi, S.A. Spatial structure and genetic variation of a mangrove species (*Avicennia marina* (Forssk.) Vierh) in the Farasan Archipelago. *Forests* **2020**, *11*, 1287. [CrossRef]
38. Zhang, Y.; Zhang, J.-W.; Yang, Y.; Li, X.-N. Structural and comparative analysis of the complete chloroplast genome of a mangrove plant: *Scyphiphora hydrophyllacea* Gaertn. f. and related Rubiaceae species. *Forests* **2019**, *10*, 1000. [CrossRef]
39. Yao, Y.-F.; Liao, W.-B.; Song, X.-Y.; Li, C.-S. Current status and conservation of mangrove resources in Tielu Harbor, Sanya, Hainan (in Chinese). *Mar. Sci. Bull.* **2010**, *29*, 150–155.
40. Tao, L.-P.; Huang, S.-M. Research on Mangrove Resources and Communities in Sanya (in Chinese). *Nat. Sci. J. Hainan Univ.* **2004**, *22*, 70–74.
41. Arnaud-Haond, S.; Teixeira, S.; Massa, S.I.; Billot, C.; Saenger, P.; Coupland, G.; Duarte, C.M.; Serrão, E.A. Genetic structure at range edge: Low diversity and high inbreeding in Southeast Asian mangrove (*Avicennia marina*) populations. *Mol. Ecol.* **2006**, *15*, 3515–3525. [CrossRef]
42. Geng, Q.F.; Lian, C.L.; Tao, J.M.; Li, S.Q.; Hogetsu, T. Isolation and characterization of 10 new compound microsatellite markers for a mangrove tree species, *Avicennia marina* (Forsk.) Vierh. (Avicenniaceae). *Mol. Ecol. Notes* **2007**, *7*, 1208–1210. [CrossRef]

43. Islam, M.S.; Lian, C.; Kameyama, N.; Wu, B.; Hogetsu, T. Development of microsatellite markers in *Rhizophora stylosa* using a dual-suppression-polymerase chain reaction technique. *Mol. Ecol. Notes* **2004**, *4*, 110–112. [CrossRef]
44. Shinmura, Y.; Wee, A.K.S.; Takayama, K.; Meenakshisundaram, S.H.; Asakawa, T.; Onrizal, O.; Adjie, B.; Ardli, E.R.; Sungkaew, S.; Malekal, N.B.; et al. Isolation and characterization of 14 microsatellite markers for *Rhizophora mucronata* (Rhizophoraceae) and their potential use in range-wide population studies. *Conserv. Genet. Resour.* **2012**, *4*, 951–954. [CrossRef]
45. Raymond, M.; Rousset, F. GENEPOP (Version 1.2): Population genetics software for exact tests and ecumenicism. *J. Hered.* **1995**, *86*, 248–249. [CrossRef]
46. Peakall, R.; Smouse, P.E. GENALEX 6: Genetic analysis in Excel. Population genetic software for teaching and research. *Mol. Ecol. Notes* **2006**, *6*, 288–295. [CrossRef]
47. Szpiech, Z.A.; Jakobsson, M.; Rosenberg, N.A. ADZE: A rarefaction approach for counting alleles private to combinations of populations. *Bioinformatics* **2008**, *24*, 2498–2504. [CrossRef] [PubMed]
48. Slate, J.; Marshall, T.; Pemberton, J. A retrospective assessment of the accuracy of the paternity inference program CERVUS. *Mol. Ecol.* **2000**, *9*, 801–808. [CrossRef] [PubMed]
49. Piry, S.; Luikart, G.; Cornuet, J.-M. Computer note. BOTTLENECK: A computer program for detecting recent reductions in the effective size using allele frequency data. *J. Hered.* **1999**, *90*, 502–503. [CrossRef]
50. Geng, Q.; Lian, C.; Goto, S.; Tao, J.; Kimura, M.; Islam, M.S.; Hogetsu, T. Mating system, pollen and propagule dispersal, and spatial genetic structure in a high-density population of the mangrove tree *Kandelia candel*. *Mol. Ecol.* **2008**, *17*, 4724–4739. [CrossRef]
51. Hardy, O.J.; Vekemans, X. SPAGeDi: A versatile computer program to analyse spatial genetic structure at the individual or population levels. *Mol. Ecol. Notes* **2002**, *2*, 618–620. [CrossRef]
52. Vekemans, X.; Hardy, O.J. New insights from fine-scale spatial genetic structure analyses in plant populations. *Mol. Ecol.* **2004**, *13*, 921–935. [CrossRef]
53. Ritland, K. Extensions of models for the estimation of mating systems using n independent loci. *Heredity* **2002**, *88*, 221–228. [CrossRef]
54. Thomas, E.; Jalonen, R.; Loo, J.; Boshier, D.; Gallo, L.; Cavers, S.; Bordács, S.; Smith, P.; Bozzano, M. Genetic considerations in ecosystem restoration using native tree species. *For. Ecol. Manag.* **2014**, *333*, 66–75. [CrossRef]
55. Salas-Leiva, D.E.; Mayor-Durán, V.M.; Toro-Perea, N. Genetic diversity of the black mangrove (*Avicennia germinans* L.) in Colombia. *Aquat. Bot.* **2009**, *91*, 187–193. [CrossRef]
56. Salas-Leiva, D.E.; Mayor-Durán, V.M.; Toro-Perea, N. Genetic diversity of black mangrove (*Avicennia germinans*) in natural and reforested areas of Salamanca Island Parkway, Colombian Caribbean. *Hydrobiologia* **2009**, *620*, 17–24. [CrossRef]
57. Millán-Aguilar, O.; Manzano-Sarabia, M.; Nettel-Hernanz, A.; Dodd, R.S.; Hurtado-Oliva, M.Á.; Velázquez-Velázquez, E. Genetic diversity of the black mangrove *Avicennia germinans* (L.) Stearn in Northwestern Mexico. *Forests* **2016**, *7*, 197. [CrossRef]
58. Yahya, A.F.; Hyun, J.O.; Lee, J.H.; Kim, Y.Y.; Lee, K.M.; Hong, K.N.; Kim, S.-C. Genetic variation and population genetic structure of *Rhizophora apiculata* (Rhizophoraceae) in the greater Sunda Islands, Indonesia using microsatellite markers. *J. Plant Res.* **2014**, *127*, 287–297. [CrossRef]
59. Azman, A.; Ng, K.-K.-S.; Ng, C.-H.; Lee, C.-T.; Tnah, L.-H.; Zakaria, N.-F.; Mahruji, S.; Perdan, K.; Abdul-Kadir, M.-Z.; Cheng, A.; et al. Low genetic diversity indicating the threatened status of *Rhizophora apiculata* (Rhizophoraceae) in Malaysia: Declined evolution meets habitat destruction. *Sci. Rep.* **2020**, *10*, 19112. [CrossRef] [PubMed]
60. Christensen, B.; Wium-Andersen, S. Seasonal growth of mangrove trees in southern Thailand. I. The phenology of *Rhizophora apiculata* Bl. *Aquat. Bot.* **1977**, *3*, 281–286. [CrossRef]
61. Maguire, T.L.; Saenger, P.; Baverstock, P.; Henry, R. Microsatellite analysis of genetic structure in the mangrove species *Avicennia marina* (Forsk.) Vierh.(Avicenniaceae). *Mol. Ecol.* **2000**, *9*, 1853–1862. [CrossRef] [PubMed]
62. Duke, N.C. Mangrove taxonomy, biogeography and evolution-Implications for conservation and management (An Indo West Pacific perspective). Excellence in Research Australia (ERA)-Collection Centre for Marine Studies Publication. In Proceedings of the International Conference and Exhibition on the Mangroves of Indian and Western Pacific Oceans, Kuala Lumpur, Malaysia, 21–24 August 2006; Hooi, T.K., Eong, O.J., Primavera, J.H., Eds.; Maritime Institute of Malaysia (MIMA): Kuala Lumpur, Malaysia, 2008.
63. Arbeláez-Cortes, E.; Castillo-Cárdenas, M.F.; Toro-Perea, N.; Cárdenas-Henao, H. Genetic structure of the red mangrove (*Rhizophora mangle* L.) on the Colombian Pacific detected by microsatellite molecular markers. *Hydrobiologia* **2007**, *583*, 321–330. [CrossRef]
64. Chen, X.-Y.; Lin, P.; Lin, Y.-M. Mating systems and spontaneous mutation rates for chlorophyll-deficiency in populations of the mangrove *Kandelia candel*. *Hereditas* **1996**, *125*, 47–52.
65. Jump, A.S.; Peñuelas, J. Genetic effects of chronic habitat fragmentation in a wind-pollinated tree. *Proc. Natl. Acad. Sci. USA* **2006**, *103*, 8096–8100. [CrossRef]
66. Ge, J.P.; Cai, B.; Ping, W.; Song, G.; Ling, H.; Lin, P. Mating system and population genetic structure of *Bruguiera gymnorrhiza* (Rhizophoraceae), a viviparous mangrove species in China. *J. Exp. Mar. Biol. Ecol.* **2005**, *326*, 48–55. [CrossRef]
67. Hanson, T.R.; Brunsfeld, S.J.; Finegan, B.; Waits, L.P. Pollen dispersal and genetic structure of the tropical tree Dipteryx panamensis in a fragmented Costa Rican landscape. *Mol. Ecol.* **2008**, *17*, 2060–2073. [CrossRef]

68. Ha, H.T.; Duarte, C.M.; Tri, N.H.; Terrados, J.; Borum, J. Growth and population dynamics during early stages of the mangrove *Kandelia candel* in Halong Bay, North Viet Nam. *Estuar. Coast. Shelf Sci.* **2003**, *58*, 435–444.
69. Kettenring, K.M.; Mercer, K.L.; Reinhardt Adams, C.; Hines, J. EDITOR'S CHOICE: Application of genetic diversity-ecosystem function research to ecological restoration. *J. Appl. Ecol.* **2014**, *51*, 339–348. [CrossRef]

diversity

MDPI

Article

Phylogeography and Genetic Structure of Sand Dune Specialist *Stilpnolepis centiflora* (Asteraceae) in Northwest China Revealed by Molecular Data

Xiaojun Shi [1,*], Xiaolong Jiang [2], Hongxiang Zhang [3] and Juan Qiu [1]

1 College of Life Sciences, Xinjiang Agricultural University, Urumqi 830052, China; xjqiujuan@163.com
2 The Laboratory of Forestry Genetics, Central South University of Forestry and Technology, Changsha 410004, China; xiaolongjiang1@gmail.com
3 State Key Laboratory of Desert and Oasis Ecology, Xinjiang Institute of Ecology and Geography, Chinese Academy of Sciences, Urumqi 830011, China; zhanghx561@ms.xjb.ac.cn
* Correspondence: shixj@xjau.edu.cn

Abstract: *Stilpnolepis centiflora* is an endemic annual herb in the Asteraceae family found across five sand deserts in Northwest China. We aimed to investigate the genetic structure of *S. centiflora* and attempt to link species evolution with desert formation during the Pleistocene era. We used sequence data from nuclear and chloroplast genes to investigate genetic diversity among 28 populations. We analyzed sequence data using network analysis, spatial analysis of molecular variance (SAMOVA), and a Mantel test. We then used a molecular clock to place the genetic patterns in a temporal framework and tested for signals of expansion using neutrality tests and by determining mismatch distributions. Six distinct haplotypes and 31 ribotypes were identified. Significant chloroplast DNA population subdivision was detected (G_{ST} = 0.952; N_{ST} = 0.976), but only moderate nrDNA subdivision (G_{ST} = 0.360; N_{ST} = 0.579) was detected. SAMOVA revealed four diverging groups of related haplotypes, coinciding with the boundaries of deserts. Molecular dating suggests that the clades representing different deserts diverged from 1.2 to 0.20 Ma, concordant with the Kun-Huang Movement of Qinghai Tibet Plateau uplift and a glacial event (Naynayxungla) during the Middle–Late Pleistocene. The disjunction of *S. centiflora* among different deserts was correspondingly reflected in the examined genetic traits with consistent spatiotemporal evolution between species and deserts. Therefore, the evolutionary dynamics of *S. centiflora* appear to have been driven by geological movement and climate change. The patterns described here are potentially useful to conservation biologists and may serve as a model for other sand-obligate organisms found in the deserts of Northwest China.

Keywords: chloroplast DNA; ITS; *Stilpnolepis*; allopatric divergence; genetic structure

Citation: Shi, X.; Jiang, X.; Zhang, H.; Qiu, J. Phylogeography and Genetic Structure of Sand Dune Specialist *Stilpnolepis centiflora* (Asteraceae) in Northwest China Revealed by Molecular Data. *Diversity* **2022**, *14*, 104. https://doi.org/10.3390/d14020104

Academic Editor: Michael Wink

Received: 22 November 2021
Accepted: 29 January 2022
Published: 31 January 2022

1. Introduction

Historically, geographic and climatic events have had a strong influence on the genetic diversity of species [1]. In the Northern Hemisphere, most biogeographical studies on plants have shown that those currently inhabiting formerly glaciated areas retreated to the southern glacial refugia, and then expanded to their modern ranges after the last glacial maximum (LGM) [2–4]. However, the impact of these climatic oscillations was different on different continents. In the desert regions of Northwest China, lacking continental glaciers, glacial periods had the primary effect of intensifying aridity. How did desert plants respond to these conditions? Since desert plants tolerate arid conditions, the climatic oscillations might have affected their survival less [5,6]. On the other hand, the mobilization and accumulation of sand were greatly enhanced. Thus, the distribution of desert plants might have increased or diminished during climatic oscillations because of the expansion or shrinkage of desert sand dunes [7,8]. In addition, deserts served as geographical barriers and thus could have promoted species differentiation in desert organisms [9–13].

Large sand deserts extend from northwestern to northern China, including the great Taklimakan Desert and the five deserts considered in this study. These deserts may have played an important role in the speciation and evolution of plants in the arid regions of Northwest China [14]. Desert formations resulted from the effects of an intensely arid climate on Quaternary paleo-eolian sands [15–17], presumably related to the uplift of the Tibetan Plateau [18–20] during the period when large-scale sand dunes expanded [21–23].

Phylogeographical analyses can trace the influence of paleoenvironments and climate change on species distribution and population demography [24]. Previous phylogeographical studies conducted in the arid regions of Northwest China provide an understanding of the spatiotemporal diversification patterns associated with environmental and climatic change; examples include *Gymnocarpos przewalskii* [25], *Reaumuria soongarica* [7], *Nitraria sphaerocarpa* [8], and *Juniperus sabina* [14]. These studies demonstrated that climate oscillations have profound effects on the evolutionary processes of native desert species, resulting in allopatric divergences [14], regional range expansion [7], and the contraction or fragmentation of population distribution [8]. However, little is known about how desert species in Northwest China responded to past geological changes, especially during the formation and development of deserts. In addition, most studies have focused on shrub species instead of herbs, which are better suited for depicting some aspects of plant evolution due to their short life cycle.

The annual herb *Stilpnolepis centiflora* belongs to Anthemideae Cass. (Asteraceae) [26]. *Stilpnolepis* is considered a monospecific genus based on research about its life history, geographical distribution, and pollen morphology [27,28]. *S. centiflora* is an endemic species with disjunct distribution between five deserts (Figure S1) [27], primarily found on mobile sand dunes and flat sand sheets between dunes. Its flowers are pollinated by a wide variety of insects such as bees and flies, and the seeds are usually dispersed by gravity. Due to the uplift of the QTP and climate change during the Quaternary period, the geographical and natural environments varied dramatically in the arid land region in Northwest China, especially in desert regions. *S. centiflora*, as an endemic desert plant species, has historically experienced many environmental and geographic changes; hence, we hypothesized that the present distribution of *S. centiflora* is the consequence of a series of geomorphological adjustments in the region.

In the present study, we amplified and sequenced two maternally inherited chloroplast DNA (cpDNA) markers (psbA-trnH, trnQ-rps16) as well as the biparentally inherited internal transcribed spacer (nrITS) region. The data were combined to detect intraspecific divergence and possible hybridization and introgression events [29,30]. Our objective for the present study was to understand the level of genetic variation, population structure, and genetic divergence of S. centiflora using cpDNA and ITS markers.

2. Materials and Methods

2.1. Sampling Methods

Two hundred eighty *S. centiflora* individuals were collected from 28 populations distributed across five deserts, covering most areas where the species is found distributed in China. The sample consisted of four populations (populations 1–4) from the Kubuqi Desert, four populations (populations 5–8) from the Mu Su Desert, four populations (populations 9–12) from the Badain Jaran Desert, eight populations (populations 13–20) from the Ulan Buh Desert, and eight populations (populations 21–28) from the Tengger Desert (Table S1). In each population, whole fresh plant organs with flowers/fruits were collected as voucher specimens, and young leaf samples were collected from 10 individuals and dried in silica gel. When sampling, a distance of at least 30 m was maintained between individuals within the same population to increase the likelihood of sampling interindividual variation. We used three species as external groups (*Elachanthemum intricatum*, *Artemisia montana*, and *Artemisia frigida*). *Elachanthemum intricatum* was sampled and used as an outgroup in the subsequent analyses. The target sequences of the other two outgroup species were extracted from the chloroplast genome sequence downloaded from GenBank: *Artemisia montana* (KF887960.1) and *Artemisia frigida* (JX293720.1). Voucher specimens were collected and deposited in the Herbarium of Xinjiang Institute of Ecology and Geography, Chinese Academy of Science (XJBI).

2.2. Laboratory Procedures

DNA was extracted using a modified CTAB protocol [31]. Firstly, we amplified the two chloroplast (cpDNA) genes *psb*A-*trn*H and *trn*Q-*rps*16 intergenic spacers using primers and cycling conditions described by Shaw et al. [32,33]. We used a total of 30 cycles of denaturation at 94 °C for 30 s, annealing at 52 °C for 30 s, and elongation at 72 °C for 90 s, and a final elongation at 72 °C for 10 min. Secondly, we amplified the nuclear internal transcribed spacer (ITS) in 250 individuals, using forward (ITS1) and reverse primer (ITS4) for amplification for both internal transcribed spacers and the 5.8S gene [34]. Amplification products were purified from 1.5% low-melting agarose gels, and the desired PCR fragment was recovered with a UNIQ-10 kit (Shanghai SBS, Biotech Ltd., Shanghai, China) according to the manufacturer's recommendations. Sequencing reactions were conducted on the recovered PCR fragment using the forward or reverse primers of the amplification reactions and the DYEnamic ET Terminator Kit (Amersham Pharmacia Biotech, Cambridge, UK), followed by sequencing with an Applied Biosystems 3730 Capillary DNA DNA Analyzer (Shanghai Sangon Biological Engineering Technology & Services Co., Ltd., Shanghai, China). DNA sequences were edited using SeqMan (Lasergene, DNASTAR Inc., Madison, WI, USA) and initially aligned with Clustal X 1.81 [35].

2.3. CpDNA and ITS Sequence Analysis

For cpDNA, we concatenated the two chloroplast gene fragments used in the analysis. For ITS, haplotypes of heterozygous individuals were reconstructed with the PHASE algorithm implemented in DnaSP 5.10 [36], using a recombination model with no assumption about rate variation and an initial estimate of 0.0004. The Markov chain Monte Carlo (MCMC) method was run for 1000 iterations with a burn-in of 100, a thinning interval of 1, and an output probability threshold of 0.9. This method is a reliable way to infer differences in alleles in heterozygotes and, therefore, a suitable alternative to cloning and omitting unresolved genotypes for phylogeographical analysis [37]. Identical haplotypes for cpDNA and phased nrDNA alleles were collapsed using DNASP 5.10. Newly identified sequences were submitted to GenBank under the accession numbers MF416962-MF417003 (Tables S2 and S3).

Phylogenetic trees were constructed using the neighbor-joining (NJ) method implemented in MEGA version 11.0 [38]. The NJ analysis incorporated Kimura's 2-parameter model of DNA evolution [39]. The relationships between cpDNA haplotypes and ribotypes for ITS were estimated with Network version 4.6.0.0 using the median-joining method [40,41].

Nucleotide diversity (π), haplotype diversity (H), and the number of segregating sites (S) were calculated from both cpDNA and ITS data using Arlequin 3.5 [42]. Isolation-by-distance analysis was conducted using a Mantel test implemented in Alleles in Space (AIS) [43]. Mantel's test explicitly tests the plausibility of an isolation-by-distance scenario in AIS to analyze the relationships between genetic and geographic distances between sampling localities.

Genetic differentiation among populations was evaluated using the G_{ST} and N_{ST} coefficients implemented in Permut CpSSR v.2.0 [44] based on 2000 random permutations of haplotypes across populations. If N_{ST} is significantly higher than G_{ST}, then genealogically closely related haplotypes tend to occur together within populations, providing evidence for phylogeographical structure.

We used spatial analysis of molecular variance (SAMOVA 2.0) to partition the populations into genetically and geographically homogeneous groups [45]. Data from cpDNA and nrDNA were separately analyzed using 100 simulated annealing processes by varying K (number of groups) from 2 to 10. The best K value was selected according to when F_{CT} reached a plateau. In addition, we performed Bayesian analysis of population structure as implemented in Structure version 2.2 to infer the most likely number of population genetic clusters (K) in the cpDNA dataset [46]. K ranged from 1 to 10, with 10 replicates performed for each K using a burn-in period of 2×10^5 and MCMC of 5×10^4. For this analysis, a "no-admixture model" and independent allele frequencies were chosen. The most likely K value was determined based on the Delta K statistic according to Pritchard et al. [47] using Structure Harvester [48]. To investigate the level of genetic variation among and within groups, hierarchical analysis of molecular variance (AMOVA) was performed using multiple categorical variables (populations, haplotypes, sampling locations, clade, and lineage) [49]. AMOVA was conducted using Arlequin v.3.5 and 1000 random permutations.

Bayesian Evolutionary Analysis Sampling Trees (BEAST) v.1.8.2 [50] was used to reconstruct haplotype gene trees and simultaneously estimate the divergence times between haplotypes. We used a constant-size coalescent tree prior and GTR substitution model in the analysis. As there is no fossil record of *Stilpnolepis*, we adopted a substitution rate method. The cpDNA substitution rates of most angiosperm species were estimated to vary between 1 and 3×10^{-9} substitutions per site per year (s/s/y). Due to the uncertainty of the rates, we used normal distribution priors with a mean of 2×10^{-9} and an SD of 6.080×10^{-10} within the 95% distribution range to estimate divergence times [51]. Although molecular clock estimates vary and, in most cases, only provide crude estimates of divergence times, they can provide insight into the approximate timing of divergences. Hence, we interpreted our molecular clock findings with an appropriate level of caution. The posterior distributions of the parameters in the MCMC analyses were approximated with 10 million steps in each analysis and sampled every 1000 generations. Convergence of the parameters sampled was checked with the program Tracer v.1.5 to examine the highest effective sampling size values (ESSs > 200) for all parameters [52]; the burn-in steps were discarded to estimate the posterior probability distribution of divergence time at the relevant node. FigTree 1.3.1 [53] was used to display the sampled trees.

The demographic history of the predicted groups based on phylogenetic analysis of haplotypes was assessed for all 28 populations; only 4 populations from Badain Jaran Desert and 8 populations from Tengger Desert used three methods implemented in Arlequin v. 3.5. Firstly, Tajima's D [54] and Fu's Fs statistics [55] were used to explore evidence for demographical expansions using a null distribution of 10,000 coalescent simulations. Significantly negative values indicate that population expansion has occurred. Secondly, mismatch distributions [56] were calculated to test for signals of demographic expansion with populations undergoing exponential growth expected to show a smooth unimodal mismatch distribution curve [57]. The significance of sum of squared deviations (SSD) and raggedness indices was determined by bootstrap resampling (10,000 replicates). Similarly, the HRag significance was determined for SSD with 1000 parametric bootstrap replicates. When an expansion model could not be rejected, we estimated the expansion time (t) as $t = \tau/2u$, where τ is calculated as the time to expansion in mutational units and u is the mutation rate per generation for the whole sequence. The u is equal to μgk, where μ is the substitution rate in substitutions per site per year (s/s/y) and k is the two-cp sequence length. The generation time (g) was estimated to be one year. The substitution rate range of the two combined cpDNA-IGS regions was perceived as the minimum and maximum mutation rates of 1.0×10^{-9} s/s/y and 3.0×10^{-9} s/s/y [58].

3. Results

3.1. Genetic Variation Based on Plastid Sequence

The total alignment length of the two chloroplast regions (*psb*A-*trn*H, *trn*Q-*rps*16) surveyed across 280 individuals from 28 populations of *S. centiflora* was 1345 bp, including nine substitutions that were represented by six haplotypes (H1–H6) (Table S4). The haplotype compositions for each population are presented in Table 1. Although the cpDNA data revealed high haplotype diversity (h = 0.794) across all 28 populations, only populations 9 and 10 sampled from Badain Jaran Desert had two haplotypes each (H5 and H6) with a haplotype diversity of 0.36 and 0.20, respectively. Nucleotide diversity was computed across all populations, and populations 9 and 10 were significantly low, varying from 0.0003 to 0.002. The remaining 26 populations only had a single haplotype with a haplotype and nucleotide diversity of zero. Three of the six haplotypes (H1, H4 and H5) were shared between two or three of the five deserts. H1 was shared in two of the populations sampled from Ulan Buh Desert and four of the populations from Tengger Desert, whereas H4 was shared among five of the populations from Ulan Buh Desert and one population from Tengger Desert. H5 was common among the five populations sampled from Badain Jaran Desert, one population from Ulan Buh Desert, and three populations from Tengger Desert. H2 and H3 were specific to Kubuqi Desert and Mu Us Desert, respectively, whereas H5 and H6 were observed together in two of the populations collected from Badain Jaran Desert (Table 1; Figure 1). The within-population gene diversity (H_S) was significantly lower (0.020) than total gene diversity (H_T) (0.820, Table 2).

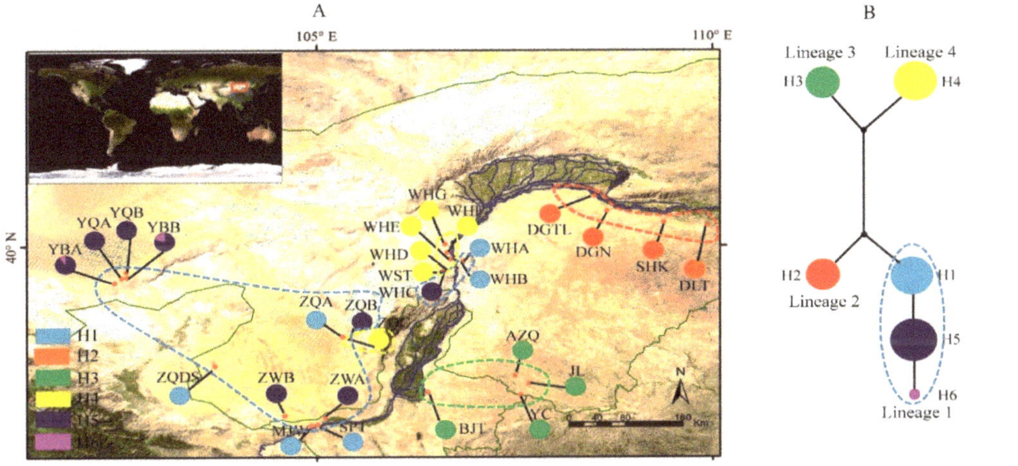

Figure 1. (**A**) Geographic distribution of cpDNA haplotypes (H1–H6) detected in the 28 populations of *Stilpnolepis centiflora*. The colored dashed curve lines delimitate the population groups (I–IV) resulting from SAMOVA and network analysis. Blue: lineage I; red: lineage II; green: lineage III; yellow: lineage IV. (**B**) The size of circles corresponds to the frequency of each haplotype. The small black circles on the branches represent hypothetical missing haplotypes.

The total gene diversity H_T (0.820) was significantly higher than the average within-population diversity H_S (0.020); therefore, both G_{ST} (0.976) and N_{ST} (0.986) were high (Table 2). Around 92.9% of the variation was attributed to genetic differentiation among the five desert regions; between-population variation accounted for just over three-quarters (98.6%) of the total variation, indicating strong differentiation and less gene flow between regions and populations (Table 3). Within each of the five regions, between-population differentiation was similarly low (Table 3). Overall, these results strongly indicate that haplotypes are geographically structured across the distribution range of the species.

Table 1. Summary of the 28 *Stilpnolepis centiflora* populations' sampling information and genetic diversity based on chloroplast DNA (cpDNA) and internal transcribed spacer (ITS) data. See Table S4 for details of the different haplotypes and ribotypes.

Code/Location	Latitude/Longitude (N/E)	cpDNA				ITS			
		N	Haplotype	h	π	N	Ribotypes	h	π
Total		280		0.7944	0.0020	250		0.8140	0.0030
Kubuqi Desert									
1 SHK	39.64°/106.60°	10	H2	0	0	22	R1, R2, R3, R4	0.6623	0.0028
2 DGN	40.71°/108.51°	10	H2	0	0	20	R1, R2, R5, R6, R7	0.6632	0.0034
3 DGTL	40.49°/108.67°	10	H2	0	0	18	R1, R2, R5, R6, R8	0.7451	0.0026
4 DLT	40.28°/109.93°	10	H2	0	0	18	R1, R2, R8	0.5817	0.0026
Mu Us Desert									
5 AZQ	40.33°/109.39°	10	H3	0	0	16	R2, R8, R9, R10, R13	0.8250	0.0020
6 BJT	38.05°/107.68°	10	H3	0	0	12	R6, R14,	0.3030	0.0025
7 YC	37.93°/106.41°	10	H3	0	0	14	R1, R8, R10	0.4835	0.0021
8 JL	37.46°/105.01°	10	H3	0	0	20	R1, R8, R9, R10, R11, R12, R13	0.6895	0.0025
Badain Jaran Desert									
9 YBA	39.35°/102.34°	10	H5 H6	0.3556	0.0005	12	R5, R6, R7, R17	0.7143	0.0012
10 YBB	39.55°/102.53°	10	H5 H6	0.2000	0.0003	14	R6, R7, R17	0.4394	0.0007
11 YQA	39.56°/102.60°	10	H5	0	0	24	R5, R6, R7, R17	0.4312	0.0007
12 YQB	39.64°/102.58°	10	H5	0	0	24	R5, R6, R7, R17, R31	0.7391	0.0014
Ulan Buh Desert									
13 WHA	39.78°/106.86°	10	H1	0	0	14	R5, R6, R15, R17	0.6923	0.0012
14 WHB	39.78°/106.85°	10	H1	0	0	20	R5, R6, R15, R17	0.6105	0.0010
15 WHC	39.64°/106.63°	10	H5	0	0	20	R5, R6, R7, R15, R24	0.6263	0.0013
16 WHD	39.64°/106.6°	10	H4	0	0	20	R5, R6, R7, R17, R24	0.4421	0.0008
17 WHE	39.91°/106.66°	10	H4	0	0	16	R6, R7, R24	0.6583	0.0016
18 WHF	39.82°/106.69°	10	H4	0	0	16	R6, R17	0.5000	0.0007
19 WHG	40.03°/106.63°	10	H4	0	0	16	R6, R17	0.1250	0.0002
20 WST	38.16°/107.51°	10	H4	0	0	12	R6, R15, R16, R17	0.5606	0.0009
Tengger Desert									
21 ZQA	38.71°/105.33°	10	H1	0	0	22	R5, R6, R7, R17, R29, R30	0.7619	0.0015
22 ZQB	38.69°/105.40°	10	H5	0	0	20	R5, R6, R7, R17, R23, R27, R28	0.6895	0.0017
23 ZWA	37.57°/105.10°	10	H5	0	0	16	R6, R7, R17, R23, R24	0.7667	0.0017
24 ZWB	37.59°/104.6°	10	H5	0	0	20	R6, R7, R17, R25	0.7105	0.00140
25 ZQC	38.55°/105.35°	10	H4	0	0	10	R6, R7, R17, R26	0.7333	0.0013
26 MJW	37.89°/107.58°	10	H1	0	0	12	R7, R17	0.5455	0.0015
27 SPT	37.45°/104.93°	10	H1	0	0	28	R6, R7, R17, R18, R19, R20, R21	0.8042	0.0018
28 ZQDS	38.30°/103.72°	10	H1	0	0	22	R5, R6, R7, R17, R22	0.7792	0.0015

N, h and π refer to number of individuals, haplotype diversity, and nucleotide diversity within populations, respectively.

Table 2. Estimation of gene diversity (H_S, H_T) and gene differentiation (G_{ST}, N_{ST}) across all populations of *Stilpnolepis centiflora*.

Date	H_S	H_T	G_{ST}	N_{ST}
cpDNA	0.020 (0.014)	0.820 (0.018)	0.976 (0.018)	0.986 (0.010)
ITS	0.617 (0.031)	0.817 (0.037)	0.244 (0.042)	0.485 (0.053)

H_S, average gene diversity within populations; H_T, total gene diversity; G_{ST}, interpopulation differentiation; N_{ST}, number of substitution types. Values are means (± SE in parentheses).

Table 3. Hierarchical analysis of molecular variance (AMOVA) based on chloroplast DNA (cpDNA) and internal transcribed spacer (ITS) data of 280 individuals from 28 *Stilpnolepis centiflora* populations.

Source of Variation	cpDNA *d.f.*	cpDNA PV (%)	cpDNA Fixation Index	ITS *d.f.*	ITS PV (%)	ITS Fixation Index
Among populations	27	98.6	F_{ST} = 0.986	27	46.68	F_{ST} = 0.467
Within populations	252	1.4		470	53.32	
Total	279			497		
Five deserts						
Among deserts	4	92.9	F_{SC} = 0.841 **	4	35.17	F_{SC} = 0.240 **
Among populations within deserts	23	5.97	F_{ST} = 0.989 **	23	15.58	F_{ST} = 0.508 **
Within populations	252	1.13	F_{CT} = 0.929 **	470	49.25	F_{CT} = 0.352 **
Total	279			497		
Four lineages						
Among lineages	3	91.73	F_{SC} = 0.879 **			
Among populations within lineages	24	7.27	F_{ST} = 0.990 **			
Within populations	252	1	F_{CT} = 0.917 **			
Total	279					

d.f., degrees of freedom; F_{ST}, correlation within populations relative to the total; PV, percentage of variation; SS, sum of squares; VC, variance components. **, $p < 0.01$; 1000 permutations.

The spatial genetic structure analysis based on cpDNA using SAMOVA showed a sharp increase in F_{CT} values from K = 2 to K = 10 and then reached a plateau at K > 7 (Figure S2), which suggests four possible groups. The groups obtained using Structure revealed the substantial phylogeographic patterns and additionally allowed recovering information about the hierarchical relationships among the groups of populations. When K = 2, individuals from the TGL-BDJL group were separated from other populations. When K = 4, further substructuring was observed that corresponds to the TGL-BDJL, WLBH, MWS, and KBQ geographic groups (Figure S3). A subsequent analysis was conducted to corroborate the structures detected with this first analysis, excluding TGL-BDJL population groups from the dataset to determine if any additional substructure could be detected. In the second analysis, the highest peak at K = 3 corresponded to MWS, WLBH, and KBQ geographic groups, which confirmed the genetic substructure detected by the SAMOVA and network analysis (Figures S2 and S3). The topology of the corresponding grouping configuration resembled that obtained from the NJ analyses (Figure 2). As shown in Figures 1A,B and 2, the four groups (a–d) were almost completely allopatric and corresponded to different desert regions (a: Badain Jaran Desert–Tengger Desert; b: Kubuqi Desert; c: Mu Us Desert; d: Ulan Buh Desert).

Three levels of hierarchical AMOVA (Table 3) conducted on cpDNA revealed significant great genetic differentiation among the groups (F_{ST} = 0.990, $p < 0.001$). Overall, these results strongly indicate that haplotypes are geographically structured across the distribution range of the species. Mantel test showed a moderate but significant correlation (r = 0.304, $p < 0.05$) between genetic differentiation and geographic distance among populations (Figure S4).

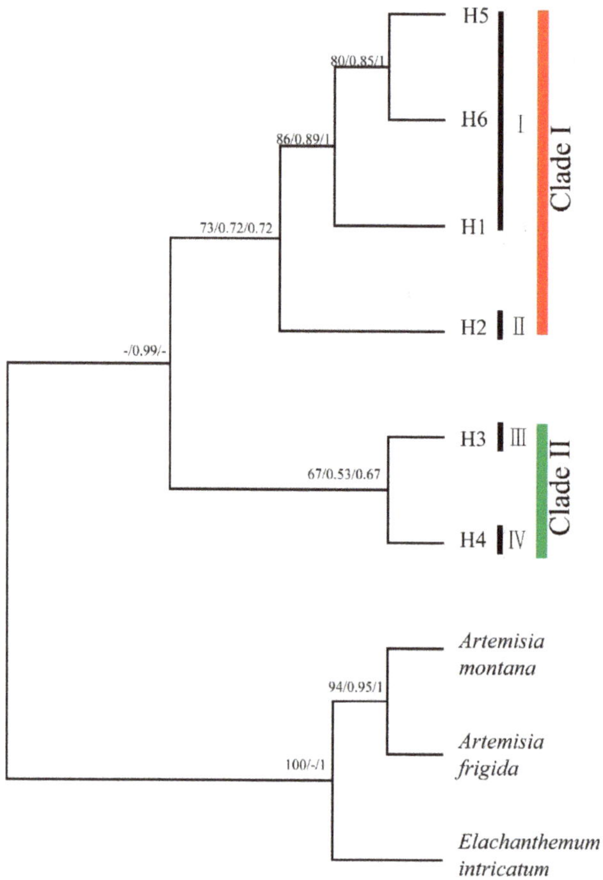

Figure 2. Phylogenetic relationships of the identified haplotypes of *Stilpnolepis centiflora* using *Elachanthemum intricatum*, *Artemisia montana*, and *Artemisia frigida* as outgroups by neighbor-joining method. Neighbor-joining bootstrap values are shown above branches.

3.2. Genetic Variation, Ribotype Distribution, Population Structure, and Phylogenetic Relationships between Ribotypes Based on ITS Sequence

Ribotype reconstruction of ITS sequences in PHASE resulted in highly supported ribotype pairs ($p > 0.90$) for 250 individuals (500 sequences in total). Although the ITS commonly represents a family of genes, sequences did not show extremely high levels of polymorphism, and manual alignment was straightforward. The aligned ITS data set (714 bp) yielded 31 ribotypes from the 500 sequences (Table S5). Among the 31 determined ribotypes, 7 nuclear ribotypes (R1, R2, R5, R6, R7, R8, R17) were relatively common (Figure 3). R6 was widely distributed in the five deserts; R5 and R7 were shared in WLBH, TGL, BDJL, and KBQ; R17 was shared in TGL, WLBH, and BDJL; R1, R6, and R8 were shared in KBQ and MWS. R24 was shared in WLBH and TGL. Of the 23 rare remaining ribotypes, 6 were restricted to MWS (R9, R10, R11, R12, R13, R14), 12 were restricted to TGL (R18, R19, R20, R21, R22, R23, R25, R26, R27, R28, R29, R30), 2 were restricted to KBQ (R3 and R4), 2 were restricted to WLBH (R15 and R16), and 1 was restricted to BDJL (R31) (Table 1). Ribotype diversity computed within populations varied from 0.125 in the WHG population to 0.825 in the AZQ population with an overall mean value of 0.814 across the 28 populations. Nucleotide diversity within populations varied from 0.0002 in the WHG population to 0.0034 in the DGN population with an overall mean value of 0.003 across the 28 populations (Table 1). Within-population gene diversity (0.617) was lower than total gene diversity (0.817) (Table 2).

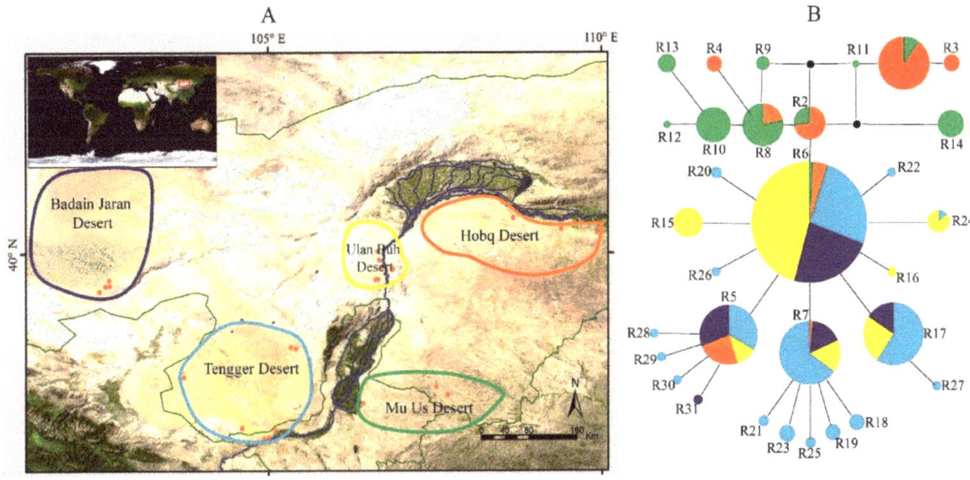

Figure 3. (**A**) The geographic distribution of the 28 *Stilpnolepis centiflora* populations in the deserts of northwestern China. Each desert is marked using different colors. (**B**) The networks of 31 ribotypes of *S. centiflora*. The size of circles corresponds to the frequency of each ribotype. Small black circles on the branches represent hypothetical missing ribotypes. Ribotypes are color-coded based on deserts: Purple: Badain Jaran; blue: Tengger; yellow: Ulan Buh; orange: Kubuqi; green: Mu Us.

For the ITS data, the phylogenetic relationships were reconstructed using NJ methods. Most of the 31 ribotypes did not form a well-supported clade (Figure S5), and the ribotype network (Figure 3B) contained the same relationships as the phylogenetic trees (Figure S5). The genealogical analysis of nuclear haplotypes (Figure 3B) showed that haplotype R6 differed from the others by one to five mutation steps. According to the ribotype network, there was no apparent association of ribotypes with geography. Compared with cpDNA data, *S. centiflora* showed lower population differentiation, with G_{ST} values of 0.244 for ITS data. The SAMOVA analysis of the ITS data did not show strong geographic patterning compared with analyses of the cpDNA sequences. Nonetheless, AMOVA revealed that differences among the four cpDNA groups (a–d identified) accounted for 39.68% of the total nrDNA variation compared with 14.14% among populations and 48.18% within populations (Table 3). The Mantel test based on ITS data revealed a significant isolation-by-distance pattern ($r = 0.2876$, $p < 0.001$) (Figure S4).

3.3. Demographic History

The neutrality tests did not support the occurrence of population expansions for all populations, Clade I, and Clade II (Figure 2). The mismatch distribution consisted of a double-peak curve (Figure 4A–C), which suggests that all populations, Clade I, and Clade II did not experience demographic expansion. The nonsignificant SSD statistics, HRag value, neutrality test, and multimodal mismatch distribution (Table 4) all suggest the *S. centiflora* population sizes remained relatively stable for long periods in the past (Figure 4; Table 4). However, the single-peak curve (Figure 4D) in the mismatch distribution for Lineage I is indicative of demographic expansion. Based on the corresponding τ values, assuming a substitution rate of between 1.0×10^{-9} and 3.0×10^{-9} s/s/y (see above), we estimated the possible expansion of *S. centiflora* in the Tengger–Badain Jaran Desert regions may have occurred between 0.08 and 0.26 Ma. The mismatch analysis and neutrality test were not conducted for Lineages I, II, and III since they each comprised only one haplotype.

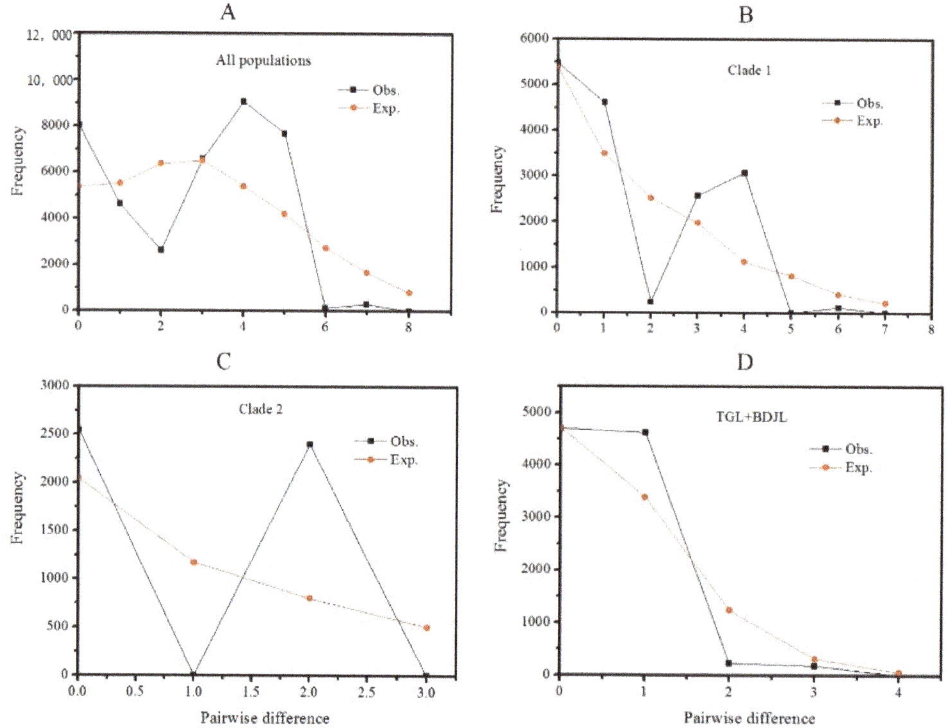

Figure 4. Mismatch distribution analysis for cpDNA data for all 28 populations (**A**), Clade 1 (**B**), Clade 2 (**C**), and Lineage I (**D**).

Table 4. Results of neutrality tests and mismatch distribution analysis for Clade 1, Clade 2, Lineage I, and the overall populations based on cpDNA.

Populations	τ	SSD (*p* Value)	HRag (*p* Value)	Tajima's D (*p* Value)	Fu's Fs (*p* Value)
Overall	4.379	0.034 (0.110)	0.064 (0.220)	1.956 (0.970)	5.793 (0.952)
Clade 1	4.326	0.043 (0.280)	0.136 (0.320)	1.119 (0.893)	4.028 (0.935)
Clade 2	2.742	0.183 (0.120)	0.736 (0.040)	2.322 (0.993)	4.467 (0.965)
Lineage I	0.703	0.026 (0.020)	0.204 (0.010)	0.105 (0.616)	1.296 (0.740)

HRag, Harpending's raggedness index; SSD, sum of squared deviations.

3.4. Phylogeny-Based Estimations of Divergence Times

The average divergence times of *S. centiflora* are shown in Figure 5. The initial divergence was estimated at approximately 1.2 Ma (95% HPD: 0.5–2.1 Mya) and the last divergence was estimated at approximately 0.2 Ma (95% HPD: 0–0.5 Mya) based on an assumed substitution rate of 1.52×10^{-9} s/s/y in cpDNA. Thus, the species divergence occurred in the Middle–Late Pleistocene.

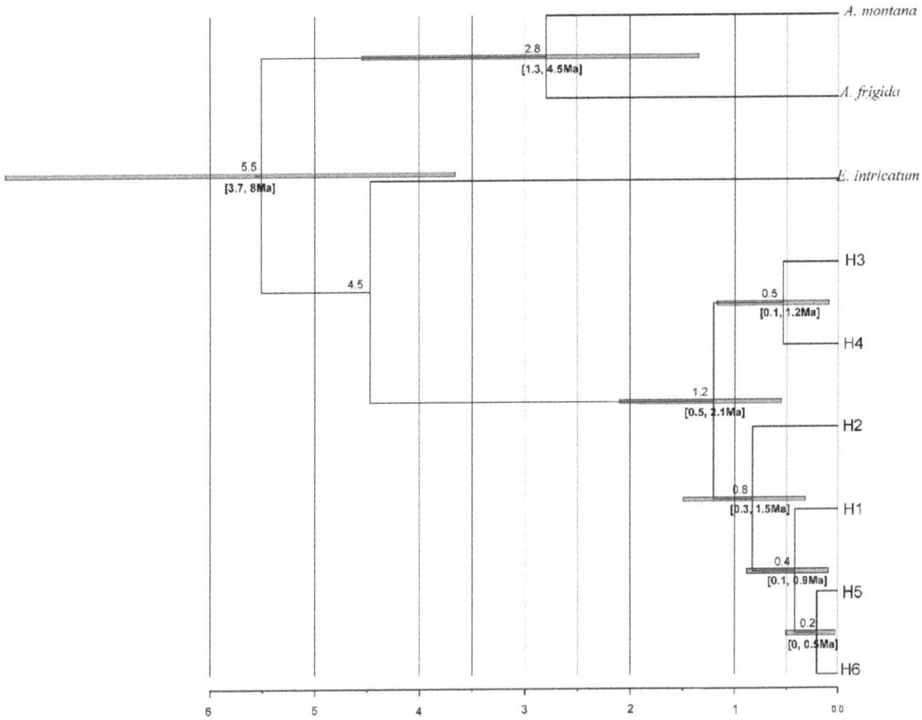

Figure 5. Bayesian divergence time estimates of *S. centiflora* based on the combined cpDNA data from two plastid gene markers (*trn*Q-*rps*16 and *psb*A-*trn*H). Blue bars on the nodes indicate 95% posterior credibility intervals.

4. Discussion

4.1. Genetic Diversity and Genetic Differentiation

Our results demonstrate that the 28 populations of *S. centiflora* have a high level of total haplotype diversity (H_T = 0.820) (Table 1). High cpDNA diversity has also been reported in several other endemic or relict species in Northwest China (*J. sabina* with H_T = 0.577 [14], *R. soongarica* with H_T = 0.607 [7], *G. przewalskii* with H_T = 0.849 [25], and *N. sphaerocarpa* with H_T = 0.887 [8]). A possible explanation for the high diversity in *S. centiflora* is its long evolutionary history, which may have allowed the accumulation of genetic variation. Moreover, the wide geographical range of this species across a geologically dynamic region provides ample opportunities for isolation, drift, and mutation.

A high level of cpDNA diversity was detected at the species level among populations (G_{ST} = 0.976); however, it was low (H_S = 0.020) within populations. Yet, the G_{ST} (0.976) of *S. centiflora* is higher than that of other angiosperm species (mean value of G_{ST} = 0.637 [59]). Low genetic diversity within populations and high genetic differentiation among populations indicate a lack of gene flow between populations for different reasons. Firstly, *S. centiflora* may have undergone long-term habitat fragmentation and geographic isolation among populations. This possibility has been interpreted as a consequence of strong bottlenecks or genetic drift associated with small effective population sizes for maternally inherited markers [60]. Secondly, high genetic differentiation among populations may also be due to limited gene flow through seeds associated either with geographical distance or mating system (either self-fertilization or clonal propagation, or both) [61]. This clearly applies for *S. centiflora,* which is hindered by geographic separation and physical barriers between deserts. In this manner, a high biologic heterogeneity may arise in desert landscapes. AMOVA revealed a marked differentiation between populations from different desert regions. Network and phylogenetic analyses also showed that ecological boundaries between the deserts limited gene flow in the species.

4.2. Identifying a Contact Zone between Deserts

Genetic divergence between lineages indicates a history of isolation. It is possible that the glacial cycling of the Plio-Pleistocene likely had a profound influence on the biogeographical history of species in this region. The continually shifting climates associated with Quaternary glacial oscillations would have dramatically affected diversification patterns within *S. centiflora*. During the Plio-Pleistocene, the climatic conditions in the deserts and steppes of Northwest China continuously varied. Steppes currently separate deserts; e.g., the Ordos grassland/steppe is located between the Kubuqi and Mu Us deserts. However, during the last glacial maximum of the Quaternary (approximately 18,000 years ago), the desert expansion in Northwest China would have also promoted desert species expansion [7]. Data regarding *S. centiflora* growing in different deserts provide strong evidence for the complex evolutionary history of this species.

Recent genetic research on the patterns of Quaternary contraction and expansion of many forest species in East China during glacial cycling has identified many refugia where multiple species survived the glacial maxima and contact zones where the same species from the different refugia met in the postglacial periods [62,63]. How can refugia and contact zones be distinguished? Generally, refugia have been identified by high levels of genetic diversity and private haplotypes within species. These haplotypes may not participate in the recolonization process and cannot be found elsewhere. Most importantly, haplotypes in refugia often have relatively close genetic relationships. However, haplotypes in the contact region genetically diverge. According to the phylogeny tree and haplotype network of cpDNA, the six chloroplast haplotypes found in our study can be divided into four lineages (I, II, III, IV) with high bootstrap supports (Figures 1A,B and 2). However, we could not detect any phylogeographic structure in N_{ST}/G_{ST} (Table 2), suggesting that the lineage admixture may exist. We found high genetic diversity, many haplotypes and ribotypes, and the occurrence of heterozygosity in the zones between the Tengger and Ulan Buh deserts. In addition, divergent haplotypes from Lineages III and IV are sympatrically distributed in these regions, suggesting that the region between the Tengger and Ulan Buh deserts is a contact zone into which different lineages dispersed from multiple refugia. Interestingly, Lineages II and III appear to be distributed strictly in an allopatric fashion; i.e., there was no contact zone between the two deserts. The most likely reason is that the impact of Quaternary climate change differed between the eastern (Kubuqi and Mu Us deserts) and western deserts (Badain Jaran, Tengger, and Ulan Buh deserts) [17]. The development and enlargement of mobile sand dunes were conducive to *S. centiflora* expansion, which resulted in haplotype sharing between the western deserts. However, the eastern deserts were less affected by Quaternary climate change than the western deserts. Therefore, semifixed or fixed sand dunes in the eastern deserts are not conducive to species expansion. Additional research must be conducted in the central Ordos area between the Kubuqi and Mu Us deserts to confirm this hypothesis and determine whether these two lineages come into contact.

4.3. nrDNA and cpDNA Variation

For *S. centiflora*, due to the arid desert environment, long distances between the five deserts, and consequent limitations on seed-mediated gene flow and pollinator movement, population discontinuities should be described by significant nrDNA and cpDNA variation. However, only cpDNA variation reflected significant genetic barriers. For the nrDNA ITS data, the AMOVA analysis showed that variation within populations was greater than that among populations (Table 3). The permutation test also revealed a significantly higher N_{ST} value than G_{ST} value in the ITS data but not in the cpDNA data (Table 2). The cpDNA data showed no shared haplotypes among the four groups or phylogeographical structure, which likely occurred because of the discontinuous distribution of common haplotypes (e.g., H1 and H5) across different geographic regions; however, nrDNA had both. The inconsistency between the two datasets may be due to their different modes of inheritance and dispersal among populations (e.g., maternal vs. bipaternal, seeds vs. both seeds and pollen), faster substitution rates, and concerted evolution of nrITS sequences [64,65]. Therefore, nrDNA ribotypes represent gene flow by both seeds and pollen in *S. centiflora*, whereas the cpDNA haplotypes represent gene flow by seeds, which are more restricted among the four groups [58,66]. Our data suggest limited seed-mediated gene flow, but extensive pollen movement between desert ranges. From our result, we conclude that the ecological transition regions between deserts, as soft barriers, play distinct roles in limiting the gene flow when considering seed vs. pollen.

4.4. Relationship between Allopatric Divergence of S. centiflora *and the Evolution of Deserts*

Many studies had shown that the dramatic climate events of the Pleistocene era were critical to the formation and development of desert flora in Northwest China [7,8,25,67]. It is especially attractive to focus on the Pleistocene's glacial history and the formation of sandy habitats as key factors promoting adaptive divergence *S. centiflora*, which is endemic to sand dunes. Six cpDNA haplotypes were found in the 28 populations across the entire geographic distribution of *S. centiflora*. These haplotypes uniquely belonged to four distinct lineages in the phylogenetic tree (Figure 2), which is consistent with their respective distribution in four separate regions: Badain Jaran–Tengger, Ulan Buh, Kubuqi, and Mu Us deserts. Although this species shows allopatric distribution patterns, it is less geologically isolated. Interestingly, populations from four regions of *S. centiflora* in Northwest China showed significant genetic differentiation, especially at the cpDNA marker, with different chloroplast haplotypes fixed in each region. Similar allopatric divergence and fragmentation have been reported for *J. sabina* [14], *G. przewalskii* [25], and *N. sphaerocarpa* [8] in Northwest China.

We hypothesized that the five sand deserts in our study originated during the Quaternary period [16,17]. The sand deserts developed due to the mobilization of paleo-eolian sand, which likely occurred due to climate cooling and aridification driven by the marked Kun-Huang Movement (1.2, 0.8, 0.6 Ma) of QTP lift during the Quaternary [68], and climate oscillation during the Naynayxungla glacial period (1.2–0.6 Ma) [6,69,70]. Evidence from eolian sand, sediments, and fossil pollen indicate that the majority of the deserts' formation and expansion occurred at roughly 1.0 Ma [16,17]. Regarding the relationship between haplotype distribution and deserts, the history of desert development serves as a background and key to understanding the evolution of *S. centiflora*. Our molecular clock results (Figure 5) dated the separation of *S. centiflora* clades at 0.2~1.2 Ma during the Pleistocene. The timing entirely coincides with the known formation and development of the deserts, which implies that the evolution of *S. centiflora* may have been synchronous with their formation. The haplotypic genetic traits can be inferred to have originated in conjunction with the formation and development of the deserts since 1.8 Ma. In particular, starting from 1.2 Ma during the Quaternary, the Kun-Huang Movement of QTP lift and climate oscillation during the Naynayxungla glacial period were possible evolutionary driving forces. Geological and climate change events played important roles in sand desert formation in Northwest China because of lowered temperatures and increased aridity. Although Northwest China remained unglaciated, this period of intense climatic instability might have caused the fragmentation of species distributions, due to reductions in suitable habitats within the region, consequently leading to genetic differentiation among isolated region populations. In addition, gene flow among *S. centiflora* populations would have been readily interrupted due to its low seed dispersal capacity and the ecological barriers between deserts. Several animal and plant species have been studied that exhibited intraspecific diversifications that partly correspond with regional uplift and Quaternary glaciations of the Middle and Late Pleistocene [71–73].

5. Conclusions

The present phylogeographic study of *S. centiflora* based on cpDNA and ITS sequences reveals the influence of complex geological and climatic events on patterns of diversification and distribution in this endemic desert species. The major diversification of *S. centiflora* that occurred 1.2 Ma is likely related to geological movement and climate change. Furthermore, Pleistocene glacial cycles may have been important for structuring the four lineages found and creating the geographical contact zone between the WLBH and TGL deserts. Our results highlight a substantial contribution to our knowledge of past vegetation and climate dynamics in the arid land in Northwest China. However, further phylogeographic studies of *S. centiflora* are needed using whole genome sequence or genome-wide markers to address the conflicting results that we reported based on cpDNA and ITS sequences.

Supplementary Materials: The following supporting information can be downloaded at: https://www.mdpi.com/article/10.3390/d14020104/s1, Figure S1: Map of sites for the species *Stilpnolepis centiflora* in the arid lands of Northwest China. Location details are given in Table 1 and the reference [37], Figure S2: Correlation between the F statistics and grouping number (K = 2–14) from the SAMOVA results based on cpDNA, Figure S3: Inference of K, the most probable number of clusters, and the proportion of genetic clusters, using Structure software. Analysis was performed based on cpDNA analysis of 280 samples of *Stilpnolepis centiflora*. (a) Proportion of genetic clusters at K = 2 and K = 4 for each of the 280 *Stilpnolepis centiflora* individuals. The smallest vertical bar represents one individual. The assignment proportion of each individual into population clusters is shown along the *y*-axis. (b) The first analysis was conducted based on cpDNA analysis of 280 samples of *Stilpnolepis centiflora*, at K = 2; the log-likelihood value of the data (Delta K) as a function of K was calculated over ten replicates. (c) The second analysis was conducted excluding populations from the TGL-BDJL group, at K = 3; the log-likelihood value of the data (Delta K) as a function of K was calculated over ten replicates. Figure S4: Scatterplots representing relationships between genetic distance and geographic distance of the species based on cpDNA (A) and nrDNA (B), Figure S5: Phylogenetic NJ-tree of ribotypes, Table S1: Summary of 28 populations used in the present study, including voucher information, Table S2: The haplotype sequences of the two chloroplast DNA fragments of 280 individuals from 28 populations of *S. centiflora* from five deserts in Northwest China used in analysis with corresponding GenBank reference numbers for each haplotype, Table S3: The ribotype sequences based on the ITS fragment from 250 individuals of *S. centiflora* from five deserts in Northwest China used in analysis with corresponding GenBank reference numbers for each ribotype, Table S4: Summary of the six haplotypes (H1–H6) based on aligned sequences of the two chloroplast DNA fragments of 280 individuals and 28 populations of *S. centiflora* from five deserts in Northwest China, Table S5: Summary of the 31 ribotypes (R1–R31) based on aligned sequences of ITS fragments from 250 individuals and 28 populations of *S. centiflora* sampled from five deserts in Northwest China.

Author Contributions: Conceptualization, X.S.; methodology, X.S., X.J. and H.Z.; investigation, data curation, X.S., X.J. and J.Q.; formal analysis, X.S. and X.J.; resources, X.S., X.J., H.Z. and J.Q.; writing—original draft preparation, X.S.; writing—review and editing, X.S., X.J., H.Z. and J.Q. All authors have read and agreed to the published version of the manuscript.

Funding: This research was financially supported by the National Natural Science Foundation of China (31960229, 31760139, 31760060) and Excellent Youth Foundation of Natural Science Foundation of Xinjiang Uygur Autonomous Region of China (2021D01E21).

Institutional Review Board Statement: Not applicable.

Data Availability Statement: Sequences are available on GenBank (see Supplementary Tables S2 and S3).

Acknowledgments: We are grateful to Carol C. Baskin, Jerry M. Baskin, Stewart C. Sanderson and the anonymous referee for their helpful comments and careful English correction.

Conflicts of Interest: The authors declare no competing interests.

References

1. Manel, S.; Holderegger, R. Ten years of landscape genetics. *Trends Ecol. Evol.* **2013**, *28*, 614–621. [CrossRef] [PubMed]
2. Avise, J. *Phylogeography: The History and Formation of Species*; Harvard University Press: Cambridge, MA, USA, 2000.
3. Hewitt, G.M. The genetic legacy of the Quaternary ice ages. *Nature* **2000**, *405*, 907–913. [CrossRef] [PubMed]
4. Hewitt, G.M. Genetic consequences of climatic oscillations in the Quaternary. *Philos. Trans. R. Soc. B Biol. Sci.* **2004**, *359*, 183–195. [CrossRef]
5. Wang, S.M.; Zhang, X.; Li, Y.; Zhang, L.; Xiong, Y.C.; Wang, G. Spatial distribution patterns of the soil seed bank of *Stipagrostis pennata* (Trin.) de Winter in the Gurbantonggut Desert of north-west China. *J. Arid. Environ.* **2005**, *63*, 203–222. [CrossRef]
6. Zheng, B.; Xu, Q.; Shen, Y. The relationship between climate change and Quaternary glacial cycles on the Qinghai–Tibetan Plateau: Review and speculation. *Quat. Int.* **2002**, *97–98*, 93–101. [CrossRef]
7. Li, Z.-H.; Chen, J.; Zhao, G.-F.; Guo, Y.-P.; Kou, Y.-X.; Ma, Y.-Z.; Wang, G.; Ma, X.-F. Response of a desert shrub to past geological and climatic change: A phylogeographic study of *Reaumuria soongarica* (Tamaricaceae) in western China. *J. Syst. Evol.* **2012**, *50*, 351–361. [CrossRef]
8. Su, Z.; Zhang, M. Evolutionary response to Quaternary climate aridification and oscillations in north-western China revealed by chloroplast phylogeography of the desert shrub *Nitraria sphaerocarpa* (Nitrariaceae). *Biol. J. Linn. Soc.* **2013**, *109*, 757–770. [CrossRef]
9. Riddle, B.R.; Honeycutt, R.L. Historical biogeography in North-American arid regions—An approach using mitochondrial-dna phylogeny in grasshopper mice (genus *Onychomys*). *Evolution* **1990**, *44*, 1–15.
10. Riddle, B.R.; Hafner, D.J.; Alexander, L.F. Comparative Phylogeography of Baileys' Pocket Mouse (Chaetodipus baileyi) and the Peromyscus eremicus Species Group: Historical Vicariance of the Baja California Peninsular Desert. *Mol. Phylogenetics Evol.* **2000**, *17*, 161–172. [CrossRef]
11. Riddle, B.R.; Hafner, D.J.; Alexander, L.F. Phylogeography and Systematics of the Peromyscus eremicus Species Group and the Historical Biogeography of North American Warm Regional Deserts. *Mol. Phylogenetics Evol.* **2000**, *17*, 145–160. [CrossRef]
12. Leache, A.D.; Mulcahy, D.G. Phylogeny, divergence times and species limits of spiny lizards (Sceloporus magister species group) in western North American deserts and Baja California. *Mol. Ecol.* **2007**, *16*, 5216–5233. [CrossRef]
13. Jaeger, J.R.; Riddle, B.R.; Bradford, D.F. Cryptic Neogene vicariance and Quaternary dispersal of the red-spotted toad (*Bufo punctatus*): Insights on the evolution of North American warm desert biotas. *Mol. Ecol.* **2005**, *14*, 3033–3048. [CrossRef]
14. Guo, Y.-P.; Zhang, R.; Chen, C.-Y.; Zhou, D.-W.; Liu, J.-Q. Allopatric divergence and regional range expansion of *Juniperus sabina* in China. *J. Syst. Evol.* **2010**, *48*, 153–160. [CrossRef]
15. Dong, G.R.; Li, B.S.; Gao, S.Y.; Wu, Z.; Shao, Y.J. Discovery of Quaternary ancient eolian sands and its significance in the Ordos Plateau. *Chin. Sci. Bull.* **1983**, *4*, 998–1001.
16. Dong, G.R.; Chen, H.Z.; Jin, J.; Wang, G.Y. Resent advancing on Quaternary research in desert area, North China. *Prog. Geo.* **1991**, *6*, 29–32.
17. Dong, G.; Li, S.; Li, B.S.; Wang, Y.; Yan, M.C. A preliminary study on the formation and evolution of deserts in China. *J. Desert Res.* **1991**, *11*, 23–32.
18. Wang, Y.; Li, S.; Wang, J.H.; Yan, M.C. The uplift of the Qinghai-Xizang (Tibetan) Plateau and its effect on the formation and evolution of Chinese deserts. *Arid Zone Res.* **1996**, *13*, 20–24.
19. Zheng, H.B.; Powell, C.M.; Butcher, K.; Cao, J.J. Late Neogene loess deposition in southern Tarim Basin: Tectonic and palaeo-environmental implications. *Tectonophysics* **2003**, *375*, 49–59. [CrossRef]
20. Sun, J.; Zhang, L.; Deng, C.; Zhu, R. Evidence for enhanced aridity in the Tarim Basin of China since 5.3Ma. *Quat. Sci. Rev.* **2008**, *27*, 1012–1023. [CrossRef]
21. Fang, X.; Lü, L.; Yang, S.; Li, J.; An, Z.; Jiang, P.; Chen, X. Loess in Kunlun Mountains and its implications on desert development and Tibetan Plateau uplift in west China. *Sci. China Ser. D Earth Sci.* **2002**, *45*, 289–299. [CrossRef]
22. Sun, J. Source Regions and Formation of the Loess Sediments on the High Mountain Regions of Northwestern China. *Quat. Res.* **2002**, *58*, 341–351. [CrossRef]
23. Yang, X.; Rost, K.T.; Lehmkuhl, F.; Zhenda, Z.; Dodson, J. The evolution of dry lands in northern China and in the Republic of Mongolia since the Last Glacial Maximum. *Quat. Int.* **2004**, *118–119*, 69–85. [CrossRef]
24. Hoskin, C.J.; Higgie, M.; McDonald, K.R.; Moritz, C. Reinforcement drives rapid allopatric speciation. *Nature* **2005**, *437*, 1353–1356. [CrossRef]
25. Ma, S.; Zhang, M. Phylogeography and conservation genetics of the relic *Gymnocarpos przewalskii* (Caryophyllaceae) restricted to northwestern China. *Conserv. Genet.* **2012**, *13*, 1531–1541. [CrossRef]
26. Shi, Z.; Humphries, C.J.; Gilbert, M.G. Asteraceae (Compositae). In *Flora of China*; Wu, Z.Y., Raven, P.H., Eds.; Science Press: Beijing, China; Missouri Botanical Garden Press: St. Louis, MO, USA, 2011; Volume 20–21, pp. 758–759.
27. Zhao, Y.Z. The area and floristic grographic element of *Stilpnolepis centiflora*. *J. Inn. Mong. Univ.* **1996**, *27*, 662–663.
28. Pellicer, J.; Hidalgo, O.; Garcia, S.; Garnatje, T.; Korobkov, A.A.; Valles, J.; Martin, J. Palynological study of Ajania and related genera (Asteraceae, Anthemideae). *Bot. J. Linn. Soc.* **2009**, *161*, 171–189. [CrossRef]
29. Liu, Y.; Yang, S.-X.; Ji, P.-Z.; Gao, L.-Z. Phylogeography of *Camellia taliensis* (Theaceae) inferred from chloroplast and nuclear DNA: Insights into evolutionary history and conservation. *BMC Evol. Biol.* **2012**, *12*, 92. [CrossRef]

30. Turchetto-Zolet, A.C.; Cruz, F.; Vendramin, G.G.; Simon, M.F.; Salgueiro, F.; Margis-Pinheiro, M.; Margis, R. Large-scale phylogeography of the disjunct Neotropical tree species *Schizolobium parahyba* (Fabaceae-Caesalpinioideae). *Mol. Phylogenet Evol.* **2012**, *65*, 174–182. [CrossRef]
31. Cullings, K. Design and testing of a plant-specific PCR primer for ecological and evolutionary studies. *Mol. Ecol.* **1992**, *1*, 233–240. [CrossRef]
32. Shaw, J.; Lickey, E.B.; Beck, J.T.; Farmer, S.B.; Liu, W.; Miller, J.; Siripun, K.C.; Winder, C.T.; Schilling, E.E.; Small, R.L. The tortoise and the hare II: Relative utility of 21 noncoding chloroplast DNA sequences for phylogenetic analysis. *Am. J. Bot.* **2005**, *92*, 142–166. [CrossRef]
33. Shaw, J.; Lickey, E.B.; Schilling, E.E.; Small, R.L. Comparison of whole chloroplast genome sequences to choose noncoding regions for phylogenetic studies in angiosperms: The tortoise and the hare III. *Am. J. Bot.* **2007**, *94*, 275–288. [CrossRef] [PubMed]
34. White, T.J.; Burns, T.; Lee, S.; Taylor, J. *Amplification and Direct Sequencing of Fungal Ribosomal RNA Genes for Phylogenetics*; Academic Press: New York, NY, USA, 1990.
35. Thompson, J.D.; Gibson, T.J.; Plewniak, F.; Jeanmougin, F.; Higgins, D.G. The CLUSTAL_X windows interface: Flexible strategies for multiple sequence alignment aided by quality analysis tools. *Nucleic Acids Res.* **1997**, *25*, 4876–4882. [CrossRef] [PubMed]
36. Librado, P.; Rozas, J. DnaSP v5: A software for comprehensive analysis of DNA polymorphism data. *Bioinformatics* **2009**, *25*, 1451–1452. [CrossRef] [PubMed]
37. Garrick, R.C.; Sunnucks, P.; Dyer, R.J. Nuclear gene phylogeography using PHASE: Dealing with unresolved genotypes, lost alleles, and systematic bias in parameter estimation. *BMC Evol. Biol.* **2010**, *10*, 118. [CrossRef]
38. Tamura, K.; Stecher, G.; Kumar, S. MEGA11: Molecular Evolutionary Genetics Analysis Version 11. *Mol. Biol. Evol.* **2021**, *38*, 3022–3027. [CrossRef]
39. Kimura, M. A simple method for estimating evolutionary rates of base substitutions through comparative studies of nucleotide sequences. *J. Mol. Evol.* **1980**, *16*, 111–120. [CrossRef]
40. Bandelt, H.J.; Forster, P.; Rohl, A. Median-joining networks for inferring intraspecific phylogenies. *Mol. Biol. Evol.* **1999**, *16*, 37–48. [CrossRef]
41. Cassens, I.; Mardulyn, P.; Milinkovitch, M.C. Evaluating intraspecific "Network" construction methods using simulated se-quence data: Do existing algorithms outperform the global maximum parsimony approach? *Syst. Biol.* **2005**, *54*, 363–372. [CrossRef]
42. Excoffier, L.; Lischer, H.E.L. Arlequin suite ver 3.5: A new series of programs to perform population genetics analyses under Linux and Windows. *Mol. Ecol. Resour.* **2010**, *10*, 564–567. [CrossRef]
43. Miller, M.P. Alleles In Space (AIS): Computer software for the joint analysis of interindividual spatial and genetic information. *J. Hered.* **2005**, *96*, 722–724. [CrossRef]
44. Pons, O.; Petit, R.J. Measuring and Testing Genetic Differentiation with Ordered versus Unordered Alleles. *Genetics* **1996**, *144*, 1237–1245. [CrossRef]
45. Dupanloup, I.; Schneider, S.; Excoffier, L. A simulated annealing approach to define the genetic structure of populations. *Mol. Ecol.* **2002**, *11*, 2571–2581. [CrossRef]
46. Evanno, G.; Regnaut, S.; Goudet, J. Detecting the number of clusters of individuals using the software STRUCTURE: A simulation study. *Mol. Ecol.* **2005**, *14*, 2611–2620. [CrossRef]
47. Pritchard, J.K.; Stephens, M.; Donnelly, P. Inference of population structure using multilocus genotype data. *Genetics* **2000**, *155*, 945–959. [CrossRef]
48. Earl, D.A.; Vonholdt, B.M. Structure Harvester: A website and program for visualizing STRUCTURE output and implementing the Evanno method. *Conserv. Genet. Resour.* **2012**, *4*, 359–361. [CrossRef]
49. Excoffier, L.; Smouse, P.E.; Quattro, J.M. Analysis of Molecular Variance Inferred from Metric Distances among DNA Haplotypes—Application to Human Mitochondrial-DNA Restriction Data. *Genetics* **1992**, *131*, 479–491. [CrossRef]
50. Drummond, A.J.; Rambaut, A. BEAST: Bayesian evolutionary analysis by sampling trees. *BMC Evol. Biol.* **2007**, *7*, 1–8. [CrossRef]
51. Wang, J.-F.; Gong, X.; Chiang, Y.-C.; Kuroda, C. Phylogenetic patterns and disjunct distribution in *Ligularia hodgsonii* Hook. (Asteraceae). *J. Biogeogr.* **2013**, *40*, 1741–1754. [CrossRef]
52. Rambaut, A.; Drummond, A.J. Tracer v. 1.5. 2009. Available online: http://tree.bio.ed.ac.uk/software/tracer/ (accessed on 25 March 2014).
53. Rambaut, A. FigTree v. 1.3.1. 2009. Available online: http://tree.bio.ed.ac.uk/software/FigTree/ (accessed on 5 May 2012).
54. Tajima, F. Statistical method for testing the neutral mutation hypothesis by DNA polymorphism. *Genetics* **1989**, *123*, 585–595. [CrossRef]
55. Fu, Y.-X. Statistical Tests of Neutrality of Mutations Against Population Growth, Hitchhiking and Background Selection. *Genetics* **1997**, *147*, 915–925. [CrossRef]
56. Slatkin, M.; Hudson, R.R. Pairwise comparisons of mitochondrial DNA sequences in stable and exponentially growing populations. *Genetics* **1991**, *129*, 555–562. [CrossRef]
57. Schneider, S.; Excoffier, L. Estimation of past demographic parameters from the distribution of pairwise differences when the mutation rates very among sites. Application to human mitochondrial DNA. *Genetics* **1999**, *152*, 1079–1089. [CrossRef]
58. Wolfe, K.H.; Li, W.-H.; Sharp, P.M. Rates of nucleotide substitution vary greatly among plant mitochondrial, chloroplast, and nuclear DNAs. *Proc. Natl. Acad. Sci. USA* **1987**, *84*, 9054–9058. [CrossRef]

59. Petit, R.J.; Duminil, J.; Fineschi, S.; Hampe, A.; Salvini, D.; Vendramin, G.G. Comparative organization of chloroplast, mitochondrial and nuclear diversity in plant populations. *Mol. Ecol.* **2005**, *14*, 689–701. [CrossRef]
60. Birky, C.W., Jr.; Fuerst, P.; Maruyama, T. Organelle gene diversity under migration, mutation, and drift: Equilibrium expectations, approach to equilibrium, effects of heteroplasmic cells, and comparison to nuclear genes. *Genetics* **1989**, *121*, 613–627. [CrossRef]
61. Petit, R.J.; Aguinagalde, I.; de Beaulieu, J.-L.; Bittkau, C.; Brewer, S.; Cheddadi, R.; Ennos, R.; Fineschi, S.; Grivet, D.; Lascoux, M.; et al. Glacial Refugia: Hotspots but Not Melting Pots of Genetic Diversity. *Science* **2003**, *300*, 1563–1565. [CrossRef]
62. Qiu, Y.-X.; Fu, C.-X.; Comes, H.P. Plant molecular phylogeography in China and adjacent regions: Tracing the genetic imprints of Quaternary climate and environmental change in the world's most diverse temperate flora. *Mol. Phylogenetics Evol.* **2011**, *59*, 225–244. [CrossRef]
63. Liu, J.Q.; Sun, Y.S.; Ge, X.J.; Gao, L.M.; Qiu, Y.X. Phylogeographic studies of plants in China: Advances in the past and directions in the future. *J. Syst. Evol.* **2012**, *50*, 267–275. [CrossRef]
64. Alvarez, I.; Wendel, J.F. Ribosomal ITS sequences and plant phylogenetic inference. *Mol. Phylogenet Evol.* **2003**, *29*, 417–434. [CrossRef]
65. Lorenz-Lemke, A.P.; Muschner, V.C.; Bonatto, S.L.; Cervi, A.C.; Salzano, F.M.; Freitas, L.B. Phylogeographic inferences concerning evolution of Brazilian *Passiflora actinia* and *P. elegans* (Passifloraceae) based on ITS (nrDNA) variation. *Ann. Bot.* **2005**, *95*, 799–806. [CrossRef] [PubMed]
66. Schaal, B.A.; Hayworth, D.A.; Olsen, K.M.; Rauscher, J.T.; Smith, W.A. Phylogeographic studies in plants: Problems and pro-spects. *Mol. Ecol.* **1998**, *7*, 465–474. [CrossRef]
67. Meng, H.-H.; Zhang, M.-L. Diversification of plant species in arid Northwest China: Species-level phylogeographical history of Lagochilus Bunge ex Bentham (Lamiaceae). *Mol. Phylogenetics Evol.* **2013**, *68*, 398–409. [CrossRef] [PubMed]
68. Li, J.J.; Shi, Y.F.; Li, B.Y. *Uplift of the Qinghai-Xizang (Tibet) Plateau and Global Change Lanzhou*; Lanzhou University Press: Lanzhou, China, 1995.
69. Shi, Y.F.; Zheng, B.X.; Li, S.J. Last glaciation and maximum glaciation in Qinghai-Xizang (Tibet) plateau. *J. Glaciol. Geocryol.* **1990**, *12*, 1–15.
70. Zhou, S.; Li, J. The sequence of Quaternary glaciation in the Bayan Har Mountains. *Quat. Int.* **1998**, *45–46*, 135–142. [CrossRef]
71. Yang, S.-J.; Yin, Z.-H.; Ma, X.-M.; Lei, F.-M. Phylogeography of ground tit (*Pseudopodoces humilis*) based on mtDNA: Evidence of past fragmentation on the Tibetan Plateau. *Mol. Phylogenetics Evol.* **2006**, *41*, 257–265. [CrossRef]
72. Qi, D.; Guo, S.; Zhao, X.; Yang, J.; Tang, W. Genetic diversity and historical population structure of *Schizopygopsis pylzovi* (Teleostei: Cyprinidae) in the Qinghai?Tibetan Plateau. *Freshw. Biol.* **2007**, *52*, 1090–1104. [CrossRef]
73. Jin, Y.T.; Brown, R.P.; Liu, N.F. Cladogenesis and phylogeography of the lizard *Phrynocephalus vlangalii* (Agamidae) on the Tibetan plateau (vol 17, pg 1971, 2008). *Mol. Ecol.* **2008**, *17*, 3033–3034.

diversity

MDPI

Article

Species Abundance Distributions Patterns between Tiankeng Forests and Nearby Non-Tiankeng Forests in Southwest China

Linjuan Huang [1,2,3,†], Hao Yang [1,2,3,†], Xiaofei An [1,2,3], Yanmei Yu [1,2,3], Linlan Yu [1,2,3], Gui Huang [1,2,3], Xinyu Liu [1,2,3], Ming Chen [1,2,3] and Yuegui Xue [1,2,3,*]

1 Key Laboratory of Ecology of Rare and Endangered Species and Environmental Protection (Ministry of Education), Guangxi Normal University, Guilin 541004, China; hglinjuan@163.com (L.H.); yanghao_henry@gxnu.edu.cn (H.Y.); anxiaofei2021@163.com (X.A.); yuyanmei2021@163.com (Y.Y.); yulinlan119@163.com (L.Y.); hgisgood@163.com (G.H.); liuxinyu_0806@163.com (X.L.); chenming087674@163.com (M.C.)
2 Guangxi Key Laboratory of Landscape Resources Conservation and Sustainable Utilization in Lijiang River Basin, Guangxi Normal University, Guilin 541004, China
3 College of Life Sciences, Guangxi Normal University, Guilin 541004, China
* Correspondence: xueyuegui@126.com
† These authors contributed equally to this work.

Abstract: Identifying the species abundance distributions (SADs) in Tiankeng forests is crucial for restoring and managing degraded karst ecosystem, whereas previous studies rarely explored the differences and response of vegetation dynamics to environmental variations. The species composition and SADs of the inner and outer fringe areas of Tiankeng forest and nearby non-Tiankeng forest were compared in Southwest China. Six models were adopted to compare SADs of three habitats. Kolmogrov–Smirnov (K–S) test was selected to compare the discrepancy between the simulated and observed SAD patterns. The Akaike Information Criterion (AIC) test was adopted to compare the models, and the best model was indicated by the lowest AIC value. The results showed that (1) the species dispersal from the inside of Tiankeng forests to the nearby non-Tiankeng forests is limited, while species have unlimited dispersal from nearby non-Tiankeng forests to the inside of Tiankeng forests via the fringe of Tiankeng forests. (2) Species abundance, species rarity, richness, and species accumulation rate in the Tiankeng forests were significant in non-Tiankeng forests ($p < 0.05$), and most species in inner Tiankeng forests originated from nearby non-Tiankeng forests. (3) Based on the criterion of K-S values, all models have passed the K–S test ($p > 0.05$), which indicated that niche processes and neutral process worked together in the maintenance of community species diversity, the community in study area is a niche-neutral continuum. (4) Considered the lowest AIC value, the neutral ($\triangle_{mean}$ AIC = 1.3) models performed better than the niche ($\triangle_{mean}$ AIC = 22.7) models and statistical ($\triangle_{mean}$ AIC = 2.7) in the Tiankeng forest, while the statistical models performed better than the niche and neutral models in the non-Tiankeng forests. The results suggested that the main driving force of Tiankeng forests is the neutral process. The negative terrain in Tiankeng restricted the species dispersal due to topographic constraints. However, the species dispersal from the nearby non-Tiankeng forests could promote the species succession in the inner Tiankeng. Therefore, we propose that nearby non-Tiankeng forests should be emphasized for protecting the biodiversity of Tiankeng forests.

Keywords: species abundance distribution (SAD); neutral process; Tiankeng forests; negative terrain; plant refuge

Citation: Huang, L.; Yang, H.; An, X.; Yu, Y.; Yu, L.; Huang, G.; Liu, X.; Chen, M.; Xue, Y. Species Abundance Distributions Patterns between Tiankeng Forests and Nearby Non-Tiankeng Forests in Southwest China. *Diversity* **2022**, *14*, 64. https://doi.org/10.3390/d14020064

Academic Editors: Hong-Hu Meng and Yi-Gang Song

Received: 28 November 2021
Accepted: 12 January 2022
Published: 19 January 2022

1. Introduction

Understanding the mechanism for the maintenance of species diversity in communities has become an issue of ecological concern [1]. At present, two major theories have been advanced to explain the rules of community assembly mechanisms. Niche theory proposes that species diversity is affected by deterministic processes, such as environment filtering, competition, and exclusion within species [2–4], while neutral theory emphasizes the effect of random processes, such as diffusion limitation on the composition and distribution of species, distance decay, and species turnover rate [5–7]. Species abundance distribution (SAD), the basic proportional abundance of species combined in an ecological community, is one of the important ways to encapsulate the characteristics of community diversity [8,9].

Since the 1940s, all kinds of models have been used to verify mechanistic rules to explain the structure of ecological communities, and reveal the effects of various ecological processes in SADs patterns [10,11]. There are three types of models, namely statistical, niche, and neutral. Statistical models, such as Fisher's log-series and Preston log-normal, describe the complexity of community composition using mathematics [12,13]. Niche models, such as the Broken stick and Niche preemption, focus on the connection of niche theory and SAD to infer species competition or repulsion [14–16]. While the neutral models, such as Zero-sum polynomials [16,17] and Volkov model [18], are used in testing for random species abundance patterns and emphasizing random effects on community construction [8]. Different models can only demonstrate a part of community ecological process. Therefore, many scholars mostly adopt a combination of multiple models to analyze the SAD pattern in different objects, including alpine meadows [19], tropical forests [20], subtropical evergreen-deciduous broadleaved mixed forests [21]. However, the current research on forest communities of SAD models in negative terrain habitats is still unclear.

Tiankeng is a kind of karst negative terrain with a huge volume [22,23]. The Tiankeng forest is located in the Tiankeng. The Tiankeng forest is divided into two areas and each have their own unique features. At the top of Tiankeng is the fringe area, a complex ecotone between the nearby non-Tiankeng forest and Tiankeng forests was influenced by some human disturbance. The Tiankeng forest interior is located on the negative terrain of the Tiankeng and is surrounded by the fringe of Tiankeng forest. Here, the topography is steeped and sloped, and there are adequate hydrothermal conditions and less human interference [23,24]. The nearby non-Tiankeng forest, located in a positive terrain, is a well-preserved karst forest near Tiankeng [25–27]. From an existing vegetation distribution perspective, the non-Tiankeng forest vegetation reflects the adaptation of species to limestone geology and human disturbance, while the Tiankeng forest reflects species independent evolution in a natural negative terrain [27,28]. At present, most studies on Tiankeng forests focus on geological features [29], landscape value [22], biodiversity [24], plant community structure [23], interspecific association [30], functional traits, and species adaption to the environment [25]. However, only a few studies have focused on the effects of nearby non-Tiankeng forests, especially for the species dispersal or invasion [24,27].

The Tiankeng and nearby non-Tiankeng forests are unique vegetation distributed on karst landforms. Combining the SADs pattern in Tiankeng forests and nearby non-Tiankeng forests not only benefits us to reveal the maintenance mechanism of species diversity in fragmented forests, but also helps us to suggest vegetation restoration and reconstruction in ecologically sensitive areas. In this study, we selected three habitats, the inside of Tiankeng forests, the fringe of Tiankeng forests and the nearby non-Tiankeng forests as the study area A total of six sampling plots were established at a scale of 40 m × 40 m. Based on the composition characteristics of woody plant, the variations in important value of dominant species, species empirical cumulative distribution functions (ECDF) curve, species-abundance curve, Jaccard similarity index, and six SAD's models fitting (Fisher's log-series, Preston log-normal, Broken stick, Niche preemption, Zero-sum polynomial, and Volkov model) have been utilized to address the following questions for three habitats (the inside of Tiankeng forests, the fringe of Tiankeng forests, and the nearby non-Tiankeng forests): (1) What are their difference in species composition; (2) What are their SADs patterns; (3) What are their main driving forces for the maintenance of community diversity?

2. Materials and Methods

2.1. Study Area

This study was carried out in the Dashiwei Tiankeng Group branch, located at 24°30′ N to 25°03′ N and 106°10′ E to 106°51′ E, in Leye, Guangxi, southwest China. The study area has a big difference in elevation, higher in the southwest (Guizhou plateau) and lower in east, north, and west (Guangxi basin). The local climate is mid-subtropical monsoon. There is a rainy season from May to October. The annual average rainfall is 1400mm, rainfall accounts for 85% of the mean precipitation, annual relative humidity is 85%, and mean average temperature is 16.6–23.0 °C. Hydrological conditions are superior, with the first-level tributary of the Hongshui river–Bailang underground river system. In the middle of the "S"-shaped fold of the Bailang underground river, the river channels are intricate and unique, forming many special peak clusters, depressions, and Tiankeng negative terrain. Soils are developed from limestone weathering and are alkaline acidic clay. The Tiankeng forest was distributed mainly on the Tiankeng negative terrain. The study area is evergreen and deciduous broad-leaved mixed forest with average altitude of 1200~1266 m [22–25,27].

2.2. Survey and Sampling

To explore SADs between Tiankeng and nearby non-Tiankeng forest, a reconnaissance survey and vegetation data collection in Dashiwei Tiankeng Group was carried out from 10 August to 20 November 2019. Because most of the Tiankeng forests are steep and difficult to reach, the number of plots in each area was determined by the size of forests, community structure, vegetation, and topography conditions [25]. Sampling methods followed Ma [8] and Zhang [31]. Three habitat types (the inside of Tiankeng forest, the fringe of Tiankeng forest, and the nearby non-Tiankeng forest) from six sampling sites were selected within the study area based on differences in location of phytocoenoses in Tiankeng. Furthermore, the study area for each type of habitat was 3200 m^2, and each sampling site was divided into 16 plots at a scale of 10 m × 10 m.

The inside of Tiankeng forest includes Liuxing Tiankeng (LX) and Dacao Tiankeng (DC). Liuxing Tiankeng is of 890 m × 320 m and a depth of 210–290 m, which is the second largest Tiankeng in the Dashiwei group, and Dacao Tiankeng is a general Tiankeng of 250 m × 140 m with a depth of 62–108 m. In the inner area of both Tiankengs are reachable, natural, and contain abundant species [22]. The fringe of Tiankeng forest contains the Luojia (LJ) and Shenmu Tiankeng (SM). The Luojia Tiankeng is 140 m × 100 m and 71–128 m deep, and Shenmu Tiankeng is 300 m × 270 m with a depth of 186–234 m. Between Luojia and Shenmu Tiankeng of the outer fringe area, there is human interference and rich species diversity [22]. The nearby non-Tiankeng also includes two sampling sites, one (LA) is near the village and other one (SW) is near the farmland; between them are well protected karst forests with complex communities. For our research, plot names are coded firstly by location (see above), and secondly as either inside of Tiankeng forest (-TK), in the fringe of Tiankeng forest (-BY), or the nearby karst non-Tiankeng forest (-FS) (Figure 1). In each quadrat, the abundance, diameter at breast height, basal diameter, and cover of all woody species with a diameter at breast height ≥1 cm were recorded.

2.3. Data Analysis

All the original data were recorded and preserved in the key laboratory of ecology of rare endangered species and environmental protection, Guangxi Normal University, in order to preliminarily identify the plant scientific names [31]. Plant family, genus, and species were identified by the online Flora of China (http://www.iplant.cn/foc/, accessed on 9 December 2021) and Wei et al. [32] for all the collected specimens. Species abundance comes from the number of individuals of each species recorded in each sampling site. The dominant species were determined by the higher importance value (P) [33]. Species importance value was calculated from the relative frequency (*RF*), relative abundance (*RA*), and relative dominance (*RD*) [34] as:

$$P = [(RF + RA + RD)/3] \times 100\% \quad (1)$$

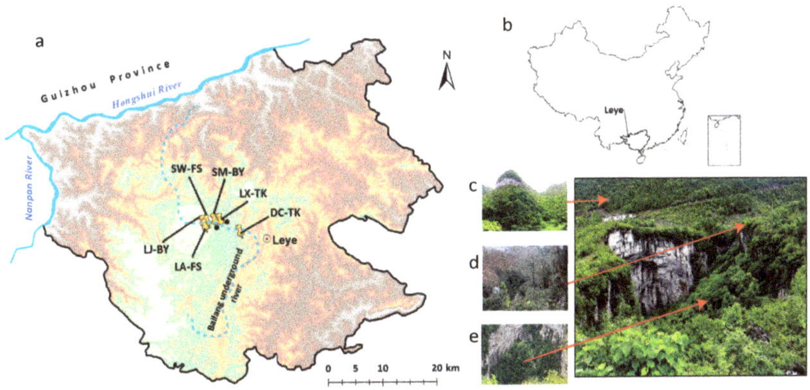

Figure 1. Map of the study area including the plot location. Sample sites in Dashiwei Tiankeng Group, Leye, China (**a**,**b**). Location of three habitats (**c**–**e**), the nearby non-Tiankeng forest (**c**), the fringe of Tiankeng forest (**d**), and the inside of Tiankeng forest (**e**). Maps are created with ArcMap 10.5. Data map source from the Ministry of natural resources, PRC (http://www.mnr.gov.cn/, figure number: GS(2016)1595, accessed on 23 November 2021), Geospatial data cloud, Chinese Academy of Sciences (http://www.gscloud.cn/, accessed on 23 November 2021).

We used two curves (species ECDF and species abundance curve) to visually observe the SADs pattern in six sampling sites, and a Kruskal–Wallis test (K–W) was used to compare the SAD's difference in the 'dplyr' packages of R software 4.0.0. (R Core Team, Vienna, Austria) [35,36].

The alpha diversity indices, such as species richness (S_p) and Shannon–Wiener index (H') were assessed for the species, the formula [37] is as follows:

$$S_P = S \quad (2)$$

$$H' = -\sum P_i ln P_i \quad (3)$$

where S represents the number of species and P_i represents the ratio of the abundance of the i-th species to the total abundance.

The beta diversity indices, such as Jaccard index (C'), were assessed for the species to compare the similarity in two communities. The formula [38] is as follows:

$$C' = c/(a + b - c) \quad (4)$$

where a and b are the number of species in the two communities and c is the number of species in common in the two communities.

2.4. Models Selection

Six models were selected to simulate the SAD pattern, including two statistical models (Fisher's log-series model and Log-normal model), two neutral models (Zero-sum polynomials model and Volkov model), and two niche models (Niche preemption model and Broken stick model) [1,8]. The details are as follows.

(1) Fisher's log-series model

This model was proposed by Fisher et al. [39]. It uses a negative binomial distribution to describe the relationship between the number of species and the individuals of species.

This model excludes species without individual numbers [40]. The calculation formula [39] is as follows:

$$S(n) = \alpha X^n/n \quad (5)$$

where α ($\alpha > 0$) is the species diversity index of the community, similar to species richness, n is species abundance, S is the number of species, and X is a constant (Tends to 1) [16].

(2) Log-normal model

This model assumes that the logarithm of the number of species in the community conforms to the normal distribution, and the abundance of the i-th species is recorded as A_i. The calculation formula [13] is as follows:

$$A_i = e^{log(\mu)+log(\varphi)}, i = 1,2,3, \ldots, S. \quad (6)$$

where μ is the mean of the normal distribution, δ is the standard deviation of the normal distribution, and φ is the normal deviation.

(3) Broken stick model

In this model, a broken stick with a length of 1 is represented as the total niche in the community, and it is given to S species [14]. The competitiveness and taxonomic status of each species are basically similar, and they exist in the community at the same time [15]. If the total number of individuals in the community is J, the abundance A_i of the i-th species is:

$$A_i = J/S, i = 1,2,3, \ldots, S. \quad (7)$$

(4) Niche preemption model

The model assumes that species 1 first occupies k parts of the niche in the community, and species 2 occupies the remaining $k(1 - k)^2$ parts, and so on, until the remaining resources cannot sustain the survival of another species [15]. If A_1 is the abundance of the most dominant species predicted by the model, the abundance A_i of the i-th species is:

$$A_i = A_1 (1 - k)^{i-1}, i = 1,2,3, \ldots, S. \quad (8)$$

(5) Zero-sum polynomials model

This model assumes that the species abundance distribution at a certain point is derived from the random drift of the neutral composite community. This model includes two parameters, the number of individual species (J) at the sampling point and the fundamental biodiversity number (θ) [17]. Therefore, the number S of species with an abundance of n at a sampling point in the composite community is expressed by the formula:

$$S(n) = \int_0^J f_{n,i}(y)(1 - y/J)^{\theta-1} dy \times \theta/n \quad (9)$$

$$f_{n,\delta} = exp(-y/\delta)y^{n-1}/\tau(n)\delta^n \quad (10)$$

(6) Volkov model

When the model fits the SAD of the community, compared to the composite community zero-sum polynomial model, the migration rate (m) is increased, and this migration coefficient is assumed to be constant from the composite community to the local community [18]. According to this model, the number S of species with an abundance of n in a local community can be expressed as:

$$S(n) = \theta[J!/n - (J - n)][\tau(\gamma)/\tau(J + \gamma)]exp(-y^\theta/)\gamma dy \int_0^\gamma \tau(n + y)\tau(J - n + \gamma - y)/\tau(1 + \gamma)\tau(\gamma - y) \quad (11)$$

$$\Gamma(Z) = \int_0^\infty t^{Z-1} e^{-t} \quad (12)$$

$$\gamma = m(J-1)/1-m \quad (13)$$

2.5. Fitting Tests for Models

We performed K-S test based on 999 times bootstrap simulation to verify the fitting results between the observed and simulation values from six models. When the $p < 0.05$, this test indicates that the model has not passed the test, otherwise, the model is accepted [41]. The goodness-of-fit of the model was determined by the lowest AIC value. If the AIC value of a model is more than 2 ($\triangle$AIC > 2) and lower than others, then this model is considered significantly better than other models [42]. These basic statistical analyses were conducted using the 'sads' and 'Matching' packages in R software 4.0.0. (R core team, Vienna, Austria) [36,42].

3. Results

3.1. Species Composition of Tiankeng Forests and Nearby Non-Tiankeng Forests

A total of 3599 individuals of woody plants were recorded in the study sites and the species composition of woody plant was rich and complex. There were 20 families, 21 genera, and 31 species in DC-TK; 24 families, 27 genera, and 32 species in LX-TK; 27 families, 27 genera, and 48 species in LJ-BY; 26 families, 26 genera, and 50 species in SM-BY; 30 families, 30 genera, and 47 species in LA-FS; and 30 families, 31 genera, and 41 species in SW-FS. If we compare the importance value of dominant species at different sites, the dominant species in the inner area of Tiankeng forests is rarely distributed in the nearby non-Tiankeng forests, while the dominant species nearby non-Tiankeng forests is frequently distributed in the inner and the fringe of Tiankeng forest (Table 1).

Table 1. The importance value of dominant species at different sites of Dashiwei Tiankeng Group.

Species	The Inside of Tiankeng Forests		The Fringe of Tiankeng Forests		Nearby Non-Tiankeng Forests	
	DC-TK	LX-TK	LJ-BY	SM-BY	LA-FS	SW-FS
Lindera glauca (Sieb. et Zucc.) Bl.	29.82%	9.27%	-	-	-	-
Schefflera guizhouensis C. B. Shang	21.44%	4.85%	-	-	-	-
Miliusa sinensis Finet et Gagnep.	11.29%	-	-	-	-	-
Machilus chinensis (Champ. ex Benth.) Hemsl.	10.69%	-	3.88%	-	-	-
Handeliodendron bodinieri (Lévl.) Rehd.	4.90%	12.32%	6.05%	2.18%	2.91%	3.38%
Choerospondias axillaris (Roxb.) B. L. Burtt & A. W. Hill	5.72%	10.97%	19.61%	-	3.81%	4.84%
Lindera pulcherrima var. *Hemsleyana* (Diels) H.P.Tsui	1.29%	9.90%	5.78%	-	2.97%	1.10%
Ilex macrocarpa Oliv.	-	-	10.83%	-	4.79%	9.99%
Itea macrophylla Wall. ex Roxb.	-	-	10.05%	-	-	-
Machilus glaucifolia S. K. Lee & F. N. Wei	-	-	8.97%	15.39%	-	-
Celtis sinensis Pers.	1.10%	8.61%	2.67%	-	11.70%	7.84%
Rhaphiolepis indica (Linnaeus) Lindley	-	1.73%	1.64%	-	3.15%	12.54%
Illicium simonsii Maxim.	-	-	-	-	10.82%	-
Cinnamomum bodinieri Lévl.	-	-	2.47%	-	1.05%	10.45%
Litsea rotundifolia var. *Oblongifolia* (Nees) Allen	-	1.62%	9.78%	-	1.26%	-
Itea yunnanensis Franch.	-	8.20%	8.82%	-	5.83%	9.61%

3.2. SADs Pattern in Tiankeng and Non-Tiankeng Forests

There was a significant difference in SADs patterns (Kruskal-Wallis chisq = 16.798, $p < 0.05$) between the Tiankeng and non-Tiankeng forests (Figure 2a). The species accumulation rate was highest in the fringe of Tiankeng forests and lowest in the inside of Tiankeng forests (Figure 2b). The Jaccard index showed that the species similarity between the fringe and inside of Tiankeng forests was the highest (0.36), followed by the non-Tiankeng forests, and the fringe of Tiankeng forests (0.29), and between the inside of Tiankeng forests and the non-Tiankeng forests was the lowest (0.27) (Figure 2c).

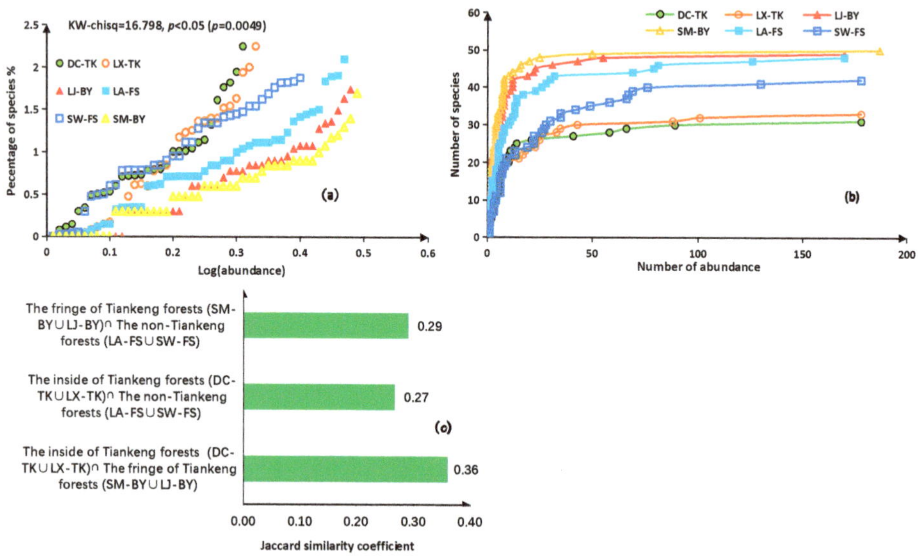

Figure 2. ECDF curve (**a**), species-abundance curve (**b**), and Jaccard similarity index (**c**) at different sites of Dashiwei Tiankeng Group.

3.3. Goodness-of Fit of SADs Models in Tiankeng and Non-Tiankeng Forests

We observed differences in the test values at the 40m×40m scale between forests with different plots. K-S tests indicated that there were no significant differences between the simulated and actual SAD values and all models have passed the K-S test ($p > 0.05$) between Tiankeng and non-Tiankeng forests (Figure 3). The AIC test of goodness-of-fit results showed that the neutral models ($\triangle_{mean}$ AIC = 1.3, AIC weights $_{mean}$ = 26.0%) performed better than the niche ($\triangle_{mean}$ AIC = 22.7, AIC weights $_{mean}$ = 0.1%) and statistical models ($\triangle_{mean}$ AIC = 2.7, AIC weights $_{mean}$ = 24.1%) in the Tiankeng forest, while the statistical models ($\triangle_{mean}$ AIC = 2.4, AIC weights $_{mean}$ = 20.0%) performed better than the niche ($\triangle_{mean}$ AIC = 11.2, AIC weights $_{mean}$ = 15.0%) and neutral ($\triangle_{mean}$ AIC = 3.5, AIC weights $_{mean}$ = 15.1%) models in the non-Tiankeng forests. According to the results of neutral theory parameters and the species diversity (Table 2), the fundamental biodiversity number (θ), species immigration rate (m), species richness, and Shannon–Wiener index was the highest in the fringe of the Tiankeng forest and the lowest was in the inside the Tiankeng forests.

Table 2. Neutral theory parameter and species diversity.

	Neutral Theory Parameter			Species Diversity Index	
	Volkov Model		Zero-Sum Polynomials Model		
Sites	θ	m	θ	Species richness	Shannon-Wiener index
DC-TK	8.31	0.35	6.71	31	3.81
LX-TK	7.3	0.4	7.24	32	3.64
LJ-BY	16.65	0.59	14.28	48	4.63
SM-BY	15.34	0.62	12.44	50	4.40
LA-FS	13.34	0.41	11.12	47	4.33
SW-FS	14.75	0.42	11.86	41	4.25

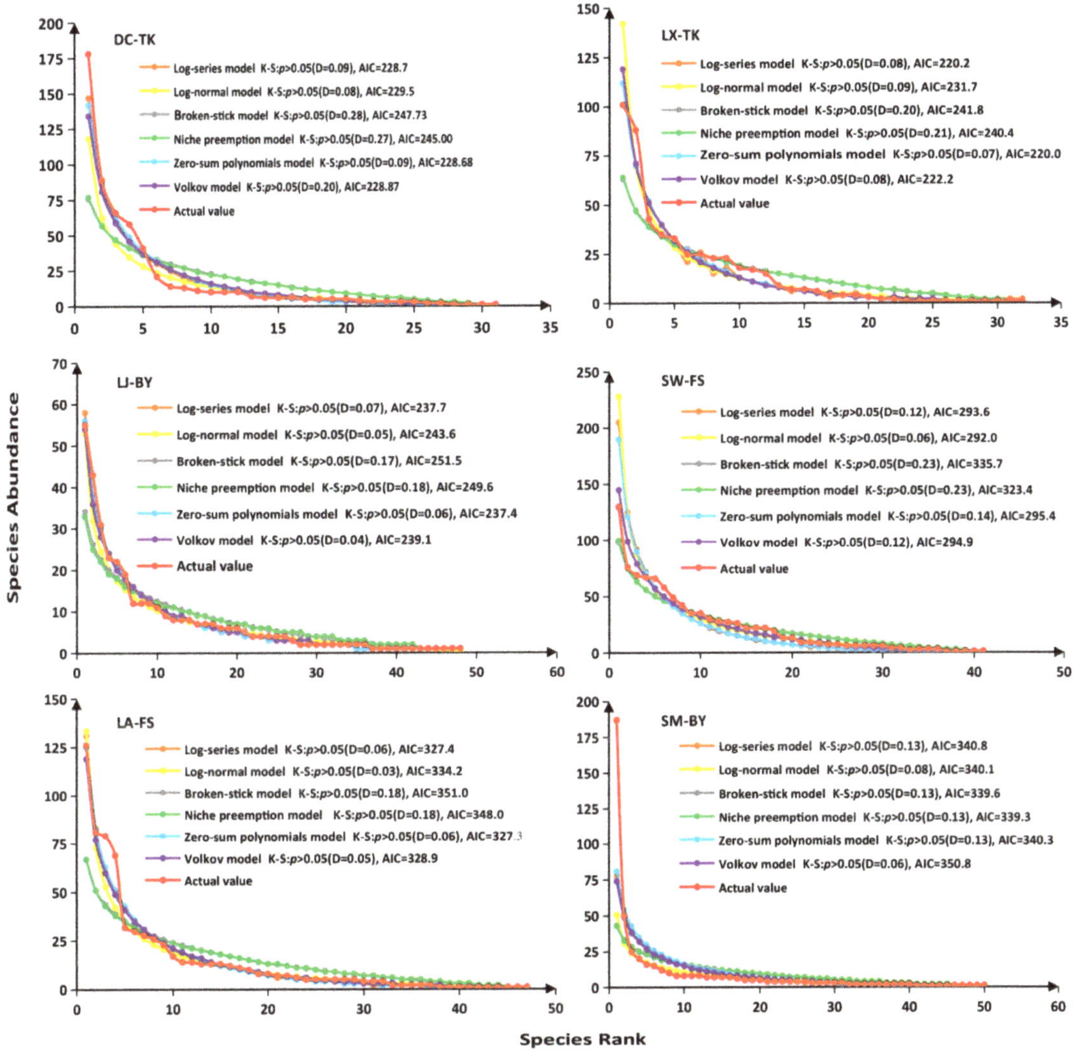

Figure 3. The SADs and models fitting for woody plants at the 40 m × 40 m scale in Tiankeng and non-Tiankeng forests. The actual values are the observed abundance.

4. Discussion

4.1. Species Composition of Tiankeng Forest and Nearby Non-Tiankeng Forests

The species dispersal plays an important role in the structure, composition, and succession of vegetation [1]. The species dispersal from the inside of Tiankeng forests to the nearby non-Tiankeng forests is limited, while species have unlimited dispersal from nearby non-Tiankeng forests to the inside of Tiankeng forests via the fringe of Tiankeng forests. The unique negative topography of Tiankeng caused a huge surface drop that further influenced the species dispersal [26]. The natural geographical barrier caused by towering cliffs and the huge depth limited the species dispersal from the inside of Tiankeng forests to the nearby non-Tiankeng forests, hindering the evolution of species-specific communication. As a result, the species composition of Tiankeng forests was changed with vegetation landscape fragmentation. However, the internal enclosed environment of Tiankeng could be conducive to the independent evolution of species in the Tiankeng forests, which provided habitats for more precious and primitive species [24]. Instead, species have unlimited dispersal from nearby non-Tiankeng forests to the inside of Tiankeng forests. Some of the common species (*Choerospondias axillaris* B. L. Burtt & A. W. Hill and *Rhaphiolepis indica* Lindley) in the nearby non-Tiankeng forests were found in the Tiankeng forests to become dominant or rare species.

It is obvious that the phenomenon of single family, genera, and species were detected in study sites of Dashiwei Tiankeng Group. The results indicate that the population in the study area was a mosaic system formed by multiple local populations. Meanwhile, there are functional exchanges between the various populations in three habitats. Disturbance promoted the increase of species diversity, and community ecotones controlled the exchange of species between different ecosystems [43]. The fringe of the Tiankeng forest was an ecocritical place of species dispersal, while the nearby non-Tiankeng forest was a regional species pool for the inside of Tiankeng forests. Fragmented habitats in Dashiwei Tiankeng Group increase the probability of random extinction of vulnerable species in Tiankeng forests. However, the species dispersal between habitat patches can also rebuild a new population in those unoccupied spaces [44]. Therefore, we should protect the nearby non-Tiankeng forests for maintenance of the ecosystem in Tiankeng forests.

4.2. SADs Pattern in Tiankeng and Non-Tiankeng Forests

There were significant differences in the SAD pattern between the inner and outer fringe area of Tiankeng forest and nearby non-Tiankeng forest. Species accumulation in the fringe of Tiankeng forest was the largest, and most species in inner Tiankeng forests originated from nearby non-Tiankeng forests via the fringe area. On the one hand, the fringe of Tiankeng forest is surrounded by non-Tiankeng forest and encircles the inside of the Tiankeng forest [30]. The non-Tiankeng forest is usually distributed in the foothills, ridges, and slopes near villages and farmlands, and is greatly affected by human activities which enhance alien species to the Tiankeng forests [22,45]. On the other hand, the inside of the Tiankeng forest is located at the bottom of the negative topographic structure, and this benefits species accumulation.

The Tiankeng formation can be tracked back to the Paleocene [46]. At that time, species in this area would freely migrate. Later, the Tiankeng and the Tiankeng forests gradually evolved and formed along with the geological movement. By now, with global warming, nearby non-Tiankeng forests were suffering soil loss, water stress, and degradation, some species could not adapt to their stress environment and took refuge in the Tiankeng negative terrain [27,47]. Sufficient water and heat condition, stable growing environment, and tundra organic-rich soil in Tiankeng provide a favorable place for the heterogeneity and diversity of species [24,27]. Therefore, by comparing the SADs difference between Tiankeng and non-Tiankeng forests, we can determine that the Tiankeng forest is a plant refuge formed by surrounding species that have adapted to the environmental changes. One of the reasons is that the negative topographic structure promotes the accumulation of species, and the other is that the external environment deteriorates and species are attracted by the hot and humid environment of Tiankeng.

4.3. Goodness-of Fit of SADs Models in Tiankeng and Non-Tiankeng Forests

SAD is usually described by various fitting curves to reflect the influence of different ecological processes [1,6]. Our SAD results show that the neutral model has the smallest AIC value in Tiankeng forests, while the statistical model has the smallest AIC value in the non-Tiankeng forests. Therefore, in Tiankeng forests, the main driving force is the neutral process, and in non Tiankeng forests, it is the niche and neutral processes. Why is there a difference? One reason may be related to community structure. In the early stages of growth, young trees are mainly affected by the neutral process and less affected by the niche process because they have weak biological characteristics and inter-specific competitiveness, and are vulnerable to the dangers of surface environment filtering and randomness [48]. As the species gradually grow, they not only enhance their biological characteristics, but also become increasingly affected by environmental filtering and the interaction between organisms. Finally, when species mature, the main community driving force is the niche process [31]. Tiankeng forest is an isolated natural secondary forest, the majority of plant individuals are in the growth period, and species relationship is mainly driven by heterogeneity of the habitat [27,30]. The negative topographic structure can protect the species from external interference, and strongly limits their dispersal from inside to outside in Tiankeng. Therefore, the neutral process plays an important role in the structure of the Tiankeng forest community, and this is one of the important points to incite Tiankeng forests being a plant refuge. However, the niche process is still evident with lesser influence [30]. Non-Tiankeng forests are near mountain secondary forests, of higher density, and with more younger trees than Tiankeng forests [22]. Additionally, their inter specific community relationships are more competitive and there is a fierce niche overlap and differentiation [22,30]. Thus, Tiankeng forests are more vulnerable to the neutral processes, while the non-Tiankeng forests are more readily affected by niche and neutral processes.

Another reason is lower human interference, and stronger erosion is a feature of negative terrain, compared to more unstable, vulnerable non-Tiankeng forest terrain [29]. The Dashiwei Tiankeng group is formed under the combined action of flowing water and gravity erosion [22]. The Tiankeng forest is situated in negative terrain and connected with the underground river, while the nearby non-Tiankeng forest is located in a downslope of karst rock mountain. This is the reason for which the Tiankeng forest suffered more hydraulic erosion and gravity erosion than non-Tiankeng. Therefore, as long as the topographical differences exist, the advantages in Tiankeng forests will continue. Additionally, this also raises conjecture, as the nearby non-Tiankeng forests were likely Tiankeng forests many years ago.

There are differences in aspects of species rarity, richness, migration rate, and accumulation rate between Tiankeng and nearby non-Tiankeng forests, but the ecological, niche, and neutral processes still work together in SADs. Some argued that niche and neutral processes are not mutually exclusive community forces, and that they are more like two endpoints in a continuous pattern, with a gradual transition phase between them [49]. In other words, any community may be able to find a corresponding position on the "niche-neutral continuum". Therefore, there is a trade-off between the niche and the neutral processes in community; when influence of the niche process declines, it is accompanied by an increase in the role of the neutral process [50]. Additionally, the trade-off relationship between the niche and neutral processes is decided by the community structure, species characteristics, environmental changes, and the outside interference.

Both Tiankeng forests and non-Tiankeng forests are karst forests, but the former, as a small patch of karst forest in the southwest of China, is under the control of nearby non-Tiankeng forests through the Tiankeng forest fringe area. As an ecotone, the outer fringe area of Tiankeng forests is affected by the nearby non-Tiankeng forests [24] and then have an influence on the inner area of Tiankeng forest. For example, *Manglietia aromatica* Dandy, a representative refuge plant of Tiankeng forest, has only ever existed in Tiankeng forest and has never been found in nearby non-Tiankeng forests of the Dashiwei Tiankeng Group. Therefore, it appears that species have successfully migrated and established in the Tiankeng forest interior.

Fierce competition among species has always existed in the non-Tiankeng forest, and the degree of niche differentiation was high, but the Tiankeng forest on negative terrain has stable environmental conditions and is good at collecting soil, reserving water, and heat [30]. Therefore, some species that prefer wet and warm environments in non-Tiankeng forest will be attracted by the environmental conditions and continue to move to the negative terrain. Subsequently, with species migration and evolution, Tiankeng forest species will not only maintain their own characteristics, but also have similarities with nearby non-Tiankeng forest communities, illustrating that the local species composition is controlled by the regional community.

Different patterns of SAD may indicate specific environmental conditions. It is doubtful whether there would be other ecological processes that respond to the species co-occurrence. This is because each forests species has its own specific role as a result of evolutionary processes [41]. Moreover, multiple-mechanism models may have better results when fitting the same data. The model fitting results based solely on the patterns of species abundance cannot fully verify the ecological mechanism behind it. Therefore, we suggested that the neutral process was important for the SAD pattern in Tiankeng forests, a further analysis needs to be carried out in combination with environmental factors and spatial patterns.

5. Conclusions

The comparing of SADs between Tiankeng and non-Tiankeng forests is essential for its management, conservation, and sustainable utilization in Dashiwei Tiankeng Group. Assessing the role of biodiversity in Tiankeng forests could help in understanding climate change, approaches to the wise use of sustainable resources, preserving biological diversity, and exploring the community maintenance mechanism. For example, we found that the species dispersal from the inside of Tiankeng forests to the nearby non-Tiankeng forests is limited, while species has unlimited dispersal from nearby non-Tiankeng forests to the inside of Tiankeng forests via the fringe of Tiankeng forests. Moreover, some species in Tiankeng forest tend to become rarer and more precious. Therefore, this study's results show that the neutral process incites Tiankeng forests to be a plant refuge throughout the fitting of niche and neutral models with nearby non-Tiankeng forests. Finally, the SAD pattern was related to the environment, climate, topography, and human activities. Tiankeng forests will remain its own unique ecosystem functions and ecological position if we enhance the protection of the nearby non-Tiankeng forest.

Author Contributions: L.H., H.Y., X.A., Y.Y., L.Y., G.H., X.L., M.C. and Y.X. conceived research idea, designed the research and collected field data. L.H., X.A. and L.Y. analyzed the data, L.H. and H.Y. wrote the first draft. L.H., H.Y., X.A., Y.Y., L.Y., G.H., X.L., M.C. and Y.X. All authors have read and agreed to the published version of the manuscript.

Funding: This work was supported by the National Natural Science Foundation of China (No. 31960047), the Guangxi Key Research and Development Program(No. AB21220057), the Research Funds of The Guangxi Key Laboratory of Landscape Resources Conservation and Sustainable Utilization in Lijiang River Basin, Guangxi Normal University (No. LRCSU21Z0101), and the Key Laboratory Research Foundation of Ecology of Rare and Endangered Species and Environmental Protection (Ministry of Education), Guangxi Normal University (No. ERESEP2019Z04).

Institutional Review Board Statement: Not applicable.

Informed Consent Statement: Not applicable.

Data Availability Statement: Data recorded in the current study are available in all Tables and Figures of the manuscript.

Acknowledgments: We are grateful to the workers at the Leye-Fengshan Global Geopark (China), who gave permission for, and contributed to, this research in Dashiwei Tiankeng Group. We also thank the locals of Shuaba village, Leye, China.

Conflicts of Interest: The authors declare that they have no known competing financial interest or personal relationships that could have influenced the work reported in this paper.

References

1. Dong, K.; Hao, G.; Yang, N.; Zhang, J.; Ding, X.; Ren, H.; Shen, J.; Wang, J.; Jiang, L.; Zhao, N.; et al. Community assembly mechanisms and succession processes significantly differ among treatments during the restoration of *Stipa grandis–Leymus chinensis* communities. *Sci. Rep.* **2019**, *9*, 16289. [CrossRef] [PubMed]
2. Schmitt, S.; Tysklind, N.; Derroire, G.; Heuertz, H.; Herault, B. Topography shapes the local coexistence of tree species within species complexes of Neotropical forests. *Oecologia* **2021**, *196*, 389–398. [CrossRef] [PubMed]

3. Germain, R.M.; Mayfield, M.; Gilbert, B. The 'filtering' metaphor revisited: Competition and environment jointly structure invasibility and coexistence. *Biol. Lett.* **2018**, *14*, 20180460. [CrossRef] [PubMed]
4. Magura, T.; Lövei, G.L.; Tóthmérész, B. Conversion from environmental filtering to randomness as assembly rule of ground beetle assemblages along an urbanization gradient. *Sci. Rep.* **2018**, *8*, 16992. [CrossRef] [PubMed]
5. Yin, D.; Ye, D.; Cadotte, M.W. Habitat loss-biodiversity relationships are influenced by assembly processes and the spatial configuration of area loss. *For. Ecol. Manag.* **2021**, *496*, 119452. [CrossRef]
6. Matthews, T.J.; Whittaker, R.J. Neutral theory and the species abundance distribution: Recent developments and prospects for unifying niche and neutral perspectives. *Ecol. Evol.* **2014**, *4*, 2263–2277. [CrossRef] [PubMed]
7. Kusumoto, B.; Kubota, Y.; Baselga, A.; Gómez, C.; Odríguez, O.C.; Shiono, T. Community dissimilarity of angiosperm trees reveals deep-time diversification across tropical and temperate forests. *J. Veg. Sci.* **2021**, *32*, e13017. [CrossRef]
8. Ma, K.M. Advances of the Study on Species Abundance Pattern. *Chin. J. Plant Ecol.* **2003**, *3*, 412–426. Available online: https://www.plant-ecology.com/CN/10.17521/cjpe.2003.0060 (accessed on 15 September 2021).
9. Hidasi, N.J.; Bini, L.M.; Siqueira, T.; Cianciaruso, M.V. Ecological similarity explains species abundance distribution of small mammal communities. *Acta Oecol.* **2020**, *102*, 103502. [CrossRef]
10. Waldock, C.; Stuart-Smith, R.D.; Albouy, C.; Cheung, W.; Edgar, G.J.; Mouillot, D.; Tjiputra, J.; Pellissier, L. A quantitative review of abundance-based species distribution models. *Ecography* **2022**, *1*, e05694. [CrossRef]
11. Thierry, E.H. Stochastic species abundance models involving special copulas. *Phys. A.* **2018**, *490*, 77–91. [CrossRef]
12. Latimer, A.; Wu, S.; Silander, G. Building statistical models to analyze species distributions. *Ecol. Appl.* **2006**, *16*, 33–50. [CrossRef] [PubMed]
13. Takolander, A.; Hickler, T.; Meller, L.; Caneza, M. Comparing future shifts in tree species distributions across Europe projected by statistical and dynamic process-based models. *Reg. Environ. Chang.* **2019**, *19*, 251–266. [CrossRef]
14. He, F.; Tang, D. Estimating the niche preemption parameter of the geometric series. *Acta Oecologica* **2008**, *33*, 105–107. [CrossRef]
15. Dallas, T.; Hastings, A. Habitat suitability estimated by niche models is largely unrelated to species abundance. *Glob. Ecol. Biogeogr.* **2018**, *27*, 1448–1456. [CrossRef]
16. Barabás, G.; Andrea, R.; Rael, R.; Meszéna, G.; Ostling, A. Emergent neutrality or hidden niches. *Oikos* **2013**, *122*, 1565–1572. [CrossRef]
17. Petr, D.; Karolina, T.; Tereza, K. Linking species abundance and overyielding from experimental communities with niche and fitness characteristics. *J. Ecol.* **2019**, *107*, 178–189. [CrossRef]
18. Volkov, I.; Banavar, J.R.; Hubbell, S.P.; Maritan, A. Neutral theory and relative species abundance in ecology. *Nature* **2003**, *424*, 1035–1037. [CrossRef] [PubMed]
19. Wang, X.Z.; Ellwood, M.; Ai, D.; Zhang, R.Y.; Wang, G. Species abundance distributions as a proxy for the niche-neutrality continuum. *J. Plant Ecol.* **2018**, *11*, 445–452. [CrossRef]
20. Villa, P.M.; Martins, S.V.; Rodrigues, A.C.; Alice, C.R.; Nathália, V.H.; Michael, A.C.; Ali, A. Testing species abundance distribution models in tropical forest successions: Implications for fine-scale passive restoration. *J. Ecol. Eng.* **2019**, *135*, 28–35. [CrossRef]
21. Feng, G.; Huang, J.; Xu, Y.; Li, J.; Zhang, R. Disentangling environmental effects on the tree species abundance distribution and richness in a subtropical forest. *Front. Plant Sci.* **2021**, *12*, 367. [CrossRef] [PubMed]
22. Zhu, X.W. *Leye Tiankeng*; Guangxi Science and Technology Publishing House: Nanning, China, 2018. (In Chinese)
23. Lin, Y. Species Diversity of Karst Tiankeng Forest in Dashiwei Tiankengs, Guangxi. Master Thesis, College of life Sciences, Guangxi Normal University, Guilin, China, 2005. (In Chinese)
24. Pu, G.; Lv, Y.; Dong, L.; Zhou, L.; Huang, K.; Zeng, D.; Mo, L.; Xu, G. Profiling the bacterial diversity in a typical Karst Tiankeng of China. *Biomolecules* **2019**, *9*, 187. [CrossRef]
25. Zhu, S.F.; Jiang, C.; Shui, W.; Guo, P.P.; Zhang, Y.Y.; Feng, J. Vertical distribution characteristics of plant community in shady slope of degraded Tiankeng talus: A case study of Zhanyi Shenxiantang in Yunnan, China. *J. Appl. Ecol.* **2020**, *31*, 1496–1504. [CrossRef]
26. Hong, S.K.; Song, I.J.; Wu, J. Fengshui theory in urban landscape planning. *Urban Ecosyst.* **2007**, *10*, 221–237. [CrossRef]
27. Su, Y.Q.; Tang, Q.M.; Mo, F.Y.; Xue, Y.G. Karst Tiankengs as refugia for indigenous tree flora amidst a degraded landscape in southwestern China. *Sci. Rep.* **2017**, *7*, 4249. [CrossRef] [PubMed]
28. Chen, J.; Lin, W.; Zhang, Y.; Dai, Y.; Chen, B. Village Fengshui Forests as Forms of Cultural and Ecological Heritage: Interpretations and Conservation Policy Implications from Southern China. *Forests* **2020**, *11*, 1286. [CrossRef]
29. Wan, F.; Zhang, X.; Li, T. Stability analysis of Jiguanshan Tunnel construction under karst tiankeng. *IOP Conf. Ser. EES* **2021**, *769*, 32–54. Available online: https://schlr.cnki.net/Detail/doi/GARJ2021_1/SIPD0DEE5CEC6C628C2AF275723D00351265 (accessed on 8 December 2021). [CrossRef]
30. Huang, L.J.; Yu, Y.M.; An, X.F.; Yu, L.L.; Xue, Y.G. Interspecific association of main woody plants in Tiankeng forests of Dashiwei Tiankeng Group, Guangxi. *Guihaia* **2021**, *41*, 695–706. (In Chinese) [CrossRef]
31. Zhang, S.Z. Study on the Maintenance Mechanisms of Species Diversity in the Matural Old Growth Tropical Forests on the Hainan Island, China. Ph.D. Dissertation, Chinese Academy of Forestry, Beijing, China, 2017. (In Chinese)
32. Wei, Y.; Wen, F.; Fu, L.; Xin, Z.; Jiang, R.; Pan, B.; Hua, F.; Mao, S.; Li, S.; Ma, H.; et al. *The Distribution and Conservation Status of Native Plants in Guangxi, China*; China Forestry Publishing: Beijing, China, 2018.
33. Avolio, M.; Forrestel, E.J.; Chang, C.C.; La Pierre, K.J.; Burghardt, K.T.; Smith, M.D. Demystifying dominant species. *New Phytol.* **2019**, *223*, 1106–1126. [CrossRef] [PubMed]

34. Skeen, J.N. An extension of the concept of importance value in analyzing forest communities. *Ecology* **1973**, *54*, 655–656. [CrossRef]
35. Hadley, W.; Romain, F.; Lionel, H.; Kirill, M. Dplyr: A Grammar of Data Manipulation. R Package Version 1.0.0. 2020. Available online: https://CRAN.R-project.org/package=dplyr (accessed on 9 December 2021).
36. R Core Team. *R: A Language and Environment for Statistical Computing*; R Foundation for Statistical Computing: Vienna, Austria, 2020; Available online: https://www.R-project.org/ (accessed on 8 December 2021).
37. Keylock, C.J. Simpson diversity and the Shannon-wiener index as special cases of a generalized entropy. *Oikos* **2005**, *109*, 203–207. [CrossRef]
38. Fletcher, S.; Islam, M. Comparing sets of patterns with the Jaccard index. *Asian-Australas. J. Anim. Sci.* **2018**, *22*, 1–17. [CrossRef]
39. Fisher, C.K.; Mehta, P. The transition between the niche and neutral regimes in ecology. *Proc. Natl. Acad. Sci. USA* **2014**, *111*, 13111–13116. [CrossRef] [PubMed]
40. Alonso, D.; Mckane, A.J. Sampling Hubbell's neutral theory of biodiversity. *Ecol. Lett.* **2004**, *7*, 901–910. [CrossRef]
41. Ramesh, T. *Trends and Changes in Hydroclimatic Variable*; Elsevier: Cambridge, MA, USA, 2018.
42. Paulo, I.P.; Murilo, D.M.; Andre, C. Maximum Likelihood Models for Species Abundance Distributions. R Packages Version 0.4.2. 2018. Available online: http://piLaboratory.github.io/sads (accessed on 8 December 2021).
43. Muluneh, M.; Feyissa, M.; Wolde, T. Effect of forest fragmentation and disturbance on diversity and structure of woody species in dry Afromontane forests of northern Ethiopia. *Biodivers. Conserv.* **2021**, *30*, 1753–1779. [CrossRef]
44. Bertassello, L.; Bertuzzo, E.; Botter, G.; Jawitz, J.W.; Aubeneau, A.F.; Hoverman, J.T.; Rinaldo, A.; Rao, P.S.C. Dynamic spatio-temporal patterns of metapopulation occupancy in patchy habitats. *R. Soc. Open. Sci.* **2021**, *8*, 201309. [CrossRef] [PubMed]
45. Chen, B.X.; Coggins, C.; Minor, J.; Zhang, Y.Q. Fengshui forests and village landscapes in China: Geographic extent, socioecological significance, and conservation prospects. *Urban For. Urban Green.* **2018**, *31*, 79–92. [CrossRef]
46. Shen, H.T.; Kimikazu, S.; Qi, M.; Masumi, M.; Tetsuya, M.; Seiji, H.; Takahashi, T.; Honda, M.; Sueki, K.; He, M.; et al. Exposure age dating of Chinese Tiankengs by 36Cl-AMS. *Nucl. Instrum. Methods B* **2019**, *459*, 29–35. [CrossRef]
47. Yang, G.; Peng, C.; Liu, Y.; Dong, F.Q. Tiankeng: An ideal place for climate warming research on forest ecosystems. *Environ. Earth Sci.* **2019**, *78*, 46. [CrossRef]
48. Wang, Y.; Lamontagne, J.M.; Lin, F.; Yuan, Z.Q.; Ye, J.; Wang, X.G.; Hao, Z. Similarity between seed rain and neighbouring mature tree communities in an old-growth temperate forest. *J. For. Res.* **2019**, *31*, 2435–2444. [CrossRef]
49. Niccolo, A.; Jorge, H.; Carlos, A.P.; Tommaso, B.; Amos, M.; Samir, S. Neutral and niche forces as drivers of species selection. *J. Theor. Biol.* **2019**, *483*, 109969. [CrossRef]
50. Pinsky, M.L. Species coexistence through competition and rapid evolution. *Proc. Natl. Acad. Sci. USA* **2019**, *116*, 2407–2409. [CrossRef] [PubMed]

diversity

MDPI

Article

Homogenized Phylogeographic Structure across the Indo-Burma Ranges of a Large Monoecious Fig, *Ficus altissima* Blume

Jian-Feng Huang [1,2], Clive T. Darwell [3] and Yan-Qiong Peng [1,2,*]

[1] CAS Key Laboratory of Tropical Forest Ecology, Xishuangbanna Tropical Botanical Garden, Chinese Academy of Sciences, Mengla 666303, China; huangjianfeng@xtbg.ac.cn
[2] Center of Plant Ecology, Core Botanical Gardens, Chinese Academy of Sciences, Mengla 666303, China
[3] National Center for Genetic Engineering and Biotechnology (BIOTEC), 113 Thailand Science Park, Pahonyothin Road, Khlong Luang 12120, Thailand; ctdarwell@gmail.com
* Correspondence: pengyq@xtbg.ac.cn

Abstract: As well as bountiful natural resources, the Indo-Burma biodiversity hotspot features high rates of habitat destruction and fragmentation due to increasing human activity; however, most of the Indo-Burma species are poorly studied. The exploration of plants closely associated with human activity will further assist us to understand our influence in the context of the ongoing extinction events in the Anthropocene. This study, based on widely and intensively sampled *F. altissima* across Indo-Burma and the adjacent south China ranges, using both the chloroplast psbA-trnH spacer and sixteen newly developed nuclear microsatellite markers (nSSRs), aims to explore its spatial genetic structure. The results indicated low chloroplast haplotype diversity and a moderate level of nuclear genetic diversity. Although limited seed flow was revealed by psbA-trnH, no discernible phylogeographic structure was shown due to the low resolution of cpDNA markers and dominance of an ancestral haplotype. From the nSSRs data set, phylogeographic structure was homogenized, most likely due to extensive pollen flow mediated by pollinating fig wasps. Additionally, human cultivation and human-mediated transplanting further confounded the analyses of population structure. No geographic barriers are evident across the large study range, with *F. altissima* constituting a single population, and extensive human cultivation is likely to have had beneficial consequences for protecting the genetic diversity of *F. altissima*.

Keywords: *Ficus*; Indo-Burma; genetic structure; human cultivation; nSSRs

Citation: Huang, J.-F.; Darwell, C.T.; Peng, Y.-Q. Homogenized Phylogeographic Structure across the Indo-Burma Ranges of a Large Monoecious Fig, *Ficus altissima* Blume. *Diversity* **2021**, *13*, 654. https://doi.org/10.3390/d13120654

Academic Editor: Mario A. Pagnotta

Received: 8 November 2021
Accepted: 1 December 2021
Published: 9 December 2021

1. Introduction

As one of the 34 identified global biodiversity hotspots [1], Indo-Burma, covering Burma, Thailand, Laos, Cambodia, and Vietnam, as well as parts of southern China, northeast India, and Bangladesh, harbours great diversity in both plants and animals. The endemic species' ecological and evolutionary trajectories were strongly influenced by plate tectonics, climatic oscillations, river system dynamics, sea level fluctuations, shifting coastlines, and human activity [2–5]. However, the biodiversity of the Indo-Burma region is severely threatened by factors such as human population growth, deforestation and habitat conversion, resource exploitation, pollution, and global warming [6–8]. Moreover, our knowledge about the underlying genomic structures (e.g., population structure, phylogeography, cryptic diversity) among Indo-Burma flora and fauna is extremely poor, mainly due to political instability and lack of infrastructure during long periods of the 20th century in many parts of this region [9,10].

In recent years, the conditions that prevented the researchers from entering large parts of Indo-Burma to explore the biodiversity have been improved along with the general development in this region. A few phylogeographic studies were conducted among

Indo-Burma fauna, including flukes [11,12], insects [13], fish [14,15], amphibians [16], reptiles [5], rodents [17,18], birds [9], and mammals [19]. For example, studies that focused on habitat-dependent freshwater animals revealed complex drainage re-alignments and sea-level fluctuation histories in the Indo-Burma region [5,15,20]. Drainage has also played an important role in shaping the phylogeography of plant species with limited pollen dispersal capability [10]. The population genetic investigation of the endemic species *Dalbergia cochinchinensis* Pierre and *D. oliveri* Gamble ex Prain represents the first detailed analysis of landscape genetics for tree species within Indo-Burma, although these two plants do not cover the entire Indo-Burma region [10]. Nonetheless, our knowledge about the genetic diversity and structure of Indo-Burma species is still extremely poor, especially for plants, which also impedes biodiversity conservation programs in this region.

The arrival of the Anthropocene has seen the transformation of ecosystems according to human use requirements [2,21]. Understanding how human activities affect the genetic diversity and structure of plants and animals is critical from both a scientific and policy perspective. Human-related threats, such as habitat destruction and fragmentation, invasive species, resource exploitation, and other indirect effects of human activities, often cause catastrophic biodiversity decline [22,23]. However, there are also some afforestation and ornamental plant species that are widely cultivated and play an integral role in the development of human civilization. Although their natural populations could have been reduced due to human-related threats, their genetic diversities may have been preserved due to extensive cultivation by humans. Few studies have focused on plants that are closely associated with humans, particularly those that inhabit badly degraded Indo-Burma forests. Studying the phylogeographic patterns of such species will be particularly beneficial to the ongoing biodiversity conservation of Indo-Burma flora and provide insight regarding the impact of human activity and whether it may degrade or augment genetic diversity.

With more than 800 species, *Ficus* L. (Moraceae) occurs pan-globally in both tropical and subtropical biomes. Coupled with their ecological significance as keystone species, and their typically species-specific relationship with co-evolved pollinating fig wasps (Hymenoptera, Chalcidoidea, Agaonidae) [24,25], the genus *Ficus* has long fascinated biologists. Many members of *Ficus* are also closely associated with humans, and often traditionally used as sources of medicines and food, as ornamental trees, religious plants, lac insect hosts (that exude useful lac gum products), fodder, fuel, hedges, or enclosures [26]. Among them, *F. altissima* is commonly known as the council tree and is frequently planted in cities, villages, or temples, both as ornamental and sacred plants [27]. The tree is also one of the recorded hosts of lac insect, and whose leaves and bark can be used as skin-disease treatments [28]. It is a long-lived perennial and hemiepiphytic tree (over 40 m tall), which can occupy a patch of about 300 m^2 (and thus almost constitute a single ecosystem [28,29]). *F. altissima*, as a keystone species, occurs naturally and widely in forests of tropical Asia at low densities and produces figs throughout the year in synchronous crops with asynchrony between trees [27,30].

Ficus altissima is a good example of a tree associated with many species and that is widespread in Indo-Burma and closely associated with human activity. Here, we present a phylogeographic study of *F. altissma* at a broad-scale across Burma, Thailand, Cambodia, Vietnam, and southern China. Based on an intensive sampling strategy across Indo-Burma, working with both chloroplast and nuclear DNA markers, we aimed to (1) investigate the level and distribution of genetic diversity of *F. altissima* within and among populations, as well as between populations collected from introduced and native areas; and (2) explore the population genetic structure of *F. altissima* within Indo-Burma. It will contribute to our understanding of the historical phylogeography of this keystone forest species of the Indo-Burma hotspot, as well as the influence on it from human activities.

2. Materials and Methods

2.1. Sample Collection

Ficus altissima populations were sampled in 37 locations in southern China, Burma, Thailand, Vietnam, and Cambodia (Figure 1, Table A1), yielding 267 individuals. The Fuz, Gan, Nid, Pan, Xia, and Yib populations were collected from areas outside its native range [31,32] and considered as introduced populations by human. However, within the native range, populations featuring disturbed habitats cannot be distinguished as either natural or cultivated. Populations such as Don and Chi, sampled from cities, may originate from their respective nearby wild source populations, rather than introduced from distant areas. The sample size for each population ranged from 1 to 23 individuals, according to limits of the population (Table A1).

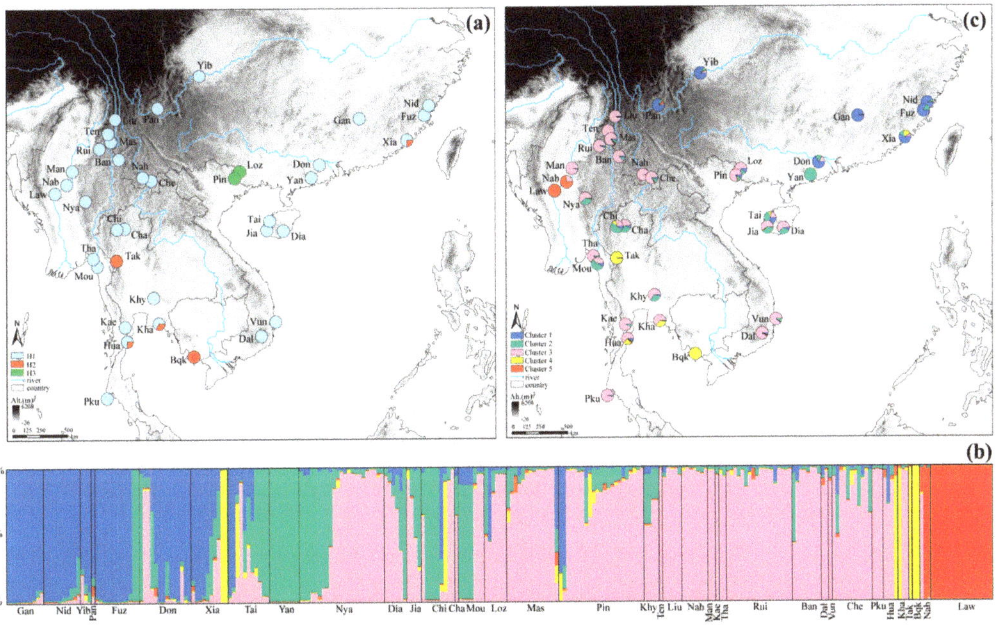

Figure 1. Sample locations and genetic structure of *F. altissima* in Indo-Burma and adjacent southern China: (**a**) geographical distribution of cpDNA haplotypes; (**b**) bar plots of the membership probabilities of *F. altissima* individuals to the different clusters from the STRUCTURE analysis at $K = 5$ based on 16 nSSRs; (**c**) geographical distribution of the genetic clusters detected in STRUCTURE, and the pie charts represent the assignment values of the admixed clustering analysis performed in STRUCTURE for all individuals.

2.2. DNA Extraction and Chloroplast DNA

Total genomic DNA of *F. altissima* was extracted from silica-gel dried material using the Plant Genomic DNAKit (Tiangen Biotech, Beijing, China). Ten chloroplast (cpDNA) intergenic regions, including ndhF-rpl32, psbA-trnH, psbB-psbF, psbC-trnS, psbJ-petA, trnD-trnT, trnF-trnV, trnL-trnF, trnQ-rps16, and trnS-trnG [33–35], were tested for amplification and sequence variation. Some of these markers are recognised as informative from earlier phylogenetic and phylogeographic research of *Ficus* plants [35–38]. To test the ten candidate cpDNA markers, one or two individuals were randomly selected from each sampled population for PCR amplification using universal primer pairs under the conditions described by Shaw et al. [33,34] and Vieira et al. [35].

Each fragment was bidirectionally sequenced by the Beijing Genomics Institute (Shenzhen, China). All forward and reverse strands were edited and assembled using the program Sequencher 4.5 (GeneCodes, Ann Arbor, MI, USA). The sequences were aligned with Clustal X [39] and then adjusted manually using BioEdit 7.0.9.0 [40]. Ultimately, only psbA-trnH showed variation in preliminary tests and was selected for the PCR amplification of all individual samples.

2.3. nSSRs

In preliminary experiments, five nuclear DNA-based markers were tested, including ETS, ITS, GBSSI, G3pdh, and ncpGS [41,42]. However, all of them showed heterozygous sequencing peaks, suggesting clone contamination, and were subsequently abandoned. Alternatively, polymorphic nSSRs were developed. One sample of the population Nah was selected to construct the DNA library following the manufacturer's protocols of the KAPAHyper prep kit (KAPA Biosystems, Wilmington, MA, USA). The sequencing processes were performed by Microread Genetics Incorporation (Beijing, China) on HiSeq 2500 (Illumina, San Diego, CA, USA), using the corollary reagents from Illumina and with paired read lengths of 2 × 150 base pairs and yielding 3.71 G raw bases. Raw data were then quality filtered by trimming the adapter sequences and by removing reads with quality scores < 20. High-quality reads were assembled into 15,532 contigs. The software MIcroSAtellite [43] was used to identify contigs with nSSRs, and Primer 3.0 [44] was used to design microsatellite primers using the default settings. Eighty primer pairs were selected as candidates for amplification and polymorphism testing. Finally, 16 of the 80 tested primer pairs were found to produce repeatable amplicons with high polymorphism and were used for genotyping. The Multiplex PCR amplification was conducted in 10 μL volumes mixed with 2.0 μL of 5 × Buffer, 0.3 μL of dNTP (10 mM each), 1 μL of DNA (100 ng), 0.4 μL of primer mixture (5 μM), and 0.1 μL of Taq polymerase (5 U), based on the following the conditions: initial denaturation at 95 °C for 8 min, 30 cycles at 94 °C for 30 s, at 56 °C for 30 s, at 72 °C for 30 s, and at 60 °C for 30 min for the final extension. Post-PCR products were analysed by capillary electrophoresis on an ABI 3730XL DNA analyser (Applied Biosystems, Foster City, CA, USA) with the GeneScan 500 ROX Size Standard. Microsatellite fragment sizes were determined using GeneMapper version 3.2.

2.4. Confirming Usability of nSSRs

Micro-checker v.2.2.3 [45] was used to examine the presence of any genotypic errors due to stuttering, large allele dropout, and null alleles. Linkage disequilibrium (LD) and Hardy–Weinberg equilibrium (HWE) were tested by GENEPOP 4.7.0 [46]. Significance levels were corrected using the Bonferroni correction for multiple tests.

2.5. Genetic Diversity

The cpDNA haplotypes were distinguished using DnaSP v5 [47] on the basis of nucleotide and indel differences. Molecular diversity indices, including the number of haplotypes (H) as well as nucleotide (π) and haplotype (H_D) diversity, were calculated using DnaSP v5. For the nSSRs data, classical indices of genetic diversity were estimated using GenAIEx 6.5 [48]. The inbreeding coefficient (F_{IS}) was calculated by FSTAT 2.9.3 [49]. The polymorphism information content (PIC) for each nSSRs was calculated with Cervus v3.0.7 [50,51]. Subsequently, the mean number of alleles (N_a) and private alleles (PA_r) per locus, observed (H_O), and expected heterozygosity (H_E) of the 37 sampled populations based on nSSRs were used to show the geographic pattern by using the inverse distance weighted (IDW) interpolation function implemented in ArcGIS 10.3 (ESRI, Redlands, CA, USA). IDW assumed that points close to each other are more relevant than those that are more distant and is weighted more closely to the predicted position than the farther distances [52].

2.6. Genetic Structure

Genetic differentiation based on cpDNA and nSSRs data were evaluated by analysis of molecular variance (AMOVA) performed in Arlequin v3.5 [53], with 10,000 permutations used for tests of significance. For nSSRs data, a pattern of isolation by distance (IBD) was assessed using a Mantel test in GenAlEx 6.5 with 9999 permutations to evaluate the correlations between pair-wise genetic ($F_{ST}/(1 - F_{ST})$) and geographic distance. The geographic distances between pairwise populations were calculated using the program Geographic Distance Matrix Generator v1.2.3 [54] based on their decimal degree coordinates. To estimate the interpopulation genetic affinity of studied populations, principal component analysis (PCoA) was conducted with GenAIEx 6.5 based on Euclidean distance and an UPGMA phylogenetic tree was constructed by MEGA 6.06 [55] based on Nei's genetic distance. To further understand the clustering patterns, the Bayesian clustering of individual genotypes was investigated using STRUCTURE v2.3.4 [56]. We employed a model with admixture, with a burn-in period of 100,000 and a run length of 1,000,000 iterations varying K from $K = 1$ to $K = 10$. For each value of K, ten runs were done. The STRUCTURE HARVESTER online program [57] was used to detect the optimal K value using Evanno method [58]. CLUMPP 1.1.2 [59] was used to summarize the membership coefficients into clusters.

2.7. Population Dynamics

For cpDNA, the population expansion hypothesis was tested using neutrality and mismatch distribution tests. For the neutrality test, Tajima's D, considering the frequency of mutations [60], and Fu's F_S [61], based on cpDNA haplotype distribution, were conducted. The mismatch distribution test is used to assess whether the observed distribution of the pairwise differences matches the expectations under the sudden demographic expansion and the spatial-demographic expansion models. These analyses were conducted in Arlequin v3.5.

The recent demographic bottlenecks were tested for the populations with more than ten individual samples using the program BOTTLENECK 1.2.02 [62] with two recommended evolutionary models for nSSRs data [63,64]: the stepwise mutation model (SMM) and the two-phase mutation model (TPM). Under the TPM, the proportion of SMM and IAM (infinite alleles model) were set with default values that 70% of the mutations were assumed to occur under the SMM, and 30% of the mutations were assumed to occur under IAM. For each mutational model, 10,000 replicates were performed. After a reduction in the effective population size, the allele diversity is expected to drop more rapidly than their heterozygosities drop [65]. Thus, the heterozygosity of the population is larger than expected considering the number of alleles found. Sign test and the allele frequency distribution mode shift analysis [66] in BOTTLENECK 1.2.02 were performed to determine the population genetic reduction signatures characteristic of recent reductions in effective population size.

3. Results

3.1. cpDNA Genetic Diversity

Of the ten tested cpDNA intergenic regions, only psbA-trnH (356 bp) showed sequence variation in *F. altissima*. Of the 233 obtained sequences, only three haplotypes (GenBank accession numbers: MT291815–MT291817) were detected with two indels and two single nucleotide polymorphisms, suggesting limited use at the population level. At the population level, H_D and π ranged from 0.000 to 0.667 and (0.000 to 0.495) $\times 10^{-2}$, respectively. Only three populations, Hua, Kha, and Xia, showed a haplotype polymorphism possessing two haplotypes, H1 and H2 (Table A1, Figure 1a). Haplotype H3 was derived from the ancestral common haplotype H1 by a single 13-bp indel. Haplotype H2 is different from H1 by two transversions and a 2-bp indel.

3.2. Usability of nSSRs and Genetic Diversity

Sixteen polymorphic nSSRs were successfully developed based on high-throughput sequencing and microsatellite genotype data for all 267 individual samples were obtained. They displayed relatively high polymorphism across *F. altissima* individuals. The sequences are deposited in GenBank and detailed characteristic information of these 16 nSSRs is listed in Table 1. The PIC values ranged from 0.137 for the Fal_61 locus to 0.813 for the Fal_73 with an average of 0.539. There was no evidence of stuttering errors or large allele dropout in any of the loci analysed by Micro-checker v.2.2.3. In turn, null alleles were detected at five loci in a few populations (Table A2). Across the 16 loci and 37 studied populations, 136 microsatellite alleles were identified, corresponding to 8.5 alleles per locus and ranging from 5 to 12 alleles for individual loci. A few pairs of loci among a few populations exhibited linkage disequilibrium. Some populations deviated from HWE in a few loci, and a deficiency of heterozygotes was observed by GENEPOP in these populations, which may be due to the presence of null alleles (Table A2). The genetic diversity parameters for *F. altissima* populations are summarized in Table A1, with an observed species-level heterozygosity (H_O) of 0.483, expected heterozygosity (H_E) of 0.576, and a species inbreeding coefficient (F_{IS}) of 0.163. For each population, the H_O, H_E, and F_{IS} ranged from 0.063 to 0.563, 0.031 to 0.580, and −1.000 to 0.451, respectively. In turn, the number of private alleles (N_p), mean number of alleles (N_a), and private alleles per locus (PA_r) ranged from 1 to 3, 1 to 5.125, and 0.000 to 0.188, respectively. Populations Hua, Pin, and Xia each have three private alleles and showed the highest PA_r values. However, no private alleles were found in 22 of the 37 sampling localities. Most of the loci showed negative or small positive F_{IS} values, which is consistent with the prevalence of outcrossing in monoecious fig species. The values of N_a, PA_r, H_O, and H_E are geographically displayed in Figure 2. The regions with high levels of genetic diversity were scattered across the Indo-Burma hotspots and adjacent southern China. Unexpectedly, the six introduced populations showed similar levels of genetic diversity with the populations collected from the native range, and many populations collected from disturbed urban areas, such as Mas, Rui, Don, Pin, and Chi, also showed high levels of genetic diversity (Figures 1 and 2, Table A1).

Table 1. Sixteen nSSRs developed for *F. altissima*.

Locus	Primer Sequences	Repeat Motif	GenBank Accession No.	PIC
Fal_2	F: CCTGTGGGAGAGTTTGAAGG R: CTTGCTGCACGAATCTGCT	(CGT)6	MN255360	0.487
Fal_9	F: GAGTACATGCAAATGCCTCG R: CTCAGCAGCAACGAAAGATG	(TTTA)5	MN255361	0.561
Fal_11	F: GACCTGTTGGAGGAGATTGC R: TCATGGGCCACTTATCCTTC	(GCG)5	MN255362	0.753
Fal_14	F: CGATCCTTATCCTCTGCTCG R: GCACGCAATTTGAACGAAC	(AAG)10	MN255363	0.740
Fal_28	F: TCAGAATTGGAACGAGGGAC R: GCAGGGACTTCTTCTCTGACC	(TTTG)7	MN255364	0.492
Fal_31	F: CGATCACCACGAGCTACTGA R: TGGCGCATGATAAGTTTGAG	(GTCT)5	MN255365	0.803
Fal_34	F: CCAACTAGCCACACTTTGGA R: TGGCACAATTGACCTCAGAA	(AATCCC)5	MN255366	0.582
Fal_41	F: CTCTTGGATACCGAGTCCGA R: GGACTGAACTGCTGTCATGTG	(TTTG)5	MN255367	0.434
Fal_45	F: TCGAAATCGGATACTCCTCG R: CATGAAGCTTGAGCATTGGA	(TTTTAT)5	MN255368	0.583
Fal_46	F: GCCACGACATCACATCATTA R: TCAGCTTACCTTATTGGCCG	(ACAT)6	MN255369	0.444
Fal_48	F: ATGTGCCAAACCCAGAACTC R: CAACCTAGCTCTCGGAGGTG	(AAAC)5	MN255370	0.764
Fal_50	F: GCCCATCTGGTGACTGAAAC R: CGTGTGCATGCTTCATCTCT	(AAT)7	MN255371	0.351
Fal_61	F: TGGGCTCGTGACTGACTAGA R: ATGTGGGGACGGCCTCTT	(TTG)6	MN255372	0.137
Fal_62	F: CACGTGGTGGCTATGTTCTG R: GCTACGGTTTATTTGCGGTG	(TTA)7	MN255373	0.215

Table 1. *Cont.*

Locus	Primer Sequences	Repeat Motif	GenBank Accession No.	PIC
Fal_73	F: ATCCTTTGCTTTGCTCGTGT R: CGAACCTTGCACACCCTAAT	(ATA)9	MN255374	0.813
Fal_75	F: GGATCCAAAATTGGGCAGT R: ATTCATGGAATCATGGGCAC	(AAT)8	MN255375	0.459

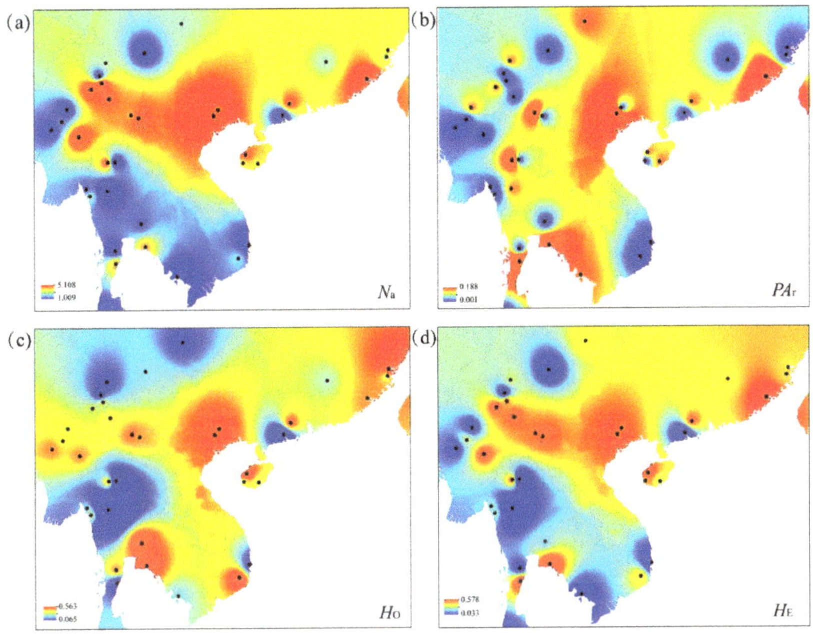

Figure 2. Genetic diversity maps of the 37 *F. altissima* populations (black dots) in Indo-Burma and adjacent southern China: (**a**) IDW interpolation of the mean number of alleles per locus (N_a); (**b**) IDW interpolation of the mean number of private alleles per locus (PA_r); (**c**) IDW interpolation of the observed heterozygosity (H_O); (**d**) IDW interpolation of the expected heterozygosity (H_E).

3.3. Population Structure

Chloroplast haplotype H1 was extremely dominant in the whole sampling area, while population Xia shared haplotype H2 with distant Thailand populations Tak, Kha, and Hua. Haplotype H3 was endemic to Guangxi of China (Loz and Pin) (Figure 1a). AMOVA indicated strong differentiation among populations in cpDNA sequences for *F. altissima* (F_{ST} = 0.958, p < 0.001) and showed that genetic variation mainly occurred between populations (Table 2).

Based on the 16 nSSRs, pairwise F_{ST} among populations ranged from −0.143 between populations Dal and Kae (979 km apart) to 0.804 between Yan and Tak (1468 km apart). Slightly negative pairwise F_{ST} (i.e., treated as zero) occurs in several population pairs, suggesting that population differentiation is negligible. Such negative estimates were converted to zero in the IBD analysis. There is no obvious correlation between genetic and geographic distance (p = 0.277, Figure 3). AMOVA tests revealed that global F_{ST} was low at 0.178 (p < 0.001), and most genetic variation (82.25%) is contained within populations, with only 17.75% genetic variation observed among populations (Table 2).

Table 2. Analysis of molecular variance (AMOVA) of the genetic diversity in *F. altissima*.

	Source of Variation	df	SS	VC	PV (%)	F_{ST}
cpDNA	Among populations	36	305.621	1.37	95.83	
	Within populations	196	11.667	0.06	4.17	
	Total	232	317.288	1.43		F_{ST} = 0.958
nSSRs	Among populations	36	554.386	0.82	17.75	
	Within populations	497	1888.006	3.80	82.25%	
	Total	533	2442.391	4.62		F_{ST} = 0.178

Notes: df, degree of freedom; SS, sum of squares; VC, variance component; PV, percentage of variation.

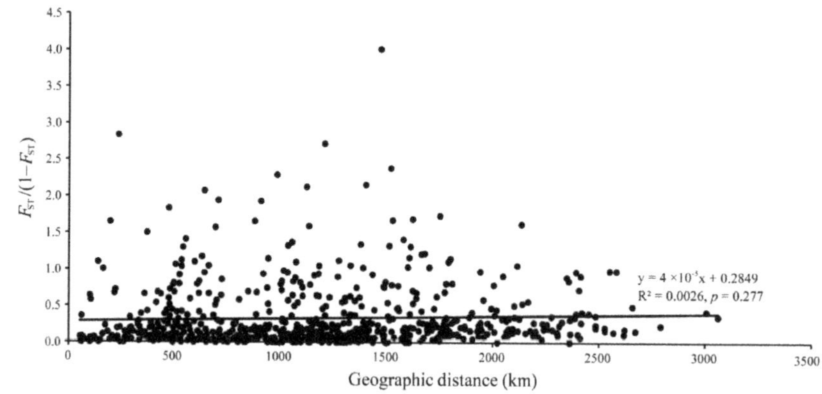

Figure 3. The regression of paired $F_{ST}/(1-F_{ST})$ vs. the geographic distances was not significant for nSSRs data.

The STRUCTURE results of the nSSRs data indicated an optimal *K* value of 5 using the delta*K* criterion [58], but this does not correspond with any obvious biological/geographical interpretation. The bar plots of the membership probabilities of *F. altissima* individuals to the different clusters using STRUCTURE analysis for *K* = 5 is shown in Figure 1b. The geographical distribution of the genetic clusters is shown in Figure 1c and pie charts represent the assignment values of the admixed clustering analysis performed in STRUCTURE for all individuals. There are no obvious boundaries defining the adjacent clusters or populations. Even geographically distant populations are highly genetically homogeneous (Figure 1b,c). The lack of a clear phylogeographic structure for *F. altissima* across the Indo-Burma range, revealed by Bayesian population structure analysis, was further validated by the PCoA (Figure 4a) and UPGMA phylogenetic analyses (Figure 4b). The different populations collected in a country also do not cluster together (Figure 4).

3.4. Population Dynamics

Neither Tajima's *D* (−0.864, $p > 0.1$) nor Fu's F_S (10.589, $p > 0.1$) vary significantly from zero, suggesting there is no evidence of recent population expansion or no recent bottlenecks [60,61]. The mismatch distribution analysis failed, as the least-square procedure to fit expected and observed mismatch distribution did not converge after 2000 steps. Based on the nSSRs data, under the TPM assumptions it was significant for five populations (Don, Fuz, Gan, Law, and Mas) while three populations (Gan, Law, and Nya) were significant under SMM ($p < 0.05$) and four populations (Gan, Law, Nid, and Xia) revealed a shifted mode of allele frequency distribution. However, only population, Law, was detected to have experienced a bottleneck effect by all of the three approaches (Table 3), and this population was genetically homogenous, as revealed by STRUCTURE analyses (Figure 1b).

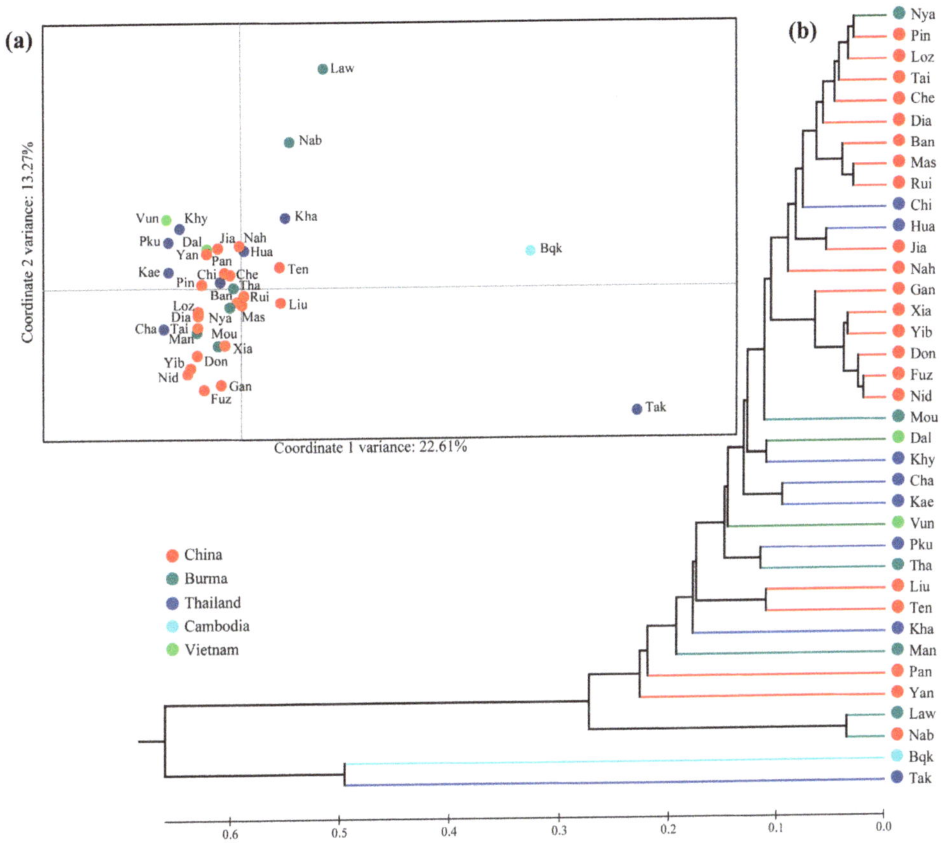

Figure 4. The interpopulation genetic affinity of 37 *F. altssima* populations based on microsatellite data: (**a**) two-dimensional scatter diagram based on principal coordinate analysis; (**b**) dendrogram based on the unweighted pair-group method with arithmetic average (UPGMA). The populations collected from different countries are marked in different colours.

Table 3. Bottleneck test of *F. altissima* populations with more than ten individual samples.

Population	Size	Model Shift	TPM	SMM
			p Value	*p* Value
Che	11	normal L-shaped distribution	0.337	0.448
Don	14	normal L-shaped distribution	0.047 *	0.336
Fuz	12	normal L-shaped distribution	0.033 *	0.245
Gan	10	shifted mode	0.010 *	0.073
Law	17	shifted mode	0.002 **	0.005 **
Mas	14	normal L-shaped distribution	0.031 *	0.0002
Nid	10	shifted mode	0.233	0.503
Nya	23	normal L-shaped distribution	0.366	0.008 **
Pin	23	normal L-shaped distribution	0.540	0.024 *
Rui	18	normal L-shaped distribution	0.355	0.078
Tai	11	normal L-shaped distribution	0.133	0.566
Xia	10	shifted mode	0.551	0.075

Notes: * Significant deviation from mutation-drift equilibrium at $p < 0.05$; ** Significant deviation from mutation-drift equilibrium at $p < 0.01$.

4. Discussion

4.1. Genetic Diversity, Population Dynamics, and Phylogeographic Patterns of Ficus altissima

The present study revealed a much lower level of cpDNA haplotype diversity in *F. altissima* than for the other published monoecious *Ficus* species [36,38,67]. To screen polymorphic markers, ten commonly used cpDNA intergenic regions in plants with the most potentially informative characters [33,34,68] were employed in our preliminary experiments, but only psbA-trnH showed sequence variation. The psbA-trnH has been commonly used in fig species [36,38,69]. However, only three psbA-trnH haplotypes were distinguished from 233 individual samples. Thus, no discernible cpDNA phylogeographic structure was revealed due to the low resolution of cpDNA regions and dominance of haplotype H1 across the whole sampling range.

Haplotypes H1 and H2 were also found in closely related fig species. For example, H1 was shared by *F. benghalensis* L., and H2 was shared by *F. binnendijkii* Miq., *F. curtipes* Corner, *F. drupacea* Thunb., *F. microcarpa* L., and *F. retusa* L., suggesting the retention of ancient polymorphisms in these closely related species. The sharing of ancestral haplotypes among widespread monoecious fig species may be a fairly common phenomenon, as observed in F. altissima and F. insipida subsp. indipida Willd [36]. These species often have large effective population sizes and long generation times, which will prolong the evolution of reciprocal monophyly [36,70]. Despite low cpDNA diversity across all samples, AMOVA revealed high interpopulation differentiation for cpDNA, suggesting that seed dispersal of *F. altissima* is limited. The derived haplotype H3, only found in Guangxi, indicates low seed dispersal as it has not dispersed into surrounding regions, or is a very recent mutation. Although there is no detailed published information, birds are probably the seed dispersal agents of *F. altissima* as it possesses a typical bird-dispersal syndrome [71]. Indeed, small non-migratory birds, including *Megalaima* Gray (Capitonidae), *Dicaeum* Cuvier (Dicaeidae), and *Pycnonotus* Boie (Pycnonotidae) species, were observed feeding on *F. altissima* trees during our field work. *F. altissima* is a hemiepiphytic tree and it exists initially as an epiphyte. The seeds of *F. altissima* germinate in the canopy of host trees [72], rather than in the soil. Hence, the germination and early seedling survival appear to be dependent on microsites provision by the host trees, such as host diameter at breast height, height and position of colonization in the canopy of host trees, crown illumination, moisture retention, and slope angle [72]. The requirements for host taxa and microsites may hinder the dispersal of *F. altissima* seeds to colonize new habitats. Short-distance seed dispersal coupled with dependency on microsites for seed development could result in limited seed flow. Nonetheless, introduced population Xia collected from Xiamen city of China shared an ancestral chloroplast haplotype H2 with the distant Thailand populations (Tak, Kha, and Hua), suggesting that human-mediated long-distance dispersal may also play a role on the current species phylogeography.

Sixteen polymorphic nSSRs for F. altissima were developed based on high-throughput sequencing and revealed a moderate level of genetic diversity. Twenty-two of the 37 populations have no private alleles, suggesting strong interpopulation pollen flow, further supported by AMOVA analyses. Extensive long-distance dispersal of pollinating wasps, aided by passive air column travel [73,74], has been documented repeatedly in monoecious figs and is ranked among the furthest known in plants [75–77]. For example, Ahmed et al. [77] showed successful pollination across distances over 160 km in monoecious *F. sycomorus* L. Meanwhile, synchronous intra-tree flowering, between tree asynchrony [78], and low densities of monoecious figs mean host-specific pollinators must travel long distances. Most populations showed negative or small positive F_{IS} values (Table A1), which is consistent with regular outcrossing associated with high dispersal events.

Rapid range expansion or a bottleneck event is not supported by assessment of Tajima's D and Fu's F_S from the cpDNA markers. However, this contrasts with the nSSRs data, where three populations (Don, Law, and Mas) were detected to have undergone bottleneck events (Table 3). However, Don is certainly a cultivated population. Mas and

Law were also collected from locations near human settlements with traces of cultivation, suggesting they experienced artificial bottlenecks.

Range expansions often result in reduced genetic diversity compared with the source populations, for example, glacial refugia compared to edge populations. Across our study range, no such significant changes in genetic diversity (e.g., haplotype diversity, observed heterozygosity, private allelic richness; Figure 2 and Table A1) based on nSSRs or cpDNA data were detected in *F. altissima*. Range expansion in obligate mutualisms involving free-living organisms is complicated by requiring the successful range extension of a pair of independently dispersing species [79,80]. Thus, the successful colonization and reproduction of *F. altissima* in a new location also depends on successful population establishment of the host-specific pollinating fig wasps, and pollinator absence often appears to limit range expansion of the *Ficus* host [80,81]. Furthermore, as a hemiepiphytic fig, germination and early seedling survival of *F. altissima* appear to be dependent on microsites provision by host trees [72]. This would be an additional impediment for the range expansion of *F. altissima*.

The origin of F. altissima has been estimated to be no earlier than the Upper Miocene [25,82], when Southeast Asia remained as a single rainforest block [83,84]. This is further supported by our finding of the widespread persistence of the ancestral H1 haplotype, across a contemporary range that is hypothesized to have undergone repeated replacement by savanna habitat during the Quaternary glaciations, with rainforest areas shrinking considerably and mostly persisting in lowland glacial refugia [85,86]. We assume that *F. altissima* survived in many lowland forests during Quaternary glaciations, but that postglacial expansions were slow. The subsequent short evolutionary history coupled with long generation times would have produced low cpDNA haplotype diversity, while genetic drift may have caused the loss of rare cpDNA haplotypes.

Indo-Burma has a complex geological and climatic history, as well as diverse habitats largely derived from the wide variation in landform, climate, and latitude [9,13,16]. Moreover, tropical forest tree species typically occur in low densities [87], and thus they may be more susceptible to genetic drift [88]. Therefore, there are both historical and ecological reasons to expect high levels of population differentiation and clear phylogeographic structure in forest trees of Indo-Burma [89,90], as observed in *D. cochinchinensis* and *D. oliveri* [10]. However, this study revealed a pattern of homogenized phylogeographic structure for *F. altissima* across Indo-Burma. As a relative generalist species, *F. altissima* is continuously distributed throughout Indo-Burma and occurs up to elevation of 2000 m. There are no obvious geographical barriers to impede the long-distance dispersal of *F. altissima*'s pollinators. In general, such postulated extensive pollen flow can buffer the effects of fragmentation and homogenize discrete populations [91,92]. A lack of phylogeographic structure for *F. altissima* in Indo-Burma appears to mainly derive from long-distance pollen dispersal. It differs from dioecious figs in which isolation by distance is often present within regions [93–95], but is consistent with patterns documented in large monoecious *Ficus* species with continuous habitat, as observed in monoecious *F. racemosa* L. [75] and *F. sur* Forssk. [96]. The notably reduced population differentiation in nuclear DNA compared to plastid cpDNA identified in *F. altissima* is also a common pattern documented in both monoecious [38,75] and dioecious fig species [94,96]. This pattern of homogenized phylogeographic structure for *F. altissima* may be observed among other Indo-Burma forest species with extensive pollen flow and limited seed transfer. However, large-scale anthropic cultivation confounds the analyses of the genetic structure of *F. altissima*.

4.2. Human Influence

Human influence on planet Earth is well-documented and considered as influential as natural processes in sculpting contemporary diversity patterns [97]. The rapid and massive loss of biodiversity directly caused (e.g., logging, hunting, fishing) and indirectly induced (e.g., pollution, infrastructure, greenhouse gas emission) by human activity in the Anthropocene is leading to the sixth mass extinction event [2,98–100]. Cultivated populations often show evidences of a reduction in genetic diversity due to artificial selection for specific quality traits [101–104]. *F. altissima* is frequently planted in its native areas or introduced into non-native areas as sacred or ornamental trees without artificial selection of specific traits; however, wild populations are being heavily impacted. It is likely to have contributed to avoiding drastic decreases in *F. altissima* genetic diversity due to declines in wild populations. For example, almost every Dai ethnic village in Xishuangbanna of China transplanted *F. altissima* from neighbouring wild resources for cultural and religious purposes [105]. Our results showed relatively high genetic diversity in many populations sampled from Dai villages (e.g., Nah, Che, Chi) and disturbed urban areas (e.g., Mas, Rui, Don, Pin). The six introduced populations also showed similar levels of genetic diversity with the populations collected from native areas, suggesting that they may originate from more than one source. This study suggests that conservation by cultivation is an effective means for protecting the genetic diversity of *F. altissima*.

It is difficult for hemiepiphytic *F. altissima* seeds to germinate and grow in urban habitats far from the natural range due to the lack of suitable host trees. Given Xiamen city is a regional center with well-established ex-patriate links to Southern Asian countries and *F. altissima* is of great ornamental value, the unexpectedly discontinuous distribution of haplotype H2 between population Xia sampled from Xiamen city and populations Tak, Kha, and Hua sampled from Thailand could be explained by long-distance human transportation, rather than by natural seed dispersal agents.

The influence of human activity also most likely explains the random and chaotic clustering revealed by nSSRs data. For example, the population Law features uniform tree age among a highly disturbed habitat close to human habitation, while population Yan is a reforestation area. The uniform genetic constitution among different individuals within these populations (Figure 1b) may originate from an individual derived germplasm. Genetic bottleneck signals were detected for all the four cultivated populations (Fuz, Gan, Nid, and Xia), suggesting artificial bottlenecks. Anthropic cultivation could have linked a naturally occurring discontinuous *F. altissima* population (e.g., due to forest fragmentation) to a continuous population in Indo-Burma and further reduced population differentiation. Therefore, the present distribution patterns and genetic structure of *F. altissima* may have been influenced by human activities. Although, it is difficult to identify the non-confounding factors that exclusively point to the impacts of human influence.

5. Conclusions

In conclusion, within Indo-Burma and south China, no clear phylogeographic structure was found in *F. altissima*, presumably as a result of extensive long-distance pollen flow unencumbered by any geographic barrier. Human cultivation and human-mediated dispersal further confound population structure, but it is also likely to have had beneficial consequences for protecting genetic diversity. This large monoecious *Ficus* species with continuous habitat constitutes a single population over a very large area. This lack of phylogeographic structure has been observed in some other plant members of Indo-Burma flora [75]. Such species will probably be highly resilient to global change as they exist over huge surfaces and have likely resisted past change due to large population ranges and a lack of restrictive adaptive specialization to localized conditions.

Author Contributions: Conceptualization, J.-F.H., C.T.D. and Y.-Q.P.; methodology, J.-F.H.; software, validation, and formal analysis, J.-F.H.; writing—original draft preparation: J.-F.H.; writing—review and editing, C.T.D. and Y.-Q.P.; visualization, J.-F.H.; supervision, Y.-Q.P.; project administration and funding acquisition, J.-F.H. and Y.-Q.P. All authors have read and agreed to the published version of the manuscript.

Funding: This research was supported by the National Natural Science Foundation of China (31800313, 32070487), Yunnan Province Applied Basic Research Project (2019FB034) and the "Light of West China" Program of the Chinese Academic of Sciences to J.-F. Huang.

Institutional Review Board Statement: Not applicable.

Informed Consent Statement: Not applicable.

Data Availability Statement: The data presented in the study are depositing in the NCBI repository, and the accession numbers are shown in the article.

Acknowledgments: The authors wish to thank Da-Rong Yang for help in collecting samples; Zhu-Zeng Huang for DNA extraction and PCR amplification; Hong-Hu Meng, Jie Gao and Finn Kjellberg for their advice during the development of this paper.

Conflicts of Interest: The authors declare no conflict of interest.

Appendix A

Table A1. Sampling information and genetic parameters of the 37 sampled *Ficus altissima* populations based on cpDNA psbA-trnH and 16 nSSRs.

Pop.	Country	Site	Lat. (N)	Long. (E)	cpDNA			nSSRs						
					Size	H_D	π (10^{-2})	Size	N_a	N_p	PA_r	H_O	H_E	F_{IS}
Yib	China	Yibin	28.623	104.418	1	na	na	3	2.563	1	0.063	0.396	0.427	0.269
Nid		Ningde	26.661	119.532	6	0.000	0.000	10	2.938	0	0.000	0.556	0.428	−0.252
Pan		Panzhihua	26.554	101.680	1	na	na	1	1.438	0	0.000	0.438	0.219	Na
Fuz		Fuzhou	26.153	119.291	7	0.000	0.000	12	2.750	0	0.000	0.469	0.435	−0.033
Liu		Nujiang	25.854	98.852	5	0.000	0.000	5	2.625	1	0.063	0.384	0.396	0.143
Gan		Ganzhou	25.850	114.928	9	0.000	0.000	10	2.563	0	0.000	0.450	0.422	−0.014
Ten		Tengchong	24.943	98.387	1	na	na	1	1.375	0	0.000	0.375	0.188	na
Xia		Xiamen	24.447	118.062	8	0.429	0.318	10	4.250	3	0.188	0.494	0.548	0.151
Mas		Mangshi	24.414	98.566	14	0.000	0.000	14	4.125	0	0.000	0.470	0.478	0.056
Rui		Ruili	23.976	97.810	18	0.000	0.000	18	4.000	1	0.063	0.465	0.518	0.130
Ban		Lincang	23.297	99.099	8	0.000	0.000	8	3.500	0	0.000	0.485	0.516	0.127
Don		Donggang	22.889	112.281	8	0.000	0.000	14	3.625	1	0.063	0.518	0.494	−0.011
Loz		Chongzuo	22.471	107.075	6	0.000	0.000	6	3.125	0	0.000	0.531	0.458	−0.069
Nah		Jinghong	22.131	100.675	7	0.000	0.000	7	3.188	2	0.125	0.509	0.474	0.008
Yan		Yangchun	22.079	111.747	3	0.000	0.000	8	1.313	0	0.000	0.313	0.156	−1.000
Pin		Pingxiang	22.056	106.736	19	0.000	0.000	23	5.125	3	0.188	0.519	0.528	0.040
Che		Mengla	21.926	101.240	11	0.000	0.000	11	3.688	0	0.000	0.494	0.509	0.076
Tai		Changjiang	19.317	109.027	11	0.000	0.000	11	3.563	2	0.063	0.542	0.512	−0.011
Jia		Ledong	18.716	108.832	4	0.000	0.000	4	2.938	0	0.000	0.469	0.451	0.104
Dia		Baoting	18.666	109.914	6	0.000	0.000	6	3.000	1	0.063	0.465	0.405	−0.057
Man	Burma	Mandalay	22.549	95.998	2	0.000	0.000	2	1.500	1	0.063	0.469	0.242	−0.875
Nab		Myotha	21.697	95.635	3	0.000	0.000	3	2.125	0	0.000	0.479	0.396	−0.011
Law		Bagan	21.128	94.851	17	0.000	0.000	17	1.563	0	0.000	0.500	0.258	−0.936
Nya		Taunggyi	20.630	96.879	23	0.000	0.000	23	4.313	0	0.000	0.500	0.515	0.052
Tha		Thaton	16.923	97.378	1	na	na	2	2.000	0	0.000	0.438	0.367	0.152
Mou		Moulmein	16.433	97.658	7	0.000	0.000	7	2.563	0	0.000	0.469	0.369	−0.197
Cha	Thailand	Lampang	18.841	99.467	1	na	na	1	1.313	0	0.000	0.313	0.156	na
Chi		Chiengmai	18.795	98.962	5	0.000	0.000	9	3.813	2	0.125	0.500	0.495	0.048
Tak		Tak	16.791	98.919	1	na	na	1	1.000	1	0.063	0.063	0.031	na
Khy		Nakhon Nayok	14.413	101.375	4	0.000	0.000	4	2.000	0	0.000	0.563	0.352	−0.500
Kha		Rayong	12.766	101.728	3	0.667	0.495	3	3.250	2	0.125	0.563	0.580	0.229
Kae		Kaeng Krachan	12.538	99.478	1	na	na	1	1.500	0	0.000	0.500	0.250	na
Hua		Thap Sakae	11.625	99.615	4	0.500	0.371	4	3.313	3	0.188	0.365	0.540	0.451
Pku		Pkuket	7.970	98.279	3	0.000	0.000	3	1.500	0	0.000	0.438	0.247	−0.680
Vun	Vietnam	Phu Yen	12.852	109.391	1	na	na	1	1.375	0	0.000	0.375	0.188	na
Dal		Lam Dong	11.940	108.458	2	0.000	0.000	2	2.375	0	0.000	0.531	0.438	0.128
Bqk	Cambodia	Bokor	10.627	104.025	2	0.000	0.000	2	1.500	1	0.063	0.438	0.234	−0.750
Species					233	0.245	0.104	267	8.500	na	na	0.483	0.576	0.163

Note: Bold indicates the six introduced populations collected from non-native range; H_D, haplotype diversity; π, nucleotide diversity; N_a, mean number of alleles per locus; N_p, number of private alleles; PA_r, mean number of private alleles per locus; H_O, observed heterozygosity; H_E, expected heterozygosity; F_{IS}, inbreeding coefficient; na, no available data.

Table A2. Null allele frequencies for each locus at each population with more than three individual samples.

Pop.	Nuclear Microsatellites															
	Fal_2	Fal_9	Fal_11	Fal_14	Fal_28	Fal_31	Fal_34	Fal_41	Fal_45	Fal_46	Fal_48	Fal_50	Fal_61	Fal_62	Fal_73	Fal_75
Nid	no	No	No	no	no	no	no	no	no	no	no	no	no	no	no	no
Fuz	no	no	**0.279**	no	no	no	no	**0.366**	no	no	no	no	no	no	no	no
Liu	no	no	No	no	no	no	no	no	no	no	no	no	no	no	no	no
Gan	no	no	**0.304**	no	no	no	no	**0.340**	no	no	no	no	no	no	no	no
Xia	no	no	No	no	no	no	no	no	no	no	no	no	no	**0.342**	no	no
Mas	no	**no**	**0.298**	no	no	no	no	no	no	no	no	no	no	no	no	no

Table A2. *Cont.*

Pop.	Nuclear Microsatellites															
	Fal_2	Fal_9	Fal_11	Fal_14	Fal_28	Fal_31	Fal_34	Fal_41	Fal_45	Fal_46	Fal_48	Fal_50	Fal_61	Fal_62	Fal_73	Fal_75
Rui	no	**0.191**	**0.402**	no	no	no	no	no	no	no	no	no	**0.208**	no	no	no
Ban	no	no	**No**	no	no	no	no	no	no	no	no	no	no	no	no	no
Don	no	no	**0.423**	no	no	no	no	0.349	no	no	no	no	no	no	no	no
Loz	no	no	No	no	no	no	no	no	no	no	no	no	no	no	no	no
Nah	no	no	No	no	no	no	no	no	no	no	no	no	no	no	no	no
Yan	no	no	No	no	no	no	no	no	no	no	no	no	no	no	no	no
Pin	no	no	**0.117**	no	no	no	no	**0.228**	no	no	no	no	**0.143**	no	no	no
Che	no	no	**0.379**	no	no	no	no	no	no	no	no	no	no	no	no	no
Tai	no	no	**0.186**	no	no	no	no	**0.375**	no	no	no	no	no	no	no	no
Jia	no	no	No	no	no	no	no	no	no	no	no	no	no	no	no	no
Dia	no	no	No	no	no	no	no	no	no	no	no	no	no	no	no	no
Law	no	no	No	no	no	no	no	no	no	no	no	no	no	no	no	no
Nya	no	no	**0.272**	no	no	no	no	no	no	no	no	no	**0.250**	no	no	**no**
Mou	no	no	No	no	no	no	no	no	no	no	no	no	no	no	no	no
Chi	no	**0.243**	No	no	no	no	no	**0.296**	no	no	no	no	no	no	no	no
Khy	no	no	No	no	no	no	no	no	no	no	no	no	no	no	no	no

Note: no, indicates no null allele; bold font indicates the deviation from Hardy–Weinberg equilibrium ($p < 0.05$).

References

1. Mittermeier, R.A.; Gil, P.R.; Hoffman, M.; Pilgrim, J.; Brooks, T.; Mittermeier, C.G.; Lamoreux, J.; de Fonseca, G.A.B. *Hotspots Revisited: Earth's Biologically Richest and Most Endangered Terrestrial Ecoregions*; CEMEX: Mexico City, Mexico, 2004.
2. Meng, H.H.; Gao, X.Y.; Song, Y.G.; Cao, G.L.; Li, J. Biodiversity arks in the Anthropocene. *Reg. Sustain.* **2021**, *2*, 109–115. [CrossRef]
3. Voris, H.K. Maps of pleistocene sea levels in Southeast Asia: Shorelines, river systems, time durations. *J. Biogeogr.* **2000**, *27*, 1153–1167. [CrossRef]
4. Woodruff, D.S. Biogeography and conservation in Southeast Asia: How 2.7 million years of repeated environmental fluctuations affect today's patterns and the future of the remaining refugial-phase biodiversity. *Biodivers. Conserv.* **2010**, *19*, 919–941. [CrossRef]
5. Lukoschek, V.; Osterhage, J.L.; Karns, D.R.; Murphy, J.C.; Voris, H.K. Phylogeography of the Mekong mud snake (*Enhydris subtaeniata*): The biogeographic importance of dynamic river drainages and fluctuating sea levels for semiaquatic taxa in Indochina. *Ecol. Evol.* **2011**, *1*, 330–342. [CrossRef]
6. Sodhi, N.S.; Koh, L.P.; Brook, B.W.; Ng, P.K. Southeast Asian biodiversity: An impending disaster. *Trends Ecol. Evol.* **2004**, *19*, 654–660. [CrossRef] [PubMed]
7. Nijiman, V. An overview of international wildlife trade from Southeast Asia. *Biodivers. Conserv.* **2010**, *19*, 1101–1114. [CrossRef]
8. Hughes, A.C. Have Indo-Malaysian forests reached the end of the road? *Biol. Conserv.* **2018**, *223*, 129–137. [CrossRef]
9. Fuchs, J.; Ericson, P.G.; Pasquet, E. Mitochondrial phylogeographic structure of the white-browed piculet (*Sasia ochracea*): Cryptic genetic differentiation and endemism in Indochina. *J. Biogeogr.* **2008**, *35*, 565–575. [CrossRef]
10. Hartvig, I.; So, T.; Changtragoon, S.; Tran, H.T.; Bouamanivong, S.; Theilade, I.; Kjaer, E.D.; Nielsen, L.R. Population genetic structure of the endemic rosewoods *Dalbergia cochinchinensis* and *D. oliveri* at a regional scale reflects the Indochinese landscape and life-history traits. *Ecol. Evol.* **2018**, *8*, 530–545. [CrossRef]
11. Attwood, S.W. Schistosomiasis in the Mekong region: Epidemiology and phylogeography. *Adv. Parasitol.* **2001**, *50*, 87–152.
12. Thaenkham, U.; Phuphisut, O.; Nuamtanong, S.; Yoonuan, T.; Sa-Nguankiat, S.; Vonghachack, Y.; Belizario, V.Y.; Dung, D.T.; Dekumyoy, P.; Waikagul, J. Genetic differences among *Haplorchis taichui* populations in Indochina revealed by mitochondrial COX1 sequences. *J. Helminthol.* **2017**, *91*, 597–604. [CrossRef]
13. Pramual, P.; Kuvangkadilok, C.; Baimai, V.; Walton, C. Phylogeography of the black fly *Simulium tani* (Diptera: Simuliidae) from Thailand as inferred from mtDNA sequences. *Mol. Ecol.* **2005**, *14*, 3989–4001. [CrossRef] [PubMed]
14. Takagi, A.P.; Ishikawa, S.; Nao, T.; Song, S.L.; Hort, S.; Thammavong, K.; Saphakdy, B.; Phomsouvanhm, A.; Nishida, M.; Kurokura, H. Genetic differentiation and distribution routes of the bronze featherback *Notopterus notopterus* (Osteoglossiformes: Notopteridae) in Indochina. *Biol. J. Linnean Soc.* **2010**, *101*, 575–582. [CrossRef]
15. Adamson, E.A.; Hurwood, D.A.; Mather, P.B. Insights into historical drainage evolution based on the phylogeography of the chevron snakehead fish (*Channa striata*) in the Mekong Basin. *Freshw. Biol.* **2012**, *57*, 2211–2229. [CrossRef]
16. Blair, C.; Davy, C.M.; Ngo, A.; Orlov, N.L.; Shi, H.T.; Lu, S.Q.; Gao, L.; Rao, D.Q.; Murphy, R.W. Genealogy and demographic history of a widespread amphibian throughout Indochina. *J. Hered.* **2013**, *104*, 72–85. [CrossRef] [PubMed]
17. Meschersky, I.G.; Abramov, A.V.; Lebedev, V.S.; Chichkina, A.N.; Rozhnov, V.V. Evidence of a complex phylogeographic structure in the Indomalayan pencil-tailed tree mouse *Chiropodomys gliroides* (Rodentia: Muridae) in eastern Indochina. *Biocheml. Syst. Ecol.* **2016**, *65*, 147–157. [CrossRef]

18. Balakirev, A.E.; Rozhnov, V.V. Taxonomic revision of beautiful squirrels (*Callosciurus*, Rodentia: Sciuridae) from the *Callosciurus erythraeus/finlaysonii* complex and their distribution in eastern Indochina. *Raffles Bull. Zool.* **2019**, *67*, 459–489.
19. Balakrishnan, C.N.; Monfort, S.L.; Gaur, A.; Singh, L.; Sorenson, M.D. Phylogeography and conservation genetics of Eld's deer (*Cervus eldi*). *Mol. Ecol.* **2003**, *12*, 1–10. [CrossRef] [PubMed]
20. Glaubrecht, M.; Köhler, F. Radiating in a river: Systematics, molecular genetics and morphological differentiation of viviparous freshwater gastropods endemic to the Kaek River, central Thailand (Cerithioidea, Pachychilidae). *Biol. J. Linnean Soc.* **2004**, *82*, 275–311. [CrossRef]
21. McGill, B.J.; Dornelas, M.; Gotelli, N.J.; Magurran, A.E. Fifteen forms of biodiversity trend in the Anthropocene. *Trends Ecol. Evol.* **2015**, *30*, 104–113. [CrossRef]
22. Newbold, T.; Hudson, L.N.; Hill, S.L.L.; Contu, S.; Lysenko, I.; Senior, R.A.; Börger, L.; Bennett, D.J.; Choimes, A.; Collen, B.; et al. Global effects of land use on local terrestrial biodiversity. *Nature* **2015**, *520*, 45–50. [CrossRef]
23. Tilman, D.; Clark, M.; Williams, D.R.; Kimmel, K.; Polasky, S.; Packer, C. Future threats to biodiversity and pathways to their prevention. *Nature* **2017**, *546*, 73–81. [CrossRef] [PubMed]
24. Herre, E.A.; Jander, K.C.; Machado, C.A. Evolutionary ecology of figs and their associates: Recent progress and outstanding puzzles. *Annu. Rev. Ecol. Evol. Syst.* **2008**, *39*, 439–458. [CrossRef]
25. Cruaud, A.; Rønsted, N.; Chantarasuwan, B.; Chou, L.S.; Clement, W.L.; Couloux, A.; Cousins, B.; Genson, G.; Harrison, R.D.; Hanson, P.E.; et al. An extreme case of plant–insect co-diversification: Figs and fig-pollinating wasps. *Syst. Biol.* **2012**, *61*, 1029–1047. [CrossRef] [PubMed]
26. Shi, Y.; Mon, A.M.; Fu, Y.; Zhang, Y.; Wang, C.; Yang, X.; Wang, Y. The genus *Ficus* (Moraceae) used in diet: Its plant diversity, distribution, traditional uses and ethnopharmacological importance. *J. Ethnopharmacol.* **2018**, *226*, 185–196. [CrossRef]
27. Peng, Y.Q.; Duan, Z.B.; Yang, D.R.; Rasplus, J.Y. Co-occurrence of two *Eupristina* species on *Ficus altissima* in Xishuangbanna, SW China. *Symbiosis* **2008**, *45*, 9–14.
28. Khare, C.P. *Indian Medicinal Plants: An Illustrated Dictionary*; Springer: Berlin/Heidelberg, Germany, 2007.
29. Berg, C.C.; Corner, E.G.H. Moraceae–Ficus. In *Flora Malesiana*; Nooteboom, H.P., Ed.; National Herbarium Nederland: Leiden, The Netherlands, 2005.
30. Peng, Y.Q.; Compton, S.G.; Yang, D.R. The reproductive success of *Ficus altissima* and its pollinator in a strongly seasonal environment: Xishuangbanna, Southwestern China. *Plant Ecol.* **2010**, *209*, 227–236. [CrossRef]
31. Zhou, Z.K.; Gilbert, M.G. Moraceae: Genus *Ficus*. In *Flora of China*; Wu, Z.Y., Raven, P.H., Hong, D.Y., Eds.; Science Press: Beijing, China; Missouri Botanical Garden Press: St Louis, MO, USA, 2003; Volume 5, pp. 37–71.
32. Fang, J.Y.; Wang, Z.H.; Tang, Z.Y. *Atlas of Woody Plants in China: Distribution and Climate*; Higher Education Press: Beijing, China; Springer-Verlag: Berlin, Germany, 2011; Volume 1, p. 203.
33. Shaw, J.; Lickey, E.B.; Beck, J.T.; Farmer, S.B.; Liu, W.; Miller, J.; Siripun, K.C.; Winder, C.T.; Schilling, E.E.; Small, R.L. The tortoise and the hare II: Relative utility of 21 noncoding chloroplast DNA sequences for phylogenetic analysis. *Am. J. Bot.* **2005**, *92*, 142–166. [CrossRef]
34. Shaw, J.; Lickey, E.B.; Schilling, E.E.; Small, R.L. Comparison of whole chloroplast genome sequences to choose noncoding regions for phylogenetic studies in angiosperms: The tortoise and the hare III. *Am. J. Bot.* **2007**, *94*, 275–288. [CrossRef] [PubMed]
35. Vieira, F.D.A.; Santana, J.D.S.; Dos Santos, R.M.; Fajardo, C.G.; Coelho, G.D.O.; Carvalho, D.D. DNA extraction protocols and cpDNA primers to *Ficus bonijesulapensis* (Moraceae). *Caatinga* **2010**, *23*, 69–74.
36. Honorio Coronado, E.N.; Dexter, K.G.; Poelchau, M.F.; Hollingsworth, P.M.; Phillips, O.L.; Pennington, R.T. *Ficus insipida* subsp. *insipida* (Moraceae) reveals the role of ecology in the phylogeography of widespread Neotropical rain forest tree species. *J. Biogeogr.* **2014**, *41*, 1697–1709. [CrossRef] [PubMed]
37. Tnah, L.H.; Lee, S.L.; Tan, A.L.; Lee, C.T.; Ng, K.K.S.; Ng, C.H. Intraspecific classification of *Ficus deltoidea* Jack subsp. *deltoidea* (Moraceae) in Peninsular Malaysia based on morphological and molecular variations. *Biochem. Syst. Ecol.* **2016**, *67*, 119–128. [CrossRef]
38. Costa, P.C.; Lorenz-Lemke, A.P.; Furini, P.R.; Honorio Coronado, E.N.; Kjellberg, F.; Pereira, R.A. The phylogeography of two disjunct Neotropical *Ficus* (Moraceae) species reveals contrasted histories between the Amazon and the Atlantic Forests. *Bot. J. Linnean Soc.* **2017**, *185*, 272–289. [CrossRef]
39. Thompson, J.D.; Gibson, T.J.; Plewniak, F.; Jeanmougin, F.; Higgins, D.G. The CLUSTAL_X windows interface: Flexible strategies for multiple sequence alignment aided by quality analysis tools. *Nucleic Acids Res.* **1997**, *25*, 4876–4882. [CrossRef]
40. Hall, T.A. BioEdit: A user-friendly biological sequence alignment editor and analysis program for Windows 95/98/NT. *Nucl. Acids Symp. Ser.* **1999**, *41*, 95–98.
41. Emshwiller, E.; Doyle, J.J. Chloroplast-expressed glutamine synthetase (ncpGS): Potential utility for phylogenetic studies with an example from *Oxalis* (Oxalidaceae). *Mol. Phylogenet. Evol.* **1999**, *12*, 310–319. [CrossRef]
42. Rønsted, N.; Weiblen, G.; Clement, W.L.; Zerega, N.J.C.; Savolainen, V. Reconstructing the phylogeny of figs (*Ficus*, Moraceae) to reveal the history of the fig pollination mutualism. *Symbiosis* **2008**, *45*, 45–55.
43. Thiel, T.; Michalek, W.; Varshney, R.K.; Graner, A. Exploiting EST databases for the development and characterization of gene-derived SSR-markers in barley (*Hordeum vulgare* L.). *Theor. Appl. Genet.* **2003**, *106*, 411–422. [CrossRef] [PubMed]
44. Rozen, S.; Skaletsky, H. Primer3 on the WWW for general users and for biologist programmers. *Methods Mol. Biol.* **1999**, *6*, 365–386.

45. Van Oosterhout, C.; Hutchinson, W.F.; Wills, D.P.M.; Shipley, P. MICRO-CHECKER: Software for identifying and correcting genotyping errors in microsatellite data. *Mol. Ecol. Notes* **2004**, *4*, 535–538. [CrossRef]
46. Rousset, F. GENEPOP'007: A complete re-implementation of the GENEPOP software for Windows and Linux. *Mol. Ecol. Resour.* **2008**, *8*, 103–106. [CrossRef] [PubMed]
47. Librado, P.; Rozas, J. DnaSP v5: A software for comprehensive analysis of DNA polymorphism data. *Bioinformatics* **2009**, *25*, 1451–1452. [CrossRef] [PubMed]
48. Peakall, R.; Smouse, P.E. GenAlEx 6.5: Genetic analysis in Excel. Population genetic software for teaching and research—An update. *Bioinformatics* **2012**, *28*, 2537–2539. [CrossRef] [PubMed]
49. Goudet, J. *FSTAT, A Program to Estimate and Test Gene Diversities and Fixation Indices, Version 2.9.3*; ScienceOpen: Burlington, MA, USA, 2001.
50. Marshall, T.C.; Slate, J.; Kruuk, L.E.B.; Pemberton, J.M. Statistical confidence for likelihood-based paternity inference in natural populations. *Mol. Ecol.* **1998**, *7*, 639–655. [CrossRef] [PubMed]
51. Kalinowski, S.T.; Taper, M.L.; Marshall, T.C. Revising how the computer program CERVUS accommodates genotyping error increases success in paternity assignment. *Mol. Ecol.* **2007**, *16*, 1099–1106. [CrossRef]
52. Hengl, T.A. *Practical Guide to Geostatistical Mapping*; University of Amsterdam: Amsterdam, The Netherlands, 2009.
53. Excoffier, L.; Lischer, H.E.L. Arlequin suite ver 3.5: A new series of programs to perform population genetics analyses under Linux and Windows. *Mol. Ecol. Resour.* **2010**, *10*, 564–567. [CrossRef]
54. Ersts, P.J. Geographic Distance Matrix Generator (Version 1.2.3). Available online: http://biodiversityinformatics.amnh.org/open_source/gdmg (accessed on 21 July 2021).
55. Tamura, K.; Stecher, G.; Peterson, D.; Filipski, A.; Kumar, S. MEGA6: Molecular evolutionary genetics analysis version 6.0. *Mol. Biol. Evol.* **2013**, *30*, 2725–2729. [CrossRef]
56. Pritchard, J.K.; Stephens, M.; Donnelly, P. Inference of population structure using multilocus genotype data. *Genetics* **2000**, *155*, 945–959. [CrossRef] [PubMed]
57. Earl, D.A.; Vonholdt, B.M. STRUCTURE HARVESTER: A website and program for visualizing STRUCTURE output and implementing the Evanno method. *Conserv. Genet. Resour.* **2012**, *4*, 359–361. [CrossRef]
58. Evanno, G.; Regnaut, S.; Goudet, J. Detecting the number of clusters of individuals using the software STRUCTURE: A simulation study. *Mol. Ecol.* **2005**, *14*, 2611–2620. [CrossRef] [PubMed]
59. Jakobsson, M.; Rosenberg, N.A. CLUMPP: A cluster matching and permutation program for dealing with label switching and multimodality in analysis of population structure. *Bioinformatics* **2007**, *23*, 1801–1806. [CrossRef] [PubMed]
60. Tajima, F. Statistical method for testing the neutral mutation hypothesis by DNA polymorphism. *Genetics* **1989**, *123*, 585–595. [CrossRef] [PubMed]
61. Fu, Y.X. Statistical tests of neutrality of mutations against population growth, hitchhiking and background selection. *Genetics* **1997**, *147*, 915–925. [CrossRef] [PubMed]
62. Piry, S.G.; Luikart, G.; Cornuet, J.M. BOTTLENECK: A computer program for detecting reductions in the effective population size using allele frequency data. *J. Hered.* **1999**, *86*, 502–503. [CrossRef]
63. Di Rienzo, A.; Peterson, A.C.; Garza, J.C.; Valdes, A.M.; Slatkin, M.; Freimer, N.B. Mutational processes of simple sequence repeat loci in human populations. *Proc. Natl. Acad. Sci. USA* **1994**, *91*, 3166–3170. [CrossRef]
64. Luikart, G.; Cornuet, J.M. Empirical evaluation of a test for identifying recently bottlenecked populations from allele frequency data. *Conserv. Biol.* **1998**, *12*, 228–237. [CrossRef]
65. Nei, M.; Maruyama, T.; Chakraborty, R. The bottleneck effect and genetic variability in populations. *Evolution* **1975**, *29*, 1–10. [CrossRef]
66. Luikart, G.L.; Allendorf, F.W.; Cornuet, J.M.; Sherwin, W.B. Distortion of allele frequency distributions provides a test for recent populations. *J. Hered.* **1998**, *89*, 238–247. [CrossRef] [PubMed]
67. Vieira, F.D.A.; Novaes, R.M.L.; Fajardo, C.G.; Santos, R.M.D.; Almeida, H.D.S.; Carvalho, D.D.; Lovato, M.B. Holocene southward expansion in seasonally dry tropical forests in South America: Phylogeography of *Ficus bonijesulapensis* (Moraceae). *Bot. J. Linnean Soc.* **2015**, *177*, 189–201. [CrossRef]
68. Kress, W.J.; Wurdack, K.J.; Zimmer, E.A.; Weigt, L.A.; Janzen, D.H. Use of DNA barcodes to identify flowering plants. *Proc. Natl. Acad. Sci. USA* **2005**, *102*, 8369–8374. [CrossRef]
69. Li, H.Q.; Chen, J.Y.; Wang, S.; Xiong, S.Z. Evaluation of six candidate DNA barcoding loci in *Ficus* (Moraceae) of China. *Mol. Ecol. Resour.* **2012**, *12*, 783–790. [CrossRef] [PubMed]
70. Rosenberg, N.A. The shapes of neutral gene genealogies in two species: Probabilities of monophyly, paraphyly, and polyphyly in a coalescent model. *Evolution* **2003**, *57*, 1465–1477. [CrossRef] [PubMed]
71. Lomáscolo, S.B.; Levey, D.J.; Kimball, R.T.; Bolker, B.M.; Alborn, H.T. Dispersers shape fruit diversity in *Ficus* (Moraceae). *Proc. Natl. Acad. Sci. USA* **2010**, *107*, 14668–14672. [CrossRef]
72. Harrison, R.D.; Hamid, A.A.; Kenta, T.; Lafrankie, J.; Lee, H.S.; Nagamasu, H.; Nakashizuka, T.; Palmiotto, P. The diversity of hemi-epiphytic figs (*Ficus*; Moraceae) in a Bornean lowland rain forest. *Biol. J. Linnean Soc.* **2003**, *78*, 439–455. [CrossRef]
73. Compton, S.G.; Ellwood, M.D.; Davis, A.J.; Welch, K. The flight heights of chalcid wasps (Hymenoptera, Chalcidoidea) in a lowland Bornean rain forest: Fig wasps are the high fliers. *Biotropica* **2000**, *32*, 515–522. [CrossRef]

74. Harrison, R.D. Fig wasp dispersal and the stability of a keystone plant resource in Borneo. *Proc. R. Soc. Lond. B Biol. Sci.* **2003**, *270*, S76–S79. [CrossRef]
75. Bain, A.; Borges, R.M.; Chevallier, M.H.; Vignes, H.; Kobmoo, N.; Peng, Y.Q.; Cruaud, A.; Rasplus, J.Y.; Kjellberg, F.; Hossaert-Mckey, M. Geographic structuring into vicariant species-pairs in a wide-ranging, high-dispersal plant–insect mutualism: The case of *Ficus racemosa* and its pollinating wasps. *Evol. Ecol.* **2016**, *30*, 663–684. [CrossRef]
76. Nason, J.D.; Herre, E.A.; Hamrick, J.L. The breeding structure of a tropical keystone plant resource. *Nature* **1998**, *391*, 685–687. [CrossRef]
77. Ahmed, S.; Compton, S.G.; Butlin, R.K.; Gilmartin, P.M. Wind-borne insects mediate directional pollen transfer between desert fig trees 160 kilometers apart. *Proc. Natl. Acad. Sci. USA* **2009**, *106*, 20342–20347. [CrossRef]
78. Lin, S.Y.; Chou, L.S.; Di Giusto, B.; Bain, A. Sexual specialization in phenology in dioecious *Ficus benguetensis* and its consequences for the mutualism. *Bot. Stud.* **2015**, *56*, 32. [CrossRef] [PubMed]
79. Alvarez, N.; McKey, D.; Kjellberg, F.; Hossaert-McKey, M. Phylogeography and historical biogeography of obligate specific mutualisms. In *The Biogeography of Host–Parasite Interactions*; Morand, S., Krasnov, B.R., Eds.; Oxford University Press: Oxford, UK, 2010; pp. 31–39.
80. Yu, H.; Nason, J.D. Nuclear and chloroplast DNA phylogeography of *Ficus hirta*: Obligate pollination mutualism and constraints on range expansion in response to climate change. *New Phytol.* **2013**, *197*, 276–289. [CrossRef]
81. Kjellberg, F.; Valdeyron, G. Species-specific pollination: A help or a limitation to range extension. In *Biological Invasions in Europe and the Mediterranean Basin*; di Castri, F., Hansen, A.J., Eds.; Dr W Junk publishers: Dordrecht, The Netherlands, 1990; pp. 371–378.
82. Xu, L.; Harrison, R.D.; Yang, P.; Yang, D.R. New insight into the phylogenetic and biogeographic history of genus *Ficus*: Vicariance played a relatively minor role compared with ecological opportunity and dispersal. *J. Syst. Evol.* **2011**, *49*, 546–557. [CrossRef]
83. Muller, J. Palynological evidence for change in geomorphology, climate and vegetation in the Mio-Pliocene of Malesia. In *Transactions of the Second Aberdeen-Hull Symposium on Malesian Ecology*; Ashton, P., Ashton, M., Eds.; University of Hull: Hull, UK, 1972; pp. 6–16.
84. Morley, R.J. *Origin and Evolution of Tropical Rain Forests*; John Wiley and Sons Ltd.: Chichester, UK, 2000.
85. Heaney, L.R. A synopsis of climatic and vegetational change in Southeast Asia. *Clim. Chang.* **1991**, *19*, 53–61. [CrossRef]
86. Tordoff, A.W.; Bezuijen, M.R.; Duckworth, J.W.; Fellowes, J.R.; Koenig, K.; Pollard, E.H.B.; Royo, A.G. *Ecosystem Profile: Indo-Burma Biodiversity Hotspot 2011 Update*; Critical Ecosystem Partnership Fund: Arlington, VA, USA, 2012.
87. Hubbell, S.P.; Foster, R.B. Diversity of canopy trees in a neotropical forest and implications for conservation. In *Tropical Rain Forest: Ecology and Management*; Sutton, S.L., Whitmore, T.C., Chadwick, A.C., Eds.; Blackwell Scientific Publications: Oxford, UK, 1983.
88. Fedorov, A.A. The structure of the tropical rain forest and speciation in the humid tropics. *J. Ecol.* **1966**, *54*, 1–11. [CrossRef]
89. Dick, C.W.; Hardy, O.J.; Jones, F.A.; Petit, R.J. Spatial scales of pollen and seed-mediated gene flow in tropical rain forest trees. *Trop. Plant Biol.* **2008**, *1*, 20–33. [CrossRef]
90. Dick, C.W.; Heuertz, M. The complex biogeographic history of a widespread tropical tree species. *Evolution* **2008**, *62*, 2760–2774. [CrossRef]
91. Liu, M.; Zhang, J.; Chen, Y.; Compton, S.G.; Chen, X. Contrasting genetic responses to population fragmentation in a coevolving fig and fig wasp across a mainland-island archipelago. *Mol. Ecol.* **2013**, *22*, 4384–4396. [CrossRef] [PubMed]
92. Colabella, F.; Gallo, L.A.; Moreno, A.C.; Marchelli, P. Extensive pollen flow in a natural fragmented population of Patagonian cypress *Austrocedrus chilensis*. *Tree Genet. Genomes* **2014**, *10*, 1519–1529. [CrossRef]
93. Chen, Y.; Jiang, Z.X.; Compton, S.G.; Liu, M.; Chen, X.Y. Genetic diversity and differentiation of the extremely dwarf *Ficus tikoua* in Southwestern China. *Biochem. Syst. Ecol.* **2011**, *39*, 441–448. [CrossRef]
94. Liu, M.; Compton, S.G.; Peng, F.E.; Zhang, J.; Chen, X.Y. Movements of genes between populations: Are pollinators more effective at transferring their own or plant genetic markers? *Proc. R. Soc. Lond. B Biol. Sci.* **2015**, *282*, 20150290. [CrossRef] [PubMed]
95. Tian, E.; Nason, J.D.; Machado, C.A.; Zheng, L.; Yu, H.; Kjellberg, F. Lack of genetic isolation by distance, similar genetic structuring but different demographic histories in a fig-pollinating wasp mutualism. *Mol. Ecol.* **2015**, *24*, 5976–5991. [CrossRef] [PubMed]
96. Deng, J.Y.; Van Noort, S.; Compton, S.G.; Chen, Y.; Greeff, J.M. The genetic consequences of habitat specificity for fig trees in southern African fragmented forests. *Acta Oecol.* **2020**, *102*, 103506. [CrossRef]
97. Corlett, R.T. The Anthropocene concept in ecology and conservation. *Trends Ecol. Evol.* **2015**, *30*, 36–41. [CrossRef]
98. Crutzen, P.J.; Stoermer, E.F. The 'Anthropocene'. *IGBP Newsl.* **2000**, *41*, 17–18.
99. Isbel, F.; Gonzalez, A.; Loreau, M.; Cowles, J.; Díaz, S.; Hector, A.; Mace, G.M.; Wardle, D.A.; O'Connor, M.I.; Duffy, J.E.; et al. Linking the influence and dependence of people on biodiversity across scales. *Nature* **2017**, *546*, 65–72. [CrossRef] [PubMed]
100. Johnson, C.N.; Balmford, A.; Brook, B.W.; Buettel, J.C.; Galetti, M.; Guangchun, L.; Wilmshurst, J.A. Biodiversity losses and conservation responses in the Anthropocene. *Science* **2017**, *356*, 270–275. [CrossRef]
101. Chen, C.; Li, P.; Wang, R.H.; Schaal, B.A.; Fu, C.X. The population genetics of cultivation: Domestication of a traditional Chinese medicine, *Scrophularia ningpoensis* Hemsl. (Scrophulariaceae). *PLoS ONE* **2014**, *9*, e105064.
102. Eyre-Walker, A.; Gaut, R.L.; Hilton, H.; Feldman, D.L.; Gaut, B.S. Investigation of the bottleneck leading to the domestication of maize. *Proc. Natl. Acad. Sci. USA* **1998**, *95*, 4441–4446. [CrossRef]
103. Londo, J.P.; Chiang, Y.C.; Hung, K.H.; Chiang, T.Y.; Schaal, B.A. Phylogeography of Asian wild rice, *Oryza rufipogon*, reveals multiple independent domestications of cultivated rice, *Oryza Sativa*. *Proc. Natl. Acad. Sci. USA* **2006**, *103*, 9578–9583. [CrossRef]

104. Sonnante, G.; Stockton, T.; Nodari, R.O.; Becerra Velásquez, V.L.; Gepts, P. Evolution of genetic diversity during the domestication of common-bean (*Phaseolus vulgaris* L.). *Theor. Appl. Genet.* **1994**, *89*, 629–635. [CrossRef]
105. Hu, H.B. Sacred natural sites in Xishuangbanna, south-western China. In *The Importance of Sacred Natural Sites for Biodiversity Conservation*; Lee, C., Schaaf, T., Eds.; United Nations Educational, Scientific and Cultural Organization: Paris, France, 2003; pp. 127–133.

MDPI
St. Alban-Anlage 66
4052 Basel
Switzerland
www.mdpi.com

Diversity Editorial Office
E-mail: diversity@mdpi.com
www.mdpi.com/journal/diversity

www.ingramcontent.com/pod-product-compliance
Lightning Source LLC
Chambersburg PA
CBHW061459220926
51850CB00004B/126
* 9 7 8 3 7 2 5 8 0 3 9 8 9 *